$T = \frac{2\pi}{\omega} = \frac{1}{f}$

$\omega = \frac{2\pi}{T}$

INTRODUCTION TO SYSTEM DYNAMICS

J. LOWEN SHEARER
Rockwell Professor of Engineering
Pennsylvania State University

ARTHUR T. MURPHY
Professor and Dean of Engineering
Widener College

HERBERT H. RICHARDSON
Associate Professor of Mechanical Engineering
Massachusetts Institute of Technology

INTRODUCTION TO SYSTEM DYNAMICS

ADDISON-WESLEY
PUBLISHING COMPANY
READING, MASSACHUSETTS
MENLO PARK, CALIFORNIA
LONDON · AMSTERDAM
DON MILLS, ONTARIO · SYDNEY

This book is in the

ADDISON-WESLEY SERIES IN SYSTEMS AND CONTROLS

Second printing, April 1971

 LIBRARY OF CONGRESS CATALOG CARD NO. 67-13403.

ISBN 0-201-07017-0
ZAABB-MU-93210

PREFACE

This book is a unified engineering treatment of mechanical, electrical, fluid, and thermal dynamic systems; their idealized models, response, and analytical description. It is intended to meet the growing need for interdisciplinary courses dealing with systems in engineering curricula, and it is flexible enough to be suitable for a variety of such courses.

Considerable emphasis is placed on physical aspects, formulation of mathematical models, and interpretation of results in terms of the behavior of real systems along with consideration of mathematical methods. The authors believe that fundamental concepts should be well understood before sophisticated mathematical abstractions are considered, and the book is organized accordingly. The early development and model-formulation techniques are valid both for linear and nonlinear systems, and methods which are *restricted* to linear systems are deferred to the second half of the text. However, most of the examples considered throughout the text deal with linear systems.

The teacher will appreciate several features of this presentation, some of which are mentioned below: (1) Two-port elements, such as transformers and transducers, are included in both specific and general fashion. (2) For clarity, examples used in the text after Chapter 4 are generally limited to electrical and mechanical systems. This allows use of the book for courses in which fluid and thermal systems are not treated, although problems dealing with these systems are given throughout for those who wish such emphasis. (3) An entire chapter is devoted to the subject of modeling. (4) A complete discussion of ways of checking analytical work is included (Section 8–6.5). (5) The use of the analogy based on through- and across-variables (the mass-capacitance or mobility analogy) is presented, together with a discussion of the merits of the various analogies (Section 8–7). (6) Emphasis is placed on the use of the linear graph as a common diagram representing structure in all physical systems considered (as opposed to the use of electric circuit symbols to represent nonelectrical systems). (7) Linear graphs are used to formulate equations as well as to visualize the structure of systems. (8) Exponential response is used to develop the concepts of system functions, poles and zeros, and is later shown to be equivalent to the use of the Laplace transform. (9) The formulation methods used early in the book are equally applicable to both linear and nonlinear systems; those restricted to linear systems are not introduced until Chapter 13 (Section 13–1 presents a review of their development).

Interdisciplinary courses are needed to educate engineers who will be able to deal with complex systems of the present and the future, systems that heavily involve dynamic interactions of many physical phenomena and traditional disciplines. As the size, complexity and necessary performance limits of engineering systems increase, their dynamic nature becomes more important, and factors such as speed of response, transient and steady-state performance, and dynamic stability become increasingly critical. The modern engineer must have a unified background, if only to communicate with the various traditional engineering disciplines.

There are two additional advantages to courses such as are presented herein. First, a properly planned curriculum which includes such a course will avoid unnecessary duplication. This will allow the introduction of more advanced material into the undergraduate program, pressure toward which exists in each school in the face of an expanding technology. Second, this interdisciplinary material is consistent with the growing tendency of students to decide on their branch of engineering relatively late in their undergraduate program. In recognition of this tendency, many engineering schools have introduced unified curricula in which specialization in a traditional branch of engineering is done only in the senior year or not at all. *System Dynamics* serves these purposes admirably, yet it may also be used in traditional curricula in electrical, mechanical, chemical, industrial, or civil engineering.

This book in its entirety is intended for use in an introductory course in engineering to be taken by all engineering students. Such an introductory course would usually appear in the sophomore or junior year, depending on the specific curriculum design. It is assumed that the student will have had the usual background of classical physics (mechanics, electricity and magnetism, heat and sound) and mathematics through calculus. A chapter on classical solution of ordinary differential equations is included (Chapter 11), which may be used either as a review from a system-response orientation for students who have had differential equations, or as new background which will enable students who have completed calculus but not differential equations to understand the rest of the book. Chapter 9, on complex numbers, may be used in a similar way.

The book can be (and has been) used for a more advanced course in a curriculum in which traditional courses in electric circuits and dynamics preceded it. It can even be (and has been) used for a graduate course for students who have little or no background in system dynamics. This material is also particularly appropriate, and has been successfully used, as part of a program of continuing education for graduate engineers who received their degrees before undergraduate engineering curricula had been given this orientation.

As a *two-semester introductory engineering course*, the entire contents of the book would be covered essentially in the order in which they are presented. We have found that there is also time to amplify the material on the analog computer, Fourier series, and Laplace transform, and to introduce special topics such as feedback control, stability, etc., depending on the preference of the teacher. We have found it particularly valuable in the two-semester course to run an analog computer laboratory concurrently with the lecture material. The material on the analog computer presented in Chapter 7 is sufficient for the students if experienced instructors will reinforce the practice in the laboratory. If some simple concepts about the operation of the computer components are presented, the laboratory work can be started immediately (before Chapter 7 is reached).

We have found that with students who have a good background, so that a brisk pace can be maintained, the following material can be covered in a *one-semester introductory course:* Chapters 1, 2, 3 (optional), 5, 6, the first six sections of Chapter 7, Chapters 8 (omitting some of the niceties of Sections 8–6 and 8–7), 10, 12, 13, and 15, followed by appropriate material from Chapters 14, 16, 3, and the last half of Chapter 7. If the students have not had sufficient mathematical preparation, Chapters 9 and 11 should be covered, with material from 14 and 16 omitted as necessary. In order to get beyond first-order systems as early as practicable, it may be desirable to omit Chapter 3 and multiport elements at first, returning to them after Chapter 13 has been studied. This covers the essentials, omitting much of the generalization.

Some curricula may be designed for an *advanced one-semester course* in system dynamics where the students will already have a background in mathematics, dynamics, and circuits. Such a course could cover Chapters 1, 4, 6, Sections 7–1 through 7–6, and Chapters 8, 10, 12, 13, 14, 15, and 16. (Depending on the nature of the preceding course on circuits, Chapter 12 could be omitted and Chapter 14 may be treated as partial review.) It is assumed that the elementary aspects would have been covered in earlier courses, and that this course would emphasize generalization.

A course in *linear* system theory which is oriented toward analytical techniques but which retains the interdisciplinary flavor may be developed from this book by introducing Chapter 16, on Laplace transforms, immediately following Chapter 10 and then slightly modifying the presentation in Chapters 12, 13, and 14 accordingly. This can be done, since all the results obtained from the exponential-response approach (response to e^{st} input) are identical to those obtained through use of the Laplace transform as indicated in Section 16–4 (initial conditions are handled as illustrated in Section 16–7). Presumably such a presentation would be made only to students who have a good background in classical solution to differential equations, and hence Chapter 11 would be omitted.

Due to size limitations, some of the topics which were originally intended for inclusion in this book have been omitted. These are state variables, mixed systems, distributed systems approximated by lumped systems, control systems, and a more extensive treatment of stability. Many of these topics are treated in problems. An original chapter on determination of initial conditions was dropped and the material is now covered in the problems at the ends of Chapters 11 and 12. The teacher who wishes to include an introduction to state variable formulation will find that this can be done naturally after Section 8–5. The formulation methods in this book are based on the use of differential equations and lend themselves to a state variable approach for either linear or nonlinear systems.

Efforts toward the development of this book began with a workshop on system dynamics held at M.I.T. during the summer of 1961. The three authors participated in this workshop and in the initial teaching and note-writing for a course given at M.I.T. in the spring of 1962. This was made possible through the support of a Ford Foundation grant to M.I.T. for the development of undergraduate engineering education. Support for the initial workshop, note-writing, subsequent faculty time, and for a visiting professorship for one of the authors (A.T.M.) is gratefully acknowledged. Dr. Gordon S. Brown, Dean of Engineering at M.I.T., and Dr. William W. Seifert, Assistant Dean of Engineering at M.I.T. were instrumental in encouraging the development of these activities and in bringing the authors together on this project. This important stimulation is appreciated.

This book, in many revisions of its preliminary form, has been class-tested at M.I.T. for four years, PMC for three years, and Penn State for two years. During these years a number of people have materially contributed to the text through the teaching of courses, discussions, development of problems and solutions, etc. The contributions of Professors John F. Kennedy, Robert Begg, Peter Griffith, Ulrich Luscher, Henry Paynter, Dinkar S. Rane, Donald L. Wise, and Messrs. Raymond P. Jefferis, III, Robert G. Leonard, Helmut Naumann, James Simes, and David N. Wormley are noted with sincere thanks. An excellent job of typing many revisions was done by Mrs. Nancy Scott Dale, Miss Mary Swan, and Mrs. Evelyn Turner and is greatly appreciated.

The philosophy and interdisciplinary approach characteristic of this book grew out of early associations of the authors with outstanding individuals who influenced them strongly. For two of the authors (J.L.S. and H.H.R.) the M.I.T. Dynamic Analysis and Control Laboratory was a strong influence. Dr. John A. Hrones, former Director of this laboratory, was responsible for the seeds of these authors' development. The interdisciplinary influence of Dr. B. Richard Teare, former Head of the Electrical Engineering Department at Carnegie Institute of Technology, and Dr. Dennistoun W. Ver Planck, former Head of the Mechanical Engineering Department at that institution, as well as close association with their work and thinking, was responsible for the development of the other author (A.T.M.). The early encouragement, leadership, and inspiration of these three outstanding engineers and educators are acknowledged with gratitude.

The order of the authors' names is arbitrary. All authors contributed to all parts of the book through writing, revision, and discussion. In most cases, revised versions of chapters were purposely rewritten by a second or third author for maximum interplay of ideas and improvement of presentation.

February 1967 J.L.S., A.T.M., H.H.R.

CONTENTS

1 INTRODUCTION

An engineer may contribute in many ways to the purposeful activities which constitute engineering. Usually he is involved somehow in creating new ways of doing things. He may have over-all responsibility for the evolution of a complex system, in which case he needs to have an exceedingly good knowledge of creative individuals as well as of inanimate physical things; he may be a key member of a team concerned with certain aspects of a new development; he may work alone on a new idea, device, or subsystem; or he may be a sort of scout out in the field, helping others to become acquainted with a new device or system and making sure that it is operated and maintained properly.

The observations, insight, and even instincts of an engineer are continually at work on the task of gaining a better understanding of the nature of the real physical world. He does this by using scientific discoveries and theories, by employing tools of analysis such as mathematics and computing systems, and through the design and evaluation of experiments.

However, his work consists of much more than attaining this comprehension of things as they are. He also is concerned with how things would become as a result of decisions and actions which might be taken now or in the future. Usually a new objective is to be achieved which requires the creation of something which did not exist before. Sometimes the new objective is simply to attain a certain improvement in performance of a system in a short time; or, at the other extreme, a true optimum may be sought in whatever time may be required to reach it. In any event, the good engineer often thinks in terms of the elusive optimum as a frame of reference by which to gauge his work.

To plan, analyze, and decide on the course of action for an engineering task, it is frequently necessary to model physical systems in such a way that they can be analyzed—usually by mathematical means, sometimes with the aid of computers. This process of modeling corresponds to observing and understanding the real physical world. Once a reasonable model has been formulated, relatively straightforward or even routine mathematical steps often are sufficient to provide solutions to problems that may be posed by the engineer. Since the results embodied in these solutions must be related to the real physical world, interpreting them usually requires as much imagination and intuition as the modeling process does. Such interpretation is a process of modeling-in-reverse.

Nearly all the systems which an engineer may work with are *dynamic*, i.e., things are usually changing so that there is no *status quo* or lasting steady state. To be sure, many of the important facets of a design may be based on steady-state considerations, but a new device or system will fail if it cannot withstand transient peak loads, respond quickly enough to a changing input, or operate without violent oscillations when disturbed. Dynamic analysis can predict such problems before a system is built; a system analysis which does not include the effects of significant dynamic phenomena is very likely to be worthless. The large suspension bridge across the Narrows near Tacoma, Washington, which failed spectacularly in 1940 [1]* from oscillations induced by a moderate (30 mph) wind, is an example of a system designed without sufficient account of a dynamic phenomenon called "aerodynamic flutter." The tragic failure, again due to flutter, of two early Lockheed Electra turboprop airliners [2] resulted from loss of stiffness in the engine mounts after unanticipated weakening of the engine nacelle structures.

The true nature of the Electra problem was especially difficult to determine because one type of dynamic weakening (overstressing of the nacelle when flying through a storm or during a hard landing) was prerequisite for another dynamic mode of failure (flutter of the outer end of the wing). During design and flight test, the estimation of gust-loading and hard-landing effects was not good enough to predict or reveal the structural weakness at the nacelle (which even after weakening was still strong enough to withstand larger-than-normal steady loads), and the exhaustive analysis and experimental study of wing flutter during design and test had not taken into account the possibility that an escessive engine vibration, due to weakened mounts, might occur and interact with the wing-flutter vibration after a sudden gust load at one critical flight speed.

Such cases of compound failure have not occurred often in the past, but the possibilities for them to occur tend to increase as engineering systems become more complex. Thus, it is necessary to recognize all the important dynamic properties of a physical system when it is being modeled. A child learns about many kinds of dynamic phenomena from direct experience; that his hand will not be burned by a hot stove or radiator unless he touches it too long, that he must lean in the direction of the turn before and while going around a corner on a bicycle, that the time required to fill a pail of water is greater if the rate of flow of water into the pail is smaller, and that a little story told to one of his imaginative and talkative friends can be exaggerated when retold and lead to a heated fistfight. All these experiences of early years may be recalled at various times when one attempts to gain a better understanding of the important dynamic properties of physical systems.

Since many of these dynamic properties occur in analogous ways in different kinds of systems, very often one may employ these analogies to help in the study of these systems. The recognition of these analogies at an early stage of a student's engineering education can reinforce the learning process and provide perspective.

As the student gains facility with modeling and analysis, he will become able to use these techniques on engineering projects which require decisions and action. Mathematical and computer studies can often be employed very effectively to help acquire the information needed to make decisions and put a program of work into action. It is very important to note here that a single flaw in the modeling process can negate the results of any subsequent analysis and can result in the making of faulty decisions at later stages of an engineering project. Thus, in addition to the many separate facets of the modeling problem which are taken up as needed in various chapters along the way, a complete chapter of this book is devoted to the subject of modeling. The usefulness of the modeling process can and should provide a great deal of motivation to the student as he proceeds through the other courses in his curriculum which deal with the basic disciplines in engineering and science.

Definition of a dynamic system. A system is defined as a collection of matter, parts, or components which are included inside a specified, often arbitrary boundary. In a dynamic system, by definition, one or more aspects of the system change with time.

An example of a very simple dynamic system is the flow of a given amount of sand through the small restriction in an hourglass to measure time. An

* Numbers in brackets are keyed to the references at the end of each chapter.

example of a very complex dynamic system is a ballistic missile, which must be controlled and guided from the time it is delicately balanced above the thrust of its rocket motors when it is launched to the time when it reaches a target thousands of miles away on the earth or millions of miles away in space.

The boundary of a dynamic system is chosen for convenient conceptual separation of the system from its environment (i.e., from other systems with which it may interact). Interactions with the environment affect the behavior of the system. Those influences which originate outside the system and act on it so that they are not directly affected by what happens in the system are called the *inputs* to the system. The changes in the state of the system (or variations in the parameters used to describe the system) which result from the action of external influences, or inputs, are called the *outputs* of the system. For the hourglass, inversion of the glass would be the *input* to the system, and the elapsed time measured by the amount of sand which has flowed through the restriction would be the *output* of the system. Figure 1–1 illustrates graphically the concept of a system with its inputs and outputs.

Some specific examples of systems responding to inputs may help to illustrate what we mean when we talk about a system with inputs and outputs.

1) When the throttle pedal (input x) of an automobile is depressed, the motor develops more power and the forward speed (output v) increases unless the car starts up a hill at the same time (another possible input β to the system). If the throttle is maintained at its new position (x, percent of power) and the automobile is traveling on a level road ($\beta = 0$), the output (v, speed) eventually levels off at a higher value than its original value. This system with its input and output is illustrated in Fig. 1–2.

2) When the steersman on a seagoing vessel turns the wheel (input) to a new position, the heading of the ship changes (output) because of hydrodynamic side forces acting on the newly positioned rudder, and if the new position (θ, degrees) of the wheel and rudder is maintained, a condition will be achieved in which a new steady rate of change of heading ($\dot{\alpha}$, degrees per minute) will be attained. In this case we designate the heading α as the output, as shown in Fig. 1–3.

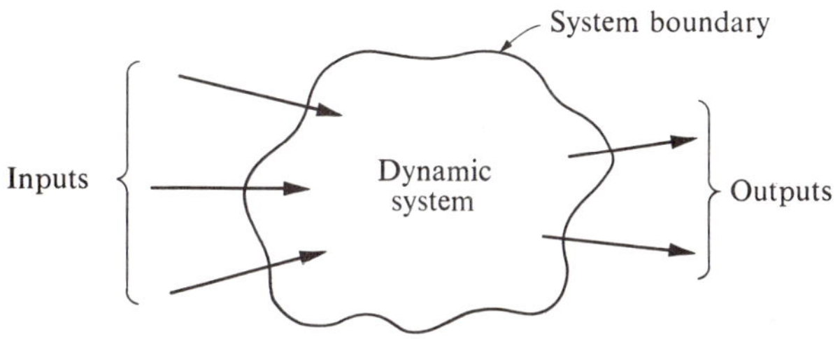

FIG. 1–1. A dynamic system with inputs and outputs.

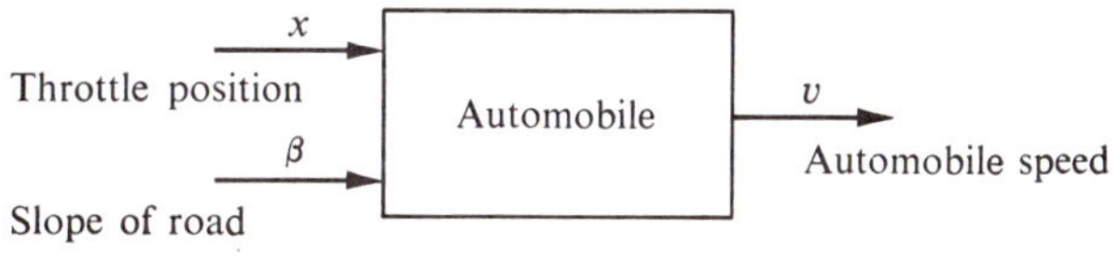

FIG. 1–2. Block diagram of moving automobile.

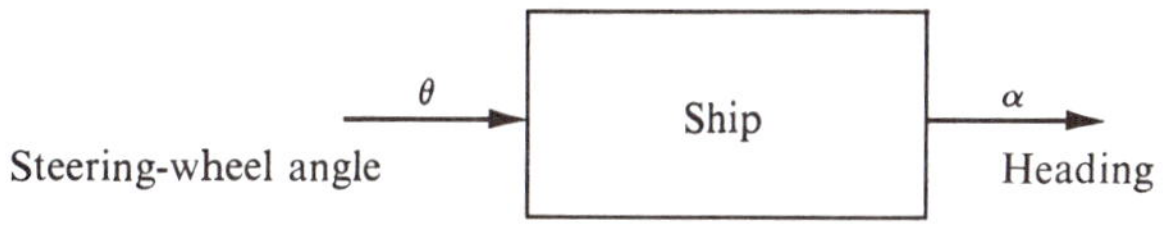

FIG. 1–3. Block diagram of steered ship.

We need to define a system and its inputs and outputs to understand cause-and-effect relationships at work in real physical systems: "If a certain change occurs in a given input, what will happen to the output or outputs?" And after we answer this question, quite naturally we follow it with: "How can the system be altered to improve the cause-and-effect relationship between input and output?" This establishment of causality in physical systems, and the study of causal relationships under conditions of unsteady operation or performance, is the central theme of this book.

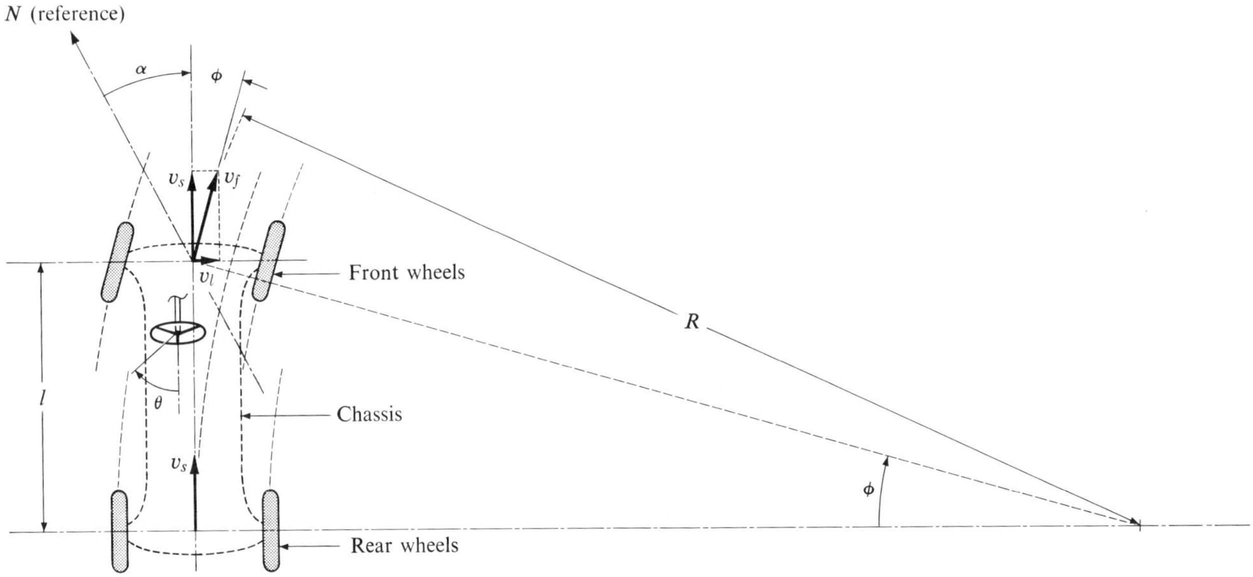

$$\phi = k\theta \qquad v_l = v_s \tan\phi \approx v_s\phi$$

$$R = \frac{l}{\tan\phi} \qquad v_f = \frac{v_s}{\cos\phi} \approx v_s$$

$$\dot{\alpha} = \frac{v_l}{l} \approx \frac{v_s\phi}{l} \qquad \alpha \approx \alpha_i + \int_0^t \frac{v_l}{l}\,dt$$

FIG. 1–4. Illustration of radius of curvature of the path an automobile follows for a constant steering angle ϕ, where k = steering gear reduction, ϕ = front wheel angle (mean), θ = steering wheel angle, and v_s = speed when following a straight path.

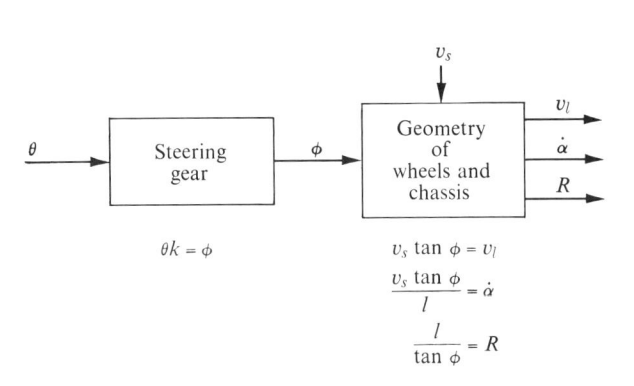

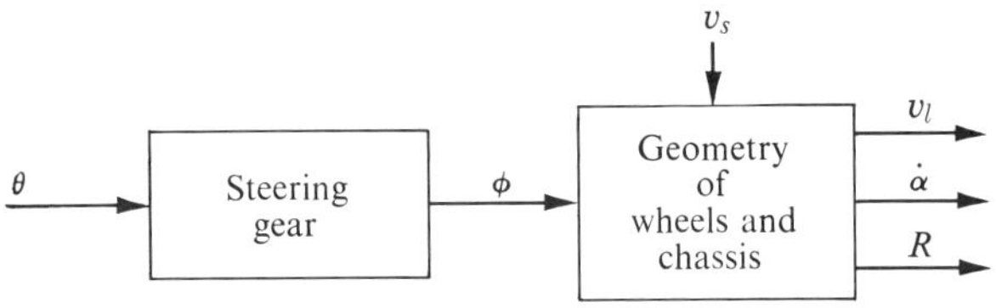

$$\theta k = \phi$$

$$v_s \tan\phi = v_l$$

$$\frac{v_s \tan\phi}{l} = \dot{\alpha}$$

$$\frac{l}{\tan\phi} = R$$

FIG. 1–5. Block-diagram representation of an automotive steering system.

Some interesting examples of more complex dynamic systems

Automotive steering system. An automotive steering system is an example of a dynamic system with which nearly everyone has first-hand experience. There are at least three different aspects of dynamic response which are of key significance in this system.

The first and most common dynamic aspect of this system has to do with how the heading of the vehicle is changed by turning the steering wheel when the vehicle is cruising along the highway at steady speed. Suddenly turning the steering wheel through a specified angle does not suddenly change the direction of motion of the car. Instead it tends to suddenly change the lateral velocity of the front end of the vehicle by an amount very nearly proportional to the change in the steering-wheel angle. This lateral velocity of the

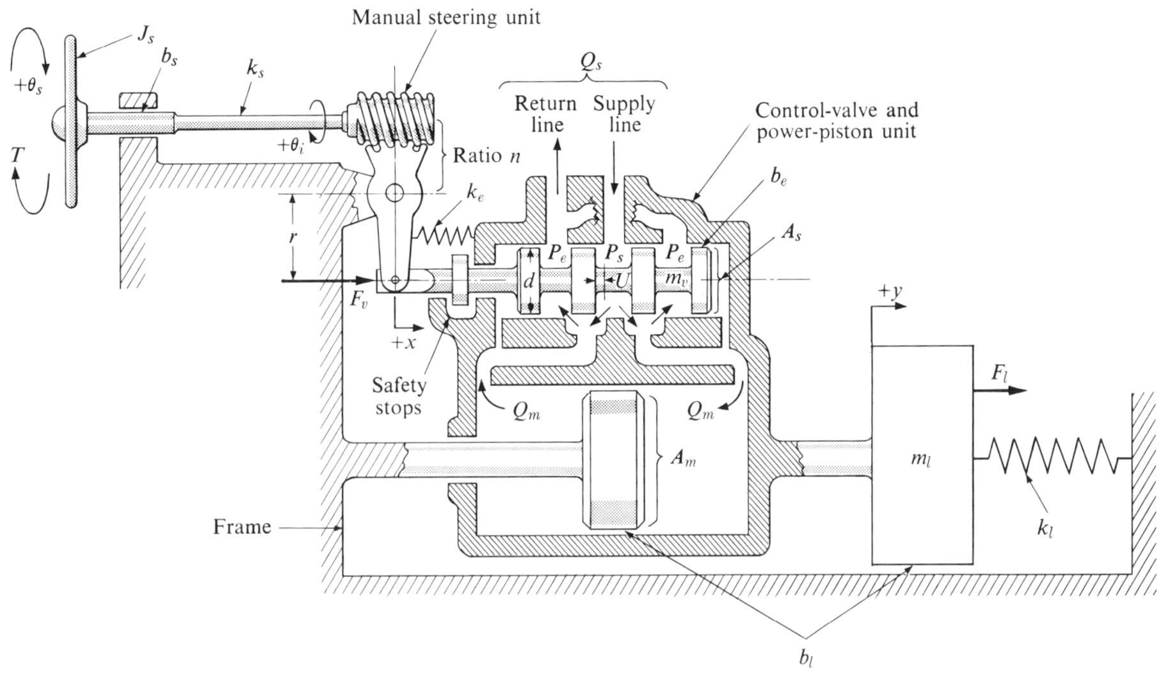

FIG. 1–6. Schematic drawing of an automotive hydraulic power steering system. (From Blackburn, Reethof, and Shearer, *Fluid Power Control*, MIT Press, 1960).

front end of the vehicle corresponds to a rate of change of the heading of the vehicle, which then follows a path of radius R (Fig. 1–4). Because the rate of change of heading is very nearly proportional to steering-wheel angle θ for small front-wheel angle ϕ, the heading α is very nearly proportional to the time integral of θ. This means that after the steering wheel has been turned it takes a finite time before the car reaches a new heading, and that unless the steering wheel is brought back to its null or neutral position, the heading of the car will continue to change. Similarly, if one is guiding a vehicle along the centerline of the highway, the distance of the vehicle from the centerline varies very nearly as the time integral of the steering wheel angle θ (for small ϕ). The essential characteristics of this steering system are shown in Fig. 1–5.

The other two dynamic aspects of an automotive steering system are considerably more complicated and will not be discussed in detail here. One has to do with a form of dynamic instability called "shimmy" which occurs when the front wheels are unbalanced or when the mechanical linkages in the system are exceedingly worn or out of adjustment. By carefully analyzing this phenomenon, one comes to realize the complexity of an automotive steering mechanism.

The other aspect has to do with returning the steering wheel to center after rounding a sharp turn. In this maneuver, the inertia of the steering wheel, acted on by the forces between the car wheels and the road, acquires considerable momentum and can cause a serious "overshoot" and "undershoot" of the intended path if the driver is not careful to keep the wheel under control as it returns to null. This is a

FIG. 1–7. Photograph of a missile-tracking antenna, ca. 1963. (Courtesy of Missile and Surface Radar Division, RCA, Moorestown, N.J.) Antenna diameter 29 ft, antenna weight 2 tons, hydraulic servo power 30 hp, tracking accuracy 0.025 mrad (approximately 1 yard in 20 miles), maximum rate 0.5 rad/sec, maximum acceleration 0.35 rad/sec^2.

case of the steering wheel wanting to respond to another input in addition to that of the driver's hands.

An additional degree of sophistication or complexity in automotive steering systems is the incorporation of a hydraulic power-steering unit into the system. A typical power-steering unit [3] consists of a valve-controlled servomotor with mechanical feedback which provides amplified power to overcome large load forces acting on the front wheels at the road surface, especially during parallel-parking maneuvers. In addition to reducing the effort required by the driver to steer the vehicle, the power steering unit also drastically changes the shimmy and return-to-center characteristics of a steering system. Thus a considerable amount of dynamic analysis has been required in the design and development of power-steering systems. A cross section of a hydraulic power-steering unit is schematically depicted in Fig. 1–6.

Weather-satellite system. A weather-satellite system is an example of a large and relatively complex dynamic system of great importance to everyone. In addition to the immediate benefits of spared losses in lives and property which are derived from the early detection of hurricanes, a weather satellite also provides basic meteorological data of great research value.

Although many articles in the popular press describe only the satellite and some of its instruments and devices, a weather-satellite *system* really is much more comprehensive. In addition to all the events involved in launching the satellite and putting it in a suitable orbit around the earth, several microwave tracking stations must be provided at different widely spaced locations around the world to provide sustained communications with the satellite throughout its useful life. Moreover, these tracking stations must be in constant communication with each other, and at least one large digital computer is required to coordinate the tracking stations and process the data received from the satellite.

Since it has only a small amount of energy available to operate its mechanical and electronic instruments, devices, and control systems, the satellite is not capable of sending strong signals to earth. To intercept the weak signals from the satellite, a large carefully focused antenna must be aimed directly at the satellite as it passes overhead and broadcasts data about what it has "seen" or detected with its scanning system. Such a *tracking antenna* (Fig. 1–7) is a rather complex dynamic system in itself. The antenna, which may be as large as 30 ft in diameter, is positioned by closed-loop azimuth and elevation drive systems comprising hydraulic servomotors, precision gear trains, electronic amplifiers, and electromechani-

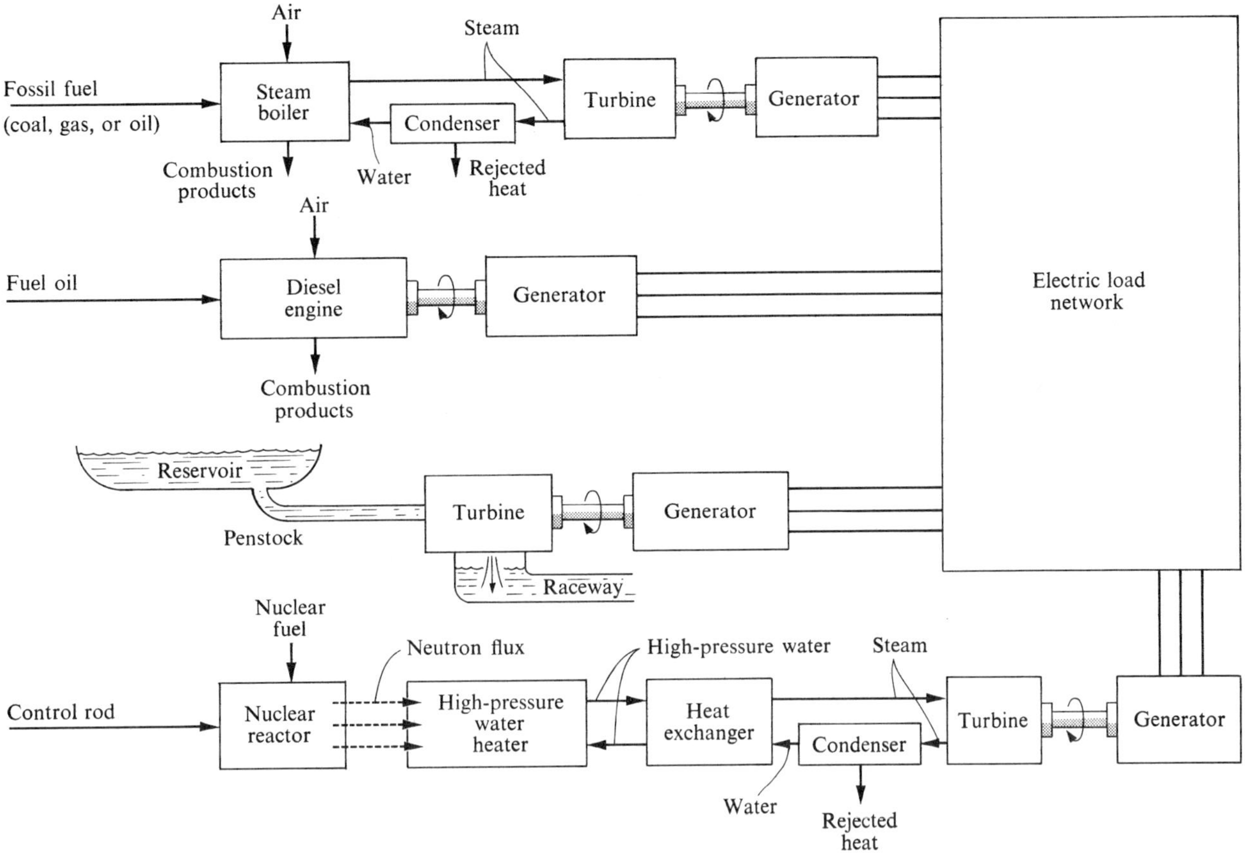

FIG. 1–8. Schematic diagram of electric-power system.

cal transducers. These closed-loop drive systems make it possible for an antenna weighing several tons to respond to an input command signal whose frequency is as high as 20 cycles per second (cps) and whose power is as little as 10^{-6} watt! The base on which the antenna is mounted and pivoted is a maze of signal wires, power cables, motors, pumps, hydraulic lines, fittings, and valves. Each component or element of the system is carefully selected or designed to execute its specific function under conditions of continually varying current, voltage, stress, strain, pressure, flow rate, or power level. Great attention is given to dynamic analysis during the design phases, and extensive dynamic system tests are made to determine whether the system is performing according to specifications. It has happened that a new system of this kind was subjected to so much dynamic testing and retesting while it was being checked out and debugged that it was literally worn out before it was ready to be shipped to the tracking-station site.

Aboard the satellite itself there are several dynamic systems or subsystems. One of these controls the satellite's attitude so that the earth-scanning device is pointed in the proper direction. Another system positions a panel or array of solar cells, which is used to generate electrical power for the electronic and mechanical devices on the satellite, to intercept the maximum incident solar energy. When a tape recorder is used to store scanned data until they can be transmitted to a tracking station, the recorder must be started, run at proper speed, and stopped by command signals beamed at the satellite from earth. The two-way microwave communication system, which amplifies and processes high-frequency electromagnetic waves on the satellite, is a dynamic system just as any radio set is.

Electric-power system. A typical electric-power system is an excellent example of a large and complex dynamic system encountered in everyday life. Such a system is dynamic in a number of respects, foremost perhaps because the electric power consists of a current and a voltage alternating sinusoidally. Such an alternating current (ac) system may distribute power throughout hundreds of square miles and contain hundreds of generators with a total power capacity of hundreds of megawatts. A single system may generate power simultaneously from several different sources, such as fossil fuels, hydraulic reservoirs, and nuclear fuels (Fig. 1–8). Each type of generating system has its own set of unique dynamic characteristics.

In a fossil-fuel energy-conversion system, the energy (heat) derived from the fuel is used to generate steam in a boiler. The steam drives a turbo-alternator (a steam turbine connected to an ac-generator) which delivers electric power to the load network. Appreciable time is required to start up a fossil-fuel system before it can deliver power at line-load levels and to change its level of power generation when it is on the line. Time lags occur in the fuel burner, the boiler, and the turbo-alternator, the most serious being that in the boiler. Thermal systems, therefore, cannot pick up sudden large increases in load. Diesel-powered generators can respond much more rapidly to sudden changes in load and are often used in parallel with turbo-alternators to provide some fast-response capability in large systems.

In regions where water power is available, hydroelectric generators are used as much of the year as possible because of their relatively low cost of operation. Many such systems must also use fossil-fuel generators to supplement water power during dry periods or peak loads. The continued development of nuclear power systems is reducing nuclear generating costs to reasonable levels, and some large electric-power systems now include one or more nuclear generators.

In addition to the dynamic-response problems caused by suddenly changing loads, other dynamics problems are caused by the interconnection of different types of generators. A generator which is stable when delivering power to a given sector can be unstable when connected to a large system. Sometimes a failure in one part of a large system overloads other parts to the extent that main circuit breakers operate and remove the whole system from the line. When the electric load is suddenly removed from a large thermal generator running at full-power level, automatic boiler-pressure and turbine-speed controls must act quickly to prevent a boiler explosion or turbine run-away. A typical thermal power generator has many automatic subsystems which control temperature, pressure, flow rate, speed, voltage, current, etc., and each of these subsystems must have suitable dynamic-response and stability characteristics.

The role of system dynamics in engineering and science. Dynamic analysis is important in many fields of engineering and in scientific, economic, and business activity. Some of these areas are:

Aerospace systems—orbit and trajectory determination, guidance, remote manipulation, attitude control, celestial navigation.

Automation—manufacturing, accounting, production processes.

Biomedical research—instrumentation of living organisms, dynamic behavior of living systems, clinical investigations.

Business systems—dynamics of a regional economy, behavior of banking systems, inventory and production control.

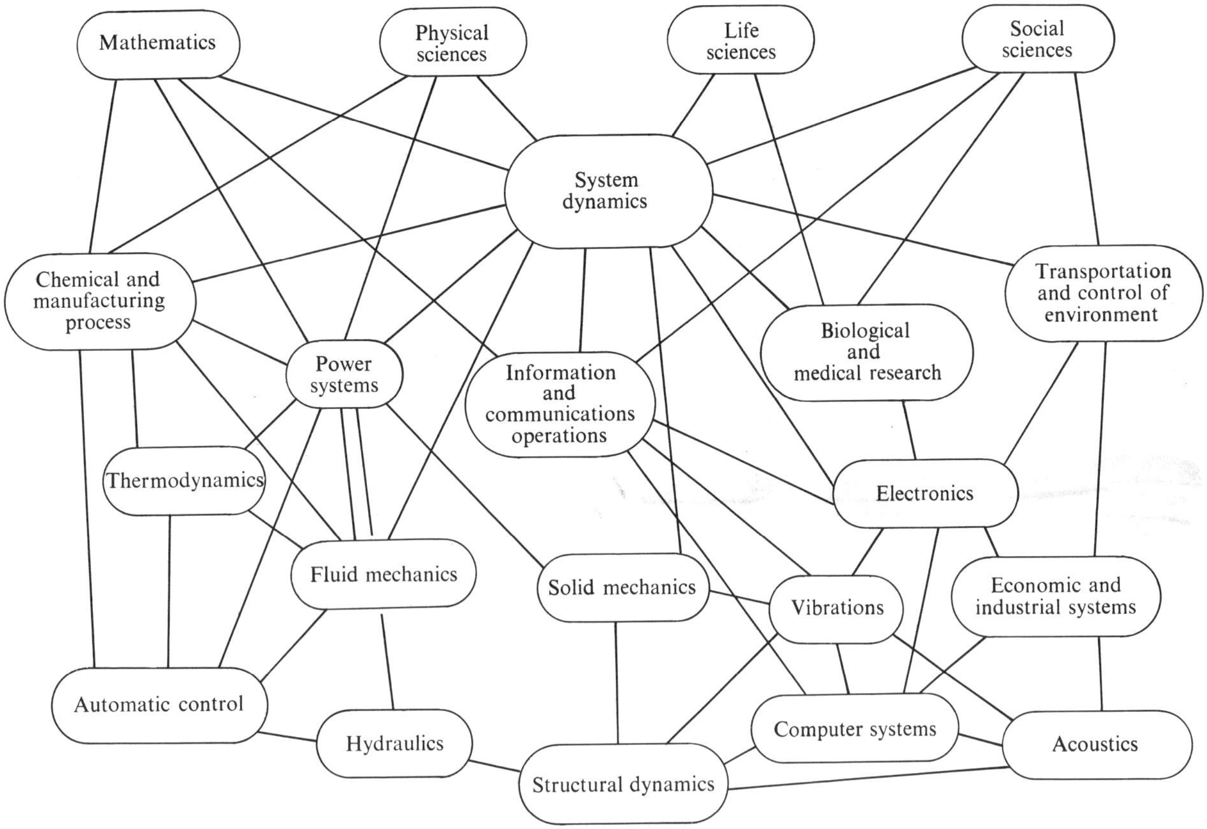

FIG. 1–9. Illustration of how system dynamics interacts with many other fields.

Communications—remote control, information transmission.

Computers and data-processing systems—solution of complex analytical problems, storage and retrieval of information, computer control of complex systems.

Defense systems—missile control, radar tracking systems, aerodynamic stability, nuclear-blast detection, weapons systems.

Power and propulsion systems—steam- and hydro-powered electric systems, control of nuclear reactors, magnetohydrodynamics, ion propulsion.

Psychology—human-operator dynamics, group dynamics, learning processes.

Structures—bridge design, dynamic behavior of buildings, flight-vehicle structures.

Transportation systems—aircraft and air-traffic control, braking systems, power steering, speed control, ship stability, water-supply systems, material distribution.

Figure 1–9 illustrates many of these fields of endeavor and some of the interactions and interdependences between them.

The references at the end of this chapter list some of the useful books about the dynamic analysis of physical systems which are good starting points for further study.

REFERENCES

1. J. P. DEN HARTOG, *Mechanical Vibrations*. New York: McGraw-Hill Book Company, Inc., pp. 398–399, 1947.
2. R. J. SERLING, *The Electra Story*. Garden City, N.Y.: Doubleday and Co., Inc., 1963.
3. J. F. BLACKBURN, G. REETHOF, and J. L. SHEARER, *Fluid Power Control*. Cambridge, Mass.: MIT Press, pp. 631–646, 1960.
4. W. A. LYNCH and J. G. TRUXAL, *Signals and Systems in Electrical Engineering, Parts 1 and 2*. New York: McGraw-Hill Book Company, Inc., 1962.
5. J. D. TRIMMER, *Response of Physical Systems*. New York: John Wiley and Sons, Inc., 1950.
6. R. L. SUTHERLAND, *Engineering Systems Analysis*. Reading, Mass.: Addison-Wesley Publishing Co., Inc., 1958.
7. H. F. OLSON, *Dynamical Analogies*. New York: D. Van Nostrand Co., Inc., 1943.
8. A. G. J. MACFARLANE, *Engineering Systems Analysis*. Reading, Mass.: Addison-Wesley Publishing Co., Inc., 1965 (originally published by G. G. Harrap and Co., Ltd., London).
9. H. E. KOENIG and W. A. BLACKWELL, *Electromechanical System Theory*. New York: McGraw-Hill Book Company, Inc., 1961.
10. S. SEELEY, *Dynamic Systems Analysis*. New York: Reinhold Publishing Corp., 1964 (also Chapman and Hall, Ltd., London).
11. W. W. HARMAN and D. W. LYTLE, *Electrical and Mechanical Networks*. New York: McGraw-Hill Book Company, Inc., 1962.
12. D. C. THORN, *An Introduction to Generalized Circuits*, Belmont, Calif.: Wadsworth Publishing Company, Inc., 1965.
13. D. K. CHENG, *Analysis of Linear Systems*. Reading, Mass.: Addison-Wesley Publishing Co., Inc., 1959.
14. H. M. PAYNTER, *Analysis and Design of Engineering Systems*. Cambridge, Mass.: MIT Press, 1961.
15. E. A. GUILLEMIN, *Theory of Linear Physical Systems*. New York: John Wiley and Sons, Inc., 1963.
16. W. J. KARPLUS and W. W. SAROKA, *Analog Methods, Computation and Simulation*. New York: McGraw-Hill Book Company, Inc., 1959.
17. M. F. GARDNER and J. L. BARNES, *Transients in Linear Systems*. New York: John Wiley and Sons, Inc., 1942.
18. B. J. LEY, S. G. LUTZ, and C. F. REHBERG, *Linear Circuit Analysis*. New York: McGraw-Hill Book Company, Inc., 1959.
19. P. E. PFEIFFER, *Linear System Analysis*. New York: McGraw-Hill Book Company, Inc., 1961.
20. R. G. BROWN and J. W. NILSSON, *Introduction to Linear Systems Analysis*. New York: John Wiley and Sons, Inc., 1962.

PROBLEMS

1–1. Consider a bicycle as a system and identify its inputs, outputs, other system variables, and its most important characteristics when it is moving under control of a rider. Note that the rider is not part of the system. Do not attempt to set up detailed equations for the system, but feel free to employ diagrams as well as words in developing your answer to this problem.

1–2. Suppose that you are aboard a ship in rough weather, trying to observe some distant target with a pair of binoculars. Take the binoculars as a system and identify their inputs, outputs, and essential characteristics as they are being trained on the target. Note that you are not part of the system to be considered.

1–3. Consider a bathroom shower having both hot and cold water valves (i.e., no automatic mixing valve) as a system and identify its inputs, outputs, and essential system characteristics.

1–4. Identify inputs, outputs, and essential system characteristics of a house which has operating windows and doors and which is being heated by a furnace that is turned on and off with a simple switch (i.e., no automatic thermostat).

1–5. A new factory starts using 200,000 gal/day of water in a community where the average daily consumption was 1,000,000 gal/day. Take the local community as a system and identify its inputs, outputs, other system variables, and essential characteristics which would enable you to study how the system might respond to the start-up of the new factory, (i.e., to the sudden increase in water consumption).

1–6. A manufacturer has just about decided to double his rate of production of automatic toasters (for bread). How would you proceed to identify the system which would be affected by his decision, including such things as effects on other manufacturers, number of people using toasters, and distributors?

1–7. Describe an elevator in a large building in terms of its system variables (including inputs and outputs) and its most important characteristics.

1–8. Find the time derivative and the time integral of each of the time functions shown in Fig. 1–10 by graphical means, sketching the derivative and integral functions immediately below the given time function as shown in the example in Fig. 1–10.

1–9. a) If the rear-axle torque for each rear wheel of an automobile varies with time as shown in Fig. 1–11, determine the forward speed of the vehicle as a function of time *assuming negligible bearing friction and negligible air resistance.* The weight of the automobile is 3000 lb and the radius of the rear wheels is 15 in.

Example

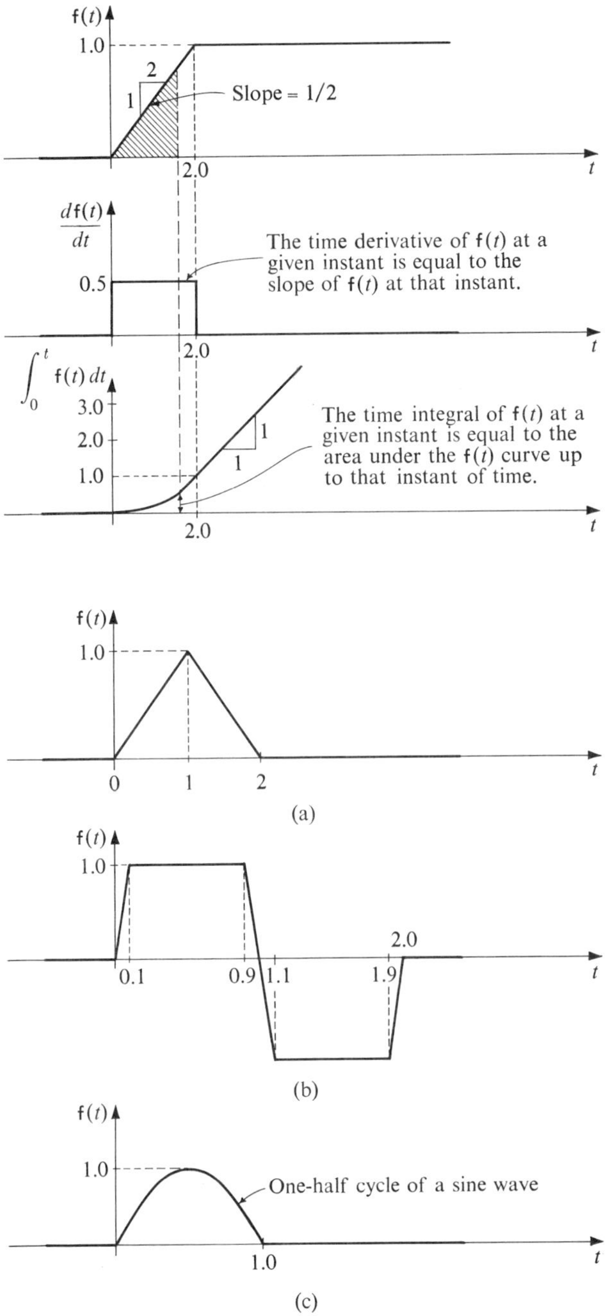

FIGURE 1–10

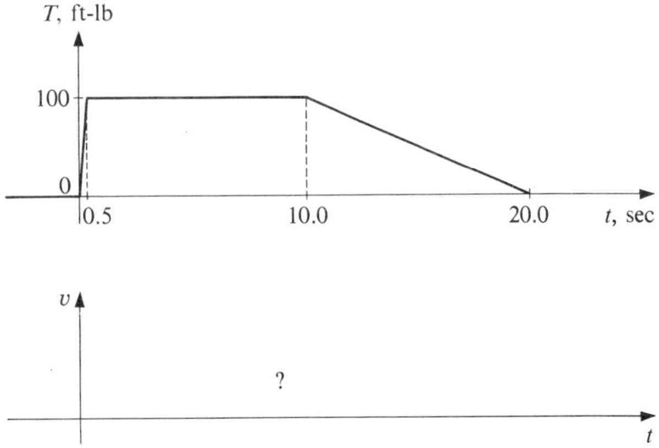

FIGURE 1–11

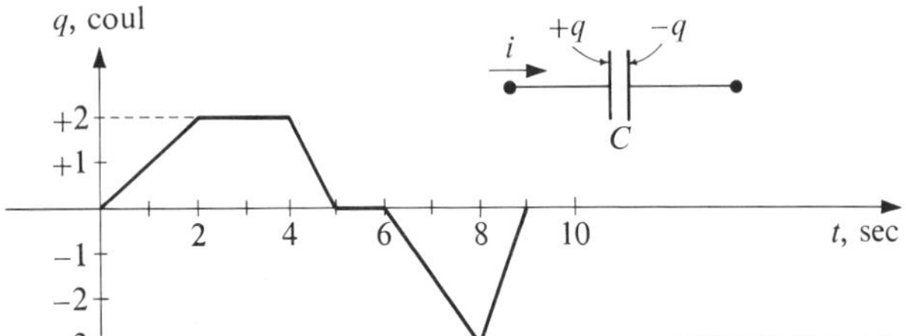

FIGURE 1–12

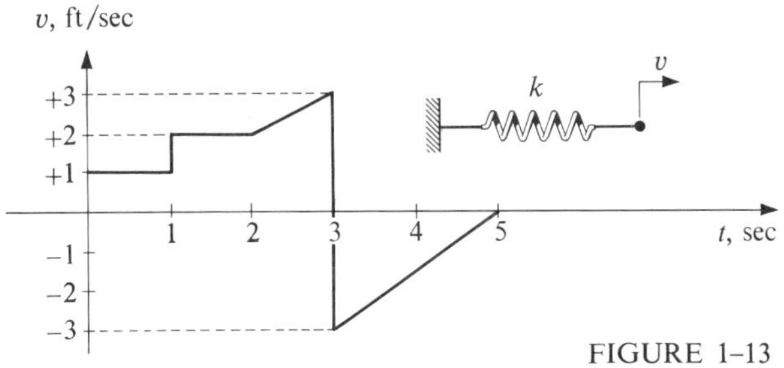

FIGURE 1–13

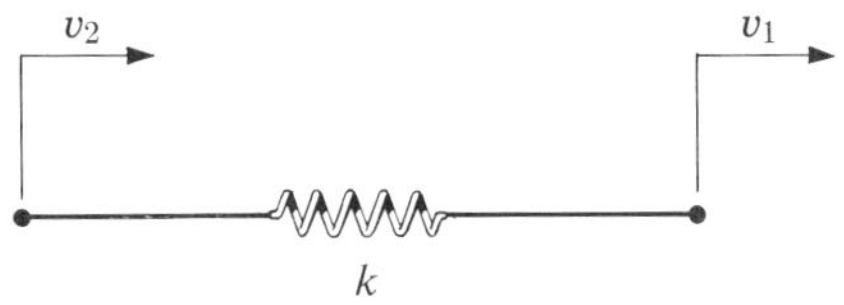

FIGURE 1–14

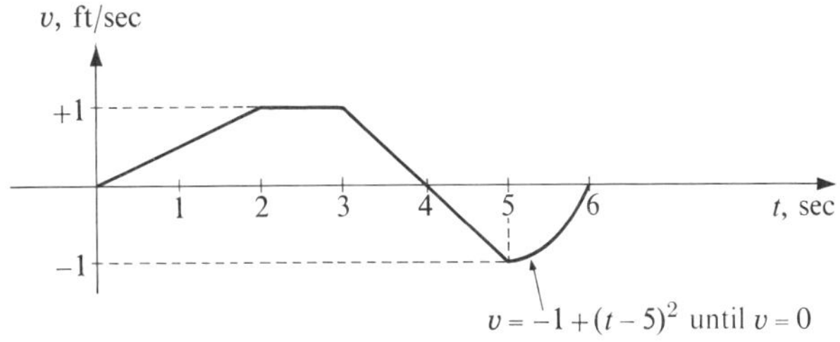

FIGURE 1–15

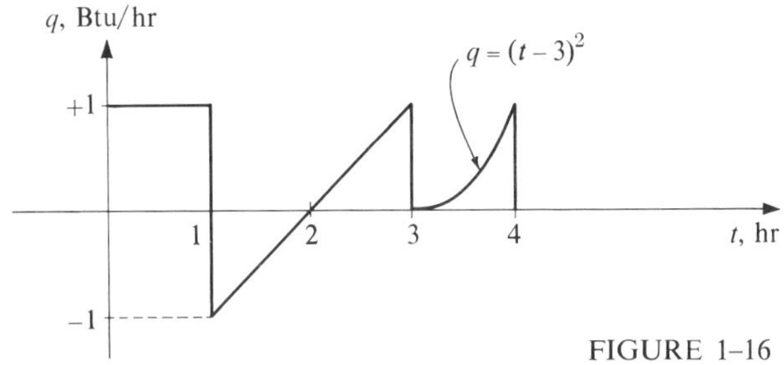

FIGURE 1–16

b) Find the distance traveled at the end of 20.0 sec.

c) Based on the answer you have obtained in part (a), do you think that allowing for bearing friction and air resistance would have resulted in an appreciably different answer? [*Hint:* Bearing friction and air resistance are approximately proportional to the square of automobile speed, and rear-axle torque required to drive the auto at a steady speed of 60 mph is 200 ft-lb.]

1–10. The graph in Fig. 1–12 is a plot of the amount of charge on the plates of a capacitor as a function of time (the number of electrons leaving the bottom plate is the same as the number entering the top plate). Plot a curve of the current i as a function of time. The current is defined as the time rate at which charge passes a cross section of the wire.

1–11. The graph in Fig. 1–13 is a plot of the velocity of one end of a spring, the other end of which is held fixed. Plot a curve of the displacement of the free end of the spring.

1–12. One end of a spring (Fig. 1–14) has the velocity (i.e., v_1) shown in Fig. 1–13, and the other end has the velocity v_2 (in ft/sec) given by

$$v_2 = 2, \quad 0 < t < 1,$$
$$v_2 = 2t, \quad 1 < t < 3,$$
$$v_2 = -6, \quad 3 < t < 5.$$

Assuming that the original length of the spring is 6 ft, plot a curve of the spring length as a function of time.

1–13. A mass has the velocity shown in Fig. 1–15. Plot the acceleration and displacement of the mass as a function of time.

1–14. The rate of heat flow into a metallic body of mass 2 slugs is shown in Fig. 1–16. The specific heat of the body (thermal energy per unit change in temperature per unit mass) is 4 Btu/slug/°R. Plot a graph of the temperature of the body as a function of time.

1–15. Given $y = f(x)$ as shown in Fig. 1–17.

a) Sketch

$$dy/dx \quad \text{and} \quad \int_0^x y\,dx \quad \text{vs. } x \text{ for } a \text{ as shown.}$$

b) Repeat (a) for a half that value.

c) Repeat (b) for the limiting case where $a \to 0$.

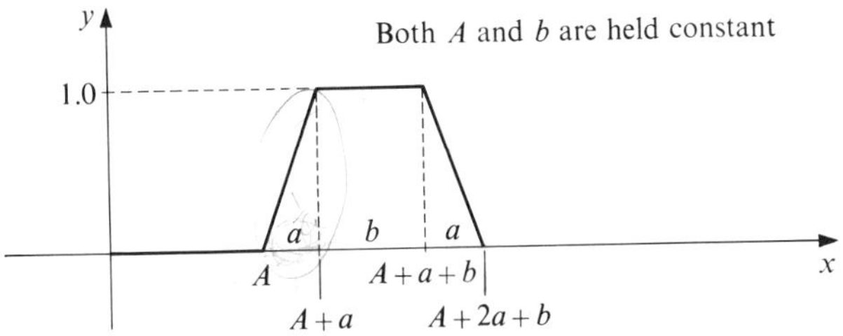

FIGURE 1–17

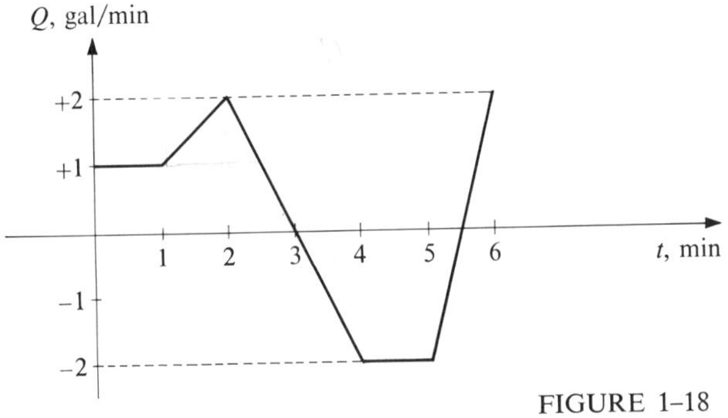

FIGURE 1–18

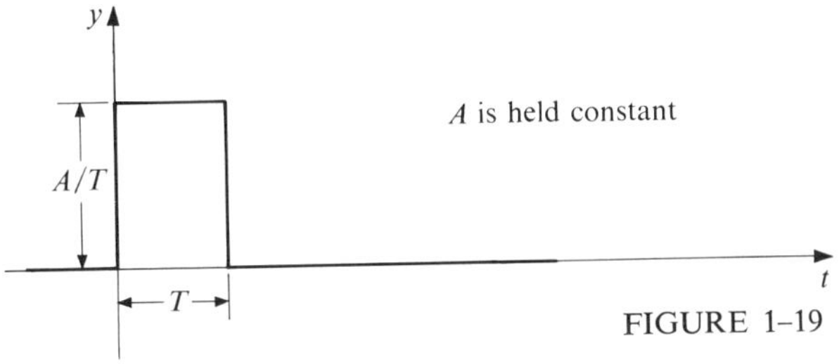

FIGURE 1–19

1–16. The equation

$$P = 2\frac{dQ}{dt} + 3Q^2 + \int_0^t Q\,dt + 10$$

relates P and Q in a particular fluid system, where P is the pressure and Q is the flow rate. If Q varies as shown in the graph of Fig. 1–18, make a sketch of P vs. time indicating significant numerical values. Part credit will be given if component curves are clearly drawn, even if the complete answer is incorrect.

1–17. a) Sketch $\int_0^t y\,dt$ vs. t for the time function y which is shown in Fig. 1–19.

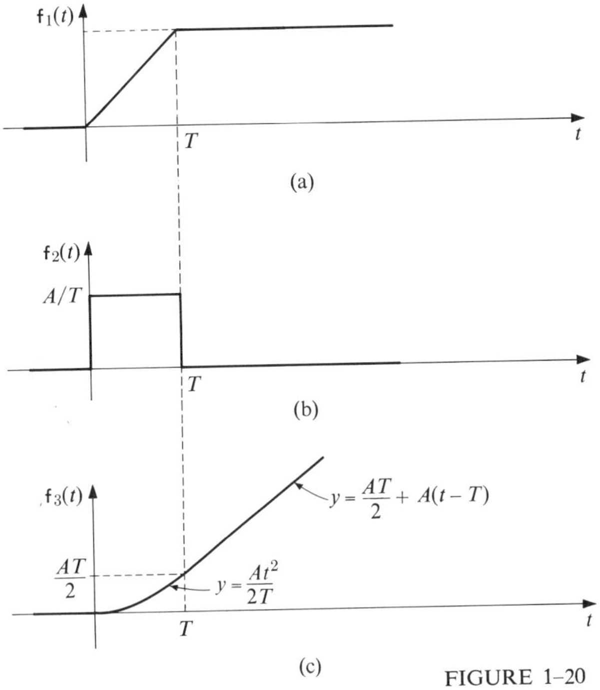

FIGURE 1–20

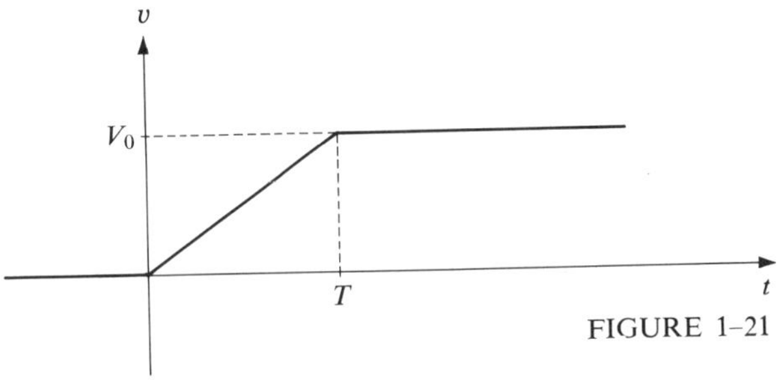

FIGURE 1–21

b) Repeat (a) for T only half as large.

c) Repeat (a) for the limiting case when $T \rightarrow 0$.

The function obtained in part (c) when $T \rightarrow 0$ is called an *impulse*. It is a function whose value goes to infinity for zero time and whose time integral is finite.

1–18. Consider the group of functions shown in Fig. 1–20.

a) How are these functions related to each other?

b) Sketch their curves as $T \rightarrow 0$. For this case they have the following special names: step function, impulse function, and ramp function.

1–19. Two variables f and v are related by the following equation:

$$f = 3v + \frac{2dv}{dt}.$$

a) If v varies as shown in Fig. 1–21, sketch a graph of f vs. t.

b) Sketch f vs. t for the limiting case where $T \to 0$.

c) If f were given rather than v, and f changed instantly from 0 to F_0, at $t = 0$, sketch a graph of v vs. t near $t = 0$ (i.e., for small values of $t > 0$) considering that at $t < 0$, $v = 0$. At the point at which f changes instantly your graph of v will either change instantly or it will not. Justify your answer with a complete argument for full credit here.

2 DYNAMIC SYSTEM ELEMENTS MECHANICAL AND ELECTRICAL

2-1. INTRODUCTION

The behavior of real physical systems which engineers must design, analyze, and understand is controlled by the flow, storage, and interchange of various forms of energy. In almost all cases, real systems are extremely complex and may involve several interacting energy phenomena, such as electromagnetic fields, heat and fluid flow, or nuclear reaction.

The analysis of a dynamic system always involves the formulation of a conceptual *model* made up of basic building-blocks that are idealizations of the essential physical phenomena occurring in real systems. An adequate conceptual model of a particular physical device or system will behave approximately like the real system. The best system model is the simplest one which yields the information necessary for engineering action or decision.

It is important that the student appreciate the advantages of simplified ideal models without oversimplifying his picture of the real physical world. An approximate answer to an engineering problem is often the best answer if the degree of approximation is known and a more precise answer is not really needed. The cost of precision is often exceedingly high, since the complexity of the conceptual models needed to represent systems usually increases very rapidly with increased requirements for accuracy. Hence efforts to attain high precision should be made only when the cost in time, money, and effort can be justified.

Of the many fundamental building blocks which the engineer must use in formulating conceptual models of particular physical systems, we shall concentrate in this chapter on simple mechanical and electrical system elements which store, dissipate, and transform energy. In subsequent chapters we will discuss fluid and thermal system elements, sources which supply energy to systems, and other idealized elements. We shall see that the many similarities between mechanical, electrical, fluid, and thermal systems will permit a unified study of their behavior.

2-2. MECHANICAL SYSTEM ELEMENTS IN TRANSLATION

2-2.1 Motion and Force

Motions of various elements in a mechanical system are nearly always associated with coexisting forces. In many instances we prefer to think of motion as resulting from the application of a force. In other instances we may prefer to think of a certain force as resulting from a given motion. In either case, interactions involving work, energy, and power occur between mechanical elements and their surroundings.

Motion is usually defined as the displacement, velocity, or acceleration of one point with respect to another. *Absolute motion* of a given point may be defined as the motion of that point relative to a fixed point in space. Because of the difficulty of providing

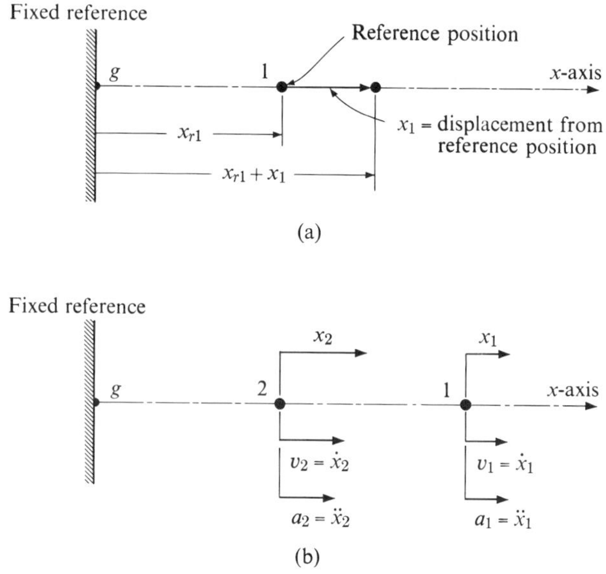

FIG. 2–1. Description of translational motion. (a) Displacement of a point relative to an initial reference position. (b) Displacements, velocities, and accelerations of two points, 1 and 2.

a real fixed point in space from which to make measurements,* our concept of absolute motion is strictly hypothetical, but nevertheless very useful.

Let us consider point 1 (Fig. 2–1a), which is assumed constrained to translation along the x-axis and is initially located a fixed distance x_{r1} (its *reference position*) from a fixed reference point. We shall use the letter g (for ground) to designate a fixed reference point. The *displacement* x_1 of point 1 is defined as the distance that point 1 moves from its reference position. This displacement is taken to be positive if it occurs in the positive x-direction. We will indicate the displacement of a point by an arrow, drawn beside the point, whose length and direction show the magnitude and direction of the displacement of the point. The point itself will usually be shown at its reference location. This convention is illustrated in Fig. 2–1(b) for two points, 1 and 2. The absolute velocity v_1 of point 1 is by definition the time derivative of x_1 (since x_{r1} is constant), and the absolute acceleration a_1 of point 1 is the time derivative of v_1:

$$v_1 = dx_1/dt, \tag{2–1}$$

$$a_1 = dv_1/dt = d^2x_1/dt^2. \tag{2–2}$$

The velocity v_2 and acceleration a_2 of point 2 are similarly related to x_2. Thus, if the displacement x of a point is completely described as a function of time, its velocity and acceleration are also determined (by differentiation). Velocities and accelerations may be shown in the same manner as displacements (Fig. 2–1b).

We shall often be concerned with the *relative* motions of two points, such as in Fig. 2–1(b). If these points are initially fixed at x_{r1} and x_{r2} and then displaced by amounts x_1 and x_2, respectively, from these reference positions, their relative displacement x_{21} is

$$x_{21} = x_2 - x_1. \tag{2–3}$$

Their relative velocities v_{21} and accelerations a_{21}, if x_2 and x_1 are changing, are

$$v_{21} = v_2 - v_1 = \dot{x}_{21}, \tag{2–4}$$

$$a_{21} = a_2 - a_1 = \dot{v}_{21}, \tag{2–5}$$

where the dots indicate differentiation with respect to time. We shall find it convenient to work primarily with velocity as the variable which describes the absolute and relative motions of points in space. In the metric (mks) system of units, velocity is measured in meters per second (m/sec), whereas in the British engineering system it is measured in inches per second (in/sec).*

For specifying the velocity (or acceleration) of a point, the location of the fixed reference point or ground is unimportant and is often not shown explic-

* All physical objects with which we are familiar are known to be in motion. The earth rotates on its axis and travels in an orbit around the sun; the sun (together with its system of planets) is believed to be traveling in the galaxy called the *Milky Way*, etc.

* The correct use of units is important in dealing with numerical calculations for dynamic systems. Appendix A gives a reference table of common units and conversion factors.

itly. It is customary to depict the absolute velocities of two points 1 and 2 as shown in Fig. 2–2.

In general, the motion of a given point will take place in three dimensions and will be described by displacement, velocity, and acceleration vectors. The preceding discussion and Eqs. (2–1) through (2–5) also apply to this general case, provided that the scalar displacements, velocities, and accelerations are replaced by general vector quantities. In this book, however, we shall deal almost entirely with one-dimensional motions which occur along prescribed paths (Fig. 2–1).

The force acting at a given point in a mechanical system is a *vector*, which has both magnitude and direction. Often this force determines a state of stress in a shaft or connecting link on which the point is located. Consider, for example, the two elements A and B connected at point 1 (Fig. 2–3a). To show the forces acting in the links and at point 1, the link must be broken (Fig. 2–3b) to establish so-called "free-body diagrams" of the two elements and of the connecting point 1. The force F acting on element B is assumed to act in the positive x-direction. According to *Newton's third law* of action and reaction, an opposite force equal to F must be exerted by element B on point 1. Also, according to *Newton's first law*, the vector sum of the forces on point 1 must be zero, and therefore element A must exert a force F to the right on point 1. Similarly, the reaction to this force on 1 must act to the left on element A (Fig. 2–3b).

In the mechanical element shown in Fig. 2–4, the forces acting on the element at points 1 and 2 are equal and opposite if the element is in *equilibrium* (i.e., if it has no acceleration). In this case, the force F may be considered to flow or be transmitted unchanged through the element.

Force will usually be expressed numerically either in newtons (n) in the metric system, or in pounds (lb) in the British engineering system. Note that one newton is equal to one kilogram·meter per second squared (1 n = 1 kg—m/sec^2).

As in the case of velocity, force is, in general, a space vector and often will not coincide in direction with the velocity vector. Thus in Fig. 2–3, force components may exist at point 1 which are perpendicular to the x-direction. In this book, however, we shall deal primarily with cases where the force components perpendicular to the direction of motion are either zero or unimportant.

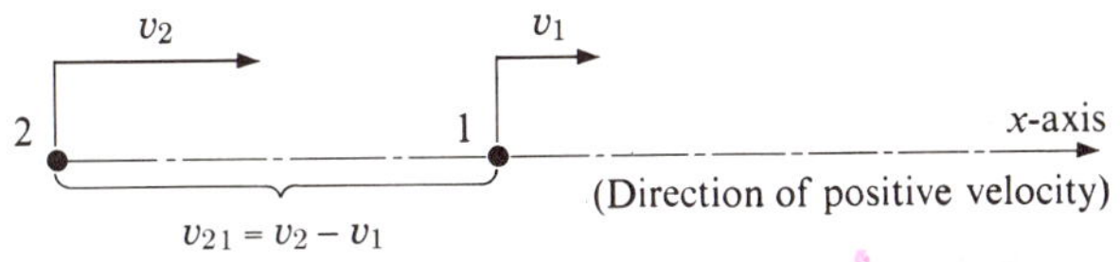

FIG. 2–2. Simplified description of translational velocity.

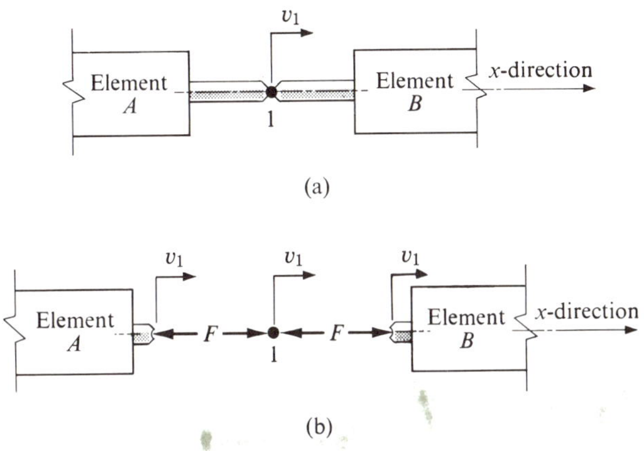

FIG. 2–3. Forces acting at a point of connection of two mechanical elements. (a) Two elements connected at point 1. (b) Free-body diagram showing force transmitted through connecting point.

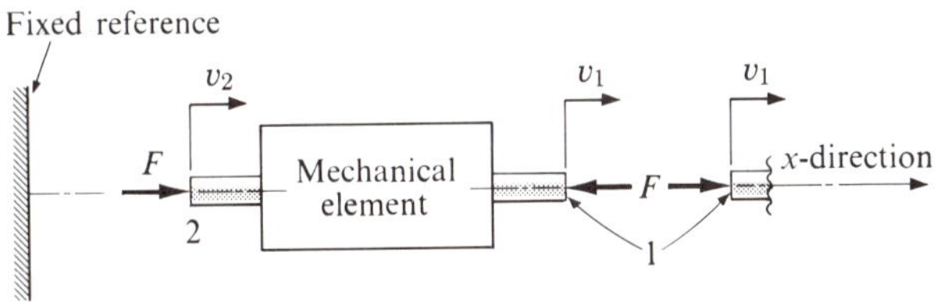

FIG. 2–4. Mechanical element with force flowing or transmitted through it.

2–2.2 Power, Work, and Energy

Power $\mathcal{P}$ is defined as the rate of flow of work or energy. The mechanical power $\mathcal{P}$ delivered to a moving point by a force which acts on that point in the direction of positive velocity is the product of the velocity and the component of force which lies in the direction of the velocity. If the force and velocity are collinear,

$$\mathcal{P} = F \cdot v, \quad \text{watts or lb·in/sec.} \tag{2-6}$$

If the force and velocity are vectors whose directions differ, then Eq. (2-6) must be interpreted as the scalar or dot product of the force and velocity vectors.

In Fig. 2-3, the power delivered to element B by element A is Fv_1, whereas in Fig. 2-4 the *net* power delivered to the mechanical element is $Fv_2 - Fv_1$, where F is the force acting on the element.

The *work* $\mathcal{W}_{ab}$ done by one mechanical element on another through a point of connection is the time integral of the power:

$$\mathcal{W}_{ab} = \int_{t_a}^{t_b} F \cdot v \, dt, \quad \text{n·m or lb·in,} \tag{2-7}$$

W = F·X

where t_a and t_b are the beginning and end of the time interval over which we desire to compute the work done.

The *mechanical energy* $\mathcal{E}$ supplied to an element is the sum of the work done on the element by all forces which act on the element.

$$\mathcal{E} = \sum_{k=1}^{n} (\mathcal{W}_{ab})_k, \quad \text{n·m or lb·in.} \tag{2-8}$$

The law of conservation of energy, or the *first law of thermodynamics*, states that the energy (of all forms) which is delivered to a system must either be stored in the system or transferred out of the system.

2-2.3 Pure Translational Mechanical Elements

In the analysis of mechanical systems,* it is convenient to make use of three idealized mechanical elements: the *mass*, the *spring*, and the *damper* (mechanical resistance, or "dashpot"). These elements represent three essential phenomena which occur in various ways in mechanical systems. Sometimes it is easy to visualize one of these phenomena predominating over the others at a given part of a mechanical system. In other instances, two or all three of these phenomena occur together in such a way that it is difficult to sort them out. This sorting out is part of the *modeling* process. The ability of the engineer to do this sorting and come up with a sufficiently good approximate model of a real system by using the right combination of these basic elements is a keystone of a large segment of engineering analysis.

* Here the term "mechanical" is used in a restrictive sense, i.e., it does not include fluid and thermal systems, which are discussed in Chapter 3.

Pure translational mass. The component parts of any mechanical system are composed of materials that have mass. Therefore, according to Newton's laws of motion, these parts will individually and collectively resist changes of velocity.

We shall begin this discussion by defining the quantity *momentum*, p, according to the following equations:

$$F = dp/dt = \dot{p} \tag{2-9}$$

or

$$p = \int_0^t F\,dt + p_0, \tag{2-10}$$

where p_0 is the momentum at $t = 0$.

In general, force and momentum are vector quantities which are related to each other in the same way that velocity and displacement are related.*

Let us consider a mass m which moves only in the x-direction (Fig. 2-5) with *absolute* acceleration a_2, velocity v_2, and displacement x_2. According to Newton's second law, the net force F which must act in the x-direction to produce motion of the center of mass of a body in the x-direction is

$$F = ma_2 = m\frac{dv_2}{dt} = m\frac{d^2x_2}{dt^2}, \tag{2-11}$$

* Some authors use the term "impulse" for the quantity p and associate momentum exclusively with mass and velocity.

where the units for the mass m are (lb·sec²/in) or (lb·sec²/ft = slugs) in the British engineering system and kilograms (kg) in the metric system.

If the motion of the mass is measured relative to a moving reference point 1, the moving reference must have a constant velocity. Since

$$x_{21} = x_2 - x_1, \qquad v_{21} = \dot{x}_{21} = \dot{x}_2 - \dot{x}_1,$$

and

$$a_{21} = \ddot{x}_{21} = \ddot{x}_2 - \ddot{x}_1,$$

it is readily seen that if $\ddot{x}_1$ is not equal to zero (i.e., if its velocity is not constant), then one cannot use a measurement of x_{21} alone (or its derivatives) to determine the absolute acceleration of point 2. It is necessary to use the *absolute* acceleration $a_2 = \ddot{x}_2$ in applying Eq. (2–11).

If we compare Eqs. (2–9) and (2–11), we see that the momentum p, according to Newton's law, is

$$p = mv_2, \qquad \text{(2–12)}$$

where m is the total mass of the body and v_2 is the velocity of its center of mass. By definition, $p = 0$ when $v_2 = 0$.

From the theory of relativity [1] we now know that this linear or proportional relationship between momentum and velocity is only valid for velocities which are small compared with the velocity of light, c. When v_2 is not small relative to c, the momentum is given by

$$p = \frac{mv_2}{\sqrt{1 - v_2^2/c^2}}, \qquad \text{(2–13)}$$

where m is the "Newtonian" mass, the mass when $v_2 \ll c$. Here the momentum is a nonlinear function of the velocity.

The particles of matter which comprise a mechanical body are not interconnected in a completely rigid manner and in general will have differing velocities. Even if a body is considered to be rigid, it may rotate and cause the velocities of its particles to differ from each other. Thus the effects of material mass are distributed in varying ways in any physical body. To separate conceptually the phenomenon of translational mass from this nonuniformity and from other nonmass effects such as friction or elastic behavior, we shall define a *pure translational mass*.

A *pure translational mass* is a mechanical element whose particles are all rigidly connected together so that they translate with identical or proportional velocities and accelerations at all times. In addition,

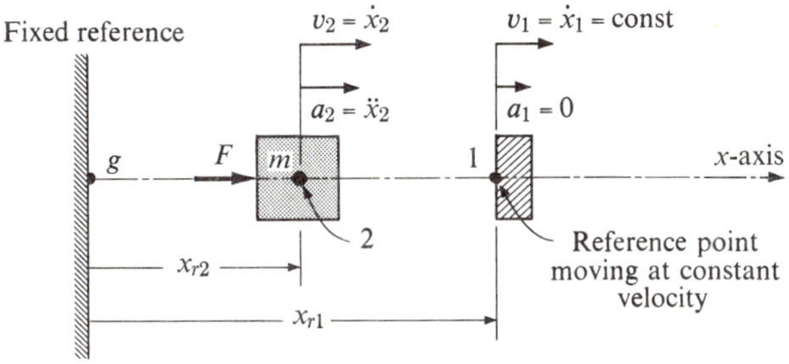

FIG. 2–5. Translational motion of a mass.

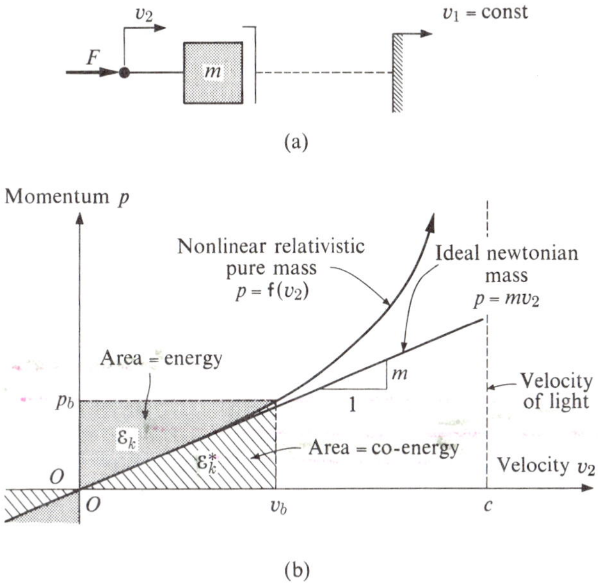

FIG. 2–6. Symbolic representation and constitutive relationships for a pure mass element. (a) Symbolic representation. (b) Constitutive relationships.

the momentum p of a pure translational mass element is by definition a single-valued function only of the absolute velocity of the element,

$$p = \mathsf{f}(v_2), \tag{2–14}$$

where f indicates a single-valued function.

The symbol for a pure translational mass element is shown in Fig. 2–6(a). The force F is the resultant force acting on the mass, v_2 is the absolute velocity of the mass, and the rectangular block represents the mass itself. To emphasize that v_2 must be measured relative to a nonaccelerating reference, the bracket and dashed line connected to the reference v_1 are used. Thus the ideal mass may be considered to have two points or *terminals* which describe its motion; the first is v_2, which describes the velocity of the mass itself, and the other is v_1, which describes the velocity of the nonaccelerating reference. Usually, we shall use a reference velocity v_1 which is equal to zero.

The behavior of a pure translational mass is completely described by the relationship between two variables, the momentum (or force) and the absolute velocity of the mass. This relationship depends only on the characteristics of the physical material which comprises the mass and on the geometry; it is therefore called a *constitutive relationship*. Figure 2–6(b) shows the constitutive relationships for a pure translational mass. For most engineering applications, the velocities are small compared with the velocity of light and momentum is a linear function of velocity. A linear pure mass will be called an *ideal mass*. For an ideal mass, Eq. (2–12) will apply, and the equation describing the mass motion, which we shall call the *elemental equation*, is then

$$F = m\frac{dv_2}{dt} = m\frac{dv_{21}}{dt}, \tag{2–15}$$

or in integral form,

$$v_{21} = \frac{1}{m}\int_0^t F\,dt + (v_{21})_0, \tag{2–16}$$

where $(v_{21})_0$ is the velocity at $t = 0$.

To be able to represent all or part of a real physical system by an ideal translational mass, that is, to "model" it as an ideal mass, it is usually necessary that all of the particles be connected together and that they move with substantially equal or proportional velocities and accelerations. Such a representation of part of a physical system will be called a *lumped ideal mass element.*

If a translating pure mass is acted on by a force, power is delivered to or removed from the mass, according to Eq. (2–6). Let us consider a pure mass, such as that represented in Fig. 2–6(a), initially at rest with $v_2 = p = 0$. Let us suppose that a force F then acts on the mass in the direction of its velocity v_2 during a time interval $t = t_a$ to $t = t_b$. An amount of work $\mathcal{W}_{ab}$ is then done on the mass, which from Eq. (2–7) is

$$\mathcal{W}_{ab} = \int_{t_a}^{t_b} Fv_2\,dt.$$

But from the definition of momentum, Eq. (2–9), $dp = F\,dt$, and the work becomes

$$\mathcal{W}_{ab} = \int_0^{p_b} v_2\,dp, \tag{2–17}$$

where p_b is the momentum at time t_b. Since the constitutive relation, Eq. (2–14), for a pure mass relates p and v_2 as indicated in Fig. 2–6(b), Eq. (2–17) is readily integrated. The integral $\mathcal{W}_{ab}$ is the area in Fig. 2–6(b) between the p-axis and the p-v_2 curve.

If then a force opposite to v_2 is applied to the mass when its momentum is p_b and its velocity is v_b, power will flow out of the mass. If the force is allowed to act until p_b decreases back to zero (remember that p is the time integral of force), an amount of work equal to $\mathcal{W}_{ab}$ is delivered by the mass to the agent which supplies the force. Thus the work done on the mass in increasing its velocity and momentum from zero to v_b and p_b, respectively, is recovered when the velocity and momentum are reduced again to zero. Therefore, the mass may be said to *store* the work $\mathcal{W}_{ab}$ as a form of internal energy associated with the velocity or momentum of the mass particles.

This energy is usually called the *translational kinetic energy*, $\mathcal{E}_k$:

$$\mathcal{E}_k = \int_{t_a}^{t_b} F v_2 \, dt = \int_0^{p_b} v_2 \, dp, \tag{2–18}$$

the shaded area in Fig. 2–6(b). The kinetic energy is zero by definition when the absolute velocity of the mass is zero. The area below the p-v_2 curve is the integral of $p\,dv_2$ and is called the kinetic co-energy, $\mathcal{E}_k^*$. For a nonlinear mass, the energy and co-energy are different, but for a linear or Newtonian mass, $\mathcal{E}_k = \mathcal{E}_k^*$.

For a Newtonian mass, $p = mv_2$, m = const, and the kinetic energy is given by the integral in Eq. (2–18):

$$\mathcal{E}_k = \tfrac{1}{2} m v_2^2 = \frac{1}{2}\frac{p^2}{m}, \quad \text{n·m or in·lb.} \tag{2–19}$$

If we think of the power-producing variables, force and velocity, we see that the kinetic energy of a mass depends directly on the velocity but only indirectly (through an integration) on the applied force. Thus we may think of a mass as an element which stores energy by virtue of its velocity. Note that the mass energy is independent of the sign of the velocity and is always positive:

$$\mathcal{E}_k \geqq 0.$$

However, the sign of the velocity may be very important to the person who wants to retrieve some stored energy, since he can do so only by slowing the mass down.

Pure translational spring. Materials undergo a certain amount of deformation when they are stressed. If a mechanical element or structure experiences a steady deformation when acted on by steady forces, it is exhibiting *compliance*, the basic characteristic of a spring. To discuss the spring characteristics of mechanical systems without the presence of other effects such as mass, we shall define a *pure translational spring* as a mechanical element of zero mass which deforms by steady amounts when loaded by steady forces. One of the most familiar mechanical devices which often approximates the behavior of a pure spring is the wire coil spring; hence this symbol is used to represent a pure spring, as shown in Fig. 2–7(a). Since by definition the pure spring has no mass, the forces acting on the element must balance. Thus, if a force F acts at point 2 of the spring, an equal and opposite force F must act at point 1. The force F may, in fact, be thought of as passing or flowing *through* the spring element.

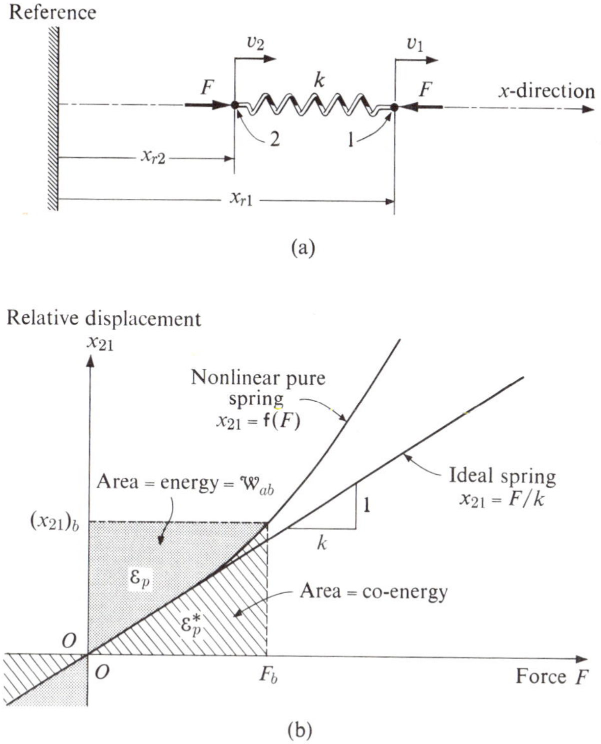

FIG. 2–7. Symbolic representation and constitutive relationships for a pure translational spring. (a) Symbolic representation. (b) Constitutive relationships.

The relationship between the velocities v_2 and v_1 of the ends of the ideal spring and the deformation or change in length of the spring is given by Eq. (2–4),

$$v_{21} = \frac{dx_{21}}{dt} = \dot{x}_{21}, \tag{2–4}$$

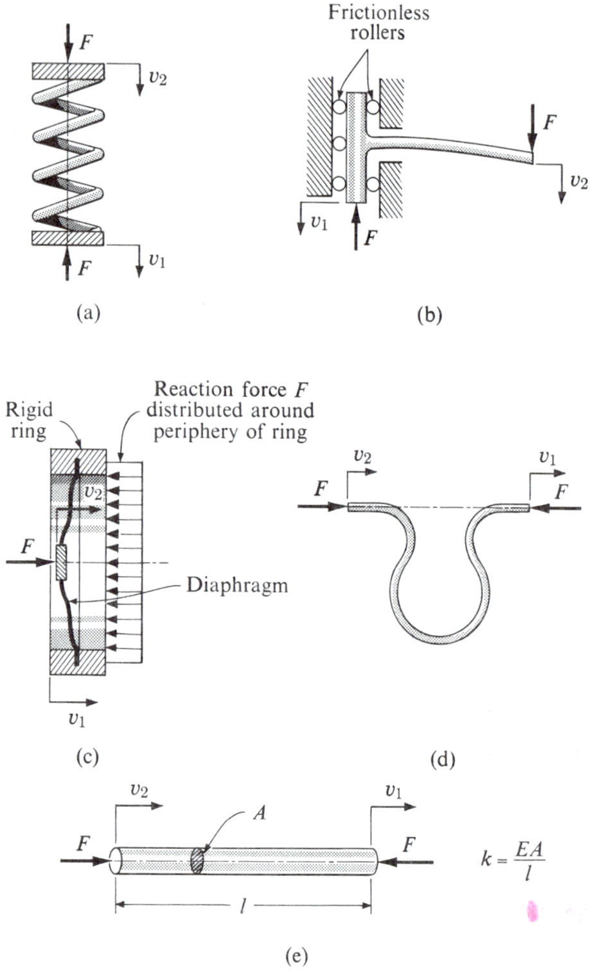

FIG. 2–8. Typical translational springs. (a) Coil spring. (b) Cantilever beam. (c) Corrugated circular diaphragm. (d) Curved bar. (e) Uniform circular bar.

or by the integral relation

$$x_{21} = \int_0^t v_{21}\,dt + (x_{21})_0, \tag{2–20}$$

where $(x_{21})_0$ is the relative displacement of points 1 and 2 at $t = 0$. The student should note the similarity between Eqs. (2–4) and (2–9) and between (2–10) and (2–20) when velocity difference and force are interchanged. The device in Fig. 2–7(a) is defined as a pure translational spring if the deformation x_{21} is a single-valued function only of the transmitted force F:

$$x_{21} = \mathsf{f}(F), \tag{2–21}$$

where f indicates a single-valued function. A linear pure spring, which we shall call an *ideal spring*, has a proportional or linear relation between deformation and force,

$$x_{21} = \frac{1}{k}F, \tag{2–22}$$

where the spring constant k has units of n/m or lb/in. Equations (2–21) or (2–22) describe the basic phenomenon of a spring and are the *constitutive relationships* for a pure translational spring. Figure 2–7(b) shows typical constitutive relations for ideal and nonlinear pure springs.

The elemental equation for an ideal spring of constant k is obtained by combining (2–22) and (2–4):

$$v_{21} = \frac{1}{k}\frac{dF}{dt}, \tag{2–23}$$

or, in integral form,

$$F = k\int_0^t v_{21}\,dt + F_0, \tag{2–24}$$

where F_0 is the force at $t = 0$. The work done by the equal and opposite forces F when a pure spring is deformed can be computed from Eq. (2–7) and the constitutive relation (2–21). Equation (2–7) combined with the velocity equation (2–4) gives

$$\mathcal{W}_{ab} = \int_{t_a}^{t_b} Fv_{21}\,dt = \int_{x_a}^{x_b} F\,dx_{21} = \int_0^{F_b} F\frac{\partial \mathsf{f}(F)}{\partial F}\,dF, \tag{2–25}$$

where F_b is the spring force at time t_b. This work integral is the area in Fig. 2–7(b) between the x_{21}-axis and the spring curve. Since Eq. (2–21) or (2–22) relates x_{21} and F, the integral (2–25) may be evaluated in terms of either the final deformation $(x_{21})_b$ or the final force F_b. As in the case of the mass element, the work done on a pure spring is retrievable by reducing the spring force to zero. Thus the spring stores the work $\mathcal{W}_{ab}$ as energy associated with the deformation of its particles. This stored energy is called the *translational potential energy*, $\mathcal{E}_p$:

$$\mathcal{E}_p = \int_{x_a}^{x_b} F\,dx_{21} = \int_0^{F_b} F\frac{\partial f(F)}{\partial F}\,dF, \tag{2–26}$$

the shaded area in Fig. 2–7(b). By definition, $\mathcal{E}_p = 0$ when the spring is undeformed ($x_{21} = 0$). The area below the F-x_{21} curve shown crosshatched in Fig. 2–7(b) is called the *potential co-energy*, $\mathcal{E}_p^*$.

For an ideal spring, $dx_{21} = (1/k)\,dF$, so that we may write

$$\mathcal{E}_p = \frac{1}{2k}F^2 = \tfrac{1}{2}k(x_{21})^2, \quad \text{n·m or in·lb.} \tag{2–27}$$

We note that the energy stored in a spring depends directly on the force transmitted through the spring, but not on the velocity difference. Hence the spring stores energy by virtue of its force. This energy, like kinetic energy, is always positive and does not depend on the sign of either the force or the spring deflection:

$$\mathcal{E}_p \geqq 0.$$

Many real physical structures of various shapes and forms, from simple uniform bars in tension to intricately shaped pieces, can often be modeled successfully as ideal translational springs. Figure 2–8 shows several mechanical elements which are frequently used in mechanisms or instruments because of their spring characteristics. The calculation of the spring constants for structures such as these may become exceedingly difficult and often requires the use of advanced methods of the theory of elasticity together with machine computation.* If the structure is actually available, the spring characteristics may, of course, be determined experimentally by applying known loads and observing the resulting deformations.

For a physical structure to be modeled as an ideal spring, the effects of mass, friction, and other influences must be negligible in the situation where the structure is employed. Often a given structure may act either as a spring or mass, for example, depending on the circumstances. We shall be concerned throughout this text with the problems of representing systems by pure elements such as springs and masses. To illustrate in a simple way the problems of modeling a structure by ideal elements, we shall consider the following example.

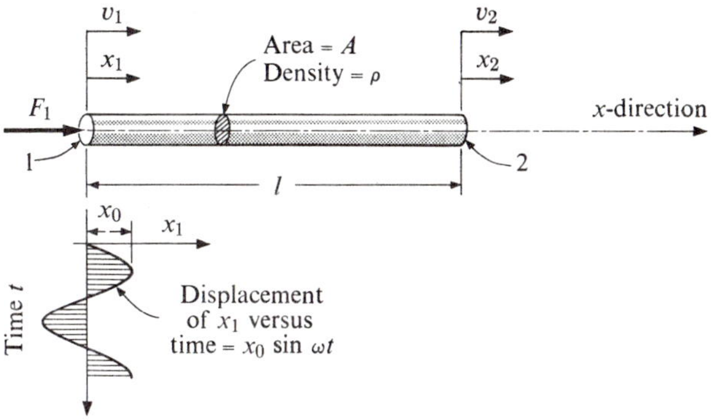

FIG. 2–9. Translational oscillation of a cylindrical rod.

Example 1. Let us suppose that a uniform cylindrical steel rod of area A, length l, and density ρ (Fig. 2–9) is constrained to translate in the x-direction. The left end, 1, of this rod is caused to oscillate sinusoidally by application of a time-varying force F_1:

$$x_1 = x_0 \sin \omega t,$$

where x_0 is the amplitude of oscillation and ω is the frequency in radians per second (rad/sec). The right end is free from force. We wish to know the conditions (approximately) under which this rod can be represented by an ideal mass.

The rod will act like an ideal mass if all its particles experience substantially equal accelerations and if

$$F_1 = m\ddot{x}_1 = m\omega^2 x_0 \sin \omega t = -m\omega^2 x_1, \tag{2–28}$$

where m is the total mass of the rod. The force F_1 applied at point 1 must drop off to zero in the x-direction, since no force exists at point 2 of the rod. However, in those parts of the rod which transmit force, deformations of the

---

* Reference 2 contains an extensive tabulation of formulas which may be used to compute the spring stiffness of various mechanical structures.

rod will occur due to spring action. If we assume the entire acceleration force to be transmitted undiminished through the rod, a uniform stress equal to F_1/A would exist in the rod. According to Hooke's law for elastic materials, a proportional strain equal to $(x_1 - x_2)/l$ would result such that

$$\frac{F_1}{A} = E\frac{(x_1 - x_2)}{l},$$

where E is Young's modulus $= 30 \times 10^6$ lb/in^2 for steel. The deformation of the bar would then be proportional to F_1, and the rod would act like an ideal spring with spring constant

$$k = \frac{F_1}{x_1 - x_2} = \frac{EA}{l}. \tag{2–29}$$

An upper limit on the bar deflection is given by Eq. (2–29) if F_1 is assumed given by Eq. (2–28). If the bar is to act like an ideal mass, however,

$$|\ddot{x}_1 - \ddot{x}_2| \ll |\ddot{x}_1|, \tag{2–30}$$

in which case the accelerations will be nearly equal everywhere. The combination of Eqs. (2–28), (2–29), and (2–30) with appropriate differentiations gives the condition

$$\omega^2 \ll k/m \tag{2–31}$$

or, since $m = \rho A l$,

$$\omega \ll (1/l)\sqrt{E/\rho}. \tag{2–32}$$

Thus if the frequency ω of oscillation of x_1 is small compared with $\sqrt{k/m}$, the rod may be assumed to act like an ideal mass.

Now let us suppose that sinusoidal displacements $x_1 = x_{01} \sin \omega t$ and $x_2 = x_{02} \sin \omega t$ are imposed at the two ends of the rod. Can you determine the conditions for the rod to behave almost like an ideal spring? Will it ever behave like an ideal mass?

Pure translational damper. When the particles of a material move relative to each other at finite velocities, their motions are resisted by forces which are proportional to or at least functions of their relative velocities. This effect is usually quite small and is normally (although not always) negligible in the so-called elastic deformation of solids, where the atoms of the material do not move very much relative to each other. In fluids, where molecules are free to slip or move continuously relative to each other in a tangential or shearing direction, forces arising from relative velocity may be substantial. The property of fluids which describes their tendency to resist a rate of change of shape is called *viscosity*. Quantitatively, *fluid viscosity* is defined as the shear stress required to produce a unity rate of shear strain of a fluid particle.

The tendency of fluid or solid materials to resist relative velocities between their particles or component parts is often called *friction*. When the forces due to relative motion are proportional to the relative velocity, the friction is usually called *viscous* or *linear* friction. Nonlinear frictional effects are, however, very common in mechanical systems.

We shall now define a third translational element, called a *pure damper*, which has no mass or spring effects but which represents only the effects of resistance to rate of deformation. Figure 2–10(a) shows the symbol which will be used for a pure translational damper.*

Since a pure damper has no mass, the forces acting on it must be balanced. Thus in Fig. 2–10(a), the force F applied at point 2 flows through the damper and out through point 1. Point 1 of the damper exerts a force on its surroundings equal to F and directed to the right. The describing equation or constitutive relationship for a pure translational damper is a single-valued function of the form

$$F = \mathsf{f}(v_2 - v_1) = \mathsf{f}(v_{21}), \tag{2–33}$$

in which $F = 0$ when $v_{21} = 0$ and the signs of F and v_{21} are the same. When the relationship between the force F and the rate of extension (or contraction) v_{21} of the damper is a straight line, as shown in Fig. 2–10(b), the damper is *ideal*, and the describing

* This symbol depicts a piston and cylinder in which fluid is forced past the piston by relative motion, thus producing resisting force.

equation is given by

$$F = b(v_2 - v_1) = bv_{21}, \tag{2–34}$$

where b is called the damping coefficient and commonly has the units n·sec/m in the metric system or lb·sec/in. in the British engineering system.

The two most common types of nonlinear pure dampers are described by the curves in Fig. 2–10(b). These are square-law damping, in which the force F is proportional to $(v_{21})^2$, and Coulomb friction damping, in which the force is dependent primarily on the direction of v_{21} and not appreciably on its magnitude. Square-law damping occurs when mechanical parts move through liquids or gases at high velocities, and Coulomb friction arises when two materials or objects are made to slide relative to each other while being held together. In this book, we shall deal chiefly with ideal or linear dampers.

When a pure damper is given a rate of extension or compression by equal and opposite forces applied at its ends or terminals, work will be done. From Eq. (2–7), the work $\mathcal{W}_{ab}$ done in a time interval t_a to t_b is

$$\mathcal{W}_{ab} = \int_{t_a}^{t_b} Fv_{21}\,dt.$$

The force may be eliminated from this integral by using Eq. (2–33):

$$\mathcal{W}_{ab} = \int_{t_a}^{t_b} v_{21}\mathsf{f}(v_{21})\,dt. \tag{2–35}$$

Unlike the spring-and-mass case, the time cannot be removed from this integral. Since by definition of a pure damper the product $v_{21}\mathsf{f}(v_{21})$ is always positive, $\mathcal{W}_{ab}$ always increases with time. This means that work can never be removed from a damper. Therefore, we say that a damper *dissipates* the work or energy delivered to it. The damper is thus fundamentally different from the spring and mass elements, which store and may give back the energy supplied to them.

The power $\mathcal{P}$ supplied to a damper is

$$\mathcal{P} = Fv_{21} = v_{21}\mathsf{f}(v_{21}).$$

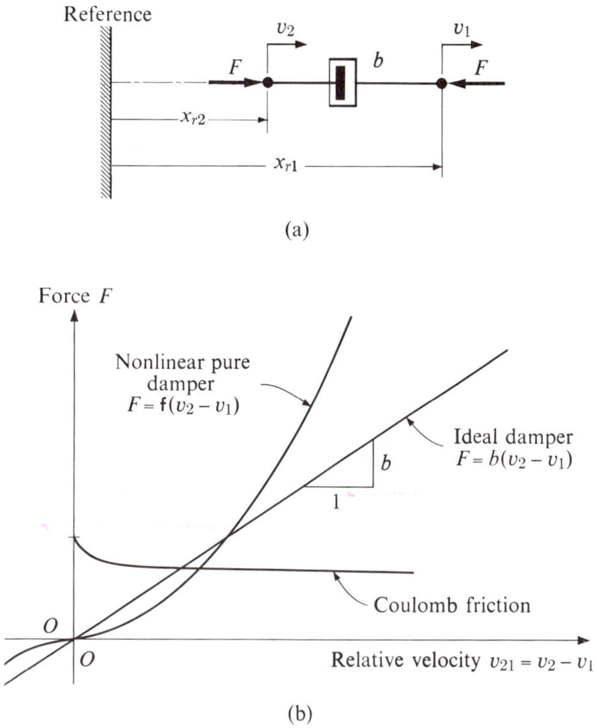

FIG. 2–10. Symbolic representation and constitutive relationships for a pure translational damper. (a) Symbolic representation. (b) Constitutive relationships.

For an ideal damper, $\mathsf{f}(v_{21}) = bv_{21}$, and

$$\mathcal{P} = bv_{21}^2 = F^2/b, \quad \text{n·m/sec} = \text{watts or in·lb/sec}. \tag{2–36}$$

This power flow is not reversible but is always positive, or into the damper. If the direction of the relative motion of the ends of the damper is reversed, the sign of the force is also reversed and power still flows into the element.

For any pure damper,

$$\mathcal{P} \geqq 0.$$

Thus, although pure springs and masses always have positive energy storage, as shown by Eqs. (2–19) and (2–27), pure dampers always have positive power flow.

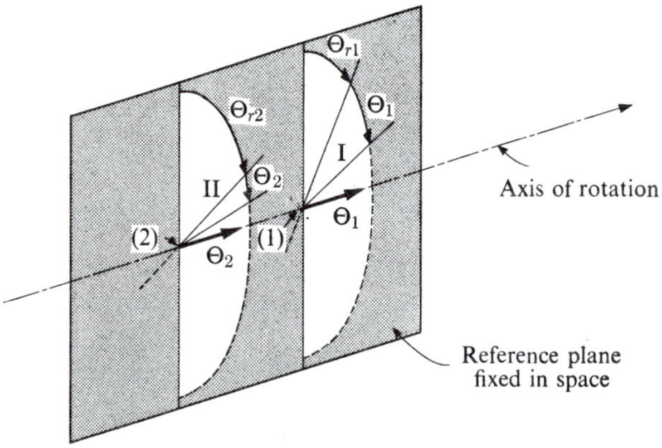

FIG. 2–11. Description of rotary motion.

Although a damper might seem at first glance to be undesirable because it only dissipates energy, it will be seen later that energy dissipation is very valuable in many dynamic systems. One of the most familiar examples of a damper is the automobile shock absorber, which is designed to dissipate energy from oscillations of the automobile mass and suspension system caused by riding over bumps in the road.

2–3. MECHANICAL SYSTEM ELEMENTS IN ROTATION

In the preceding section we considered three mechanical elements representing three essential phenomena in mechanical systems which undergo only translatory motions. Another large class of mechanical systems involves rotation about fixed or nonaccelerating axes. Many types of rotating-machinery and control-system components, such as turbines, gear trains, motors, galvanometers, and pumps, may be modeled in part by the use of three ideal rotary mechanical elements which are analogous to the translational elements already discussed. These rotary elements are *rotational mass*, or *inertia*; *torsional* or *rotational spring*; and *torsional* or *rotational damper*. The basic physical phenomena involved in each of these ideal elements is the same as for the corresponding translational element, except that in rotary systems effects take place in terms of circular motions about an axis of rotation rather than in terms of straight-line motion along an axis of translation. Also, torque, or moment of force, replaces force at points of interaction between system elements.

The unit of angular motion employed here will be the radian unless otherwise specified (the unit revolution, 2π radians, is sometimes used). The units of torque will be newton·meters (n·m) or pound·inches (lb·in.).

All the earlier comments about modeling physical systems by the use of ideal elements are equally applicable here. The concept of lumping distributed collections of matter into discrete elements is applied in the same way, and the comments and restrictions relating to each particular kind of ideal translational element apply in substantially the same fashion to each kind of rotational element.

2–3.1 Angular Motion and Torque

Absolute angular motion is described by the angle subtended between a fixed reference plane and a line drawn on the rotating element which lies in a plane perpendicular to the reference plane. The *relative* motion between two lines is the difference between the absolute motions of the lines. Thus in Fig. 2–11 the absolute displacements of lines II and I are Θ_2 and Θ_1, respectively, and the displacement of II with respect to I is

$$\Theta_{21} = \Theta_2 - \Theta_1. \tag{2-37}$$

Similarly, the relative velocity Ω_{21} of II with respect to I is

$$\Omega_{21} = \Omega_2 - \Omega_1 = \dot{\Theta}_{21}. \tag{2-38}$$

The simplified representation of angular motion shown in Fig. 2–12 will be used in the following discussion.

Torque about a given axis is defined as the resultant moment of all forces about the axis. Newton's second law may be used to show that the net torque T acting on any element having either no mass or no angular acceleration $\dot{\Omega}$ must be zero. We shall represent torque by a vector lying along the axis about which the torque acts and having a direction determined by the right-hand screw rule. Thus in Fig. 2–13, we find that, when looking from left to right, the torque at point 2 is clockwise and that at point 1 is counterclockwise. If this rotary element either has no mass or is not accelerating, the torque T flows through the element undiminished, in the same way that force flows through a translational element in equilibrium.

2–3.2 Power, Work, and Energy

The power $\mathcal{P}$ delivered by a torque T acting on a shaft rotating at speed Ω in the direction of the torque is

$$\mathcal{P} = T \cdot \Omega, \quad \text{n·m/sec} = \text{watts or lb·in/sec.} \tag{2–39}$$

If the torque and angular velocity, which are in general vectors, do not have the same direction, then (2–39) must be the dot product of the two vectors. We shall deal only with cases where T and Ω are collinear.

The work $\mathcal{W}_{ab}$ done on an element is the integral of the power:

$$\mathcal{W}_{ab} = \int_{t_a}^{t_b} T \cdot \Omega \, dt, \quad \text{n·m or lb·in.} \tag{2–40}$$

Similarly, the energy $\mathcal{E}$ delivered to an element is the sum of the work done at all points where torque and angular velocity exist.

2–3.3 Pure Rotary Mechanical Elements

Pure rotary mass or inertia. The moment of inertia J of a point mass m rotating at radius r from an axis of rotation is by definition

$$J = mr^2, \quad \text{kg·m}^2 = \text{n·m·sec}^2 \text{ or lb·in·sec}^2. \tag{2–41}$$

The moment of inertia of a number n of mass particles about an axis is

$$J = \sum_{k=1}^{n} m_k r_k^2.$$

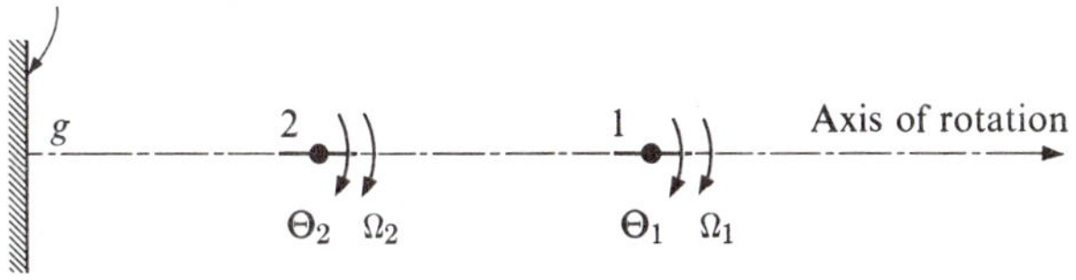

FIG. 2–12. Simplified representation of angular motion.

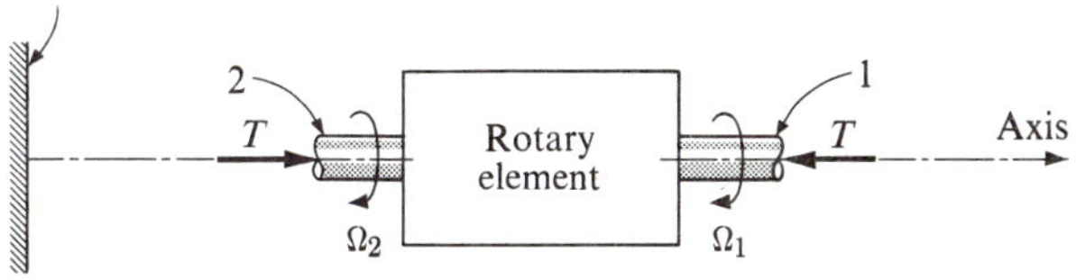

FIG. 2–13. Rotary element having torque transmitted from left to right.

Similarly, the inertia of a distributed mass is

$$J = \int r^2 \, dm, \tag{2–42}$$

where the integral is taken over the body. Equation (2–42) assumes that rotation occurs only about one axis so that all mass particles describe circular arcs as they rotate.*

We shall now define the quantity *angular momentum, h,* which is the rotational counterpart of translational momentum:

$$T = dh/dt. \tag{2–43}$$

If we define an *ideal rotational mass* as a collection or distribution of ideal translational masses rigidly connected together and constrained to rotate at Ω_2

---

* Often it is easy to compute $J_{\text{center·gravity}}$ about the center of mass of a body because of symmetry. The inertia J about a parallel axis of rotation located a distance $r_{\text{center·gravity}}$ from the center of mass is then given by

$$J = J_{cg} + mr_{cg}^2,$$

where m is the total mass and cg indicates center of gravity. This expression is known as the "parallel-axis theorem" for rotation about fixed axes.

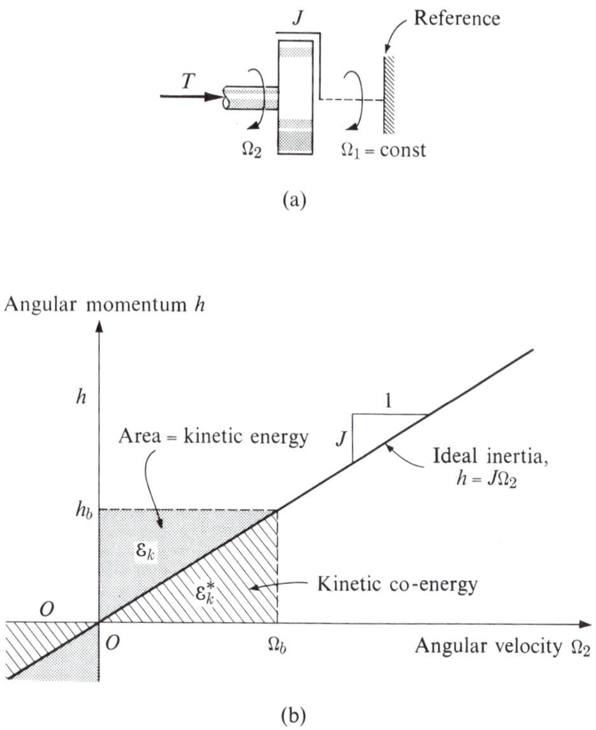

FIG. 2–14. Ideal rotary mass or inertia. (a) Symbolic representation. (b) Constitutive relationships.

about a fixed axis, we can show from Newton's second law that

$$h = J\Omega_2. \tag{2-44}$$

This result is the constitutive relation for an *ideal (linear) rotational mass*. Nonlinear pure rotary mass is not encountered in rotational systems since any rigid body will burst due to centrifugal stresses at rotary speeds where the velocities of the rotating particles are still negligible compared with the speed of light.

Figure 2–14 shows the symbolic representation and constitutive relationship for an ideal rotary mass or inertia. The elemental equation for this ideal element is obtained by combining (2–43) and (2–44):

$$T = J\frac{d\Omega_2}{dt} = J\frac{d\Omega_{21}}{dt}. \tag{2-45}$$

As with ideal mass, the velocity Ω_2 must be referred to a fixed or nonaccelerating reference, as indicated by the dashed connection in Fig. 2–14. Usually Ω_1 will be taken equal to zero.

When a pure rotary inertia, which we shall henceforth call simply a *pure inertia*, is accelerated, energy is stored in the form of rotary kinetic energy, $\mathcal{E}_k$:

$$\mathcal{E}_k = \int_0^{h_b} \Omega_2\, dh, \tag{2-46}$$

the shaded area in Fig. 2–14(b). For an ideal inertia, $h = J\Omega_2$, and Eq. (2–46) gives

$$\mathcal{E}_k = \tfrac{1}{2}J\Omega_2^2 = \frac{1}{2}\frac{h^2}{J}, \text{ n·m or in·lb.} \tag{2-47}$$

As with the translational mass, this energy can be recovered by slowing down the velocity. Rotational kinetic energy is seen to be stored by virtue of the angular velocity and to be independent of the particular value of torque. Also, from (2–47),

$$\mathcal{E}_k \geqq 0.$$

Pure rotational spring. Mechanical elements which exhibit the characteristics of a rotational spring occur in many forms. One of the most common is a torsional helical-coil spring. A schematic representation of this device (Fig. 2–15a), is used as the symbol of a pure rotary spring. The pure spring has no inertia by definition and hence transmits torque undiminished. The defining constitutive relations for a rotary spring are shown in Fig. 2–15(b). For an ideal rotary spring,

$$\Theta_{21} = \frac{1}{K}T, \tag{2-48}$$

where Θ_{21} is the relative angular deformation, which is zero when the spring is relaxed ($T = 0$), and K is the rotational spring stiffness (n·m/rad or lb·in/rad).

The angular velocities are related to Θ_{21} by

$$\Omega_{21} = \Omega_2 - \Omega_1 = \frac{d\Theta_{21}}{dt}. \tag{2-49}$$

The elemental equation for a rotary ideal spring of constant K may thus be written

$$\Omega_{21} = \frac{1}{K}\frac{dT}{dt} \tag{2–50}$$

or

$$T = K\int_0^t \Omega_{21}\,dt + T_0, \tag{2–51}$$

where T_0 is the torque at $t = 0$.

The potential energy $\mathcal{E}_p$ stored in a rotational spring as it is deformed from a relaxed state is represented by the shaded area in Fig. 2–15(b):

$$\mathcal{E}_p = \int_0^{(\Theta_{21})_b} T\,d\Theta_{21}. \tag{2–52}$$

If the spring is ideal, $d\Theta_{21} = 1/K\,dT$, and

$$\mathcal{E}_p = \frac{1}{2K}T^2 = \tfrac{1}{2}K(\Theta_{21})^2. \tag{2–53}$$

This energy is always positive, is recoverable by reducing T, and is stored by virtue of the torque rather than the angular velocity.

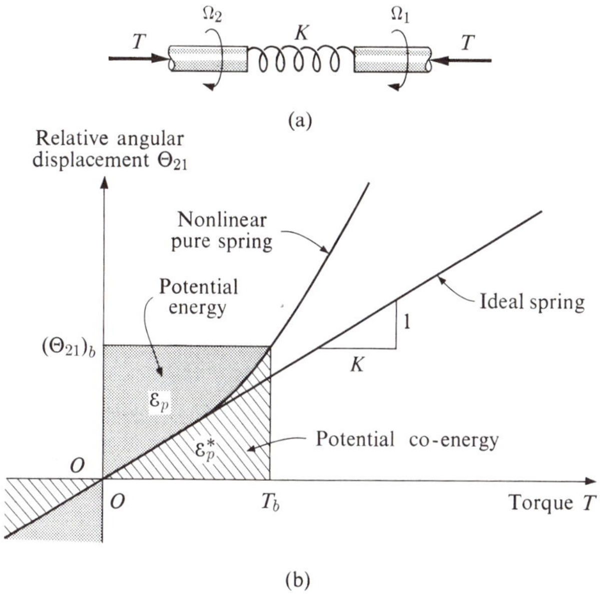

FIG. 2–15. Pure rotary spring. (a) Symbolic representation. (b) Constitutive relationships.

Pure rotational damper. Rotational damping, characterized by a resistance to the rate of relative angular motion between parts of rotating mechanical systems, is encountered in bearings, on the surface of rotating elements immersed in fluids, and in coupling devices. The fluid coupling employed in many of the automatic transmissions of automobiles is an example of a rotational damper which has been developed specifically for the purpose of introducing needed damping or energy dissipation into a system; it helps to "filter out" the jerks that occur in the forward motion of the car when the speed ratio of the transmission is automatically changed. It plays a role similar to that played by a driver with a manually operated transmission when he only partially engages the clutch to accelerate the vehicle smoothly.

The symbolic diagram for a pure rotational damper, which by definition has no mass or compliance, is shown in Fig. 2–16(a). The constitutive

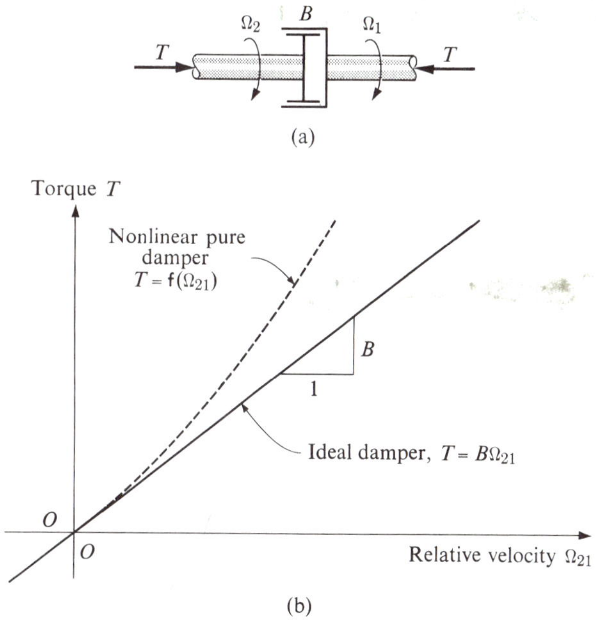

FIG. 2–16. Pure rotary damper. (a) Symbolic diagram. (b) Constitutive relationships.

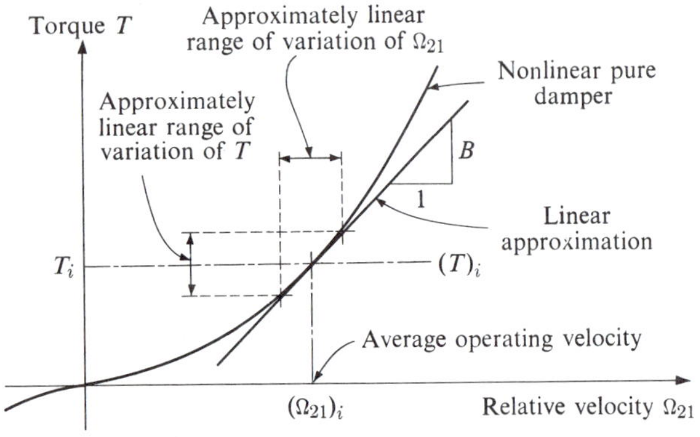

FIG. 2-17. Linear approximation of incremental behavior of nonlinear rotary damper.

relation for a pure rotary damper is

$$T = \mathsf{f}(\Omega_2 - \Omega_1) = \mathsf{f}(\Omega_{21}), \tag{2-54}$$

and for an ideal damper,

$$T = B(\Omega_2 - \Omega_1) = B\Omega_{21}, \tag{2-55}$$

where the damping coefficient B has units of n·m·sec/rad or lb·in·sec/rad.

As with the translational damper, the rotary damper dissipates energy whenever Ω_{21} is different from zero. The rate of energy flow into a linear rotary damper is

$$\mathcal{P} = B(\Omega_{21})^2 = T^2/B > 0. \tag{2-56}$$

Rotational damping which is encountered in physical systems is often nonlinear, with a characteristic like the dashed curve in Fig. 2-16(b). However, an ideal linear damper may be used to model this nonlinear characteristic if the relative speed $(\Omega_2 - \Omega_1)$ varies by only a small amount about an average value. Modeling of nonlinear characteristics in this way is called *linearization*. As an example, the local slope of the characteristic curve for the nonlinear damper shown in Fig. 2-17 is the linearized damping coefficient.

2-4. MECHANICAL TRANSFORMERS

Thus far we have dealt with two classes of mechanical system elements: those which store energy (mass and spring), and those which dissipate energy (damper). We shall now introduce a third type of element, called a *pure transformer*, which neither stores nor dissipates, but simply transforms mechanical energy. The magnitudes of torque or force and of angular or linear velocity are changed by a pure mechanical transformer in such a way that the net flow of power into the element is zero.

We will introduce the concept of a mechanical transformer by discussing a simple gear train.

2-4.1 The Simple Gear Train

The simplest and most familiar example of what we shall call a rotational transformer is the elementary gear train (Fig. 2-18). Two mating gears of different sizes are attached to two parallel shafts which in turn are free to rotate supported in a frame or housing. Gear-tooth profiles machined into the periphery of the gear disks cause the gears to behave like circular cylinders of radii r_a and r_b which roll on each other without slipping, provided that the gears are rigid and the gear teeth are geometrically perfect. These radii are called the *pitch radii* of the gears, and their magnitudes are proportional to the numbers of teeth N_a and N_b in the gears:

$$r_a/r_b = N_a/N_b = \text{gear ratio}. \tag{2-57}$$

Let us suppose that the frame in Fig. 2-18 is fixed, and that $\Omega_1 = 0$. Since the gears are assumed to roll without slipping, the arc distance $r_a\Omega_2\,dt$ turned off on gear a must equal minus the arc distance $r_b\Omega_4\,dt$ turned off on gear b. Hence

$$\frac{\Omega_4}{\Omega_2} = -\frac{r_a}{r_b} = -\frac{N_a}{N_b} = n, \tag{2-58}$$

where the negative sign indicates that the direction of Ω_4 is opposite to that of Ω_2. To determine the relationship between the torques T_a and T_b, the two gears are isolated from each other and from the frame, thus

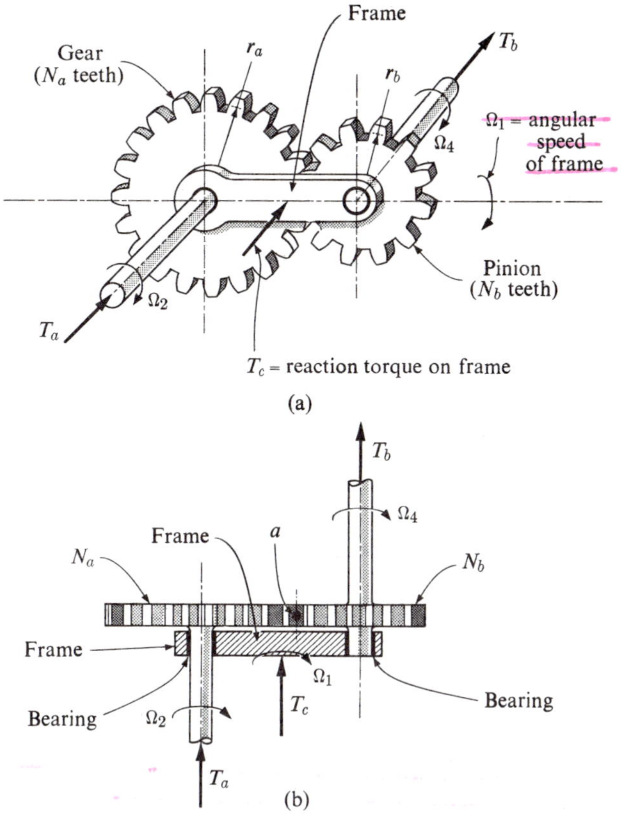

FIG. 2–18. Simple gear train; rotational transformer. (a) Perspective view. (b) Top view.

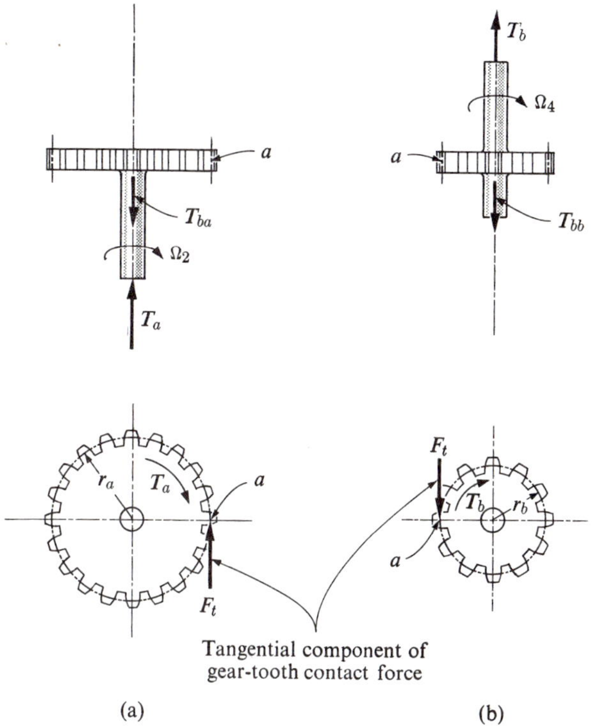

FIG. 2–19. Free-body diagrams of gear pair, showing torques and torque-producing forces. (a) Gear isolated. (b) Pinion isolated.

forming the free-body diagrams of Fig. 2–19. Only torques and torque-producing forces are shown on these free bodies, and friction between the gear-tooth surfaces is assumed to be zero. If we further assume that the gears have no rate of change of angular momentum (zero inertia or constant speed), the net torque on each gear is zero. Then

$$T_a - T_{ba} = F_t r_a, \tag{2–59}$$

$$T_b - T_{bb} = F_t r_b, \tag{2–60}$$

where T_{ba} and T_{bb} are the bearing friction torques, and F_t is the tangential component of the gear-tooth contact force. If we now assume the bearing torques to be negligible and employ the definition of n in Eq. (2–58), we obtain

$$T_b/T_a = r_b/r_a = N_b/N_a = -1/n. \tag{2–61}$$

Thus the speed is changed as n while the torque is changed as $1/n$. The net power into the gear train for $\Omega_1 = 0$ is therefore zero:

$$\mathcal{P} = T_a\Omega_2 + T_b\Omega_4 = T_a\Omega_2 + (-T_a/n)(+n\Omega_2) = 0.$$

Since the gear system is assumed to have constant angular momentum, the torque T_c shown acting on the frame in Fig. 2–18 must be equal to minus the sum of T_a and T_b, or

$$-T_c = T_a + T_b. \tag{2–62}$$

Since $\Omega_1 = 0$ in the preceding discussion, T_c does no work on the system.

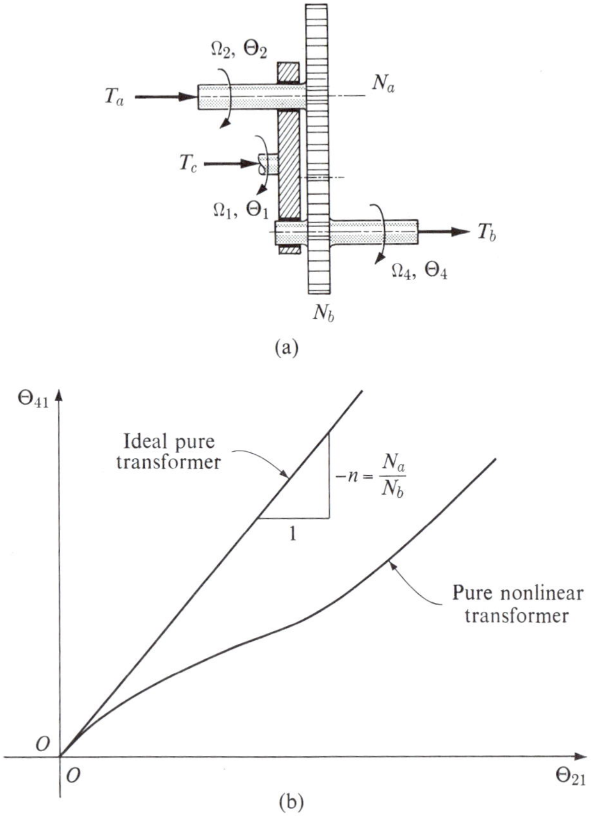

FIG. 2-20. Symbolic representation and functional relationships for a pure rotational transformer. (a) Symbolic representation. (b) Functional representation.

Now let us consider the general case in which the frame is not necessarily fixed, but may rotate with angular velocity Ω_1 (Fig. 2-18b). Equations (2-61) and (2-62) are still valid, but (2-58) must be revised. An observer standing on the frame and thus rotating at absolute velocity Ω_1 will see only the velocities of the gears relative to himself or to the frame. These relative velocities Ω_{21} and Ω_{41} are

$$\Omega_{21} = \Omega_2 - \Omega_1, \qquad \Omega_{41} = \Omega_4 - \Omega_1. \tag{2-63}$$

The observer will see that the *relative* velocities are related by the gear ratio

$$\frac{\Omega_{41}}{\Omega_{21}} = -\frac{r_a}{r_b} = -\frac{N_a}{N_b} = n. \tag{2-64}$$

If we now compute the net power flow $\mathcal{P}$ into the gear system, and employ Eq. (2-62),

$$\mathcal{P} = T_a\Omega_2 + T_c\Omega_1 + T_b\Omega_4 = T_a\Omega_{21} + T_b\Omega_{41}. \tag{2-65}$$

Assuming that Ω_2 and Ω_1 are specified and that a given torque T_a is applied to the system, we can calculate Ω_4 from Eqs. (2-63) and (2-64), and can determine T_b and T_c from Eqs. (2-61) and (2-62). The student should verify that when this is done, Eq. (2-65) becomes $\mathcal{P} = 0$.

In arriving at the preceding results for a simple gear train, we made several assumptions which the reader should review at this point. Within these assumptions we have seen that speed is changed by the gear ratio n and torque is changed by $1/n$ in such a way that the net power into the gear system is zero.

To permit this torque- and speed-transformation phenomenon to be used in the modeling of rotational mechanical systems, we shall define the *pure rotational transformer.*

2-4.2 Pure Rotational Transformers

A *pure rotational transformer* is defined as an idealized device which stores or dissipates no energy and which determines a chosen output relative angular displacement Θ_{41} (or velocity Ω_{41}) as a single-valued function of a chosen input relative angular displacement Θ_{21} (or velocity Ω_{21}). The reference angle Θ_1 (velocity Ω_1) is not necessarily constant, as discussed in Section 2-4.1. An *ideal rotary transformer* has by definition a linear relation between the input and output relative angles and velocities. Figure 2-20 shows the symbolic representation which we shall use for pure rotational transformers and indicates the describing functional relationships. In most cases the frame will be fixed and $\Omega_1 = \dot{\Theta}_1 = 0$.

There are many mechanical devices which may be represented approximately by pure or ideal rotational transformers. Belt and chain drives, cam systems, intermittent-motion devices, and linkages are in this category and are used for the following purposes:

1) To change the speed of a motor or other power source to meet the need for a lower or higher output

speed (as in the automobile transmission which couples the engine to the driving wheels).

2) To obtain a nonuniform motion from a uniform one. This is a common application in automatic machinery.

3) To achieve a mechanical advantage, that is, an increase in torque level.

Figure 2–21 illustrates schematically several devices which are designed to approximate pure rotational transformers.

2–4.3 Pure Translational Transformers

A pure translational transformer is typified by a lever, as shown in Fig. 2–22, which has no mass, is infinitely stiff, and is free from frictional effects.

The relationship between the velocities of points 4 and 2 of the lever relative to the pivot velocity is

$$v_{21}/v_{41} = -r_a/r_b = n = \text{lever ratio}, \tag{2–66}$$

where $v_{41} = v_4 - v_1$ and $v_{21} = v_2 - v_1$. Note that r_a and r_b change as Θ varies, but that $r_a/r_b = n$ remains constant. Often Θ will be limited to a small angle, so that r_a and r_b are essentially constant.

Since the massless lever has no momentum, then by definition the sum of the forces in any direction and the moments of the forces about the pivot must be zero:

$$F_a + F_b + F_c = 0, \tag{2–67}$$

$$F_b/F_a = r_a/r_b = -1/n. \tag{2–68}$$

These relationships between forces and linear velocities are the same as those between torques and angular velocities in the rotational transformer. An *ideal* linear transformer has constant lever ratio n. As in the rotary transformer, the pure translational transformer has zero net power flow. Figure 2–23 shows the symbol and the functional relationships for pure translational transformers.

How many mechanical systems can you think of which exhibit the characteristic of a pure or ideal translational transformer?

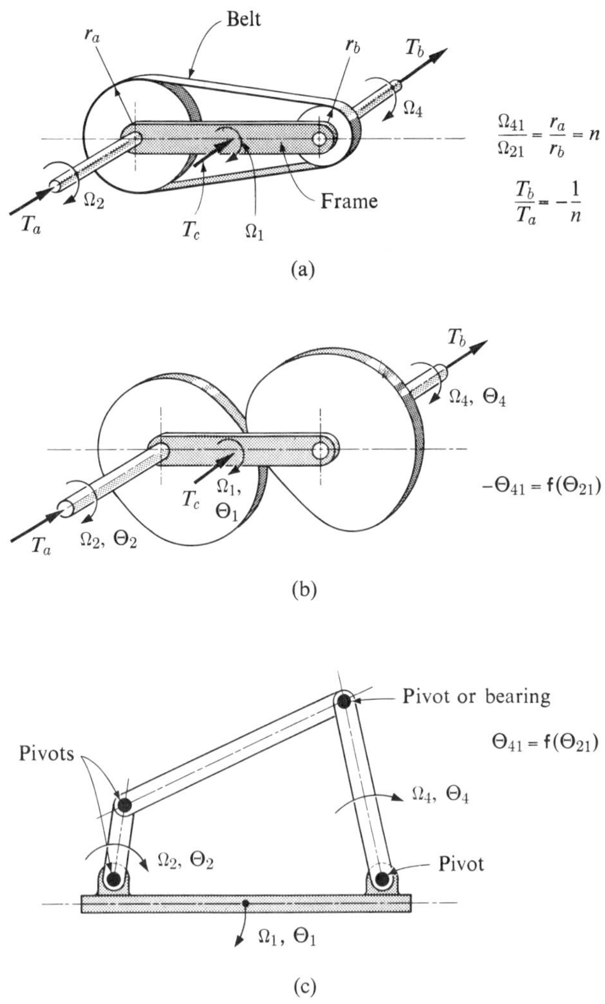

FIG. 2–21. Typical rotational mechanisms which exhibit approximately pure transformer action. (a) Belt or chain and pulley system. (b) Cam drive system; pure nonlinear transformer. (c) Four-bar linkage; pure nonlinear transformer.

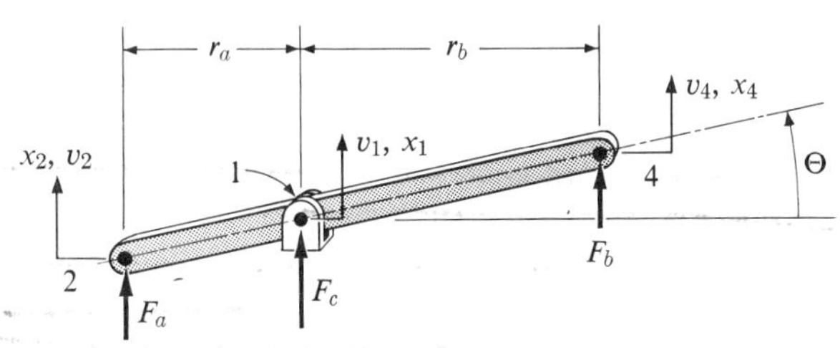

FIG. 2–22. Simple lever.

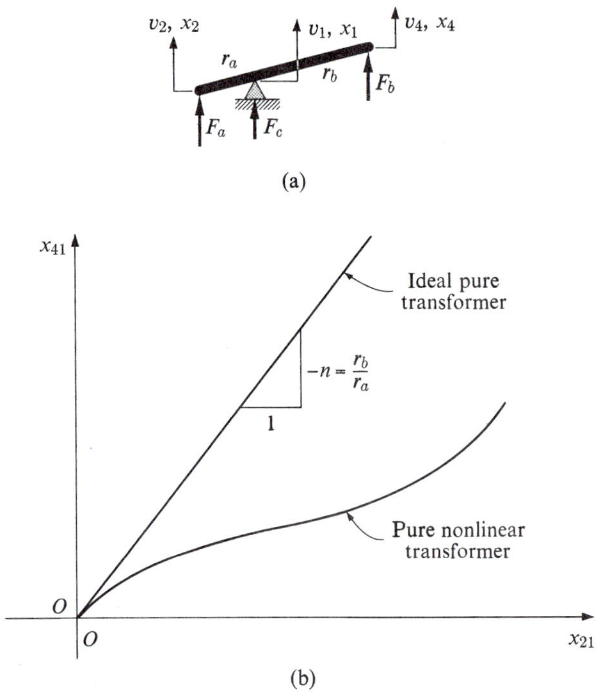

FIG. 2–23. Symbolic representation and functional relationships for a pure translational transformer. (a) Symbolic representation. (b) Functional relationships.

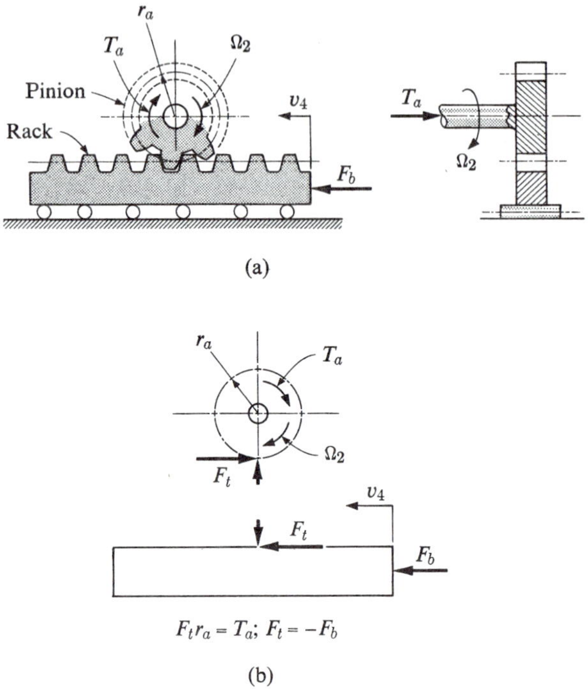

FIG. 2–24. Gear and rack system; rotary-to-translatory energy transducer. (a) Schematic diagram. (b) Free-body diagram.

2–5. MECHANICAL TRANSDUCERS

Before we conclude this discussion of mechanical system elements, we shall describe one further idealized mechanical element, called a *transducer*, which is similar to the transformer. We have seen that a pure rotational transformer changes the level of output torque and angular velocity relative to input torque and velocity while maintaining zero net power input to the transformer. Both the input and output power were in the rotational form (torque times angular velocity). Similarly, the pure translational transformer preserved power in the translational form.

A *transducer* is defined as a device which transforms energy or power in one form (e.g., torque times angular velocity) into energy or power in another form (e.g., force times lineal velocity). A *pure transducer* performs this function with no energy loss or storage, so that the net power flow into a pure transducer is zero.

Some familiar systems which exhibit the properties of a transducer are electric generators (mechanical power to electrical power), electric motors (electrical to mechanical), gas turbines (fluid to mechanical), hydraulic pumps (mechanical to fluid), and thermocouples (thermal to electrical). Several of these systems will be discussed in later chapters of this book. Here we shall introduce only the concept of transduction by discussing the rotary-mechanical to translatory-mechanical transducer.

Figure 2–24(a) shows a simple gear and rack (a rack is a gear of infinite radius) drive system. The gear is mounted, with freedom only to rotate, in a stationary frame, and the rack is free to translate horizontally relative to the frame. If the gears are geometrically perfect and rigid, the linear distance moved by the rack must equal the arc distance through which the gear pitch circle moves:

$$v_4\,dt = \Omega_2 r_a\,dt,$$
$$v_4/\Omega_2 = r_a = n = \text{transduction ratio.} \tag{2–69}$$

If we assume that the parts have zero momentum and that friction torques and forces are zero, the following equation is obtained from the two free-body diagrams of Fig. 2–24(b):

$$F_b/T_a = -1/r_a = -1/n \tag{2–70}$$

The net power $\mathcal{P}$ into the device is

$$\mathcal{P} = T_a\Omega_2 + F_b v_4$$
$$= T_a\Omega_2 + \left(-\frac{T_a}{r_a}\right)(r_a\Omega_2) = 0.$$

Hence the idealized gear and rack, with zero net power flow like the transformer, transduces torque into force and angular motion into translational motion.

A *pure* rotary-to-translatory energy transducer produces a translatory displacement (or velocity) which is a single-valued function of the rotary displacement (or angular velocity). The net power into the device is identically zero. An *ideal* rotary-to-translatory energy transducer has rotary motion proportional to translatory motion.

2–6. ELECTRICAL SYSTEM ELEMENTS

Many important problems may be solved and the behavior of many electrical systems may be understood by means of the concepts of electric circuits and idealized electric elements. We shall see that the behavior of these electric circuits and elements is very similar to the behavior of systems of the mechanical lumped elements discussed earlier in this chapter.

2–6.1 Voltage and Current

To describe mathematically idealized electric circuit elements and analyze their interactions, we must employ the concepts of voltage (electric field potential) and current (flow of electric charge). A variation of voltage at a given point in an electric system is almost always associated with a coexisting current. In some instances it is possible to think of the voltage as resulting from an imposed current; in other instances it may be preferable to think of the current as resulting from an applied voltage. As in mechanical systems, work, energy, and power are crucial factors in the study of electrical systems.

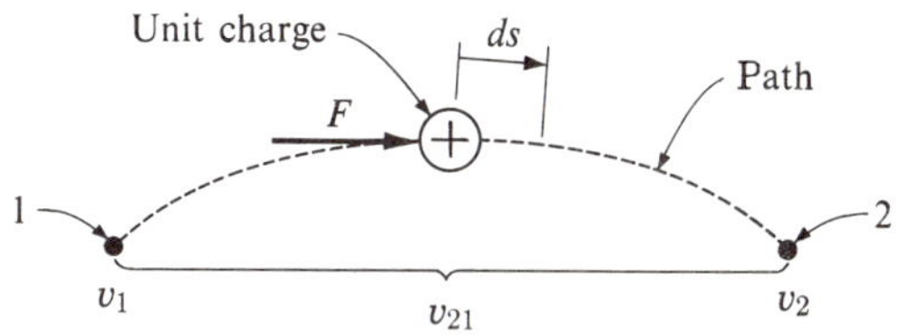

FIG. 2–25. Voltage difference between two points.

The *voltage* (or potential difference) v_{21} between two points such as 1 and 2 in Fig. 2–25 is defined as the work which would be done (or the energy required) in carrying a *unit positive charge* from one point to the other. For the concept of voltage to be applicable, it is necessary that the work determined in this way be independent of the path along which the charge is moved.* If it is necessary to do work on a positive charge (i.e., to apply a force in the direction of motion, or more precisely, if the integral of $F{\cdot}ds$ over the path is positive) as it moves from point 1 to point 2, then point 2 is said to be at a higher potential than point 1 and the voltage v_{21} is considered positive.

Since it is often not known *a priori* which of two points will be at the higher potential or voltage, it is necessary to establish an algebraic reference convention for voltage. In Fig. 2–25 the voltage difference $v_2 - v_1 = v_{21}$ is the voltage of point 2 minus that of point 1. Thus v_{21} is positive when $v_2 > v_1$, and negative when $v_2 < v_1$.

* This requirement implies that the rate of change of the magnetic field in the region of the path must be either zero or negligible.

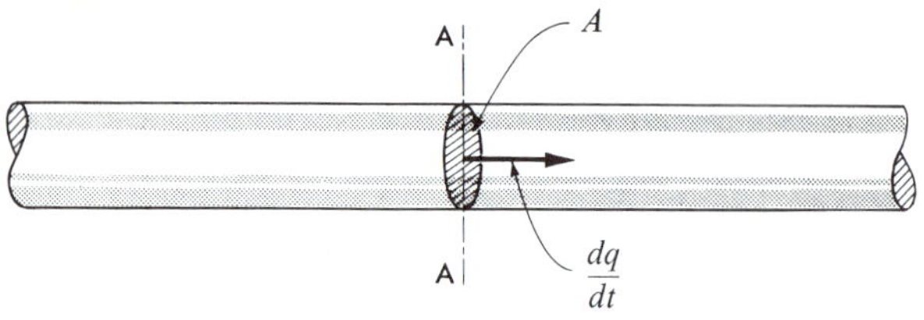

FIG. 2–26. Current flow in a wire.

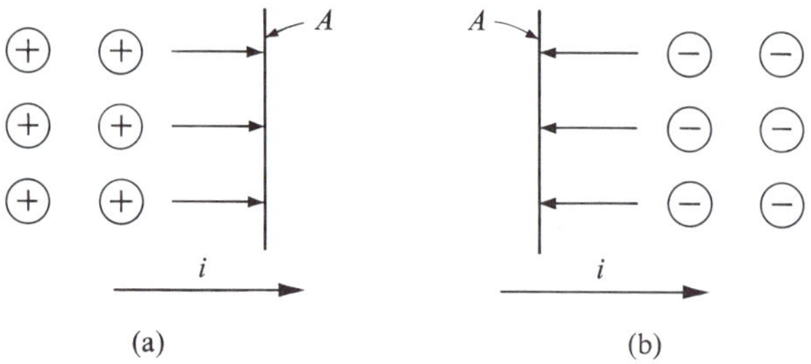

FIG. 2–27. Conventions for positive current. (a) Positive charge flowing across A from left to right. (b) negative charge flowing across A from right to left.

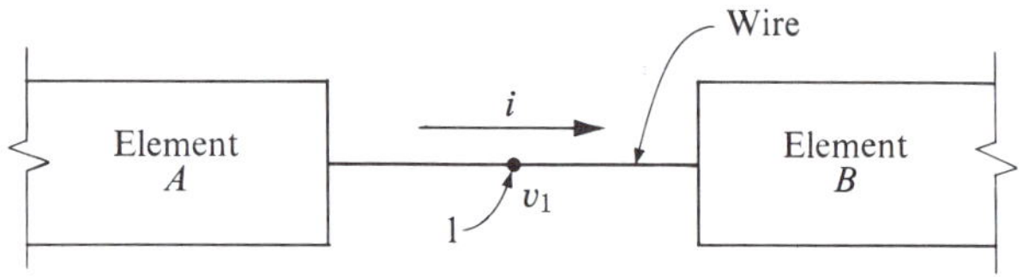

FIG. 2–28. Current flow from one element to another.

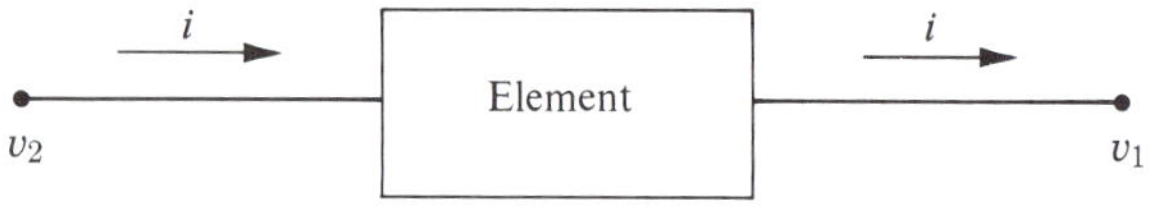

FIG. 2–29. Two-terminal electrical system element.

From the definition of voltage, the amount of work $d\mathcal{W}_{21}$ done when an infinitesimal charge dq is moved from 1 to 2 is $d\mathcal{W}_{21} = v_{21}\,dq$, or

$$v_{21} = \frac{d\mathcal{W}_{21}}{dq}, \quad \frac{\text{j}}{\text{coul}} = \text{v}. \tag{2–71}$$

One joule (one watt-second) is equivalent to 0.737 ft·lb or 8.84 in·lb. One coulomb is equal in magnitude but opposite in sign to the charge of 6.22×10^{18} electrons. Equation (2–71) is especially important because it provides an essential link between electrical and mechanical systems.

Current is defined as the rate of flow of charge across a given area, often the cross-sectional area of a wire.

The current at section A–A in Fig. 2–26 is equal to the net charge crossing area A per unit time. This current may be changing with time; therefore the measurement should be made in a very short time, infinitesimal in the limit. Thus, if charge dq crosses area A in time dt, then the current is

$$i = \frac{dq}{dt}, \quad \text{amp} = \frac{\text{coul}}{\text{sec}}. \tag{2–72}$$

This current has a direction as well as a magnitude. The direction convention commonly chosen is that direction in which a net flow of *positive* charge has occurred. Thus the two situations shown in Fig. 2–27 give a current flow from left to right.

Frequently the direction of a current will not be known *a priori*, and thus it is necessary to establish an algebraic reference for current. This reference is shown by an arrow which indicates the direction of positive current. Thus in Fig. 2–28 i is positive when the current is from left to right. When the current is from right to left, i is negative.

When two elements are connected together as shown in Fig. 2–28, the *principle of conservation of electric charge* requires that the net charge leaving element A must enter element B. This principle also requires that the charge entering an element minus the charge leaving the element must equal the change of net charge stored within the element. We shall restrict ourselves in this book to situations in which no accumulation of net charge occurs over any time period. Thus the simplest electrical system element is one which has two terminals through which charge may enter and leave the element. The charge and current going into the element must equal the charge and current leaving (Fig. 2–29). This situation is similar to the equal forces which exist at both ends of a spring or damper.

Continuity of flow of charge in a two-terminal electrical element corresponds to continuity of force in a two-terminal mechanical element. Electric elements transmit current as mechanical elements transmit force.

It is evident from the stated condition of current continuity that current can flow in an electrical element only if a *circuit* or closed path for current exists. Hence systems of interconnected electrical elements are called *electric circuits*, and the elements themselves are called *circuit elements*.

2–6.2 Power, Work, and Energy

The electrical power $\mathcal{P}$ delivered *to* a two-terminal electrical element can be computed from the definitions of voltage and current, Eqs. (2–71) and (2–72). For the element of Fig. 2–29,

$$\mathcal{P} = \frac{d\mathcal{W}_{21}}{dt} = \frac{dq}{dt}\frac{d\mathcal{W}_{21}}{dq},$$

$$\mathcal{P} = iv_{21}, \quad \text{j/sec} = \text{watts}, \tag{2–73}$$

where i is the current flowing through the element and v_{21} is the voltage difference across it.

The electrical energy $\mathcal{E}$ supplied to an element is the time integral of the power

$$\mathcal{E} = \int_{t_a}^{t_b} v_{21}i\,dt, \quad \text{j or n·m}, \tag{2–74}$$

where t_a and t_b are the beginning and end of the time interval during which power flows. If energy is delivered to an electrical system through more than one pair of terminals, the total energy supplied is the sum of the energies supplied at all the terminals.

Equation (2–73) is really equivalent to Eq. (2–6). In the mks system of units, mechanical and electrical power are both measured in watts. One watt is equal to one joule/sec or one n·m/sec. Since one horsepower (hp) is equal to 550 ft-lb/sec and one joule is equal to 0.737 ft-lb, one horsepower is equivalent to 746 watts. We shall see that there are many similarities between electrical and mechanical systems and between the equations relating system variables.

2–6.3 Pure Electrical System Elements

We have stated that in many instances mechanical systems may be modeled by idealized lumped elements which represent the essential mechanical energy storage, dissipation, and transformation phenomena. In like fashion, electrical systems may often be described, with sufficient accuracy for engineering purposes, by a similar set of idealized elements which represent essential electrical phenomena. These elements are *capacitance* (which stores energy in an electric field), *inductance* (which stores energy in a magnetic field), *resistance* (which dissipates energy), and the *transformer* (which transforms energy without dissipation).

Pure capacitance. If two pieces of conducting material are so arranged that they are separated by a dielectric material (a material in which an electric field can be established without allowing a significant flow of charge through it), an electric field is established between the conductors when charge flows into one conductor and out of the other. This electric field results in a potential difference between the two conductors which depends on the amount of charge placed on the conductors. Physical devices which exhibit this type of relation between charge and voltage are said to have *capacitance*. To permit this capacitive effect to be separated or sorted out from other electrical phenomena, we shall define a *pure capacitance*. A pure capacitance is a two-terminal electrical element in which charge q is a single-valued function of the voltage difference across the element:

$$q = f(v_{21}). \tag{2–75}$$

Figure 2–30 shows the symbol to be used for a pure capacitance. This symbol represents two parallel flat plates separated by a dielectric. The assumed positive direction of current flow is indicated by an arrow drawn beside the element. Usually the sign convention for voltage will be taken so that voltage drops in the direction of current flow. Thus in Fig. 2–30, v_{21} will usually be taken as algebraically positive.

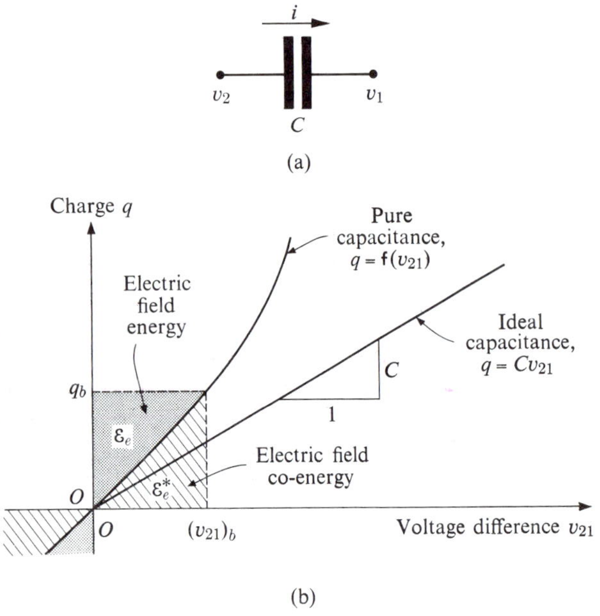

FIG. 2-30. Symbolic representation and constitutive relationships for a capacitance. (a) Symbolic representation. (b) Constitutive relationships.

Equation (2-75) is called the *constitutive relationship* for a pure capacitance. An *ideal capacitance* has charge proportional to voltage difference,

$$q = Cv_{21}, \tag{2-76}$$

where C is the capacitance of the element, having units of farads (f = amp·sec/v).

The elemental equation for an ideal capacitance of constant C is therefore

$$i = C\frac{dv_{21}}{dt}, \tag{2-77}$$

or

$$v_{21} = \frac{1}{C}\int_0^t i\,dt + (v_{21})_0, \tag{2-78}$$

where $(v_{21})_0$ is the voltage across the capacitance when $t = 0$.

In an ideal capacitance, the conductors offer zero resistance to flow of charge, the dielectric separating the conductors has infinite resistance to the flow of charge, and magnetic field effects are absent.

When charge is caused to flow into a capacitance, energy $\mathcal{E}_e$ is transferred to the element. From the power equation (2-73), the definition of current, (2-72), and the defining equation for a pure capacitance (2-75),

$$\mathcal{E}_e = \int_0^{q_b} v_{21}\,dq, \tag{2-79}$$

the shaded area in Fig. 2-30(b). If charge is now allowed to flow out of the capacitance until q_b and v_{21} return to zero, an amount of energy $\mathcal{E}_e$ will be taken *out* of the element. Hence, the energy $\mathcal{E}_e$ is *stored* by the capacitance. A capacitance stores energy in its electrostatic field; therefore $\mathcal{E}_e$ is called the *electric-field energy*. The area below the q-v_{21} curve is the integral of $q\,dv_{21}$ and is called the *electric-field co-energy*, $\mathcal{E}_e^*$. For an ideal capacitance, $\mathcal{E}_e = \mathcal{E}_e^*$.

The energy stored in an ideal capacitance is obtained from (2-79) with $q = Cv_{21}$:

$$\mathcal{E}_e = \tfrac{1}{2}Cv_{21}^2 = \frac{1}{2}\frac{q^2}{C}, \quad \text{j}. \tag{2-80}$$

The process of energy storage is reversible, and all the electrical energy stored in an ideal capacitor is retrievable in the form of electrical energy. Note that the energy stored is a function of the square of the voltage across the capacitor. It depends neither on the sign of the voltage nor on the instantaneous value of the current:

$$\mathcal{E}_e \geqq 0.$$

Most electrical components exhibit the property of capacitance to some extent. A device designed to behave primarily like a capacitance is called a *capacitor*. Figure 2-31 shows several physical components which can often be approximately represented by ideal capacitances. The capacitance of structures such as these depends only on the geometry and the material properties. For example, it may be recalled from physics that the capacitance of the closely spaced parallel-plate capacitor shown in Fig. 2–31(a) is

$$C = \epsilon A/d,$$

where C is the capacitance in farads, ϵ is the permittivity (a property) of the medium between the plates in farads/meter (f/m), A is the plate area in m^2, and d is the plate spacing in m. For air,

$$\epsilon = 8.85 \times 10^{-12} \cong 1/36\pi \times 10^{-9} \text{ f/m}.$$

In general, the calculation of capacitance for particular structures [e.g., that in Fig. 2–31(b)] is very difficult and requires advanced methods of electric-field theory. (See [3] for formulas for the capacitance of typical structures.) Of course the capacitance of an existing element can always be determined experimentally.

Pure inductance. When current flows through a conducting structure, a magnetic field is established in the space or the material around the structure. If this current is changed as a function of time, the intensity of the magnetic field will also vary with time. According to Lenz's law, this changing field will induce voltage differences in the conducting structure which will tend to oppose the changing current. The basic characteristic by which an electrical element resists with a voltage difference the rate of change of current flow through it is called *inductance*.

Let us define a quantity *flux linkage*, λ, in the following way. For any electrical element having two terminals, 1 and 2 (Fig. 2–29), by definition

$$v_{21} = \frac{d\lambda_{21}}{dt} \tag{2–81}$$

or

$$\lambda_{21} = \int_0^t v_{21}\, dt + (\lambda_{21})_0, \tag{2–82}$$

where $(\lambda_{21})_0$ is the flux linkage when $t = 0$. The unit of flux linkage is webers (= v·sec). The similarity between these equations and Eqs. (2–4) and (2–20) should be noted.

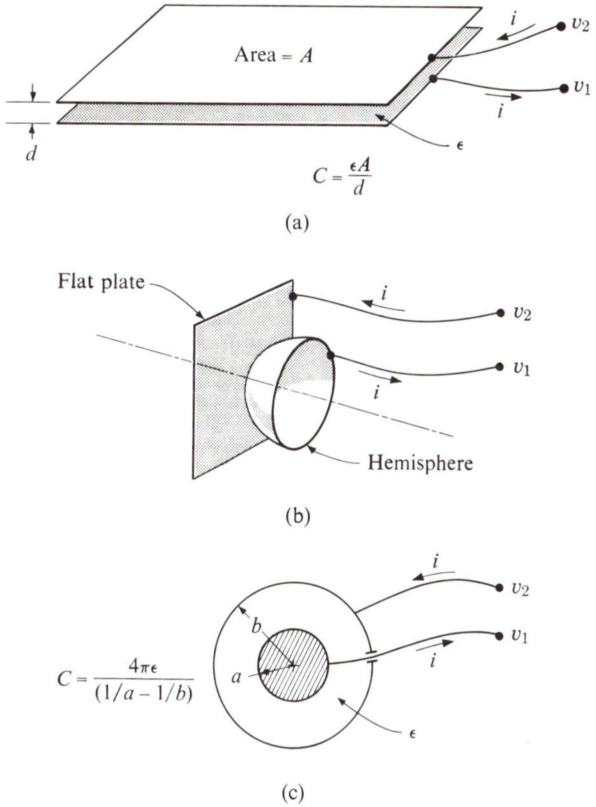

FIG. 2–31. Typical structures which exhibit capacitance. (a) Parallel-plate capacitor. (b) Plate and hemisphere. (c) Concentric spheres.

An electric element is said to be a *pure inductance* provided that the flux linkage λ_{21} associated with its terminal voltage difference is a single-valued function of the current i flowing through the element,

$$\lambda_{21} \doteq \mathsf{f}(i). \tag{2–83}$$

A linear pure inductance, which we shall call an *ideal inductance*, has flux linkage proportional to current:

$$\lambda_{21} = Li, \tag{2–84}$$

where the inductance L is measured in henries, h (henry = v·sec/amp), and λ_{21} is defined to be zero when $i = 0$. Figure 2–32(b) shows the constitutive relations (2–83) and (2–84) for pure inductances. The

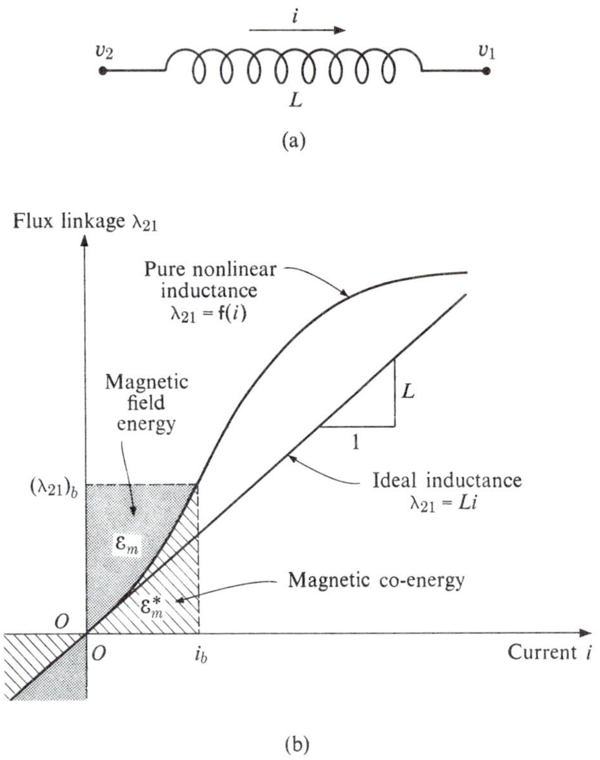

FIG. 2–32. Symbolic representation and constitutive relationships for a pure inductance. (a) Symbolic representation. (b) Constitutive relationships.

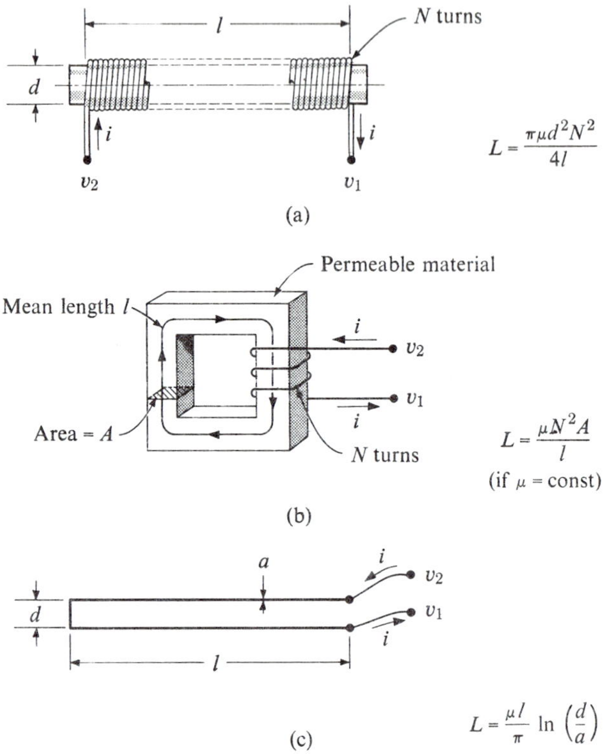

FIG. 2–33. Typical structures which exhibit inductance. (a) Long solenoid. (b) Permeable-core inductor. (c) Two long thin parallel wires with short circuit at one end.

classical example of a structure designed to exhibit inductance is a helical coil of conducting wire; therefore, the schematic diagram shown in Fig. 2–32(a) is used to represent a pure inductance.

The elemental equation for an ideal inductance of constant L follows from Eqs. (2–81) and (2–84):

$$v_{21} = L\frac{di}{dt}, \tag{2–85}$$

or

$$i = \frac{1}{L}\int_0^t v_{21}\,dt + i_0, \tag{2–86}$$

where i_0 is the current when $t = 0$.

An inductance stores energy in the magnetic field associated with the current. The electrical energy stored in a pure inductance is called the magnetic-field energy, $\mathcal{E}_m$:

$$\mathcal{E}_m = \int_{t_a}^{t_b} v_{21} i\,dt = \int_{\lambda_a}^{\lambda_b} i\,d\lambda_{21}. \tag{2–87}$$

The energy $\mathcal{E}_m$ is represented by the shaded area in Fig. 2–32(b). The area $\mathcal{E}_m^*$ below the λ-i curve is the magnetic co-energy.

If the inductance is ideal, Eqs. (2–84) and (2–87) give

$$\mathcal{E}_m = \tfrac{1}{2}Li^2 = \frac{1}{2}\frac{(\lambda_{21})^2}{L}, \quad \text{j.} \tag{2–88}$$

This energy is stored by virtue of the current and does not directly depend on the voltage difference. The

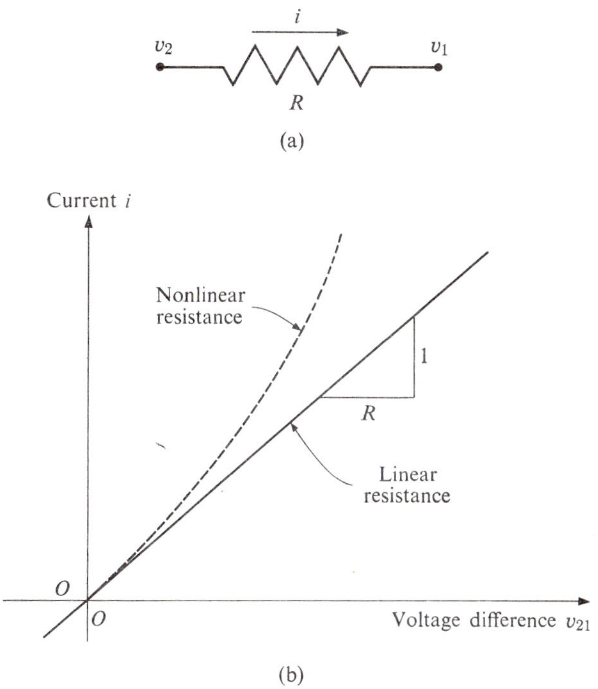

FIG. 2–34. Symbolic representation and constitutive relationships for a pure resistance. (a) Symbolic representation. (b) Constitutive relationships.

energy is always positive,

$$\mathcal{E}_m \geqq 0,$$

and any energy stored by increasing the current from i_a to i_b is recovered completely when the current is decreased to i_a.

An electrical configuration which behaves primarily like an inductance is called an *inductor*. As with capacitance, the computation of the inductance for a given structure may be difficult and will usually require the use of field theory.* It is noted that inductance is a property determined purely by geometry and material properties.

* Formulas for computing inductance and capacitance may be found in [3] and [4].

The inductance of a closely wound, long solenoid coil can be calculated in a relatively simple manner. In terms of the length l, diameter d, and number of turns N, the inductance of the coil shown in Fig. 2–33(a) is

$$L = \frac{\pi N^2 \mu d^2}{4l}, \quad \frac{\text{v·sec}}{\text{amp}} = \text{h},$$

where μ is the permeability of the material in the vicinity of the coil. For a vacuum (or very nearly for air),

$$\mu = 4\pi \times 10^{-7}, \quad \text{h/m}.$$

The inductance of a coil in air will be nearly linear, but the magnitude of the inductance is very small. This magnitude may be increased by factors up to 10^6 by the use of ferromagnetic materials such as iron, cobalt, and nickel. However, inductors which use ferromagnetic materials are nonlinear and give responses somewhat like the nonlinear curve shown in Fig. 2–32(b). The flattening of this curve as i is increased is called *saturation.*

Figure 2–33 shows several structures which may act approximately like pure inductances.

Pure resistance. All ordinary materials exhibit some resistance to the flow of electric charge. Materials in which this resistance is small are called *conductors* (copper, silver, and many other metals), and those in which the resistance is high are called *insulators* (glass, mica, and plastics, for example). We shall call this resistance to the flow of charge simply *resistance.*

A *pure resistance* is defined by a single-valued functional relationship between current and voltage difference, such that $i = 0$ when $v_{21} = 0$,

$$i = \mathsf{f}(v_{21}), \tag{2–89}$$

and the signs of i and v_{21} are the same. An *ideal resistance* is a pure linear resistance; thus its elemental equation is

$$i = \frac{1}{R} v_{21}, \tag{2–90}$$

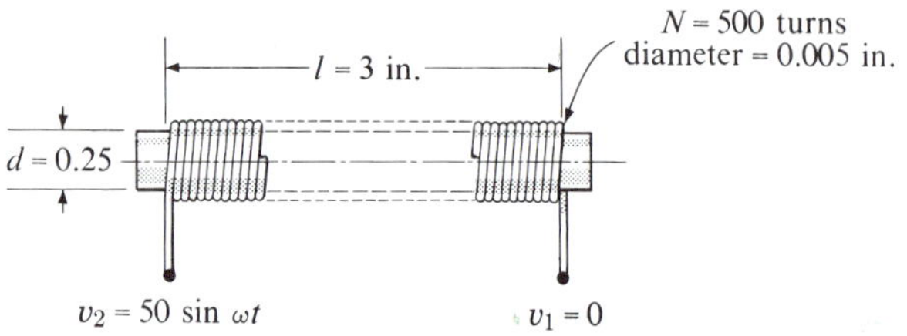

FIG. 2–35. Wire-wound resistor.

where the resistance R is measured in ohms (= v/amp). The constitutive relationships in Eqs. (2–89) and (2–90) are shown in Fig. 2–34(b). The symbol which we shall use for a pure resistance is shown in Fig. 2–34(a).

Electric and magnetic field effects (which result in capacitance and inductance) are zero in a pure resistance. It can store neither electric-field energy nor magnetic-field energy. However, electrical power is delivered to a resistance whenever current flows through it. From Eqs. (2–73) and (2–89) or (2–90),

$$\mathcal{P} = iv_{21} = v_{21}\mathsf{f}(v_{21}) \tag{2–91}$$

for a pure resistance, or

$$\mathcal{P} = \frac{1}{R}(v_{21})^2 = i^2R, \quad \text{watts} \tag{2–92}$$

for an ideal resistance. Power always flows into the resistance regardless of the signs of i and v_{21}:

$$\mathcal{P} \geqq 0.$$

The power transferred to a resistance cannot be retrieved and is said to be *dissipated.* Usually the power is transformed into thermal internal energy in the resistance. The temperature rise which accompanies the increase in internal energy usually results in flow of heat to the environment. This heat might be useful or harmful, but it is not readily converted back into electrical power. Although resistance might at first glance seem to be undesirable because of the power loss expressed by Eq. (2–92), it will be seen later that resistance can have many useful functions in systems. For instance, a typical television receiver may have 100 or more *resistors* (devices designed to exhibit primarily resistance) intentionally built into it and necessary for the function of its circuits.

The simplest form of resistor is a fine metal wire or a thin carbon film. Resistors are encountered in many shapes and forms, often designed to provide a needed resistance effect at some place in an electric system, but sometimes as an unwanted (parasitic) effect.

Example 2. A 500-turn coil of steel wire 0.005 in. in diameter is closely wound around a plastic cylinder of diameter d = 0.25 in. and length l = 3.0 in. (Fig. 2–35). A sinusoidally varying voltage difference,

$$v = 50 \sin \omega t, \text{ v.}$$

is applied to the ends of the coil, where ω is the frequency in rad/sec and t is time in seconds. We wish to find the range of frequency ω, if any, over which this coil behaves essentially like a pure resistance. The resistivity ρ (resistance times area divided by length) of the wire is

$$12 \times 10^{-6} \text{ ohm·cm.}$$

Since the permeability μ of plastic and air is about $4\pi \times 10^{-7}$ h/m, the coil inductance, L, is

$$L = \frac{\pi N^2 \mu\, d^2}{4l} = \frac{\pi(500)^2 4\pi \times 10^{-7}(0.25)^2 2.54 \times 10^{-2}}{12}$$

$$= 13.1 \times 10^{-5} \text{ h.}$$

The coil resistance R is

$$R = \frac{\rho N d}{r^2} = \frac{12 \times 10^{-6} 500(0.25)}{(0.0025)^2 2.54} = 94.5 \text{ ohms.}$$

If a voltage difference $v = v_{21}$ is applied across the terminals of this coil, part of the voltage will be used to overcome the coil resistance and part will be used to produce a rate of change of current through the coil inductance. The current i which the voltage v_{21} would produce in an ideal resistance is given by Eq. (2–90),

$$i = (1/R)v_{21} = (1/R)50 \sin \omega t, \quad \text{amp,}$$

and then $di/dt = (50\omega/R) \cos \omega t$, amp/sec.

The voltage necessary to cause this di/dt in an inductance L is given by Eq. (2–85):

$$(v_{21})_{\text{inductance}} = L\,(di/dt).$$

The coil will act like an ideal resistance, provided that the inductive voltage is negligible compared with the applied voltage. If we take "negligible" to mean less than one percent, $(v_{21})_{\text{inductance}} = 0.50$ v, then

$$(50L\omega/R)\cos\omega t < 0.50\sin\omega t.$$

This condition can never be met for all times t, since at $t = 0$, $\sin\omega t = 0$. However, if we compare only peak values of v and $(v_{21})_{\text{inductance}}$ we have a nearly ideal resistance if

$$\omega < 0.01R/L$$
$$< 7210 \text{ rad/sec or } 1150 \text{ cycles/second (cps).}$$

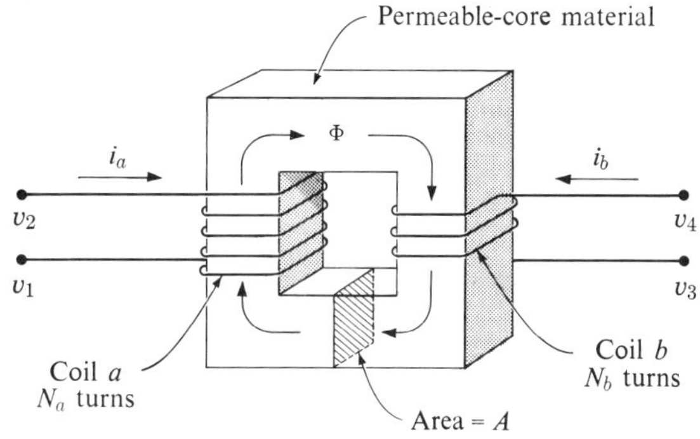

FIG. 2–36. Electrical transformer.

2–6.4 Pure Electrical Transformer

A *pure electrical transformer* is an idealized device which does not store or dissipate energy, but determines an output-voltage difference as a single-valued function of an input-voltage difference. An ideal transformer has a linear relationship between input and output voltages.

In Fig. 2–36, two coils of N_a and N_b turns are wound on a ferromagnetic core whose permeability μ is very large compared with that of air. Now suppose that a magnetic flux Φ is completely confined to the core, i.e., that no flux "leaks" out. Then all of the flux which links coil a will also link coil b. According to Faraday's law and the definition of λ_{21}, the flux linkage λ_{21} in coil a is

$$\lambda_{21} = \Phi N_a,$$

and the flux linkage λ_{43} for coil b is

$$\lambda_{43} = \Phi N_b.$$

Therefore,

$$\lambda_{43}/\lambda_{21} = N_b/N_a. \tag{2–93}$$

Since the voltage differences are equal to the rates of

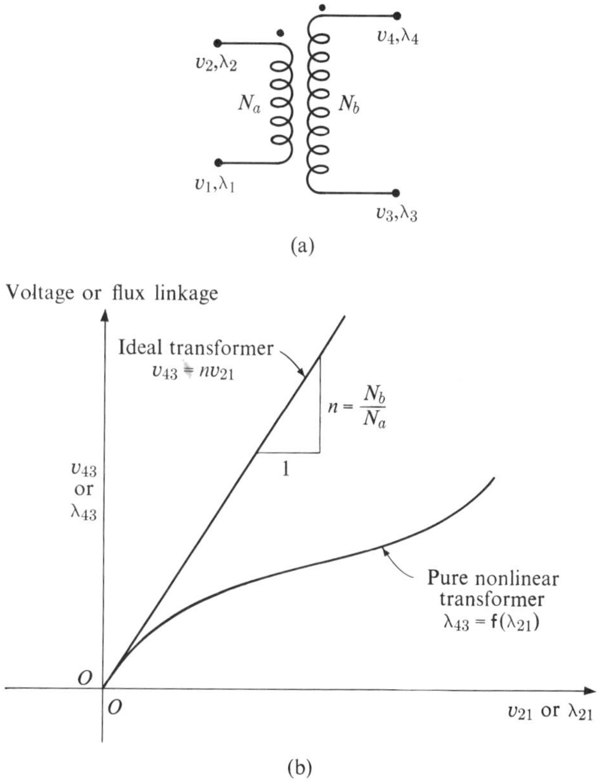

FIG. 2–37. Pure electrical transformer. (a) Symbolic representation. (b) Functional relationships.

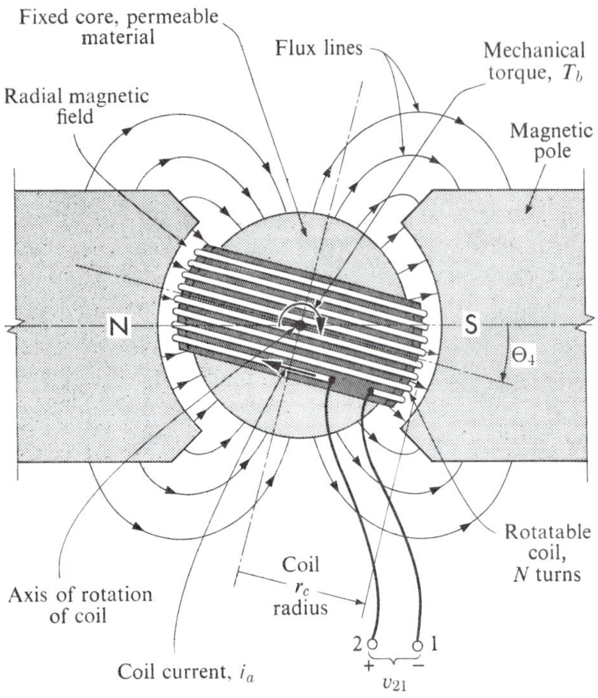

FIG. 2–38. Schematic diagram of rotary electromagnetic electrical-to-mechanical transducer.

change of flux linkage, provided that the coil resistances are negligible, it follows that

$$v_{43}/v_{21} = N_b/N_a = n = \text{turns ratio.} \tag{2–94}$$

The sign of v_{43} relative to v_{21} depends on the relative winding directions of the coils. In Fig. 2–36, v_{43} is positive when v_{21} is positive.

If the permeability μ of the core is very large, if no capacitive effects are present, and if no energy losses exist, the device shown in Fig. 2–36 has the input power to coil a plus the input power to coil b equal to zero:

$$i_a v_{21} + i_b v_{43} = 0,$$

or

$$i_b/i_a = -v_{21}/v_{43} = -1/n. \tag{2–95}$$

Equations (2–94) and (2–95) describe an *ideal electrical transformer.*

Figure 2–37(a) shows the symbol to be used for pure and ideal electrical transformers. The sign of the transformation ratio is often indicated by dots. That is, when the dotted terminal on the left coil has positive voltage relative to its opposite terminal, the dotted terminal on the right coil is positive relative to its opposite terminal.

2–6.5 Electromechanical Transducers

The concept of an energy transducer was introduced in Section 2–5. One of the most important classes of transducers is the electromechanical transducer, which converts energy from electrical to mechanical form and vice versa. Electric motors and generators, piezoelectric and electrostrictive crystals, loudspeaker coils, solenoids, and relays are all examples of electromechanical transducers.

One common electromechanical transducer is illustrated in Fig. 2–38. This device transduces electric current into mechanical torque and voltage into angular velocity, or conversely. A nearly radial magnetic field of strength B (flux per unit area in webers/m^2) is established by two opposite magnetic poles in an annular air gap formed between the pole pieces and a fixed cylindrical core of permeable material. A movable coil of N turns is mounted free to rotate about the axis of the fixed core (Fig. 2–38). When a current i_a passes through the coil, which is in the radial magnetic field, a mechanical torque T_b results from the interaction of the current and the magnetic field. This torque may be derived from the Biot-Savart law [5], and is proportional to the flux density B, the total length $2Nl$ of conductor lying in the field, the coil radius r_c, and the current i_a:

$$T_b = -2NBlr_c i_a = -\frac{1}{n} i_a. \tag{2–96}$$

The quantity n which relates T_b to i_a is called the *transduction ratio* or *electromechanical coupling constant.*

Since the magnetic field is radial, the force on the coil (which is perpendicular to both the current and magnetic field directions) is always tangential to the coil radius; hence the torque is independent of the coil angle Θ_4.

If the coil moves with angular velocity $\Omega_4 = \dot{\Theta}_4$ relative to the magnetic field, a voltage difference v_{21} will be induced across the terminals 2 and 1 of the coil. This voltage opposes the current i_a by Lenz's law and is given by Faraday's law as the rate of change of flux linkage of the coil:

$$v_{21} = N\frac{d\Phi}{dt} = 2NBlr_c\Omega_4. \tag{2–97}$$

Comparison of this equation with Eq. (2–96) gives

$$\Omega_4 = nv_{21}. \tag{2–98}$$

From Eqs. (2–96) and (2–98), the net power $\mathcal{P}$ into the device is identically zero:

$$\begin{aligned}\mathcal{P} &= v_{21}i_a + T_b\Omega_4\\ &= v_{21}i_a + \left(-\frac{1}{n}i_a\right)(v_{21}n) = 0.\end{aligned}$$

This transducer is used in the movements of most electrical meters for measuring voltage or current. The d'Arsenval galvanometer movement utilizes a rotary transducer of this type in conjunction with a rotary spring to give a displacement of a pointer on a scale proportional to the current through the transducer. The dynamic properties of a real galvanometer movement are analyzed in some detail in Chapter 7.

An analogous transducer in which the coil *translates* in a radial magnetic field is used in most loudspeakers. Here electrical energy is transduced into the mechanical motion of the loudspeaker cone and then into acoustic energy.

Other electromechanical transducers will be considered in the text and problem material of subsequent chapters. Detailed treatments of such transducers are available at both elementary [6] and advanced levels [7, 8].

2–7. SIMILARITIES, ANALOGIES, AND GENERALIZATIONS ABOUT IDEAL SYSTEM ELEMENTS

The previous sections of this chapter have dealt with a number of individual basic pure elements which are to be used as building blocks in the modeling of mechanical and electrical systems. It has undoubtedly occurred to the reader that there are a number of similarities between certain of these elements. For example, if we compare the elemental equation (2–15) for an ideal mass,

$$F = m\frac{dv_{21}}{dt} = m\frac{dv_2}{dt},$$

with the elemental equation (2–77) for an ideal capacitance,

$$i = C\frac{dv_{21}}{dt},$$

we see that the equations are similar. If we substitute force for current, mass for capacitance, and velocity for voltage (we have intentionally used the same symbol for velocity and voltage) we can obtain the mass equation from the capacitance equation. Two ideal elements related in this way are said to be *analogous*.

Why do we choose to draw comparisons between force and current and between velocity and voltage? The basis of these comparisons is the idea of *through*- and *across*-variables.

2–7.1 Through- and Across-Variables

A *variable* is a measurable characteristic of a system which may change with time. It has been seen that a lumped system element is usually described by a relationship between two physical variables, a *through*-variable, which has the same value at the two terminals or ends of the element, and an *across*-variable, which is specified in terms of a relative value or difference between the terminals.

Insight into the criteria for defining through- and across-variables can be gained by a consideration of the method by which a particular variable would be measured in the actual physical system. In fact, it is from this consideration that the names "through" and "across" arise. Forces and torques can be measured by means of a calibrated spring scale. To measure the force or torque at a point, we must *sever* the system at that point and insert the spring scale between the two resultant sections. Current is measured by an ammeter, which is a device with a coil suspended in a magnetic field and restrained by a spring. The current to be measured must flow through this coil. Thus an electrical system containing a current to be measured must be broken into at the point at which the current is to be determined, and the ammeter must be inserted between the two sections. Therefore, force, torque, and current are thought of as being applied "through" the measuring device and hence can be described by the common name *through-variable.* We shall see later that fluid flow and heat flux are also through-variables.

A voltmeter, which is used to measure voltage, has two terminals, which are connected to the points between which the voltage is to be determined. It is not necessary to make any severing entry into the system; the connections can be made without such a step. To measure angular velocity, a device is used that makes a measurement of the speed of rotation of an input shaft with respect to the speed of rotation of another part of the device. Thus a tachometer-generator measures the speed of a motor rotor with respect to the stator. The student should not be misled by normal usage where the stator is not rotating. Typical instruments for measuring translational velocity often seem at first glance to only have one point of attachment. However, closer inspection shows that they operate on a principle which involves the motion (or absence of motion) of a mass, and that the measurement is thus being made with respect to an inertial reference frame which is, in most cases, approximated by the earth. A more general velocity meter could be envisioned as a device which determines the rate of separation of two points within it, each of which is rigidly connected to the two points between which the velocity is desired. With any of these velocity meters (angular or translational) the measurement can be made by simply appropriately attaching the measuring device to the system; it is not necessary to break into the system. Thus velocity, angular velocity, and voltage are thought of as existing across two points and can be described by the common name of *across-variables.* Similarly, in fluid and thermal systems, pressure *difference* and temperature *difference* will be seen to be across-variables.

We shall use the general symbols f and v to stand for any through- and across-variables, respectively.

2-7.2 Linear-Graph Representation and Summary of Pure Elements

The two-terminal elements discussed thus far are characterized by functional relationships between their through- and across-variables. A convenient symbol for this relationship is the *linear graph* shown in Fig. 2-39(a). The two ends or terminals of this graph indicate the across-variables for the element, and the line between these terminals represents the continuity of the through-variable in the element. The term "linear" in "linear graph" means that the graph is defined by a line segment and should not be confused with "linear" as applied to the constitutive relationship for an ideal element.

The sign conventions selected for the across-variable difference and for the through-variable can be shown by a single arrowhead drawn on the graph as in Fig. 2-39(b). The graph is then said to be *oriented.* The arrow pointing from 2 to 1 means that v_2 is algebraically larger than v_1 when v_{21} is positive. It also means that f is positive when it flows from 2 to 1, i.e., when it tends to produce a positive v_{21} in the element. These two conventions require that power flow into the element when v_{21} and f are both positive.

Figure 2-39(c) is a free-body diagram of a non-mass mechanical element showing $v_2 > v_1$ and a compressive force F. The oriented graph for this element is shown in Fig. 2-39(d). The mass elements require special treatment because one terminal must

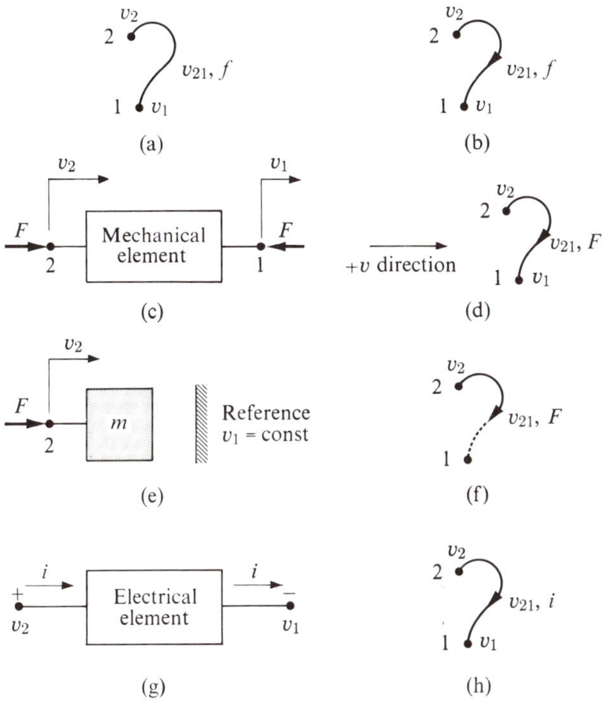

FIG. 2–39. Linear graph representation of pure mechanical and electrical system elements. (a) Linear graph. (b) Oriented linear graph. (c) Nonmass mechanical element. (d) Oriented graph. (e) Translational mass. (f) Oriented graph. (g) Electrical element. (h) Oriented graph.

be a fixed reference and the force is not physically transmitted to that terminal. Figure 2–39(e) shows a free-body diagram and 2–39(f) the oriented graph for a translational mass.

Electrical elements are represented as in Fig. 2–39(g) and (h). Here voltage drops and current flows with the arrow on the oriented graph.

Table 2–1 summarizes the ideal two-terminal mechanical and electrical elements (see page 48).

2–7.3 Power and Energy

The power flow and energy transferred from one element to another at a connecting point is summarized in Table 2–2 for the three systems studied thus far. In each case the power is the product of the through-variable and the across-variable.

TABLE 2–2

Power and Energy Transferred at a Connecting Point

Mechanical-translational	Mechanical-rotational	Electrical
$F \longrightarrow \; \leftarrow F$ (v, x)	$T \longrightarrow \; \longleftarrow T$ (Ω, Θ)	$i \longrightarrow \cdot \longrightarrow i$ (v, λ)
$\mathcal{P} = Fv$	$\mathcal{P} = T\Omega$	$\mathcal{P} = iv$
$\mathcal{E} = \int_{t_a}^{t_b} Fv\,dt$	$\mathcal{E} = \int_{t_a}^{t_b} T\Omega\,dt$	$\mathcal{E} = \int_{t_a}^{t_b} iv\,dt$

TABLE 2–3

Summary of Energy Storage and Dissipation for Ideal Elements

A-type storage	*T*-type storage	*D*-type dissipation
Mass $\mathcal{E}_k = \frac{1}{2}mv_2^2$	Spring $\mathcal{E}_p = \frac{1}{2}\frac{F^2}{k}$	Damper $\mathcal{P} = bv_{21}^2 = \frac{F^2}{b}$
Inertia $\mathcal{E}_k = \frac{1}{2}J\Omega_2^2$	Spring $\mathcal{E}_p = \frac{1}{2}\frac{T^2}{K}$	Damper $\mathcal{P} = B\Omega_{21}^2 = \frac{T^2}{B}$
Capacitance $\mathcal{E}_e = \frac{1}{2}C(v_{21})^2$	Inductance $\mathcal{E}_m = \frac{1}{2}Li^2$	Resistance $\mathcal{P} = \frac{(v_{21})^2}{R} = Ri^2$

The ideal elements have been classified as energy-storage and dissipation elements in earlier sections of this chapter. The mass, inertia, and capacitance store energy by virtue of their *across*-variables (velocity and voltage) and will be called *A-type* energy-storage units. Springs and inductance store energy by virtue of their *through*-variables and will be called *T-type* energy-storage devices. The dampers and resistance dissipate energy and will be called *D-type* elements. Table 2–3 summarizes the energy-storage and dissipation functions for the ideal mechanical and electrical elements.

We have also discussed energy transformers and transducers in which the input power is equal to out-

Ω – (ω) angular velocity

TABLE 2–1

Summary of Ideal Mechanical and Electrical Elements

Name	Symbol	Linear graph	Describing equations	Energy or power
Mass	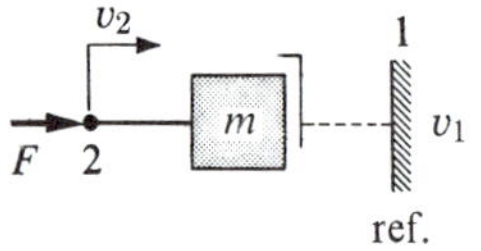	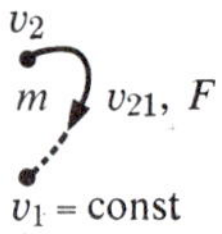	$F = m\dfrac{dv_2}{dt}$ $v_{21} = \dfrac{1}{m}\int_0^t F\,dt + (v_{21})_0$	$\mathcal{E}_k = \frac{1}{2}mv_2^2$ $\mathcal{E}_k \geqslant 0$
Damper	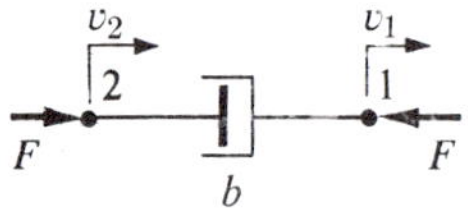	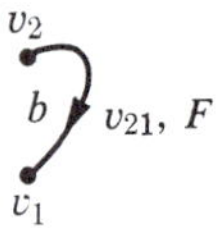	$F = bv_{21}$ $v_{21} = \dfrac{1}{b}F$	$\mathcal{P} = bv_{21}^2$ $\mathcal{P} \geqslant 0$
Spring	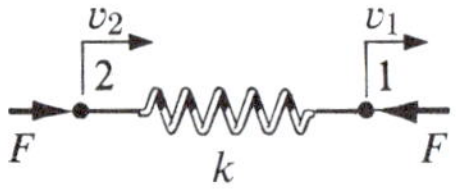	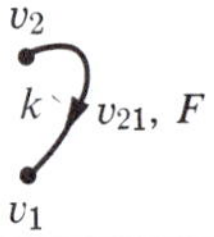	$F = k\int_0^t v_{21}\,dt + F_0$ $v_{21} = \dfrac{1}{k}\dfrac{dF}{dt}$	$\mathcal{E}_p = \dfrac{1}{2}\dfrac{F^2}{k}$ $\mathcal{E}_p \geqslant 0$
Inertia	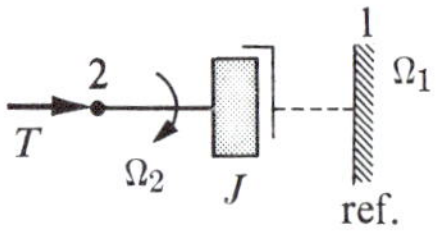	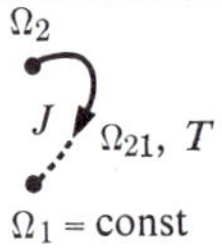	$T = J\dfrac{d\Omega_2}{dt}$ $\Omega_{21} = \dfrac{1}{J}\int_0^t T\,dt + (\Omega_{21})_0$	$\mathcal{E}_k = \frac{1}{2}J\Omega_2^2$ $\mathcal{E}_k \geqslant 0$
Rotational damper	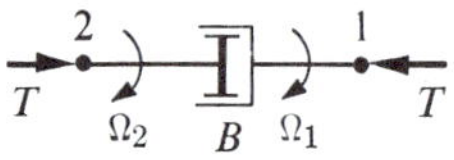	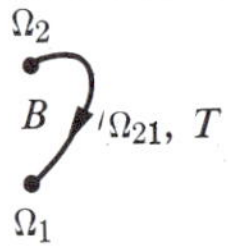	$T = B\Omega_{21}$ $\Omega_{21} = \dfrac{1}{B}T$	$\mathcal{P} = B\Omega_{21}^2$ $\mathcal{P} \geqslant 0$
Rotational spring	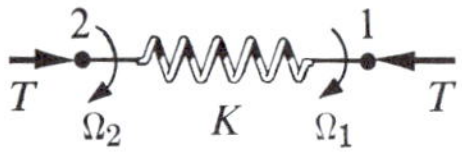	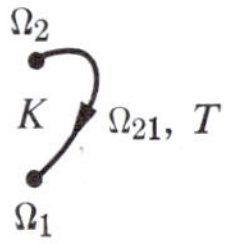	$T = K\int_0^t \Omega_{21}\,dt + T_0$ $\Omega_{21} = \dfrac{1}{K}\dfrac{dT}{dt}$	$\mathcal{E}_p = \dfrac{1}{2}\dfrac{T^2}{K}$ $\mathcal{E}_p \geqslant 0$
Capacitance	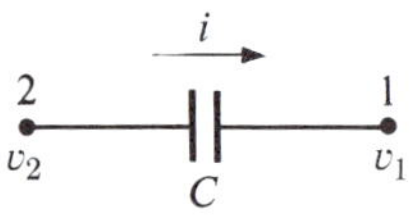	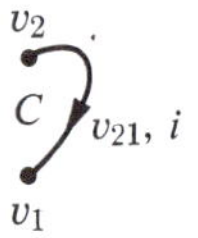	$i = C\dfrac{dv_{21}}{dt}$ $v_{21} = \dfrac{1}{C}\int_0^t i\,dt + (v_{21})_0$	$\mathcal{E}_e = \frac{1}{2}Cv_{21}^2$ $\mathcal{E}_e \geqslant 0$
Resistance	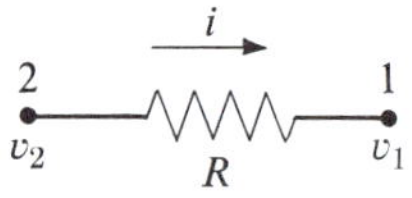	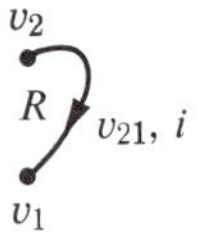	$i = \dfrac{1}{R}v_{21}$ $v_{21} = Ri$	$\mathcal{P} = \dfrac{1}{R}v_{21}^2$ $\mathcal{P} \geqslant 0$
Inductance	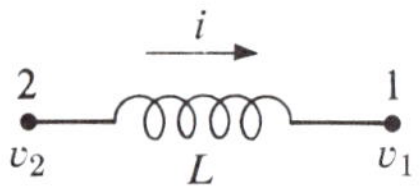	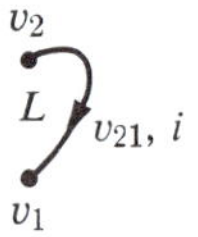	$i = \dfrac{1}{L}\int_0^t v_{21}\,dt + i_0$ $v_{21} = L\dfrac{di}{dt}$	$\mathcal{E}_m = \frac{1}{2}Li^2$ $\mathcal{E}_m \geqslant 0$

TABLE 2–4

Generalized Describing Equations for Lumped Ideal Elements

A-type elements	*T*-type elements	*D*-type elements
Mass, inertia, capacitance	Spring, inductance	Damper, resistance
$f = C\dfrac{dv_{21}}{dt}$	$v_{21} = L\dfrac{df}{dt}$	$v_{21} = Rf$
$v_{21} = \dfrac{1}{C}\int_0^t f\,dt + (v_{21})_0$	$f = \dfrac{1}{L}\int_0^t v_{21}\,dt + f_0$	$f = \dfrac{1}{R}v_{21}$
$\mathcal{E}_a = \frac{1}{2}C\,(v_{21})^2$	$\mathcal{E}_t = \frac{1}{2}Lf^2$	$\mathcal{P} = \dfrac{(v_{21})^2}{R}$
	Linear graph 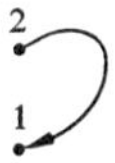 v_{21}, f	

put power. For this type of device, the product of the input through- and across-variables plus the product of the output through- and across-variables is zero.

Later we shall discuss energy *sources* which supply energy and power to dynamic systems.

2–7.4 Generalized Elemental Equations

The behavior of all the ideal mechanical and electrical elements can be described by a single set of three elemental equations for *A*-type, *T*-type, and *D*-type elements, written in terms of the generalized through- and across-variables f and v. Table 2–4 shows these general elemental equations.

The quantity C stands for mass, inertia, or capacitance; L for the reciprocal of spring constant, or inductance; and R for reciprocal damping, or resistance. When C represents mass or inertia, v_1 must be zero or constant.

We shall see in Chapter 3 that fluid and thermal system elements may also be represented by the generalized elemental equations of Table 2–4.

REFERENCES

1. H. P. Robertson, *Modern Physics for the Engineer*, L. N. Ridenour, Ed. New York: McGraw-Hill Book Company, Inc., 1954.
2. R. J. Roark, *Formulas for Stress and Strain*. New York: McGraw-Hill Book Company, Inc., 1954.
3. *Handbook of Chemistry and Physics*. Cleveland: The Chemical Rubber Publishing Company, 1961.
4. F. Langford-Smith, *Radiotron Designer's Handbook*, 4th Ed. Sydney, Australia: Wireless Press. (Distributed by Radio Corporation of America, Hanson, N.J.)
5. J. D. Kraus, *Electromagnetics*. New York: McGraw-Hill Book Company, Inc., 1953.
6. H. H. Skilling, *Electromechanics*. New York: John Wiley and Sons, Inc., 1962.
7. D. C. White and H. H. Woodson, *Electromechanical Energy Conversion*. New York: John Wiley and Sons, Inc., 1959.
8. W. P. Mason, *Electromechanical Transducers and Wave Filters*. New York: D. Van Nostrand Company, Inc., 1958.

PROBLEMS

2–1. As a passenger train leaves a station on a straight track, its velocity increases according to the function of time shown in Fig. 2–40. Thirty seconds after the train leaves, a boy at the rear of the train begins to run forward at a constant speed of 5 ft/sec relative to the train. Find and sketch versus time:

a) the absolute acceleration of the boy,

b) the absolute velocity of the boy,

c) the position of the train relative to the station,

d) the position of the boy relative to the train,

e) the position of the boy relative to the station.

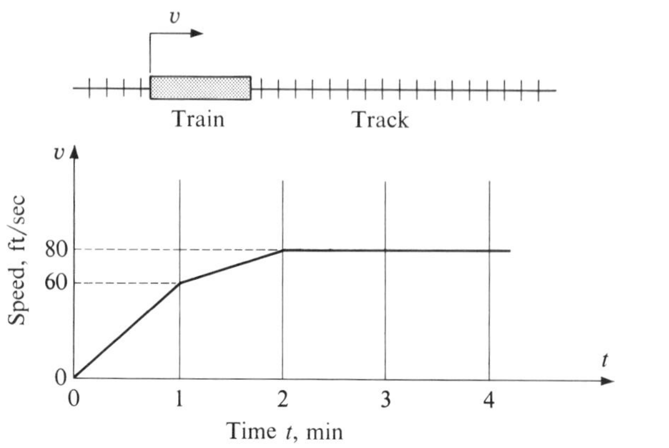

FIGURE 2–40

2–2. In Problem 2–1, find as functions of time the horizontal force that the driving wheels must produce on the train and the horizontal force that the boy must exert against the floor of the train to cause the stated motions. The train weighs 15 tons and the boy 90 lb.

2–3. In Problem 2–1, find the power delivered by the train's engine and by the boy's muscles versus time in producing the given motions.

2–4. The force applied to a point and the velocity of that point are shown graphically in Fig. 2–41. (The system itself is very complex to produce such a velocity variation. It may contain other forces within it.) The velocity reference is in the same direction as the force reference (i.e., positive force is in the same direction as positive velocity). Plot a graph of the power input to the system at this point, as well as a graph of the energy input between time zero and t as a function of t.

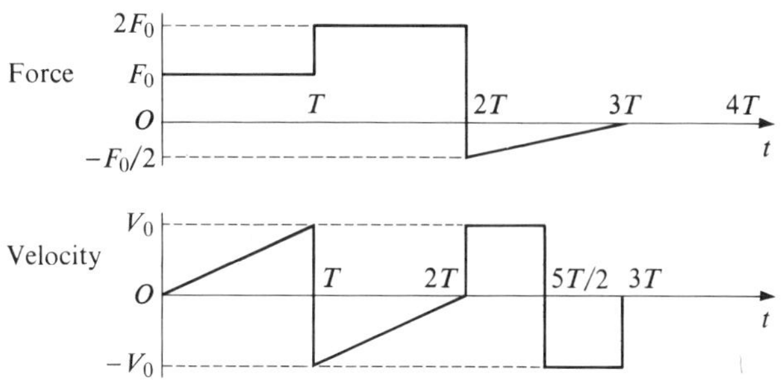

FIGURE 2–41

2–5. Power is the time rate of flow of energy. Consider a mechanical system in which a force acts through an infinitesimal distance dx (for the case we consider here, the force and the displacement are in the same direction). The (infinitesimal) amount of work done is $d\mathcal{W}$, and

$$d\mathcal{W} = F\,dx.$$

The time rate at which this work is being done, which is the time rate of flow of energy into the system, is $\mathcal{P}$:

$$\mathcal{P} = \frac{d\mathcal{W}}{dt} = \frac{F\,dx}{dt} = Fv,$$

where v is the velocity of the point at which the force is applied.

a) Check the units of $\mathcal{P} = Fv$ to show that they are consistent.

b) For an accelerating mass m moving at velocity v, show that this relation applies by showing that the time rate of increase of stored energy is equal to Fv.

c) For a spring with one end free, the force required to displace the spring an amount x from rest is proportional to x (Hooke's Law); i.e., $F = kx$, where k is the spring constant (Fig. 2–42). Determine the amount of (potential) energy stored in a stretched spring by using $\mathcal{P} = Fv$ and $F = kx$ to determine the energy input to the spring, required to stretch it an amount x. (Remember that $\mathcal{P} = d\mathcal{W}/dt$.)

2–6. The momentum p of a relativistic (nonlinear) mass is given by Eq. 2–13. Find the translational kinetic energy stored in the mass and determine the kinetic co-energy when the velocity is increased from zero to V.

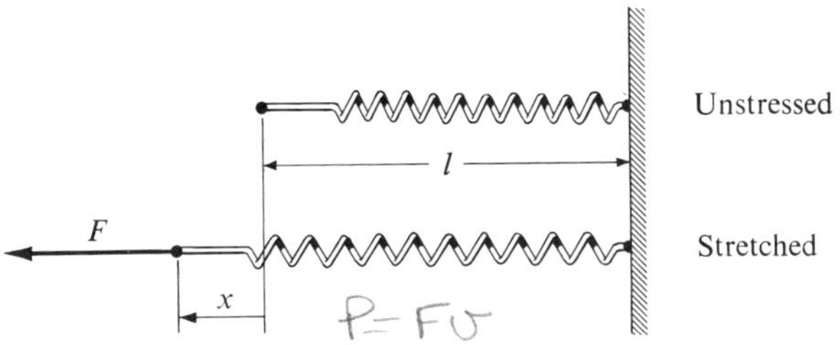

FIGURE 2–42

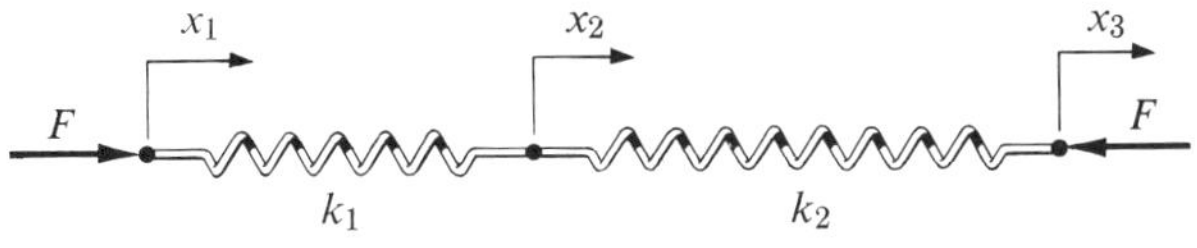

Series: $k_s = F/(x_1 - x_3)$

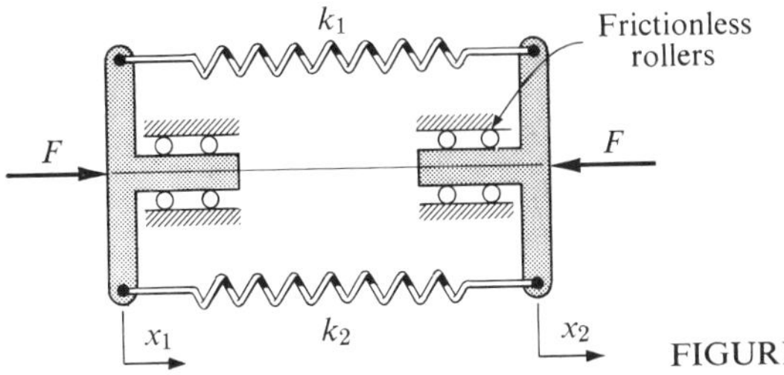

FIGURE 2–43

Parallel: $k_s = F/(x_1 - x_2)$

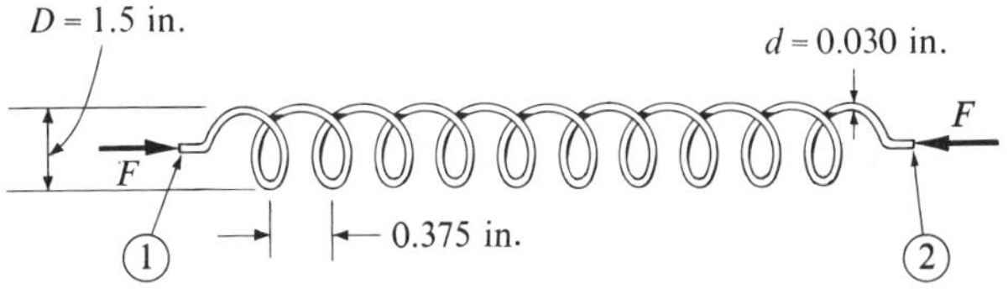

FIGURE 2–44

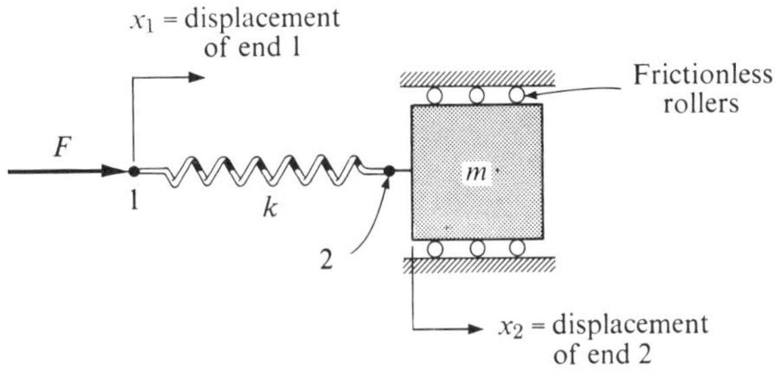

FIGURE 2–45

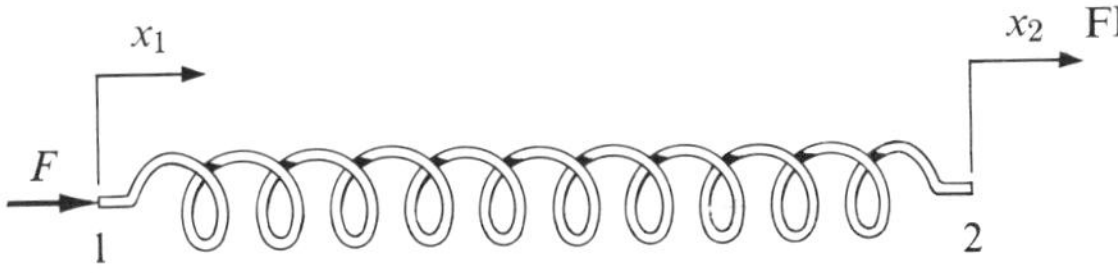

FIGURE 2–46

2–7. Two ideal springs may be interconnected to form a system of springs. The springs may be connected in such a way that their forces are equal (series) or their deflections are equal (parallel) (Fig. 2–43). Find the total stiffness k_s for each combination.

2–8. The coil of wire shown in Fig. 2–44 has a pitch diameter D (mean coil diameter) of 1.5 in., a pitch of 0.375 in./turn and is formed of wire having a diameter of 0.030 in. The coil has 10 turns.

a) The coil is loaded by two equal and opposite constant forces F acting along the centerline of the coil. By the methods of applied mechanics it can be shown that the deflection δ of end 1 of the coil relative to end 2 is approximately

$$\delta = 8D^3nF/Gd^4,$$

where n is the number of turns and G is the shear modulus. Assuming that the coil is made of steel and that $G_{\text{steel}} = 12 \times 10^6$ psi, calculate the spring constant k and determine the deflection and energy stored when a force F is applied.

b) Estimate the total mass of the coil of steel wire.

c) Under what conditions (frequencies) will the coil behave mainly as a spring?

2–9. An ideal mass is attached to end 2 of an ideal spring, and end 1 of the spring is acted on by an axial force F (Fig. 2–45). The end 1 is moved sinusoidally at frequency ω and amplitude A, $x_1 = A \sin \omega t$, where t = time.

a) For what range of frequency ω will the entire system (mass and spring) behave essentially like an ideal mass? [*Hint:* The system will behave like a mass if the spring deflection is small compared with x_1 and x_2.]

b) For what range of frequencies will the system behave like an ideal spring which has end 2 fixed?

2–10. The coil of wire described in Problem 2–8 is so supported that it can translate only in the x-direction (Fig. 2–46). End 1 is moved sinusoidally according to the

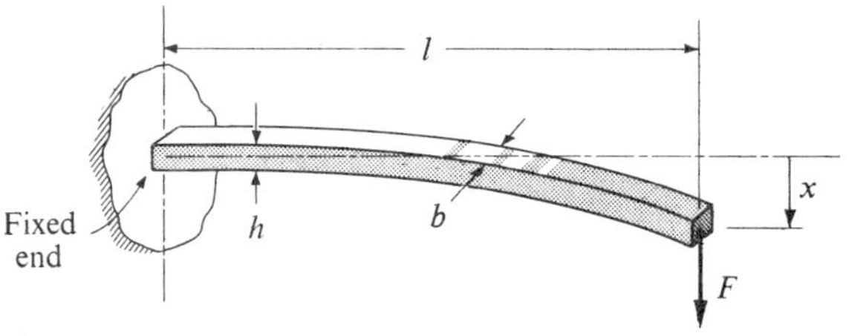

FIGURE 2–47

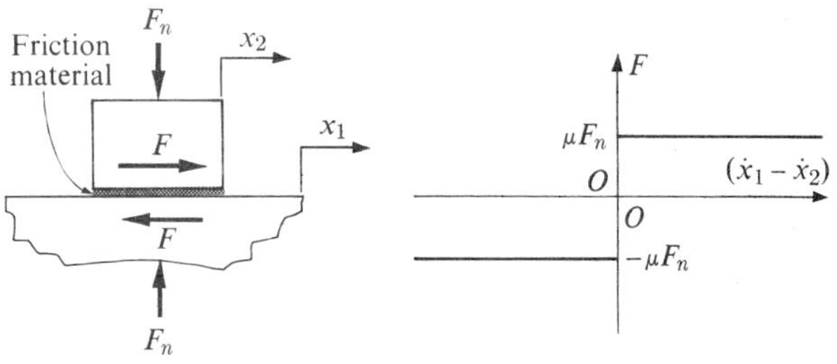

FIGURE 2–48

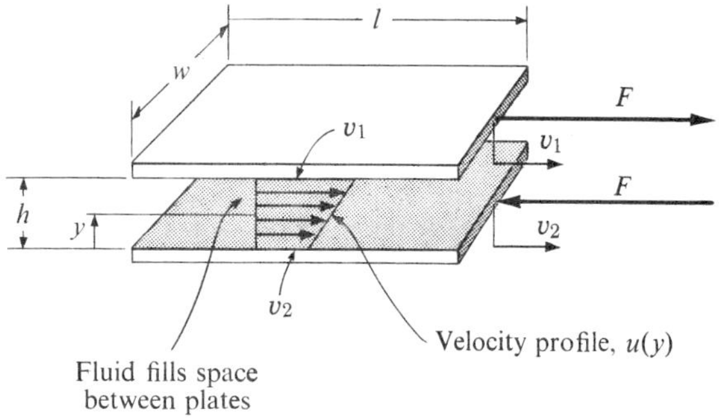

FIGURE 2–49

formula $x_1 = 0.2 \sin \omega t$, in., by applying a force F to that end.

a) Assuming that end 2 is force free, estimate the range of frequency ω over which the coil will behave essentially like an ideal mass.

b) If end 2 is *fixed*, $x_2 = 0$, and x_1 is moved sinusoidally as before, then the coil will act like an ideal spring for sufficiently low frequencies. Estimate how low the frequency must be for this to be true.

c) How could you represent the fixed-end coil for frequencies much higher than that calculated in (b)?

2–11. A rectangular cantilever beam made of metal having a Young's modulus $E = 10^7$ psi and a density of 0.1 lb/in^3 is built into a fixed wall (Fig. 2–47).

a) Calculate the spring stiffness F/x for a steady load applied at the end of the beam.

b) Calculate the mass of the beam.

c) Using the type of reasoning presented in Example 2–1 of the text and Problem 2–9, estimate the range of frequency, when x varies sinusoidally, for which the beam behaves nearly like an ideal spring.

2–12. A nonlinear damper is designed to dissipate energy during sliding of two parts in an automatic machine and thus to prevent excessive sliding motions (Fig. 2–48). The tangential force F produced between the two contacting parts by the friction material is described approximately by the Coulomb friction characteristic shown above, where μ is the coefficient of friction. The normal force F_n is constant.

The parts move sinusoidally so that

$$x_1 = a_0 \sin \omega t, \qquad x_2 = b_0 \sin \omega t.$$

a) Find and sketch versus time the relative velocity between the parts and the friction force F.

b) Derive and sketch an expression for the power flow *into* the friction device versus time.

c) Find the total energy dissipated per cycle, assuming that $\mu = 0.2$, $a_0 = 1$ in., $b_0 = 2$ in., $\omega = 200$ rad/sec, and $F_n = 10$ lb.

d) If the velocities $\dot{x}_1$ and $\dot{x}_2$ are doubled, how do the results of parts (a), (b), and (c) change?

2–13. Two parallel flat plates are placed as shown in Fig. 2–49. The small gap between the plates is filled with oil or other viscous fluid. The upper plate moves with a constant velocity v_1, and the lower plate moves with velocity v_2. If the plates are parallel and no pressure variations occur inside the gap between plates, the fluid velocity will vary linearly from v_2 at the lower plate to v_1 at the upper plate. From the definition of fluid viscosity, the shear stress τ acting on the surface of one of the plates is $\tau = \mu(du/dy)$, where μ is viscosity, u is velocity of the fluid, and y is distance measured normal to the plates.

a) Derive an expression relating the force F and the velocities v_1 and v_2. What type of ideal element represents this relationship?

b) If the velocities v become very high or if the gap h becomes large, the velocity profile $u(y)$ will become distorted and the flow between the plates will become turbulent. Under these conditions, the shear stress on the plates varies as the square of the velocity difference:

$$\tau = c(v_1 - v_2)|v_1 - v_2|.$$

Can the behavior of the plates (F vs. v_1 and v_2) still be represented by a pure element? Can it be represented approximately under any conditions by an ideal element?

2–14. A so-called "drag cup" is constructed by immersing a thin cylinder in a cup of oil of viscosity μ (Fig. 2–50).

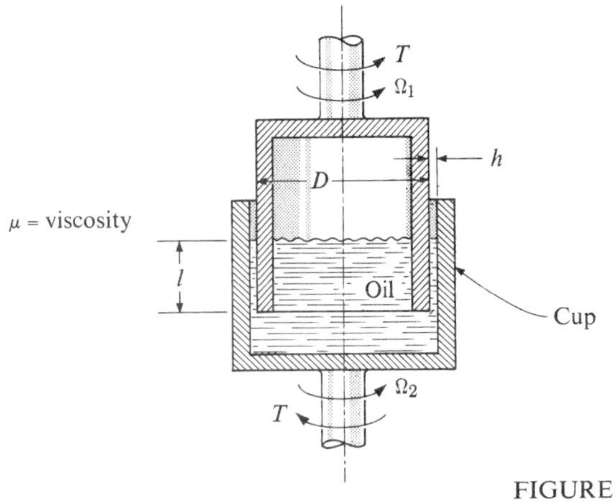

FIGURE 2–50

a) Use the definition of fluid viscosity given in Problem 2–13 to derive an expression for the function relating Ω_1, Ω_2, T; the geometry and the viscosity for the case where Ω_1 and Ω_2 are constant.

b) What ideal element could represent this functional relationship?

c) If the speeds Ω_1 and Ω_2 are not constant, what other ideal elements would be required to represent this physical device?

2–15. The landing strut for an aircraft contains a variable-area orifice which allows the hydraulic fluid to pass from one side to the other at a rate which depends on the position of the piston and the load on the strut. Figure 2–51 shows the structure and the desired deceleration-vs.-time curve. At the end of 2 sec vertical motion has stopped.

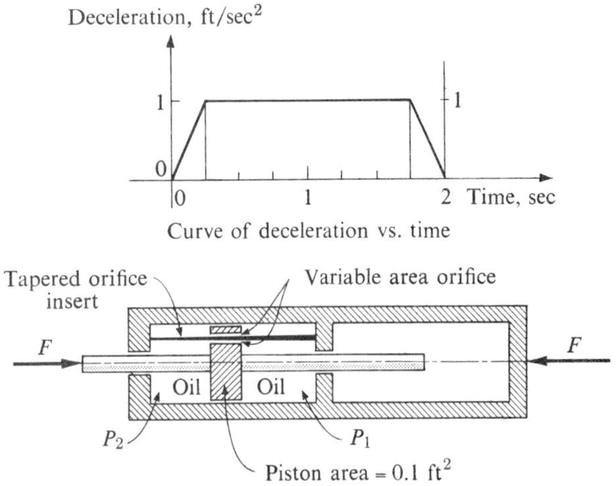

FIGURE 2–51

a) What is the force-position curve for this shock absorber? Assume that there are two wheels on the aircraft, the weight is 100,000 lb, and there is just 100,000 lb of lift acting.

b) For the strut in Fig. 2–51, what is the curve of volume-flow rate vs. position?

c) For the flat region of the deceleration-time curve, what would be the desired pressure-difference versus position curve?

d) Depending on the design of the orifice, two limits are possible:

$$\frac{Q_{\text{oil}}}{A_{\text{orifice}}} = K_1(P_2 - P_1) \qquad \text{(viscous fluids)},$$

$$\frac{Q_{\text{oil}}}{A_{\text{orifice}}} = K_2(P_2 - P_1)^{1/2} \qquad \text{(thin fluids)}.$$

In these expressions, K_1 and K_2 are functions of the geometry and fluid properties. For the flat region of the deceleration curve, discuss which characteristics are best for the orifice. What considerations would determine which characteristics would be chosen?

2–16. A simple torsion bar is formed by a cylindrical steel rod of constant diameter which is fixed to a rigid support at one end (Fig. 2–52). The stiffness K_Θ for input torque applied at the left end is

$$K_\Theta = T/\Theta = GI_p/l,$$

where G = shear modulus, I_p = polar moment of inertia* = $\pi d^4/32$, and l = length.

* For the plane section.

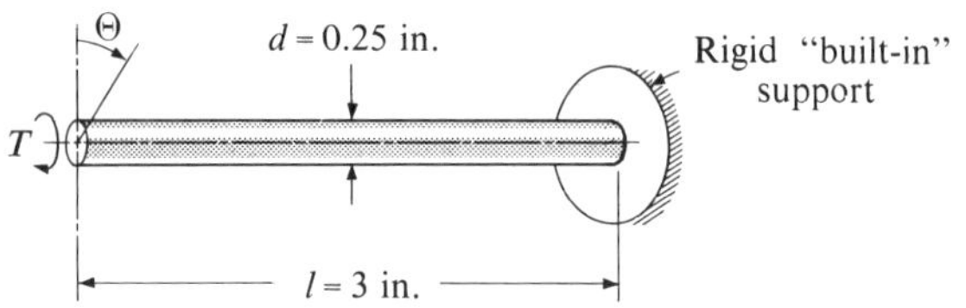

FIGURE 2–52

a) Compute the numerical value of the torsional stiffness K_Θ.

b) Compute the inertia J of the shaft.

2–17. The end of the bar in Problem 2–16 is oscillated sinusoidally:

$$\Theta = \Theta_0 \sin \omega t.$$

Estimate the range of frequency ω over which the bar will act approximately like an ideal spring.

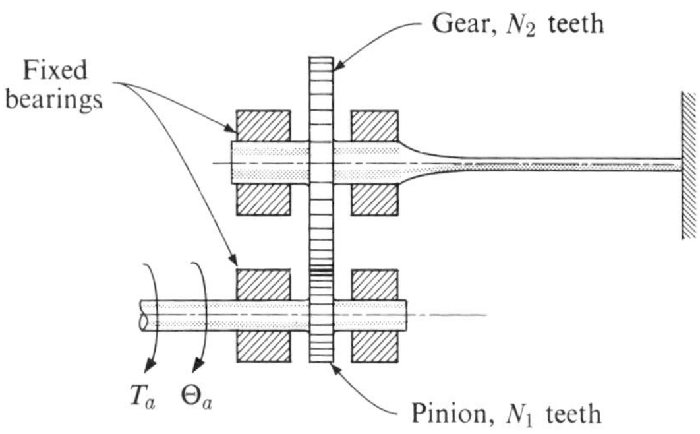

FIGURE 2–53

2–18. An ideal torsional spring is attached to the shaft of a gear which is driven by a smaller pinion gear (Fig. 2–53). A torque T_a is applied to the shaft of the pinion, causing both gears to rotate, thus twisting the spring.

a) Find the torsional spring stiffness that would be observed at the pinion shaft. This is called the stiffness "reflected" through the gear train.

b) Assuming that $K = 100$ in·lb/rad, $N_1 = 5$, $N_2 = 20$, compute the energy stored in the spring when $T_a = 100$ in·lb.

c) If the gear ratio (transformer ratio) is doubled while everything else remains the same, how does the stored energy change?

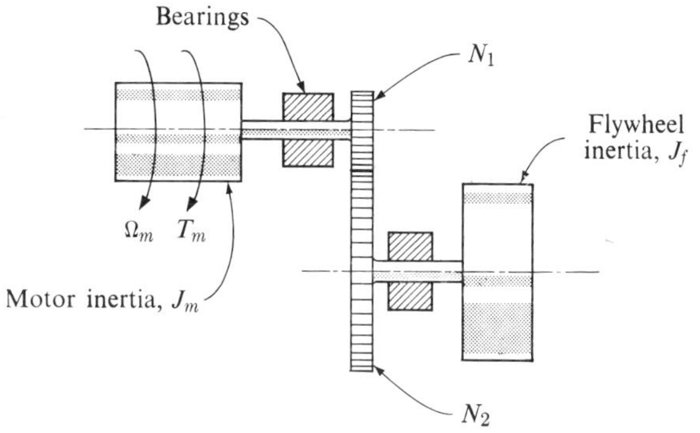

FIGURE 2–54

2–19. An electric motor drives a flywheel through a gear train (Fig. 2–54). The motor and flywheel are supported in bearings. If the stiffnesses of the various parts of the system are very high and friction is negligible, what is the relationship between the motor speed Ω_m and the motor torque T_m? What single ideal element will represent this system?

2–20. A lever (a translational transformer) is used to apply a force to an element (Fig. 2–55). Assume that the lever is massless and perfectly rigid and that only small angular deflections are allowed.

a) Determine an equivalent element which represents this system to the force F for each of the three cases (Fig. 2–56) where the element is attached to the lever.

b) Consider an interchange of the locations of applied force and element connection. Describe the difference.

c) Compare these situations to that of a rotational system where a torque is applied to each of the three ideal rotational elements through a set of gears. Define a lever ratio n, by analogy with the gear ratio.

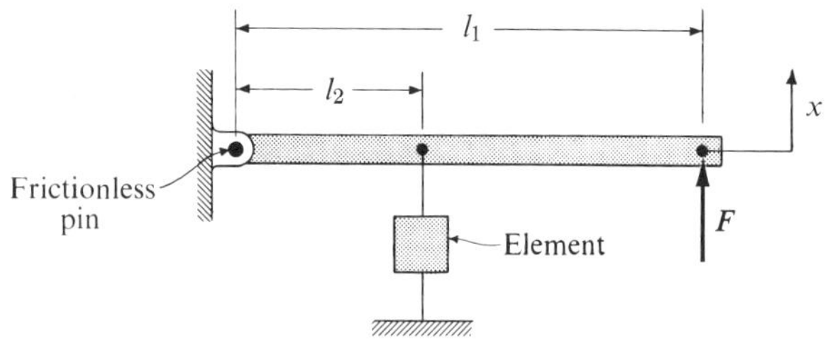

FIGURE 2–55

2–21. A torsional element is constructed by combining an ideal linear spring with a pure rotational-to-translational transducer (Fig. 2–57). The unstressed length of

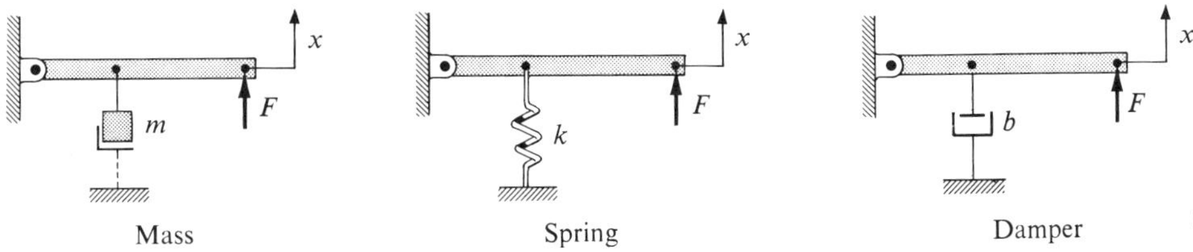

FIGURE 2–56

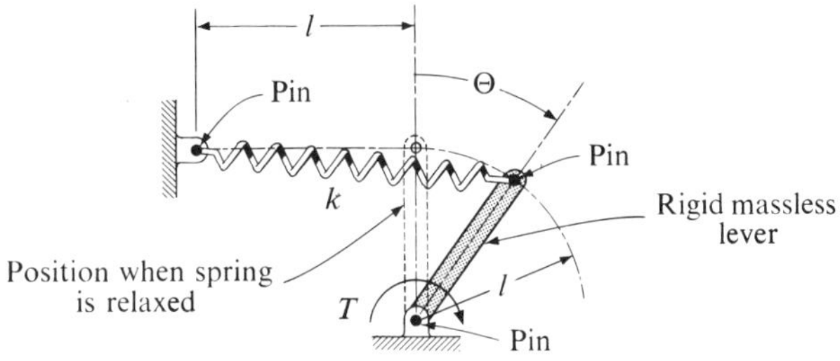

FIGURE 2–57

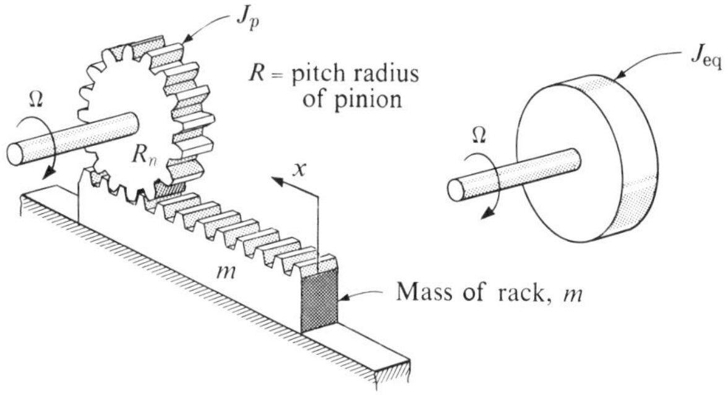

FIGURE 2–58

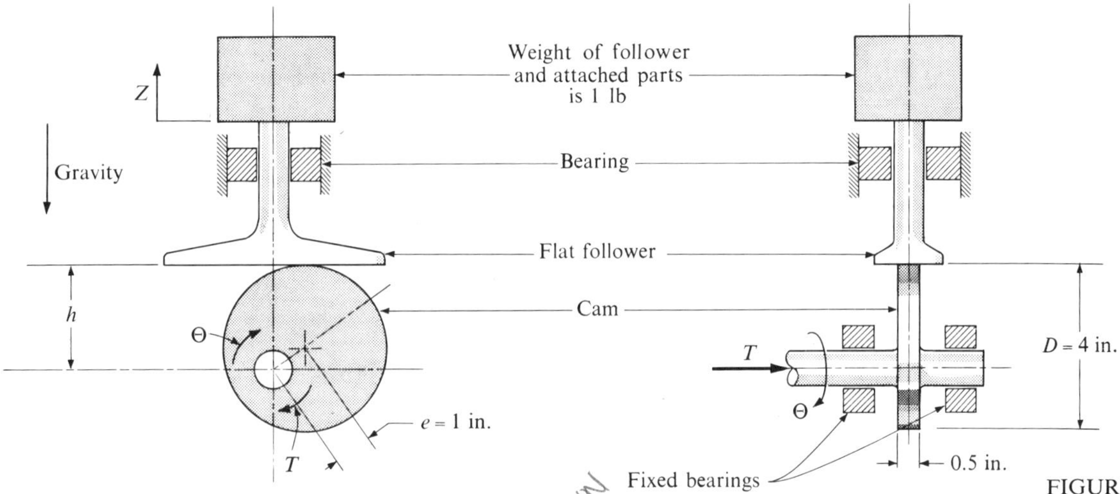

FIGURE 2–59

the spring is l, and the lever radius is also l. The lever is vertical and the spring axis is horizontal when the torque T is zero.

a) Determine and sketch an expression for T vs. Θ for $0 < \Theta < \pi/2$. What element represents this function?

b) On your sketch show graphically the area representing the energy stored when $\Theta = \pi/4$.

2–22. For each of the three ideal mechanical elements (mass, damper, and spring), determine how the velocity varies if the force is $F_0 \sin \omega t$.

2–23. What conditions must be satisfied for the rack and pinion system shown in Fig. 2–58 to be suitably modeled by an equivalent inertia J_{eq} as shown at the right? Express the equivalent inertia J_{eq} in terms of J_p and m (and anything else you think may be important).

2–24. To perform a certain function in a beer-bottle capping machine, it is necessary to achieve a reciprocating motion of amplitude Z of ± 1 in. at a cyclic rate of 1 cps. The designer of the machine has decided to use the so-called flat-follower cam system illustrated in Fig. 2–59.

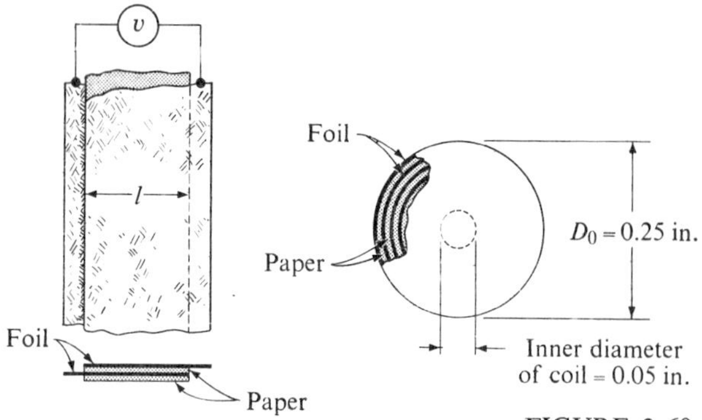

FIGURE 2–60

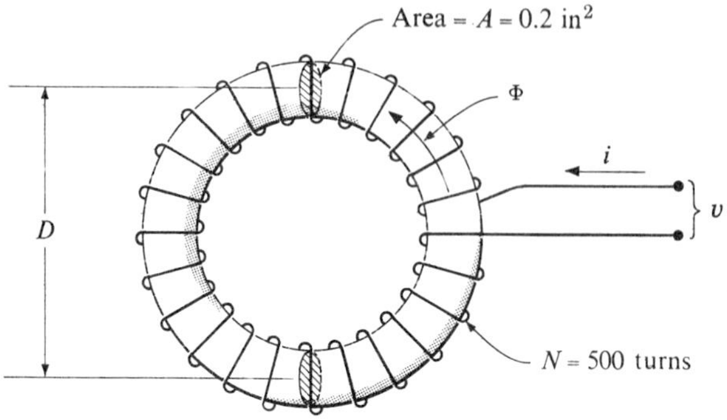

FIGURE 2–61

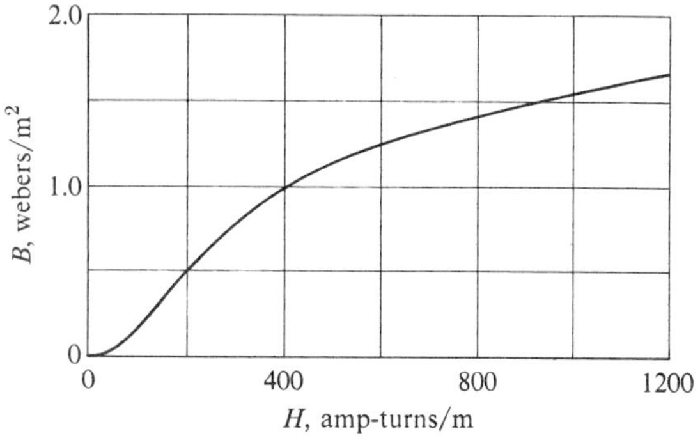

FIGURE 2–62

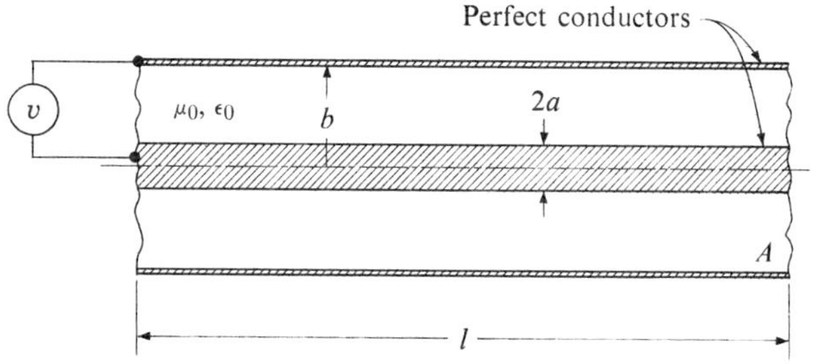

FIGURE 2–63

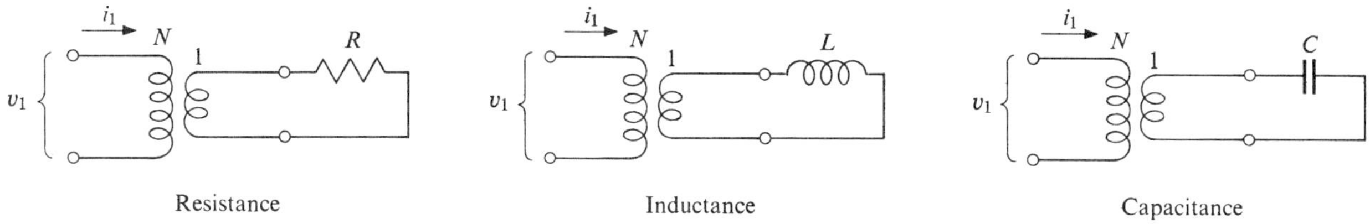

FIGURE 2–64

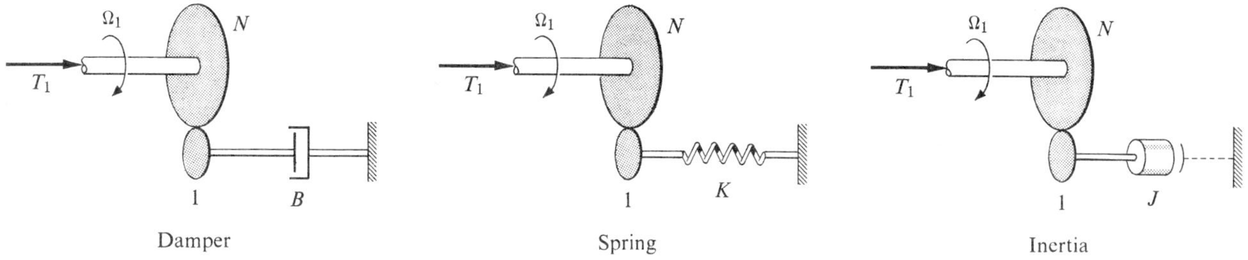

FIGURE 2–65

a) Assuming that the cam shape is a perfect circle of diameter $D = 4$ in., that the follower surface is perfectly flat, and that the parts are infinitely rigid, compute and plot output displacement Z as a function of input angular displacement Θ if Θ is varied very slowly. What type of system element does the cam system represent?

b) Estimate the maximum value of the time-varying torque T which must be supplied to the shaft if $d\Theta/dt$ is held constant at the 1 cps rate. Neglect all friction forces.

2–25. A precision parallel-plate capacitor uses mica (permittivity relative to air $= 6.0$, conductivity

$$\sigma = 10^{-15}\ (\text{ohm·m})^{-1})$$

as the dielectric material. The capacitor plates have an area of 1 in^2 and a spacing of 10 mils (0.010 in.).

a) Compute the capacitance of this capacitor.

b) Compute its resistance.

2–26. A paper dielectric capacitor is constructed by rolling up a long sandwich-like strip composed of a pair of metal foils and a pair of impregnated paper sheets (Fig. 2–60). After the roll is made, the extended ends of the foils are soldered together so that the terminal voltage v is applied across each layer of the paper dielectric. Assuming that the foil and paper thicknesses are 0.005 in. each, the width of the paper is $l = 1$ in., and the permittivity of the dielectric is 1.5 relative to air, estimate the capacitance. Compare this value with the capacitance of the unrolled strip.

2–27. For the geometry of Fig. 2–31(c), compute the resistance and capacitance between terminals 2 and 1, assuming that the dielectric has a permittivity of 1.8 relative to air and a conductivity of 5×10^{-14} (ohm·m)$^{-1}$. The radii are $a = 1$ in., $b = 1.10$ in.

2–28. An electrical transmission line consists of two parallel wires 10 in. apart extending for 10 mi. The wire is No. 6 copper, has a diameter of 0.162 in., and has a resistance of 0.4 ohm/1000 ft. Compute the inductance and resistance, assuming that the wires are connected at one end.

2–29. In the system of Problem 2–28, a sinusoidally varying current $i = 10 \sin \omega t$ amp is caused to flow through the lines. A voltage difference which is the sum of the voltage drops caused by the inductance and by the resistance will appear at the terminals. Compute the frequency at which the magnitude of the inductive voltage becomes equal to the magnitude of the resistive voltage drop.

2–30. A real air-core inductor consists of a single layer of 500 turns of No. 40 copper wire wound on a cylindrical form. (Wire diameter = 3.1 mils; resistance = 1070 ohms/1000 ft.) The length of the inductor is 5 in. and its diameter is 1 in.

a) Find the inductance of the inductor.

b) Find its resistance.

c) When changes in the resistance of the wire due to skin effect are neglected, for what range of frequency can this be considered an ideal inductor (i.e., compare voltage drop due to resistance to voltage drop due to inductance)? For what range of frequencies is it an ideal resistor?

2–31. An iron-core inductor is built by winding 500 turns of copper wire around a steel torus with a mean diameter D of 3 in. and a cross-sectional area of 0.2 in^2 (Fig. 2–61). If we assume that the coil resistance is negligible and that flux is uniform throughout the iron core, we see that Faraday's law gives

$$v = N\frac{d\Phi}{dt} = \frac{d\lambda}{dt}\ \text{volts}.$$

The flux density B in the iron is defined as Φ/A, and the magnetizing force H resulting from the current i is

$$H = \frac{Ni}{l}\ \frac{\text{amp-turns}}{\text{m}},$$

where l is the length of the magnetic path $= \pi D$. The units of B in the mks system are webers/m^2 (1 weber-turn = 1 v·sec). The B-H curve for steel as determined experimentally is shown in Fig. 2–62.

a) Determine the relationship between v and i for the inductor.

b) Compute the energy stored by the inductor when $i = 0.16$ amp.

2–32. The nature and function of an element are dependent on, among other things, the constraints of the element. The circular coaxial cable (Fig. 2–63) has inner and outer radii of a and b, respectively, and length l very large compared with b. What ideal element represents the coaxial cable, for steady-state operation, and what is the value of the associated parameter in terms of geometry and the electromagnetic properties under each of the following conditions?

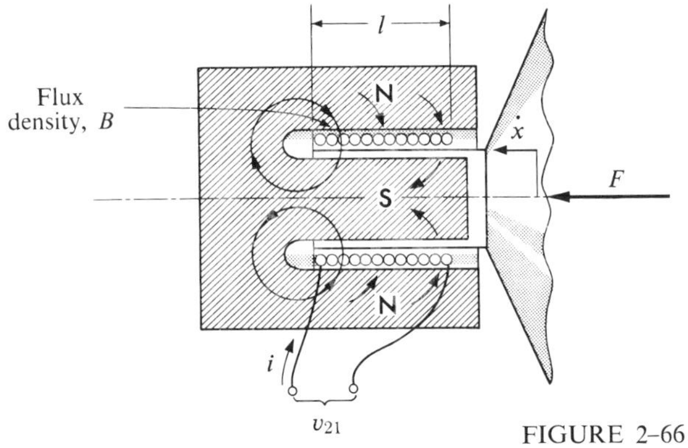

FIGURE 2–66

a) Cable open at end A and annular region occupied by a perfect vacuum. (Neglect end effects.)

b) Inner and outer cylinders connected by a perfect conducting plate at end A and annular region occupied by a perfect vacuum.

c) Annular region filled with material of conductivity σ.

d) What conditions are necessary for the system to act nearly like an ideal element in each of these cases?

2–33. Find the equivalent elements which result when each of the ideal electric elements is connected across the terminals of an ideal transformer (Fig. 2–64). Compare these results with those obtained when the ideal mechanical elements are connected to an ideal gear train (Fig. 2–65).

2–34. The terminals of a galvanometer movement such as the one described in Section 2–6.5 are connected through an ideal resistance R. Assuming that the galvanometer is represented by an ideal transducer with transduction ratio n:

a) Find the relationship between the mechanical torque and the angular velocity of the movement.

b) What ideal rotary mechanical element represents this electromechanical system?

c) Repeat (a) and (b) with a capacitance connected across the terminals of the coil.

2–35. Figure 2–66 shows a "voice-coil" movement used in most loudspeakers. A current i flowing through the coil interacts with a constant radial flux of density B webers/m^2 set up in the air gap by a permanent magnet. The coil has N turns, length l, and radius R.

Assuming that all masses are negligible and that the current i does not affect the flux density B, derive an idealized relation between the two electrical variables v_{21} and i and between the two mechanical variables $\dot{x}$ and F. Determine the transduction ratio n.

3 DYNAMIC SYSTEM ELEMENTS, FLUID AND THERMAL

3–1. INTRODUCTION

Often we can model physical systems which involve the flow of fluids and the transfer of heat by the use of idealized elements analogous to the mechanical and electrical elements discussed in Chapter 2. These elements represent the basic physical phenomena of fluid and thermal systems.

The fluid system elements will fit directly into the generalized scheme of ideal elements presented in Section 2–7. Although thermal systems are similar in many respects to mechanical and electrical systems, they show marked departures from the generalized scheme.

3–2. FLUID-SYSTEM ELEMENTS

Many engineering systems contain working fluids in motion under pressure. Some examples are power-generation and energy-conversion systems (hydroelectric and thermoelectric power plants, internal-combustion engines, jet engines, etc.) and control systems (for machine tools, automatic machinery, chemical processes, automotive equipment, aircraft, missiles, and space vehicles). In many cases, thermal effects are also involved in these fluid systems. Thus sometimes fluid mechanics and thermodynamics are both needed in dealing with fluid systems. However, the simple fluid-system elements which will be discussed here involve a minimum of thermal effects, and as idealized fluid elements they will involve only the most elementary aspects of fluid mechanics.

The types of fluid elements to be discussed are directly analogous to corresponding electrical- and mechanical-system elements. In practice, they are connected with tubes or pipes to form networks much as electrical-system elements are connected with wires to form networks. Volume rate of flow of working fluid is analogous to electric-current flow, and fluid pressure is analogous to electric potential.

The working fluid may be either a liquid or a gas, although the characteristics of pneumatic elements tend to be more highly nonlinear than corresponding hydraulic elements, so that the range of linear operation is frequently smaller for a given pneumatic device than it is for the corresponding hydraulic device.

3–2.1 Fluid Pressure and Flow Rate

Fluid pressure is the quantity which causes (or results from) fluid flow or sustains the weight of a column of fluid. We shall talk in terms of pressure differences between two points, although often one of these points will be an implied reference pressure not explicitly shown.

Any surface (real or imaginary) in a fluid can be acted on by forces which may be resolved into normal and tangential components. Pressure is defined as the normal force per unit area (normal stress) acting on the surface. The tangential force per unit area is called the shear stress. Since we shall be dealing with fluids contained in pipes, let us consider the section in Fig. 3–1 taken normal to the centerline of a pipe. On

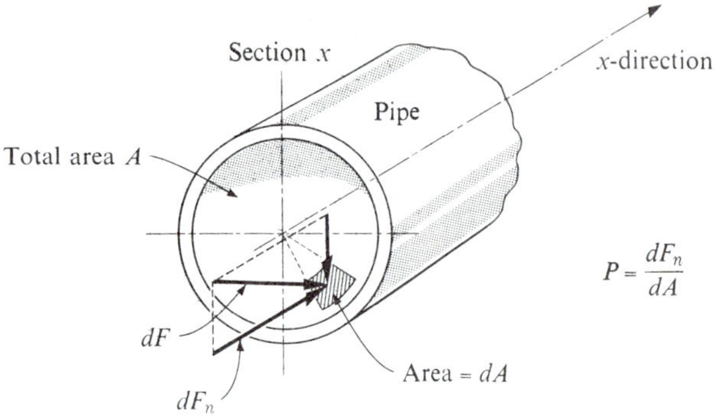

FIG. 3-1. Fluid pressure in a pipe.

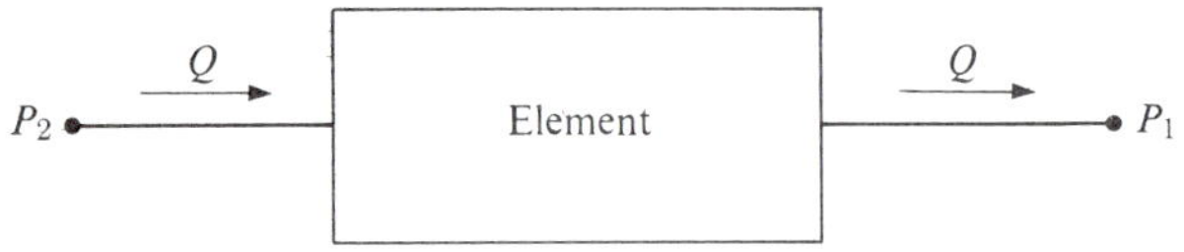

FIG. 3-2. Two-terminal fluid-system element.

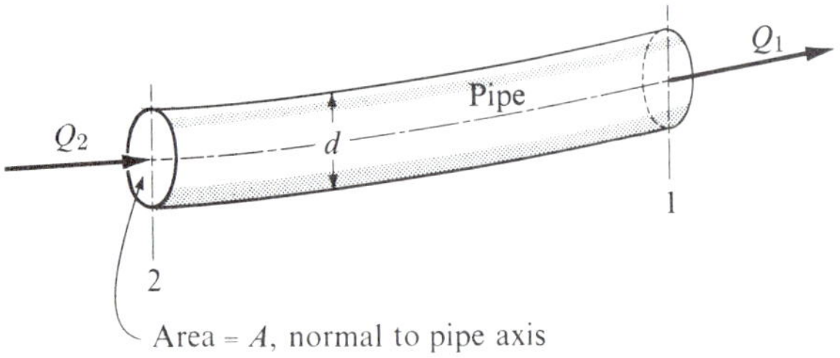

FIG. 3-3. Fluid flow in a constant-area pipe.

each infinitesimal area dA of this section an infinitesimal normal force dF_n acts. By the definition of pressure, $dF_n = P\,dA$. The total axial force on the (plane) area is the integral of these forces over the area:

$$F_n = \int_A dF_n = \int_A P\,dA. \tag{3-1}$$

For the situations we shall consider, the pressure can be assumed uniform over the area; therefore

$$F_n = PA. \tag{3-2}$$

If P varies over the cross section, the above equation defines an average pressure, which is the quantity with which we shall work in that case. Usually P will be measured in n/m^2 or lb/in^2.*

We can see that pressure has additional significance when we bring energy into consideration. If a quantity of fluid passes the area A in the direction of F_n or P, the amount of work $d\mathcal{W}$ done by the force F_n in moving this fluid through a distance dx is

$$d\mathcal{W} = F_n\,dx = PA\,dx.$$

The quantity or volume dV of fluid passing A is

$$dV = A\,dx;$$

therefore

$$P = d\mathcal{W}/dV, \qquad n/m^2 \text{ or } lb/in^2. \tag{3-3}$$

Thus pressure is the work done in passing a unit volume of fluid across the area on which P acts. In this sense the pressure is directly analogous to electrical potential, which is work per unit quantity of charge.

Pressure is considered positive when the fluid is in a state of compression. Since most fluids will not support tension without cavitation (vaporization), we shall be concerned only with positive pressures. Algebraic references for pressure differences will be used in the same way as for velocity and voltage differences. Thus in Fig. 3-2 the pressure difference $P_2 - P_1$ is positive if $P_2 > P_1$.

Fluid-flow rate is the variable which measures the quantity of fluid passing a given area per unit time. Since we shall be working with piping situations, the flow area will normally be the cross-sectional area of a pipe, as shown in Fig. 3-1.

If in time dt a quantity (volume) of fluid dV crosses area A, then the volume flow rate Q is defined as

$$Q = dV/dt, \qquad m^3/sec \text{ or } in^3/sec. \tag{3-4}$$

---

* See Appendix A.

It is seen that Q is analogous to i, the electric current, which is the rate of flow of electric charge.

If the fluid is *incompressible* and the pipe is *rigid*, the flow Q_2 entering at section 2 in Fig. 3–3 must equal the flow Q_1 leaving at section 1, and it is appropriate to speak of *the* flow $Q = Q_2 = Q_1$ *through* the pipe.

When two parts or elements of a fluid system are connected by a section of pipe (Fig. 3–4), the principle of conservation of mass states that the amount of fluid leaving element A must enter element B. This principle also requires that the net flow rate of mass into any fluid system must equal the rate of increase of mass inside the system. If the fluid is incompressible, the preceding statement also applies to volume flow rate and fluid volume.

The assumed direction of positive flow is indicated by an arrow (Fig. 3–4). To measure the flow Q in a pipe, we can sever the pipe at the section of interest and collect the flow in a container for a known time interval. The volume of the container filled in a unit time interval is the flow rate. It is evident that flow rate is a *through*-variable and hence that pressure is an *across*-variable in fluid systems. How might pressure be measured in a pipe without severing the pipe or appreciably disturbing the flow inside the pipe?

3–2.2 Power and Energy

Power, as we know, is the rate of flow of work. Hence from Eqs. (3–3) and (3–4),

$$\mathcal{P} = \frac{d\mathcal{W}}{dt} = P\frac{dV}{dt} = PQ, \qquad \text{watts or in·lb/sec.} \qquad (3\text{–}5a)$$

This power is the product of the across- and through-variables P and Q.

The net power delivered to a two-terminal element (Fig. 3–2), which has flow in equal to flow out, is

$$\mathcal{P} = Q(P_2 - P_1) = QP_{21}. \qquad (3\text{–}5b)$$

The fluid energy delivered to a fluid element is the time integral of the power.

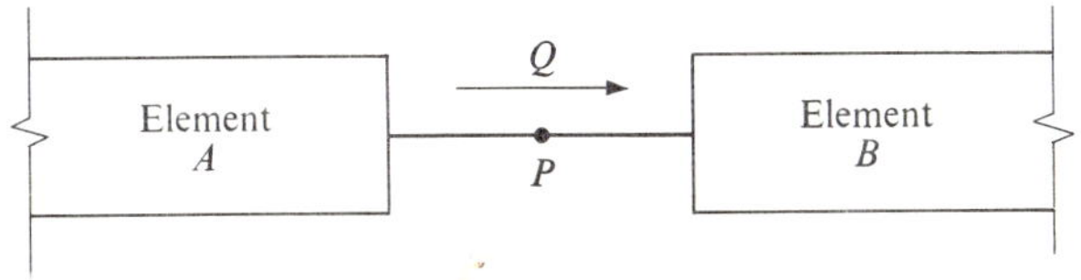

FIG. 3–4. Fluid flow from one element to another.

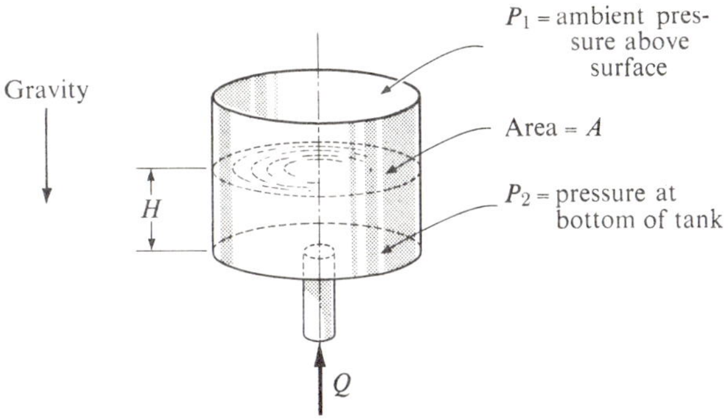

FIG. 3–5. Open fluid reservoir.

3–2.3 Pure Fluid System Elements

Fluid systems may often be represented by a series of idealized elements which characterize the mechanisms of fluid energy storage, dissipation, and transfer. These idealized elements are analogous to the electrical and mechanical elements discussed previously. *Fluid capacitance* stores energy by virtue of the pressure (across-variable), *fluid inertance* stores energy by virtue of the flow (through-variable), and *fluid resistance* dissipates energy.

Pure fluid capacitance. a) *Fluid reservoir.* A fluid element in which the energy stored is a function of pressure may be thought of as a *fluid capacitor*, analogous to an electrical capacitor, in which the stored energy is a function of voltage. One of the simplest examples of a fluid capacitor is an open tank or reservoir which is located in a gravity field and supplied with a liquid working fluid through a port or pipe at the bottom (Fig. 3–5). When a flow Q occurs

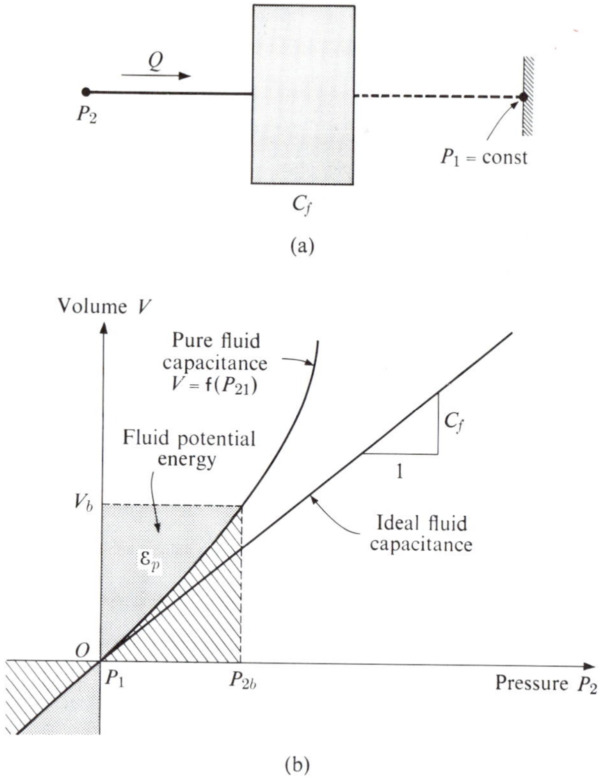

FIG. 3-6. Symbolic representation and constitutive relationships for a pure fluid capacitance. (a) Symbolic representation. (b) Constitutive relationships.

so that fluid is forced into this system, energy is stored because fluid particles in the reservoir are elevated in the gravitational field of the earth.

The law of conservation of mass (fluid-flow continuity) for the reservoir states that

$$\rho Q = \frac{d}{dt}(\rho A H),$$

where

ρ = mass density of fluid, kg/m^3 or lb·sec^2/in^4,
Q = volume flow rate, m^3/sec or in^3/sec,
A = cross-sectional area of tank, m^2 or in^2,
H = height of liquid level in tank, m or in.

If the fluid is incompressible (ρ = const) and the walls of the reservoir are rigid,

$$Q = A\frac{dH}{dt}. \tag{3-6}$$

If there are no fluid accelerations of importance (except due to gravity) and no fluid-resistance effects, the pressure difference $P_2 - P_1 = P_{21}$ between the top and bottom of the reservoir must support the weight of the liquid (we assume that P_1 is a constant ambient or atmospheric pressure):

$$P_{21} = \frac{\text{weight}}{\text{area}} = \frac{\rho g H A}{A} = \rho g H, \tag{3-7}$$

where g is the acceleration due to gravity, 9.86 m/sec^2 or 386 in/sec^2. Equations (3-6) and (3-7) are now combined:

$$Q = \frac{A}{\rho g}\frac{dP_{21}}{dt} = \frac{A}{\rho g}\frac{dP_2}{dt}. \tag{3-8}$$

The similarity between this equation and the elemental equation (2-77) for an ideal capacitance is evident.

Because of the direct analogy of the reservoir to an electrical capacitor when flow rate Q is considered analogous to current and pressure P_2 is considered analogous to voltage, the quantity $A/\rho g$ may be considered the "capacitance" C_f of the reservoir:

$$C_f = \frac{A}{\rho g}, \qquad \text{m}^5/\text{n or in}^5/\text{lb}. \tag{3-9}$$

Since the volume of fluid V in the tank is AH, Eq. (3-7) may be written

$$V = \frac{A}{\rho g}P_{21} = C_f P_{21}. \tag{3-10}$$

Noting that $Q = dV/dt$, we see that, when differentiated, Eq. (3-10) yields Eq. (3-8). Equation (3-10) is similar to the constitutive relationship for an ideal electrical capacitance, Eq. (2-76).

We shall define a *pure fluid capacitance* as an idealized fluid system element in which pressure difference P_{21} is a single-valued function of fluid volume V inside the element:

$$V = f(P_{21}). \tag{3–11}$$

An *ideal fluid capacitance* has volume directly proportional to pressure, as in Eq. (3–10).

It is evident that the fluid capacitance is an element having only one port or terminal through which flow takes place. The pressure P_2 at this port must be measured with respect to a fixed reference pressure P_1 (usually atmospheric pressure). Figure 3–6(a) shows the symbol to be used for a pure fluid capacitance. The reference of P_2 to a fixed P_1 is shown by a dashed connection to P_1 (usually thought of as "ground"); no flow takes place through the dashed connection. This symbol is a schematic representation of a tank into which there is a flow Q.

Figure 3–6(b) shows the constitutive relations for pure fluid capacitance. The reader should compare Fig. 3–6 with Figs. 2–6 and 2–14, which define the pure translational and rotational masses. The fixed reference for P here is similar to the fixed reference for velocity in the case of mass.

The elemental equations for an ideal fluid capacitance are

$$Q = C_f \frac{dP_{21}}{dt} = C_f \frac{dP_2}{dt} \tag{3–12}$$

or

$$P_{21} = \frac{1}{C_f} \int_0^t Q\,dt + (P_{21})_0, \tag{3–13}$$

where $(P_{21})_0$ is the pressure at $t = 0$.

Energy is stored in a fluid capacitance by virtue of the pressure and will be called *fluid potential energy*, $\mathcal{E}_p$. Fluid may be forced to flow into the reservoir of Fig. 3–5 by attaching a pump between a second tank or supply of fluid and the inlet to the reservoir (Fig. 3–7). The pump acts as a *source* of fluid power in this case and raises the fluid pressure from P_1 to P_2. Sources will be discussed in Chapter 4.

The power $\mathcal{P}_p$ supplied by the pump is given by Eq. (3–5b), providing that the fluid density is constant:

$$\mathcal{P}_p = Q(P_2 - P_1) = QP_{21}.$$

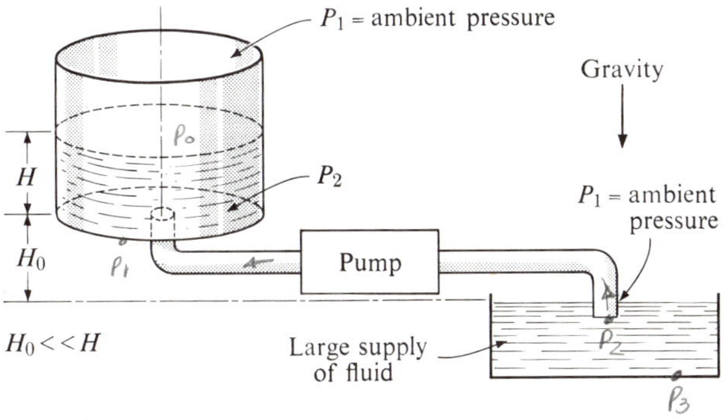

FIG. 3–7. Pump supplying fluid to a reservoir.

The energy supplied by the pump in raising the pressure in the reservoir from P_1 to P_2 is defined as the *fluid potential energy*, $\mathcal{E}_p$. When $P_1 = P_2$, $\mathcal{E}_p = 0$ by definition:

$$\mathcal{E}_p = \int_{t_a}^{t_b} \mathcal{P}_p\,dt = \int_0^{t_b} P_{21}Q\,dt = \int_0^{P_{21}} P_{21}\,dV. \tag{3–14}$$

This energy is represented by the shaded area in Fig. 3–6(b). Since the reservoir is assumed to behave like an ideal fluid capacitance, Eq. (3–14) can be integrated by introducing Eq. (3–12) in the form

$$dV = C_f\,dP_{21},$$

$$\mathcal{E}_p = \int_0^{P_{21}} C_f P_{21}\,dP_{21} = \frac{C_f (P_{21})^2}{2} = \frac{V^2}{2C_f}, \tag{3–15}$$

where Eq. (3–10) has been used to express P_{21} in terms of V.

The energy $\mathcal{E}_p$ can be recovered by allowing the fluid volume V to flow out of the reservoir. Thus $\mathcal{E}_p$ is *stored* in the reservoir. This energy is stored by virtue of the across-variable P_{21} and does not depend directly on the value of Q. Note that although P_1 is a constant ambient pressure, $\mathcal{E}_p$ depends on $P_2 - P_1$.

b) *Pressurized tank.* A second example of a system which may behave approximately like a pure fluid capacitance is a closed tank in which fluid energy is stored because of the compressibility of the fluid. Consider a closed chamber of volume $\mathcal{V}$ with rigid

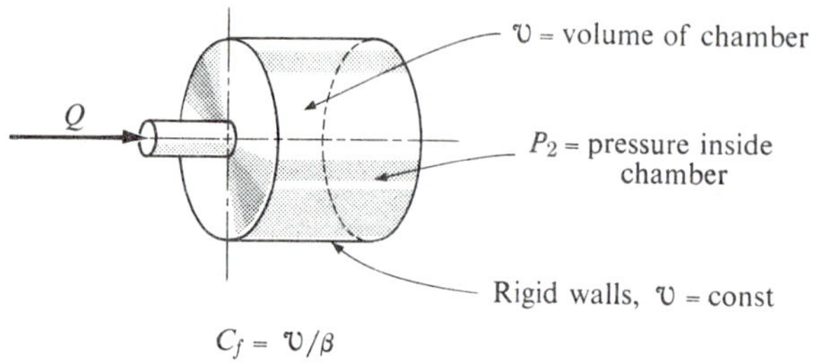

FIG. 3–8. Tank or chamber containing compressed fluid.

walls completely filled with fluid of density ρ (Fig. 3–8). In this case we shall consider the possible variation of pressure in the tank due to gravity to be negligible (or assume negligible gravity), and consider only the variation in pressure in the tank which occurs as a result of the amount of fluid which has been forced into it. The pressure will be assumed uniform (but not necessarily constant) throughout the chamber. In addition, fluid inertia and frictional effects are neglected.

The law of conservation of mass for the tank gives

$$\rho Q = \frac{d}{dt}(\rho \mathcal{V}) = \mathcal{V}\frac{d\rho}{dt}, \tag{3–16}$$

where $\mathcal{V}$ is the constant tank volume and ρ is the fluid density.

To employ tank pressure P_2 rather than the mass density ρ as a variable, we must know how P_2 and ρ are related for the fluid (i.e., its equation of state). For most *liquids*, a reasonably good first-order approximation for the equation of state is given by

$$d\rho = \frac{\rho}{\beta}\,dP_2, \tag{3–17a}$$

where β is the bulk modulus of elasticity of the liquid in n/m^2 or lb/in^2. For most *gases* (e.g., air),

$$d\rho = \frac{\rho}{nP_2}\,dP_2, \tag{3–17b}$$

where n is a constant whose value is between 1.0 and 1.4, depending on how rapidly the fluid is compressed and P_2 is the absolute fluid pressure. Thus, if the pressure variations in a gas are small, it may be considered to have a bulk modulus of elasticity $\beta \approx nP_2$.

Employing the equation of state for a fluid with bulk modulus β in Eq. (3–16) yields

$$Q = \frac{\mathcal{V}}{\beta}\frac{dP_2}{dt} = \frac{\mathcal{V}}{\beta}\frac{dP_{21}}{dt}, \tag{3–18}$$

where P_1 is the constant ambient pressure outside the tank.

The volume of fluid V introduced into the tank when its pressure rises from P_1 to P_2 is obtained by integration of Eq. (3–18):

$$V = \int Q\,dt = \mathcal{V}\int_0^{P_{21}} \frac{dP_{21}}{\beta}. \tag{3–19}$$

If β is constant, Eq. (3–19) can be integrated directly. If β depends only on the tank pressure P_2, the integral can be evaluated after substituting the expression for β into Eq. (3–19). For a liquid, Eq. (3–19) yields

$$V = \frac{\mathcal{V}}{\beta}P_{21}, \tag{3–20}$$

and the tank behaves like an ideal fluid capacitance with

$$C_f = \frac{\mathcal{V}}{\beta}. \tag{3–21}$$

Once again, when drawing the symbolic diagram for this element one must always consider one port to be connected to a constant-pressure reference (Fig. 3–6a).

c) *Other types of fluid capacitance.* Two other devices which may act like fluid capacitances are often encountered in fluid systems. These are the spring-loaded "accumulator" shown in Fig. 3–9 and the aircharged (air spring) accumulator shown in Fig. 3–10. In both of these devices, the major energy storage occurs in a type of spring: a mechanical spring, in Fig. 3–9, and an "air spring," in Fig. 3–10. In some instances, if the walls are sufficiently elastic, the tank itself can serve as the spring for storing energy, thus creating a fluid-capacitance effect.

Problems 3–2, 3–5, 3–6, and 3–7 deal with the calculation of the capacitance C_f for various systems.

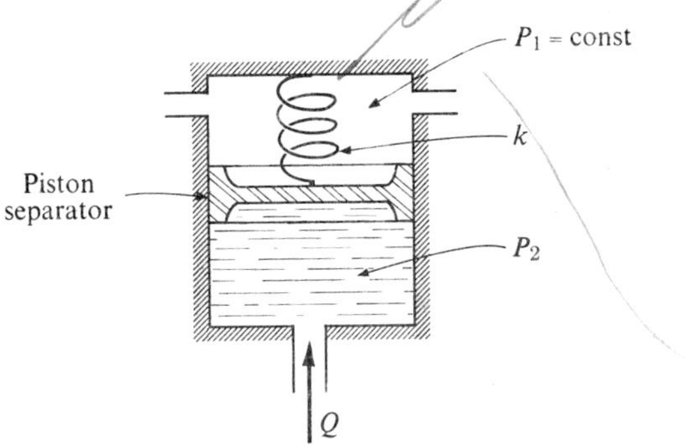

FIG. 3–9. Spring-loaded accumulator.

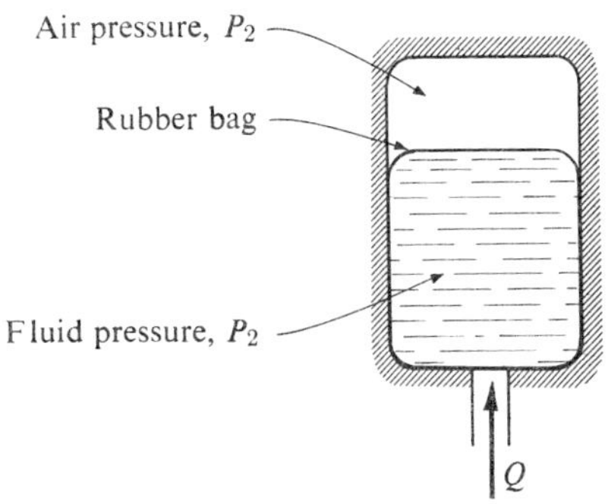

FIG. 3–10. Air-charged accumulator.

Pure fluid inertance. The fluid element whose characteristic is analogous to electrical inductance has a property which we will call *fluid inertance* because it comes from the inertia forces required to accelerate a fluid in a pipe or passage. To visualize the concept of inertance, let us consider the unsteady frictionless flow of an incompressible fluid in a nonaccelerating tube or pipe (Fig. 3–11).

If the fluid is incompressible, the flow rate at any section of the pipe is the same. If the pipe has constant area and the velocity of the fluid is uniform across any cross section of the pipe, we can say that every fluid particle has the same velocity and hence the same acceleration. The force F necessary to produce an acceleration dv/dt of the fluid mass in the pipe is

$$F = \underbrace{A(P_2 - P_1)}_{\substack{\text{Accelerating}\\ \text{force}}} = \underbrace{\rho A l}_{\substack{\text{Mass}\\ \text{of fluid}\\ \text{in pipe}}} \underbrace{\frac{dv}{dt}}_{\substack{\text{Acceleration}\\ \text{of fluid}\\ \text{mass}}}$$

The velocity v is equal to the flow rate Q divided by the area A, so that we may write

$$P_2 - P_1 = P_{21} = \frac{\rho l}{A}\frac{dQ}{dt} = I\frac{dQ}{dt}, \tag{3–22}$$

where the quantity $\rho l/A$ is the inertance, I, of the flow in the pipe. Because this element is analogous to

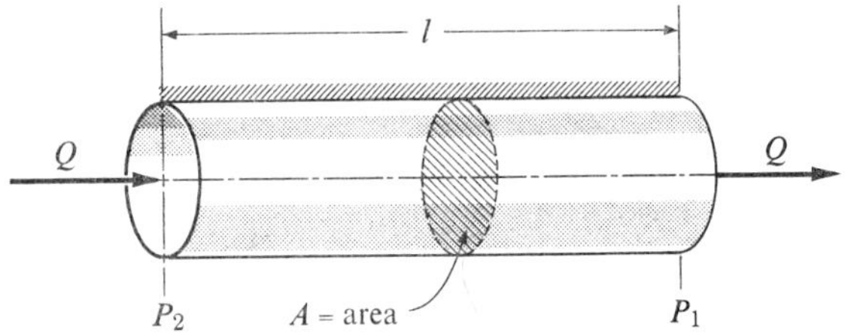

FIG. 3–11. Acceleration of fluid flow in a pipe.

an electrical inductor, the quantity $\rho l/A$ is sometimes called *fluid inductance*. Define the quantity Γ_{21}, called the *pressure-momentum* difference, according to the relations

$$P_{21} = \frac{d\Gamma_{21}}{dt}, \tag{3–23}$$

$$\Gamma_{21} = \int_{t_a}^{t_b} P_{21}\,dt + (\Gamma_{21})_0, \tag{3–24}$$

where $(\Gamma_{21})_0$ is the pressure-momentum at $t = 0$.

A *pure inertance* is defined as a fluid element in which the change of pressure momentum Γ_{21} is a single-valued function of the fluid flow Q:

$$\Gamma_{21} = f(Q). \tag{3–25}$$

An *ideal inertance* has a linear relation between Γ_{21} and Q:

$$\Gamma_{21} = IQ, \tag{3–26}$$

where Γ_{21} has units of n·sec/m^2 or lb·sec/in^2, and I

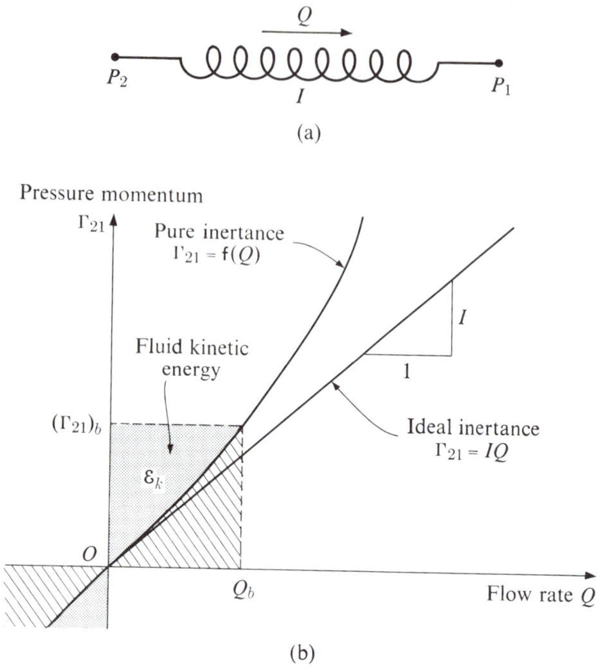

FIG. 3–12. Symbolic representation and constitutive relationships for pure inertances. (a) Symbolic representation. (b) Constitutive relationships.

has units of n·sec^2/m^5 or lb·sec^2/in^5. For constant I, the elemental equation for an ideal inertance is

$$P_{21} = I\frac{dQ}{dt} \tag{3–27}$$

or

$$Q = \frac{1}{I}\int_0^t P_{21}\,dt + Q_0, \tag{3–28}$$

where Q_0 is the flow at $t = 0$. Figure 3–12 shows the symbol and the constitutive relationships for pure and ideal inertance.

The energy stored in a pure inertance is called *fluid kinetic energy*, $\mathcal{E}_k$, since it results from the kinetic energy of the fluid mass:

$$\mathcal{E}_k = \int_{t_a}^{t_b} QP_{21}\,dt = \int_{\Gamma_a}^{\Gamma_b} Q\,d\Gamma_{21} = \int_{Q_a}^{Q_b} Q\frac{\partial f(Q)}{\partial Q}\,dQ, \tag{3–29}$$

the shaded area in Fig. 3–12. For an ideal inertance, Eq. (3–26) permits this integral to be evaluated explicitly:

$$\mathcal{E}_k = \tfrac{1}{2}IQ^2 = \frac{1}{2}\frac{(\Gamma_{21})^2}{I}. \tag{3–30}$$

We see that energy is stored in an ideal inertance by virtue of the through-variable Q. Nonlinear inertance is not usually encountered in practice because fluid flow velocities are normally very small compared with the speed of light.

In actual fluid piping, significant friction effects (discussed below) are often present along with the inertance effect, and the inertance effect tends to predominate only when the rate of change of flow rate (fluid acceleration) is relatively large. Since flow resistance in a pipe decreases more rapidly with increasing pipe area A than does inertance, it is easier for inertance effects to overshadow resistance effects in pipes of larger sizes. However, when the rate of change of flow rate is large enough, significant inertance effects are sometimes observed even in fine capillary tubes.

It is important to note that the relationships developed above hold only when the pipe is not being accelerated. If the pipe itself has an acceleration in the direction of flow through it, additional pressure-difference effects will be experienced between the ends of the pipe.

Pure fluid resistance. There are four basic types of fluid resistance that are most frequently encountered in fluid systems: (a) porous medium or "plug," (b) long capillary tube, (c) turbulent pipe flow, and (d) orifice flow. When only small pressure differences compared with the average pressure are used, the resistance characteristic for each device is of the same form for liquid and gaseous fluids. The limitation to small pressure differences reduces the otherwise important effect of compressibility which would be observed in the case of a gaseous fluid. These flows are often called incompressible simply because compressibility effects are small enough to be neglected in a great number of engineering applications.

a) *Incompressible flow through a porous plug.* D'Arcy's law for incompressible flow through a porous plug [1] is analogous to Ohm's law for an electrical resistor. In Fig. 3–13,

$$P_{21} = R_f Q, \tag{3–31}$$

where

P_{21} = pressure difference across the plug,

Q = volume flow rate,

R_f = fluid resistance.

The fluid resistance R_f depends on the geometry of the porous medium and often on the magnitude of the pressure difference. The units of R_f are n·sec/m^5 or lb·sec/m^5.

The value of R_f for a given porous plug may be estimated if its mean pore size, the plug porosity, and the viscosity of the working fluid are known [1]. Otherwise a simple experimental measurement affords a direct means of determining R_f for a given working fluid.

b) *Incompressible flow through a long capillary tube* (laminar flow). Here the Hagen-Poiseuille law applies [2] and, again, it gives a result analogous to flow of electrical current through a resistor. Figure 3–14 shows the geometry of a capillary tube.

Equation (3–31) applies also to capillary tube flow of an incompressible fluid, where R_f is given by

$$R_f = \frac{128\mu l}{\pi d^4}, \tag{3–32}$$

where

μ = absolute viscosity of the fluid, n·sec/m^2 or lb·sec/in^2,

l = length of tube, m or in.,

d = inside diameter of capillary, m or in.

The tube must be long enough in relation to its diameter so that effects of nonuniform flow near the entrance are negligible. Also, the rate of change of flow rate must be small enough that inertia effects are negligible.

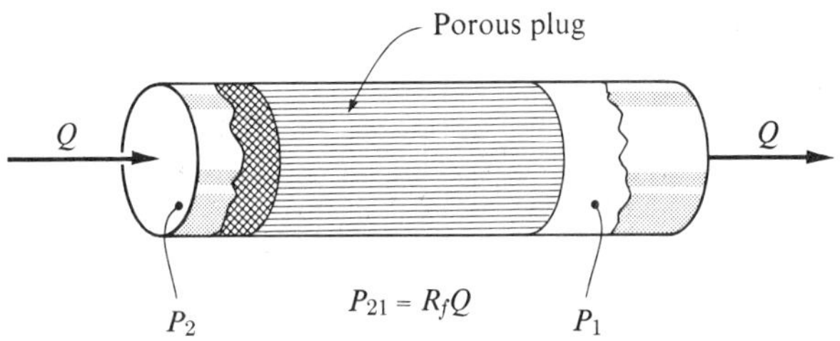

FIG. 3–13. Flow through a porous plug.

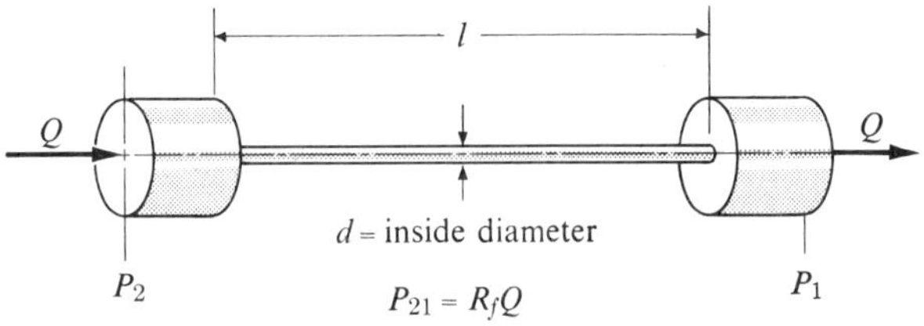

FIG. 3–14. Flow through a capillary tube.

c) *Incompressible flow through a long pipe.* When the flow rate through a tube becomes large so that the *Reynolds number*, Rey, is greater than 2000, where

$$\text{Rey} = \frac{4\rho Q}{\pi d\mu} \tag{3–33}$$

and ρ is the fluid density, the laminar capillary flow equations (3–31) and (3–32) are no longer valid. When the Reynolds number is greater than 5000, the flow is very likely to be turbulent, and the following approximate relation exists:

$$P_{21} = a_t Q|Q|^{3/4}, \tag{3–34}$$

where the bars indicate the absolute value and a_t is a constant whose value depends on the flow rate at which transition to turbulent flow occurs, the dimensions of the pipe (diameter and length), the properties of the fluid (density ρ and viscosity μ), and the roughness of the inner walls of the pipe. It is often simplest to employ a simple experimental measurement to determine a_t for a given specific pipe.

d) *Incompressible flow through an orifice.* Figure 3–15 shows an *orifice.* An approximate expression for the

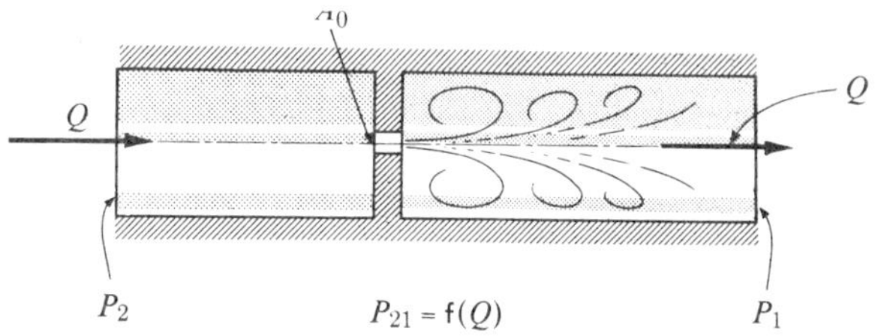

FIG. 3–15. Flow through an orifice.

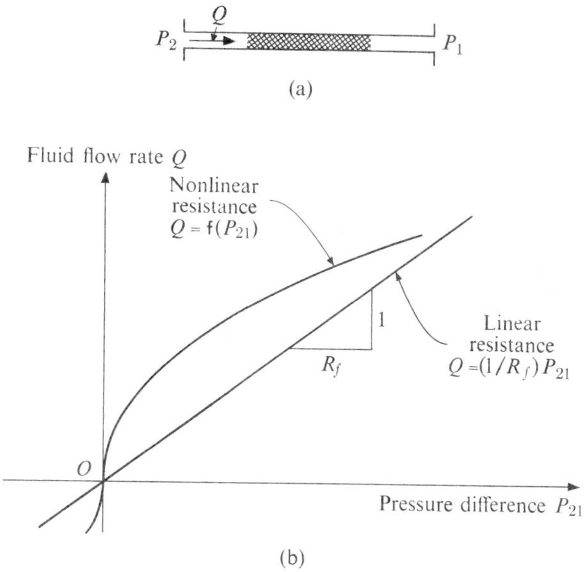

FIG. 3–16. Symbolic diagram and constitutive relationships for pure fluid resistance. (a) Symbolic diagrams. (b) Constitutive relationships.

pressure drop across an orifice is

$$P_{21} = a_0 Q|Q|, \tag{3–35}$$

$$a_0 = \rho/2C_d^2 A_0^2, \tag{3–36}$$

where C_d is a "discharge coefficient" (0.62 for a sharp-edged orifice) and A_0 is the orifice area.

We shall now define a *pure fluid resistance* by the single-valued function

$$Q = f(P_{21}), \tag{3–37}$$

where $Q = 0$ when $P_{21} = 0$, and the signs of Q and P_{21} are the same. For an *ideal fluid resistance*,

$$Q = \frac{1}{R_f} P_{21}. \tag{3–38}$$

By analogy with the electrical resistance, the symbol shown in Fig. 3–16(a) will be used sometimes to represent a pure fluid resistance. Figure 3–16(b) shows the constitutive relationships for fluid resistances.

It is important to note that Eqs. (3–31) through (3–38) hold only when the rate of change of flow through each resistance is small enough to avoid fluid inertia effects (this is analogous to avoiding inductance effects in an electrical resistor).

A discussion of fluid-resistance characteristics when fluid-compressibility effects are important (as they often are with gaseous fluids) is beyond the scope of this text. The reader who is interested in penetrating this subject more deeply may consult References [1], [2] or [5].

We see that fluid resistance dissipates fluid power. For incompressible fluids, the dissipated power $\mathcal{P}$ is

$$\mathcal{P} = P_{21}Q = R_f Q^2 = (P_{21})^2/R_f, \qquad \mathcal{P} \geqq 0. \tag{3–39}$$

Just as an electric wire can act like a resistor, capacitor, or inductor, depending on how it is used, a fluid-filled pipe can exhibit the characteristics of resistance, capacitance, or inertance, depending on how it is used. We shall discuss the problem of representing physical systems by various combinations of ideal elements in subsequent chapters.

3–2.4 Pure Fluid Transformer

A pure fluid transformer is an idealized device which is analogous to the pure mechanical or electrical transformers discussed in Chapter 2. The pure fluid transformer has zero net power flow and determines an output fluid volume V_b (integral of flow rate) which is a single-valued function of an input fluid volume V_a. In an ideal fluid transformer, these volumes are directly proportional to each other.

Figure 3–17 shows a system whose behavior approximates that of an ideal fluid transformer under certain circumstances. Two pistons, I and II, are inserted in two circular pipes of areas A_a and A_b. The pistons are assumed to fit the pipes closely enough that no significant leakage of fluid occurs past the pistons. Also, the friction forces between the pistons and their pipes or cylinders are assumed to be negligible. The pistons are connected by a rigid rod so that they have identical velocities. The chamber between the pistons is open to an ambient pressure P_1. The force acting on piston I is $(P_2 - P_1)A_a$, which must be balanced (assuming negligible momentum of the pistons) by an equal and opposite force on II, equal to $(P_4 - P_1)A_b$:

$$P_{41}/P_{21} = A_a/A_b = n = \text{area ratio} \tag{3–40}$$

where $P_{41} = P_4 - P_1$ and $P_{21} = P_2 - P_1$. The relation between the flows Q_a and Q_b is obtained by equating the velocities v_p of the two pistons:

$$v_p = Q_a/A_a = -Q_b/A_b.$$

Hence

$$Q_b/Q_a = V_b/V_a = -A_b/A_a = -1/n. \tag{3–41}$$

The relations (3–40) and (3–41) are analogous to the previously derived expressions for mechanical and electrical transformers. Can you show that the net power flow into the fluid transformer is zero? The symbol used for a pure fluid transformer is shown in Fig. 3–18. The pressures P_2 and P_4 must be referred to a common reference or "ground" pressure P_1.

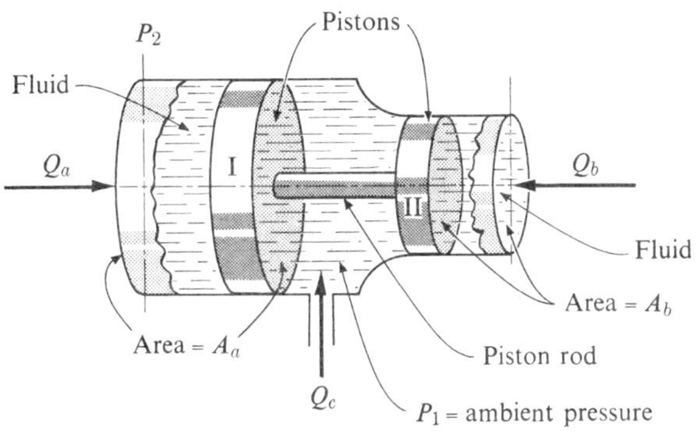

FIG. 3–17. Fluid transformer.

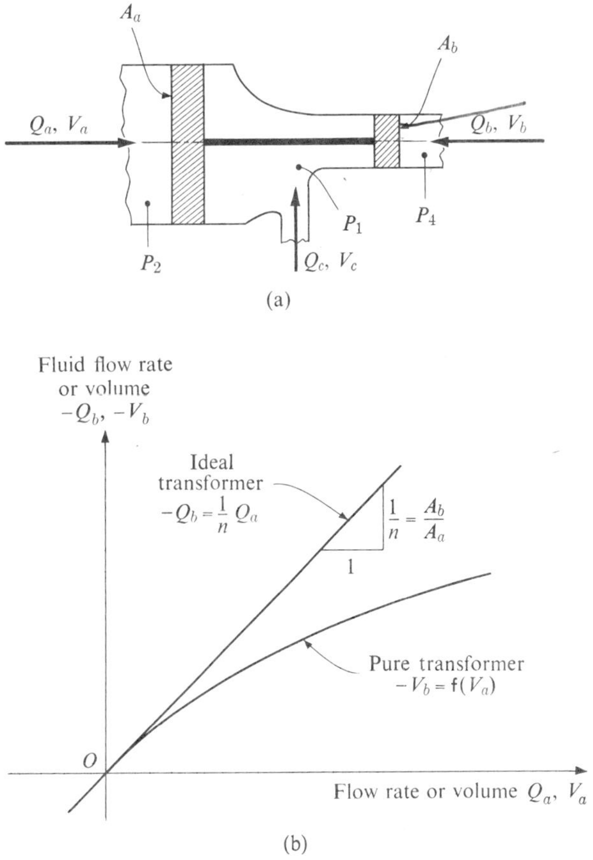

FIG. 3–18. Symbol and functional relationships for pure fluid transformers. (a) Symbol. (b) Functional relationships.

3–3. SIMPLE COUPLING OF FLUID AND MECHANICAL SYSTEMS, THE GYRATING TRANSDUCER

In Chapter 2 we introduced the idea of a *transducer* in the context of a change from mechanical rotational energy to mechanical translational energy (rack-and-pinion gear train) without storage or dissipation of energy. In that case we saw that the through-variable

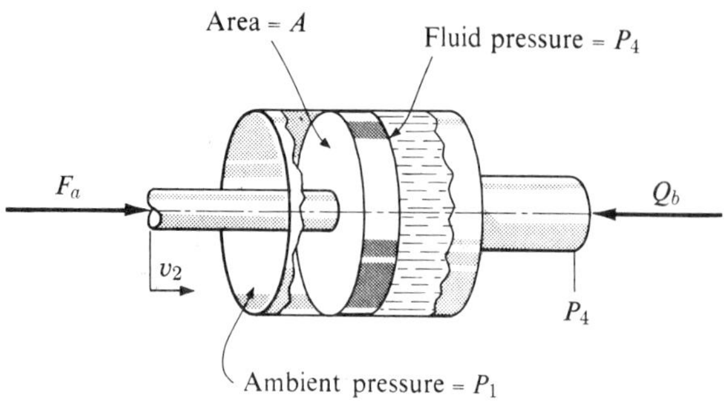

FIG. 3–19. Piston and cylinder; fluid-to-mechanical transducer.

in the rotational system (torque) was directly related to the through-variable in the translational system (force). Similarly, the across-variables were directly related.

Let us consider now the piston and cylinder illustrated in Fig. 3–19. For continuity of fluid flow, the piston area A times the piston's velocity v_2 must equal minus the flow Q_b:

$$-Q_b = Av_2. \tag{3–42}$$

A force balance on the piston gives

$$F_a = (P_4 - P_1)A = P_{41}A. \tag{3–43}$$

The mechanical power into the piston is

$$\mathcal{P}_{\text{mechanical}} = F_a v_2,$$

and the fluid power out of the cylinder is

$$\mathcal{P}_{\text{fluid}} = P_{41}Q_b.$$

Hence, using Eqs. (3–42) and (3–43), we see that the net power $\mathcal{P}$ is

$$\mathcal{P} = F_a v_2 + P_{41}Q_b = F_a v_2 + (F_a/A)(-Av_2) = 0.$$

In this device the fluid through-variable Q_b is directly related (through A) to the mechanical across-variable v_2. Similarly, P_{41}, the fluid across-variable, is proportional to F_a, the mechanical through-variable. A transducer which switches the roles of across- and through-variables will be called a *gyrating transducer*. To distinguish between this device and the transducer which transforms a through-variable in one medium to a through-variable in another medium (e.g., the rack and pinion), we shall call the latter a *transforming transducer*.

3–4. THERMAL SYSTEM ELEMENTS

The flow of heat from point to point in a thermal system is usually related to the temperatures of the system in much the same way that current is related to voltage in an electrical network, force to velocity in a mechanical system, or flow rate to pressure in a fluid system. Complex thermal systems also frequently involve fluid flow, mechanical motion, and electromagnetic field effects. However, this discussion will be limited to simple thermal systems in which only heat flow and associated energy storage are involved. We shall see that the behavior of thermal systems deviates considerably from that of the mechanical, electrical, and fluid systems considered thus far, although much similarity of behavior does exist.

3–4.1 Temperature and Heat Flow

The temperature of a substance or of a particular locality within a substance is a measure of the mean kinetic energy of the molecules. It is that quantity which is measured by a thermometer.

When the temperature of a body varies, changes occur in various physical characteristics of the body. For example, as the temperature of a metal increases, its linear dimensions will increase, its electrical resistance normally will increase, and the color of its surfaces may change. Such changes may be used to establish a quantitative measure of the temperature of the body. A mercury-bulb thermometer, for instance, measures temperature by the length of a mercury column contained in a fine glass tube. The scale of temperature in this case is obtained by dividing into equal increments the change in length of the column which occurs when the thermometer is immersed first in ice water and then in boiling water. The divisions of this interval are called *degrees*. In the Fahrenheit system (°F) there are 180° in the interval, and the temperature of ice water is taken to be 32°F. In the centigrade system (°C) the interval is divided into 100 one-degree increments, and the ice-water temperature is defined as 0°. It is evident from thermodynamics [3] that an *absolute* scale of temperature can be defined which is independent of physical characteristics of substances. This definition is arrived at through the *second law of thermodynamics*, which also shows that negative absolute temperatures are impossible. To avoid negative values, the absolute scale measures temperatures upward (positively) from zero. When this measurement is made in degrees Fahrenheit, the absolute scale is called the *Rankine* scale and the units are °R, and when it is made in degrees centigrade the scale is called the *Kelvin* scale and the units are °K. Thus the freezing point of water, ordinarily expressed as 0°C or 32°F, can also be expressed in the absolute scales as 273.2°K or 491.6°R. We shall use the symbol θ to represent temperature measured by any of the above scales.

If two bodies, 1 and 2, having different temperatures, θ_1 and θ_2, are brought into contact with each other, their temperatures will be seen to change in such a way that θ_1 will approach θ_2. In other words, one temperature will increase and the other will decrease. Thus it may be said that some quantity has been transferred or has flowed from one body to the other to produce the observed temperature changes. The quantity which flows from one body to another (or from point to point inside a single body) when a temperature difference exists is called *heat*. If the body temperature is decreased by the flow, the heat flow is considered to be out of the body. Since temperature is a measure of the mean energy of the molecules of the body, heat is a form of energy.

A quantitative measure of heat may be obtained by allowing the heat to flow into a known mass of a substance such as water. The resulting increase in temperature is a function of the amount of heat transferred. We shall use the symbol $\mathcal{H}$ for quantity of heat. The standard units of heat are the British thermal unit (Btu) and the kilogram·calorie (kg·cal). One Btu is the amount of heat $\mathcal{H}$ which must be supplied to a 1-lb mass of water to raise its temperature by 1 F°. One kilogram·calorie is the amount of heat required to raise the temperature of 1 kg of water by 1 C°. One Btu is equivalent to 0.252 kg·cal.

The flow rate of heat q is obtained by differentiating $\mathcal{H}$:

$$\mathsf{q} = \frac{d\mathcal{H}}{dt}. \tag{3–44}$$

Consideration of the methods by which temperature and heat flow could be measured reveals that temperature is an across-variable (it can be measured without breaking into the system) and that heat flow is a through-variable.

3–4.2 Thermal Power and Energy

In the systems we have studied previously, power has been given by the product of the through- and across-variables at any section of the system. In thermal systems this situation does not prevail, since heat flow (the through-variable) is *itself the thermal power*. Thus

$$\mathcal{P} = \mathsf{q}. \tag{3–45}$$

Thermal power is measured in kg·cal/sec or Btu/sec. Equation (3–45) may also be expressed in watts or ft·lb/sec by means of the equivalence of mechanical work and heat determined by the first law of thermodynamics. Thus 1 Btu/sec equals 778.26 ft·lb/sec, and 1 kg·cal/sec is 3.96 Btu/sec, or 4200 n·m/sec.

The thermal energy $\mathcal{H}$ transferred through a surface (real or imaginary) in a body is the time integral of the heat flow:

$$\mathcal{H} = \int_0^t \mathsf{q}\, dt. \tag{3–46}$$

The first law of thermodynamics tells us that for any system defined by a closed imaginary surface, the total heat transferred into the system must equal the energy transferred out of the system in other forms (such as mechanical work, electrical energy, or fluid energy) plus the increase of energy stored in the system.

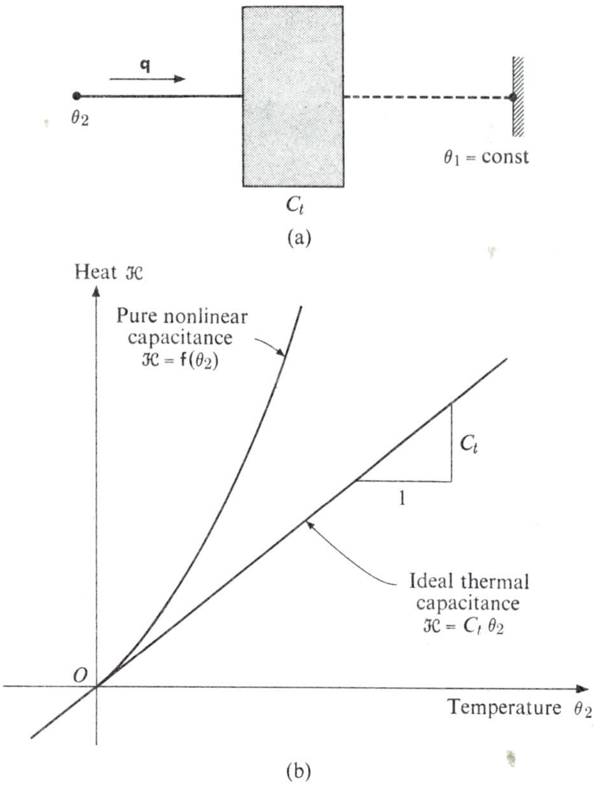

FIG. 3-20. Symbolic representation and constitutive relationships for pure thermal capacitance. (a) Symbolic representation. (b) Constitutive relationships.

3-4.3 Pure Thermal System Elements

We might expect from our earlier study of mechanical, electrical, and fluid systems that three ideal thermal elements could also be found which would represent the phenomena of thermal capacitance, thermal inductance, and thermal resistance. Thermal resistance and capacitance do in fact occur in thermal systems, but thermal inductance apparently *does not exist* in thermal systems operating at normal temperatures.

Pure thermal capacitance. The capacity of a given material or combination of materials to store internal energy by virtue of a temperature rise when it receives a net flow of heat is determined by its mass and its so-called *specific heat,* C_p. This energy-storage phenomenon may be described by a thermal capacitance. When heat flow q takes place into a system having uniform (but not necessarily constant) temperature θ_2 relative to a constant reference temperature θ_1,

$$\mathsf{q} = C_p m \frac{d\theta_2}{dt} = C_p m \frac{d\theta_{21}}{dt}, \tag{3-47}$$

where

C_p = specific heat of mass, kg·cal/kg-°C or Btu/lb-mass-°F;*

m = mass of system, kg or lb-mass.

A pure thermal capacitance by definition has a heat content $\mathcal{H}$ which is a single-valued function of its temperature θ_2. In an *ideal thermal capacitance,* $\mathcal{H}$ and θ_2 are proportional:

$$\mathcal{H} = C_t \theta_2, \tag{3-48}$$

where C_t is the thermal capacitance, in kg·cal/°C or in Btu/°F. Comparison of Eqs. (3-47) and (3-48) shows that

$$C_t = C_p m. \tag{3-49}$$

The symbol and the constitutive relations for a pure thermal capacitance are shown in Fig. 3-20. As for a fluid capacitance and ideal masses, one terminal of the thermal capacitance must be connected or referred to a reference temperature θ_1 which is zero or constant. No heat flow occurs through the reference "connection," and hence the reference is shown dashed.

The energy $\mathcal{E}_t$ stored in a thermal capacitance in the absence of work effects is simply equal to the heat transferred into it. Here again, energy does not enter

* The specific heat of a given material has the same numerical value in both of these sets of units. The specific heat depends somewhat on the constraints applied to the material during heating. For example, heating at constant pressure yields a specific heat different from that obtained by heating at constant volume.

the picture in the same way as in the case of mechanical, electrical, and fluid elements, because the stored energy is a function of only the first power of the across-variable θ_2:

$$\mathcal{E}_t = \mathcal{H} = \mathsf{f}(\theta_2). \tag{3–50}$$

For a system to be modeled by an ideal thermal capacitance, all thermal resistance effects (discussed below) should be negligible within it (usually in comparison with external resistances) so that the temperature will be uniform throughout the mass which comprises the capacitor. A mass of metal or a well-stirred tank of fluid often behave very nearly like ideal thermal capacitances.

Pure thermal inductance. Our second thermal element should be an energy-storage element which stores energy by virtue of a heat flow through it, as distinct from a thermal capacitance, which stores energy by virtue of its temperature. *There is no known thermal phenomenon at ordinary temperatures which acts in this way; thus there is no thermal element which could be called an inductance, by analogy to electrical inductance.*

A phenomenon which has been found to exist near absolute zero temperatures may indicate another type of thermal energy storage, which is yet to be understood. This phenomenon is called "second sound" to indicate that heat transmission at these low temperatures has been observed to occur in the same fashion as sound transmission (i.e., a wave phenomenon); at ordinary temperatures, this is not experienced. Investigation of various wave phenomena shows that two types of energy storage must always be present together before wave propagation can occur. If this is a general principle, then the observance of thermal wave motion at low temperatures means that the second type of thermal energy storage must exist and perhaps could be isolated.

Pure thermal resistance. All materials offer some resistance to heat flow, which is evidenced by the fact

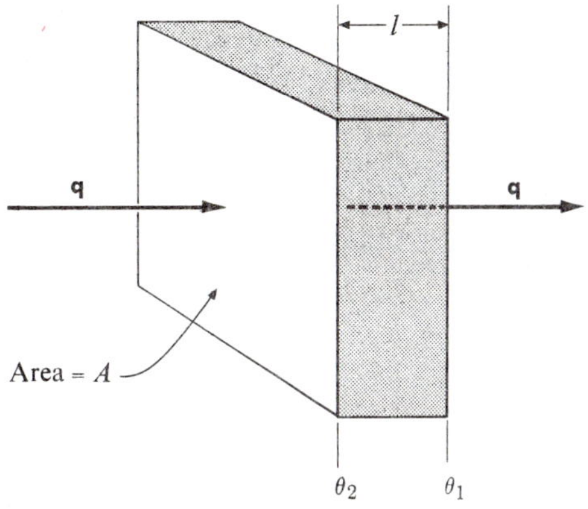

FIG. 3–21. Heat flow through a conducting material.

that the temperature drops in the direction of heat flow through material.* Even when the heat flow is in the form of radiation through space (vacuum), a negative temperature gradient is associated with the heat flow. When the material offers a large degree of resistance to heat flow, it is often called an *insulator*, and its thermal conductivity is low. A material through which heat flows relatively freely is called a *conductor* and its thermal conductivity is high.

The flow of heat through a solid material, such as the sheet illustrated in Fig. 3–21, occurs by *conduction*, that is, by the transfer of thermal kinetic energy from atom to atom on a microscopic scale within the material. This type of heat flow is described quantitatively by *Fourier's law:*

$$\mathsf{q} = \frac{\sigma_c A(\theta_2 - \theta_1)}{l}, \tag{3–51}$$

where A is the area normal to the heat flow direction, l is the length in the flow direction, and $\theta_2 - \theta_1 = \theta_{21}$ is the temperature difference where $\theta_2 > \theta_1$. The thermal conductivity σ_c is a specific property of a material, and is defined by Eq. (3–51). In the mks

* At very low temperatures, near absolute zero, certain materials may exhibit "superconductivity," in which case heat may flow with practically zero temperature difference.

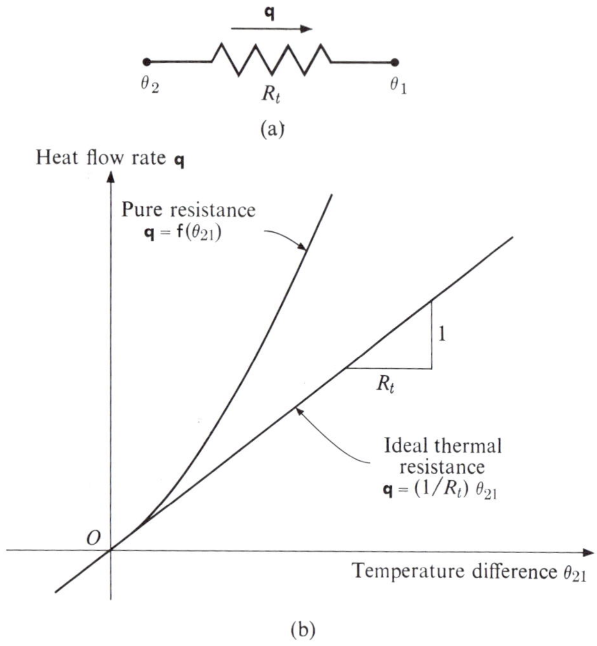

FIG. 3–22. Symbolic representation and constitutive relationships for pure thermal resistance. (a) Symbolic representation. (b) Constitutive relationships.

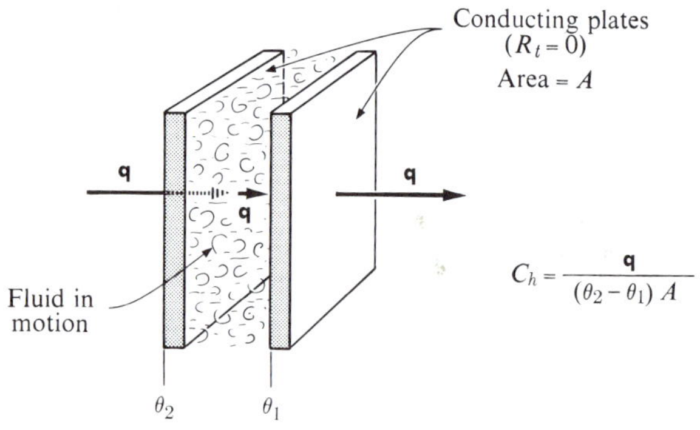

FIG. 3–23. Convective heat transfer.

system, σ_c is expressed in kg·cal/sec·m·°C; in the British engineering system, it is in Btu/sec·in·°F.

Equation (3–51) relates the through-variable q to the across-variable θ_{21} by a simple constant factor. By analogy with previous ideal system elements, we will define a *pure thermal resistance* by the following equation:

$$q = f(\theta_{21}), \tag{3–52}$$

in which $q = 0$ when $\theta_{21} = 0$, the signs of q and θ_{21} are the same, and where f is a single-valued function. An *ideal thermal resistance* has the defining equation

$$q = \frac{1}{R_t}\theta_{21}. \tag{3–53}$$

Comparison of Eqs. (3–51) and (3–53) shows that heat conduction is represented by an ideal resistance

$$R_t = \frac{l}{\sigma_c A}, \tag{3–54}$$

where

R_t = thermal resistance, $\frac{°\text{C·sec}}{\text{kg·cal}}$ or $\frac{°\text{F·sec}}{\text{Btu}}$.

A given material may be considered as a pure or ideal thermal resistance only if the amount of heat stored in it is negligible or does not change appreciably during operation. Figure 3–22 shows the symbol and the constitutive relationships for pure thermal resistances.

When two sheets of solid material are placed together with imperfect contact (which is typical), the total resistance of the combination is greater than the sum of the individual resistances of each of the sheets because of the small air space (or vacuum or oxide film in some cases) which exists at the interface. Although the thickness of the air space is very small, this extra resistance can be appreciable, because still air is a relatively poor conductor of heat. In fact, good insulating materials are usually fabricated so that they are not solid but contain a great many "dead" air spaces.

When the material through which heat is flowing is in part a fluid (liquid or gas), one may compute its resistance in much the same way as for a solid (Eq. 3–54), but *only so long as the fluid is not in motion.* When fluid motion occurs (natural convection, forced flow, etc.), the heat-transfer picture becomes very complex due to the heat-transport effects derived from the fluid motion. This heat-transfer mechanism is called *convection,* and is illustrated in Fig. 3–23. Motion, especially if it is turbulent, reduces the resistance to heat flow through a fluid—so much so, in fact, that the major thermal resistance exists in a very thin layer near the wall where fluid motion is small. This resistance is described in terms of a *film coefficient of heat transfer*.

The notion of an *over-all* coefficient of heat transfer for a given material, combination of materials, or complex physical arrangement of components in a thermal system is frequently employed, and it is defined as follows:

$$C_h = \frac{\text{rate of heat flow}}{\text{(area)(over-all temperature drop)}} = \frac{\mathsf{q}}{(\theta_2 - \theta_1)A}, \tag{3–55}$$

where C_h = coefficient of heat transfer,

kg·cal/°C·sec·m^2

or

Btu/°F·sec·in^2.

A determination of the heat-transfer coefficient for a given thermal situation (which often includes fluid-flow phenomena as well as thermal effects) is very complex and nearly always involves experimental measurements. The coefficient C_h is not constant for a given geometry, but may often be assumed constant over limited ranges. Methods of determining heat-transfer coefficients are discussed in detail in Reference [4].

Equation (3–55) describes an ideal thermal resistance R_t whose value is

$$R_t = 1/C_hA. \tag{3–56}$$

In addition to thermal conduction and convection, heat may be transferred to or from a system by *thermal radiation.* Radiation is described quantitatively by the Stefan-Boltzmann law,

$$\mathsf{q} = C_r(\theta_2^4 - \theta_1^4), \tag{3–57}$$

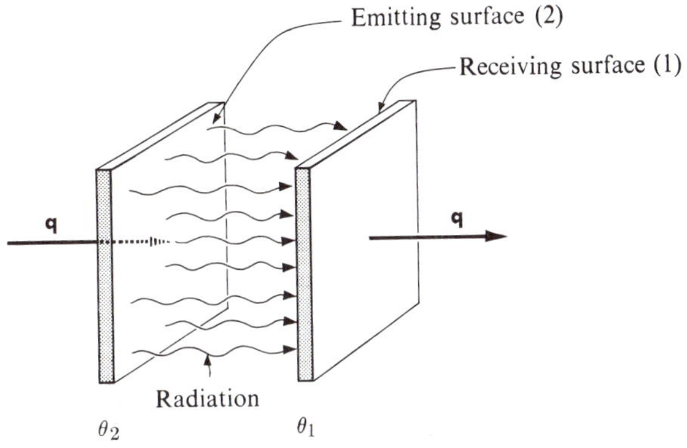

FIG. 3–24. Radiative heat transfer.

where C_r is a constant which depends on the shape, area, and surface characteristics of the body which is emanating radiation (or which is receiving radiation), and on the wavelength of the radiation. The temperatures θ_2 and θ_1 must be the *absolute temperatures* (°K or °R) of the surfaces emitting and receiving radiation, respectively. Figure 3–24, illustrates schematically radiative heat transfer. Here it is assumed that all the radiant energy emitted by surface 2 is absorbed by surface 1.

The constant C_r in Eq. (3–57), measured in kg·cal/sec·°K^4 or Btu/sec·°R^4 is very small, so that heat transfer by radiation is usually insignificant unless large temperatures are involved. As an example, for radiation from a spherical black body to a black environment which completely surrounds it, the value of C_r is given by σA_1, where σ is the Stefan-Boltzmann constant,

$$\sigma = 1.37 \times 10^{-5} \text{ kg·cal/m}^2\text{·sec·°K}^4$$

or

$$\sigma = 3.34 \times 10^{-13} \text{ Btu/in}^2\text{·sec·°R}^4,$$

and A_1 is the area of the radiating sphere in m^2 or in^2, respectively.

The expression (3–57) for heat transfer is nonlinear and does not represent a pure thermal resistance as we have defined it, because q does not depend on temperature difference. If the temperatures θ_2 and θ_1

change by only small amounts from mean values θ_{20} and θ_{10}, Eq. (3–57) may be *linearized.* Let q_0 be the heat flow when

$$\theta_2 = \theta_{20} \quad \text{and} \quad \theta_1 = \theta_{10},$$

let θ_1 be constant and $\Delta\theta_2$ be a small change (compared with θ_{20} and θ_{10}) of temperature θ_2. Then Δq is the change in heat flow,

$$\Delta q = 4C_r\theta_{20}^3\,\Delta\theta_2 = \frac{1}{R_{eq}}\Delta\theta_2, \qquad q_0 = C_r(\theta_{20}^4 - \theta_{10}^4); \tag{3–58}$$

and R_{eq} is the equivalent linear resistance when $q = q_0$.

When a thermal-resistance effect occurs due to combined radiative and convective heat transfer, it is often true that the over-all resistance is nearly linear and can be represented over the range of interest by an ideal thermal resistance [4].

A thermal resistance, unlike mechanical, electrical, and fluid resistances, dissipates no energy since the net heat flow is always zero. It also stores no energy, since no work is done during the heat-flow process in a thermal resistance (first law of thermodynamics). There is, however, an increase of the property *entropy* associated with the flow of heat through a thermal resistance (second law of thermodynamics), and hence a loss or dissipation of the available part of the energy* supplied to it at temperature θ_2.

3–4.4 Pure Thermal Transformer

A pure transformer is a device which provides changes in the magnitudes of the through- and across-variables while maintaining instantaneous power balance. Such a device is clearly impossible in thermal systems because the through-variable itself is power (heat flow). Instantaneous power balance requires that $q_{in} = q_{out}$, and therefore no transformation of q can occur.

3–4.5 Thermoelectric Transducers

There are several kinds of thermal-to-electrical transducers. The most common of these is the thermocouple, in which a voltage difference is generated between two dissimilar metals when a temperature difference is applied. This effect has been used for temperature measurement and for the generation of small amounts of electric power. When energy flows from the electrical system to the thermal system, refrigeration or cooling can be obtained. Small thermoelectric refrigerators have been made for the cooling of electric components and for use as automatically controlled constant-temperature references.

REFERENCES

1. L. Prandtl, *Essentials of Fluid Dynamics.* New York: Hafner Publishing Company, 1952.
2. J. F. Blackburn, G. Reethof, and J. L. Shearer, *Fluid Power Control.* New York: The Technology Press of M.I.T. and John Wiley and Sons, Inc., 1960.
3. J. H. Keenan, *Thermodynamics.* New York: John Wiley and Sons, Inc., 1941.
4. W. McAdams, *Heat Transmission.* New York: McGraw-Hill Book Company, Inc., 1942.
5. A. H. Shapiro, *The Dynamics and Thermodynamics of Compressible Fluid Flow*, Vols. I and II. New York: Ronald Press Company, 1953.

* If the heat flow q were put into a reversible heat engine (a thermal-to-mechanical transducer) an amount of mechanical power equal to $q(\theta_2 - \theta_1)/\theta_2$ could have been taken out of the engine.

PROBLEMS

3–1. A simple hydraulic pump (a source of fluid power) consists of two meshing gears contained in a close-fitting housing (Fig. 3–25). As the gears rotate, each tooth space captures a nearly constant amount of fluid at the inlet and carries it around to the outlet. At the meshing region the gear teeth contact and seal the outlet. Ideally, Q = constant through the device and the pressure rises from 0 to P_s.

a) Qualitatively discuss how Q is related to the gear-tooth size, the speed Ω, and the width w of the gears.

b) If there are no energy losses, how are Q, P_s, T, and Ω related?

c) The above pump is connected to a fluid system. During operation, Q = const = 10 gal/min, and the outlet pressure P_s is found to vary sinusoidally about a mean pressure of 500 psi, $P_s = 500 + 100 \sin 20\pi t$. Determine the average power and the maximum instantaneous power delivered to the system.

3–2. In a chemical-processing plant, liquid flows into a long cylinder with a stainless-steel wall 1 in. thick (Fig. 3–26). The pressure in the tank may be assumed uniform and varies from 100 to 500 psi during operation of the process. In this problem we wish to study the capacitive properties of this tank. The fluid has a bulk modulus of 150,000 psi and a density of 60 lb/ft^3; the steel has a Young's modulus of 30×10^6 psi and a Poisson's ratio of 0.3.

a) Assuming that the tank is rigid, compute the fluid capacitance of the tank, $C_f = (\int(Q_1 - Q_2)\,dt)/P = V/P$.

b) Compute the energy stored when P increases from 0 to 500 psi.

c) Assuming that the fluid is incompressible, compute C_f and compute the energy stored when P = 500 psi.

d) Combine the above to obtain the actual tank capacitance and the total energy stored when P = 500 psi.

e) Discuss the linearity of the V/P relations in (a), (c), and (d).

3–3. Consider an open reservoir being filled with water. The pressure in the tank bottom varies as shown in Fig. 3–27. Sketch curves of the flow rate Q into the tank and the height H of the water in the tank vs. time.

3–4. The flow of water into the tank of Problem 3–3 varies as shown in Fig. 3–28. Given that A = 1 ft^2, determine, as a function of time, the pressure P required at the inlet to produce this flow rate. $P = P_0$ at $t = 0$.

3–5. A cylindrical steel tank of the dimensions shown in Fig. 3–29 has a single inlet for oil flow. The tank is rigid, and the oil has a bulk modulus β = 200,000 psi and a density ρ_O = 62 lb/ft^3. The average pressure in the tank

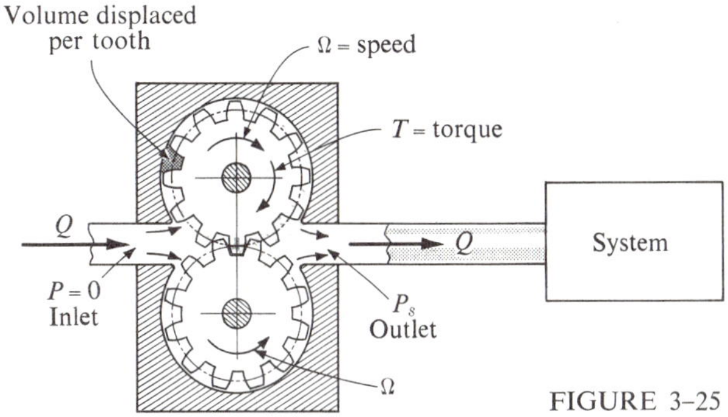

FIGURE 3–25

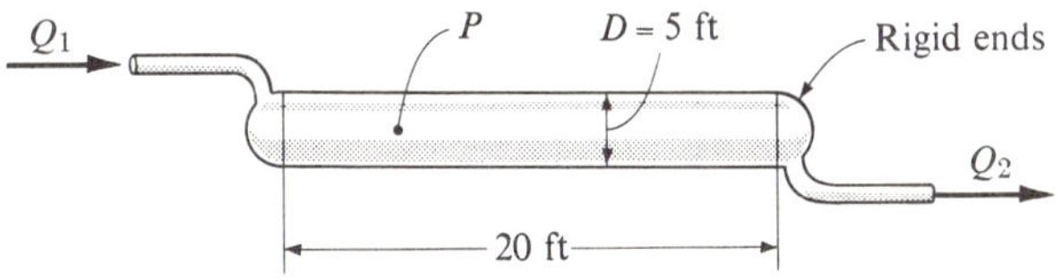

FIGURE 3–26

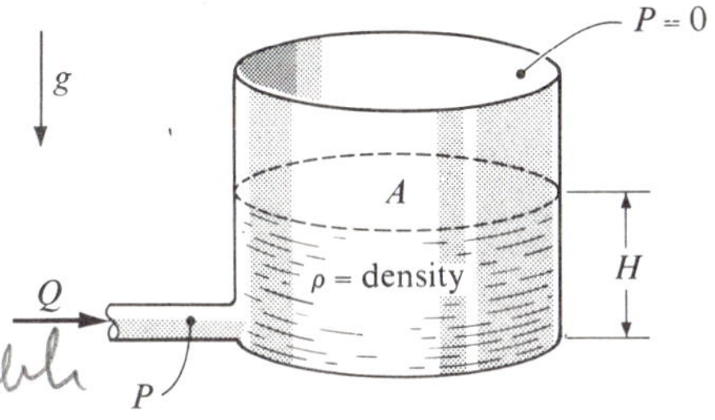

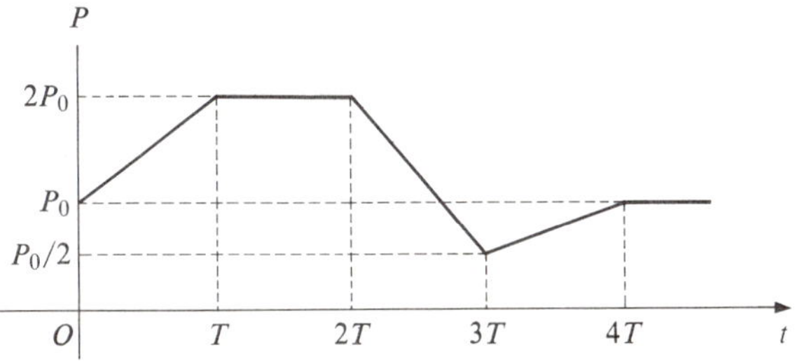

FIGURE 3–27

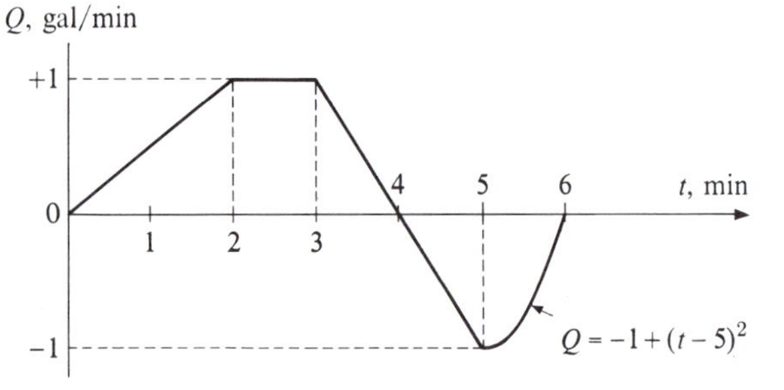

FIGURE 3–28

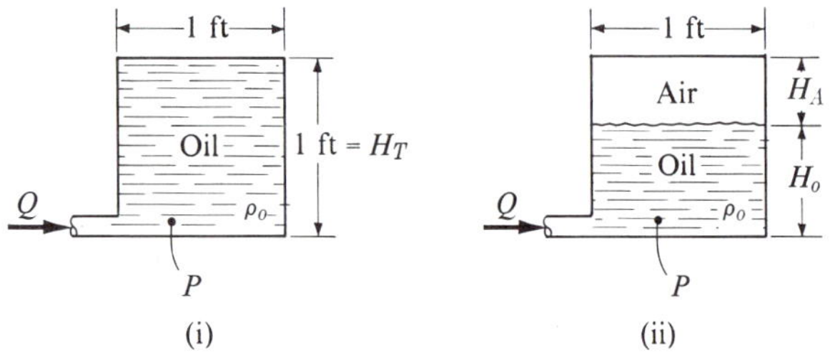

FIGURE 3–29

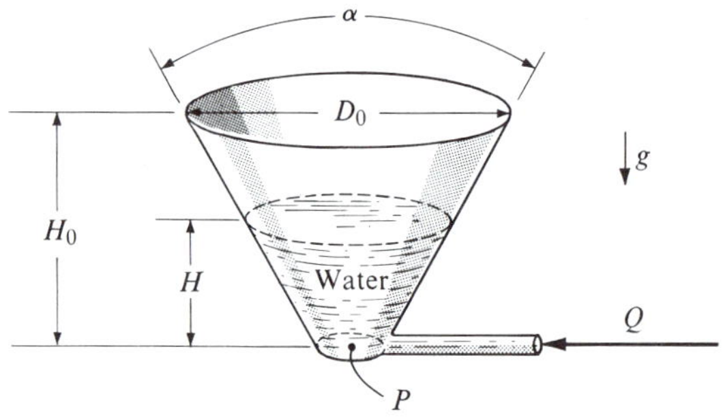

FIGURE 3–30

is P_O, and P varies slightly from this value. Gravity forces are negligible.

a) Assuming that the tank is entirely full of oil as in Fig. 3–29(i), compute the fluid capacitance.

b) In Fig. 3–29(ii) a quantity of air was accidentally left in the tank when it was filled with oil. Derive an expression for the capacitance when $P = P_O + \Delta P$, where ΔP is small in terms of H_O, H_A, P_O, etc., assuming that the air is isothermal and acts like a perfect gas.

c) If $H_A = 0.1\ H_O$ and $P_O = 200$ psi, compute the capacitances of the tanks in Fig. 3–29(i) and (ii). Comment on the effect of entrained gas bubbles on the capacitance of liquid-filled tanks.

3–6. A water storage tank has a conical shape (Fig. 3—30). Assuming that $\alpha = 30°$, $H_0 = D_0$, and the tank is rigid, compute and sketch a curve of stored volume V vs. pressure P for $0 < H < H_0$. Sketch a curve of stored energy vs. P. For small changes of P about a mean value P_0 the tank is approximated by an ideal capacitance. Sketch a curve of capacitance vs. P_0.

3–7. A fluid accumulator (Fig. 3–9) is required to have a capacitance of 20 lb/in^5 and a volume of 50 in^3. Design the device; i.e., specify piston area, spring stiffness, length and wall thickness of cylinder, and any other important parameters. Discuss qualitatively the conditions on Q, P_2, and their respective integrals and derivatives which must be met for the device to behave approximately like an ideal capacitance.

3–8. A conduit, called a *penstock*, which leads water from a dam to the inlet of a hydraulic turbine driving an electric generator has a diameter of 5 ft and a length of 700 ft. Assuming that the conduit is rigid, compute the inertance of the fluid. During full-load operation the average flow velocity in the pipe is 50 ft/sec. For this condition, find the energy stored in the fluid inertance and determine how long a 100-watt light bulb could operate from this amount of energy.

3–9. In Problem 3–8, an attempt is made to reduce the flow in the conduit to zero as a linear function of time in 30 sec, starting with full-load conditions. This is done by closing a gate valve near the turbine inlet. Estimate the pressure rise at the valve if the upstream pressure at the inlet to the conduit remains constant.

3–10. A circular tube has a gradual uniform taper, so that its internal diameter decreases linearly from D_2 at the left end to D_1 at the right. The tube is rigid and has length L.

a) Assuming that an incompressible fluid of density ρ flows from left to right in the tube, compute the fluid inertance.

b) If the fluid flows from right to left, how does the inertance change?

c) If the taper is nonuniform, how will this affect the inertance, assuming that D_2 and D_1 do not change?

3–11. A fluid having density $\rho = 0.04\ \text{lb/in}^3$ and viscosity $\mu = 5 \times 10^{-5}$ lb·sec/in^2 flows through a tube of 0.032-in. internal diameter and 1-ft length.

a) Compute the fluid resistance, assuming Hagen-Poiseuille flow.

b) Estimate the range of steady flow rates Q where (a) is valid.

c) Compute the fluid inertance.

3–12. Assuming that the flow varies sinusoidally in the tube of Problem 3–11, $Q = Q_0 \sin \omega t$, estimate the range of frequency ω over which the tube can be represented by an ideal resistance. Note that the pressure drop due to inertance must be small.

3–13. The pressure-flow relation for an orifice is given by Eqs. (3–35) and (3–36). If $P_{21} = P_0 + \Delta P_{21}$, and if $Q = Q_0 + \Delta Q$, where ΔP_{21} and ΔQ are small, there will be a nearly linear relation between ΔP_{21} and ΔQ, that is,

$$\Delta P_{21} = R_f \Delta Q.$$

This is called the *linearized* or *local* resistance near $P_{21} = P_0$ and $Q = Q_0$, and is approximately the local slope of the P_{21}-Q curve (Fig. 2–17). Determine R_f as a function of P_0 and sketch qualitatively. Comment on the resistance of an orifice at very small flow rates.

3–14. The fluid transformer (side *b*) of Fig. 3–18(a) is connected in turn to an ideal resistance, an ideal inertance, and an ideal capacitance. Find the equivalent ideal elements as observed from side *a* of the transformer. Compare these results with the results obtained when a gear train is connected successively to an ideal damper, a spring, and an inertance.

3–15. The gear pump described in Problem 3–1 may be considered to be a transducer between (fluid flow and pressure) and (rotary torque and speed).

a) Determine an appropriate relation between these four variables. State your assumptions clearly.

b) Is the device a transforming or gyrating transducer? Define and compute an appropriate transduction ratio.

c) A fluid capacitance is connected to the output of the gear pump, $Q = C_f(dPs/dt)$. Determine the relationship between torque T and speed Ω at the drive shaft of the pump. What ideal element represents this relationship?

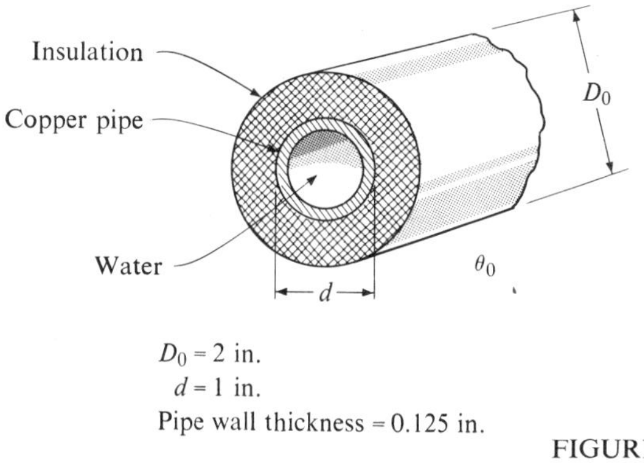

FIGURE 3–31

3–16. Figure 3–31 shows the cross section of an insulated water pipe. This pipe passes near the outside wall of a poorly heated building, and the object of the insulation is to prevent the water from freezing. The specific heats of the insulation, copper pipe, and water are 0.2, 0.093, and 1 Btu/lb·mass, and their thermal conductivities are 0.26, 242, and 0.34 Btu/hr·ft·°F, respectively.

a) Estimate the thermal resistances per unit length of the insulation and the copper.

b) Estimate the thermal capacitances per unit length of the water, the copper, and the insulation, assuming each to be at uniform temperature.

c) Compare the resistances and capacitances of the various components, and describe a simplified model of this configuration.

3–17. A thick hollow cylinder of insulation has outer diameter D_o and inner diameter D_i. If the material has conductivity σ_c, compute the thermal resistance

a) for heat flow outward, and

b) for heat flow inward.

3–18. The inner surface of a thick hollow cylinder is at temperature θ_i and has diameter D_i. The outside, of diameter D_o, is in contact with air, and there is an over-all coefficient of heat transfer C_h between this surface and the surroundings, which are at temperature θ_o.

a) Determine an expression for the total thermal resistance between θ_i and θ_o.

b) Does the resistance R_t continue to increase as more insulation is added (D_o increases)? If not, find the optimum D_o which maximizes R_t.

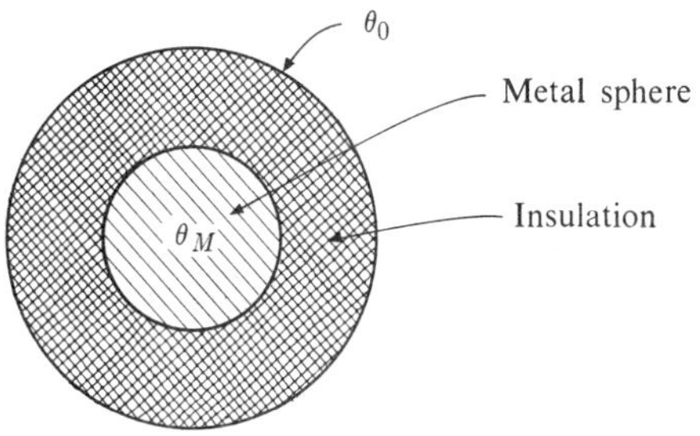

Overall thermal resistance R_t of insulation = 1/3 °F·hr/Btu
Thermal capacitance C_t = 2 Btu/°F of metal

FIGURE 3–32

3–19. An electric hot-water heater for home use is cylindrical in shape and is insulated by a 2 in.-thick layer of glass wool having a conductivity of 0.25 Btu/hr·ft·°F. The heat-transfer coefficient C_h between the outside of the tank and the air is 1 Btu/hr·ft^2·°F. The specific heat of water is 1 Btu/lb-mass·°F. The tank is 5 ft high by 2 ft in diameter.

a) Assuming that the water temperature is uniformly equal to 170°F and the average room temperature is 60°F, estimate the electrical energy in kw-hr/day required to maintain the water at constant temperature.

b) Assuming that there are 100 gal/day of hot water used and replaced by cold water at 50°F, estimate the electrical energy required per day to heat the water.

c) By adding the results of (a) and (b), estimate the total energy per day required. Is this estimate high or low? Why?

3–20. The thermal system in Fig. 3–32 consists of insulation wrapped around a large mass of metal. The system is placed in an oven and heated so that the temperature of the metal (which had been constant at room temperature) increases linearly by 100°F in 1 hr and then remains constant. Determine θ_o, the temperature of the outside of the insulation vs. time, and sketch.

3–21. A block of copper whose mass is 2 slugs is surrounded with insulation. The property of this insulation is that the temperature difference across it is proportional to the heat flow through it, so that a temperature difference of 10°F produces a heat flow of 10 Btu/hr. The system is initially at room temperature and then the outside of the insulation is suddenly raised 100 F° above room temperature.

Discuss this situation in detail. Make sketches of the change of temperature of the copper with time as well as the amount of heat flow into the system. Explain your sketches and *make some estimate of the time it takes* for the copper to heat up.

4 GENERALIZATION OF DYNAMIC SYSTEM ELEMENTS

4–1. INTRODUCTION

Chapters 2 and 3 dealt with many of the essential physical phenomena which determine the dynamic behavior of mechanical, electrical, fluid, and thermal systems. We have seen that there are many striking similarities between the basic elements of these several classes of physical systems. In this chapter we shall summarize the system elements discussed thus far and generalize their properties in terms of energy storage, dissipation, and transformation. A uniform terminology and symbolism applicable to all of the physical systems considered will be developed. In addition, we shall introduce the concepts of pure energy sources, which can supply energy to dynamic systems, pure gyrators, and pure transmission elements. Finally, the idea of power modulation will be presented. By proper use of these idealized lumped elements, we will be able to model any physical system, linear or nonlinear. The extent to which the behavior of the lumped model represents the actual behavior of the real physical system depends on the sophistication and completeness of the model and on the degree of insight and judgment with which the model is formulated.

4–2. GENERALIZED VARIABLES, POWER AND ENERGY

In Chapters 2 and 3 we saw that the behavior of system elements is determined by certain functional relationships between two variables which we have called the *through*-variable and the *across*-variable. The through-variable has the same value at each end or terminal of the element, and the across-variable is specified in terms of a relative value or difference between the terminals. We shall use the symbols f and v to indicate *any* physical through- and across-variables, respectively.

f = generalized through-variable,

v = generalized across-variable.

These two variables may be expressed as the time derivatives of the *integrated through-variable* h and the *integrated across-variable* x, respectively.

$$f = dh/dt, \tag{4–1}$$

$$v = dx/dt. \tag{4–2}$$

Table 4–1 lists the through- and across-variables f and v and their respective integrals h and x for the five physical systems we have considered.

The power flow $\mathcal{P}$ into an element or system through two points 1 and 2 which have a common through-variable f and an across-variable difference v_{21} is in general

$$\mathcal{P} = fv_{21}, \tag{4–3}$$

and the energy $\mathcal{E}$ transferred is the time integral of the power. Thus, during time interval $t_a \to t_b$,

$$\mathcal{E} = \int_{t_a}^{t_b} \mathcal{P}\, dt = \int_{t_a}^{t_b} fv_{21}\, dt. \tag{4–4}$$

TABLE 4–1

Through and Across-Variables for Physical Systems

System	Through-variable f	Integrated through-variable h	Across-variable v	Integrated across-variable x
Mechanical-translational	Force F	Translational momentum p	Velocity difference v_{21}	Displacement difference x_{21}
Mechanical-rotational	Torque T	Angular momentum h	Angular velocity difference Ω_{21}	Angular displacement difference Θ_{21}
Electrical	Current i	Charge q	Voltage difference v_{21}	Flux linkage λ_{21}
Fluid	Fluid flow Q	Volume V	Pressure difference P_{21}	Pressure-momentum Γ_{21}
Thermal	Heat flow q	Heat energy $\mathcal{H}$	Temperature difference θ_{21}	Not used in general

The only exception to these energetic relations is the thermal system, in which power is the through-variable itself (heat flow q) and energy is the integrated through-variable or the amount of heat transferred (Chapter 3). As discussed in Section 4–7, other physical systems exist in which through- and across-variable concepts are useful but in which the product of these two variables cannot be interpreted as power.

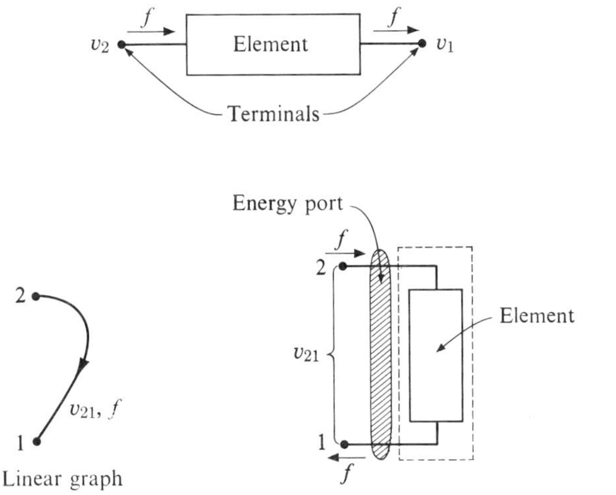

FIG. 4–1. Two-terminal or single energy-port element, showing through- and across-variables.

4–3. TWO-TERMINAL OR SINGLE ENERGY-PORT ELEMENTS

An element, or a system composed internally of many elements, which is described (Fig. 4–1) by the relation between a single through-variable f and a single across-variable difference v_{21}, is called a *two-terminal* element or system. Since energy can flow into or out of this system only by virtue of f and v_{21} (their product in most cases), the system is often called a *single energy-port* or a *one-port* system. Because the elemental behavior prescribes a relation between f and v_{21}, the energy flow to or from a given single-port element is determined by either f or v_{21}.

We shall now discuss the various classes of single-port elements that exist in physical systems.

4–3.1 Pure *T*-Type Energy Storage, Generalized Inductance

Pure translational and rotary springs, pure inductance, and pure fluid inertance are all defined by a single-valued relationship between their through-variables f and their integrated across-variables x. For an element having terminals 2 and 1,

$$x_{21} = \mathsf{f}(f) \tag{4–5}$$

where the reference for x_{21} is such that $x_{21} = 0$ when $f = 0$ and f is a single-valued function. An element described by Eq. (4–5) will be called a *generalized inductance*. If the element is *ideal* (linear), Eq. (4–5) becomes a proportionality

$$x_{21} = Lf \tag{4–6}$$

or, if L is constant,

$$v_{21} = \frac{dx_{21}}{dt} = L\frac{df}{dt}, \tag{4–7}$$

where L is called the *generalized inductance*. For ideal springs, L is the reciprocal stiffness or compliance; for ideal inductances, L is the inductance; and for ideal fluid elements, L is the inertance.

The energy $\mathcal{E}$ supplied to a pure element defined by Eq. (4–5) is

$$\mathcal{E} = \int_{t_a}^{t_b} fv_{21}\,dt = \int_0^f f\,dx_{21}. \tag{4–8}$$

This energy is a function only of the elemental equation (4–5) and the final value of the through-variable. Energy is thus stored by virtue of the through-variable, and these elements are called *T-type energy storage elements*, as already explained in Section 2–7.2.

For an ideal T-type element, the stored energy is

$$\mathcal{E} = \tfrac{1}{2}Lf^2 = \frac{1}{2}\frac{(x_{21})^2}{L}. \tag{4–9}$$

The first section of Table 4–2 summarizes the pure T-type elements studied in Chapters 2 and 3. In general the coil symbol indicates a T-type element. Note that there is no inductive thermal element.

4–3.2 Pure A-Type Energy Storage, Generalized Capacitance

Pure translational and rotary masses, as well as pure electrical, fluid, and thermal capacitances, are defined by a single-valued function of the form

$$h = \mathsf{f}(v_{21}), \tag{4–10}$$

where h is the integrated through-variable and is defined to be zero when v_{21} is zero. In all cases except the electrical capacitance, the elements we have considered that are described by Eq. (4–10) must have one terminal attached to a constant through-variable reference, so that $dv_1/dt = 0$. An element which behaves according to Eq. (4–10) will be called a *generalized capacitance*. For any ideal pure capacitance,

$$h = Cv_{21}, \tag{4–11}$$

and, if C is constant,

$$f = dh/dt = C\frac{dv_{21}}{dt} \tag{4–12}$$

where C is the *generalized capacitance*. For ideal masses, C is the mass or moment of inertia. For ideal electrical, fluid, and thermal capacitance, C is the electrical, fluid, or thermal capacitance, respectively.

The energy supplied to a generalized capacitance is

$$\mathcal{E} = \int_{t_a}^{t_b} v_{21}f\,dt = \int_0^{v_{21}} v_{21}\,dh. \tag{4–13}$$

Equation (4–10) substituted into (4–13) permits $\mathcal{E}$ to be expressed as a function only of v_{21}, the final state of the across-variable. Energy is therefore stored as a function of v_{21}, and this type of element is therefore called an *A-type energy storage element*.

For ideal (linear) A-type elements,

$$\mathcal{E} = \tfrac{1}{2}Cv_{21}^2 = \frac{h^2}{2C}. \tag{4–14a}$$

Note that this equation does not apply to the thermal capacitance, for which

$$\mathcal{E} = \int_{t_a}^{t_b} \mathsf{q}\,dt = \mathcal{H} = C_t\theta_2. \tag{4–14b}$$

The second section of Table 4–2 summarizes the pure A-type elements we have considered. The symbol for the electrical capacitance will be used to indicate a generalized capacitance. The student should keep in mind that when a generalized capacitance is used to represent a nonelectrical capacitance, one terminal must have a constant (usually assumed to be zero) across-variable.

TABLE 4–2

Summary of Pure and Ideal System Elements

Type of element	Physical element	Linear graph	Diagram	Constitutive relationship	Energy or power function	Ideal elemental equation	Ideal energy or power
T-type energy storage; $\mathcal{E} \geq 0$; f, v_2, x_2, v_1, x_1	Translational spring	2, 1, k, v_{21}, F	F, F, v_2, x_2, v_1, x_1	$x_{21} = \mathrm{f}(F)$	$\mathcal{E} = \int_0^F F\,dx_{21}$	$v_{21} = \frac{1}{k}\frac{dF}{dt}$	$\mathcal{E} = \frac{1}{2}\frac{F^2}{k}$
	Rotational spring	2, 1, K, Ω_{21}, T	T, T, Ω_2, Θ_2, Ω_1, Θ_1	$\Theta_{21} = \mathrm{f}(T)$	$\mathcal{E} = \int_0^T T\,d\Theta_{21}$	$\Omega_{21} = \frac{1}{K}\frac{dT}{dt}$	$\mathcal{E} = \frac{1}{2}\frac{T^2}{K}$
Pure: $x_{21} = \mathrm{f}(f)$; $\mathcal{E} = \int_0^f f\,dx_{21}$ — Ideal: $x_{21} = Lf$; $\mathcal{E} = \frac{1}{2}Lf^2$	Inductance	2, 1, L, v_{21}, i	i, i, v_2, λ_2, v_1, λ_1	$\lambda_{21} = \mathrm{f}(i)$	$\mathcal{E} = \int_0^i i\,d\lambda_{21}$	$v_{21} = L\frac{di}{dt}$	$\mathcal{E} = \frac{1}{2}Li^2$
	Fluid inertance	2, 1, I, P_{21}, Q	Q, Q, P_2, Γ_2, P_1, Γ_1	$\Gamma_{21} = \mathrm{f}(Q)$	$\mathcal{E} = \int_0^Q Q\,d\Gamma_{21}$	$P_{21} = I\frac{dQ}{dt}$	$\mathcal{E} = \frac{1}{2}IQ^2$
A-type energy storage; $\mathcal{E} \geq 0$; f, h, v_2, v_1	Translational mass	2, 1, m, v_{21}, F	F, p, $v_1 = \text{const}$, v_2	$p = \mathrm{f}(v_2)$	$\mathcal{E} = \int_0^{v_2} v_2\,dp$	$F = m\frac{dv_2}{dt}$	$\mathcal{E} = \frac{1}{2}mv_2^2$
	Inertia	2, 1, J, Ω_{21}, T	T, h, $\Omega_1 = \text{const}$, Ω_2	$h = \mathrm{f}(\Omega_2)$	$\mathcal{E} = \int_0^{\Omega_2} \Omega_2\,dh$	$T = J\frac{d\Omega_2}{dt}$	$\mathcal{E} = \frac{1}{2}J\Omega_2^2$
	Electrical capacitance	2, 1, C, v_{21}, i	i, q, i, v_2, v_1	$q = \mathrm{f}(v_{21})$	$\mathcal{E} = \int_0^{v_{21}} v_{21}\,dq$	$i = C\frac{dv_{21}}{dt}$	$\mathcal{E} = \frac{1}{2}Cv_{21}^2$
Pure: $h = \mathrm{f}(v_{21})$ — Ideal: $h = Cv_{21}$	Fluid capacitance	2, 1, C_f, P_{21}, Q	Q, V, $P_1 = \text{const}$, P_2	$V = \mathrm{f}(P_2)$	$\mathcal{E} = \int_0^{P_2} P_2\,dV$	$Q = C_f\frac{dP_2}{dt}$	$\mathcal{E} = \frac{1}{2}C_fP_2^2$
Pure: $\mathcal{E} = \int_0^{v_{21}} v_{21}\,dh$ — Ideal: $\mathcal{E} = \frac{1}{2}Cv_{21}^2$	Thermal capacitance	2, 1, C_t, θ_{21}, q	$\mathrm{q}, \mathcal{H}$, $\theta_1 = \text{const}$, θ_2	$\mathcal{H} = \mathrm{f}(\theta_2)$	$\mathcal{E} = \int_0^{\theta_2} \mathrm{q}\,dt = \mathcal{H}$	$\mathrm{q} = C_t\frac{d\theta_2}{dt}$	$\mathcal{E} = C_t\theta_2$
D-type energy dissipators; $\mathcal{P} \geq 0$; f, v_2, v_1	Translational damper	2, 1, b, v_{21}, F	F, F, v_2, v_1	$F = \mathrm{f}(v_{21})$	$\mathcal{P} = Fv_{21}$	$F = bv_{21}$	$\mathcal{P} = bv_{21}^2$
	Rotational damper	2, 1, B, Ω_{21}, T	T, T, Ω_2, Ω_1	$T = \mathrm{f}(\Omega_{21})$	$\mathcal{P} = T\Omega_{21}$	$T = B\Omega_{21}$	$\mathcal{P} = B\Omega_{21}^2$
	Electrical resistance	2, 1, R, v_{21}, i	i, i, v_2, v_1	$i = \mathrm{f}(v_{21})$	$\mathcal{P} = iv_{21}$	$i = \frac{1}{R}v_{21}$	$\mathcal{P} = \frac{1}{R}v_{21}^2$
Pure: $f = \mathrm{f}(v_{21})$; $\mathcal{P} = v_{21}\mathrm{f}(v_{21})$ — Ideal: $f = \frac{1}{R}v_{21}$; $\mathcal{P} = \frac{1}{R}v_{21}^2 = Rf^2$	Fluid resistance	2, 1, R_f, P_{21}, Q	Q, Q, P_2, P_1	$Q = \mathrm{f}(P_{21})$	$\mathcal{P} = QP_{21}$	$Q = \frac{1}{R_f}P_{21}$	$\mathcal{P} = \frac{1}{R_f}P_{21}^2$
	Thermal resistance	2, 1, R_t, θ_{21}, q	q, q, θ_2, θ_1	$\mathrm{q} = \mathrm{f}(\theta_{21})$	$\mathcal{P} = \mathrm{q}$	$\mathrm{q} = \frac{1}{R_t}\theta_{21}$	$\mathcal{P} = \frac{1}{R_t}\theta_{21}$
Energy sources; $\mathcal{P} \gtrless 0$; $\mathcal{E} \gtrless 0$	*A*-type across-variable source	2, 1, v	$+$, v, $-$, v_2, v_1	$v_{21} = \mathrm{f}(t)$	$\mathcal{P} = fv_{21}$		
	T-type through-variable source	2, 1, f	f, v_2, v_1	$f = \mathrm{f}(t)$	$\mathcal{P} = fv_{21}$		

Nomenclature

$\mathcal{E}$ = energy, $\mathcal{P}$ = power

f = generalized through-variable, F = force, T = torque, i = current, Q = fluid flow rate, q = heat flow rate

h = generalized integrated through-variable, p = translational momentum, h = angular momentum, q = charge, V = fluid volume displaced, $\mathcal{H}$ = heat

v = generalized across-variable, v = translational velocity, Ω = angular velocity, v = voltage, P = pressure, θ = temperature

x = generalized integrated across-variable, x = translational displacement, Θ = angular displacement, λ = flux linkage, Γ = pressure-momentum

L = generalized ideal inductance, $1/k$ = reciprocal translational stiffness, $1/K$ = reciprocal rotational stiffness, L = inductance, I = fluid inertance

C = generalized ideal capacitance, m = mass, J = moment of inertia, C = capacitance, C_f = fluid capacitance, C_t = thermal capacitance

R = generalized ideal resistance, $1/b$ = reciprocal translational damping, $1/B$ = reciprocal rotational damping,
R = electrical resistance, R_f = fluid resistance, R_t = thermal resistance

4–3.3 Pure *D*-Type Energy Dissipators, Generalized Resistance

Pure translational and rotary dampers, and electrical, fluid, and thermal resistances are defined by the single-valued function

$$f = \mathsf{f}(v_{21}), \tag{4–15}$$

where the function is such that $f = 0$ when $v_{21} = 0$ and the signs of f and v_{21} are always alike. Any element that obeys Eq. (4–15) under all circumstances is called a *generalized resistance*. If the resistance is ideal,

$$f = \frac{1}{R} v_{21}, \tag{4–16}$$

where R is the *generalized resistance* of the element. The resistance of ideal translational and rotational dampers is the reciprocal of their damping coefficients. For electrical, fluid, and thermal elements the ideal resistances are equal to R.

The power $\mathcal{P}$ supplied to a generalized resistance is

$$\mathcal{P} = f v_{21} = v_{21}\mathsf{f}(v_{21}). \tag{4–17}$$

Since the signs of f and v_{21} are always alike, in accordance with the above definition, $\mathcal{P}$ is always positive and power always flows *into* the resistance. Hence the generalized resistance dissipates energy and is thus called a *D*-type energy dissipator.

For an ideal resistance,

$$\mathcal{P} = f^2 R = v_{21}^2/R. \tag{4–18}$$

Table 4–2 shows the *D*-type elements for the mechanical, electrical, fluid, and thermal systems. The electrical resistance symbol is used to represent a generalized resistance.

4–3.4 Pure *T*-Type and *A*-Type Energy Sources

We have thus far investigated only those two-terminal elements which store or dissipate energy. We shall now introduce the concept of a two-terminal "source." By "source" we mean a device which is capable of delivering energy continuously to a system. The energy which a source delivers is provided by an energy reservoir external to the system under consideration. Let us consider the simple system of a mass suspended by a spring, the other end of which is attached to the ceiling of a room (Fig. 4–2). If a person pushes on the mass, one could consider that the force he exerts on the mass is a source for the spring-mass system. Any energy which is delivered to the system comes from within the person. The person and his biological equipment are not taken to be part of the system; only the force exerted is of interest. (At some later time one might wish to redefine the system to include the person; then the sources would be different, such as the quantity of food eaten, etc.)

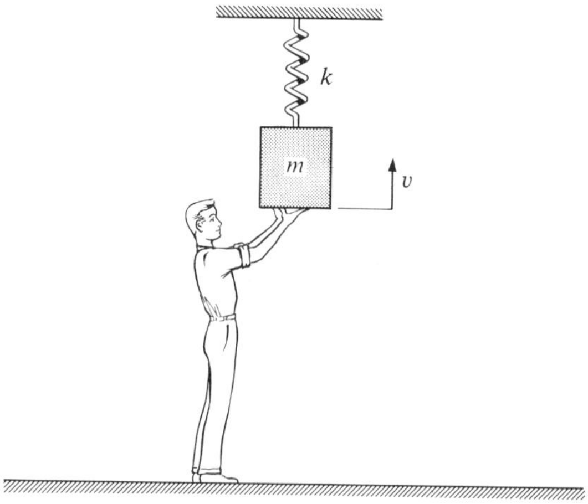

FIG. 4–2. A person acting as a source to a spring-mass system.

In the above example, one could have assumed that the person applied a *motion*, v, to the mass instead of a force. Then the source would be the motion of the mass. Some thought will reveal that no human being is likely to be able to exert either a *prescribed* force or a *prescribed* velocity independent of the system characteristics. For instance, if the person in Fig. 4–2 were to try to impose a sudden change of velocity on the spring-mass system, he would be unable to do so since such a change would require infinite force. Similarly, if the person tries to impose a

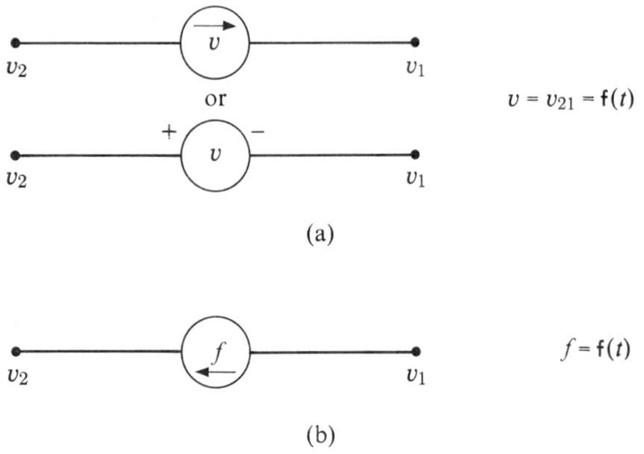

FIG. 4–3. Ideal *A*- and *T*-type source elements. (a) *A*-type source. (b) *T*-type source.

specified force, the system will move at varying velocities, thereby making it difficult for the person to vary the force in the specified fashion. Thus for this system the "source" is a complicated one in which the relationship between the force applied to the mass and the velocity of the mass is complex.

It is useful in system analysis to define idealized sources in much the same way that the idealized energy storage and dissipation elements were defined. Two limiting cases are considered, one in which the source across-variable is a specified function of time, and one in which the through-variable is a specified function of time. The former is called an ideal *A-type source* and the latter an ideal *T-type source;*

$v_{21} = \mathsf{f}(t)$, *A*-type source,

$f = \mathsf{f}(t)$, *T*-type source,

where f indicates specified functions.

The symbols to be used for ideal sources are shown in Fig. 4–3. In Fig. 4–3(a), v_{21} is a specified function of time independent of the through-variable f which the source must supply. The arrow indicates that when v_{21} is numerically positive, v_2 is algebraically larger than v_1. Thus the arrow points in the direction of across-variable drop. In many textbooks, especially those dealing with electric circuits, this convention is shown by attaching plus and minus signs to the two terminals of the source (Fig. 4–3a). In Fig. 4–3(b), f is

TABLE 4–3

Physical Systems that Approximate Ideal Sources

System	Source medium	Type of source
Battery	Electrical voltage	*A*-type
Hydraulic pump	Fluid flow	*T*-type
Melting ice	Liquid temperature	*A*-type
Torque motor	Torque	*T*-type
Ocean depth	Pressure	*A*-type

a specified function of time independent of the value of v_{21} across the terminals and in the direction of the arrow.

Although most sources are used to permit energy to be supplied to systems, we shall define the sign of the conjugate variable for a source (f for an *A*-type source or v_{21} for a *T*-type source) to be positive when power flows into the source. Any source may absorb as well as supply power and energy. For any source, the flow of power *into* the source is

$$\mathcal{P} = fv_{21}. \tag{4–19}$$

Thus, if f and v_{21} are of opposite sign,

$$\mathcal{P} < 0, \quad \text{and} \quad \mathcal{E} < 0.$$

Although $\mathcal{P}$ may be negative for energy storage elements (when energy is being removed from the element) only sources can supply power and energy continuously over an extended period of time. Note that $\mathcal{E}$ can never be negative for energy storage elements. Table 4–2 includes the two ideal source elements. We shall discuss sources further in Chapter 6; however, several physical situations which give rise to approximately ideal sources are listed in Table 4–3.

4–3.5 Summary of Single-Port Lumped Elements

The various single-port elements we have studied are summarized in Table 4–2. The generalized relations between the through- and across-variables for the *A*-type, *T*-type, and *D*-type elements may be sum-

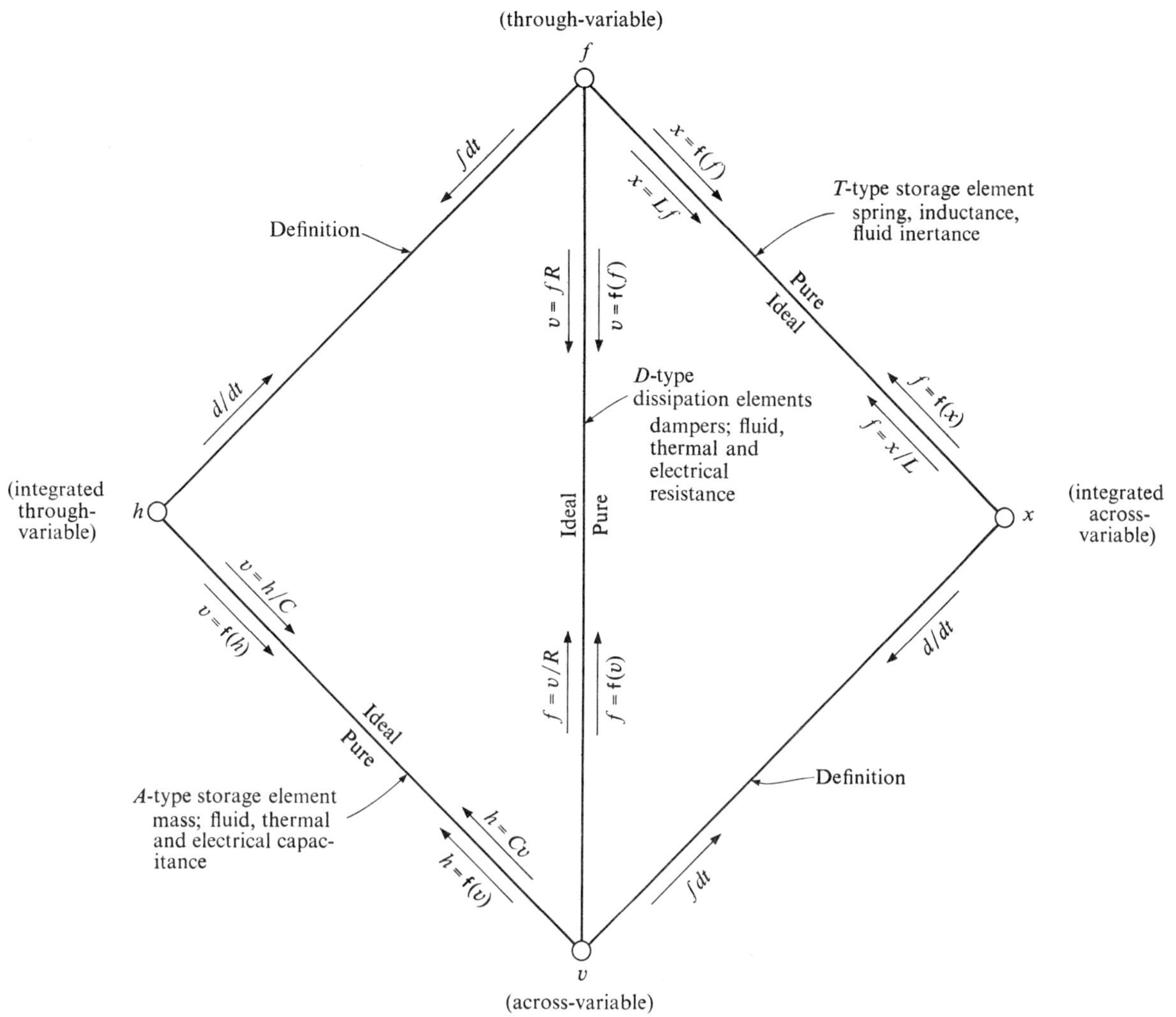

FIG. 4–4. Tetrahedron of state for single-port elements (adapted from Reference 1).

marized by means of a *tetrahedron of state* (adapted from Reference [1]) where the A- and T-variables are said to determine the *state* of the element. Figure 4–4 shows the relations between these variables. The through-variable f and the integrated through-variable h are related by definition and integration or differentiation. The across-variable v and the integrated across-variable x are similarly related. The branch connecting f and x represents all T-type elements (nonlinear as well as ideal), and the branch connecting v and h represents all A-type elements. The diagonal connecting f and v represents all D-type elements or energy dissipators. The source elements in which the through-variable f or the across-variable v are specified functions of time permit energy to be supplied to the various elements summarized by Fig. 4–4.

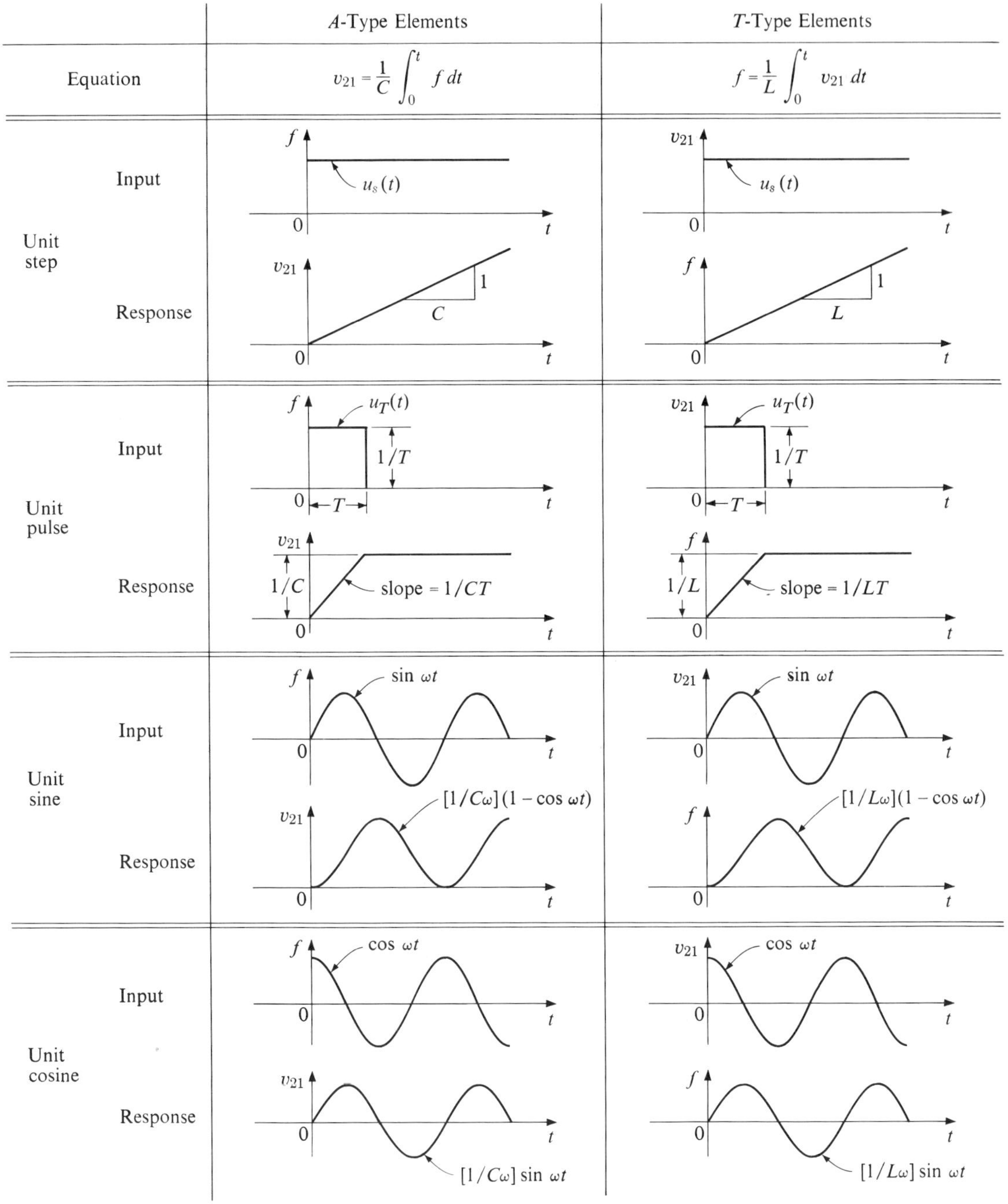

FIG. 4–5. Summary of dynamic response characteristics of ideal A-type and T-type elements.

4–3.6 Response of Ideal System Elements to Simple Source Functions

We can derive additional insight into the nature of the ideal two-terminal system elements by considering their behavior when one of their variables is changed by means of a simple source function. The variable controlled by the source is called the *input* variable to the element. The resulting conjugate variable is called the *output* or *response* variable.

First let us consider a source function which is a sudden change of a variable from one constant value to another. This is called a *step change* in the variable. When the source or input function is a sudden change from zero to a unit amplitude it is called a *unit step*, $u_s(t)$, and the response of an element (or system) to this input is called the *unit step response*. If one takes into account energy and power, it is evident that it is not physically possible to suddenly (i.e., in zero time) change the across-variable of an A-type element or the through-variable of a T-type element. For example, if one were to accomplish a sudden change in v_{21} of an A-type element, the derivative of v_{21} with respect to time at the instant of the sudden change would be infinite and this would, according to Eq. (4–12), require f to be infinite. Moreover, suddenly changing v_{21} would mean a sudden change in the energy stored in the generalized capacitance, which in turn means that it would have been necessary to transfer energy at an infinite rate into the element at the time of the sudden change in v_{21}. In other words, a source with infinite power capability would have been required.

On the other hand, it is quite reasonable physically to impose a sudden change of the through-variable f on an A-type element or of the across-variable v_{21} on a T-type element. Figure 4–5 shows the responses of ideal A- and T-type elements to unit step inputs of the through- and across-variables, respectively.

When one step input is followed a short time later by another step of equal amplitude but opposite sign, a rectangular pulse is created. A unit pulse, $u_T(t)$ is a pulse having unit area or time integral. Thus if a unit pulse has T duration, its height must be $1/T$. The unit pulse and the responses of ideal A- and T-type elements to f and v_{21} inputs, respectively, are also shown in Fig. 4–5. By letting T approach zero, one can approach the case of a sudden change in the across-variable for an A-type element or in the through-variable for a T-type element. In the limit as $T \rightarrow 0$, the unit pulse is called a *unit impulse*, and the resulting response, the *unit impulse response*. Although a true impulse is not physically possible because it requires infinite power transfer, it is often possible to bring about a pulse of sufficiently short duration that it appears as an impulse to the system to which it is applied. Systems frequently can be tested in this manner. An example is testing the dynamic-response characteristics of a machine frame by striking it with a hammer.

Another simple input or source function which is frequently encountered is the sinusoid. The responses of A-type and T-type ideal elements to f and v_{21} sinusoids which begin at $t = 0$ are obtained by integration, as shown in Fig. 4–5. Conversely, the responses of A-type and T-type elements to v_{21} and f sinusoids can be obtained by differentiation.

The response characteristics of the ideal dissipative elements are exceedingly simple. They respond instantaneously in a proportional fashion to any kind of change of one of their variables. A step change in the f-variable results in a step change in the v_{21}-variable, and vice versa. This instantaneous response is possible because there is no energy storage involved in the operation of these elements.

4–3.7 Linear Graph Representation of Single-Port Elements

The pure system elements listed in Table 4–2 are characterized by two terminals which as a pair allow energy transfer between the element and its surroundings and by a functional relationship between their through- and across-variables. A convenient general symbol for any two-terminal element is the *linear graph*, shown in Fig. 4–6(a). The ends or terminals of the graph indicate the across-variables for the element, and the line between the terminals represents

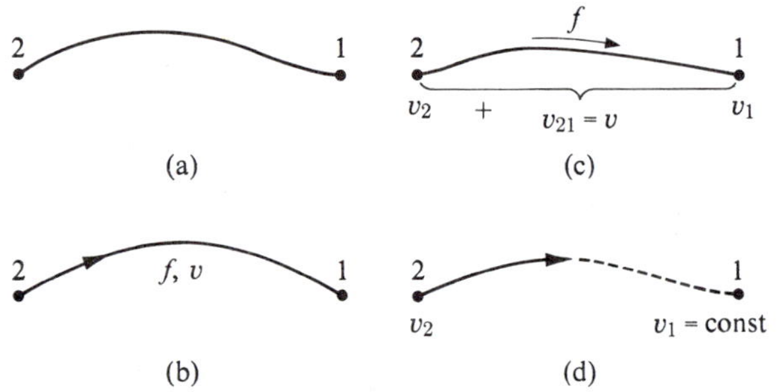

FIG. 4–6. Linear graph representation of two-terminal elements. (a) Linear graph. (b) Oriented linear graph. (c) Sign conventions implied by oriented linear graph in (b). (d) Linear graph for nonelectrical A-type element.

the continuity of the through-variable in the element. The term "linear" in linear graph means that the graph is defined by a line segment and should not be confused with "linear" as applied to the constitutive relationship for an ideal element. The sign conventions for the through- and across-variables can be incorporated into the graph by adding an arrow on the graph, thus forming an *oriented linear graph*, as shown in Fig. 4–6(b). The arrow implies that f is positive in the direction of the arrow and that v_{21} is positive when $v_2 > v_1$. The arrow in Fig. 4–6(b) therefore establishes the two sign conventions illustrated in Fig. 4–6(c).

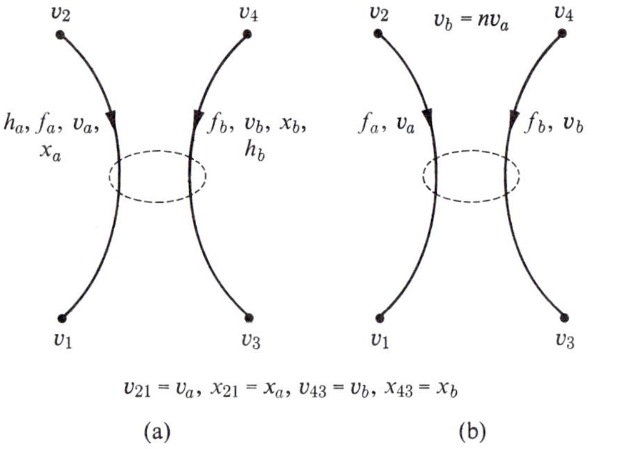

FIG. 4–7. Oriented linear graph symbols for pure and ideal transformers. (a) Pure transformer, $x_b = f(x_a)$. (b) Ideal transformer, $v_b = nv_a$; $f_b = (-1/n)f_a$.

With the exception of the electrical capacitance, all the A-type elements we have considered must have one terminal at a constant or zero value of the across-variable. The through-variable then does not flow into the constant across-variable terminal. The linear graph of elements of this type may be drawn as shown in Fig. 4–6(d). Table 4–2 shows the linear graph representation of all the pure system elements.

The advantages of linear graphs as symbols for pure system elements will be seen in Chapter 8, where their use in equation formulation is discussed.

4–4. TWO-ENERGY-PORT TRANSFORMATION ELEMENTS

Any device or system which is characterized by a set of relationships between *two* sets of through- and across-variables is called a *four-terminal* (or three-terminal if two terminals are common), a *two-terminal pair*, or a *two-energy-port system*. Such a device can exchange energy with its surroundings at each pair of terminals.

In Chapters 2 and 3 we saw that pure transformers and transducers were four-terminal devices which provided transformation of the through- and across-variables while maintaining instantaneous continuity of power. We shall now review these devices and extend our considerations to include other possible energy-transformation elements.

4–4.1 Pure Transformers

Pure transformers, discussed in Chapters 2 and 3, provide a ratio of through- and across-variables in a particular physical medium such that the total power flow into the device is identically zero. Rotary gear trains, levers, linkages, electrical transformers, and fluid pistons of differential area are examples of physical devices whose behavior is approximated in many instances by that of a pure transformer.

A *generalized pure transformer* may be defined and represented by the linear graph symbol shown in Fig. 4–7(a). The oriented graphs 2–1 and 4–3 imply that v_a is positive when $v_2 > v_1$, that v_b is positive when $v_4 > v_3$, and that f_a and f_b are positive in the directions of the arrows. The dashed circle enclosing the two graphs in Fig. 4–7 implies transformer coupling (across-variable v_a transforms into across-variable v_b and through-variable f_a transforms into through-variable f_b). We have defined pure transformers as devices in which the integrated across-variables or in some cases the integrated through-variables for the input and output terminals are related by single-valued functions,

$$x_b = \mathsf{f}(x_a) \tag{4–20}$$

or

$$h_b = \mathsf{f}(h_a), \tag{4–21}$$

and the net power flow $\mathcal{P}$ is zero:

$$\mathcal{P} = f_a v_a + f_b v_b = 0. \tag{4–22}$$

An ideal transformer has a proportionality between x_a and x_b or between h_a and h_b, and hence between v_a and v_b and also between f_a and f_b:

$$v_b = n v_a, \tag{4–23}$$

$$f_b = -\frac{1}{n} f_a. \tag{4–24}$$

The constant n is called the *transformation ratio*. Table 4–4 summarizes the types of transformers we have discussed and lists the ratios n for the ideal devices. Note that we have *defined* n by Eq. (4–23). Therefore, if our reference conventions and the nature of the physical device are such that v_a is actually negative when v_b is positive, n will be a negative number. This is illustrated by the lever shown in Table 4–4. Here the negative value of n means that when v_{21} is positive (2 moves *up* relative to 1), v_{41} must be negative (4 moves *down* relative to 1). Similarly the negative n in the gear train of Table 4–4 means that when Ω_{21} is clockwise, Ω_{41} is counterclockwise. How could a gear train having a *positive* transformation ratio be formed?

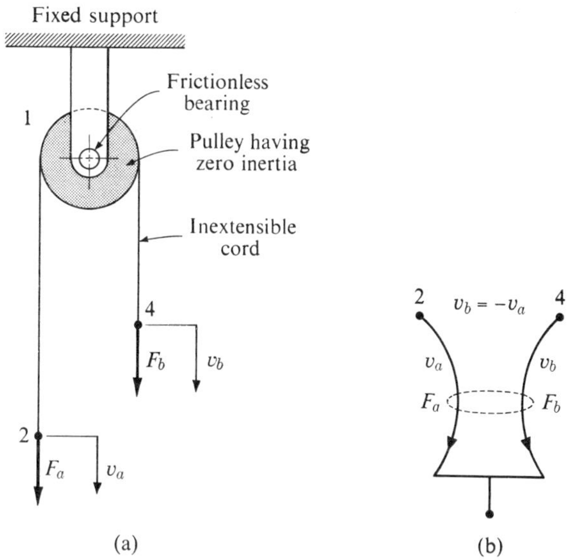

FIG. 4–8. A mechanical transformer having a transformation ratio $n = -1$ and its linear graph. (a) System. (b) Linear graph.

Mechanical transformers having $n = -1$ often appear in physical systems, where they act only to change the direction of a velocity or force. A simple pulley (Fig. 4–8a) is one example of such a direction reversal. The linear graph for this system is shown in Fig. 4–8(b). The transformation ratio n (see Table 4–4) may depend on the ratio of a physical constant of side a to a physical constant of side b or vice versa, depending on the nature of the transformer. Note also in Table 4–4 that the pure differential-piston fluid transformer is considered to relate the integrated through-variables V_a and V_b, whereas the other

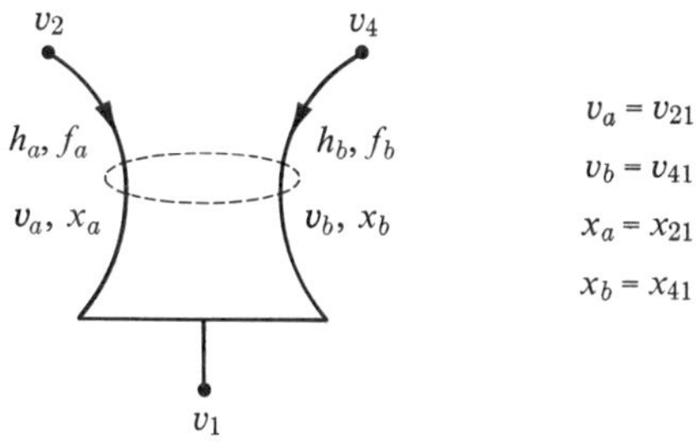

FIG. 4–9. Linear graph for mechanical and fluid transformers.

TABLE 4–4. Pure and Ideal Transformers

System	Symbol	Pure transformer	Ideal transformer	Transformation ratio
Mechanical translation (lever)	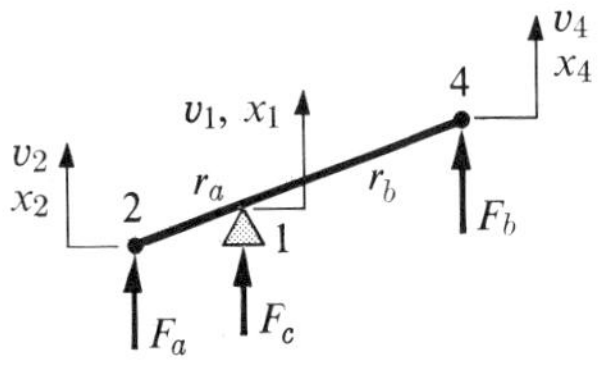	$x_{41} = f(x_{21})$	$v_{41} = nv_{21}$ $F_b = -\frac{1}{n} F_a$	$n = -\frac{r_b}{r_a}$ lever ratio
Mechanical rotational (gears)	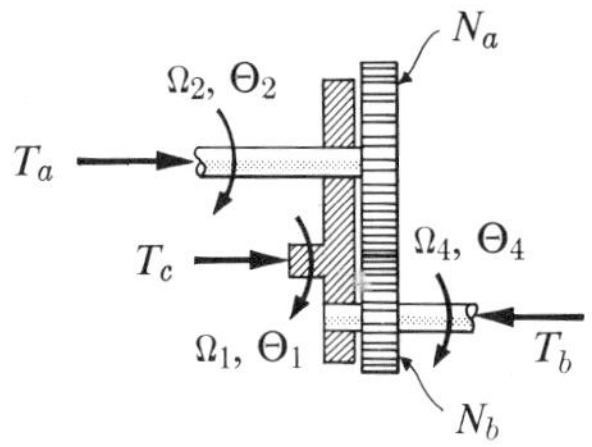	$\Theta_{41} = f(\Theta_{21})$	$\Omega_{41} = n\Omega_{21}$ $T_b = -\frac{1}{n} T_a$	$n = -\frac{N_a}{N_b}$ gear ratio
Electrical (magnetic)	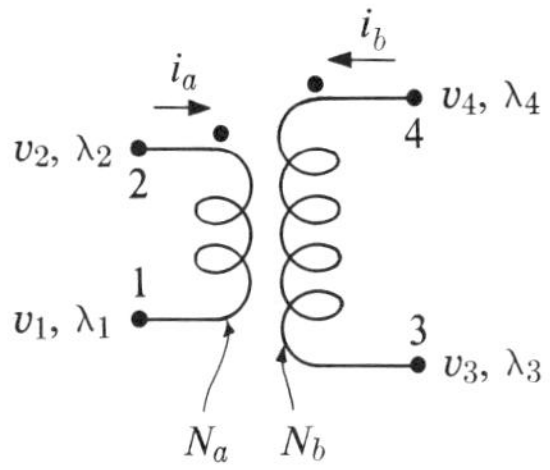	$\lambda_{43} = f(\lambda_{21})$	$v_{43} = nv_{21}$ $i_b = -\frac{1}{n} i_a$	$n = \frac{N_b}{N_a}$ turns ratio
Fluid (differential piston)	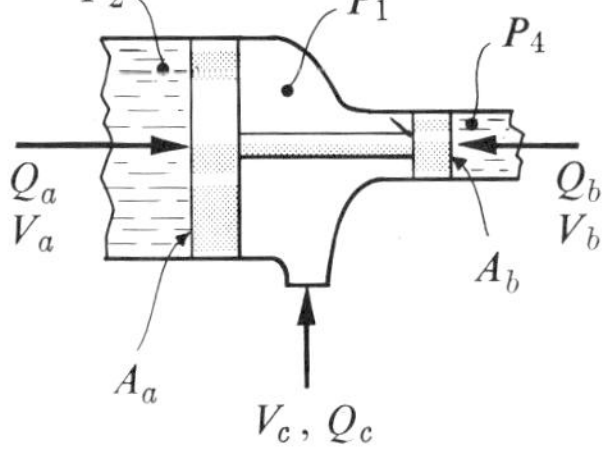	$V_b = f(V_a)$	$P_{41} = nP_{21}$ $Q_b = -\frac{1}{n} Q_a$	$n = \frac{A_a}{A_b}$ area ratio

devices listed relate the integrated across-variables for the two pairs of terminals. With the exception of the electrical transformer, all of the systems shown in Table 4–3 must have the terminals 1 and 3 (see Fig. 4–7) common. The linear graph for these elements should thus be drawn as in Fig. 4–9 (p. 91).

In thermal systems the through-variable itself is power, and hence a device having constant power and a transformation of the through-variable magnitude is not possible, since it would violate energy conservation.

4–4.2 Pure Gyrators

The transformers just discussed *transform* an across-variable into an across-variable of different magnitude and the corresponding through-variable into a through-variable of different magnitude. Another device is possible which, in a given physical medium, changes a through-variable into an across-variable and an across-variable into a through-variable. Such a device is called a *gyrator*. A *pure gyrator* accomplishes this change with instantaneous power continuity (i.e., no energy storage or loss). The linear graph of Fig. 4–10 represents a generalized pure gyrator (the dashed reverse loop indicates gyration). Physical devices which perform gyration are far less common than those which perform transformation. In mechanical systems gyroscopes produce steady rates of rotation about one axis in response to steady applied torques about another axis and thus act in part as gyrators. The so-called Hall effect in electrical systems involves gyration of a current into a voltage difference by the action of a constant magnetic field.

An ideal gyrator is defined by the relations

$$v_b = rf_a, \tag{4–25}$$

$$f_b = -\frac{1}{r}v_a, \tag{4–26}$$

where r is called the gyration ratio or gyrational resistance and has the units of an across-variable over a through-variable (e.g., v/amp = ohms).

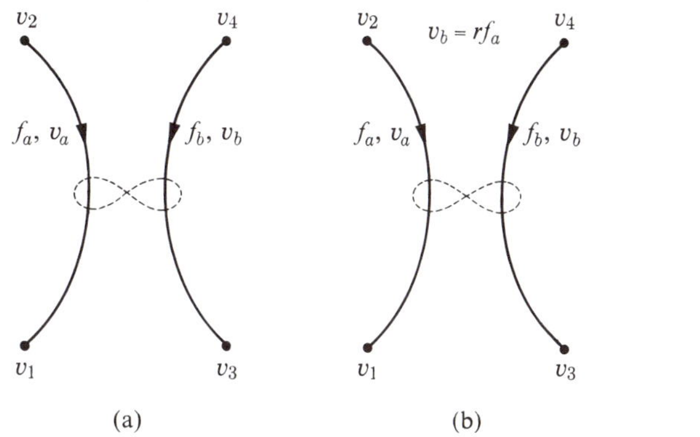

FIG. 4–10. Oriented linear graph for pure and ideal gyrators. (a) Pure gyrator. (b) Ideal gyrator.

From Eqs. (4–25) and (4–26) we have

$$\mathcal{P} = f_a v_a + f_b v_b = 0. \tag{4–27}$$

4–4.3 Pure Transducers

The idea of a pure transducer, which changes energy from one physical medium to another without energy loss or storage, has been introduced in Chapters 2 and 3. We have seen that two types of transducers are possible: the *transforming transducer* and the *gyrating transducer*.

a) *Transforming transducers.* A transforming transducer *transduces* a through- (or across-) variable in one physical medium into a through- (or across-) variable in another medium, and its linear graph is the same as that of a transformer. For example, the gear and rack discussed in Section 2–5 is a rotational-to-translational transforming transducer since it couples translational velocity to rotational velocity and force to torque.

Electric motors and generators transduce current into torque, and conversely. Thermocouples are thermal-to-electrical transforming transducers which relate temperature difference to voltage difference. The reader should try to list other examples of this kind of transducer.

7:00

b) *Gyrating transducers.* The gyrating transducer *transduces* a through- (or across-) variable in one physical medium into an across- (or through-) variable in another medium, and its linear graph is the same as that of a gyrator. The piston-and-cylinder discussed in Section 3–3 is an example of a fluid-to-mechanical gyrating transducer. Piezoelectric crystals are used for microphones, phonograph pickups, and in various electromechanical measuring instruments to produce a voltage dependent on applied force, and they therefore exhibit the characteristic of an electromechanical gyrating transducer. Certain fluid machines also act as gyrating fluid-to-mechanical transducers.

The describing relations and linear graphs for pure and ideal transducers are identical to those presented in Sections 4–4.1 and 4–4.2, except that the through- and across-variables v_a and f_a for terminal pair 2–1 are for one physical medium, whereas the variables v_b and f_b for terminal pair 4–3 are for a different physical medium. The power flow $\mathcal{P}$ for a pure transducer is identically zero:

$$\mathcal{P} = f_a v_a + f_b v_b = 0. \tag{4–28}$$

In applying Eq. (4–28) to transducers, we must be careful to express $f_a v_a$ and $f_b v_b$ in the same units. For instance, if f_a and v_a are electric current (amperes) and voltage (volts), $f_a v_a$ must be in watts. If f_b and v_b are force and velocity, the units of f_b and v_b must be chosen so that $f_b v_b$ is also in units of watts for (4–28) to be numerically satisfied. In this case f_b could be expressed in newtons and v_b in m/sec, giving a product with units of watts.

The transduction ratio n for an ideal transforming transducer can be expressed by Eqs. (4–23) and (4–24), where now n has the units of the across-variable in medium b divided by the across-variable in medium a. The ratio n will be numerically equal in Eqs. (4–23) and (4–24) if a consistent set of units (e.g., mks) is used for all variables.

Similarly, the transduction ratio r for an ideal gyrating transducer can be expressed by Eqs. (4–25) and (4–26) where r has units of the across-variable in medium b divided by the through-variable in medium a.

4–4.4 Pure Transmitters; Energy Transmission Elements

A final two-energy-port energy transformation device is one which transmits energy in a particular physical form over a distance from one location to another. A pure transmitter accomplishes this with no energy losses or storage. This element will not be discussed further in this book, but is introduced here only to complete the discussion of energy transformation devices. Physical systems that behave in part like energy transmitters include fluid pipelines, transmission shafts, and electrical power lines. Real transmission systems always involve energy storage, time or transport delay, and some dissipation, and a detailed understanding of their behavior requires a study of wave propagation and diffusion phenomena. Nevertheless, in this book we shall approximate or model real two-port transmitters by lumped systems (e.g., shafts by springs, pipes by fluid resistors, etc.).

4–5. OTHER MULTIPORT ELEMENTS; MODULATORS

We have discussed in some detail special classes of one- and two-port systems. All the essential physical phenomena of energy supply, storage, and dissipation are described by the family of single-port elements summarized in Table 4–2. Single-port systems, characterized by the presence of only one pair of terminals, of more complicated types are possible. In many cases such systems may be represented by interconnected combinations of pure source, energy storage, and dissipative elements.

The basic phenomena of energy transformation, transduction, and transmission are represented by the pure two-port elements discussed in Section 4–4. In general, two-port elements may also include energy sources, energy storage, and dissipation. A real gear-pair having inertia, elastic deflections, and friction losses is one example of a real two-port system which has energy transformation, storage, and dissipation.

Complex systems may have many energy ports or places at which energy may be exchanged with the environment, and hence may be called *multiports*.

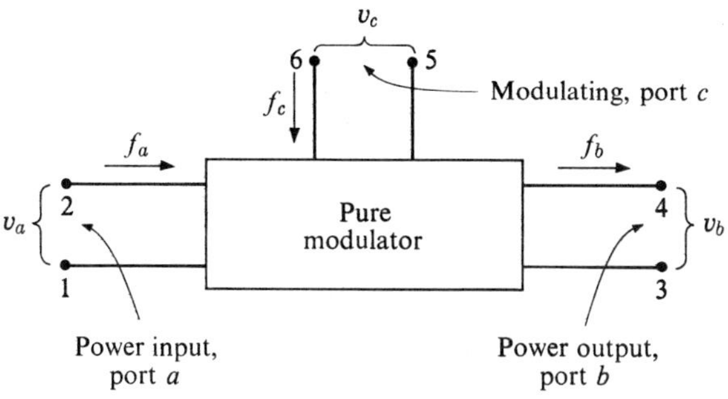

FIG. 4–11. Schematic diagram of pure modulator.

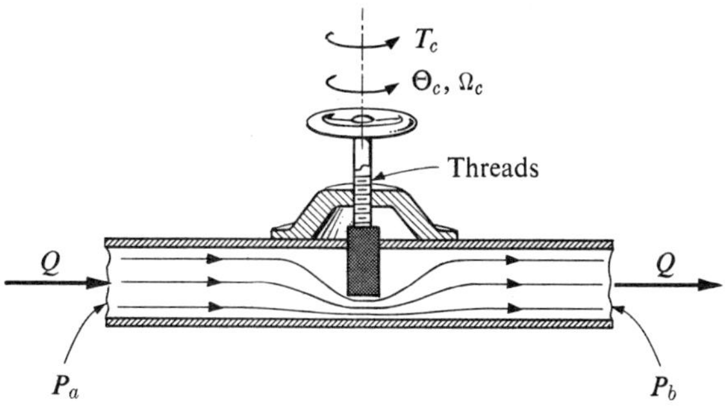

FIG. 4–12. Fluid modulator; variable-area valve.

One basic class of *three-port* elements of great importance will now be introduced before we terminate this discussion of dynamic system elements. These elements are called *modulators*. A *pure modulator* is a device which stores no energy, in which the flow of energy at two ports can be varied or controlled by an energy input at the third port. Figure 4–11 illustrates this situation schematically. A real modulator is necessarily a dissipative device in which the total power $\mathcal{P}_a + \mathcal{P}_c$ entering the system through ports a and c exceeds the power $\mathcal{P}_b$ leaving the system. Often the modulation power $\mathcal{P}_c$ will be much smaller than $\mathcal{P}_a$ or $\mathcal{P}_b$. The physical media at ports a and b are normally the same and may differ from that at port c. When $\mathcal{P}_c$ is considered the input and $\mathcal{P}_b$ (the modulated power) the output of the system, the modulator is called an *amplifier*.

One of the simplest examples of a pure modulator is a fluid valve (similar to a water faucet). Fluid power in the form of a flow rate Q at pressure P_a enters the valve in Fig. 4–12. The restricting orifice in the valve reduces the fluid pressure and the flow leaves the valve at a lower pressure P_b. The input and output fluid powers are

$$\mathcal{P}_a = QP_a,$$

$$\mathcal{P}_b = QP_b < \mathcal{P}_a.$$

The modulating input is a mechanical rotation of the valve shaft which requires power

$$\mathcal{P}_c = \Omega_c T_c \ll \mathcal{P}_a.$$

This condition may not hold when the valve is nearly closed and $\mathcal{P}_a$ is very small. The valve acts like a variable fluid resistance, R_f, with respect to the fluid powers $\mathcal{P}_a$ and $\mathcal{P}_b$:

$$P_a - P_b = R_f Q,$$

where R_f is a function of the valve shaft rotation. Although modulators are in general three-port elements, in the case of incompressible flow the valve-type fluid modulator is actually a *two-port* element; the first port being associated with $(P_a - P_b)Q$ and the second with the mechanical power $\Omega_c T_c$. However, with compressible gases, even the very same valve must be considered a three-port device.

Other devices which act like modulators include vacuum tubes and transistors, fluid couplings, clutches, variable resistors or potentiometers, switches, and relays.

In the analysis of many of these real modulators which are operated in conjunction with an energy source, the concept of a *pure dependent source* is useful. A pure dependent source is an ideal source of the type discussed in Section 4–3.4 in which the source variable (f or v) is a function of a second, independent, variable. In an ideal dependent source, this function is a proportionality. Figure 4–13 illustrates dependent through-variable and across-variable sources. One could also, of course, have dependent sources which depend on another through-variable rather than on an across-variable, as shown in Fig. 4–13. In these pure dependent sources, the variable which modulates

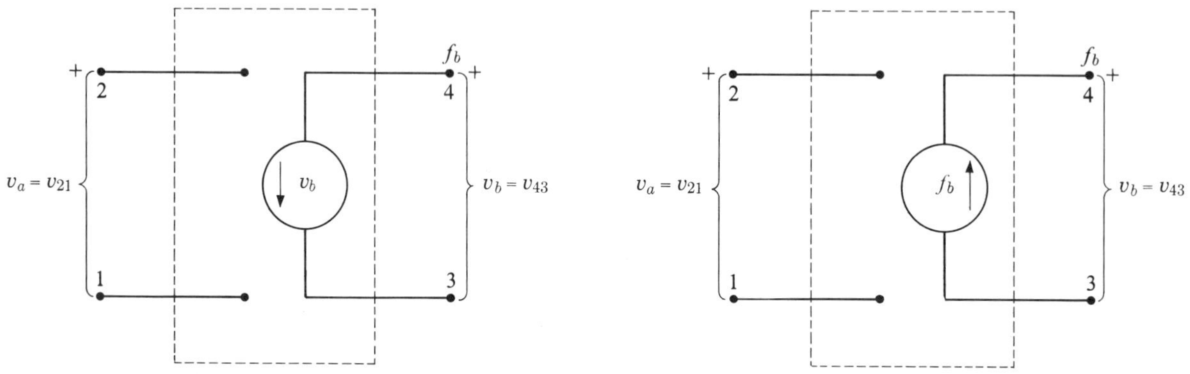

$v_b = \mathbf{f}(v_a)$ . . . . pure . . . . $f_b = \mathbf{f}(v_a)$

$v_b = \mu v_a$ . . . . ideal . . . . $f_b = g v_a$

(a) (b)

FIG. 4-13. Dependent sources. (a) Dependent across-variable source. (b) Dependent through-variable source.

the source is sometimes assumed to have a conjugate variable of zero so that no energy input is required at the modulating energy port (terminal pair 2–1 in Fig. 4–13).

The linear graphs of Fig. 4–14 may be used to represent pure dependent across- and through-variable sources. The dependent source is similar to a transformer or gyrator except that the dependent source does not obey a zero-net-power condition.

Vacuum tubes, transistors, fluid valves, electric motors, fluid machines, strain gauges, and many other

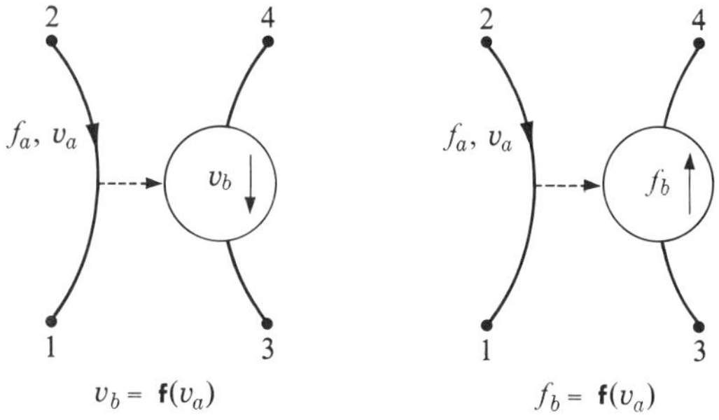

FIG. 4–14. Linear graphs of dependent sources.

physical devices may often be adequately modeled by means of dependent sources. The manner in which this modeling is done will be brought out in the text and problem material of subsequent chapters.

4–6. ENERGY AND POWER RELATIONSHIPS FOR PURE SYSTEM ELEMENTS

The pure system elements discussed in this chapter may be classified as energy storage elements, $\mathcal{E} \geq 0$; energy dissipators, $\mathcal{P} > 0$; energy sources, $\mathcal{P} \lessgtr 0$; energy transformers, $\mathcal{P} = 0$; and energy modulators. Lumped-element models of any physical system can be formulated by suitable interconnections of these idealized elements. In Chapter 5 we shall begin to see how these interconnections are made and how the behavior of the resulting dynamic system can be determined. Chapter 6 will discuss how various pure elements are selected to permit the formulation of a proper system model.

The diagram in Fig. 4–15 summarizes the energetic relations for pure system elements.

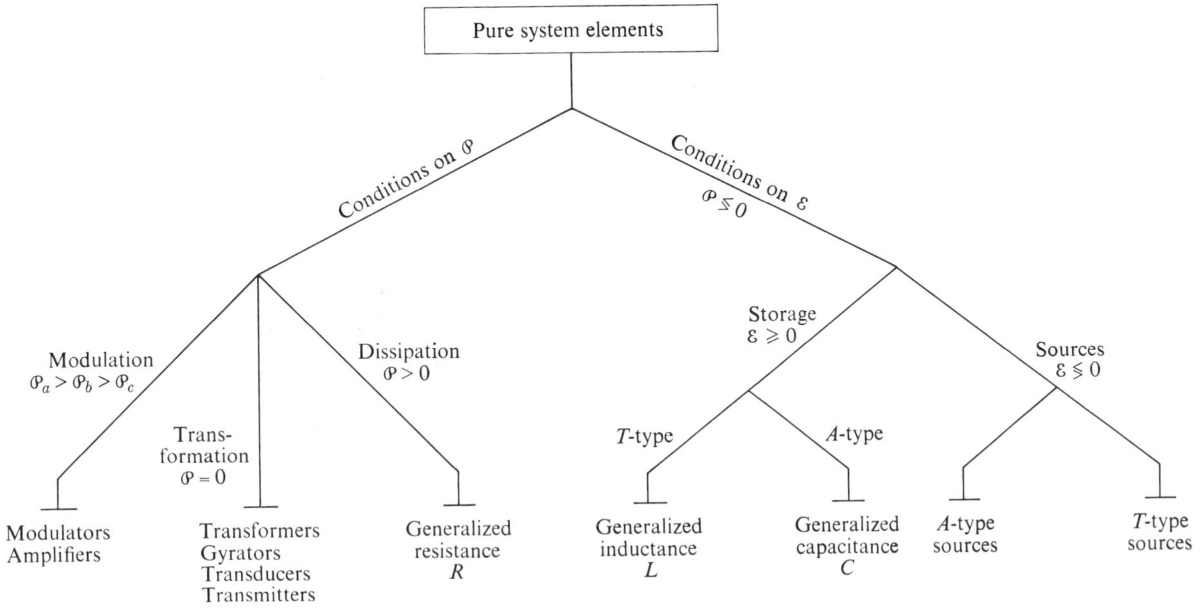

FIG. 4–15. Energetic relations for pure system elements.

4–7. EXTENSION OF IDEALIZED ELEMENT CONCEPTS TO NONENERGETIC SYSTEMS

Although the majority of the dynamic systems considered in this book are those in which interactions between various energy-source, storage, transformation, and modulation elements control system behavior, it should be mentioned that the techniques developed in subsequent chapters may also be applied to other systems involving such areas as business and economic systems, traffic flow, and population growth.

In traffic flow, for instance, a through-variable equal to the rate at which cars pass a given point may be defined. The across-variable might then measure the proclivity of the drivers to reach their destinations. The product of traffic flow and proclivity could not, however, be interpreted as power. Lines of traffic can be said to possess attributes similar to generalized inductance since the through-variable (flow rate) cannot be changed instantaneously. Similarly, capacitance exists in parking areas and other places where numbers of vehicles may accumulate. Resistance to traffic flow occurs at constrictions in the roadways such as at the entrances and exits to turnpikes, narrow tunnels, bridges, and intersections. By attributing quantitative values to these various parameters and effects, it has been possible in some cases to model and predict the behavior of traffic flow. A discussion of the problems of representing the dynamics of traffic is contained in Reference 2.

The use of dynamic-system concepts to study and control the operation of such systems has only begun to be exploited and will undoubtedly develop rapidly in the years to come.

REFERENCES

1. H. M. Paynter, *Analysis and Design of Engineering Systems.* Cambridge, Mass.: The M.I.T. Press, 1961.
2. F. A. Haight, *Mathematical Theories of Traffic Flow.* New York: Academic Press, 1963.
3. A. J. G. MacFarlane, *Engineering Systems Analysis.* Reading, Mass.: Addison-Wesley Publishing Co., Inc., 1965.

PROBLEMS

4–1. On one sheet of paper, sketch a complete cycle of $y = A \sin(\omega t + \phi)$ for

a) $y_1: A = 1, \quad \omega = 10\pi, \quad \phi = 0;$

b) $y_2: A = 1, \quad \omega = 30\pi, \quad \phi = 0;$

c) $y_3: A = 1, \quad \omega = 10\pi, \quad \phi = \pi/4;$

d) $y_4: A = \frac{1}{2}, \quad \omega = 10\pi, \quad \phi = -\pi/2.$

4–2. Referring to Problem 4–1, sketch on one graph (by graphical addition and subtraction)

a) $y_1 + y_2$, b) $y_1 - y_2$.

4–3. Referring to Problem 4–1, sketch on one graph (by graphical addition and subtraction)

a) $y_1 + y_3$, b) $y_1 - y_3$, c) $y_1 - y_3 + y_4$.

4–4. We have seen that power is in general equal to the product of the through- and across-variables (except in thermal systems). In many situations it is desirable to determine the average power when the power is periodic and thus repeats itself cyclically with a period T. By average power, $\mathcal{P}_{\text{ave}}$, we mean that quantity which when multiplied by the time of one period T will yield the correct net energy transferred during the period:

$$\mathcal{P}_{\text{ave}} T = \mathcal{E}_T.$$

From the definition of power, the energy transferred is

$$\mathcal{E}_T = \int_0^T \mathcal{P}\, dt$$

and thus,

$$\mathcal{P}_{\text{ave}} = \frac{1}{T} \int_0^T \mathcal{P}\, dt.$$

Once $\mathcal{P}_{\text{ave}}$ is computed for one period T, the total energy transferred in an integral number of periods is obtained by multiplying $\mathcal{P}_{\text{ave}}$ by the total time. If the time of interest involves many periods and if the time interval is not an exact multiple of the period, a negligible error is made.

a) In many cases the variables are sinusoidal:

$$f = F_0 \sin \omega t, \qquad v = V_0 \sin(\omega t + \phi).$$

Determine the power fv and sketch a curve of fv and f, on the same scale (take $\phi = 45°$).

b) Use trigonometric identities to show

$$\mathcal{P} = fv = (F_0 V_0/2)[\cos \phi - \cos(2\omega t + \phi)].$$

Check your sketch of (a) with this equation.

c) What are the periods of f, v, and $\mathcal{P}$?

d) Since $\int (y_1 + y_2)\, dt = \int y_1\, dt + \int y_2\, dt$, determine $\mathcal{P}_{\text{ave}}$ by integrating $\mathcal{P}$ as two terms.

e) What is the integral of any sinusoid $A \sin(\omega t + \phi)$ over one period?

f) Suppose f and v are given as $f = F_0 \sin(\omega t + \alpha)$, $v = V_0 \sin(\omega t + \alpha + \phi)$; show that $\mathcal{P}_{\text{ave}}$ is independent of α.

4–5. In comparing mechanical and electrical one-port elements, we have chosen to take force and current as analogous based on the idea of through- and across-variables. This is called the *mobility analogy* and is quite arbitrary. Force and voltage difference could also be made analogous if these quantities were considered to be mechanical and electrical *efforts*. The velocity and current would then be considered to be *flows*. Construct a table like Table 4–2 for mechanical translational and electrical elements, assuming that force and voltage, and velocity and current, are compared. Elements which store energy by virtue of their efforts may be called E-type elements, those which store energy by virtue of their flow F-type, etc.

4–6. A square-law damper is described by the equation $F = C_d|v|v$, where $C_d = 2\ \text{lb}\cdot\text{sec}^2/\text{ft}^2$ and v is the velocity difference. Sketch this function carefully on a sheet of graph paper.

a) A sinusoidal force $F_0 \sin \omega t$ is applied. Sketch v vs. time for $F_0 = 2$, and 18 lb.

b) A sinusoidal velocity $v = V_0 \sin \omega t$ is applied. Sketch F vs. time for $V_0 = 1$, 2, and 3 ft/sec.

c) Estimate the average power dissipated in (a) when $F_0 = 8$ and in (b) when $V_0 = 2$. What can you conclude about the response of a nonlinear damper to a sinusoidal input function?

4–7. A unit impulse $u_i(t)$ is the limit of a unit pulse (see Fig. 4–5) as the duration of the pulse approaches zero, $u_i(t) = \lim_{T\to 0} u_T(t)$. Determine the following:

a) the response of an ideal mass to a unit impulse of force;

b) the response of an ideal capacitance to a unit impulse of current;

c) the response of an ideal spring to a unit impulse of velocity;

d) generalize the above to find the response of any A-type energy storer to a unit impulse of the through-variable and of any T-type energy storer to a unit impulse of the across-variable.

4–8. Two ideal capacitances C_1 and C_2 having initial voltages v_1 and v_2 are suddenly connected in parallel with each other so they have equal voltages after connection.

a) Find the total energy stored in C_1 and C_2 before connection.

b) Find the charges on C_1 and C_2 after connection.

c) Determine the total energy in C_1 and C_2 after connection.

d) What happens to the current when the connection is made?

e) What physical quantity is conserved during the interaction of C_1 and C_2?

4–9. Repeat Problem 4–8, assuming that two masses m_1 and m_2 having initial momenta p_1 and p_2 suddenly collide and stick together.

4–10. Two ideal springs are initially constrained or clamped by holding point a (Fig. 4–16). Spring k_1 is compressed by x_1 and k_2 is stretched by x_2 from their unstressed lengths. At a certain instant, the clamp is suddenly removed.

a) Find the force exerted by the clamp before release.

b) Find the total energy stored in the springs before release.

c) Determine the force in the springs after release.

d) Find the total energy stored after release.

e) How does the displacement of point a vary when the clamp is released?

f) What is conserved during the release?

4–11. Repeat Problem 4–10 for two inductances, L_1 and L_2, initially carrying currents i_1 and i_2 and having flux linkages λ_1 and λ_2, which are suddenly connected (Fig. 4–17). Substitute current for force and flux linkage λ for displacement in parts (a), (c), and (e).

4–12. A sinusoidal force $F = F_0 \sin \omega t$ acts on an ideal mass m.

a) Find the resulting velocity $v(t)$. What is the phase angle between these sinusoids?

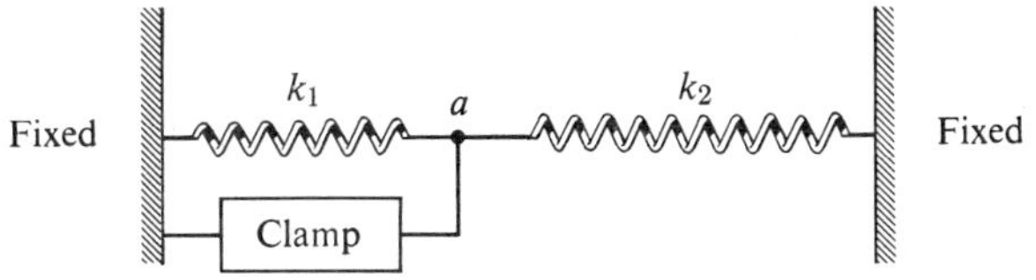

FIGURE 4–16

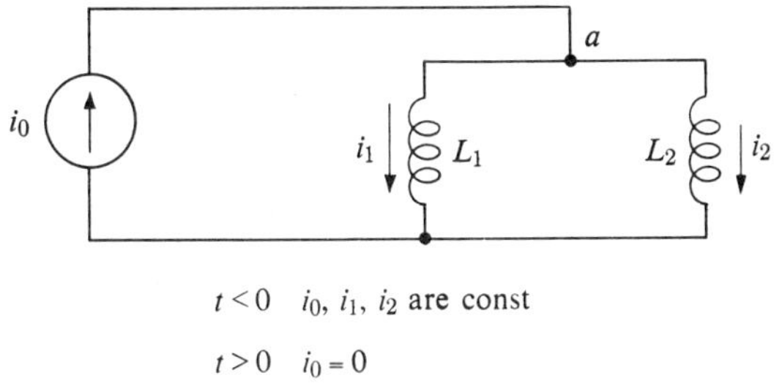

FIGURE 4–17

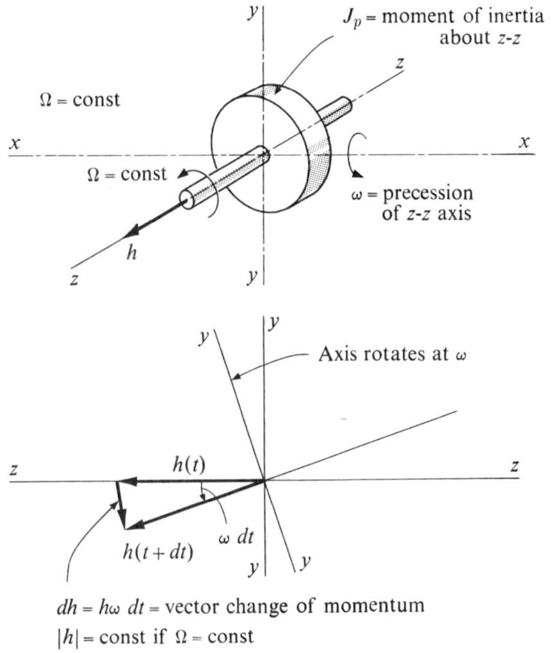

FIGURE 4–18

b) Sketch graphs of the amplitude of the $v(t)$ sinusoid and phase angle between F and v vs. frequency.

c) On cartesian coordinates cross plot F vertically vs. v horizontally. This is called a *Lissajous figure*. Show curves for two values of ω.

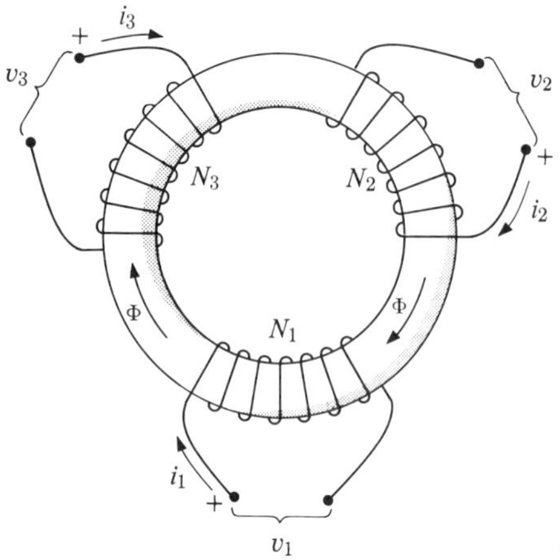

FIGURE 4–19

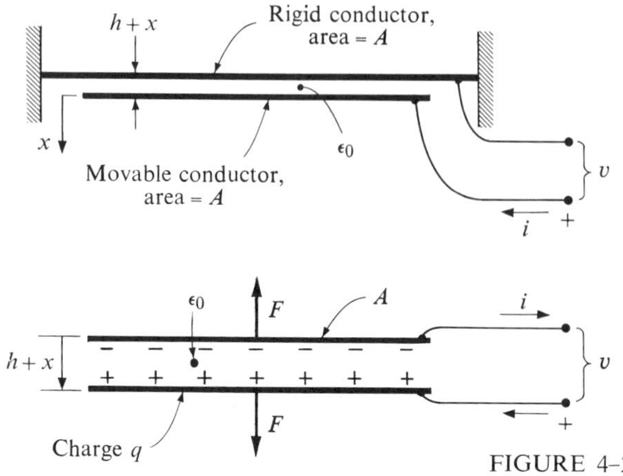

FIGURE 4–20

4–13. Repeat Problem 4–12, assuming that the sinusoidal force is applied to an ideal spring k.

4–14. Repeat Problem 4–12, assuming that the sinusoidal force is applied to an ideal damper b.

4–15. Consider a wheel rotating at constant angular velocity Ω about its polar axis (Fig. 4–18). The angular momentum h is a vector ΩJ_p lying along the axis. In three dimensions the vector rate of change of h equals the vector torque on the body. If the wheel is precessed at constant speed about axis x–x, the angular momentum will change due to the directional change of h and a torque will be produced about the y–y axis.

a) Show that if $|h| = \Omega J_p$ is constant, $dh/dt = h\omega$ due to rotation of the h-vector.

b) Find the direction and magnitude of the torque. The x-, y-, z-axes precess with the wheel at ω. This torque may be used as an indication of the angular rate ω. Devices which measure rate in this way are called *rate gyroscopes.*

c) Considering the x- and y-axes as energy ports, show that under steady conditions this device behaves like an ideal gyrator. Compute the gyrational resistance.

4–16. A toroidal air core is wound with three windings having N_1, N_2, and N_3 turns (Fig. 4–19). The toroid has mean radius R and sectional area A. Air has permeability μ_0.

a) Assuming that all flux is contained in the core, and that resistive and capacitive effects are zero, determine the relations between v_1, v_2, v_3, i_1, i_2, and i_3.

b) Show that your result describes a two-winding transformer as analyzed in the text if $i_3 = 0$.

4–17. An idealized representation of a capacitance loudspeaker is afforded by a parallel-plate capacitor whose plates are capable of moving (Fig. 4–20). When a charge is produced on the plates by an applied voltage, the electric attractive forces produce a net force $-F$ on the plates. The plate motion generates sound waves.

a) Assuming fixed plates, $x = 0$, compute the electrical energy stored in the capacitance as a function of v and repeat as a function of q.

b) Find the capacitance as a function of $h + x$, ϵ_0, and A.

c) Find the change in electrical stored energy if the plate separation increases a small amount to $h + dx$ with voltage held constant.

d) Repeat (d) with charge held constant, i.e., with $i = 0$.

e) For a small increase of gap from h to $(h + dx)$, assume that energy is conserved and equate the mechanical work done, $F\,dx$, to the change in stored electrical energy at constant charge, and thereby compute F as a function of h, A, v, and ϵ_0.

f) Repeat (e) with voltage constant and show that the same result is obtained. Note that electrical work

$$vi\,dt = v\,dq$$

is added through the terminals.

g) Use the previous results and conservation of energy to find the relations between F, v, i, and v. What type of ideal device do these relations describe?

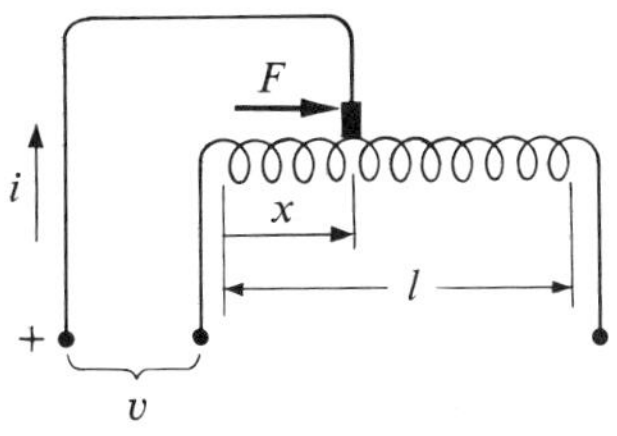

N coils, total length = *l*
maximum resistance = *R*

FIGURE 4–21

4–18. A potentiometer consists of a wound coil of resistive wire and a mechanical slider which can be moved along the coil to vary the number of coils in the resistor (Fig. 4–21). A Coulomb friction force, F = const in magnitude, acts between the slider and the coils when the slider is moving. Determine the relations between the mechanical variables F and x (or dx/dt) and the electrical variables v and i. Show that the device acts like a modulator.

5 ANALYSIS OF ELEMENTARY DYNAMIC SYSTEMS

2:15

5-1. INTRODUCTION

To this point in the development of the subject matter of dynamic systems, we have discussed at some length the pure lumped elements and their use in approximating real devices. We have seen that these elements may be characterized as energy (1) storers, (2) dissipators, (3) sources, (4) transmitters, (5) transformers, and (6) modulators. However, knowledge of the elements alone is not sufficient to investigate the behavior of dynamic systems. The objective of this chapter will be to orient the reader to the complete problem of analysis of a dynamic system. We will consider elementary examples, and we will also use methods which will be the subject of detailed consideration in later chapters. Thus, in a sense, for motivational purposes as well as for an introduction to elementary problems and methods of analysis, this chapter is a preview of the rest of the book.

When the engineer is confronted with a dynamic-system problem, he must first *recognize* that the problem is one to which he can apply the principles developed in this book. This may seem obvious, and of course, all the problems with which the student will deal in *this* book will be of this nature. This may be misleading, however, since the practicing engineer is faced with many types of situations and has a wide range of subject matter at hand to apply to any given situation. His education has hopefully given him the perspective with which to judge the problem at hand and to categorize it properly. Lack of such perspective may be one of the major stumbling blocks to many practicing engineers. The integration of subject matter in this book is designed to allow the engineer to recognize that a large class of problems can be considered in a similar and systematic fashion.

The engineer must then *define* the system to be considered (should the ambient-temperature variation be considered an input? is the inertia of a connecting shaft important in this situation? etc.). He then may describe the system by means of the various dynamic system elements which we have considered. His next step is to investigate the energetic interactions between these elements when they are interconnected and then excited by some signal. The engineer's most important (but frequently overlooked) job is to establish a *mathematical model* of the system to be analyzed. This involves the identification and idealization of individual system components as well as identification and idealization of their interconnection. This chapter will describe some elementary interconnections and the laws governing the relations between system variables under these interconnections.

The mathematical statement of the governing relationships between system variables is called the *formulation problem.* Interconnection of the elements imposes constraints on the variation of system variables, and the convenient way of specifying these constraints is by a mathematical statement of the way in which the various *through-variables* are related and the way in which the various *across-variables* are related. The elemental equations then relate the through- and across-variables for each individual element. This package of equations is a complete mathematical description of the system.

To investigate the *dynamic behavior* of the system, we must solve this set of equations. The input and outputs are selected and a single differential equation, called the *system equation*, relating each output and input must be determined. The initial state of the system must also be specified by a set of *initial conditions*. The system differential equation must be solved for the output response under the specified input signal and initial conditions. There are numerous approaches to the *solution* of the system equation, and the choice of a method will depend on the problem at hand. Briefly, these methods are (1) graphical; (2) numerical, with the possible use of a digital computer; (3) operational block diagram of the system with the eventual use of an analog computer; and (4) a purely mathematical solution. This chapter will describe the advantages of each of these approaches.

Following this, we then consider certain results concerning the dynamic behavior of elementary systems. Our methods are also elementary, and later chapters will generalize these methods to establish more efficient tools for solution. The results developed here, however, will serve to indicate the types of behavior of interest and the reasons for detailed later consideration.

After obtaining a solution for the response, the engineer has another major job. The tasks of initial modeling and final analyzing or designing are functions which clearly distinguish the engineer from the mathematician. The engineer must *check* his solution (is it correct dimensionally? does it correspond to physical reality? does it check for simplified situations which can be easily analyzed? etc.). His interest in the whole matter of dynamic systems is to eventually *analyze* to determine whether a certain performance is obtained and is satisfactory, or more generally, he must *design* the system so that it will meet certain performance specifications. The process of designing is usually an iterative analysis. The results of one analysis point toward a change in the system which may improve the performance; another analysis is done on the new system to determine performance, etc. The design aspects distinguish the engineer from the scientist. The engineer is ultimately interested in building a system which will perform a useful function for the benefit of mankind.

In summary, our approach to a dynamic system problem can be enumerated as follows:*

1) define the system and its components,
2) formulate the mathematical model,
3) determine the system equations,
4) solve for the desired output,
5) check the solution,
6) analyze or design.

5–2. MODELING

After the system has been defined, a mathematical model must be established for it. No real physical system dynamically behaves exactly according to any of the equations which describe the ideal elements. A coil of wire may act *primarily* like a spring or *primarily* like a mass, depending on the nature of the forces applied to it and on the type of geometrical constraints to which it is subjected; it will never behave *exactly* like an ideal mass or spring. The ideal elements are mathematical fictions which approximate the behavior of certain physical devices only under specialized conditions. If several of the ideal elements are combined, however, an ideal dynamic system is formed which may represent the behavior of a real physical system over a wider range of conditions. For example, the coil of wire mentioned above may be adequately represented, for low-frequency motions, by an ideal spring in combination with an ideal mass. The combination of ideal elements which is intended to represent the behavior of a physical system is called the *model* of the system. The degree to which the behavior of the ideal model corresponds to the behavior of the physical system represented by the model is a function of the experience, skill, and engineering judgment of the modeler.

* See Reference 1 for a general statement of a professional method of problem solving. Our approach here, although limited to dynamic system problems, follows this method closely.

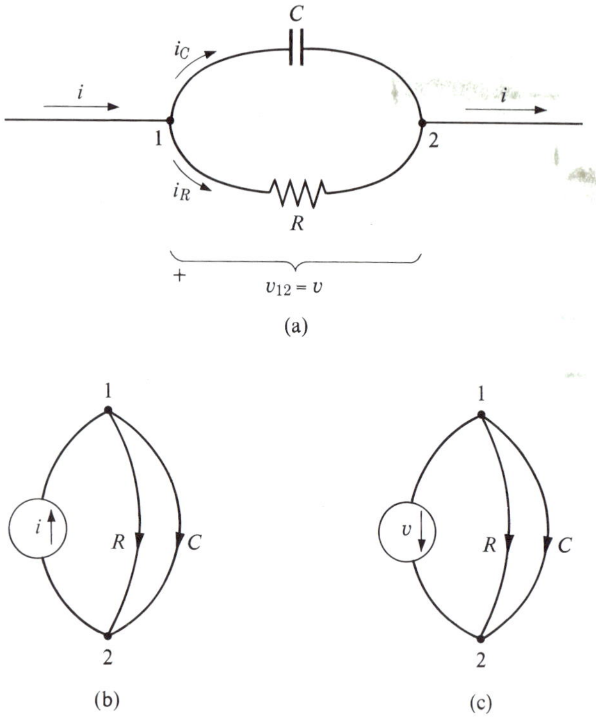

FIG. 5–1. Parallel combination of resistance and capacitance. (a) Electric circuit. (b) Linear graph with current source. (c) Linear graph with voltage source.

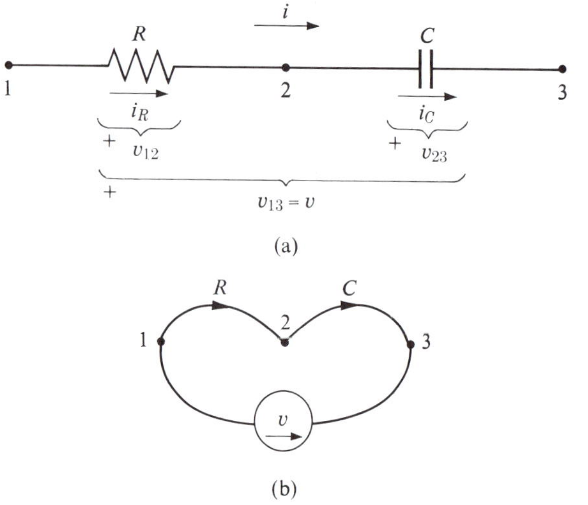

FIG. 5–2. Series combination of resistance and capacitance. (a) Electric circuit. (b) Linear graph with voltage source.

Usually, the model must represent a compromise between its complexity and the degree of accuracy required in the predicted behavior of the physical system.

The importance of this process cannot be overemphasized. Here is where the engineer must make assumptions about a real-world situation to make it mathematically tractable and yield a solution with a reasonable effort. *The usefulness of the solution is directly dependent on how closely the model represents the behavior of the physical system.* Chapter 6 gives detailed consideration to the modeling problem; however, to appreciate whether a particular model would be appropriate, we must understand the behavior of the idealized elements (and simple combinations of them). Therefore, this chapter will introduce these ideas by considering the interconnection of idealized elements.

5–3. FORMULATION OF SYSTEM EQUATIONS

We will now discuss the simplest of ideal dynamic systems, those consisting of only two ideal elements. To minimize complexity at this stage, we will limit our discussion of these elements to one energy storer and one energy dissipator. Systems with two or more energy storers have more complex differential equations and will be considered later. Section 5–6.2 will formulate the equation for a system with two elements which store energy.

5–3.1 Interconnection of Two Ideal Elements

Resistance and capacitance. Let us suppose that a dynamic system is formed by combining two ideal elements; e.g., an ideal capacitance and an ideal resistance. There are only two possible ways to achieve this combination:

a) *Common voltage.* This arrangement (Fig. 5–1) is ordinarily called a *parallel* connection because the current i entering the system divides between two parallel paths. The voltages at the terminals of the two elements are equal. The figure shows the circuit and linear-graph representation for both a voltage- or current-source input.

b) *Common current.* Figure 5–2 shows the alternative way of connecting a resistance and capacitance. This is called a *series* connection since the same current $i = i_R = i_C$ flows through both elements. The voltage $v = v_{13} = v_1 - v_3$ divides between the two elements. The linear graph shows the system driven by a voltage-source input.

Mass and damper. Similarly, two possible dynamic systems can be constructed from the combination of an ideal mass and ideal damper.

a) *Common velocity.* The mass and viscous damper shown schematically in Fig. 5–3(a) might at first glance seem to be connected in series. However, more careful observation reveals that because they have a common velocity $v_{1g} = v$, and because the net force accelerating the mass is the difference between the applied force F and the force in the dashpot, they are actually a pair of elements acting in parallel, as is clearly shown in the linear-graph representation. The applied force divides between the mass and dashpot. It is important to note the similarities and dissimilarities between the parallel connection of a resistance and capacitance in Fig. 5–1 and the parallel connection of a mass and damper in Fig. 5–3. The physical schematic diagram for the mechanical system (which shows approximately how it is constructed) does not reveal the elemental aspects of the system as well as the linear graph does. This simple system readily shows the desirability of using linear graphs to portray the structure of mechanical systems.

Because the linear graphs are identical, it seems that the systems are analogous to each other. Force in the mechanical system is analogous to current in the electrical system, and similarly velocity is analogous to voltage. However, it must be determined whether the characteristics of the elements are analogous before one can say that the systems are true analogs to each other. Since the resistance and damper are analogous and the capacitance and mass are analogous (Table 2–4), the analogy is complete; a later section will deal more completely with analogies.

There is an important aspect in which the two systems differ, however. In the mechanical-system linear graph, one "terminal" of the mass element must be referred or "connected" to a nonaccelerating reference or ground (the dashed line in Fig. 5–3b). Actually, there is no physical connection between the mass and ground, and no force is transmitted from the mass to ground in the manner that the damper transmits force to the ground by means of its physical connection to ground. The only force felt by the ground is the damper force. This difficulty does not exist with the electrical system since the full current i leaves the vertex at 2.

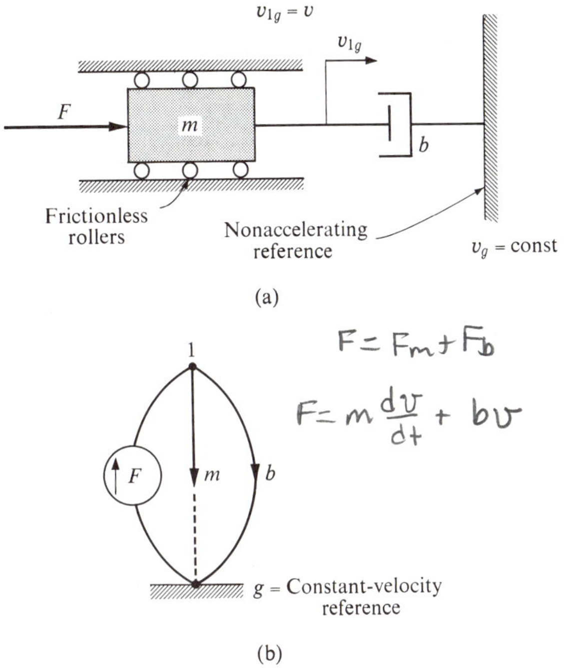

FIG. 5–3. Parallel combination of mass and damper. (a) Physical schematic diagram. (b) Linear graph with force source.

b) *Common force.* The mass and viscous damper shown schematically in Fig. 5–4(a) have the same force and constitute a series connection of these elements, as shown in Fig. 5–4(b). The velocity v_1 divides between the damper and the mass. Note the similarity to the electrical circuit in Fig. 5–2. Note also the dashed "connection" to ground of the mass element in the linear graph. The ground feels no force through this "connection."

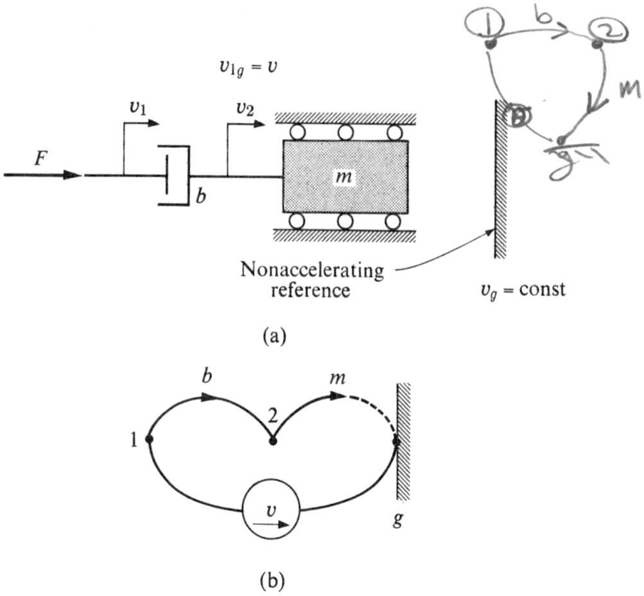

FIG. 5–4. Series combination of mass and damper. (a) Physical schematic diagram. (b) Linear graph with velocity source.

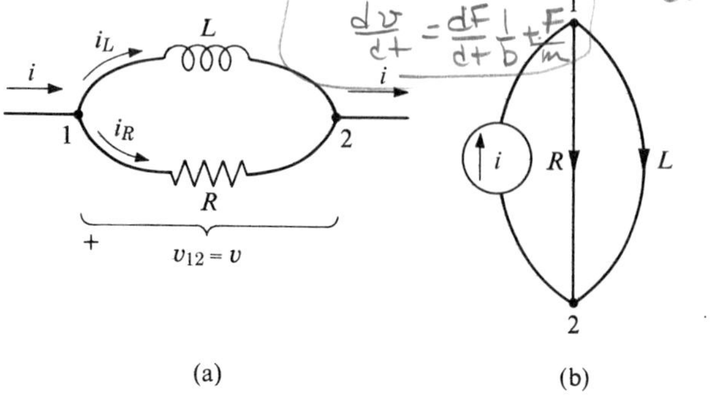

FIG. 5–5. Parallel combination of inductance and resistance. (a) Electric circuit. (b) Linear graph with current source.

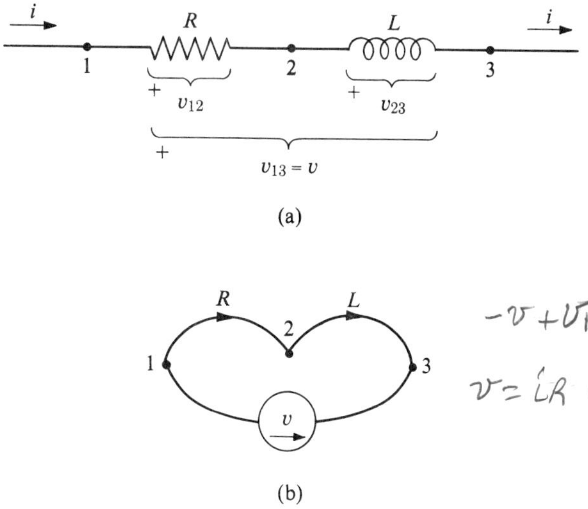

FIG. 5–6. Series combination of inductance and resistance. (a) Electric circuit. (b) Linear graph with voltage source.

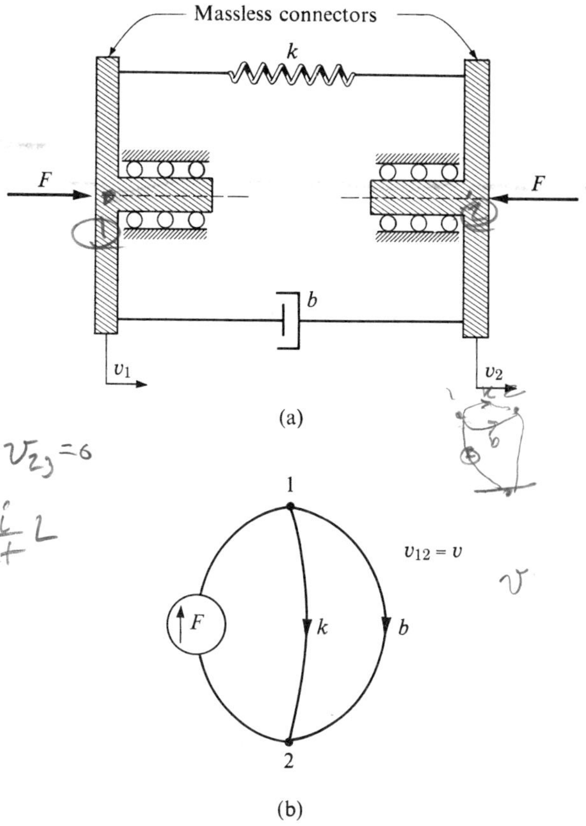

FIG. 5–7. Parallel combination of spring and damper. (a) Physical schematic diagram. (b) Linear graph with force source.

A series of simple systems may also be formed by connecting a resistance and an inductance or a damper and a spring. These systems are shown in Figs. 5–5 to 5–8.

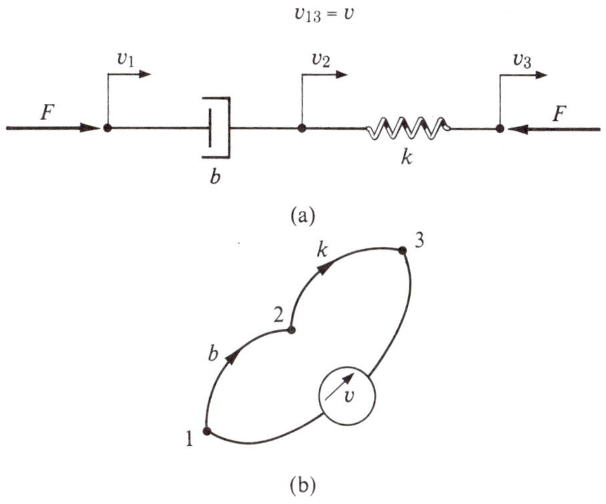

FIG. 5-8. Series combination of spring and damper. (a) Physical schematic diagram. (b) Linear graph with velocity source.

5-3.2 Governing Relationships for Dynamic Systems

The unsteady behavior of dynamic systems is governed by the way in which energy is handled by the system. If one can describe mathematically all the energetic actions and interactions in a given system, he has described the dynamic behavior of the system. The problem of recognizing all these energetic actions and interactions may be very difficult in some systems, and it is helpful to have a systematic means of recognizing and assembling all the important phenomena when making a dynamic analysis of a given system.

This is accomplished by using simple idealized system elements whose individual elemental equations are known. In addition to the elemental description, there are two important conditions which must be satisfied when the elements are combined or connected together. These two conditions will be called *compatibility* and *continuity*. Thus the system analysis will be carried out by systematic use of:

a) equations describing elemental behavior,

b) compatibility (path) conditions applied to across-variables,

c) continuity (vertex) conditions applied to through-variables.

The elemental equations have already been discussed in Chapters 2, 3, and 4, and in each case the storage or dissipation of energy was revealed for the respective elements. Thus the elemental equations contain information about energetic actions within the elements. The compatibility and continuity conditions provide a means of systematically dealing with the energetic interactions which occur between the system elements.

The *compatibility* requirement is established by the manner in which the dynamic elements are connected, and it results in a relation among the various across-variables. For example, when two elements are connected in parallel, as in Figs. 5-1, 5-3, 5-5, and 5-7, compatibility requires equal voltages or velocities at the points where the elements are connected. In a series connection, Figs. 5-2, 5-4, 5-6, and 5-8, compatibility requires that the voltage or velocity difference across both elements be equal to the sum of the voltage or velocity differences across the individual elements. In a mechanical system, the concept of compatibility means that the *geometric constraints* placed on the motions of the elements are expressed. In an electrical circuit, compatibility requires that the sum of all voltage drops around any closed circuit be zero. For electrical circuits, the compatibility requirement is called *Kirchhoff's voltage law* [2]. In fluid and thermal systems, compatibility is a statement of the fact that pressure and temperature are scalar quantities and must add as one moves from point to point.

Continuity implies that charge is conserved in an electrical circuit or that force is conserved in a mechanical network. Thus, the continuity requirement results in a relationship among the various through-variables. In the system shown in Fig. 5-3, for example, continuity requires that the total force F applied to point 1 must equal the sum of the forces F_m and F_b applied to the mass and damper, respectively. This condition is recognized as *Newton's law* for equilibrium of forces at a point. For a point which has no mass, the sum of all forces acting must be zero, or any force transmitted into the point must also be transmitted away from the point. In electrical systems, continuity requires that charge be conserved and, therefore, that the sum of all currents flowing into any point or node in an electrical circuit must sum to zero. In Fig. 5-1, for example, continuity requires that the total current i introduced at vertex 1 be equal to the sum of the currents i_C and i_R which flow into the capacitance and resistance, respectively. In electrical circuits, the continuity of charge at any node is called *Kirchhoff's current law*. In fluid systems, continuity is an expression of the *law of conservation of matter*. In thermal systems, continuity is an expression of the *law of conservation of energy*.

The compatibility and continuity requirements may also be thought of as dealing with *path* and *vertex* conditions, respectively. In Fig. 5-1, for example, we may consider points 1 and 2 as *vertices* (where the *net* current flow into each vertex is zero) and the connections between 1 and 2 as paths (and the voltage difference is the same for each path between 1 and 2). The linear graph (Fig. 5-1b or c) then provides a graphical statement of the compatibility and continuity relationships for the circuit of Fig. 5-1(a). The graph similarly provides these statements for the systems of Figs. 5-2 through 5-8.

As discussed previously, the arrows on the linear graph serve to establish reference conditions for all variables. For instance, in Fig. 5-1, a positive numerical value of i_R means that the current is actually flowing from 1 toward 2 through the resistance. If i_R is numerically negative, then the current is actually flowing from 2 toward 1. Similar statements can be made concerning all the other variables shown on the figures. The directions shown are only the reference directions; the actual directions can be the way shown or the opposite way and are distinguished by the sign of the numerical value of the variable.

The vertex condition for Fig. 5-1 states that

$$i = i_R + i_C$$

at vertices 1 and 2, and the path condition states that the voltage difference across path R equals the voltage difference across path C, or that the sum of all voltage drops around the *closed path* $R \rightarrow C$ must be zero. The linear graph provides the means of visualizing the compatibility and continuity relationships for the system. Linear graphs, compatibility, and continuity will be discussed in more detail in Chapter 8.

Note that the statements of continuity and compatibility or the construction of the linear graph for the circuit of Fig. 5-1 in no way involve the nature of the elements included in the circuit but depend only on the way in which the elements are interconnected. Thus, the linear graphs for the inductance-resistance system of Fig. 5-7 and the spring-damper system of Fig. 5-3 are identical to the linear graph for the system in Fig. 5-1.

The *elemental behavior* requirement refers to the relationship established between voltage and current or between force and velocity by the physical characteristics of the element in question. In an ideal spring, for example, the force must be proportional to the relative displacement of its two ends. The elemental behavior may not always be described by simple ideal elements but may involve nonlinearity, multivalued functions, and so forth. However, a knowledge of the elemental relationships, when combined with the compatibility and continuity requirements, permits formulation of the equations of a dynamic system.

It is well to point out that these three requirements for system behavior (continuity, compatibility, and elemental behavior) are not limited to dynamic systems. In the mechanics of deformable bodies, one should recognize continuity as force equilibrium, compatibility as geometrical compatibility, and the elemental behavior as the stress-strain relationships [3]. In electromagnetic-field theory, one should recognize continuity as the law of conservation of charge, compatibility as the potential concept for conservative fields, and the elemental behavior as the constitutive relations for conductors, dielectrics, or magnetic materials [4].

5-3.3 Determination of System Equations

The equations for the behavior of the circuit shown in Fig. 5-1 may now be obtained by satisfying the conditions discussed in the preceding section.

a) *Compatibility.* This requirement is satisfied by setting the voltage at the terminals of the resistance equal to that of the capacitance:

$$v_{12} = v = v_R = v_C. \quad (5\text{-}1)$$

b) *Continuity.* Current is conserved:

$$i = i_R + i_C. \quad (5\text{-}2)$$

c) *Elemental behavior.* For the ideal capacitance,

$$i_C = C\frac{dv_C}{dt} = C\frac{dv}{dt}, \quad i = i_R + i_C \quad (5\text{-}3)$$

and for the ideal resistance, $i = \frac{v}{R} + C\frac{dv}{dt}$

$$i_R = v_R/R = v/R. \quad (5\text{-}4)$$

Equation (5-1) has already been used by substituting v for v_R and v_C in Eqs. (5-3) and (5-4). Equations (5-2), (5-3), and (5-4) are so simple that there is little difficulty in combining them to obtain the *system equation*. This will not be so simple for more complex systems, however, and a considerable portion of the subject matter of dynamic systems is devoted to ways of performing this combination and methods of simplifying the problem. If many compatibility, continuity, and elemental equations must be combined, then confusion results if some orderly pattern is not followed. In Chapter 8 we will present some orderly approaches and consider some systems where the difficulty of dealing with a large number of simultaneous differential equations will be apparent. This difficulty is so great that methods have been developed which considerably simplify this problem. These simplifications are discussed in Chapters 13, "Response of Linear Systems," and 14, "Simplification of Linear System Analysis."

For the simple equations we have here, it is easy to obtain the system equation by substituting Eqs. (5-3) and (5-4) into (5-2), which gives the system equation for Fig. 5-1:

$$C\frac{dv}{dt} + \frac{v}{R} = i. \quad (5\text{-}5)$$

Equation (5-5) is a first-order differential equation involving the variable i and the variable v and its first time derivative dv/dt. The equation is linear since no powers or products of the variables or their derivatives appear in the equation, and the equation has constant coefficients (C and $1/R$). The equation relates voltage v to current i and vice versa, and it is obvious that if one of these quantities is specified as a known time function (i.e., if either i or v is a source), the other may be calculated from a solution of Eq. (5-5). The specified time function is an independent variable, which we call an *input* to the dynamic system, and the other function is a dependent variable or output, called the *response* of the system to the given input.

If the circuit is driven by a voltage source, i.e., v is taken as the input, the output current i is very easily determined by Eq. (5-5) and differentiation. For example, if v is given by the function shown in the graph in Fig. 5-9(a), $C = 2$, and $R = 1$, the graphical addition of Fig. 5-9(b), (c), and (d) yields the current i. The reader should study this graphical development carefully since it gives considerable insight into the performance of dynamic systems. Note that the current in the resistance, i_R, varies as the voltage; however, the current in the capacitance is not of this nature at all, but in fact, varies as the derivative of the voltage (note that i_C is zero when i_R is maximum, for instance). The sum of these two currents will in general be a complex function which will depend heavily on the values of R and C.

On the other hand, if the circuit is driven by a current source, i.e., i is the input quantity, the linear first-order differential equation must be "solved" to determine the output voltage v. This is a more difficult problem than that just considered and, in fact, Section 5-4 and a large part of the remainder of this book is devoted to this problem and the understanding of its solution. Thus it is apparent that the behavior of the system depends on the nature of the system itself and also on the nature of the input to the system. Before we discuss the performance of this system further, it might be well to return to Section 5-3.1 and

$i = \frac{v}{R} + C\frac{dv}{dt}$

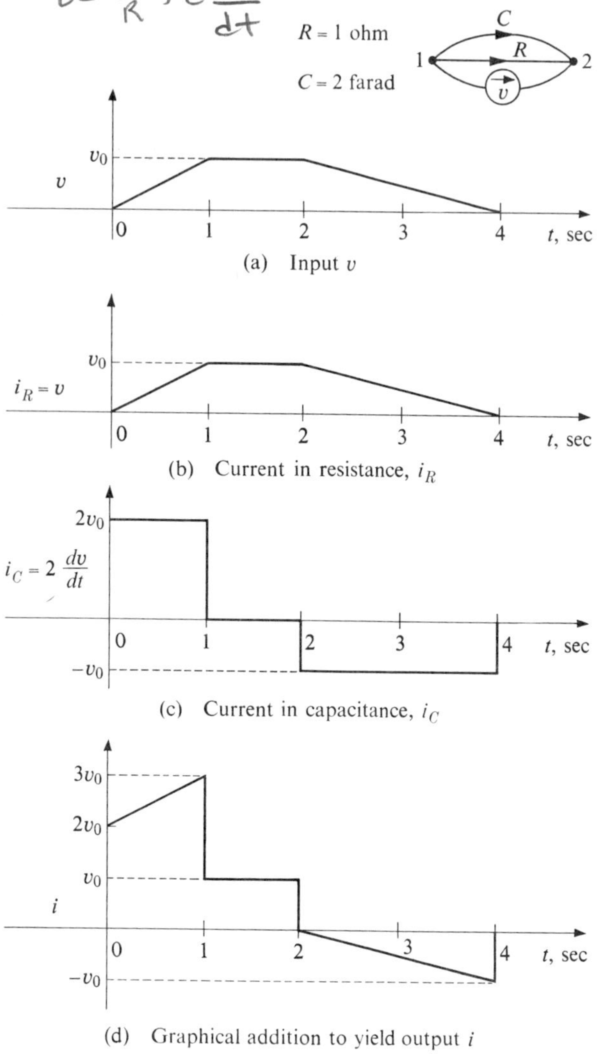

FIG. 5–9. Solution of Eq. (5–5) for v input.

formulate the differential equations for the other simple systems described.

For the series resistance-capacitance circuit of Fig. 5–2 and its linear graph of Fig. 5–2(b), the equations are given below.

a) *Compatibility:*

$$v_{13} = v = v_{12} + v_{23} = v_R + v_C. \tag{5–6}$$

b) *Continuity:*

$$i = i_R = i_C. \tag{5–7}$$

c) *Elemental behavior:*

$$i_R R = iR = v_R, \tag{5–8}$$

$$i_C/C = i/C = dv_C/dt. \tag{5–9}$$

Again, these simple equations are readily combined. When Eq. (5–6) is differentiated with respect to time and combined with Eqs. (5–7), (5–8), and (5–9), we have

$$R\frac{di}{dt} + \frac{i}{C} = \frac{dv}{dt}, \tag{5–10}$$

which is the system equation for Fig. 5–2.

For the mechanical system of Fig. 5–3, the conditions are:

a) *Compatibility.* The velocity v of the mass is the same as the velocity difference across the damper.

b) *Continuity:*

$$F = F_m + F_b. \tag{5–11}$$

c) *Elemental behavior:*

$$F_m = m\frac{dv}{dt}, \tag{5–12}$$

$$F_b = bv. \tag{5–13}$$

Combining Eqs. (5–11), (5–12), and (5–13), we obtain the system equation,

$$m\frac{dv}{dt} + bv = F. \tag{5–14}$$

The similarity between this equation and Eq. (5–5) should be noted.*

---

* Note that Eqs. (5–5) and (5–14) are identical if i is replaced by F, C by m, and R by $1/b$. These two systems are therefore said to be analogous, as indicated earlier. If the behavior of one of these systems is known for a given set of conditions, the behavior of the other is also known for the corresponding set of conditions. The use of analogies will be treated more completely in Section 5–5.3.

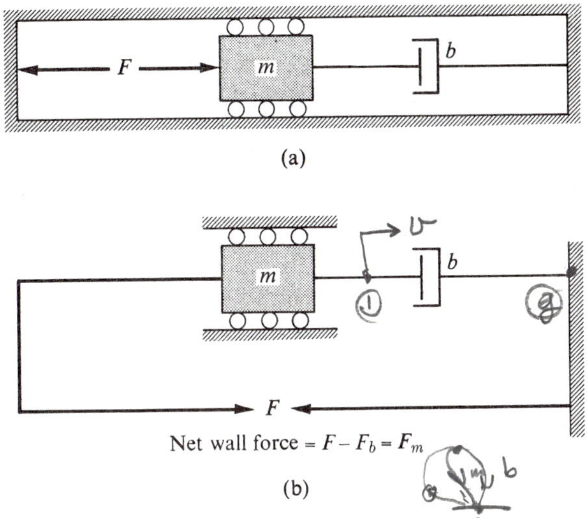

FIG. 5–10. Illustration of effect of complete system at reference.

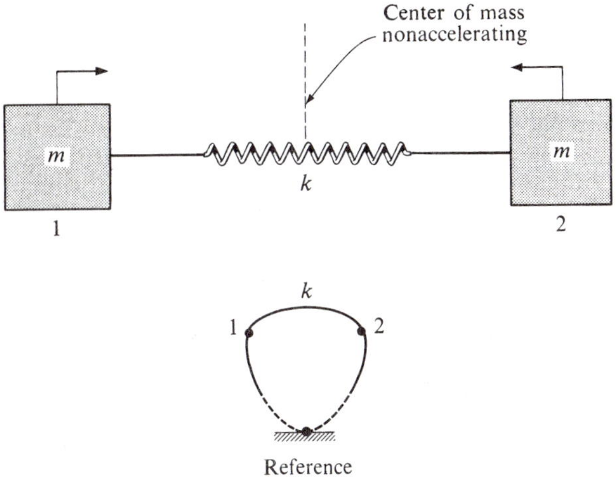

FIG. 5–11. Closed mechanical system with no external forces and no necessity for earth connection.

A basic difference should be noted between analogous electrical and mechanical systems when one of the mechanical elements is a mass. It was seen that the velocity of the mass must be referred to a nonaccelerating reference, and a discontinuity of force actually occurs across the mass element. In Fig. 5–3(b), for instance, the force F_m acting on the mass is not actually transmitted "through" the mass to the nonaccelerating reference g. The use of a dashed line for the mass branch linear graph is employed to emphasize that no actual force is transmitted through that path to the reference (ground). This is obvious from the original physical schematic, Fig. 5–3(a).

However, if one considers that the force source, F, is a two-terminal source (as discussed in Chapter 4), then the conditions at the other terminal of this source should also be investigated. It has been assumed that this other terminal is also connected to a nonaccelerating reference. Two possibilities are shown in Fig. 5–10. It can easily be seen that these are equivalent situations, since the force F exerted to the left on the left-hand wall in Fig. 5–10(a) results in a force F exerted to the left on the right-hand wall since this force is transmitted by the rigid and nonaccelerating surrounding reference structure. The total force exerted on the wall in Fig. 5–10(b) (as well as on the entire structure in Fig. 5–10(a)) is $F - F_b$ to the left. Since continuity at the connection point of b and m has demanded $F = F_b + F_m$, we conclude that the net force $F - F_b$ is equal to F_m to the left. If a fictitious force, F_m, is considered to be transmitted to the reference through the dashed branch of the linear graph, then continuity holds at the reference vertex. That is, if we demand that the sum of the forces equal zero at the reference, then we obtain $F = F_m + F_b$ at this point as well as at the connection point of the mass and damper. This is an additional reason for considering the mass to be a two-terminal element and showing the fictitious force to be transmitted to the reference; the linear graph illustrates this extremely clearly.

The true situation should not be overlooked, however, and there is an actual unbalance of force equal to F_m acting on the reference, in the opposite direction to the force acting on the mass. In the usual engineering situation, the reference is the earth and the unbalanced force will tend to accelerate the earth. Since the mass of the earth is so large, this acceleration can usually be considered negligible, and therefore the earth is a satisfactory approximation to a nonaccelerating reference. The idealized situation then is that the reference point has infinite mass and therefore has absolutely zero acceleration; the earth is a sufficiently good approximation to this idealization.

It should be recognized that the systems under discussion are closed systems (with the earth included as part of the system) and no external forces are acting. It will be recalled from physics (dynamics) that the center of mass of such a system will have zero acceleration. This principle presents another way to view this situation: The center of mass of the system is essentially at the center of the earth since the earth is so massive compared to the mass m; therefore the nonaccelerating reference can be taken as the center of the earth with good approximation, and if the earth is considered rigid, then the attachment point can also be considered a nonaccelerating reference.

This viewpoint also presents another way of providing a closed system with no external forces which contains a nonaccelerating reference without the use of an earth connection. It is only necessary to add another mass m to the system which will experience a force of magnitude F_m which will accelerate it in the direction opposite to that experienced by the first mass. Figure 5–11 shows such a situation for a common system having no external forces.

Continuing with the development of the system equations for the other two-element systems, we can, by analogy, obtain the equation for the system of Fig. 5–4 from the equation for the system of Fig. 5–2. Replacing C by m, i by F, and R by $1/b$ in Eq. (5–10), we obtain

$$\frac{1}{b}\frac{dF}{dt} + \frac{F}{m} = \frac{dv}{dt}. \tag{5–15}$$

The reader should check this result by application of compatibility, continuity, and elemental equations.

The reader should apply compatibility, continuity, and elemental equations to the remaining two-element systems and obtain the following results. For the parallel resistance-inductance circuit of Fig. 5–5 with $v_{12} = v$,

$$\frac{1}{R}\frac{dv}{dt} + \frac{v}{L} = \frac{di}{dt}. \tag{5–16}$$

For the series resistance-inductance circuit of Fig. 5–6 with $v_{13} = v$,

$$L\frac{di}{dt} + Ri = v. \tag{5–17}$$

For the parallel spring-damper system of Fig. 5–7 with $v_{12} = v$,

$$b\frac{dv}{dt} + kv = \frac{dF}{dt}. \tag{5–18}$$

For the series spring-damper system of Fig. 5–8 with $v_{13} = v$,

$$\frac{1}{k}\frac{dF}{dt} + \frac{F}{b} = v. \tag{5–19}$$

Comparison of Eqs. (5–17) and (5–19) and of Eqs. (5–16) and (5–18) shows that the inductance L is analogous to the compliance $1/k$ of the spring, and the resistance R corresponds to the inverse of the mechanical damping coefficient, $1/b$, when current is considered the analog of force.

All the preceding equations describing the relation between force and velocity or current and voltage have a common characteristic. All are linear first-order differential equations with constant coefficients. Each of the systems described in Figs. 5–1 through 5–8 is composed of one energy-storage element in combination with an energy-dissipation element. Systems of this type, which are described by a first-order equation, are called *first-order systems*.

Another approach to the analysis of systems of this kind is to account for all the energy in the system directly, rather than using the elemental equations and the compatibility and continuity conditions. As an example, the system shown in Fig. 5–1 may be analyzed as follows.

The energy stored in the system (in the capacitance) is given by

$$\mathcal{E}_s = Cv^2/2. \tag{5–20}$$

The rate of energy dissipation in the system (in the resistance) is given by:

$$d\mathcal{E}_d/dt = i_R^2 R = v^2/R. \tag{5–21}$$

The rate at which energy is received by the system (electrical power from the connecting leads) is given by:

$$d\mathcal{E}_i/dt = iv. \tag{5–22}$$

The principle of conservation of energy (as applied in this case in which no matter is being converted to energy) states that

power supplied = power into storage
+ power dissipated,

or

$$d\mathcal{E}_i/dt = d\mathcal{E}_s/dt + d\mathcal{E}_d/dt. \tag{5–23}$$

Substitution from Eqs. (5–20), (5–21), and (5–22) yields

$$iv = Cv\frac{dv}{dt} + v^2/R.$$

Dividing all terms by v and rearranging gives

$$C\frac{dv}{dt} + \frac{v}{R} = i, \tag{5–24}$$

which agrees with Eq. (5–5). In this instance, the energy method is about as simple as the network method. This is not always true, however, especially in work with more complex systems. In addition to providing a systematic means of analyzing a system, the network method involves the use of a network diagram or linear graph which exposes the structure of the system and reveals analogies which exist between different systems.

Another method which involves the use of energies and coenergies is called *Hamilton's principle* and allows the formation of the LaGrange equations which have been applied most often in classical dynamics. These methods can be extended to all types of systems, and the interested reader is referred to Reference 5, which completely develops and applies this approach.

5–3.4 Initial Conditions

Before considering the solution of system equations, we should point out that the system equation alone is not sufficient to obtain a solution. For instance, repeating the equation for the parallel resistance-capacitance system of Fig. 5–1,

$$C\frac{dv}{dt} + \frac{v}{R} = i, \tag{5–5}$$

we see that if i is a prescribed function of time, then the solution consists of determining v as a function of time. To do this, we must know the value of v at $t = 0$ (for physical reasons, a value of v is usually known at $t = 0$; however, mathematically it is only necessary to know v at any finite time). The specification of this value describes the state of this system before the application of the current source i. By giving this value, we are essentially stating the amount of energy stored in the system initially (since $\mathcal{E}_e = \frac{1}{2}Cv^2$).

Let us consider a situation for which $v = 0$ before the current source is applied (zero initial energy storage). Since $i_C = C(dv/dt)$, the voltage v cannot change instantly at $t = 0$ unless i_C were infinite (since an instantaneous change in v means that dv/dt would be infinite at that time). Unless the current source is infinite, i_C will not be infinite; therefore, v will not change instantly when a finite current source is applied.

To properly discuss the initial conditions for all the systems which have been presented, it will help if we categorize the system equations. It can be noted that there are two different types of system equations which have been derived. In one case, the right-hand side contains the input itself, whereas in the other case, the right-hand side contains the derivative of the input. The equations are grouped in Table 5–1 in this manner.

Let us consider the systems in the left-hand column for which the general form of the system equation is

$$a_1\frac{dx_r}{dt} + a_0x_r = b_0x_f. \tag{5–25}$$

For these cases it is noted that the system output is the variable by virtue of which energy is stored. The *RC*- and *bm*-systems contain *A*-type elements, and the across-variable is the output. The *RL*- and *bk*-systems contain *T*-type elements, and the through-variable is the output.

For all the situations in the left-hand column of Table 5–1, an argument similar to that made above for the parallel *RC*-circuit will yield similar results.

TABLE 5–1

System Equations and Initial Conditions

Finite input: Systems initially relaxed, $x_r = 0$ at $t = 0-$

General form $a_1 \frac{dx_r}{dt} + a_0 x_r = b_0 x_f$ $x_r = 0$ at $t = 0+$	General form $a_1 \frac{dx_r}{dt} + a_0 x_r = b_1 \frac{dx_f}{dt}$ $x_r = \frac{b_1}{a_1} x_f \Big\|_0$ at $t = 0+$
Parallel *RC* $C \frac{dv}{dt} + \frac{v}{R} = i$ $v = 0$ at $t = 0+$	Series *RC* $R \frac{di}{dt} + \frac{i}{C} = \frac{dv}{dt}$ $i = \frac{v\|_0}{R}$ at $t = 0+$
Parallel *bm* $m \frac{dv}{dt} + bv = F$ $v = 0$ at $t = 0+$	Series *bm* $\frac{1}{b} \frac{dF}{dt} + \frac{F}{m} = \frac{dv}{dt}$ $F = bv\|_0$ at $t = 0+$
Series *RL* $L \frac{di}{dt} + Ri = v$ $i = 0$ at $t = 0+$	Parallel *RL* $\frac{1}{R} \frac{dv}{dt} + \frac{v}{L} = \frac{di}{dt}$ $v = Ri\|_0$ at $t = 0+$
Series *bk* $\frac{1}{k} \frac{dF}{dt} + \frac{F}{b} = v$ $F = 0$ at $t = 0+$	Parallel *bk* $b \frac{dv}{dt} + kv = \frac{dF}{dt}$ $v = \frac{F\|_0}{b}$ at $t = 0+$

That is, if the system is relaxed (no energy stored) before the input is applied (called $t = 0$), then the output will not change instantly when a finite input is applied; therefore the output is zero just after (called $t = 0+$) the application of a finite input. The various results on initial conditions are also shown in Table 5–1. The reader should check these conclusions by using a physical argument (like that used for the *RC*-circuit) on each system.

The situation presented by those systems listed in the right-hand column is different, however. The general form of the system equation there is

$$a_1(dx_r/dt) + a_0 x_r = b_1(dx_f/dt). \tag{5–26}$$

For these cases it is noted that the equations all have a derivative of the input and that if the input changes instantly (a suddenly applied source), this term will be infinite and one should expect the output to change also. In all cases, the output is *not* the variable by virtue of which energy is stored. The *RC*- and *bm*-systems contain *A*-type elements, but the through-variable is the output. The *RL*- and *bk*-systems contain *T*-type elements, but the across-variable is the output.

For these systems, although it is still true that the variable by virtue of which energy is stored cannot change instantly, the initial conditions must specify another variable. For example, in the series *RC*-circuit we have $v_C = 0$ at $t = 0+$, but the initial condition must specify i at $t = 0+$. Using the compatibility equation (5–6) and the resistance elemental equation (5–8), we have

$$v = v_R + v_C = iR + v_C,$$

$$v_C = 0 \quad \text{at} \quad t = 0+,$$

$$i = \frac{v|_0}{R} \quad \text{at} \quad t = 0+$$

where

$$v|_0 = v \quad \text{at} \quad t = 0+.$$

A similar reasoning process can be followed for the remaining systems, and the results given in Table 5–1 are obtained. Note that *this type of performance is found for the systems in the left-hand column for other outputs.* For instance, if in the parallel *RC*-circuit, the current in *C* is selected as the output, an equation like those in the right-hand column is obtained. Likewise, if in the series *RC*-circuit, the voltage across *C* is chosen as an output, an equation like those in the left-hand column is obtained.

It is important to emphasize that the reader should not attempt to memorize all the foregoing results for either system equations or initial conditions. They have been listed in tabular form for convenience of discussion only. The results are sufficiently simple that they can be easily derived when needed. The reader should appreciate the development and physical reasoning behind each result.

The information essential to the development of the above arguments is the behavior of the various elements when a sudden change occurs somewhere in the system. We saw earlier that the voltage across a capacitance cannot change instantly, and a similar argument will show that the *current* in an *inductance* also cannot change instantly. The following summarizes the elemental behavior with respect to sudden changes.

a) The across-variable v of a generalized capacitance (mass and electrical, fluid and thermal capacitances) cannot change instantly without the application of an infinite through-variable f. This element can therefore be described quantitatively as one which tends to oppose *changes* in v.

b) The through-variable f of a generalized inductance (spring, inductance, and inertance) cannot change instantly without the application of an infinite across-variable v. This element therefore tends to oppose *changes* in f.

c) The generalized resistance (damper and electrical, fluid, and thermal resistance) is neutral to *changes* in its variables. The energy-storage elements and sources in the system will determine whether any instantaneous changes will occur in these variables.

5–4. SOLUTION OF SYSTEM EQUATIONS

To this point we have considered various types of systems, the determination of the system equations, and initial conditions. We have seen that if the input to the system is the variable which appears on the left-hand side of the system equation given in (5–25) or (5–26), then the determination of the output is relatively straightforward, as was shown in Fig. 5–9. We will now consider methods of determining the solution to the more difficult problem presented when the input is the variable on the right-hand side of the system equation.

All the methods we will consider here are used by workers in the field. In the remainder of this book, however, we will deal with the analog computer and mathematical methods. The graphical and numerical methods are presented for completeness and will also serve to introduce the student to the solution of differential equations and the physical significance of this solution. It will be seen that each method will have its advantages and disadvantages, and the engineer must develop the judgment, through experience, to make the proper choice for any particular problem.

It is first necessary to define what is meant by a solution. The system equation relates the output of the system to the input. Upon specification of the type of input (i.e., what time variation the input has), it is then desired to determine what time variation the output has. A *solution* to the system equation, then, gives the output as an explicit function of time which corresponds to the given input function.

The following methods will be presented primarily in terms of a single equation, that for the parallel *RC*-circuit, which is repeated here:

$$C\frac{dv}{dt} + \frac{v}{R} = i; \qquad v = 0 \quad \text{at} \quad t = 0+. \tag{5–5}$$

It will sometimes simplify the discussion to rewrite this equation in the following form:

$$\tau\frac{dv}{dt} + v = v_f; \qquad v = 0 \quad \text{at} \quad t = 0+, \tag{5–27}$$

where $v_f = Ri$ (volts) and $\tau = RC$ (seconds).

5–4.1 Graphical Solution

To perform a graphical solution of a differential equation, we must separate the variables in such a manner that the resulting equation can be written in the form of integrals. The integrals are then evaluated graphically by using the concept that the integral is the area under the curve of the integrand. The area is obtained by plotting the curve and counting squares.

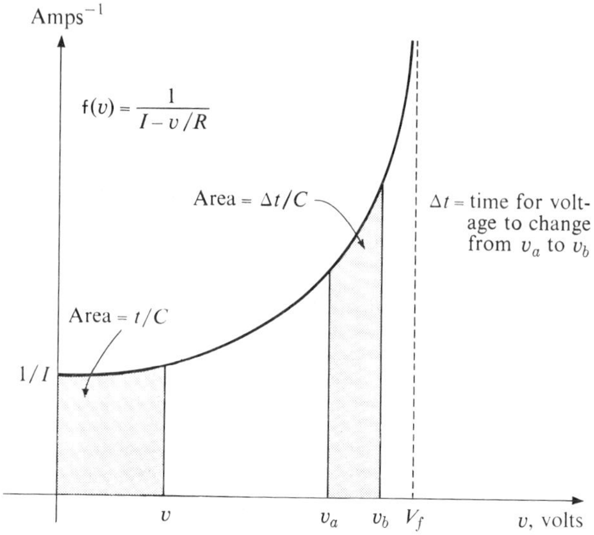

FIG. 5–12. Method for graphical solution of Eq. (5–5).

This process will be illustrated with the sample equation (5–5). Although (5–27) (which would give a more general solution) could be used, the use of graphical solutions in general requires the assignment of numerical values to all quantities. Hence we will work with (5–5) in that fashion.

Let us consider the situation where a suddenly applied constant current source is the input to the system:

$$i = 0 \qquad t < 0,$$
$$i = I \qquad t > 0.$$

$C\frac{dv}{dt} + \frac{v}{R} = i$

Such an input is called a *step* function and is a common disturbance. Since v_f is proportional to i, it will also change in this manner.

It is desired to determine $v(t)$ for $t > 0$, given that $v = 0$ at $t = 0+$ and $i = I$ for $t > 0$. The equation can then be written

$$C\frac{dv}{dt} + \frac{v}{R} = I,$$

or

$$C\frac{dv}{dt} = I - \frac{v}{R}. \tag{5–28}$$

Since the left-hand side of (5–28) is a function only of v (and not of t), it is possible to separate the variables by multiplying through by the differential and dividing through by $I - v/R$. Thus

$$\frac{dv}{I - v/R} = \frac{dt}{C},$$

and

$$\int_0^v \frac{dv}{I - v/R} = \int_0^t \frac{dt}{C}, \tag{5–29}$$

where the information that $v = 0$ when $t = 0$ has been used for the lower limits. It should be noted that if i was a function of t, rather than a constant, it would not have been possible to separate the variables, and the following graphical solution could not be used.

The right-hand side of (5–29) can be integrated by inspection and becomes t/C. The left-hand side of (5–29) can be interpreted as the area under a curve of $f(v) = 1/(I - v/R)$. Then

$$\int_0^v f(v)\,dv = t/C, \tag{5–30}$$

where

$$f(v) = \frac{1}{I - v/R}.$$

By choosing values of v between 0 and $V_f = IR$ and computing $f(v)$ for each of these values, we can plot a curve of $f(v)$ vs. v (Fig. 5–12). The integral in Eq. (5–30), then, is the area under this curve between 0 and any point v of interest. Equation (5–30) states that this area is equal to t/C. Therefore, once the area is known for any v, the time t at which this v occurs can be easily computed. If a number of (v, t) points are found, the response can be presented by a plot of v vs. t.

A sample plot is shown on Fig. 5–12. The area from 0 to v is equal to t/C, where t is the time measured from $t = 0$ for v to change from 0 to v. Another area is shown between v_a and v_b, from which the time increment for the voltage to change between these values can be found. This process is equivalent to integrating between the limits of v_a and v_b in Eq. (5–29).

A number of things should be noted from Fig. 5–12.

a) It will take longer for a given voltage change to occur as time passes, or conversely, a given time increment will result in a smaller voltage change as time passes. This can be seen from Fig. 5–12, where the two areas shown are about equal but the change in voltage from v_a to v_b is much smaller than the change in voltage from 0 to v.

b) The area under the curve from 0 to V_f is infinite. This shows that it will take an infinite time for v to reach V_f after the application of a constant current source. The truth of this result can be seen from Eq. (5–28), which shows that the closer v is to V_f, the smaller is the rate at which v is changing (dv/dt), and therefore, v can never reach V_f. Physically, this means that all the current starts flowing through the capacitance, and thus builds up the charge and voltage on the capacitance quickly. However, the current in the resistance then increases, which must result in a decrease of the capacitance current, since the sum of the two currents is equal to the constant input current. When the capacitance current decreases, the rate at which charge and voltage on the capacitance increase is reduced. The closer v gets to V_f, the more nearly the resistance current equals the input and the smaller the capacitance current becomes. The capacitance will, therefore, never get completely charged up to V_f.

The required areas are most easily found by counting squares on fine graph paper. It may be necessary to plot expanded-scale curves over some regions. Figure 5–13 shows the final result of v vs. t as actually determined by the graphical calculation.

The example used is one in which it is not necessary to use graphical techniques, since the integral in Eq. (5–30) can be evaluated mathematically, and this is done in Section 5–4.4. This example was used here, however, to give a feeling for the meaning of the solution of this equation.

However, the graphical solution may be applied to nonlinear equations which do not lend themselves to mathematical solution. For instance, the following equations can be solved by the same procedures just described, since the variables are separable.

$$a\left(\frac{dv}{dt}\right)^m + bv^n = c, \qquad (5\text{–}31)$$

$$n = \text{any constant}, \quad m = \text{any constant},$$

or

$$\frac{dv}{[(c - bv^n)/a]^{1/m}} = dt,$$

$$a\left(\frac{dv}{dt}\right)^m + \mathsf{f}(v) = c, \qquad (5\text{–}32)$$

$\mathsf{f}(v)$ = any function, and can be given graphically

or

$$\frac{dv}{[(c - \mathsf{f}(v))/a]^{1/m}} = dt.$$

The disadvantages of this method of solution are

a) can be applied only to variables-separable equations,

b) accuracy limited by graph size,

c) tedious.

5–4.2 Numerical and Digital-Computer Solution

The availability of the high-speed electronic digital computer in recent years has caused a rebirth of

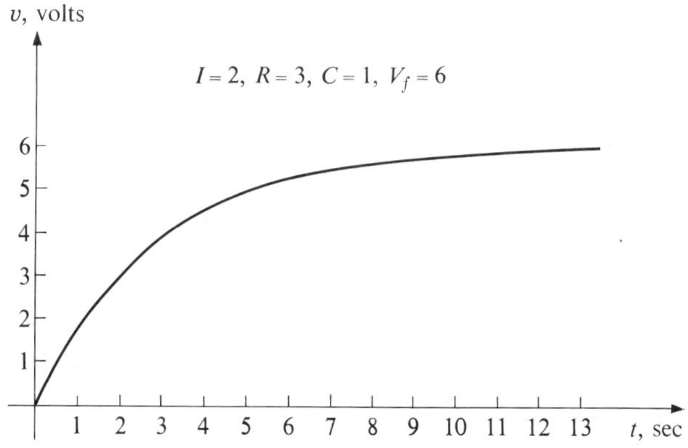

FIG. 5–13. Response by graphical solution.

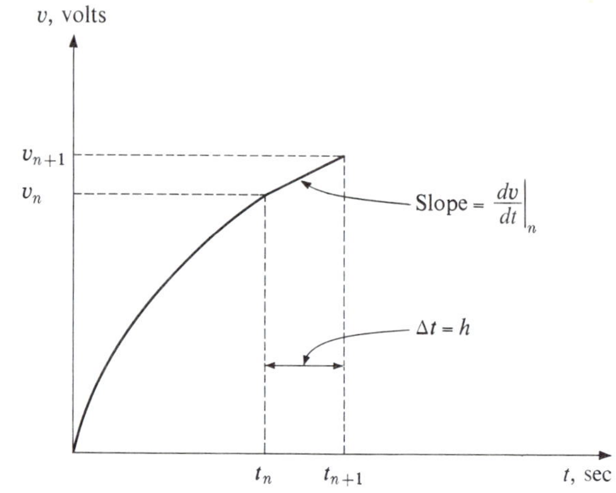

FIG. 5–14. Approximation by straight-line segments.

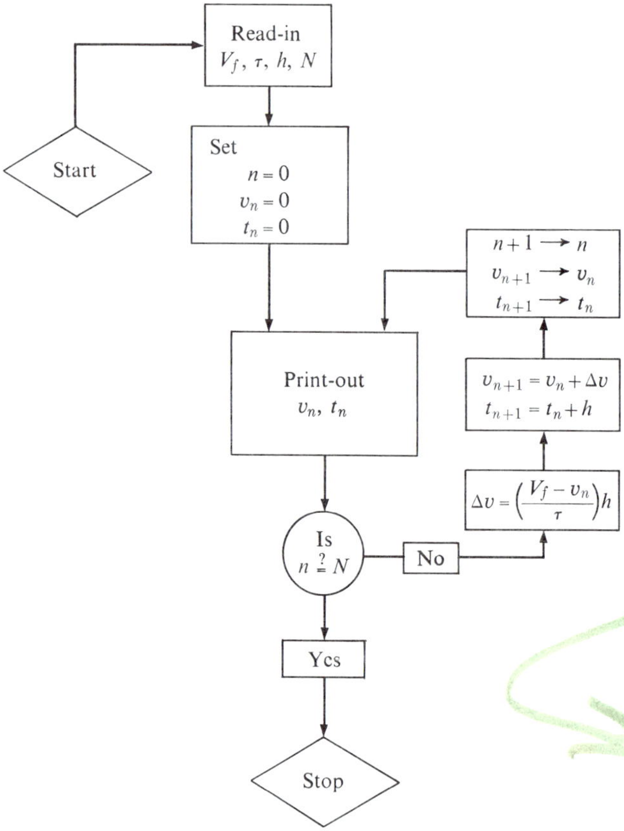

FIG. 5–15. Digital computer diagram for N increments.

numerical methods. Although any of these methods may be used with the slide-rule and/or the desk calculator, the digital computer has made them more attractive since the tedium has been eliminated (replaced by programming), and also, since answers can be obtained rapidly, problems which are so difficult that they would never be undertaken with the desk calculator can be done in a relatively short time on the digital computer.

The theory and application of numerical methods are highly developed, and the interested reader is referred to Reference 6. The approach which will be used here is most elementary but will also serve to give significance to the meaning of a solution of the differential equation.

Let us consider Eq. (5–27), rewritten in revised form as

$$\frac{dv}{dt} = \frac{V_f - v}{\tau}. \tag{5–33}$$

If v is known at any t, then dv/dt at that t may be determined by simple substitution. If it is considered that the slope will stay approximately constant for short time intervals, the curve of v vs. t can be considered to be made up of short segments of straight lines (Fig. 5–14). Extending the straight line over a short time interval will determine another value of v. The new value of v may be substituted into Eq. (5–33) to compute a new slope, and then the process can be repeated.

This process may be described in the following way. Suppose that v at some t, say t_n, is known; this v is called v_n. The slope at this point is then

$$\left.\frac{dv}{dt}\right|_n = \frac{V_f - v_n}{\tau}, \tag{5–34}$$

and the next v is v_{n+1},

$$v_{n+1} = v_n + \left.\frac{dv}{dt}\right|_n \Delta t, \tag{5–35}$$

where $\Delta t = t_{n+1} - t_n = h$ is some small increment of time. Substituting Eq. (5–34) into (5–35) yields the

5-4 | SOLUTION OF SYSTEM EQUATIONS

TABLE 5-2
Accuracy of Numerical Solution, Values of $v(t)/V_f$ *h = some small increment of time Δt*

t/τ	Exact $h/\tau = 0$	$h/\tau = 0.01$	$h/\tau = 0.1$	$h/\tau = 0.5$	t/τ	Exact $h/\tau = 0$	$h/\tau = 1.5$	$h/\tau = 2$	$h/\tau = 3$
1.0	0.632	0.634	0.651	0.750	1.5	0.778	1.500	—	—
2.0	0.865	0.866	0.879	0.938	2.0	0.865	—	2.000	—
3.0	0.950	0.951	0.958	0.984	3.0	0.950	0.750	—	+3.000
4.0	0.982	0.982	0.985	0.997	4.0	0.982	—	0.000	—
5.0	0.993	0.993	0.995	0.999	4.5	0.989	1.125	—	—
6.0	0.998	0.998	0.999	1.000	6.0	0.998	0.938	2.000	−3.000
					7.5	0.999	1.031	—	—
					9.0	1.000	0.984	—	+9.000
							converging oscillations	constant oscillations	diverging oscillations

Reasonable | Unreasonable

recursion formula

$h = t_{n+1} - t_n$

$$v_{n+1} = v_n + \left(\frac{V_f - v_n}{\tau}\right) h \quad (5\text{-}36)$$

Equation (5-36) may be used repeatedly on a desk calculator or may be programmed for a digital computer. The solution will then be computed as a series of points of t_n and v_n. Although little can be said here concerning the operation of a digital computer, it is of interest to show the process of computation on a flow diagram as in Fig. 5-15. This process will cause the basic computation to be repeated N times. The "printout" will be a table of v_n and t_n.

The accuracy of this method will obviously depend on how small a time increment $\Delta t = h$ is chosen. If h is too large, poor accuracy will be obtained (or maybe a meaningless solution if h is very large). If h is very small, accuracy will be high, but a large number of computations will be required to obtain the full solution, thus taking more computer time and costing more money.* The engineer must decide what accuracy is necessary and adjust h accordingly. The references on numerical methods listed at the end of this chapter investigate the accuracy of various approaches in depth. A good "experimental" way of determining whether one has sufficient accuracy is to perform the solution for an h which is believed to be sufficiently small. Then the solution is repeated with h taken as one-half the earlier value. If the values of v do not change by more than the accuracy requirement, then the first h was sufficiently small. If they

---

* There is a limit to the improvement of accuracy, since "round-off" errors are introduced by the dropping of the least significant digits when the number of digits is limited. See Reference 6 for a complete discussion.

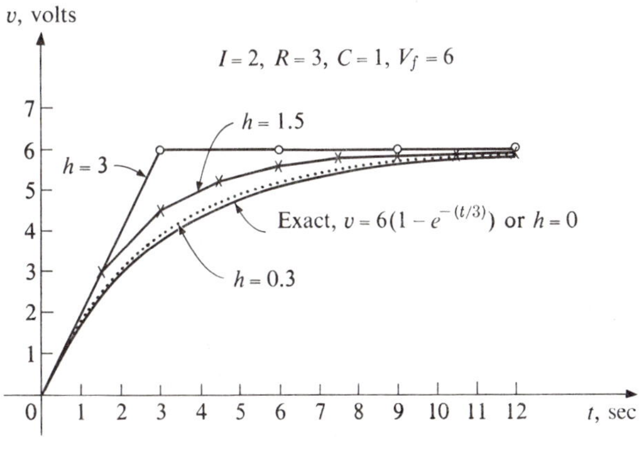

FIG. 5-16. Response by numerical solution.

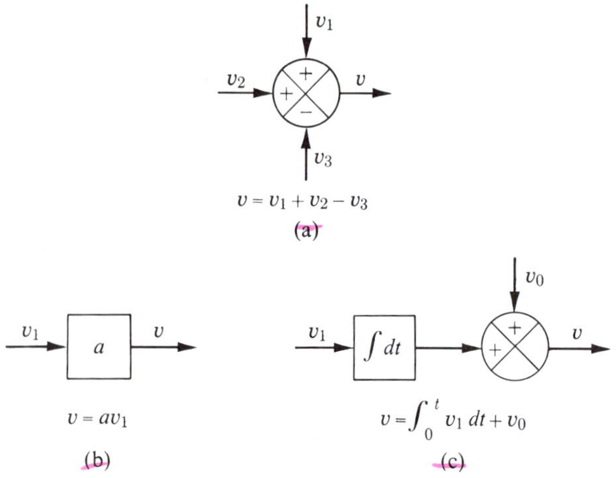

FIG. 5–17. Some operational blocks. (a) Summation. (b) Multiplication. (c) Integration.

do change, then repeat the process with h cut in half again, etc.

Figure 5–16 shows the results of numerical solution for various choices of h. It is easily seen that h must be small (compared with the time it takes the system to have a significant change in output) to obtain an accurate solution, that is, h must be small compared to τ. Table 5–2 shows the results of the numerical calculation for various choices of h/τ. It should be observed that large values of h/τ give completely unreasonable results. (See Prob. 5–18.)

5–4.3 Operational Block Diagram and Analog Computer Solution

Operational block diagrams are useful for representing the interrelationships which may occur in a dynamic system. An *operational block* is a symbol which represents a particular mathematical operation. Figure 5–17 shows the operational block representations of addition, multiplication by a constant, and integration with respect to time. For integration, the use of an initial condition in the manner shown makes it possible to integrate from a given starting time at which the output of the integrator itself is set at zero.

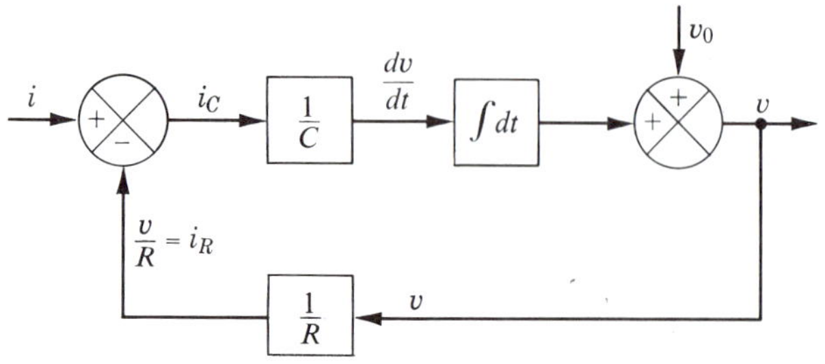

FIG. 5–18. Operational block diagram for parallel RC-circuit.

When operational block diagrams are used, the equations are best left in fundamental form. The strength of this approach is that it shows the physical significance of the various inputs and outputs. Thus we will work with Eq. (5–5).

An operational block diagram for Eq. (5–5) which represents the resistance-capacitance circuit of Fig. 5–1 can be constructed in the following way. Equation (5–5) is first rewritten with the highest derivative by itself on one side of the equation:

$$C\frac{dv}{dt} = i - \frac{v}{R}. \tag{5–37}$$

The quantity $C(dv/dt)$, which is equal to i_C, is obtained by summing i and $-v/R$. Then dv/dt is obtained by division of $C(dv/dt)$ by C, and v is obtained by integration of dv/dt. The value of v thus obtained is used to form v/R, which is used in the summation. Figure 5–18 shows the complete block diagram, which relates the output v to the input current i.

Each of the systems listed in the left-hand column of Table 5–1 can be represented by the block diagram of Fig. 5–18, provided that the proper substitutions are made for the constants R and C and the variables i and v.

The series RC-circuit is represented by a block diagram (Fig. 5–19). The reader should derive this from the Eqs. (5–6), (5–7), (5–8), and (5–9) [rather than use Eq. (5–10), since the input appears as a derivative there]. All the systems in the right-hand column of Table 5–1 can be represented by Fig. 5–19, provided that the proper substitutions are made.

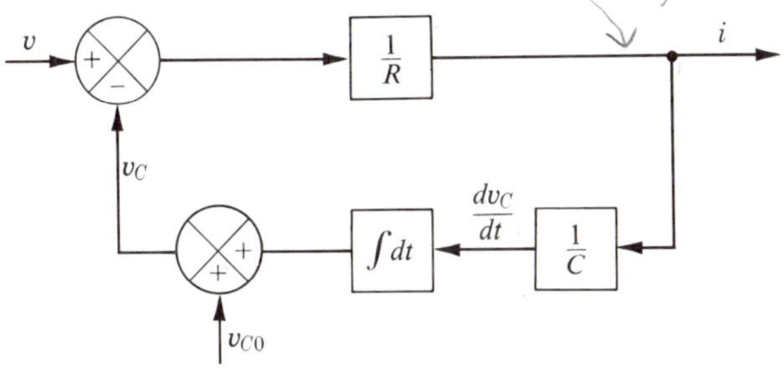

FIG. 5–19. Operational block diagram for series *RC*-circuit.

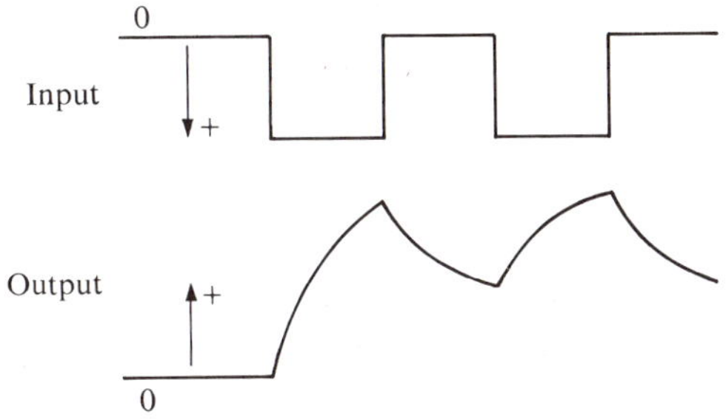

FIG. 5–20. Analog computer results.

The various blocks in the operational diagram can usually be related to the physical phenomena which take place in the actual system. For example, the summation in the block diagram of Fig. 5–18 represents the continuity equation for the circuit, the integration represents the buildup of voltage on the capacitor due to influx of current, and the quantity v/R represents the current flow through the resistor. Examination of the block diagram shows that for a constant input current i, the voltage v will build up continuously until the current v/R through the resistance just equals the input current. At this point dv/dt will be zero and v will thereafter remain constant.

In addition to its usefulness for visualization of system performance, the block diagram is particularly noteworthy because it is possible to electronically perform all the operations indicated. A machine which makes it possible to easily perform these operations is called an *electronic analog computer*. Chapter 7 will discuss this computer in some detail, but it can be briefly described here as a device in which all the variables are represented by electrical voltages. Amplifiers, potentiometers, and electrical elements are used to perform the indicated operations.

The input is caused to vary in the same way as the system input (by means of an electronic signal generator), and the output is recorded on an oscillograph or cathode-ray oscilloscope. This method of solution is particularly suited for situations in which the input is a complex function or for which a number of parameter (R and C in our examples) values need to be investigated. Nonlinear systems may also be investigated with facility since it is possible to perform nonlinear operations (such as multiplication of two variables) electronically.

For the parallel *RC*-circuit of Fig. 5–18, the results shown in Fig. 5–20 were obtained on an analog computer. The input was a square wave, and the corresponding output is shown. It is possible to mathematically determine this result for this system; however, even in the case of a simple *RC*-circuit, the calculations would be lengthy for an input as complex as a square wave.

5–4.4 Mathematical Solution

A mathematical solution is in general most desirable since it allows a straightforward investigation of the effect of different values of system parameters. However, the mathematical approach is often intractable since it requires an inordinate amount of effort. In many cases (nonlinear, particularly) there is no known approach for solutions in general, and each situation must be treated in a special way. Many nonlinear problems cannot be solved analytically (but will yield to analog, numerical, or graphical methods).

In the case of linear differential equations, particularly with constant coefficients, however, the mathematical approach has been most successful. The mathematical theory of linear systems is highly developed and allows one to make great generalizations. The results of linear analysis allow for extended understanding of nonlinear phenomena also, and thus linear theory forms the basis for system analysis. A major part of the remainder of this book is devoted to the appreciation of the response of linear systems. In this chapter, simple approaches will be used to introduce the reader to the types of results of interest.

Let us consider the prototype equation (5–27),

$$\tau \frac{dv}{dt} + v = v_f; \qquad v = 0 \quad \text{at} \quad t = 0+, \tag{5–27}$$

for the situation of a suddenly applied constant current source (step function):

$$\begin{aligned} i &= 0, \quad v_f = 0, && \text{for} \quad t < 0, \\ i &= I, \quad v_f = IR = V_f, && \text{for} \quad t > 0. \end{aligned} \tag{5–38}$$

For this case the variables may be separated (in the same fashion as in Section 5–4.1) to give

$$\tau \frac{dv}{dt} = V_f - v, \qquad \text{or} \qquad \frac{dv}{V_f - v} = \frac{dt}{\tau}$$

and

$$\int_0^v \frac{dv}{V_f - v} = \int_0^t \frac{dt}{\tau} = \frac{t}{\tau}. \tag{5–39}$$

The left-hand side of Eq. (5–39) may be integrated to give a natural logarithm:

$$-\ln (V_f - v)\Big|_0^v = \frac{t}{\tau},$$

or

$$\ln V_f - \ln (V_f - v) = \frac{t}{\tau}$$

and

$$\ln \left(\frac{V_f - v}{V_f}\right) = -\frac{t}{\tau}. \tag{5–40}$$

Hence

$$\frac{V_f - v}{V_f} = e^{-t/\tau},$$

or

$$v = V_f(1 - e^{-t/\tau}), \tag{5–41}$$

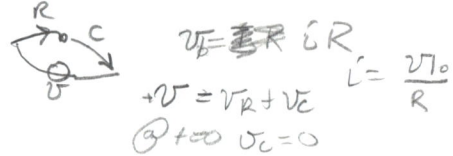

which is the desired solution giving v as a function of t. This is the equation used to plot the exact solution in Fig. 5–16.

It is also easy to use this method to obtain a solution for the other type of differential equation which describes the systems in the right-hand column of Table 5–1. Taking the series RC-circuit as an example, we have

$$R \frac{di}{dt} + \frac{i}{C} = \frac{dv}{dt}; \qquad i = \frac{v|_0}{R} \quad \text{at} \quad t = 0+. \tag{5–42}$$

For a step input in v (a suddenly applied constant voltage source, such as a battery connected to the circuit by the closing of a switch),

$$\begin{aligned} v &= 0 \qquad t < 0, \\ v &= V \qquad t > 0, \end{aligned} \tag{5–43}$$

and for $t > 0$,

$$R \frac{di}{dt} + \frac{i}{C} = 0; \qquad i = \frac{V}{R} \quad \text{at} \quad t = 0+,$$

or

$$\tau \frac{di}{dt} + i = 0; \qquad i = \frac{V}{R} \quad \text{at} \quad t = 0+; \qquad \tau = RC. \tag{5–44}$$

Proceeding as before, separating the variables, we have

$$\int_{V/R}^i \frac{di}{i} = -\int_0^t \frac{dt}{\tau} = -\frac{t}{\tau},$$

$$\ln i\Big|_{V/R}^i = \ln i - \ln \frac{V}{R} = \ln \left(\frac{iR}{V}\right) = -\frac{t}{\tau},$$

or

$$i = \frac{V}{R} e^{-t/\tau}, \tag{5–45}$$

which is the desired solution for the current in the series RC-circuit for a step voltage input. The significance of the solutions given in Eqs. (5–41) and (5–45) will be discussed in Section 5–5.1.

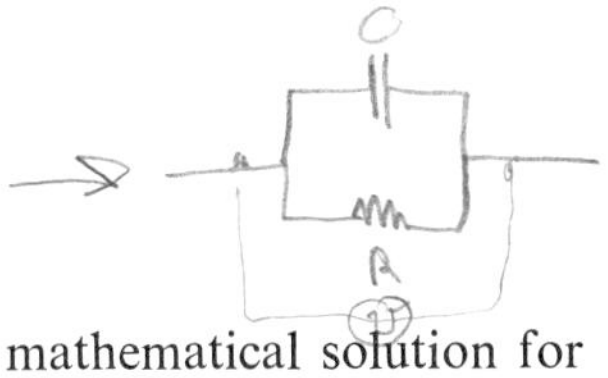

Let us now consider a mathematical solution for another type of input of interest—the sinusoid. If a steady sinusoidal excitation has been applied to the parallel RC-circuit for a long time, then it would be expected that the system would be in a steady state; i.e., the system variables would be periodic and varying at the same frequency as the input.

In Eq. (5–27), let

$$v_f = V_f \sin \omega t = IR \sin \omega t. \tag{5–46}$$

Then Eq. (5–27) can be written

$$\tau \frac{dv}{dt} + v = V_f \sin \omega t. \tag{5–47}$$

The initial condition is not required since we have assumed that the input was applied a long time before $t = 0$ and the output has settled to a steady-state sinusoidal variation.

This equation cannot be solved by the separation of variables used for the step input because the right-hand side is now a function of t. However, since we have stipulated that the output is in a steady state, it is reasonable to assume that v will also vary sinusoidally at the same frequency (this is a general conclusion that can be applied to any linear system, as we will see in Chapter 11). It can also be seen that a simple sine function for the output will not satisfy Eq. (5–47) since the derivative would then be a cosine; neither will a cosine function satisfy Eq. (5–47). Thus the assumed solution will be

$$v = A \sin \omega t + B \cos \omega t, \tag{5–48}$$

where A and B are constants. Substituting Eq. (5–48) into the differential equation (5–47) gives

$$\tau A\omega \cos \omega t - \tau B\omega \sin \omega t + A \sin \omega t + B \cos \omega t = V_f \sin \omega t. \tag{5–49}$$

Equation (5–49) can hold for all time t if and only if the complete coefficients of the sine and cosine terms both vanish,

$$\tau A\omega + B = 0, \qquad -\tau B\omega + A = V_f. \tag{5–50}$$

Therefore, solving Eqs. (5–50) for A and B gives

$$A = \frac{V_f}{1 + \tau^2\omega^2}, \qquad B = \frac{-\tau\omega V_f}{1 + \tau^2\omega^2}, \tag{5–51}$$

and the assumed solution for v is a valid one for these values of A and B; therefore

$$v = \left(\frac{V_f}{1 + \tau^2\omega^2}\right) \sin \omega t - \left(\frac{V_f\tau\omega}{1 + \tau^2\omega^2}\right) \cos \omega t. \tag{5–52}$$

The sine and cosine functions in Eq. (5–52) may be combined into a single sine function by means of the trigonometric identity

$$D \sin (\theta - \phi) = D\,[\sin \theta \cos \phi - \cos \theta \sin \phi], \tag{5–53}$$

where

$$\theta = \omega t,$$

$$D \cos \phi = \frac{V_f}{1 + \tau^2\omega^2} = A,$$

$$D \sin \phi = \frac{\tau\omega V_f}{1 + \tau^2\omega^2} = -B.$$

The significance of these relations can be seen from the triangle of Fig. 5–21. Hence

$$\tan \phi = \tau\omega, \qquad D = \frac{V_f}{\sqrt{1 + \tau^2\omega^2}}, \tag{5–54}$$

and finally,

$$v = \frac{V_f}{\sqrt{1 + \tau^2\omega^2}} \sin (\omega t - \phi); \qquad \phi = \tan^{-1} (\tau\omega), \tag{5–55}$$

which is the desired solution for a sinusoidal input.

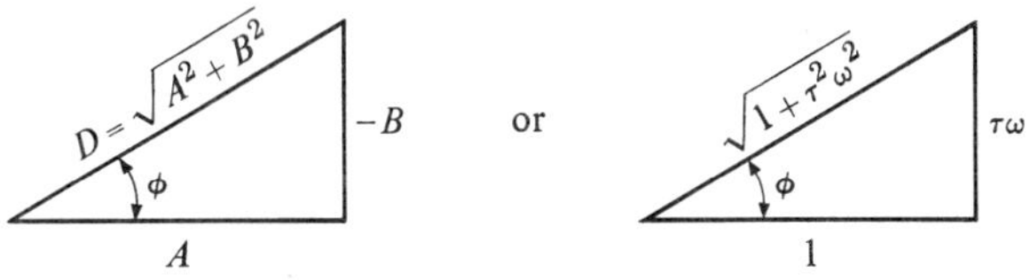

FIG. 5–21. Triangle used to combine sine and cosine functions.

TABLE 5–3
Summary of Solution Methods

$$C\frac{dv}{dt}+\frac{v}{R}=i,\quad v=0 \text{ at } t=0+$$

Graphical	Numerical	Analog	Mathematical
$i=I,\ t>0;\ \frac{t}{RC}=\int_0^v \frac{dv}{IR-v}$ 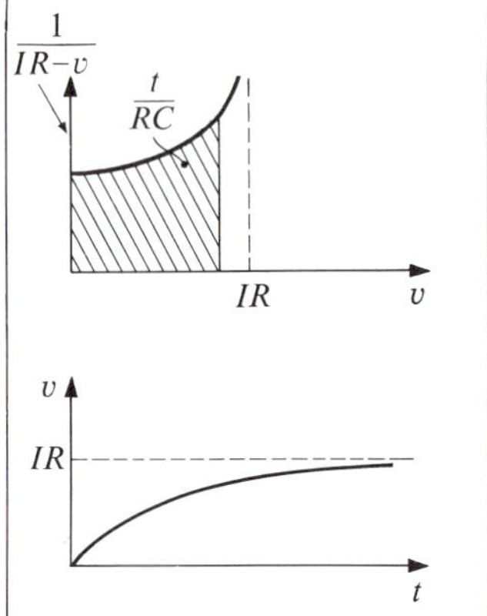	$i=I,\ t>0$ $v_{n+1}=v_n+\left(\frac{IR-v_n}{RC}\right)\Delta t$ 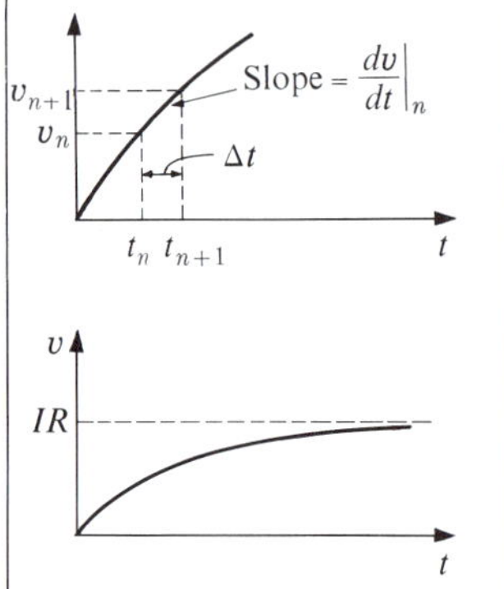	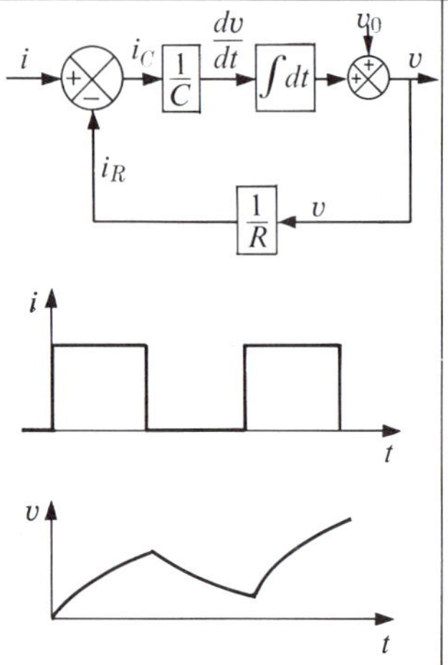	$i=I,\ t>0$ $\frac{t}{RC}=\int_0^v \frac{dv}{IR-v}$ $v=IR(1-e^{-t/RC})$ 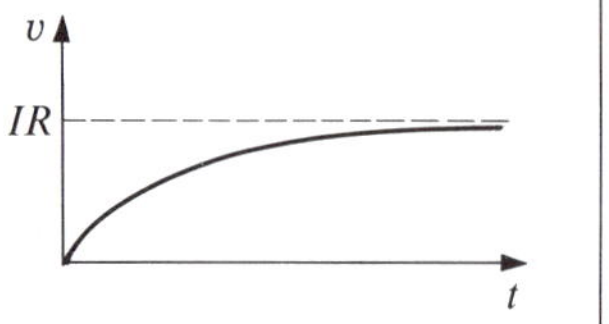
Advantages Handles nonlinear cases	Advantages Handles nonlinear cases Accurate if take small Δt Can be programmed for digital computer	Advantages Retains physical significance Handles nonlinear cases Handles difficult inputs Easy to investigate parameter variation	Advantages General solution Gives insight to general behavior
Disadvantages Accuracy limited by graph size Must be able to separate variables Must re-solve to investigate parameter variation	Disadvantages Must re-solve to investigate parameter variation	Disadvantages Accuracy limited by components	Disadvantages Cannot in general be used for nonlinear Limited to simplified situations

As we will see, the mathematical solution allows one to tell by simple inspection what the effect of a change in τ would be. It is suggested that the reader use the method just presented to solve Eq. (5–42) for a sinusoidal input.

5–4.5 Summary of Solution Methods

Table 5–3 summarizes the various solution methods which have been discussed and lists some advantages and disadvantages of each for comparison. The analog computer solution will be treated in Chapter 7 and other mathematical solutions will be discussed in the remainder of the book.

5–5. DYNAMIC BEHAVIOR OF LINEAR FIRST-ORDER SYSTEMS

The solutions for the prototype linear first-order system have been obtained. In this section we will investigate the significance of these solutions to familiarize the reader with the dynamic behavior of systems. The types of behavior investigated here are those which are of general interest and will serve to set the stage for the investigation of more complex systems throughout the remainder of the book.

The generality of these solutions will be shown by considering the various systems in the section on analogs and duals. Finally a brief introduction to the nature of stability will be given.

5–5.1 Transient Response to Step Input

The step, or sudden change, input from one constant to another, is of considerable interest in system analysis. Not only do many physical phenomena occur approximately in this fashion, but any arbitrary input function may be represented by a series of small successive changes, as shown in Fig. 5–22.

Figure 5–23 shows a unit step input which occurs at time $t = 0$. The amplitude of the step is unity, and the function is represented by $u_s(t)$. Steps of magnitude different from unity are represented by an amplitude times the unit step, i.e., $Au_s(t)$.

The step function $u_s(t)$ is described mathematically by

$$\begin{aligned} u_s(t) &= 0 \qquad t < 0, \\ u_s(t) &= 1 \qquad t > 0. \end{aligned} \tag{5–56}$$

Consider the parallel RC-circuit for which the system equation is Eq. (5–27):

$$\tau \frac{dv}{dt} + v = v_f; \qquad v = 0 \quad \text{at} \quad t = 0+, \tag{5–27}$$

$$v_f = IR \text{ (volts)}; \qquad \tau = RC \text{ (seconds)}.$$

With a step input, we have

$$v_f = V_f u_s(t) \tag{5–57}$$

and

$$\tau \frac{dv}{dt} + v = V_f u_s(t) \qquad v = 0 \quad \text{at} \quad t = 0+. \tag{5–58}$$

The solution for this case has been found and is given in Eq. (5–41), which is repeated here:

$$v = V_f(1 - e^{-t/\tau}). \tag{5–41}$$

When the step occurs, the voltage across the capacitance is zero and does not change instantly, the current through the resistance is zero, with the entire input current going through C. From Eq. (5–58), the initial rate of change of v is

$$\left.\frac{dv}{dt}\right|_0 = \frac{V_f}{\tau} = \frac{IR}{RC} = \frac{I}{C}, \tag{5–59}$$

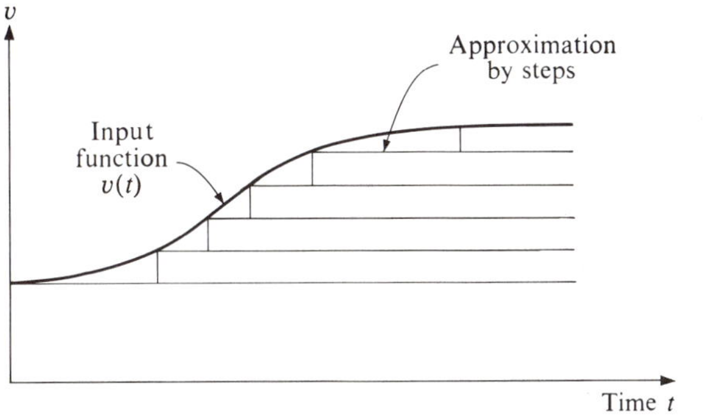

FIG. 5–22. Representation of a function by successive step changes.

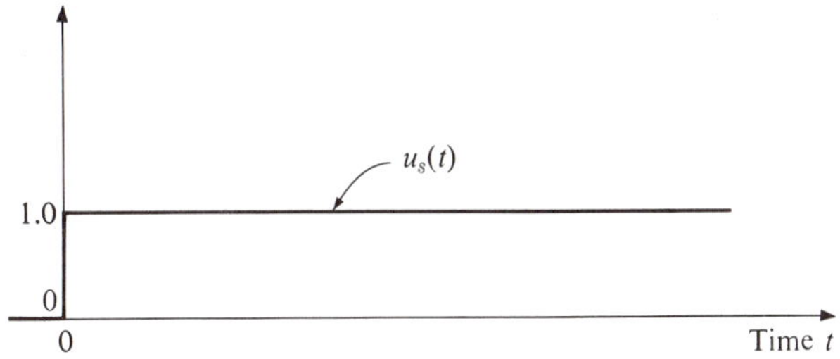

FIG. 5–23. Unit step function.

which agrees with the elemental equation for the capacitance when the entire input current is through C. As the voltage v increases, less current goes through C and more through R. Finally, the output v approaches a steady value so that $dv/dt = 0$. Equation (5–58) then gives the final value of v:

$$v_{t\to\infty} = V_f = IR, \tag{5–60}$$

which agrees with Eq. (5–41) as $t \to \infty$. This final value corresponds to the physical situation where the current in the capacitance is zero and all the input current goes through the resistance.

One can sketch the response curve without having the complete solution as given in Eq. (5–41) since the initial slope is V_f/τ, the final value is V_f, and Eq. (5–58) dictates that the slope must decrease as v increases:

$$\frac{dv}{dt} = \frac{V_f - v}{\tau}. \tag{5–61}$$

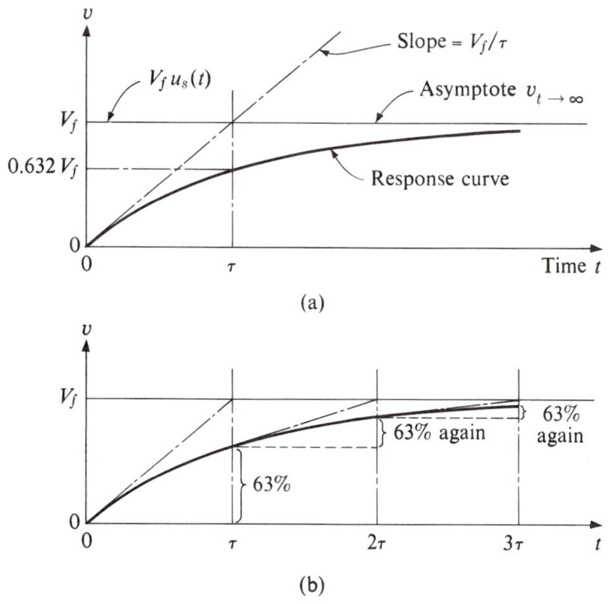

FIG. 5–24. System response. (a) Response to step function. (b) Construction of response curve.

Consequently the slope decreases monotonically from V_f/τ at $t = 0$ to zero at $t = \infty$.

Note that v is proportional to V_f at any time. This is a characteristic of a *linear* system. If an input which is twice as large is used, the resulting output at any time is doubled. The time response v/V_f is determined by the single parameter τ, which is called the *time constant* of the system. This value measures how fast the system responds to a disturbance; if τ is large, the system is sluggish and responds slowly; if τ is small the system responds rapidly. The time constants of dynamic systems will range from hours (for large thermal systems) to fractions of microseconds (for small electrical systems). The definite significance of the time constant is seen from Table 5–4, where the ratio of output to final value is shown for various times. It is seen from this table and Fig. 5–24(a), that although it takes an infinite time for the output to reach exactly the final value, it gets very close to the final value in a time equal to a few time constants. The value of four time constants ($t = 4\tau$) is frequently used as the time at which

TABLE 5–4

Output of First-Order Step-Input Systems at Various Times

t	$v/V_f = 1 - e^{-t/\tau}$
0	0
τ	0.632
2τ	0.865
3τ	0.950
4τ	0.982
5τ	0.993
6τ	0.998

the variation is essentially over, since it is seen from Table 5–4 that less than 2% of the change remains to occur. *It should be remembered that the output of a first-order system is within 2% of the final value in four time constants*, and this time will be used as a standard of the time it takes for a change to occur.

It is easy to sketch an exponential response by using the following steps: (a) the initial slope is such that if the curve continued at that slope, it would reach the final value in a time τ; (b) the actual change in time τ is about $\frac{2}{3}$ the total change (actually 63.2%); (c) the slope at τ is again that which would cause the curve to reach its final value in another time τ from the value at one τ; (d) the actual change is again about $\frac{2}{3}$ the remaining change, etc. This process is continued for about 4τ. The constant percent change for constant time increments is a property of the exponential function. Figure 5–24(b) shows this sketching.

Since all the variables of our elementary systems are described by first-order equations, they will all vary in exponential fashion but some will go up and others down, depending on the variable and its meaning in the system. Figure 5–25 shows the variation of all system variables for the parallel *RC*-circuit and the series *RC*-circuit for step inputs. It will be recalled that the solution for the series *RC*-circuit was obtained in Eq. (5–45). The reader should check all these responses, using the compatibility, continuity, and elemental equations and the concept of exponential response of first-order systems as exemplified by the time constant.

5-5 | DYNAMIC BEHAVIOR OF LINEAR FIRST-ORDER SYSTEMS   127

Parallel RC, Current input	Series RC, Voltage input
Input: step I i_C decay i_R rising exponential v rising exponential to IR	Input: step V v_R decay v_C rising exponential i decay from V/R
$i = i_C + i_R$ $i_R = v/R$ $i_C = C\dfrac{dv}{dt}$ $\tau = RC$	$v = v_C + v_R$ $i = v_R/R$ $i = C\dfrac{dv_C}{dt}$ $\tau = RC$

FIG. 5-25. Response of RC-circuits to step inputs.

$v = v_R + v_C$

$\dfrac{dv}{dt} = \dfrac{dv_R}{dt} + \dfrac{dv_C}{dt}$

$\dfrac{dv}{dt} = \dfrac{di}{dt}R + \dfrac{i}{C}$

$v_0 = IR; \ \dfrac{dv}{dt} = 0$

$\dfrac{di}{i} = -\dfrac{dt}{RC} = -\dfrac{dt}{\tau}$

$\ln(i)\Big|_{V/R}^{i} = -\dfrac{t}{\tau}$

$\ln(i) - \ln(V/R) = -\dfrac{t}{\tau}$

$\ln\left(\dfrac{iR}{V}\right) = -t/\tau$

$i = \dfrac{V}{R} e^{-t/\tau}$

$i = \dfrac{V}{R} e^{-t/\tau}$

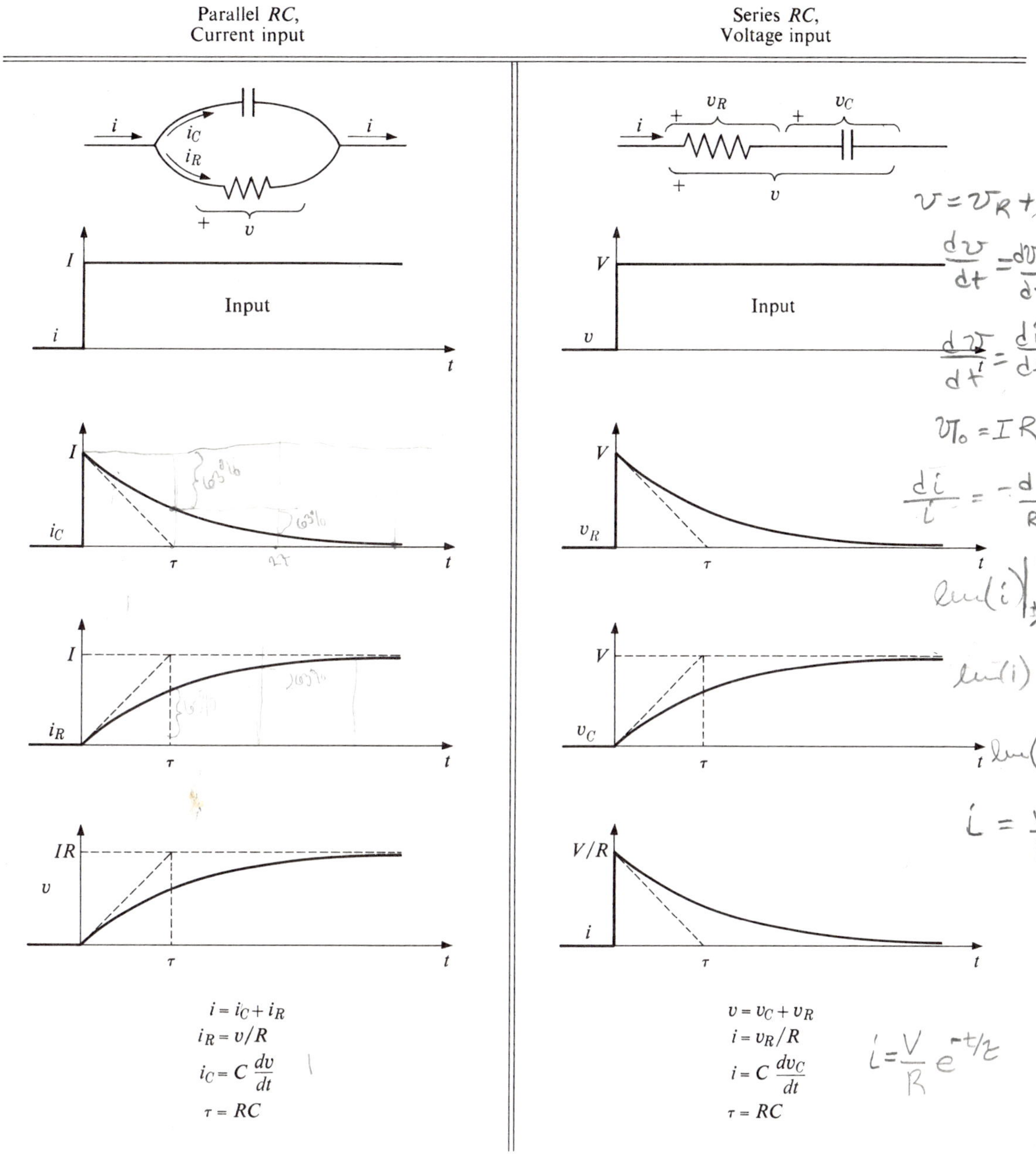

We have seen that applying a step change in the input of a system results in a time response which reveals considerable information or data about the system. Thus the response to a step input, sometimes referred to as *indicial response* (for a unit step) may be used to dynamically characterize the system. It does not tell of what elements the system consists, but from the point of view of an observer concerned only with the input and output, it does give all the necessary information about the system. In fact, some workers in the field tend to think about different systems primarily in terms of their indicial responses.

5–5.2 Steady-State Response to Sinusoidal Input

Many inputs to systems change periodically. Examples are reciprocating engines, diurnal temperature variation, ac (alternating current) electrical power, etc. It is thus of interest to investigate the response of dynamic systems to periodic inputs. The sinusoid is the simplest periodic function, and in fact, it will be shown in Chapter 10 that any periodic function may be represented by a Fourier series which is the sum of a number of sinusoids (over a given time interval, any time function may also be so represented). For these reasons, measuring the responses of systems to sinusoids has become an important means of getting a feel for the system performance. Of interest is the way in which the system will respond to high and low frequencies. (Since we are dealing with linear systems, the output amplitude will be proportional to the input amplitude, but the constant of proportionality will depend on the frequency and element parameters.) It is important to realize that when a sinusoidal input is first applied, the output will not vary periodically at first. It will take some time for the system to respond, as was seen in the previous section. After about four time constants have passed, a first-order system will then be essentially in a steady state, where the output will also be varying sinusoidally. It is this steady-state response which will be investigated here. Figure 5–26 shows how the response proceeds from transient to steady-state.

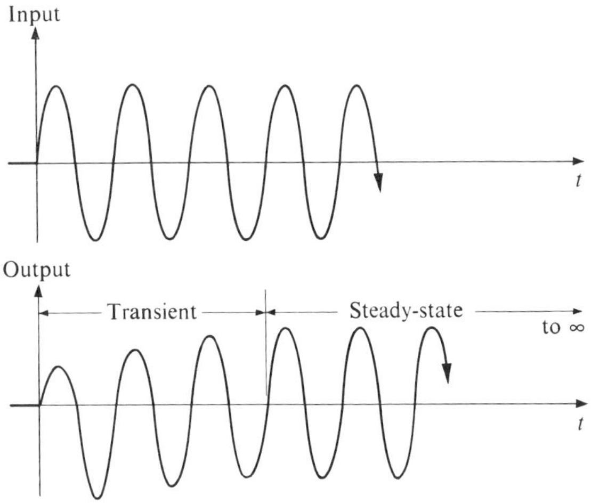

FIG. 5–26. Complete response of system to sinusoid.

The steady-state response of the parallel *RC*-circuit to a sinusoidal input has been obtained in Section 5–4.4. For convenience, these results are repeated here. The system equation was

$$\tau \frac{dv}{dt} + v = V_f \sin \omega t, \tag{5–47}$$

and the response is

$$v = V_f \frac{1}{\sqrt{1 + \tau^2\omega^2}} \sin (\omega t - \phi), \qquad \phi = \tan^{-1} \omega\tau. \tag{5–55}$$

A general equation for a sinusoid can be written as in Eq. (5–62) and is shown graphically in Fig. 5–27.

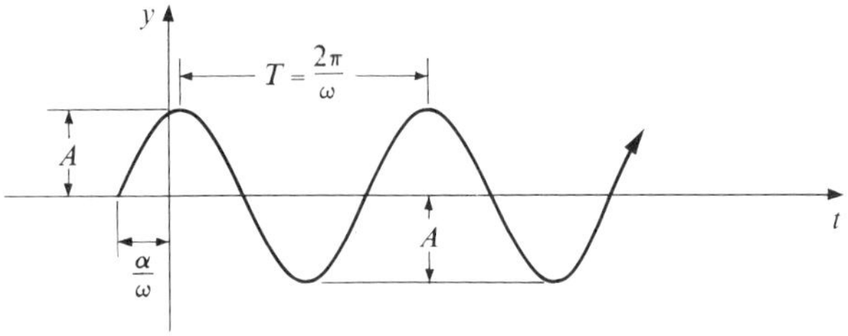

FIG. 5–27. A general sinusoid.

$$y = A \sin (\omega t + \alpha). \tag{5–62}$$

The following definitions will be used:

A = the amplitude,
ω = the angular frequency (rad/sec),
α = the phase angle (rad),
$T = 2\pi/\omega$ = the period (sec),
$f = 1/T = \omega/2\pi$ = the frequency (cps).

For the output v, then:

a) The frequency is the same as the input frequency (likewise for the angular frequency and period).

b) The amplitude is $V_f/\sqrt{1 + \tau^2\omega^2}$; the amplitude of the input is V_f.

c) The phase angle is $-\tan^{-1}(\omega\tau)$; the phase angle of the input is zero. The output *lags* the input by the angle ϕ, since the output peaks after the input peaks. The phase angle between the output and input is $-\phi$, where the minus sign indicates that the output lags the input.

Figure 5–28 shows the input sinusoid v_f and the output sinusoid v as functions of time for various values of $\omega\tau$. For small $\omega\tau$, i.e., low frequency or small time constant, the phase angle ϕ is very small and the output/input amplitude ratio is nearly unity, as may be seen from Eq. (5–55). Under these conditions, the output is almost *in phase* with the input, and the amplitude of the output is almost equal to the amplitude of the input. This corresponds physically to the fact that almost all the input current is going through R and relatively little through C. A solution for the current through C would give this result. However, we can find the current through C without having to solve another differential equation, since

$$i_C = C\frac{dv}{dt} = \frac{\omega C V_f}{\sqrt{1 + \omega^2\tau^2}} \cos(\omega t - \phi),$$

or (5–63)

$$i_C = \frac{\omega\tau I}{\sqrt{1 + \omega^2\tau^2}} \sin(\omega t + 90 - \phi).$$

We see that i_C is nearly zero for small $\omega\tau$. Figure 5–28 shows the output v for both large and small values of $\omega\tau$.

The response of various parts of the system to different frequencies is of considerable importance since it shows how these parts of the system will respond to fast and slow changes. This information is usually given in the form of *frequency response* curves of the variation of output/input amplitude ratio and phase as a function of frequency. The frequency response for v is shown in Fig. 5–29. We can see that for applied frequencies such that $\omega\tau \gg 1$, the output amplitude will be essentially zero, whereas for frequencies such that $\omega\tau \ll 1$, the output will be essentially equal to the input, independent of frequency in this range.

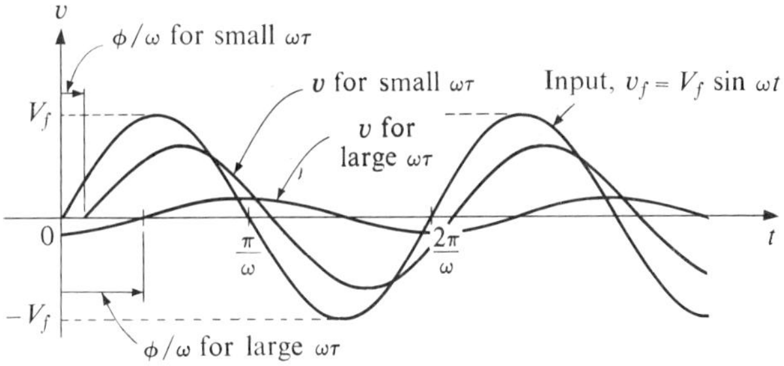

FIG. 5–28. Response of a first-order system to a sinusoidal input.

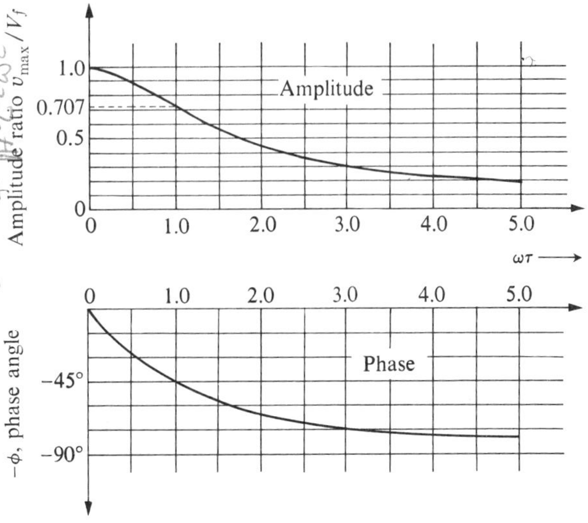

FIG. 5–29. Amplitude and phase response of a first-order system to a steady sinusoidal input.

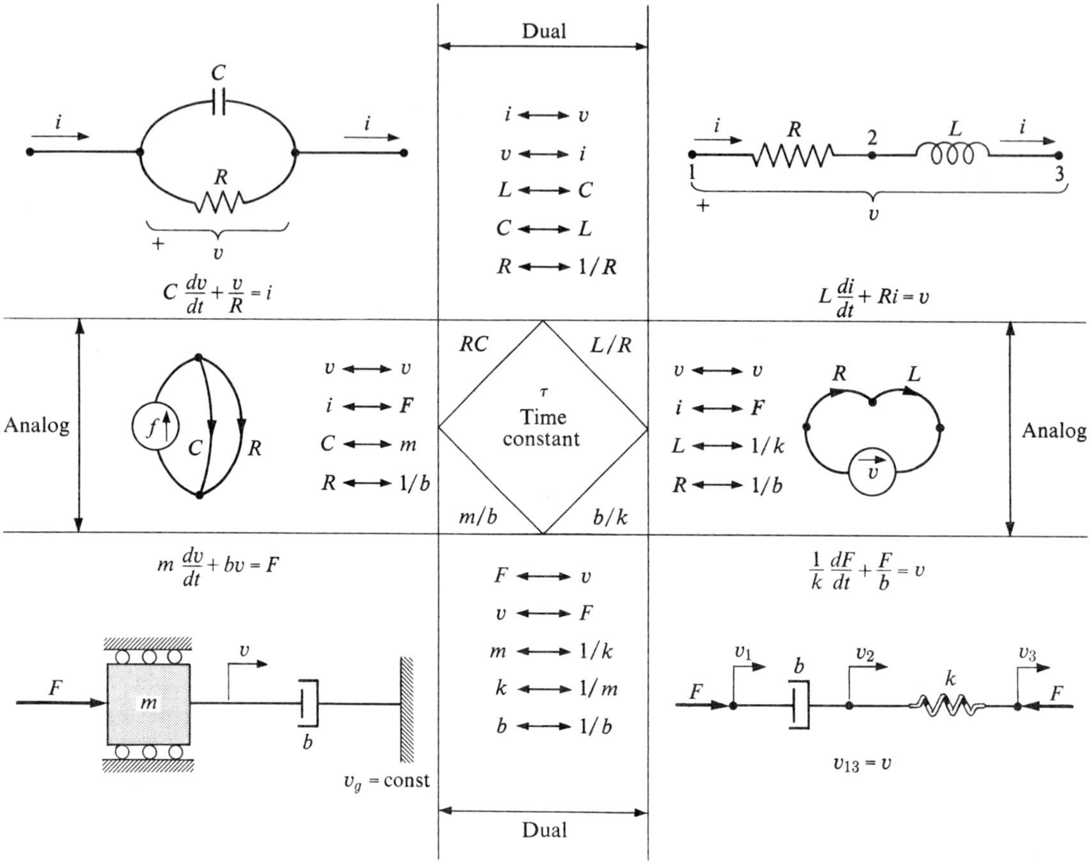

FIG. 5–30. Analogs and duals for *RC*-parallel prototype.

For more complex systems, such as third-order or higher, the determination of the frequency response by the method used in Section 5–4.4 becomes cumbersome. More efficient approaches have been developed, which will be discussed in Chapters 13 and 15.

5–5.3 Analogs and Duals

We have noted that the equations describing different physical systems are similar. This similarity means that the results determined for one system can be extended to the others which have similar equations. The knowledge we have gained of one system, then, is of broader significance than might have first seemed possible. Understanding of a mechanical system leads to an understanding of the analogous electrical system, and vice versa. Thus the *RC*- and *bm*-systems are analogous, and so are the *RL*- and *bk*-systems. These systems are called *analogs* of each other. The analog concept has been exploited most fruitfully (see Reference 7).

In addition, there is a similarity between two different electrical systems such as the parallel *RC*- and the series *RL*-circuits. These networks are called *duals*. The current in one behaves like the voltage in the other, and vice versa. A series connection is the dual of a parallel connection. Figure 5–30 shows the

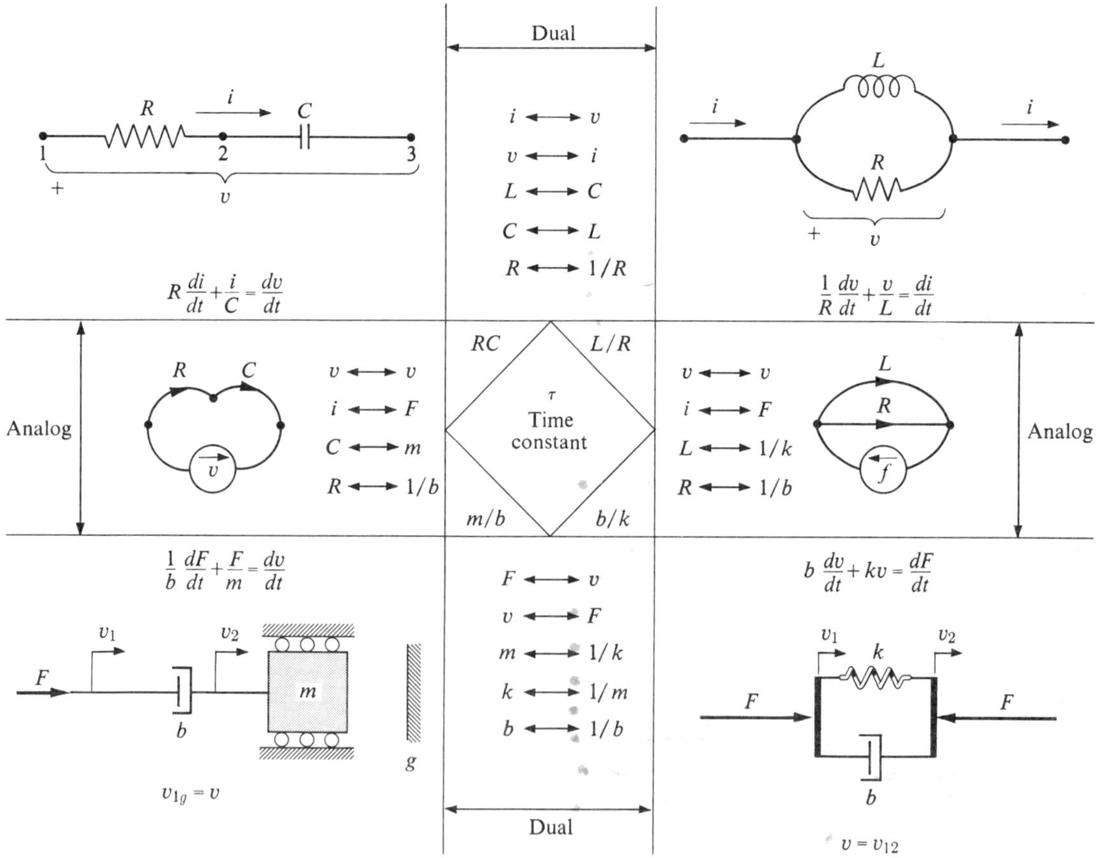

FIG. 5–31. Analogs and duals for *RC*-series prototype

relationship between the various systems which are listed in the left-hand column of Table 5–1, and Fig. 5–31 treats the systems listed in the right-hand column of Table 5–1.

5–5.4 Stability

A system is in equilibrium if it can persist indefinitely in its present state so long as it remains undisturbed. A state of equilibrium is *stable* if the system, when disturbed slightly and released, returns to its initial equilibrium state. In this section, we will investigate the stability and instability of simple first-order dynamic systems.

Repeating our prototype example, Eq. (5–27), we have

$$\tau \frac{dv}{dt} + v = v_f. \tag{5–27}$$

If v_f is constant, $v_f = V_f$, then the only *equilibrium* state of this system is $v = V_f$, for only then can v remain indefinitely at one value, since only for this case will $dv/dt = 0$.

To determine if this equilibrium is *stable*, we should determine what happens if the voltage v is initially different from its equilibrium value. Suppose, for instance, that $v = 0$. The response of this system to this disturbance has been studied already in Section 5–5.1, where it was found that the voltage v rises exponentially from zero to $v = V_f$ according to Eq. (5–41):

$$v/V_f = 1 - e^{-t/\tau}.$$

Provided that τ is a positive quantity, v will tend to approach the equilibrium point V_f if v is initially different. We conclude therefore that this system is *dynamically stable* if its time constant τ is positive. Similar conclusions will also apply for all the first-order systems in Table 5–1, since all are governed by the same type of linear first-order differential equation. Note also that for all these systems the physical constants m, b, k, L, C, and R are always positive. Hence dynamic instability cannot occur in any first-order system composed of the ideal mechanical and electrical elements which have been discussed. All these ideal elements have the common characteristic that they either store or dissipate energy; they never can supply energy which has not previously been stored in the elements. For this reason these elements are called *passive elements*, and a network containing only passive elements is called a *passive network*. *No passive first-order system can be unstable, since its time-constant is always positive.*

To examine the behavior of an unstable first-order system, let us consider a thermal system consisting of a piece of wire through which there is an electric current which is fixed regardless of the wire resistance (the current is determined by other elements in the electrical circuit). It is well known that the resistance of metal increases with temperature in a nearly linear fashion. If we assume that this linear relation holds over the range of temperatures of interest, then the electrical power converted to heat power in the wire is

$$q_{\text{gen}} = I^2R_0(1 + \alpha\theta),$$

where I is the current in the wire, θ is the temperature of the wire above room temperature, R_0 is the wire resistance at room temperature, and α is a constant determined by the wire material.

Let us model this wire as a body whose temperature is uniform at θ and which loses heat to the atmosphere at a rate proportional to the temperature above ambient (room) temperature (see Section 3–4.3). The wire will also store heat energy as a thermal capacitance. Applying continuity to this system, we have:

heat power generated electrically
= heat power lost to the atmosphere
+ heat power into storage,

or

$$I^2R_0(1 + \alpha\theta) = C_h\theta + C_t\frac{d\theta}{dt},$$

where

C_h = over-all coefficient of heat transfer, and

C_t = thermal capacitance.

This equation may be rewritten in the following form:

$$C_t\frac{d\theta}{dt} + (C_h - I^2R_0\alpha)\theta = I^2R_0,$$

or (5–64)

$$\tau\frac{d\theta}{dt} + \theta = \frac{I^2R_0}{(C_h - I^2R_0\alpha)},$$

where

$$\tau = \frac{C_t}{(C_h - I^2R_0\alpha)} = \text{time constant.}$$

The only possible equilibrium condition for this system is

$$\theta_{\text{eq}} = \frac{I^2R_0}{(C_h - I^2R_0\alpha)},$$

since only for this value is $d\theta/dt$ equal to zero.

Let us consider that the wire is initially at room temperature and the current is suddenly turned on. Equation (5–64) applies, and the solution for this temperature as a function of time is

$$\theta = \theta_{\text{eq}}(1 - e^{-t/\tau}),$$

since Eq. (5-64) is of the same form as Eq. (5-27), which was solved earlier.

For $C_h > I^2R_0\alpha$, τ is positive and the temperature variation is as shown in Fig. 5-32(a). This situation corresponds physically to the condition in which the heat loss to the atmosphere per unit change of temperature is greater than the increase in heat generated per unit change of temperature. However, if the increase in heat generated per unit change of temperature is greater than the heat lost to the atmosphere per unit change of temperature, then $C_h < I^2R_0\alpha$ and τ is negative (θ_{eq} also is negative). Under these conditions a rise in temperature will cause more heat to be generated than can be lost, so that the temperature will rise at an ever-increasing rate and the system is unstable. The temperature will rise until (a) the wire melts or (b) the model we have used is no longer accurate. The solution for this case is

$$\theta = \theta_{\text{eq}}(1 - e^{+t/|\tau|}),$$

and the temperature variation is shown in Fig. 5-32(b).* Note that in the unstable case for a first-order system the output tends monotonically toward infinity; i.e., no oscillations are present.

The stability criterion for a *linear first-order* dynamic system can be stated as follows.

Stability criterion. A linear first-order dynamic system is stable if and only if its time constant is positive.

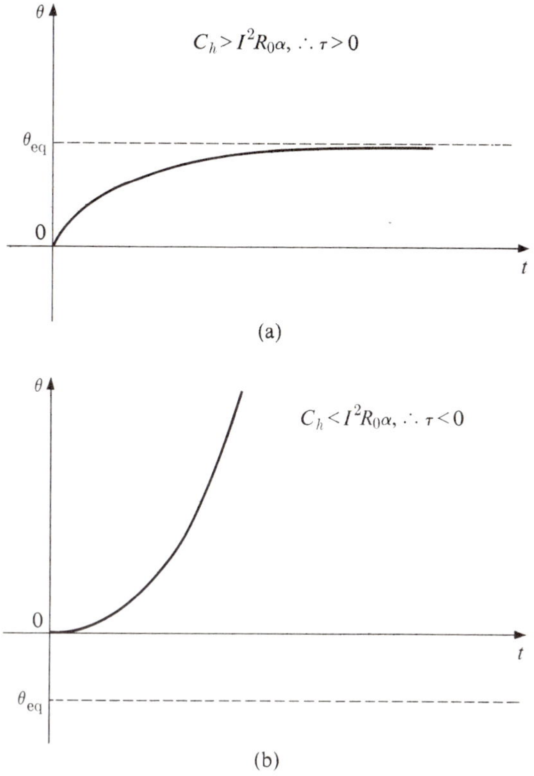

FIG. 5-32. Temperature variation of wire. (a) Stable. (b) Unstable.

5-6. DYNAMIC SYSTEMS WITH MORE THAN TWO ELEMENTS

In the preceding section, systems composed of only two ideal elements have been studied. All of these were governed by first-order differential equations which had a single term on the right-hand side (see Table 5-1). Systems with more elements will result in more complex equations.

5-6.1 First-Order Systems

If we consider systems with only one energy-storage element, the equations will still be of first order, but can, in general, have two terms on the right-hand side. The most general linear first-order equation relating two variables x_r and x_f is of the form

$$a_1\frac{dx_r}{dt} + a_0x_r = b_1\frac{dx_f}{dt} + b_0x_f, \tag{5-65}$$

where a_1, a_0, b_1, and b_0 are constants determined by the system element parameters.

* This instability will tend to occur with wires of small radius (which would have high resistance and low surface area for heat loss). This may be a partial explanation for the "exploding wire" phenomenon, which has been of some scientific interest.

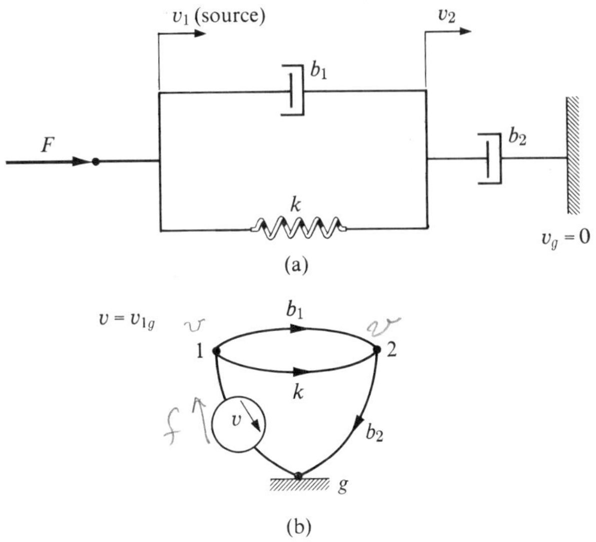

FIG. 5-33. Spring-damper system. (a) Physical schematic diagram. (b) Linear graph for velocity input.

To obtain an equation of this form we need at least *three* ideal elements. For example, let us consider the spring-damper systems shown in Fig. 5-33, where $v = v_{1g}$ is considered the system input. This system consists of a damper and spring in parallel, connected in series with a second damper. Note that only one energy-storage element is involved. For this system we have the following equations.

a) compatibility:

$$v = v_{12} + v_2. \tag{5-66}$$

b) continuity:

$$F_{b2} = F_{b1} + F_k = F. \tag{5-67}$$

c) elemental behavior:
for damper b_1,

$$F_{b1} = b_1 v_{12}, \tag{5-68}$$

for spring k,

$$\frac{dF_k}{dt} = k v_{12}, \tag{5-69}$$

and for damper b_2,

$$F = b_2 v_2. \tag{5-70}$$

Combining Eqs. (5-66) through (5-70) to eliminate all forces gives the following relationship for output v_2:

$$(b_1 + b_2)\frac{dv_2}{dt} + k v_2 = b_1 \frac{dv}{dt} + kv. \tag{5-71}$$

Using Eq. (5-70), one may then find the system equation when F is considered the output:

$$(b_2 + b_1)\frac{dF}{dt} + kF = b_2\left(b_1 \frac{dv}{dt} + kv\right). \tag{5-72}$$

Two time constants may be used to describe this dynamic system:

$$\tau_1 = \frac{b_1 + b_2}{k}, \qquad \tau_a = \frac{b_1}{k}. \tag{5-73}$$

Equation (5-72) then may be written in the following form:

$$\tau_1 \frac{dF}{dt} + F = b_2\left(\tau_a \frac{dv}{dt} + v\right). \tag{5-74}$$

This equation will not be solved here, but it can be seen that since both v and dv/dt are present on the right-hand side, the solution should present characteristics in common with each of the prototype equations investigated earlier in Table 5-1.

5-6.2 Second-Order Systems

If we allow two independent energy-storage elements to appear in a system, the system equation will be of second order; i.e., its highest derivative will be a second derivative.

Consider the circuit shown in Fig. 5-34 where the input is e and the output of interest is v. For this system, we have the following.

a) compatibility:

$$e = v_R + v. \tag{5-75}$$

b) continuity:

$$i_1 = i_2 + i_3. \tag{5–76}$$

c) elemental equations:
for the resistance R,

$$i_1 = \frac{v_R}{R}, \tag{5–77}$$

for the inductance L,

$$v = L\frac{di_2}{dt}, \tag{5–78}$$

for the capacitance C,

$$i_3 = C\frac{dv}{dt}. \tag{5–79}$$

Substituting these equations into (5–76) and substituting $v_R = e - v$ from (5–75), we obtain an equation which contains only v and e:

$$\frac{e - v}{R} = \frac{1}{L}\int_0^t v\,dt + i_{20} + C\frac{dv}{dt}.$$

Differentiating this equation, we obtain the final system equation:

$$C\frac{d^2v}{dt^2} + \frac{1}{R}\frac{dv}{dt} + \frac{v}{L} = \frac{1}{R}\frac{de}{dt}. \tag{5–80}$$

The most general second-order equation would be of the form

$$a_2\frac{d^2x_r}{dt^2} + a_1\frac{dx_r}{dt} + a_0x_r = b_0x_f + b_1\frac{dx_f}{dt} + b_2\frac{d^2x_f}{dt^2} \tag{5–81}$$

for x_f input and x_r output.

The step-function response of a second-order equation is more difficult and requires a general consideration of solution of differential equations as presented in Chapter 11. The sinusoidal response can be found by the methods of Section 5–4.4. It will be seen that basically different phenomena can occur in a second-order system, since the two energy-storage elements can shift the energy back and forth between them, causing oscillations in the system. For a step input of e, the output v may vary in the ways shown in Fig. 5–35, depending on the value of R, for example. These various responses will be the subject of Chapter 12.

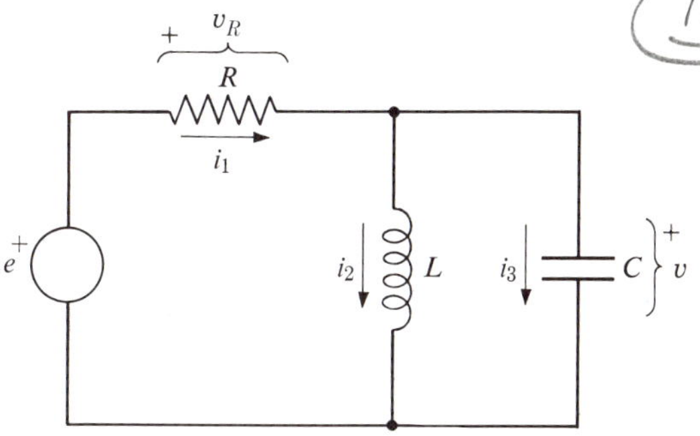

FIG. 5–34. Second-order system.

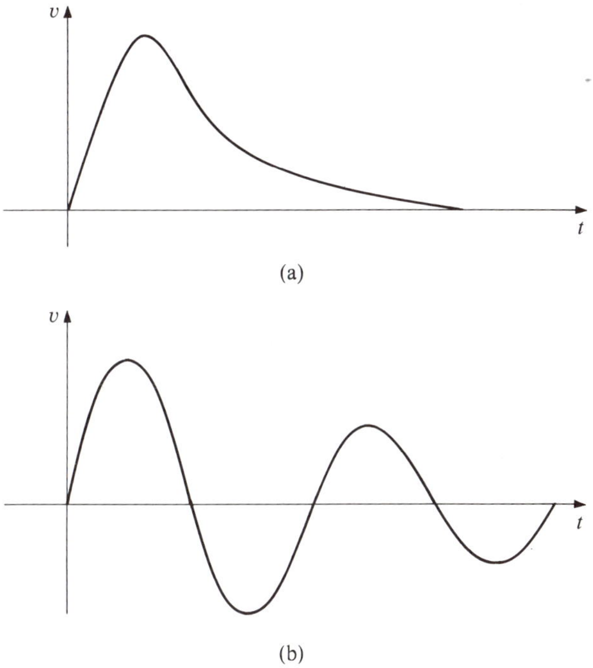

FIG. 5–35. Behavior of second-order system. (a) Small R. (b) Large R.

REFERENCES

1. D. W. VER PLANCK and B. R. TEARE, JR., *Engineering Analysis*. New York: John Wiley and Sons, Inc., 1954.
2. E. A. GUILLEMIN, *Introductory Circuit Theory*. New York: John Wiley and Sons, Inc., 1953.
3. S. H. CRANDALL and N. C. DAHL, *Introduction to the Mechanics of Solids*. New York: McGraw-Hill Book Company, Inc., 1959.
4. R. M. FANO, L. J. CHU, and R. B. ADLER, *Electromagnetic Fields, Energy and Forces*. New York: John Wiley and Sons, Inc., 1960.
5. S. H. CRANDALL, *A Unified Approach to Dynamics via Hamilton's Principle*. Cambridge, Mass.: School of Engineering, Massachusetts Institute of Technology, 1962.
6. R. W. HAMMING, *Numerical Methods for Scientists and Engineers*. New York: McGraw-Hill Book Company, Inc., 1962.
7. W. J. KARPLUS and W. W. SOROKA, *Analog Methods*, 2nd Ed. New York: McGraw-Hill Book Company, Inc., 1959.

PROBLEMS

5–1. Indicate which of the two physical systems in Fig. 5–36 is a "common velocity" connection and which is a "common force" connection. Explain in detail. Show the linear graph of each.

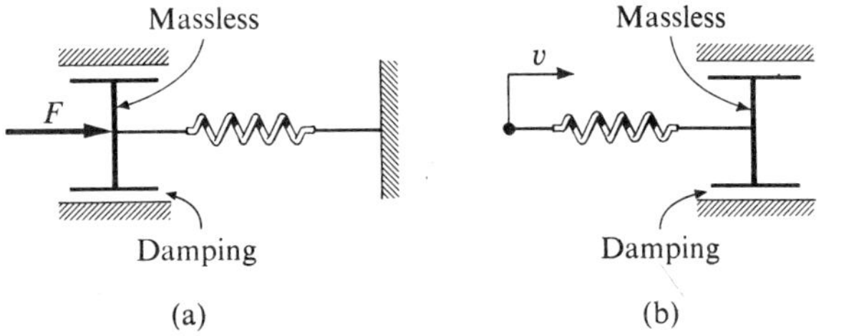

FIGURE 5–36

5–2. A tank has three pipes through which water can flow in and out. Figure 3–37 shows Q_1 and Q_2 (flow rates in gal/min) as well as the quantity of water in the tank. Determine the rate of flow *out of* the third pipe.

5–3. Derive the system differential equation for the parallel *RC*-circuit shown in Fig. 5–1 for an output of the capacitance current, i_C, where the input is i. Compare this with Eq. (5–5) for the output v.

5–4. Apply compatibility, continuity, and elemental equations to obtain the system equations for

a) the parallel *RL*-circuit of Fig. 5–5 to obtain Eq. (5–16),

b) the series *RL*-circuit of Fig. 5–6 to obtain Eq. (5–17).

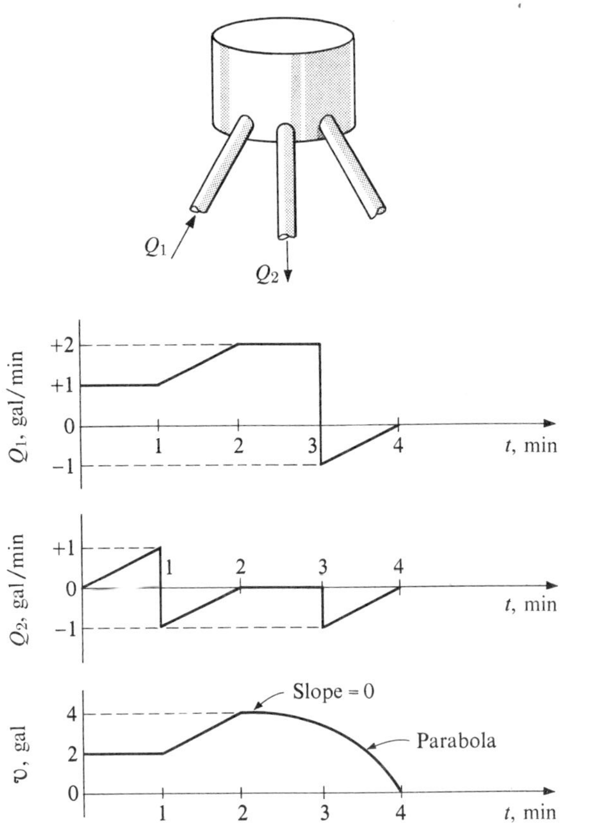

FIGURE 5–37

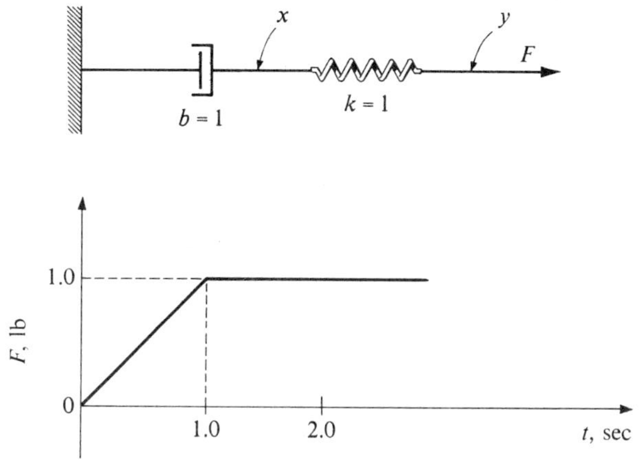

FIGURE 5–38

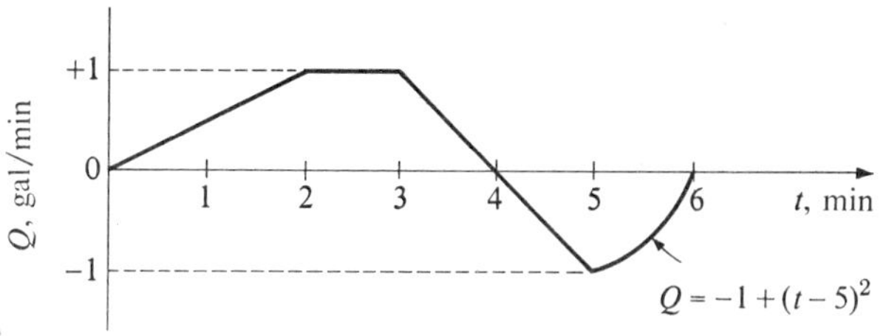

FIGURE 5–39

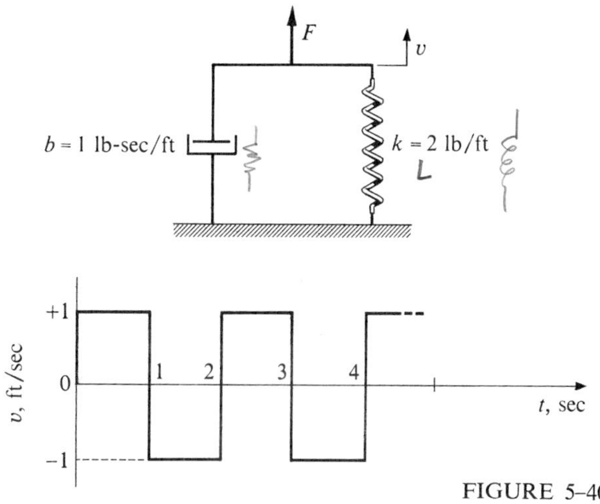

FIGURE 5–40

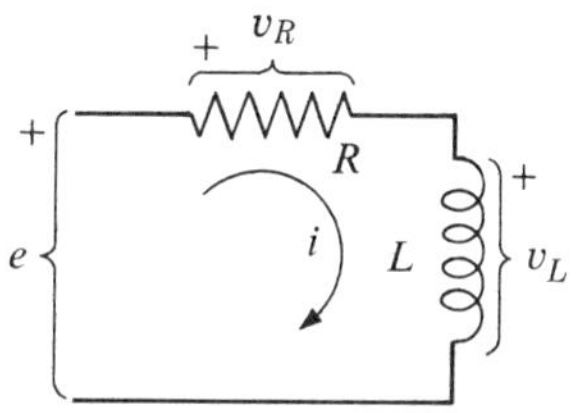

FIGURE 5–41

5–5. Apply compatibility, continuity, and elemental equations to obtain the system equations for

a) parallel bk-system of Fig. 5–7 to obtain Eq. (5–18),

b) series bk-system of Fig. 5–8 to obtain Eq. (5–19).

5–6. Derive the system differential equation for the series bk-system shown in Fig. 5–8 for the output velocity of the spring v_2 (with $v_3 = 0$) where the input is v_1. Compare this with the Eq. (5–19) for the output F.

5–7. A force F is applied to a mechanical translational system (Fig. 5–38).

a) Sketch the velocity variation at point x.

b) Sketch the velocity variation at point y.

c) Sketch the time variation of power input.

d) How much energy is stored in the system at $t = 2$?

e) Draw an electrical analog.

5–8. A tank of area 1 ft^2 has a long thin pipe leading into it. The flow characteristics of such a pipe demand that the flow rate through the pipe be proportional to the pressure drop across the pipe, i.e., it is a linear fluid resistance. This particular pipe causes a 10-psi pressure drop at a flow of 20 gal/min. Determine the pressure required at the open end of the pipe to give the flow of water into the tank, as shown in Fig. 5–39.

5–9. Assuming that a square wave of velocity is applied to the system in Fig. 5–40, sketch graphs of the force in the damper, the force in the spring, and the total force necessary to be applied to the system. The spring is unstressed when the square wave starts at $t = 0$.

5–10. The circuit in Fig. 5–41 is composed of ideal elements $R = 10$ ohms and $L = 0.1$ h

a) Assuming that the current i is

$$i = 0, \qquad t < 0,$$
$$i = 2e^{-50t}, \qquad t > 0,$$

sketch i. Determine equations for and sketch v_R, v_L, and e.

(Continued.)

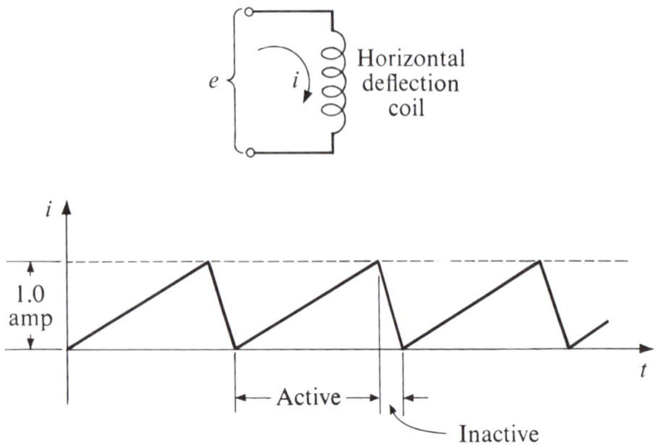

FIGURE 5–42

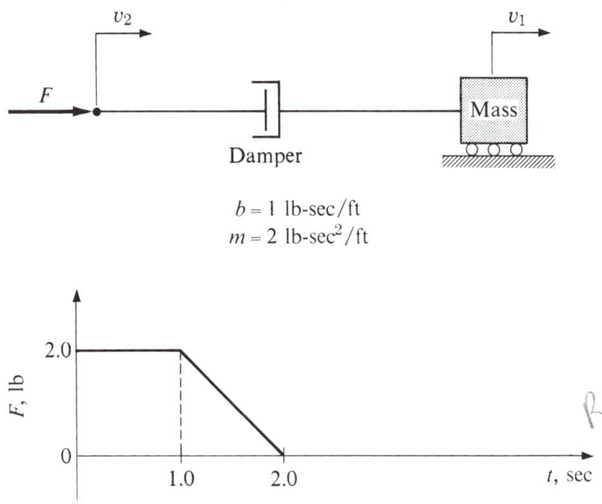

FIGURE 5–43

b) Assuming that the current i is

$$i = 0, \qquad t < 0,$$
$$i = 10(1 - e^{-100t}), \qquad t > 0,$$

sketch i. Determine equations for and sketch v_R, v_L, and e.

c) Assuming that the current i is

$$i = 0, \qquad t < 0,$$
$$i = 10 \sin 50\pi t, \qquad t > 0,$$

sketch i. Determine equations for and sketch v_R, v_L, and e. Express e in the form $A \sin(\omega t + \phi)$.

5–11. In the cathode ray tube used in television receivers, the electron beam is ordinarily deflected magnetically. To form the raster of lines on the face of the tube, the beam must be simultaneously deflected both vertically and horizontally at different rates. We will here consider only the horizontal deflection (Fig. 5–42).

The current through the deflection coil (which causes the magnetic field) is required to vary as the sawtooth wave shown, to obtain a linear sweep. A typical deflection coil has a resistance of 50 ohms and an inductance of 6 mh. The frequency must be 15,750 cps, and the upward slope causes the beam to move progressively from left to right while the downward slope returns the beam to the left side of the tube (during the return, or inactive period, the beam is shut off so that it causes no image on the screen). The inactive period is $\frac{1}{15}$ the total period.

a) Determine how the voltage e must vary with time to give the desired current (show a sketch of e vs. time, stipulating *your* reference for e, labeling significant magnitudes in volts). (Once this variation is known, an electronic circuit can be devised to give the desired output.)

b) What is the peak power the source must deliver? (The source is e.)

c) What is the average power output of the source?

d) Account for the power output of the source at each instant. (That is, how much power is going into heat and how much into stored energy in the field of the inductance at each instant?)

5–12. Given: the mechanical-translational system in Fig. 5–43 with the force applied and at rest at $t = 0$.

a) Sketch a graph of the velocity v_1 vs. time.

b) Sketch a graph of the velocity v_2 vs. time.

c) Determine the energy stored in the system at $t = 1.0$ sec.

d) Determine the total energy input to the system between $t = 0$ and $t = 1.0$. Explain why this differs from (c).

e) Draw the analogous electrical circuit. What is analogous to the velocity v_2?

5–13. A torque $T_1(t)$ applied to the system in Fig. 5–44 causes an angular velocity $\Omega_1(t)$ as shown. The spring is unstressed at $t = 0$.

a) How much work is done by the torque?

b) How much energy is stored in the system for $t > 2$? Explain why the result differs from (a). (Continued.)

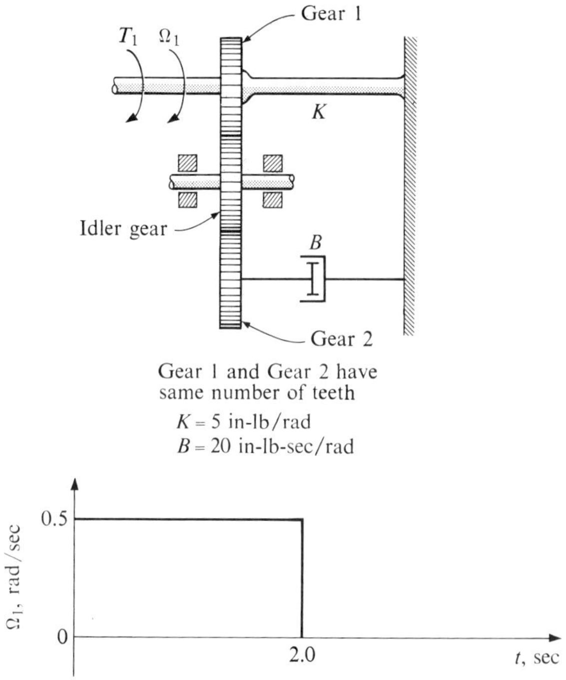

FIGURE 5–44

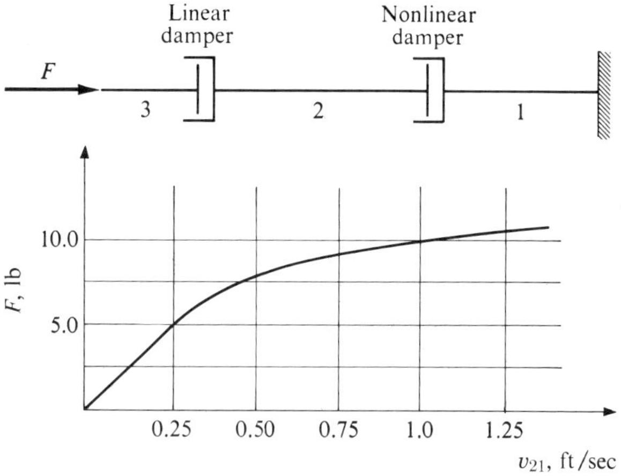

FIGURE 5–45

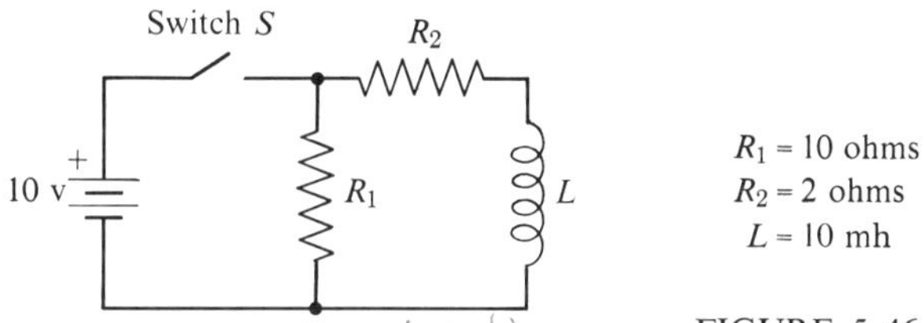

FIGURE 5–46

c) Show the mechanical translational analog of this system.

5–14. a) The mechanical system in Fig. 5–45 consists of two dampers in series. One damper has linear characteristics with $b = 10.0$ lb/fps. The other damper is nonlinear and has the characteristics shown in the graph. Find the velocity of point 3 for the following values of F: $F = 5$ lb, $F = 10$ lb, and $F = 8$ lb. Using these points and the origin, plot F vs. v_{31} for this system.

b) Assuming that both dampers were linear with

$$b_1 = 10.0 \text{ lb/fps} \quad \text{and} \quad b_2 = 20 \text{ lb/fps},$$

find the velocity of point 3 for $F = 5$ lb and $F = 10$ lb. Plot.

c) Note that the velocity in (b) for $F = 10$ lb is double the velocity for $F = 5$ lb. This is an example of the *principle of superposition* which holds for linear systems. Does this work for case (a)?

5–15. For the system in Fig. 5–46 the voltage has been applied for some time when the switch S is suddenly opened. Determine the value of the current through R_1 at the instant *after* the switch is opened. Recall the behavior of the various elements with respect to sudden changes as discussed in Section 5–3.4. What is the current through R_1 the instant *before* the switch is opened?

5–16. For the graphical integration considered in Section 5–4.1, perform the actual calculations by plotting and counting squares to obtain the result given in Fig. 5–13. The numerical values of the parameters are given in the figure. Use increments of v of 0.5 up to $v = 5.0$. You will note that the accuracy will be reduced in the range from 5 to 6 because of the rapid change in the function. Plot an expanded scale graph from $v = 5.0$ to 6.0 and use 0.1-increments of v between 5.0 and 6.0.

5–17. Solve the following differential equation by graphical means.

$$\left(\frac{dx}{dt}\right)^2 + 10x = 15, \qquad x = 0 \quad \text{at} \quad t = 0.$$

Compare the result with that obtained in Problem 5–20.

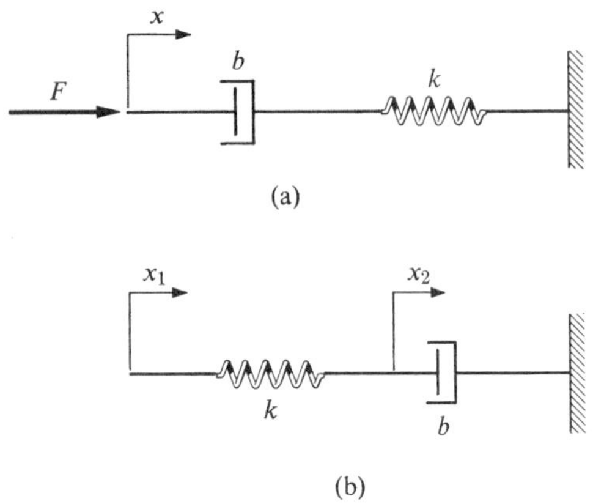

FIGURE 5–47

5–18. The example used to illustrate the numerical solution of Section 5–4.2 is so simple that the numerical process can be analyzed and a closed-form solution (of the finite-difference equation) can be obtained (this cannot usually be done). This solution can be used to show the effect of the size of h on the accuracy.

The process of Eq. (5–36) is

$$v_{n+1} = v_n + \left(\frac{V_f - v_n}{\tau}\right) h = V_f \frac{h}{\tau} + v_n \left(1 - \frac{h}{\tau}\right).$$

a) Show that v_n can be written as a series:

$$v_n = V_f \frac{h}{\tau} \sum_{k=0}^{n-1} \beta^k,$$

where $\beta = (1 - h/\tau)$.

b) The above series is known as a *geometric* series, and its sum is known in closed form. Show that the sum is

$$\sum_{k=0}^{n-1} \beta^k = \frac{1 - \beta^n}{1 - \beta}.$$

[*Hint:* S_n = Sum of n terms $= 1 + \beta + \beta^2 + \cdots + \beta^{n-1}$. Also $S_{n+1} = S_n + \beta^n$. Express βS_n in terms of S_{n+1} and solve for S_n.]

c) Substitute the result of (b) into (a) and show that

$$v_n = V_f \left[1 - \left(1 - \frac{h}{\tau}\right)^n\right].$$

This result may be used to determine v_n without stepping off all the previous calculations. Note that $nh = t$.

d) Consider the cases $h < \tau$, $h = \tau$, $\tau < h < 2\tau$, and $h > 2\tau$ and comment on the form of the solution as determined from the magnitude and sign of the second term in (c) as n increases. Compare with the results of Table 5–2.

e) The exact solution can be obtained from (c) by taking the limit as $h \to 0$ and using the definition of $e = 2.71828$, where

$$e = \lim_{x \to 0} (1 + x)^{1/x}.$$

Perform this limiting process to obtain Eq. (5–41).

5–19. Apply Eq. (5–36) 10 times for $h/\tau = 0.1$, $V_f = 1.0$, $v_0 = 0$ and compare the result with that obtained from the "exact numerical solution" given in part (c) of Problem 5–18. Any differences are due to round-off errors.

5–20. Solve the following differential equation numerically:

$$\left(\frac{dx}{dt}\right)^2 + 10x = 15, \qquad x = 0 \quad \text{at} \quad t = 0.$$

Compare the result with that obtained in Problem 5–17.

5–21. Prepare an operational block diagram for the series RL-circuit of Fig. 5–6 for an output of v_2 ($v_3 = 0$) and an input of v_1.

5–22. Prepare an operational block diagram for the parallel bk-system of Fig. 5–7 for an output of F_k and an input of F.

5–23. The spring and damper system shown in Fig. 5–47(a) is to be driven with a force source F. We are interested in how the displacement x is related to the force input.

a) What are the initial values of x and $\dot{x}$ immediately after a force F is suddenly applied to the dashpot?

b) Assuming that the force F is held constant after it has been applied, what are the final values x and $\dot{x}$ which will result after a long period of time?

c) Derive the differential equation relating x to F and the differential equation which relates $\dot{x}$ to F.

d) Sketch the responses x and $\dot{x}$ as functions of time to a unit step change in F. (Continued on next page.)

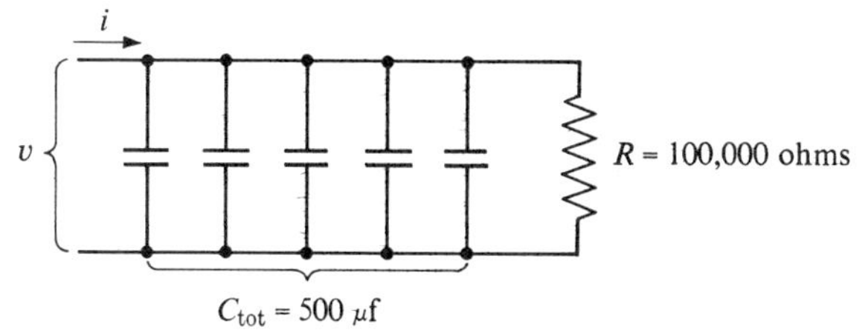

FIGURE 5–48

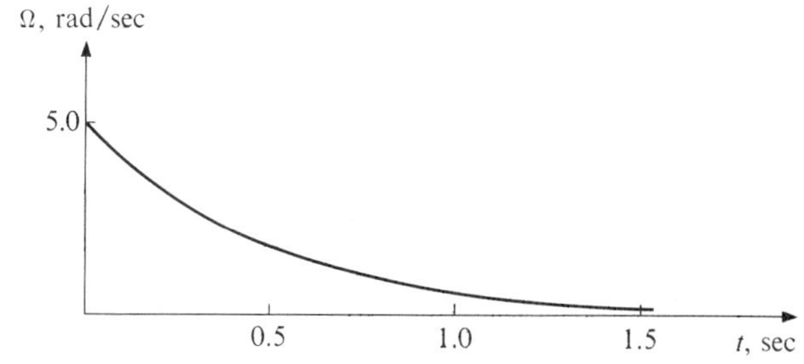

FIGURE 5–49

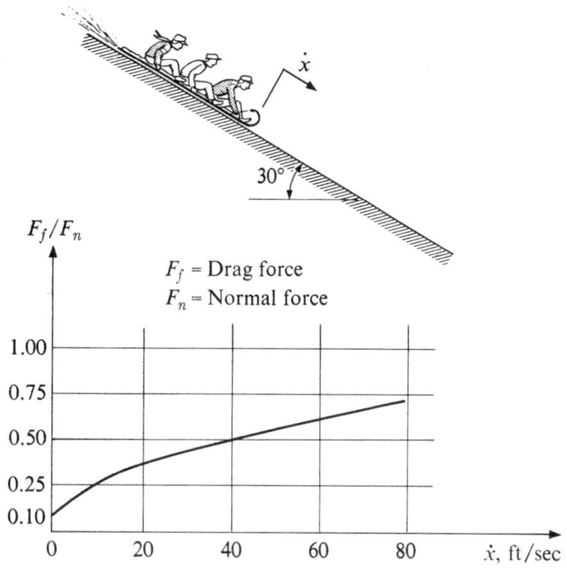

FIGURE 5–50

e) Let the spring and dashpot system be rearranged and driven with a displacement source x_1, as shown in Fig. 5–47(b). Discuss the similarities and dissimilarities between this system and the previous system.

5–24. The circuit in Fig. 5–48 is sometimes used in connection with photoflash bulbs. The capacitors are charged in parallel and discharged (through the bulb) in series.

a) How long will it take to charge the bank of capacitors from zero to 150 v with a current of 2 ma when the leakage resistor R is connected as shown?

b) What is the maximum voltage which can be attained with this scheme? What is the stored energy associated with this charge?

c) How long would it have taken to reach 150 v if the leakage resistor R had not been connected?

5–25. Estimate the rotational viscous friction coefficient B of the rotational damper which is constructed as shown in Problem 2–14 from the following experimental information. One shaft is held while the other is rotated at a speed of 5.0 rad/sec and then released. The plot in Fig. 5–49 is obtained by attaching a tachometer generator whose voltage output is read on a recorder. The inertia J of the rotating parts of the apparatus is found to be 0.01 lb·in·sec^2.

5–26. How long does it take the loaded toboggan (Fig. 5–50) to reach 95% of its terminal velocity? The frictional drag on a toboggan sliding on snow is also given. The total weight of the toboggan with passengers is 300 lb.

5–27. Consider the mechanical system shown in Fig. 5–4. Suppose that the input velocity v has been the constant value v_0 for a long time and is suddenly changed to zero at $t = 0$. What is the value of v_2 at $t = 0+$? Solve for v_2 as a function of time and sketch the resulting response curve. What is the time constant?

5–28. The capacitor C in the circuit shown in Fig. 5–51 is charged to 10 v when the switch is thrown. Derive the differential equation for the capacitor voltage. Solve for this voltage as a function of time, and sketch a plot of it.

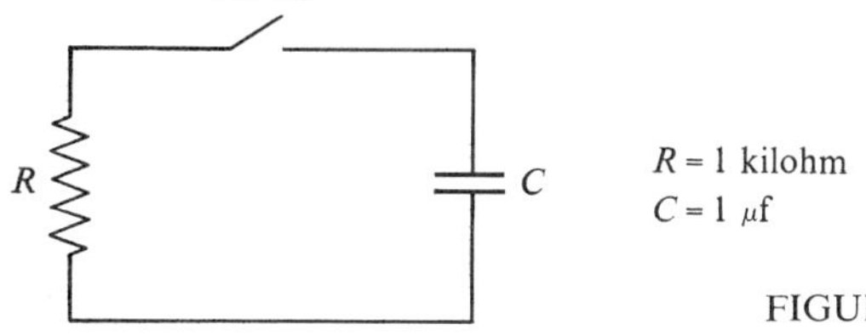

FIGURE 5–51

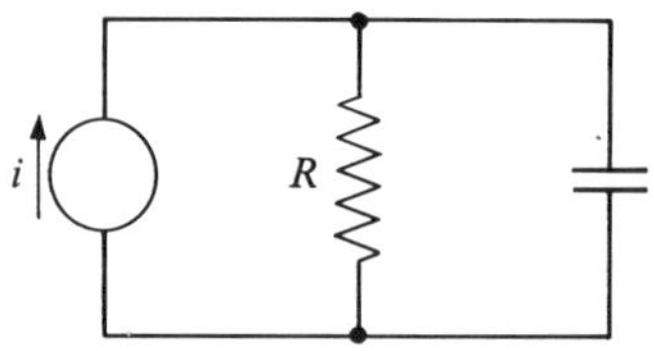

FIGURE 5–52

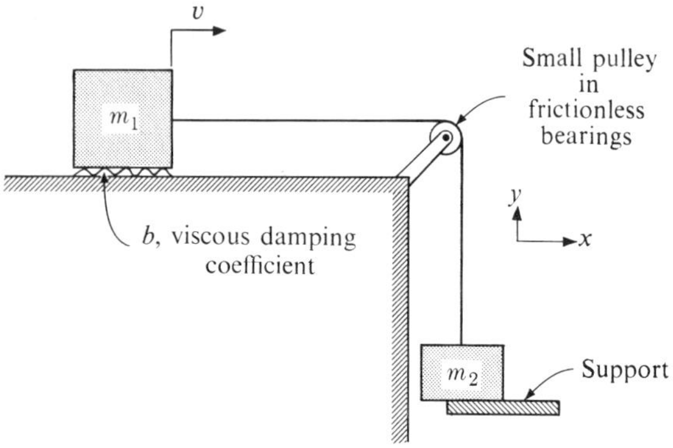

FIGURE 5–53

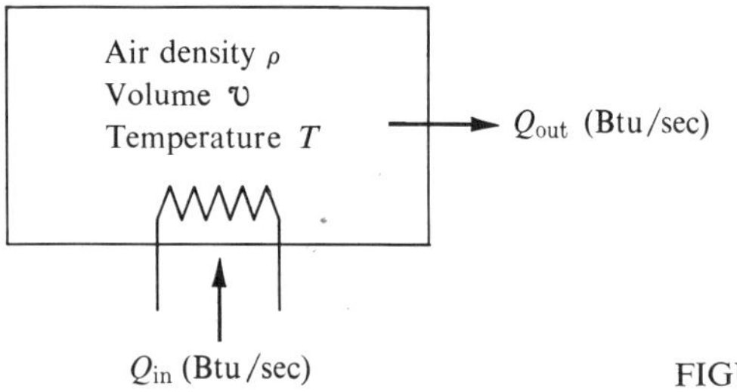

FIGURE 5–54

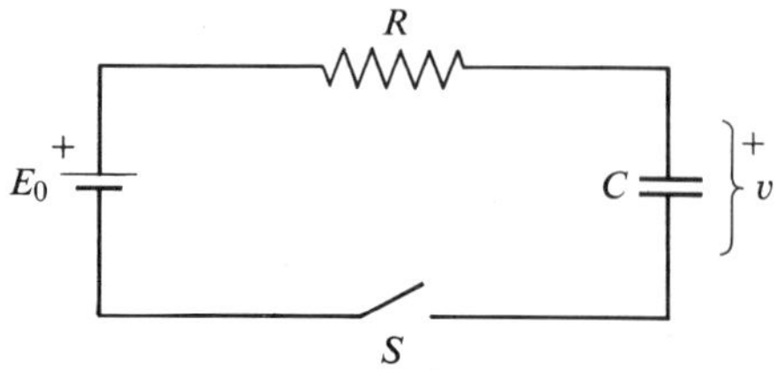

FIGURE 5–55

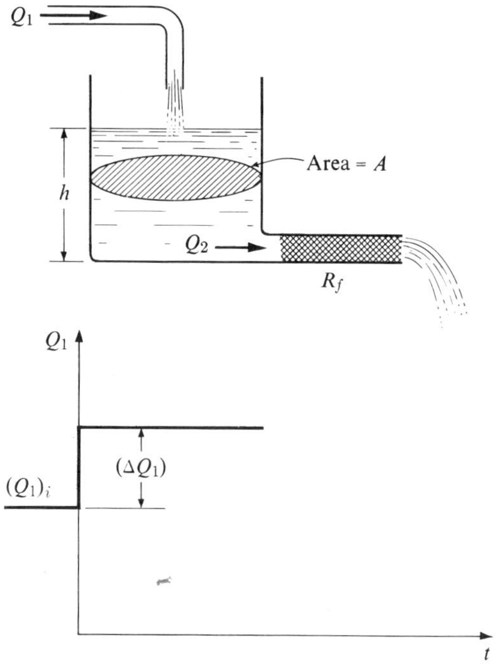

FIGURE 5–56

5–29. Given: the electric circuit in Fig. 5–52, where the voltage across C is zero at $t = 0$.

a) Derive a differential equation for the voltage across the capacitor in terms of the current source i.

b) If the current i is a step function which changes from 0 to I_0 at $t = 0$, determine whether or not the voltage across C changes instantly at $t = 0$. Why?

c) Determine the initial (an instant after $t = 0$) value of the time rate of change of the voltage across the capacitor for the step-function current input.

d) What is the value of the voltage across C a long time after the step function is applied?

e) Sketch a curve of the voltage across C vs. time.

5–30. A system consisting of two masses m_1 and m_2 connected by a flexible but inelastic cable (i.e., cable stiffness in tension is very large) is orginally held in equilibrium by having a support under the mass m_2 (Fig. 5–53). This problem involves determining the velocity v as a function of time after the support is suddenly removed at $t = 0$.

a) Draw a simple lumped-parameter model of this system involving translational elements which move in only the x-direction. Include an input to this model which is equivalent to removing the support in the real system.

b) Find v as a function of time after the support is removed. Make a clearly labeled sketch of this system response, showing all its salient features. (Continued.)

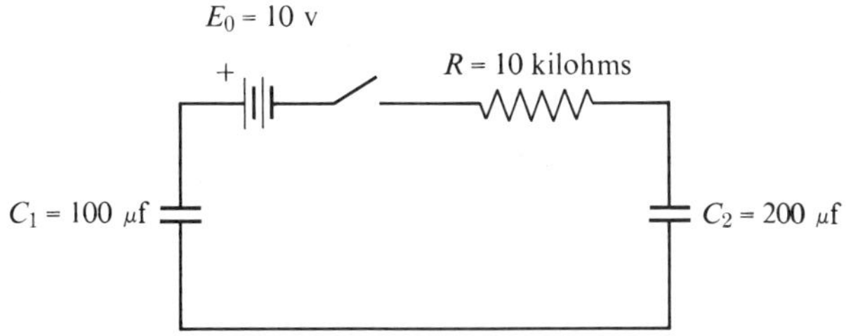

Initial voltage across C_1 and C_2 is zero.

FIGURE 5–57

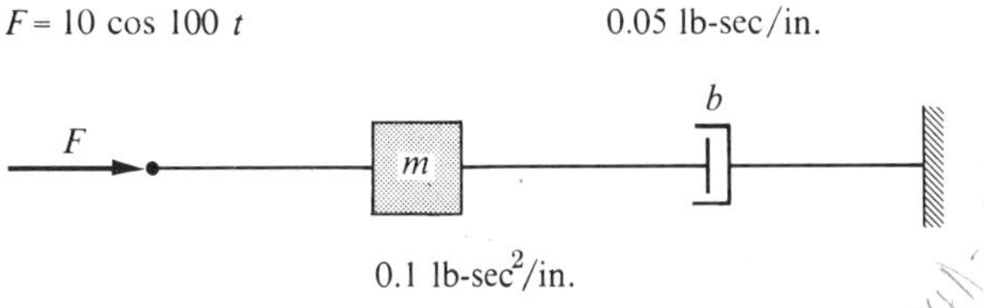

Initial velocity of mass is 2 in/sec.

FIGURE 5–58

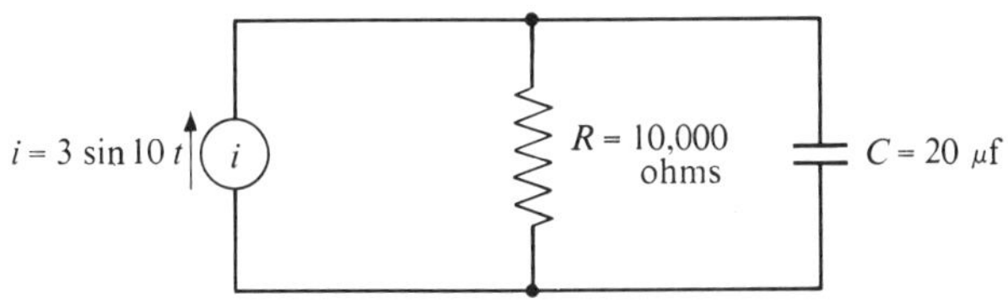

FIGURE 5–59

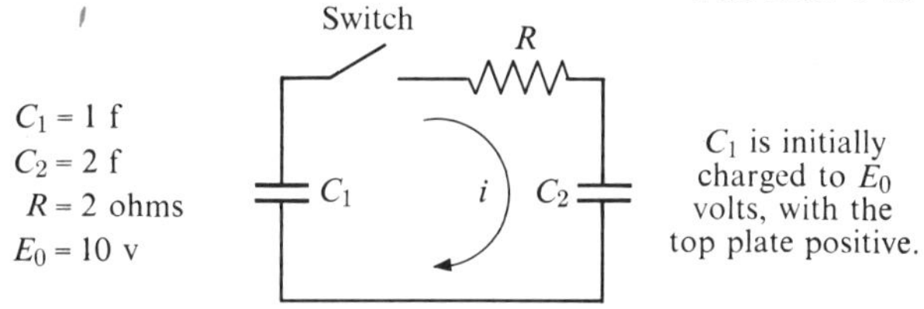

FIGURE 5–60

c) Explore qualitatively the differences in system response which would be obtained if each of the following changes were made in the system (one at a time):

1) the mass m_2 is increased by a factor of 5.

2) a frictionless pulley of inertia J and radius R is used instead of the original pulley.

5–31. Write the differential equation for each of the following systems, and obtain the time solution in each case.

a) A room has an initial temperature of T_1 when there is a sudden drop in outside temperature to T_0. The specific heat of air is c. The heat capacity of the building can be neglected. Assume that $Q_{\text{out}} = K(T - T_0)$. Solve for T. (See Fig. 5–54).

b) Current flows in the circuit (Fig. 5–55) when the switch is closed. Solve for v. Initially, $v = 0$.

5–32. a) Develop the differential equation relating the level h of the liquid in the open tank to the rate of inflow Q_1 into the tank, as shown in Fig. 5–56. The flow resistance R_f is linear.

b) Determine and sketch the response in the level h and the flow Q_2 to a step change in Q_1 at $t = 0$. The system is to be assumed to be initially in a steady state with $Q_1 = (Q_1)_i = (Q_2)_i$ before the step change in Q_1 occurs.

5–33. Find and sketch the voltage v_2 across C_2 (Fig. 5–57) as a function of time after the switch is closed.

5–34. Find and sketch the steady-state velocity of the mass (Fig. 5–58) as a function of t.

5–35. Find and sketch the steady-state current in C (Fig. 5–59) as a function of time.

5–36. In the circuit in Fig. 5–60, capacitor C_1 is initially charged to E_0 v. Capacitor C_2 is initially uncharged. The switch is then closed at $t = 0$.

a) Determine the energy stored initially in C_1.

b) Determine the initial rate of energy flow out of C_1 after the switch is closed.

c) Determine the final energy stored in C_1 after the system settles down.

d) Determine the final energy stored in C_2.

e) Determine a differential equation for the current i.

f) Sketch curves of the time variation of i, the voltage across C_1, and the voltage across C_2.

g) Draw the analogous fluid system and stipulate the initial state of the fluid system which would be analogous to the initial state of the electrical system. (If you are having trouble with (c), (d), and (f), it may help to think of the circuit behavior in terms of the fluid problem.)

h) Repeat g for the analogous thermal system.

i) Repeat g for the analogous mechanical translational system. (Continued.)

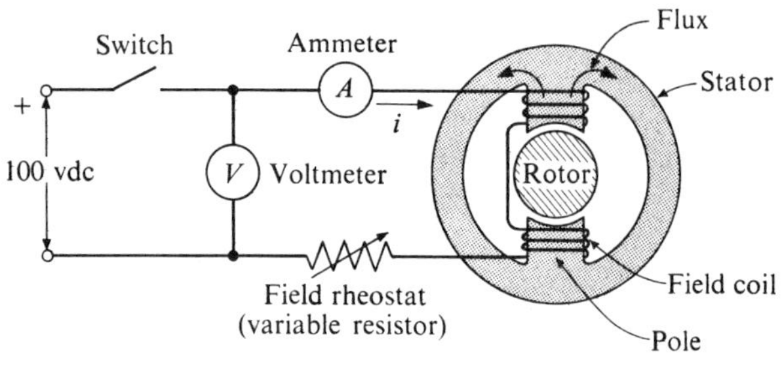

FIGURE 5–61

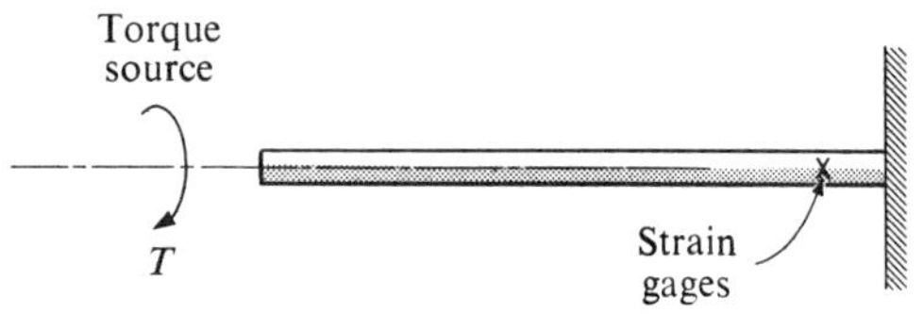

FIGURE 5–62

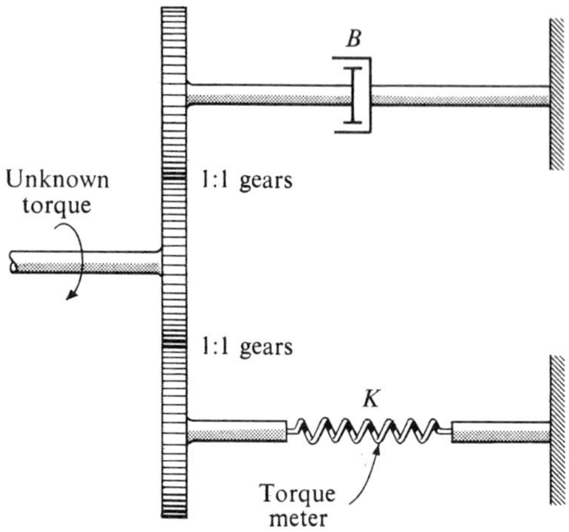

FIGURE 5–63

j) Repeat *g* for the analogous mechanical rotational system.

5–37. In a direct-current (dc) motor, the *field* flux which interacts with current in the *armature* (or rotor) to produce torque is set up by coils wound about the iron structure of the motor stator. The motor shown (Fig. 5–61) has two *poles*, and the current *i* supplied to the coils may be influenced by a variable resistor called a *field rheostat*. The coils may be approximated as a linear inductance of 5 h in series with the resistance of the coils.

A test is run on the field circuit of the motor using an ammeter in series and a voltmeter across the line. The switch is closed and after some time the steady readings of the voltmeter and ammeter are 100 v and 2 amp, respectively. The switch is then suddenly opened, instantaneously disconnecting the source from the circuit.

The ammeter has negligible resistance (i.e., behaves like a short circuit) and a full-scale range of 5 amp. The voltmeter has a resistance of 200,000 ohms and a full-scale range of 200 v.

a) Draw the equivalent circuit diagram for the electrical system using ideal electrical elements.

b) Determine the combined resistance of the field rheostat and field coils.

c) Sketch the time variation of voltage across the voltmeter and of current through the ammeter after the switch is opened. Show voltage and current levels and label time scales.

d) Find the initial rate of energy dissipation when the switch is opened and indicate quantitatively where in the circuit this energy is lost.

e) Discuss the effects of this test on the meters. Suggest an improved test procedure.

5–38. A torque meter is made from a circular shaft which is rigidly mounted at one end and has strain gages attached to it (Fig. 5–62). A static calibration test (with known constant *T* applied) yields a spring constant of 15.0 ft·lb/rad. The magnitude of the applied torque is time-dependent.

In a certain application it is desired to measure a sinusoidally varying torque while filtering out some high frequency components of the source, which are called *noise*. To filter out the noise, the system in Fig. 5–63 is proposed. In the following the inertia of the rotating parts may be neglected.

a) What value of *B* should be used so that for frequencies between 0 and 30 cps the torque is measured with an accuracy of 2%?

b) For very high frequencies, what is the approximate ratio of the applied torque to the measured torque?

c) Assuming that the noise is also sinusoidal with an amplitude that never exceeds 20% of the amplitude of the torque that you want to measure, what minimum multiple of the test-signal frequency must the noise frequency be if the noise in the output is not to exceed 2% of the desired output?

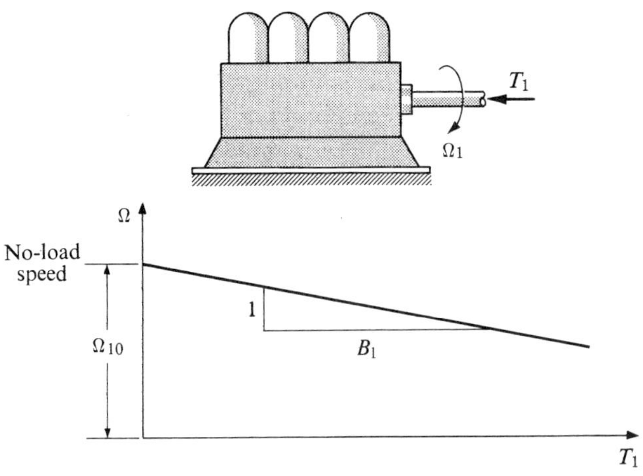

FIGURE 5–64

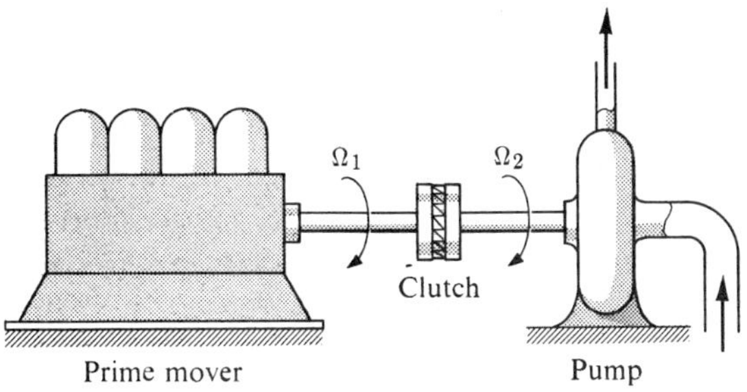

FIGURE 5–65

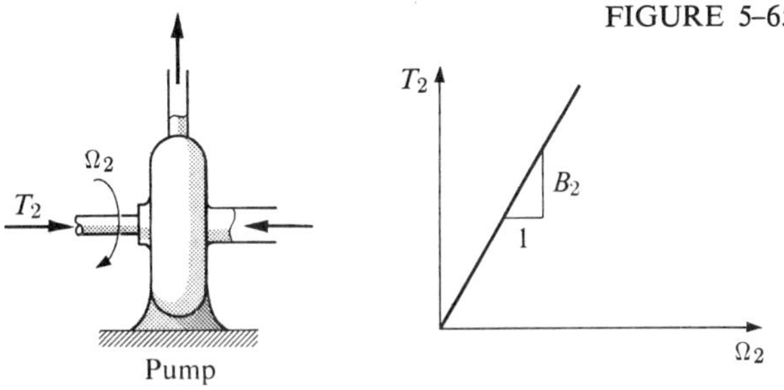

FIGURE 5–66

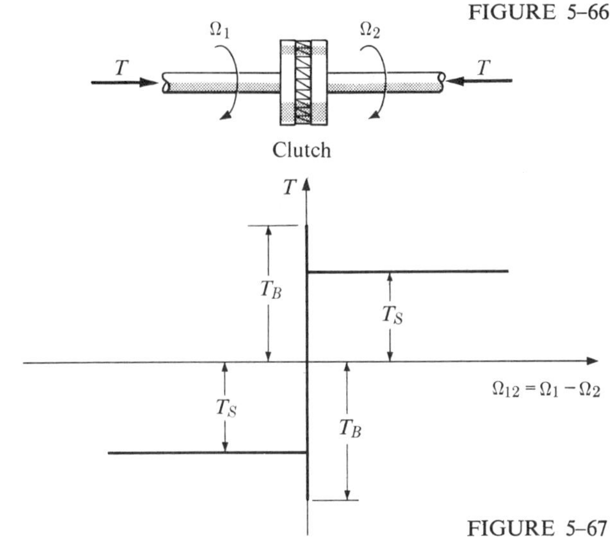

FIGURE 5–67

5–39. A diesel engine prime mover having the steady-state speed-vs.-torque characteristic shown in Fig. 5–64 is to be connected through a static-friction clutch as shown in Fig. 5–65 to a centrifugal pump having a steady-state torque-vs.-speed characteristic such as that shown in Fig. 5–66. The clutch consists of two parallel plates which are forced to rub against each other by axial forces (not shown) when the clutch is engaged. When it is disengaged, the clutch transmits zero torque (i.e., the plates move freely relative to each other.) When it is engaged, the clutch acts as a rigid connection (i.e., a shaft) unless the applied torque exceeds the breakaway level (Fig. 5–67). When the applied torque reaches the value T_B, the clutch will begin to slip, and the transmitted torque drops to T_S. Thus the slipping clutch is a constant-torque device.

a) If the effective inertias of the moving parts of the prime mover and the pump are J_1 and J_2, respectively, draw a linear graph of this system using lumped elements driven by an appropriate source.

b) When there is relative motion between the two halves of the clutch, what does the linear graph look like (i.e., what element describes the clutch)?

c) When there is *no* relative motion between the two halves of the clutch, what does the linear graph reduce to?

d) If the clutch is suddenly engaged with the prime mover initially running at speed Ω_{10} with the pump at rest, determine the responses of the shaft speeds Ω_1 and Ω_2 and the clutch torque T. Make sketches of these responses vs. time. Consider all possible ranges of the system parameters and make separate sketches if basically different phenomena can occur for different parameter values.

e) If $\Omega_{10} = 3000$ rpm,

$B_1 = 0.5$ ft·lb/rpm, $\quad B_2 = 0.1$ ft·lb/rpm,

$T_B = 650$ ft·lb, $\quad T_S = 500$ ft·lb,

$J_1 = 1.0$ ft·lb·sec^2, $\quad J_2 = 10$ ft·lb·sec^2,

determine the responses Ω_1, Ω_2, and T and sketch to scale.

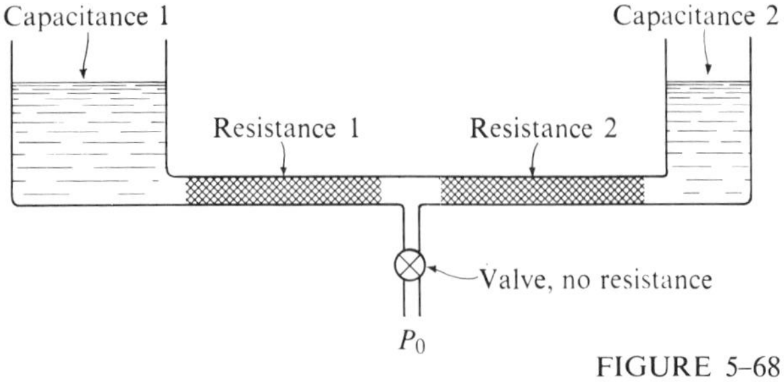

FIGURE 5–68

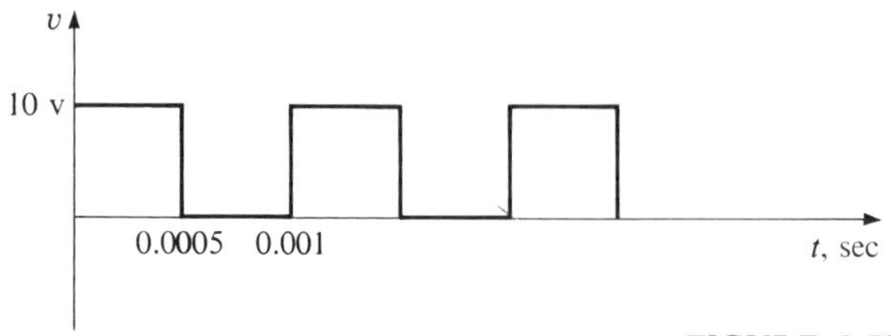

FIGURE 5–70

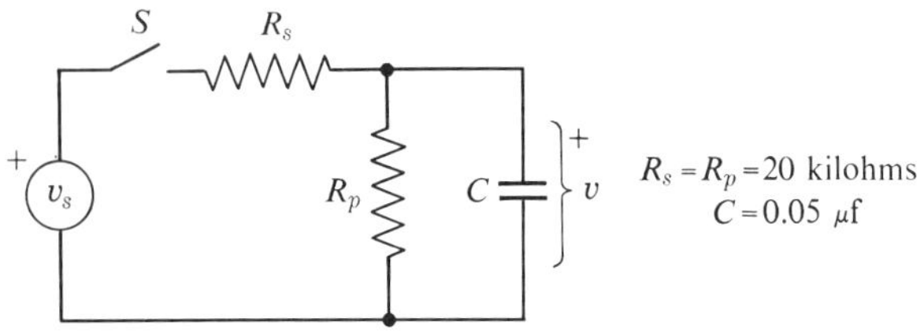

FIGURE 5–69

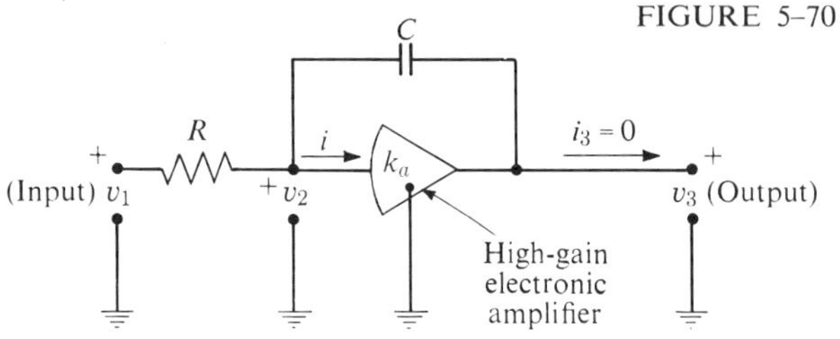

FIGURE 5–71

5–40. For the system of Fig. 5–68, determine and sketch how the system reacts when

a) the valve is suddenly opened and a constant pressure is applied at the valve (the tanks had been previously partially filled and had been at rest for some time);

b) the valve is suddenly closed at a time when the water level in the two tanks is different.

5–41. An electrical circuit is built as in Fig. 5–69. By means of an electronic circuit the switch is turned on and off periodically at a frequency of 1 kc. Thus the voltage v_s as a function of time is a square wave (Fig. 5–70). Prior to $t = 0$, the circuit is relaxed; i.e., there is no stored energy.

a) Find the voltage v and the current in the capacitor just after $t = 0$ ($t = 0+$).

b) Find v as a function of time for $0 < t < 0.001$ sec and sketch.

c) After a long time, a periodic variation of v will be established at the switching frequency. Sketch *qualitatively* the shape of this voltage function.

d) If the voltage v_s is doubled, how do the above results change?

e) If the resistances are doubled, how will the results change?

5–42. The circuit diagram (Fig. 5–71) is a simplified version of an electronic integrator of the kind used in the analog computer. To a good approximation the amplifier characteristic is $v_3 = -k_a v_2$, where the amplifier gain k_a is a very large number (between 10^4 and 10^8). The current i flowing into the amplifier at its input is so small (of the order of 10^{-10} amp) that it may be assumed to be zero at all times.

Develop the differential equation relating v_3 to v_1 and show the conditions under which it may be considered to be a good integrator. [*Hint:* It may be helpful to find the step response to see how nearly this system behaves like a true integrator.]

5–43. It is proposed to use the circuit in Fig. 5–72 as an integrating circuit for a vibration analyzer. This circuit is sometimes used when requirements are not so rigid as to necessitate the use of an operational amplifier, as was considered in Problem 5–42. The vibration analyzer transducer puts out a voltage e which is proportional to velocity in the range of frequencies above 500 cpm (note *per minute*). To measure displacement, this velocity is to be integrated in the proposed circuit.

a) Determine the value required for the product RC so that the circuit will perform to the amplitude accuracy required up to a frequency of 40,000 cpm. Assume that the integrator must be accurate to 3%.

b) What is the magnitude of the output voltage e_0 at both ends of the frequency range? Determine the gain K of the amplifier which would be required to bring the voltage level back up to that of e_0 at the lowest frequency.

c) Verify that $\phi \approx 90°$.

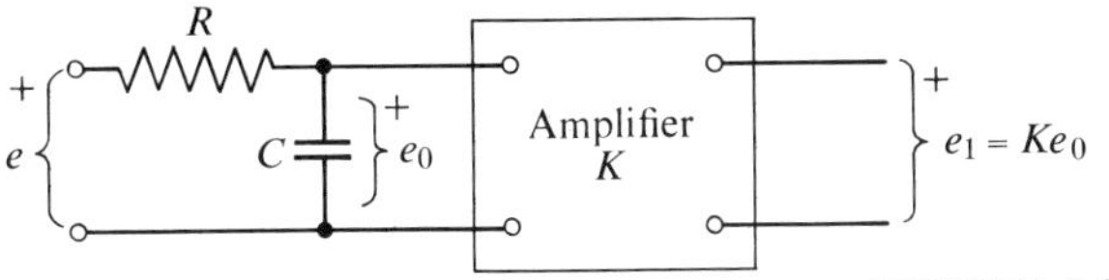

FIGURE 5–72

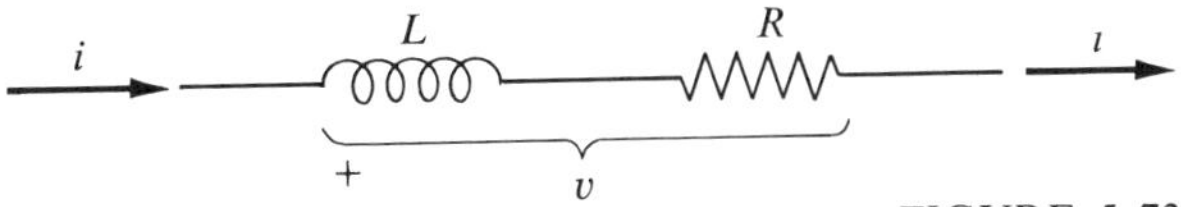

FIGURE 5–73

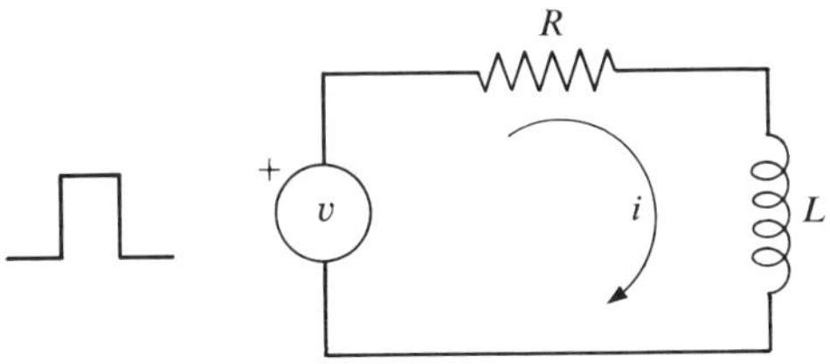

FIGURE 5–74

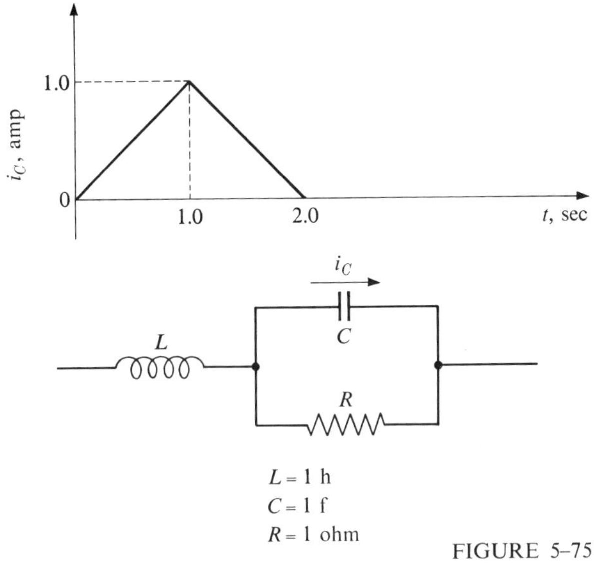

FIGURE 5–75

5–44. It is proposed that a coil of wire be modeled by an in-series system of lumped inductance and lumped resistance (Fig. 5–73). Assuming that $v = V_0 \sin \omega t$, determine the frequency at which the current equals $0.707 = 1/\sqrt{2}$ its value at $\omega = 0$. Compare this frequency to the frequency at which the coil looks approximately like a pure inductance. Compare the current values also.

5–45. For the parallel RC-circuit (Fig. 5–1) which has a suddenly applied sinusoidal input current $i = \text{Im}[I\,e^{i\omega t}] = A \sin \omega t + B \cos \omega t$, the response is as shown in Fig. 5–26. Estimate how long it takes the transient to die out. Does this time depend on the frequency of the sinusoid? Is it possible that no transient will occur? Under what conditions would it not occur?

5–46. We have seen that the current through an inductance cannot change instantly unless an infinitely high voltage is applied to it. For the circuit in Fig. 5–74, a very high voltage is applied for a very short time. Sketch the rise and fall of current in the circuit. What would happen as the voltage amplitude became infinitely large?

5–47. Indicate which of the following first-order equations are stable.

a) $a\dfrac{dx}{dt} - bx = y,$ y-input, x-output

b) $m\dfrac{dv}{at} + bv = F,$ F-input, v-output

c) $-a\dfrac{dx}{dt} - bx = y,$ y-input, x-output

d) $-a\dfrac{dx}{dt} + bx = y,$ y-input, x-output

e) $a\dfrac{dx}{dt} + bx = -y,$ y-input, x-output

f) $a\dfrac{dx}{dt} - bx = y,$ x-input, y-output

5–48. The growth of the world population may be modeled. The population is controlled by the birth rate and the death rate from all causes. Assuming that these rates are constants (not true in the real world), write a differential equation whose solution is the population. Solve this equation and sketch the result. Consider the cases where the birth rate is greater than the death rate, and vice versa.

5–49. In the electric circuit in Fig. 5–75, the current through the capacitor, i_C, is specified as shown in the graph. The capacitor is uncharged at $t = 0$. With *your* reference conditions shown on *your* diagram, determine and present graphically with significant numerical values,

a) the voltage across C,

b) the current in R,

c) the current in L,

d) the voltage across L, (Continued.)

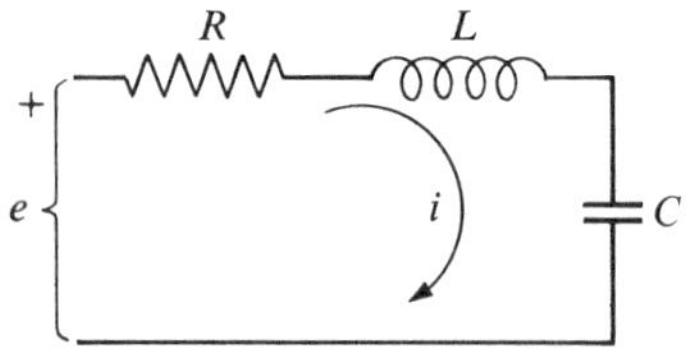

$R = 1/2$ ohm
$L = 1$ h
$C = 2$ f
C is uncharged at $t = 0$

FIGURE 5–76

Given i as shown

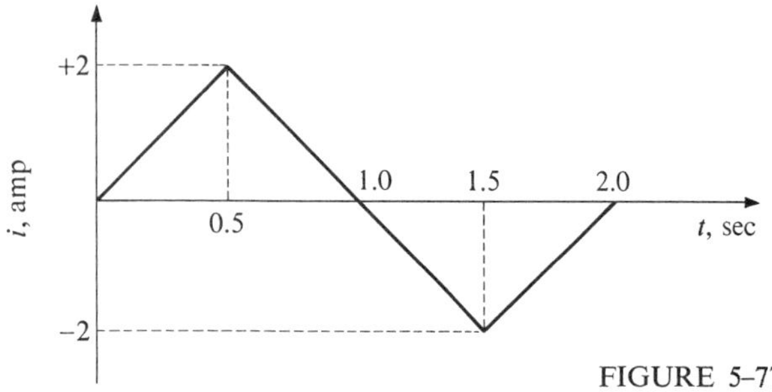

FIGURE 5–77

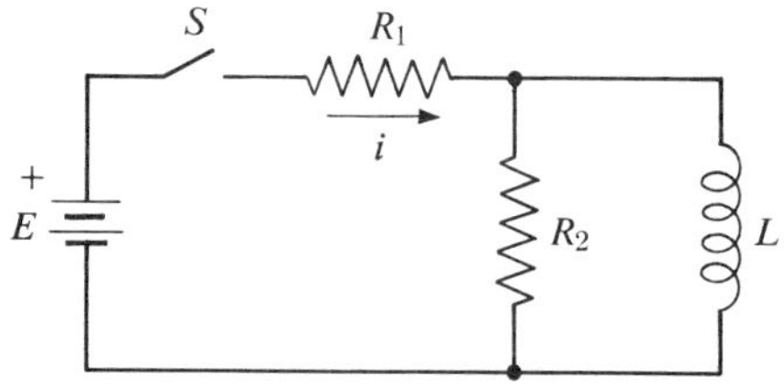

E = const, circuit relaxed at $t = 0-$

FIGURE 5–80

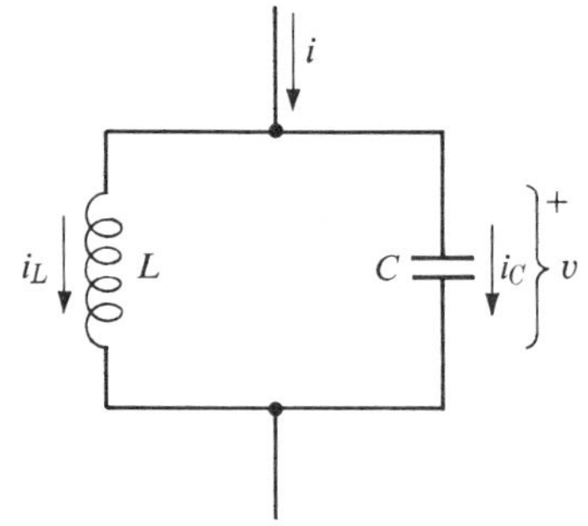

FIGURE 5–83

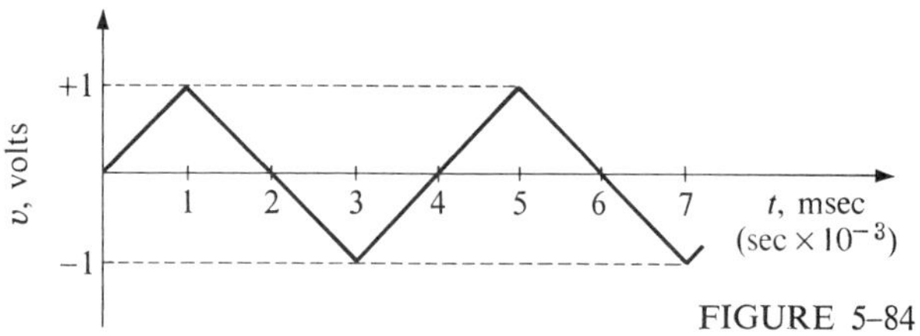

FIGURE 5–84

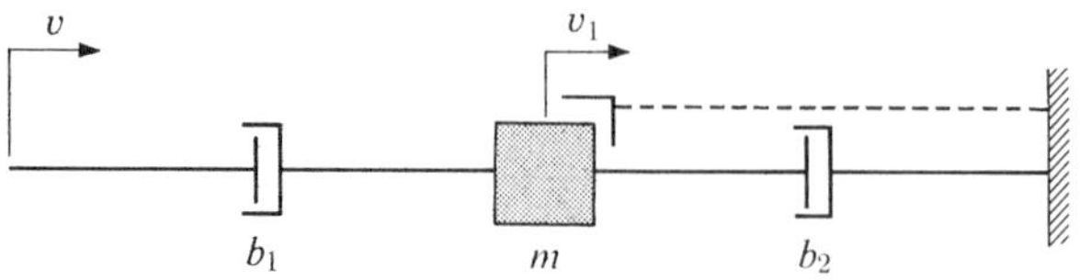

FIGURE 5–78

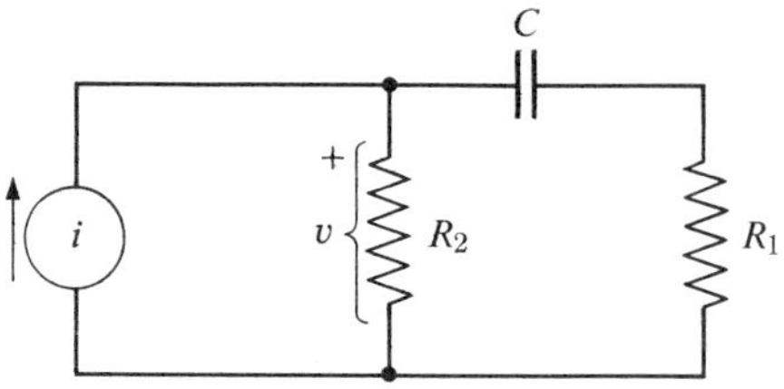

FIGURE 5–79

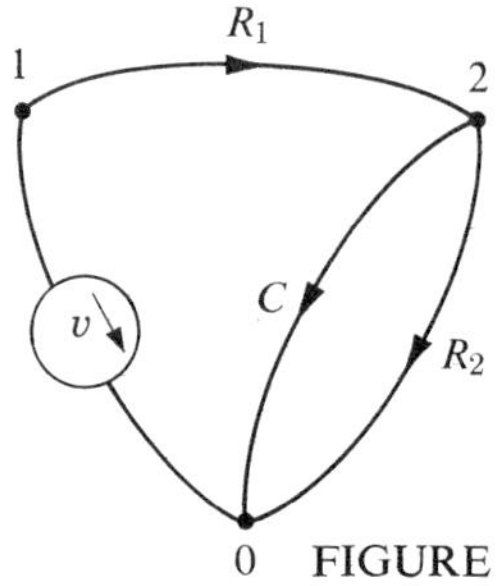

FIGURE 5–81

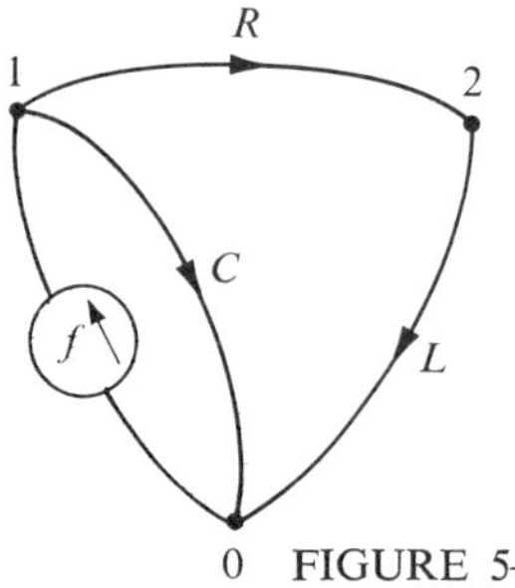

FIGURE 5–82

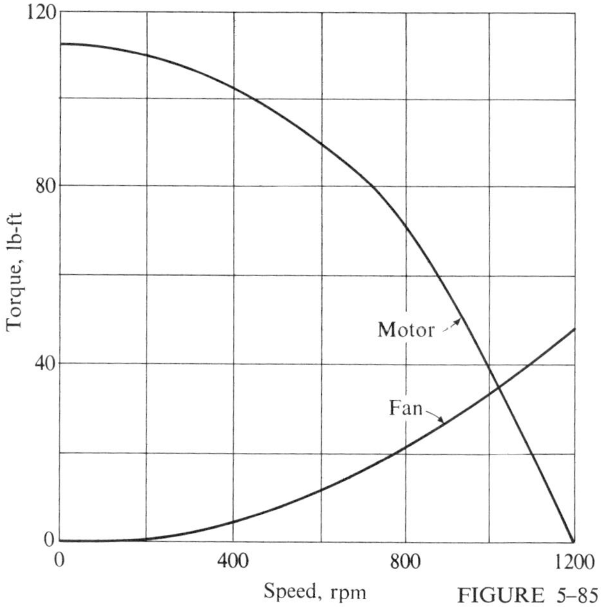

FIGURE 5–85

e) Draw the mechanical translational system which is analogous to the electrical circuit (force-current analogy).

f) Draw the fluid system which is analogous to the electrical circuit (flow rate-current analogy).

h) Draw the thermal system which is analogous to the electrical circuit (heat flux-current analogy).

5–50. For the circuit shown in Fig. 5–76:
a) Given i as in Fig. 5–77, determine the voltage e.

b) Determine the voltage e when

$$i = 0, \qquad t < 0,$$
$$i = 10 \sin 2t, \qquad t > 0.$$

c) Determine the voltage e when

$$i = 0, \qquad t < 0,$$
$$i = 2e^{-3t}, \qquad t > 0.$$

d) Given the current in (a), find the energy stored in L and C at $t = 1.5$ sec.

e) Find the energy which has been dissipated in R between $t = 0$ and $t = 0.5$ sec for the current given in (a).

5–51. Derive a differential equation for the velocity of the mass v_1 in terms of the velocity source v (Fig. 5–78).

5–52. The electrical system of Fig. 5–79 which is powered by a current source i is given:

a) Find the differential equation for v, the voltage across the resistor R_2.

b) Check the dimensions of this equation.

c) Check that it gives the correct result for $R_2 = 0$ and $R_1 = 0$ separately.

5–53. For the circuit in Fig. 5–80, the switch is closed at $t = 0$.

a) What is the time constant?

b) What is the initial value (at $t = 0+$) of each current?

c) What is the final value of each current?

d) Derive the differential equation for the current i.

5–54. Derive the system equation for the system in Fig. 5–81 for v_2-output and v-input.

5–55. Derive the system equation for the system in Fig. 5–82 for f_L-output and f-input.

5–56. The circuit in Fig. 5–83 is composed of ideal elements $L = 0.1$ h, $C = 10\,\mu$f and $i_L = 0$ at $t = 0$.

a) Assuming that the voltage v is a triangular wave as in Fig. 5–84, sketch i_L, i_C, and i.

b) Assuming that the voltage v is

$$v = 0, \qquad t < 0,$$
$$v = 10e^{-1000t}, \qquad t > 0,$$

sketch v. Determine equations for and sketch i_L, i_C, and i.

c) Assuming that the voltage v is

$$v = 0, \qquad t < 0,$$
$$v = 10 \sin 1000t, \qquad t > 0,$$

sketch v. Determine equations for and sketch i_L, i_C, and i.

5–57. For the circuit shown in Fig. 5–34, assuming that $e = 0$ (short circuit) and that the capacitance has an initial charge, discuss the system response for various values of R. Let $R \to \infty$, then let $R \to 0$. Explain what happens to the stored energy.

5–58. Consider the system in Fig. 5–33. Determine the differential equation for (a) F_k and (b) F_{b1}.

5–59. Consider the solution to Eq. (5–74) which is for the system shown in Fig. (5–33). If v is a step function, dv/dt is infinite at the moment of the instantaneous change. Discuss why this result is significant with respect to Eq. (5–74). What are the initial conditions on F? Solve this equation for $F = \mathsf{f}(t)$ and sketch.

5–60. For the circuit shown in Fig. 5–34, determine the system equation for i_1. Compare the result with Eq. (5–80) for v. Which terms of Eq. (5–81) are present on the right-hand side? Why?

5–61. A three-phase 10-hp wound-rotor induction motor is coupled to a centrifugal fan. The torque-speed curves of the motor and the fan are given in Fig. 5–85. The combined inertia of the motor rotor and fan blades is 1.26 slug·ft^2. Determine the time it takes for the system to come up to speed after the motor is energized if it is started at full voltage. Plot a curve of speed vs. time.

6 MODELING OF PHYSICAL SYSTEMS

6–1. CONCEPT OF A MODEL OF A PHYSICAL SYSTEM

The objective of an engineering analysis of a physical system is prediction of its behavior, often when the system is being designed. Since all real systems are extremely complex when viewed in detail, an "exact" analysis of any system is never possible. Therefore simplifying assumptions concerning the properties of the system components must be made, thus reducing the system to an idealized version whose behavior approximates the significant behavior of the physical system. In many cases these assumptions involve neglecting effects which are clearly negligible. Many of these effects are in fact neglected as a matter of course without a clear statement of the implied assumption. For instance, the effect of mechanical vibration on the performance of an electrical circuit is ordinarily not considered; however, it can be seen that there are situations in which vibrations might have a serious effect. If the vibrational magnitude were such that it caused a significant change in the spacing between capacitor plates, quite different currents would be observed in the circuit than would be predicted in an analysis which neglected this vibration. As another example, the induced voltage (called "pickup") in an electrical circuit due to surrounding time-varying electromagnetic fields (radio waves, radiation from fluorescent lights, etc.) is ordinarily not considered in a mathematical analysis. However, in many electronic circuits of high effective resistance, this effect is frequently sufficient to cause faulty operation. (In fact, shielded rooms, within which experiments are performed, are often necessary for sensitive high-frequency measurements.) Still another example is the usual neglect of the gravitational pull of nearby masses compared to the pull of the gravity of the earth. Likewise, the effect of the position of the moon and planets is not ordinarily considered in determining the motion of a body on earth, but it is clear that these positions will affect slightly the forces exerted on any mass on the earth.

These examples concern factors which are completely negligible in many situations of interest. However, in most situations there are slightly more important effects which will still be neglected frequently to define a problem so that it can be handled mathematically without inordinate complexity. It is this type of assumption which is most often faced. Sometimes these effects will be neglected in a first attack on a problem so that the engineer can get a quick feeling for the predominant effects. Then, in a more precise analysis, the more complex factors will be considered. In other cases, the complicating effects are considered sufficiently small that the essential behavior of the system is exhibited by the simplified version. In many cases, the importance of a complicating factor cannot be determined without recourse to experiment.

The process by which a physical system is simplified to obtain a mathematically tractable situation is called *modeling*. The resulting simplified version of the real system is called the *mathematical model*, or simply the *model*, of the system. The degree to which the behavior of the model represents that of the actual system clearly depends greatly on the validity of the various assumptions made in arriving at the model. The experience, skill, and judgment of the engineer are thus crucial factors in successful modeling. *A mathematical analysis is useless, however clever and complete, if it is based on a model which does not closely represent the behavior of the physical system under consideration.* On the other hand, undue complexity in formulation of system models is also to be avoided. First, the cost of completing the analysis of a system model increases very rapidly with its degree of complexity, and second, the use of an excessively complicated model will often obscure (with a mass of nonessential detail) predominant or first-order effects. The *best* model for any system is one that provides, with minimum complexity, the information necessary for engineering action.

Often there is an advantage in building up a system model by successive degrees of complexity until all important aspects of its behavior have been included. This process usually begins by representing components of the system with the pure system elements discussed in Chapters 2, 3, and 4.

6–2. MODELING OF SYSTEM COMPONENTS BY PURE SYSTEM ELEMENTS

6–2.1 Primary Lumped-Element Properties

Component parts of systems may often be identified which behave almost like the pure system elements discussed in Chapters 2, 3, and 4. For example, for predicting the flightpath of a space vehicle around the earth, the vehicle can usually be assumed to be an ideal mass. Resistors used in electronic circuits are designed to act like ideal resistances as closely as possible. Similarly, tanks of gas often behave primarily like pure (usually not ideal) fluid capacitances.

It should be clear from our considerations in previous chapters that the *primary* property of a component as a lumped element is determined by its behavior at low rates-of-change (approaching zero) of its through- and across-variables. If we think of steady sinusoidal variations of these variables, low rates of change correspond to low frequencies of oscillation. In addition, the primary property depends on the constraints applied to the component. The lumped-element behavior of a system component can therefore be determined by considering the case in which the through- and across-variables are constant (constant, of course, includes the special case of zero value). These primary properties are illustrated in Fig. 6–1, for the case of a mechanical component. Similar illustrations may be given for other physical systems. Consider the three electrical configurations shown in Fig. 6–2. (These are discussed from the field point of view in Reference 1.)

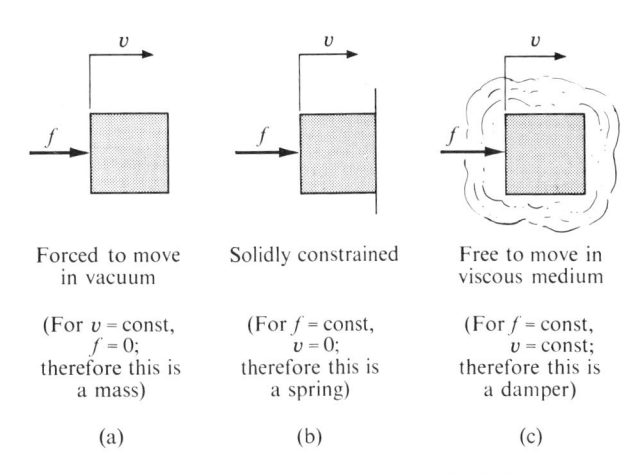

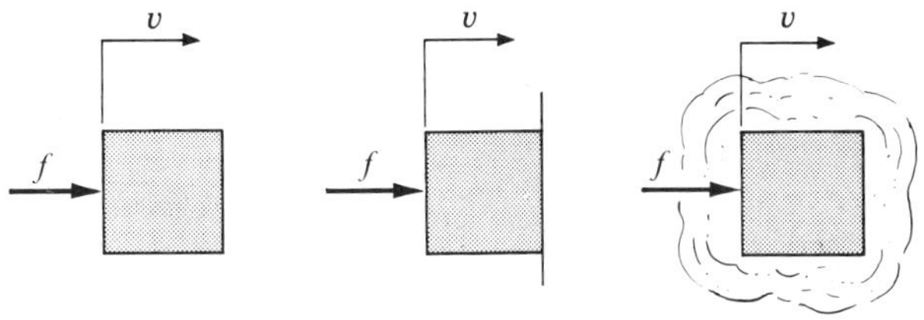

(a)	(b)	(c)
Forced to move in vacuum	Solidly constrained	Free to move in viscous medium
(For v = const, $f = 0$; therefore this is a mass)	(For f = const, $v = 0$; therefore this is a spring)	(For f = const, v = const; therefore this is a damper)

FIG. 6–1. Elemental behavior of a mechanical element as a function of its constraints.

A qualitative application of the principles of electromagnetic field theory shows that, for other than the static case, these structures exhibit properties in addition to their primary properties indicated in Fig. 6–2. For instance, for case (a), if the voltage is time-varying, a time-varying electric field exists between the plates. Again from field theory, using Ampere's law, as generalized by Maxwell to include displacement current, we see that a time-varying electric field induces a magnetic field which links the electric field. Thus there is a storage of magnetic energy in this "capacitive" structure when time varia-

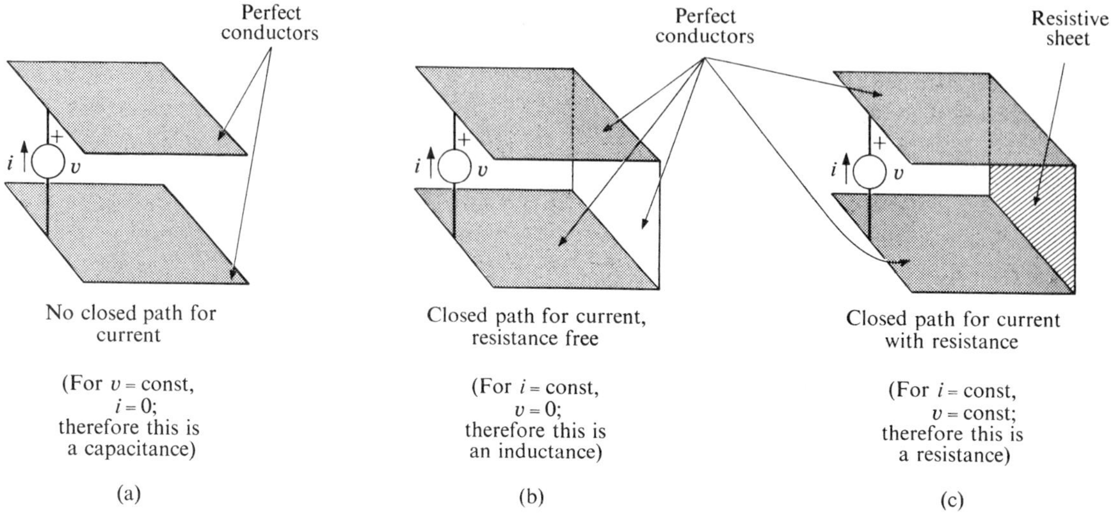

FIG. 6–2. Elemental behavior of an electrical structure as a function of its constraints.

tions exist. Only for sufficiently slow variations will this inductive effect be negligible. Likewise, for case (b), if the current is time varying, then a time-varying magnetic field exists. By Faraday's law, this induces a time-varying electric field which links the magnetic field, and thus there is a storage of electric energy as well as magnetic energy in the structure and its surroundings when time variations exist. Thus the "inductor" also exhibits the properties of a capacitor for sufficiently fast variations.

For case (c), it is easily seen from the similarity of the structure with the other two cases that both electric and magnetic fields (and therefore these respective energy storages) will exist even for constant voltage and current. Therefore, when time variations are present, this case will exhibit inductive and capacitive effects, along with the resistance. These effects, called *parasitic effects*, may be quite negligible over a wide frequency range for appropriate values of the resistance. However, for very high resistance values the capacitive effect will predominate over the inductive effect; it can also predominate over the resistance effect at rather low frequencies if the resistance is high enough. For low resistance, corresponding statements can be made concerning the predominance of the inductive effect.

We should, of course, also realize that perfect conductors do not exist, so that every closed circuit will exhibit a resistive behavior in the static case. Inductors are ordinarily made of coils of wire which, of course, have resistance. Most common magnetic and dielectric materials also have energy losses.

It should be noted here, if only to show the true complexity of the real physical world, that there are situations for which voltage and current cannot even be meaningfully defined. In such an event, the only recourse is a complete field approach to the problem by use of Maxwell's equations. Full appreciation of the limits of applicability of lumped-parameter electric-circuit theory can be obtained only by a thorough understanding of field theory.

6–2.2 Parasitic or Secondary Elements

Parasitic or secondary elements represent effects which always inherently accompany the primary behavior of a component as a lumped element. The term "parasitic" is used when the secondary effect is undesirable. For instance, the presence of resistance in a coil of wire intended to be an inductance would be called a *parasitic element*. In the discussion of the three structures in Fig. 6–2, these structures were seen to

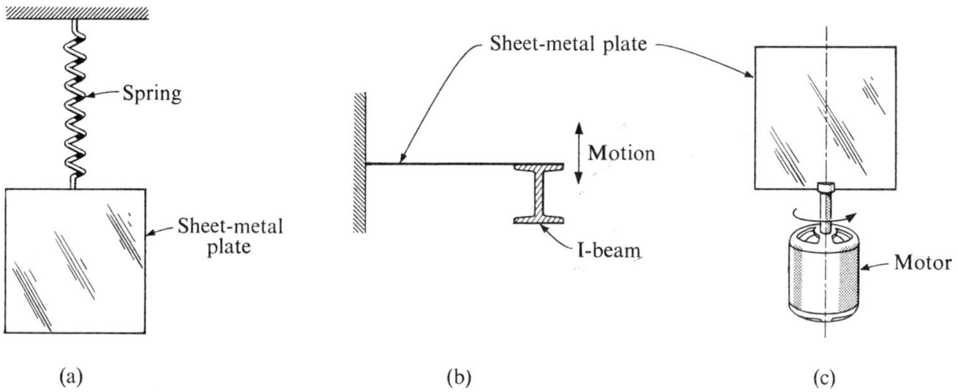

FIG. 6–3. Metal plate acting primarily as (a) a mass, (b) a spring, and (c) a damper.

possess parasitic effects in addition to their primary elemental behavior. The following example is provided to make the idea of parasitic or secondary effects clearer.

Let us consider a plate of metal. It is impossible to tell whether, as an unspecified system, this plate could be considered a mass, a spring, a damping mechanism, or even some combination of all three. Then again, it might be used simply as a lever (acting like a mechanical transformer) or even as a base plate to be bolted to the floor. There is no reason to restrict this list of applications to purely mechanical considerations. This plate could also be a microwave reflector, part of an electrostatic shield for a transformer, a workpiece to be heated in an induction heater, or an antenna for a paddlewheel satellite. The list could be continued indefinitely. Limiting the consideration to mechanical systems, we can easily see that this plate perhaps should be treated like a "mass" in many cases. However, it also has elasticity and thus could serve as a "spring" either in bending, tension, or torsion. Additionally, if it were to move through the air broadside at high speeds, then its drag or "damping" properties would be predominant. The systems illustrated in Fig. 6–3 show these uses in simplified situations.

It should be clear, however, that in each of these situations the other properties of the metal plate, besides the one of primary importance, will also be present. In case (b) for instance, the mass of the metal plate and the damping due to its motion through the air would be called parasitic elements. Whether these parasitic effects are important depends on the other parameters of the system and the conditions under which it is operated. In case (b), if the mass of the I-beam is sufficiently *large* compared to the mass of the metal plate, and the motion in the vertical plane takes place at *slow* rates of change, then only the "springiness" of the plate need be considered. It is to be noted that the italicized comparatives in the previous sentence are typical of conditions which must exist for parasitic effects to be neglected. The reader should review Examples 1 (Section 2–2.3) and 2 (Section 2–6.3) in Chapter 2.

6–2.3 Nonlinearity

Nonlinear effects, which frequently occur in physical devices, may affect the behavior of a component as a lumped element. Let us consider, for example, a mechanical device whose expected primary behavior is that of a spring. The actual force-displacement relation for the device might take any of the forms shown in Fig. 6–4 when the load is slowly applied and released. In case (a) we would say that the component acts like a pure but nonlinear spring since force is apparently a single-valued function of displacement. Case (b) cannot be represented exactly by a spring because the displacement depends on the history or previous values of the force as well as on the present

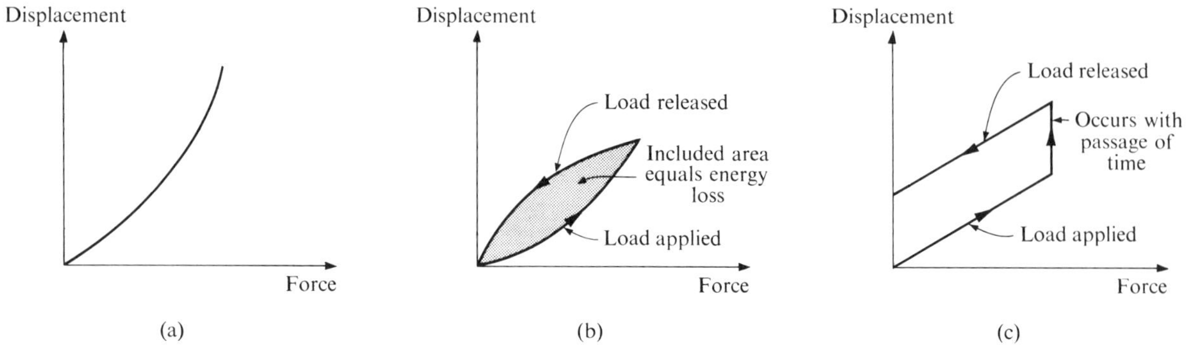

FIG. 6–4. Common types of nonlinearity.

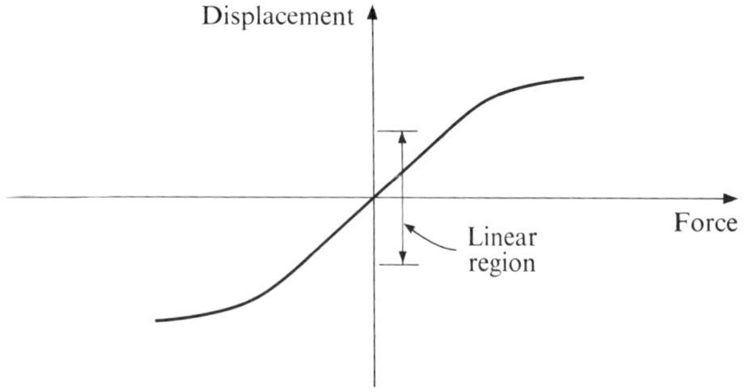

FIG. 6–5. Automobile air-spring characteristics.

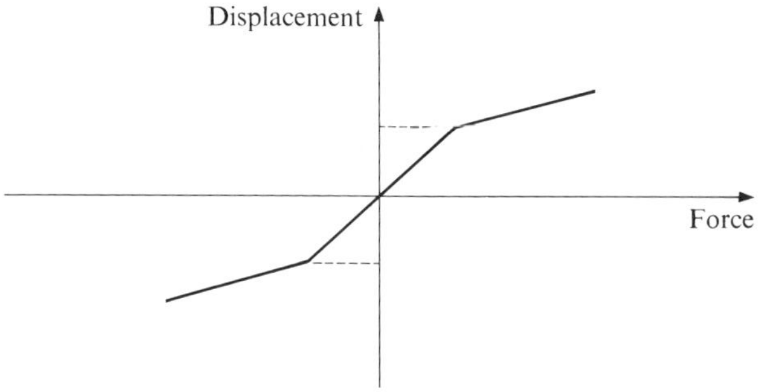

FIG. 6–6. Piecewise linear approximation to the air-spring characteristics shown in Fig. 6–5.

magnitude of the force. This type of behavior is often called *hysteresis* and is of great importance in inductors, which have magnetic energy storage in ferromagnetic materials (in this case the function relating flux linkage λ and current i would be of interest). The hysteresis effect involves energy loss, as is apparent from Fig. 6–4(b), since the energy is the integral of the force-displacement curve or the area included in the force-displacement loop.

The behavior illustrated in Fig. 6–4(c) involves a continued displacement at constant force as time passes and is commonly called "creep." The creep phenomenon occurs in plastic materials such as nylon and in metals at high temperatures. The response of soils to applied loads such as the weight of buildings also exhibits creep. In some instances, it is possible to represent creep as a secondary or parasitic damping element in the manner discussed in Section 6–2.2.

So long as the foregoing nonlinear effects are small, we may be willing to consider that the device is sufficiently described by Hooke's law, which describes force as being proportional to displacement. In other words, we might approximate the behavior by that of an ideal spring.

The reason for approximating system characteristics by linear relations is that the analysis of linear systems is much simpler than that of nonlinear systems. However, if the nonlinear effects are large or if we are concerned with their influence on the behavior of a system, approximation by linear relations would lead to results of little value.

In some cases, nonlinearity is purposely designed into a device to obtain specified performance. An example of this is the air spring now used on many automobiles. A typical air-spring characteristic is shown in Fig. 6–5. For small displacements the spring is almost linear. As the displacement increases, the spring becomes "stiffer." This tends to reduce the "swells" when riding over ruts and holes. An analysis which attempted to compare the performance of an air spring with that of coil springs and failed to take this nonlinearity into account would obviously be invalid. However, the engineer might wish to use a "piecewise" linear approximation (Fig. 6–6).

Other examples of this type of nonlinearity commonly occur with (1) damping mechanisms, where at low velocities the drag force is proportional to velocity, whereas at high velocities the drag force becomes proportional to the square of the velocity (actually, more generally this occurs at low and high fluid-flow Reynolds' numbers, respectively (Eq. 3–32); (2) inductors with cores of magnetic material where the material "saturates" when the magnetic field gets large, and thus makes the inductance high at low values of current and smaller at high values of current; (3) nonlinear resistors such as vacuum tubes and semiconductor devices; and (4) heat loss due to convection and/or radiation, as described in Section 3–4.3.

For the above types of nonlinearity, the nonlinear function may be "linearized" or approximated by a linear function, provided that the excursions of the variables are sufficiently small. The following example illustrates this technique (see also Fig. 2–17 and Eq. 3–58).

A certain carbon resistor used in an electronic amplifier has the curve of current vs. voltage-difference shown in Fig. 6–7. Physically this behavior occurs because the resistivity of carbon decreases with increasing temperature. As the current through the resistor increases, its temperature rises because of the conversion of electrical power into heat, thus causing its resistance to become smaller. If the excursions of i and v_{21} lie between points a and b on the curve of Fig. 6–7, the curve can be represented approximately by a straight line over this interval. This line may be drawn tangent to the curve at v_i, i_i or it may be drawn parallel to the chord which passes through points a and b. In the latter case illustrated in Fig. 6–7, we can write

$$v_{21} - v_i = R_i(i - i_i), \tag{6–1}$$

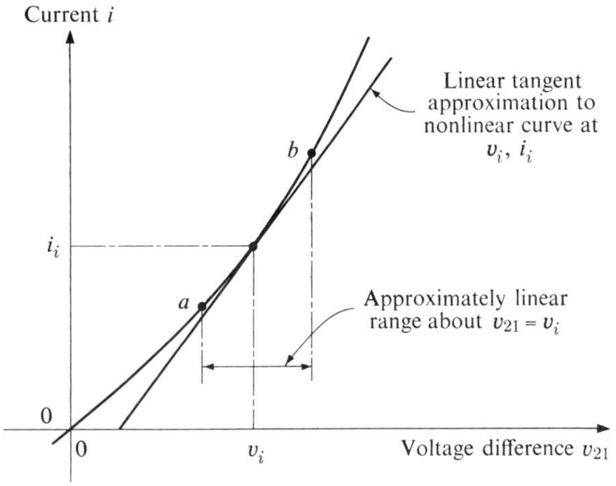

FIG. 6–7. Linearization of nonlinear electrical resistor characteristics about a quiescent operating point.

or letting the symbol Δ mean a small change in the indicated variable from its initial or average value,

$$\Delta v_{21} = R_i \Delta i, \tag{6–2}$$

where R_i, called the *incremental resistance*, is the slope of the $v_{21} - i$ curve at $v_{21} = v_i$:

$$R_i = (dv_{21}/di)_{v_i} = \Delta v_{21}/\Delta i.$$

This quantity can be computed if an equation for the curve is available, or it can be measured if the curve is presented graphically.

Some caution is necessary in working with linearized expressions such as (6–1) and (6–2) when energy and power are considered. For example, the power dissipated in the resistance is

$$\begin{aligned} P &= v_{21}i = (v_i + R_i \Delta i)(i_i + \Delta i) \\ &= v_i i_i + R_i i_i \Delta i + v_i \Delta i + R_i(\Delta i)^2. \end{aligned}$$

The first term is the power dissipated when $\Delta i = 0$, and the following three terms equal the *incremental power*, or change in power, due to Δi. Note that this power is not equal to the power one would compute by applying the power equation (Eq. 2–73) to Eq. (6–2). The latter power, $R_i(\Delta i)^2$ is in fact only a vanishingly small part of the incremental power as $\Delta i \to 0$.

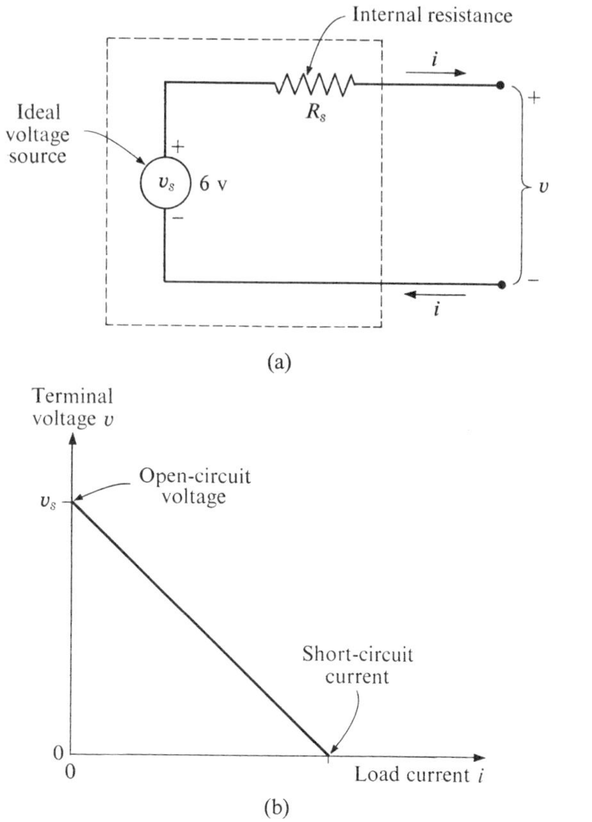

FIG. 6–8. Two-element lumped-parameter model of an electric battery. (a) Battery model. (b) Source characteristic.

6–3. MODELING OF COMPLETE SYSTEMS

Nearly every physical system of interest will be sufficiently complex that its behavior cannot be adequately described by any single pure system element or nonlinear function. However, by subdividing the system into an appropriate number of interconnected linear or nonlinear system components, we can obtain an adequate system model. This model typically will be composed of dynamic system elements of the types summarized in Chapter 4, interconnected so that they satisfy the structural conditions of compatibility and continuity. When the number of elements used is finite, we have a lumped-parameter or reticulated (network-like) model. As the number tends to infinity we approach a continuous or distributed-parameter model. The selection of an appropriate model for a physical system requires insight and engineering judgment, qualities which will develop with experience.

6–3.1 Representation by Finite Systems of Pure Elements

Many systems can be adequately modeled by interconnected combinations of pure system elements. To illustrate the process of model formulation by means of lumped-parameter system elements, let us consider two examples.

Example 1. Modeling of electrical sources

We have defined ideal sources in Chapter 4 as elements that provide specified across- or through-variables. For instance, an across-variable source provides a specified across-variable independent of the value of the through-variable. No actual system is likely to be capable of providing this ideal situation, and hence real sources often cannot be modeled by ideal source elements alone. An adequate representation requires a model composed of at least two elements. Let us consider the electrical source.

An example of an electrical source is a battery. Battery outputs are measured in volts, v. A specification of 6 v does not mean, however, that the battery will hold its terminal voltage at 6 v regardless of the load current i flowing out of it. The 6-v rating is only the voltage when i is zero, called the *open-circuit voltage*. A simplified view of the battery may be obtained if one assumes that it can be represented by a chemical process which produces 6 v independently of the current drawn, and an "internal" resistance due to the electrolyte resistivity. A two-element model of the battery then consists of an ideal voltage source connected, as in Fig. 6–8(a), in series with a pure resistance. If the currents drawn are not too large compared to the short-circuit current (i when $v = 0$ in Fig. 6–8), the resistance will be nearly ideal for many types of batteries. Figure 6–8(b) shows the "source characteristic" curve given by the two-element model of the battery. If the battery is used to supply an electrical circuit in which high-frequency variations of current or terminal voltage are required, the simple model of Fig. 6–8(a) may become invalid, and other system elements would be required (e.g., inductance and capacitance) for the battery to be modeled adequately.

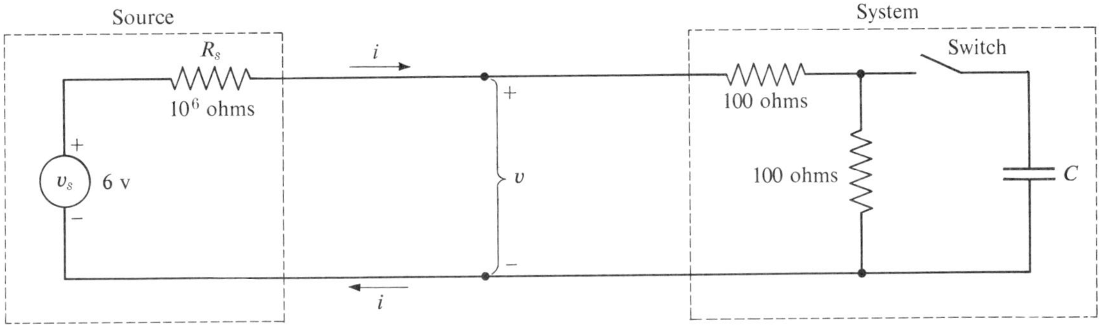

FIG. 6–9. Battery with large internal resistance supplying a low-resistance system.

However, under certain conditions, the battery may be assumed to act like an ideal source. To show that the type of source to be assumed depends on the system to which the battery is attached, let us consider the battery to be connected to the system depicted in Fig. 6–9. Suppose that R_s is 10^6 ohms (1 megohm). In this situation, the resistance in series with the source is so large that the current into the system is essentially constant. To see this, let us assume that i is given by the value it would have if the system were replaced by a short circuit (i.e., $i = 6\text{ v}/1\text{ megohm} = 6\,\mu\text{a}$). With the system as shown, the voltage v would vary between $1200\,\mu\text{v}$ ($200 \times 6 = 1200$ when C is fully charged and looks like an open circuit) and $600\,\mu\text{v}$ ($100 \times 6 = 600$ when C is uncharged and looks like a short circuit). In either case, however, this voltage is so small compared to 6 v that the entire 6 v is essentially applied across the 1-megohm resistance. Thus this is a source which is best described by saying that it puts out a *constant current* of $6\,\mu\text{a}$ regardless of the voltage. (It should be noted that v itself changed by a factor of 2 in the two cases calculated above.) However, if the resistances in the system were of magnitude comparable to one megohm, then the above statement concerning the constant-current characteristic of the source would be inappropriate. Also, if the resistances in the system were very *large* compared with 1 megohm, then we could easily show that the battery could appropriately be modeled as an ideal voltage source.

Example 2. Modeling of an automobile suspension system

For a second example, let us consider the problem of modeling an automobile and its suspension, with the objective of predicting the vertical motions of the vehicle and passengers. The objective of the modeling process and subsequent analysis might be the optimum design of the spring and shock-absorber system.

Figure 6–10(a) is a sketch of the automobile. As the vehicle moves along the road, the vertical variations x_1 and x_2 in the road height act as velocity or displacement sources to the tires. The forces caused by this input will impart vertical motion to the vehicle.

The simplest possible model of this system is shown in Fig. 6–10(b), where the vehicle is assumed to behave like a single lumped ideal mass moved vertically by the road displacement. This displacement, which is time-varying because of the forward velocity V of the car, might be assumed equal to the $x(t)$ seen by an observer moving along the road at speed V. If the road is bumpy and the vehicle has a large forward velocity, this model would predict extremely high vertical loads and accelerations and would also predict that the car would bounce away from the road. (Can you see why these effects are predicted by the model?) The objective of the suspension system is to reduce vertical accelerations to "comfortable" levels.

An improved model which gives some indication of the influence of the suspension is illustrated in Fig. 6–10(c). Here k represents the automobile springs and b is the effective damping coefficient of the shock absorbers (actually both these elements could be nonlinear, but we might assume them to be linear in a first analysis). Although this model is still very much simplified, it would allow us to predict approximately the motion x_c and the forces in the springs and shock absorbers. A much better model, particularly when the automobile speed is high or the intervals between bumps are very short, is afforded by including the wheel and axle mass m_w and the tire stiffness k_t, as in Fig. 6–10(d).

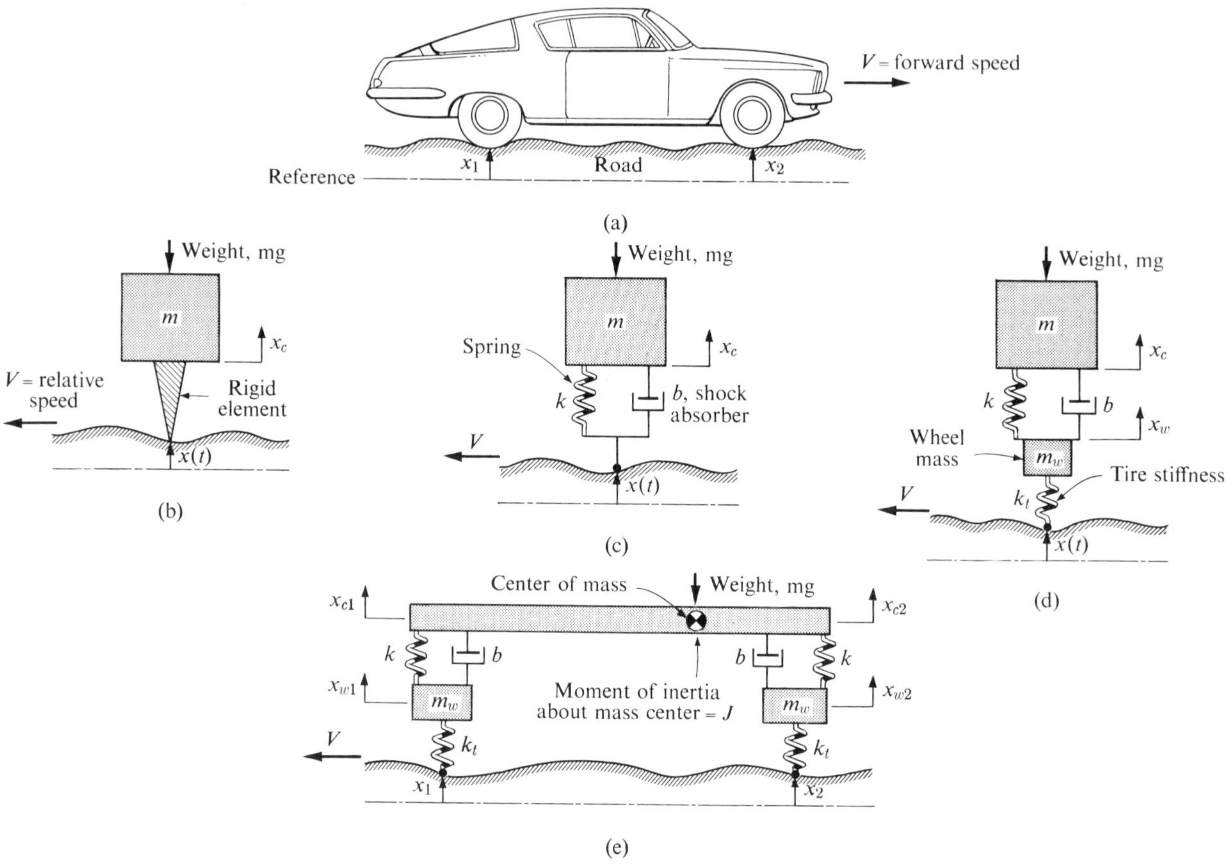

FIG. 6–10. Successive models of automobile and suspension system. (a) Automobile and suspension system. (b) Lumped mass model. (c) Mass, spring, and damper model. (d) Model including tire stiffness and wheel and axle mass. (e) Two-dimensional model.

All the foregoing models provide some information about the system behavior. If we were interested in a good quantitative estimate of system performance, including rocking of the automobile body, it would be necessary to go to more complex models, such as the one shown in Fig. 6–10(e). Here the auto body is assumed to be a rigid body of mass m and moment of inertia J about its center of mass.* The presence of two sets of tires, wheels, springs, and shock absorbers is admitted in this model. We could formulate more complex models which included nonlinearity of the system elements or which included three-dimensional motions of the automobile.

* It may be recalled from physics that the motion of any rigid body can be considered a translational motion of the center of mass (caused by the resultant force acting through the center of mass) plus a rotational motion about the center of mass (caused by the resultant moment about the center of mass).

In the process of generating increasingly complex and increasingly complete models of a system, the engineer should always be searching for the simplest model which will provide the qualitative and quantitative information which his particular problem requires.

6–3.2 Representation of Distributed Systems

Although many physical systems can be adequately modeled by suitable structures or networks of lumped elements, all such systems are substantially continuous. That is, in the earlier examples, it should be clear that each infinitesimal portion of a physical body exhibits both elastic and massive properties; i.e., the mass and elasticity are *distributed* properties. Likewise, in a structure in which time-varying electromagnetic effects occur, both magnetic and electric energy will be stored at each point in space; thus the inductance and capacitance are also *distributed* parameters. Similar observations may be made regarding any of the other pure system parameters we have discussed.

In some cases, a realistic model of a system requires detailed consideration of these distributed effects. In other words, the various energy storage and dissipation effects cannot be sorted out or isolated from each other and then considered as interconnected discrete lumped elements. Two examples of phenomena which must be modeled by the consideration of distributed effects are the propagation of pressure, or sound waves in solids and fluids, and the transmission of electrical power over long distances by means of wires. If a small pressure change is impressed at one end of a long fluid pipeline, this pressure is not felt immediately at the other end, but propagates through the fluid at a finite rate called *the speed of sound*. The moving pressure disturbance (and of course the associated flow) in the pipe is called a *wave*. Such waves are in part responsible for the effect called "water hammer" which you may have noticed as a rattle in water pipes at home when a faucet is suddenly closed. This wave phenomenon, including the rate of advance of the wave, depends on the distributed fluid capacitance and inertance. Similarly, when electric power is transmitted over long lengths of wire, wave propagation and reflection effects occur which can only be explained by considering the distributed capacitance and inductance (and often resistance) of the wires.

For many distributed systems, the energy storage and dissipation effects can be sorted out on a microscopic or infinitesimal level, resulting in a model composed of an infinity of infinitesimal lumped elements. The mathematical equations for such models will be *partial* differential equations,* whereas the equations for systems of finite lumped elements are *ordinary* differential equations.

It should be made clear at this point that any detailed investigation of distributed systems is beyond the scope of this book. The reader is referred to the many texts in the various fields of elasticity, fluid mechanics, heat and mass transfer, and electromagnetic fields for the background prerequisite to such investigations.

6–4. ADVANTAGES OF THE STUDY OF IDEALIZED MODELS

The analysis of any physical system must be based on a simplified model conceived by the engineer and formulated on the basis of his insight and judgment. Although we will spend a great deal of time on the *analysis* of system models, the student should recognize the crucial importance of the modeling process. The ability to perform satisfactory modeling will grow with experience; the work here will form only a foundation for such growth.

The reader may wonder at this stage whether there is any value in a study of lumped-parameter systems which are obviously greatly idealized. The answer is clearly "yes," for at least three reasons. First, a large number of systems behave very nearly like reasonable idealized models. Slight deviations are a small price to pay for the tremendous simplification of analysis afforded by a lumped-parameter model. The warnings implied by the discussions contained in this chapter are intended to caution the reader to be constantly alert to possible significant deviations from his model. These statements should not be interpreted as indicating that the model will be inappropriate more often than not.

Second, the study of idealized models gives a powerful insight into the nature of phenomena in dynamic systems and thus forms a desirable, if not

* See Chapter 11 for the definitions of partial and ordinary differential equations.

absolutely necessary, background for consideration of the more complicated situations (nonlinear and/or distributed) which have been referred to in this chapter. Thus the student should approach the situations to be considered in this book from the viewpoint of determining what *would* happen *if* the real system were actually as described by the model. This information, plus further knowledge, gained experimentally or by more complicated analysis, of the deviations of the system behavior from the simple model frequently will form a substantial basis for engineering action.

Third, to a large degree the mathematical techniques to be learned in dealing with lumped-parameter linear systems are also directly applicable to distributed systems, and many nonlinear methods of analysis make extensive use of linear theory. Thus the background acquired here will be of continuous value in other areas and in more advanced work.

REFERENCES

1. R. M. FANO, L. J. CHU, and R. B. ADLER, *Electromagnetic Fields, Energy, and Forces.* New York: John Wiley and Sons, 1960.

2. R. M. ADLER, L. J. CHU, and R. M. FANO, *Electromagnetic Energy Transmission and Radiation.* New York: John Wiley and Sons, 1960.

3. J. P. DEN HARTOG, *Mechanical Vibrations.* New York: McGraw-Hill Book Company, Inc., 1947.

4. J. F. BLACKBURN, G. REETHOF, and J. L. SHEARER, *Fluid Power Control.* Cambridge, Mass.: The M.I.T. Press, 1959.

5. F. A. MCCLINTOCK and A. S. ARGON, *An Introduction to the Mechanical Behavior of Materials.* Reading, Mass.: Addison-Wesley Publishing Co., Inc., 1966.

6. W. P. MASON, *Electromechanical Transducers and Wave Filters.* New York: D. Van Nostrand Company, Inc., 1958.

PROBLEMS

6–1. The drag cup described in Problem 2–14 is intended to exhibit the primary characteristic of an ideal damper. Unfortunately, the cups will necessarily have inertia, and hence, at best, parasitic mass effects will be present.

a) Make a lumped-element model of this system which includes the cup inertias. Draw the network diagram and the linear graph.

b) Derive, for $\Omega_2 = 0$, a differential equation for Ω_1 vs. the applied torque T.

c) For steady sinusoidal applied torque, $T = T_0 \sin \omega t$, find the frequency range ω over which the amplitude of Ω_1 can be predicted within 5% when the effect of inertia is neglected.

6–2. A parallel-plate capacitor has been found by measurement to have a capacitance of 10 $\mu\mu$f and a leakage resistance of 1 megohm. This device is attached in series with a 50,000-ohm resistor and a voltage source $v(t)$. At $t = 0$, when the capacitance is uncharged, $v(t)$ suddenly increases from zero to a constant value of 10 v.

a) First neglect the leakage resistance (assume it to be infinite); that is, assume that the capacitor is ideal. Solve for the capacitor voltage vs. time.

b) Repeat part (a) but including the leakage resistance. Sketch the solution and compare with part (a). Determine the magnitude of the largest error that is made by neglecting the leakage resistance.

6–3. The wire-wound coil described in Example 2 of Chapter 2 is connected across an ideal voltage source which is initially at 0 v. At $t = 0$ the voltage is suddenly increased to 50 v and held constant.

a) Assuming that the coil is an ideal inductance, solve for the current in the coil and sketch a plot of your result.

b) Repeat (a) using the resistance-inductance model developed in Chapter 2 and sketch.

c) Over what time interval, if any, can the current-time function be predicted within 1% if the coil is considered only as an ideal inductance?

6–4. Make a lumped-element model of the capacitor described in Problem 2–25. A sinusoidal voltage

$$v = v_0 \sin \omega t$$

is impressed across the capacitor. Find the frequency at which the amplitude of the current through the resistance is 5% of the total current amplitude to the capacitor. Determine the phase angle between applied voltage and total current at this frequency. Discuss the behavior of this device at very low and very high frequencies.

6–5. Make a lumped-element model of the fluid flow in the tube of Problem 3–11 for the range where Hagen-Poiseuille flow is valid. Draw the linear graph. Assuming that the pressure drop across the tube P varies sinusoidally, $P = P_0 \sin \omega t$,

a) solve for the steady sinusoidal variation of flow;

b) determine the range of frequency ω over which the amplitude of the sinusoidal flow can be predicted within 5% by assuming that the tube is an ideal fluid resistance.

6–6. Make a lumped-element model of the thermal system described in Problem 3–16 and draw the linear graph. The water is stationary within the pipe.

a) If only the thermal resistance of the insulation and the thermal capacitance of the water are considered, write the differential equation relating water temperature to the temperature of the outside surface of the insulation.

b) Estimate the time for the water to reach the freezing point if the outside temperature suddenly drops to $-20°$F from an initially steady value of 60°F.

6–7. For the fluid system in Fig. 6–11, we have

$$C_f = 1000 \text{ in}^5/\text{lb}, \qquad R_1 = 0.005 \text{ lb·sec/in}^5,$$
$$R_2 = 0.001 \text{ lb·sec/in}^5.$$

a) Determine a differential equation relating the pressure at the tank P_t to the system flow Q.

b) Consider that the system flow Q has been 10^4 in^3/sec for some time and then suddenly doubles. Make sketches of the variation of (1) the flow in R_1, (2) the flow into the tank, and (3) the tank pressure P_t as a function time.

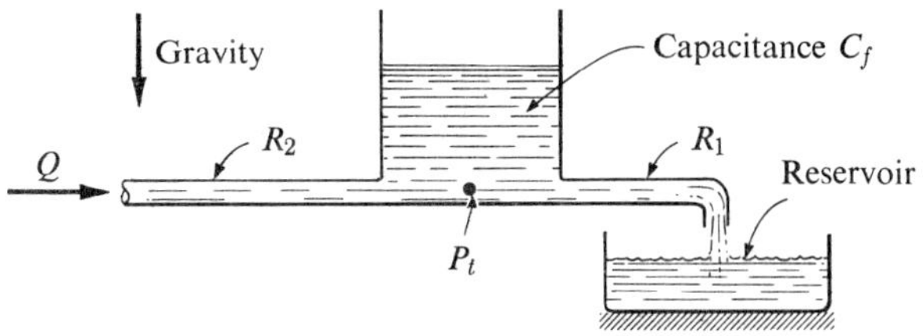

FIGURE 6–11

Tank area = 1 ft^2
Pipe characteristic, $\Delta P = KQ$
$K = 500$ lb-sec/ft^2

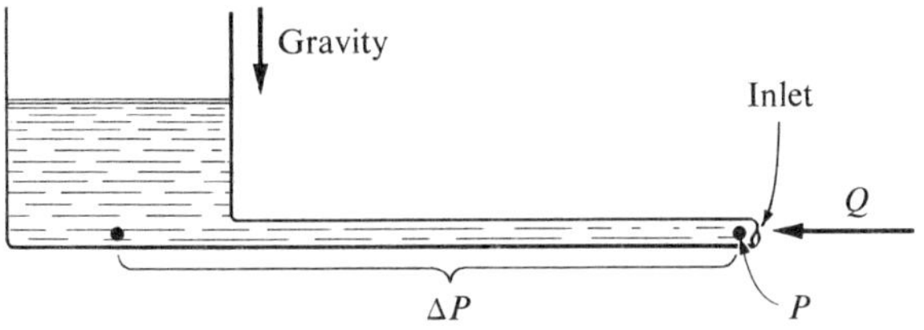

FIGURE 6–12

Explain these curves on the basis of the differential equation in (a) and any auxiliary equations you used to obtain (a).

6–8. A fluid-flow system containing a tank and connecting pipe (through which water flows) has the following properties: tank area = 1 ft^2; pipe characteristic, $\Delta P = KQ$, and $K = 500$ lb·sec/ft^5. The tank initially contains 2 ft^3 of water when the inlet pressure P suddenly doubles in value. See Fig. 6–12.

a) Make a lumped-element model of this system.

b) Determine a differential equation relating the system flow rate to the inlet pressure P.

c) Determine a differential equation relating the height of water in the tank to the inlet pressure.

d) Make sketches of the variation of the flow rate and water height vs. time when the system is subjected to the above pressure change. Explain all discontinuities in slopes and magnitude in terms of the differential equations given in (b) and (c).

e) Compute the final volume of water in the tank.

6–9. A fluid-flow system is arranged as shown in Fig. 6–13. The tank and pipe have the properties given in Problem 6–7.

a) Make a lumped-element model of the system.

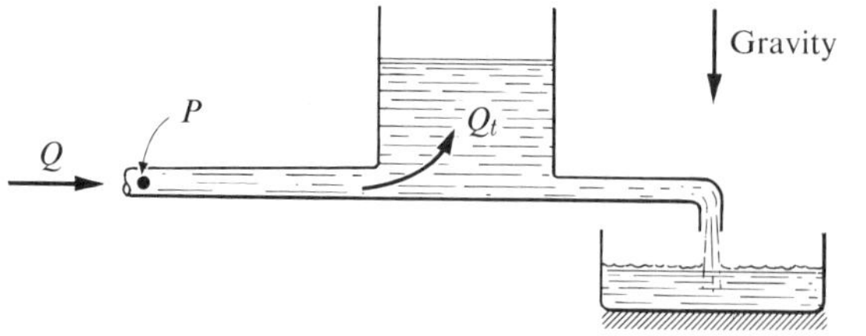

FIGURE 6–13

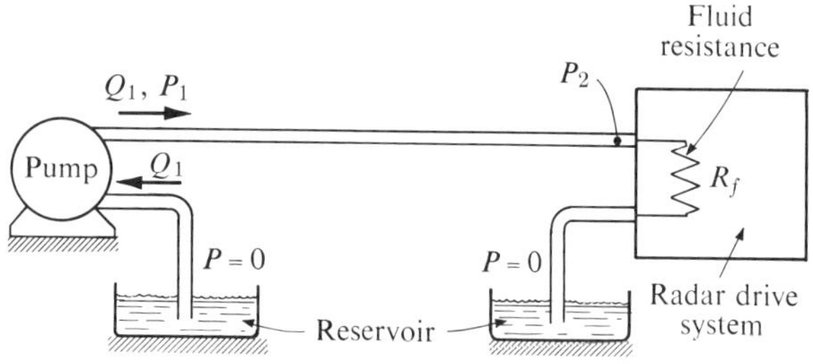

FIGURE 6–14

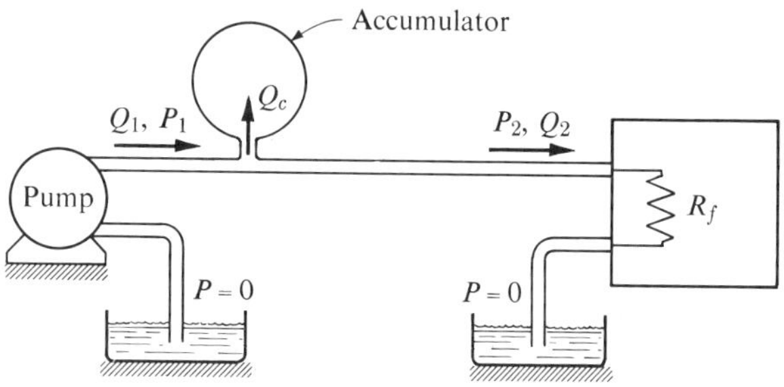

FIGURE 6–15

b) Determine a differential equation relating the tank inlet pressure P to the flow Q which enters the entire system.

c) Determine a differential equation relating the flow Q_t into the tank to the system flow Q.

d) Consider that the system flow has been at such a value that the tank has had a constant 20 ft^3 of water in it. The flow then changes to three times this value very suddenly. Make sketches of the variation of (1) the flow in the pipe, (2) the flow into the tank, and (3) the pressure at the tank bottom as functions of time. Explain these curves on the basis of the differential equations given in (b) and (c).

6–10. A volume of copper is heated by induction heating. (Induction heating is a method based on electromagnetic induction of circulating currents; we may assume that it is simply a way to produce a given (uniform) rate of heat-energy generation inside the copper.) The copper is surrounded with insulation and the outside of the insulation is held at a uniform temperature by a water bath (say at 32°F). We assume that (1) the copper is at a uniform temperature throughout and (2) the insulation has negligible heat capacity compared to the copper; the mass of copper is 2 slugs, the insulation characteristic is $\mathbf{q} = k\,\Delta\theta$, and $k = 1$ Btu/hr·°F.

a) Determine a differential equation relating the temperature of the copper to the rate of heat generation within the copper.

b) Determine a differential equation relating the rate of heat storage in the copper to the rate of heat generation in the copper.

c) If the rate of heat generation has been at a value which makes the copper have a temperature of 50°F for some time and suddenly becomes four times that value, sketch curves of the variation of the copper temperature and the rate of heat storage in the copper vs. time. Explain these curves based on the equations in (a) and (b).

6–11. The source for an automatic hydraulic radar-antenna drive is a positive-displacement pump (flow source) Fig. 6–14. A steel tube of inside diameter D and length L connects this pump to the drive system. The drive may be modeled here by an ideal fluid resistance, R_f. The fluid has density ρ, viscosity μ, and bulk modulus β.

a) Assuming that the pipe is rigid and the fluid is incompressible ($\beta = \infty$), propose and draw a lumped-element model of the system. If Q_1 is suddenly changed, what will be the maximum value of P_1? Comment on the validity of the model for this case.

b) Propose an improved model which includes fluid compressibility ($\beta \neq \infty$). Express the values of the lumped parameters in terms of D, L, ρ, μ, and β.

c) To reduce pulsations of flow and pressure P_2 and Q_2 at the drive system, it is proposed to place a fluid accumulator near the pump (Fig. 6–15). The accumulator is described by the equation $Q_c = C_f(dP_1/dt)$.

Using the assumptions of part (a), draw a lumped-element model of the new system including the accumulator.

If Q_1 varies sinusoidally at frequency ω about an average value Q_0, for what range of ω will the time variation of P_2 be less than 5% of the average value of P_2?

6–12. Consider the fluid piston system in Fig. 6–16. List the conditions which must hold for this device to behave like an ideal fluid transformer.

a) If the piston assembly has mass, its effect on the fluid system will be to add inertance to the fluid system. This inertance could be placed on either side of an ideal transformer to model the system. Compute the values of inertance to be used in each case (i.e., inertance shown on either side), given that the piston has mass m and draw the fluid linear graph.

b) Assuming that the piston is massless but the connecting rod is elastic [i.e., that the rod is a spring (k)], repeat part (a) but introduce capacitance.

c) If the piston is not perfectly sealed, fluid will leak between points 2 and 1 and between 4 and 1. Assuming that the leakages are represented by ideal fluid resistances, make a lumped fluid model of the *real* fluid transformer including leakage, mass, and elasticity. Draw the linear graph.

6–13. Figure 6–17 shows a single pair of spur gears having input shaft 1 and output shaft 2. It is desired to make a lumped-element model of this real mechanical transformer which includes the inertias J_1 and J_2, the elasticity of the gear teeth, and the rotational friction on the bearings.

a) Find the ideal relation between T_1, T_2, Ω_1, and Ω_2 when the gears are rigid, frictionless, and massless.

b) If $\Omega_2 = 0$ and a steady torque T_1 is applied, a proportional rotation Θ_1 of shaft 1 will occur due to tooth deflection, $\Theta_1 = T_1/K$. Assuming that $\Omega_1 = 0$ and a steady T_2 is applied to shaft 2, estimate Θ_2. Draw a linear graph which will describe this model, in which bearing friction and inertia are not considered.

c) If the gears are assumed rigid and frictionless, but have inertia, an input torque T_1 is reduced by the amount necessary to accelerate gear 1 before it is transmitted to gear 2. This torque is then further reduced by the inertia of gear 2 before it appears at the output 2. Write the equations relating T_1, T_2, Ω_1, and Ω_2 for this case and represent them by a linear graph.

d) Repeat part (c) considering only bearing friction; i.e., assume that the gears are rigid and massless. Draw the appropriate linear graph.

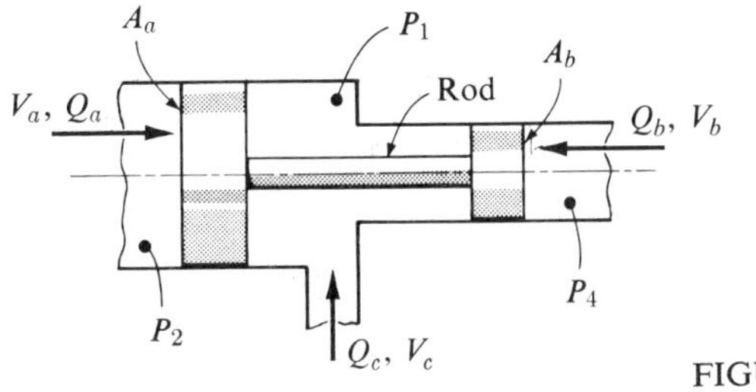

FIGURE 6–16

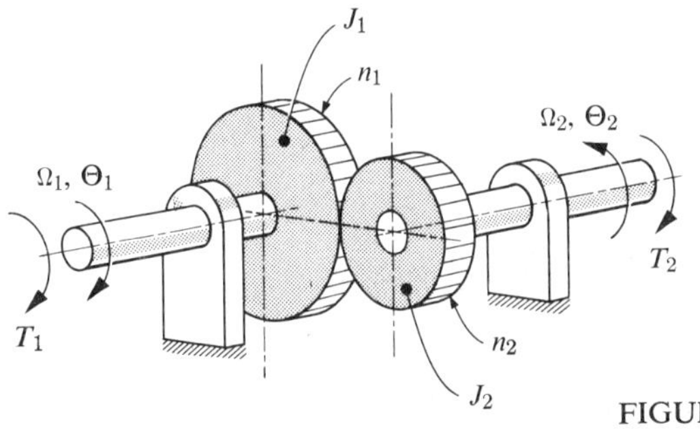

FIGURE 6–17

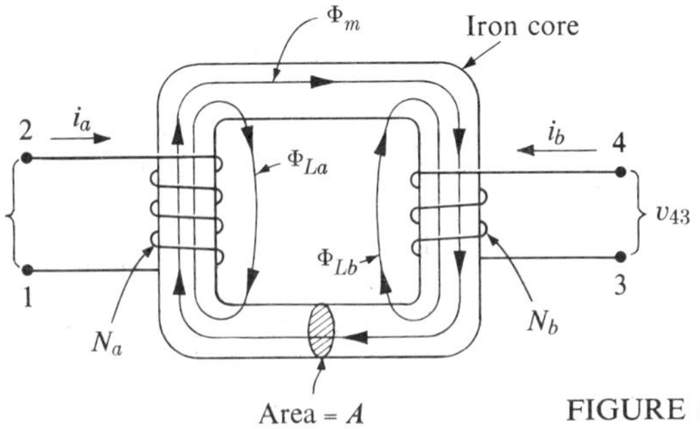

FIGURE 6–18

e) Finally, assemble the results of parts (a) through (d) to derive the equations relating T_1, T_2, Ω_1, and Ω_2 for the real mechanical transformer. Draw the linear graph and label each branch.

6–14. A more realistic model of an electrical transformer than that discussed in Section 2–6.4 includes the effects of winding resistance and leakage flux (Fig. 6–18). Summing voltage drops due to resistance and induced voltage, we see that

$$v_{21} = i_a R_a + N_a \frac{d\Phi a}{dt},$$

$$v_{43} = i_b R_b + N_b \frac{d\Phi_b}{dt},$$

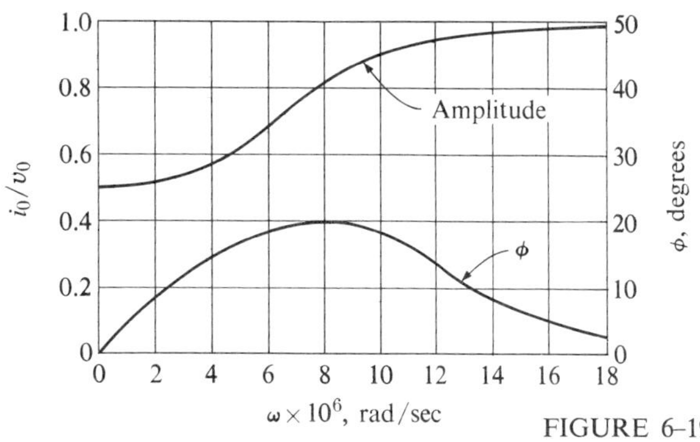

FIGURE 6–19

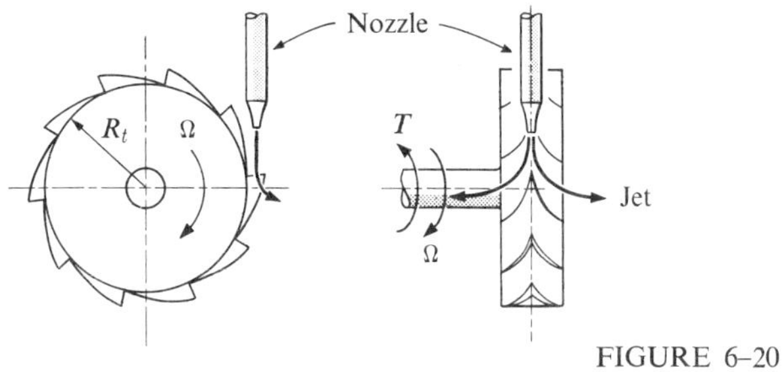

FIGURE 6–20

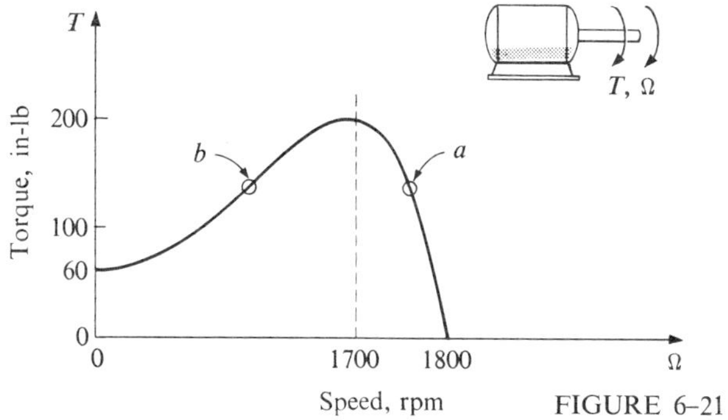

FIGURE 6–21

where $\Phi_a = \Phi_m + \Phi_{La}$, $\Phi_b = \Phi_m + \Phi_{Lb}$. Ampère's law for any closed flux line enclosing N ampere-turns is

$$\oint \frac{B}{\mu} \cdot dl = Ni,$$

where B is the flux density, μ is the permeability, and dl is an element of length along the path.

As a first approximation, assume that the permeability of the iron core is very large compared with that of air (it is 10,000 to 100,000 times larger). Then

$$\mu_0 N_a i_a = \Phi_{La} \int \frac{dl}{A} = C_a \Phi_{La},$$

$$\mu_0 N_b i_b = \Phi_{Lb} \int \frac{dl}{A} = C_b \Phi_{Lb},$$

where C_a and C_b are geometrical factors. For the mutual flux Φ_m,

$$N_a i_a + N_b i_b \approx 0, \qquad \text{since } \mu \gg 1 \text{ for the core.}$$

a) From the above relations, determine the equations relating v_{21}, v_{43}, i_a, and i_b. Represent these equations by a linear graph which includes an ideal transformer, two ideal resistances, and two ideal leakage inductances. Find expressions for the transformation ratio and for the leakage inductances.

b) If the permeability of the core were small enough that $N_a i_b \neq N_b i_b$, a so-called magnetizing inductance would appear in the electrical network and in the linear graph. Letting

$$\mu_i(N_a i_a + N_b i_b) = \Phi_m \int \frac{dl}{A} = C_m \Phi_m,$$

derive the governing equations and draw the linear graph for this case.

6–15. A two-terminal electrical system is tested by applying a steady sinusoidal voltage $v_0 \sin \omega t$ across the terminals and observing the current which flows through the system. This current $i(t)$ is found to be substantially sinusoidal, $i = i_0 \sin (\omega t + \phi)$. When ω is varied, i_0 and ϕ vary as in Fig. 6–19. Find a lumped-element electrical system which will exhibit this type of behavior (three ideal elements are required). Specify your best choice of element values and sketch the amplitude and phase characteristic for your equivalent system.

If the system were excited by a current source and the voltage were observed, would your equivalent system necessarily behave in the same way as the unknown physical system? Can you invent a situation where it would not?

6–16. A source of mechanical power for doing mechanical work, driving other mechanical devices or electrical generators, is afforded by the hydraulic or gas turbine, which is supplied from a source of pressurized fluid. One of the

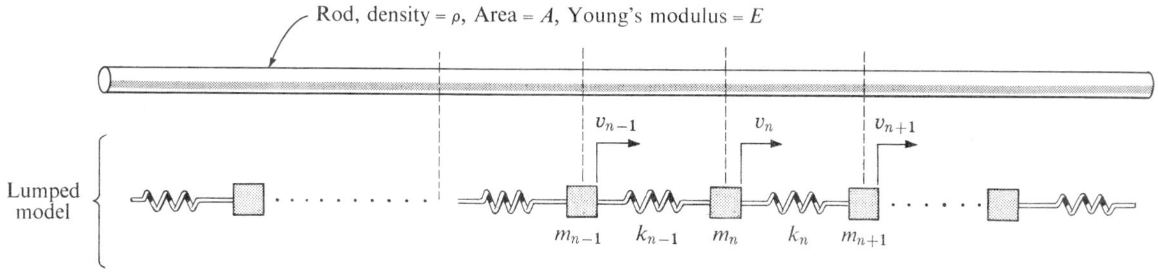

FIGURE 6–22

simplest of turbines is the impulse turbine shown in Fig. 6–20.

Fluid of density ρ is accelerated to a velocity v in a converging nozzle. The tangential force F exerted by the fluid jet on the periphery of the turbine wheel equals the rate of change of the linear momentum of the fluid particles from the jet. Assume that the jet has tangential velocity v at the nozzle exit and that it leaves the turbine wheel with no tangential velocity *relative to the wheel.*

a) For a nozzle area A, determine the torque T vs. speed Ω curve for various constant speeds.

b) Propose a simple lumped-element mechanical system which would represent this characteristic. This is a real mechanical source.

c) Modify your model in (b) to account for turbine-wheel inertia and draw the linear graph.

6–17. The steady-state torque-speed curves of electric motors are often highly nonlinear. A typical induction (ac) motor curve is shown in Fig. 6–21.

Assuming that the motor is loaded with a large average torque so that it initially operates at point a, propose a linear model which would represent the torque-speed curve in the neighborhood of a. Show how you would determine the parameters in this model.

If operation were to occur near point b, how would the ideal elements required in the model change?

6–18. The motor in Problem 6–17 is used to drive a fan whose torque varies as the square of its rotary speed. The input shaft to the fan requires 100 in·lb at 1800 rpm. The inertia of the fan measured at the input shaft is 1 lb·in^2, and the motor inertia is 0.2 lb·in^2. The connecting shafts are rigid.

a) Estimate the speed of operation of the fan.

b) Determine a lumped-element model which will permit the variations of fan speed to be studied for small variations about the steady-state speed.

c) Derive a differential equation for the small variations in speed resulting from the application of a small external torque to the fan-and-motor system.

6–19. A lumped-element model (Fig. 6–22) of a long uniform rod for extensional motions is to be made by dividing the rod into small segments, each of which has mass and flexibility.

a) Assuming that the rod is divided into n equal segments, find the stiffness k_n of each segment and the mass m_n of each segment.

b) Derive a differential equation for the motion of the nth mass in terms of the motions of its immediate neighbors.

c) Qualitatively discuss the accuracy of the lumped model as the number of segments is increased from one to many.

7 OPERATIONAL BLOCK DIAGRAMS AND ANALOG COMPUTER SOLUTIONS

7-1. INTRODUCTION

Block diagrams can be of great help in understanding dynamic system behavior because they reveal cause-and-effect relationships and interactions within the system that are not so readily evident from schematic diagrams or linear graphs alone.

The response of a dynamic system to a stimulus or input usually occurs in such a way that certain parts of the system, being more quickly affected by the stimulus, respond earlier than other parts. Reasoning on a physical basis, one can visualize or anticipate the relative quickness with which different parts of a system will respond to an input by thinking in terms of the transfer, storage, and dissipation of energy.

Let us consider, for instance, the response of a system consisting of two inertias and a damper to a step change in torque T as shown in Fig. 7-1. The first inertia J_1 initially feels the full effect of the suddenly applied torque. The torque transmitted through the damper B is initially zero (assuming the system is initially at rest). The inertia J_1 accelerates, and its velocity Ω_1 starts to increase as it receives energy from the source supplying the torque T. In the meantime the inertia J_2 does not immediately respond because the torque transmitted by the damper B is initially zero. As time goes by, Ω_1 increases, the torque transmitted by the damper B increases, the inertia J_2 experiences an increasing acceleration, and its velocity Ω_2 eventually begins to increase, storing energy that is transmitted to it by the damper. At the same time the torque in the damper B reacts on the inertia J_1, tending to reduce its acceleration and subsequently to reduce the rate at which energy is stored in J_2.

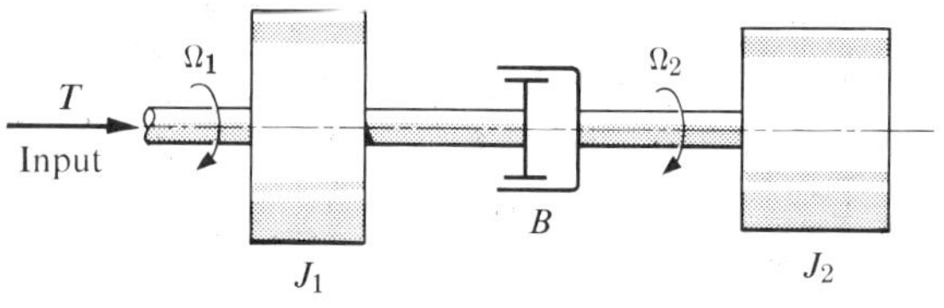

FIG. 7-1. Two rotational inertias coupled by rotational damper.

This qualitative sort of physical reasoning is especially helpful in gaining a qualitative insight into the behavior of relatively simple systems. However, what is needed is a more precise, systematic way of expressing the manner in which different parts of a system respond to a given input. Operational block diagrams fulfill this need admirably.

An operational block diagram provides a symbolic representation of all the basic system equations in one picture or "road map" of the system. All inputs and all outputs are seen at once. The relative quickness with which different parts of the system respond to a given input is immediately evident, and all the significant cause-and-effect relationships are revealed at a glance. Furthermore, formulation of a suitable operational block diagram of a system is the first step in preparing a given problem for machine computation. The use of computing machines in the study of dynamic systems becomes imperative as systems grow larger or more complex. The use of computers has removed much of the drudgery of analysis without eliminating the need for mathematical equations. Also, it has made it possible to obtain solutions for some problems, notably with nonlinear systems, that are not susceptible to closed-form mathematical solution.

7–2. OPERATIONAL ELEMENTS OR BLOCKS

The operational elements or blocks that are used in operational block diagrams correspond to the elementary mathematical operations that are used in mathematical analysis. The simplest of these, conceptually, is perhaps the operation of multiplying a variable x_1 by a constant coefficient, as shown in Fig. 7–2, to obtain another variable x_2. In addition to the fact that the operational block in Fig. 7–2 satisfies or expresses the mathematical equation $x_2 = ax_1$, it also prescribes a causal relationship, namely that x_2 results from x_1. It says that x_1 is an input, or independent variable, and x_2 is an output, or dependent variable, so far as this operational block is concerned. This block *is not capable of accepting x_2 as an input and giving x_1 as an output in the form $x_1 = (1/a)x_2$*. Thus it is necessary to know that one wishes to express x_2 as resulting in a causal way from the occurrence of x_1. If one wished instead to express x_1 as resulting in a causal way from x_2, then it would be necessary to use a different operational block, the one shown in Fig. 7–3.

Another simple mathematical operation is that of summation, or bringing together a number of variables by addition or subtraction to form a new variable which is equal to their sum (Fig. 7–4). Again it should be noted that a particular causal relationship is prescribed by this block diagram. In other words, it is necessary to decide ahead of time which variables are to be considered inputs. Note that the operational block can have only one output. If only two of the other three variables can be set up as inputs, then the other two must come out together as a single output

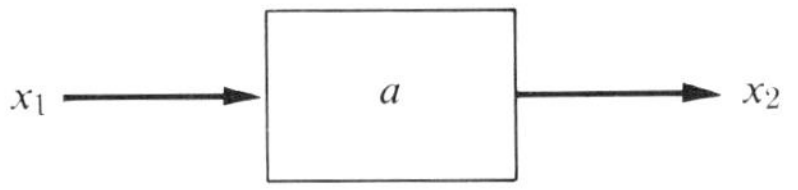

$x_1 \cdot a = x_2$

FIG. 7–2. Block diagram of coefficient operation.

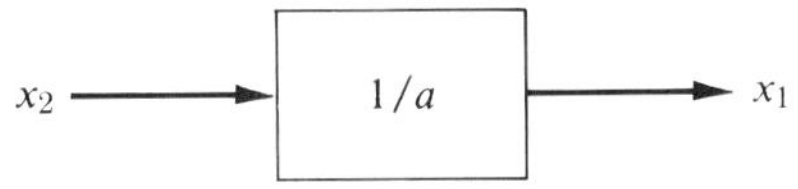

$x_2 \cdot \frac{1}{a} = x_1$

FIG. 7–3. Block diagram of inverse coefficient operation.

Diagram:

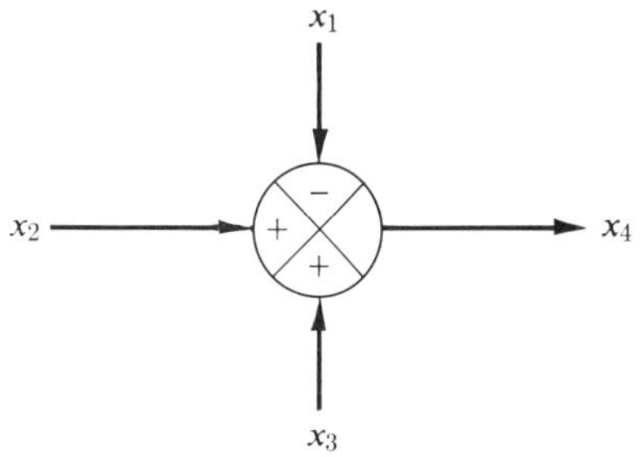

Equation: $-x_1 + x_2 + x_3 = x_4$

FIG. 7–4. Block diagram of summation operation.

Diagram:

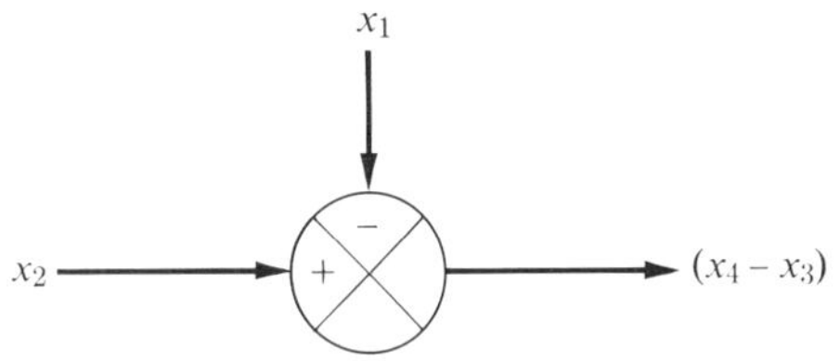

Equation: $-x_1 + x_2 = (x_4 - x_3)$

FIG. 7–5. Summation with alternative causal relationship having only two inputs.

(Fig. 7–5). Other particular prescribed causal relationships for *the same mathematical relationship* are shown in Fig. 7–6. Thus Figs. 7–4, 7–5, and 7–6 are different ways of expressing the same equation. It is seen that whereas the mathematician can express the

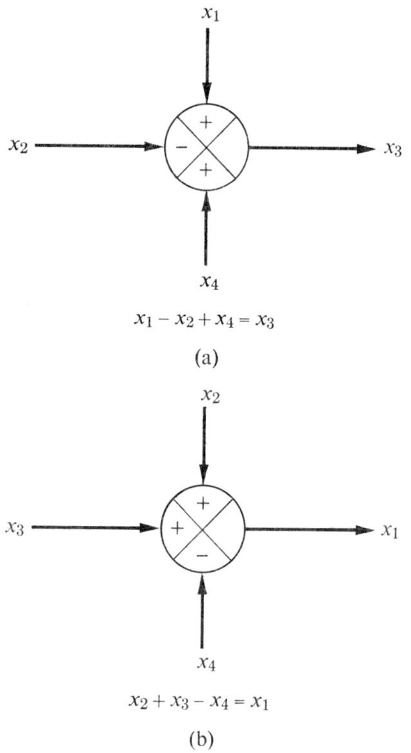

FIG. 7–6. Summation with other causal relationships.

Diagram:

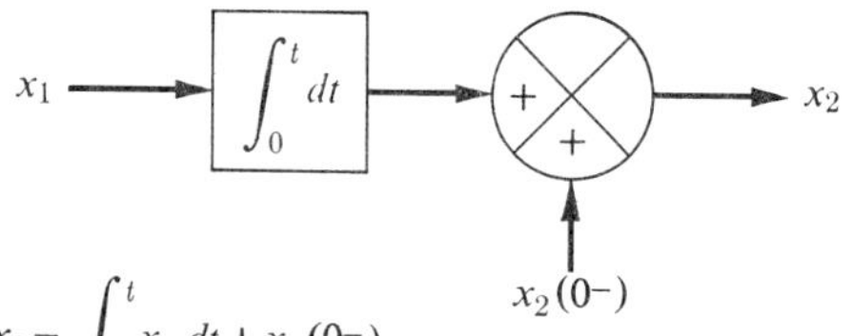

Equation: $x_2 = \int_0^t x_1\, dt + x_2(0-)$

FIG. 7–7. Block diagram of integration operation.

Diagram:

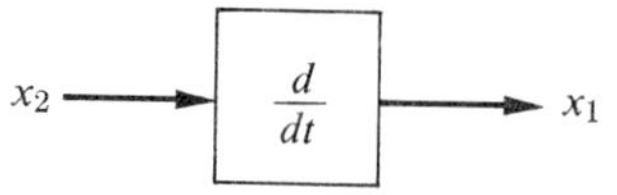

Equation: $dx_2/dt = x_1$

FIG. 7–8. Block diagram of differential operation.

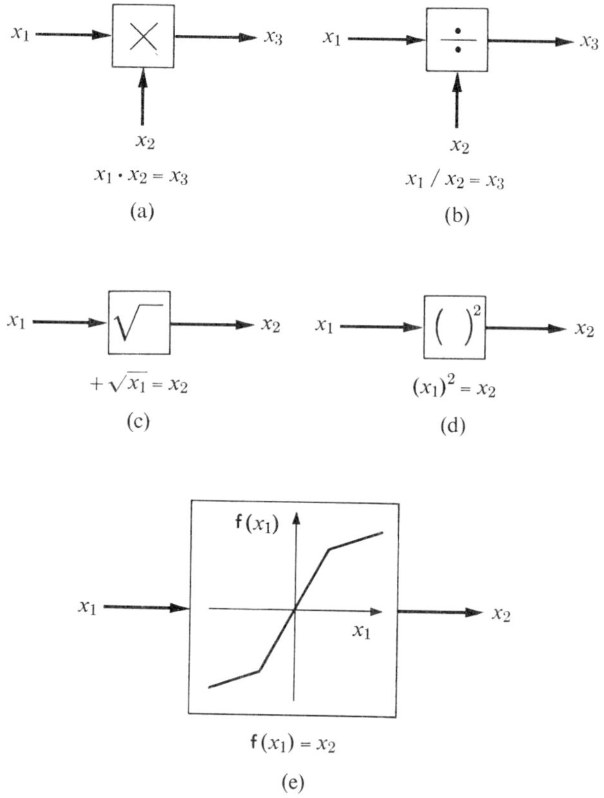

FIG. 7–9. Block diagram of some nonlinear operation.

relationship between x_1, x_2, x_3, and x_4 without regard to which variables are dependent and which are independent, an engineer constructing an operational block diagram must be very conscious of causality and must make decisions about dependency of variables as he proceeds from block to block in building up a complete operational block diagram of a system.

A third mathematical operation which is important in the construction of block diagrams of dynamic systems is integration with respect to time. Although the equations which are usually derived to describe dynamic systems are differential equations consisting of terms which express derivatives of the system variables with respect to time, the mathematical operation of differentiation is seldom used in operational block diagrams.* By careful planning and establishment of causal relationships between system variables, we can replace each differentiation with respect to time by a corresponding integration with respect to time. The formation of a variable as the time integral of another variable, or simply the operation of integration, is shown in Fig. 7–7. The causal relationship between x_1 and x_2 cannot be reversed with this operational block; *one cannot introduce x_2 as an input on the right and have x_1 come out on the left as the derivative of x_2.* Occasionally, when it is absolutely necessary to employ the operation of differentiation, it is represented as shown in Fig. 7–8.

All the operational elements described above are linear. Block diagrams of a number of nonlinear operations which are employed in the analysis of nonlinear systems are shown in Fig. 7–9. Although very little attention will be devoted to nonlinear systems in this book, it should be noted that nonlinearities pose no serious problems in the formulation of operational block diagrams—with the result that computers are frequently used to solve nonlinear problems.

Diagram:

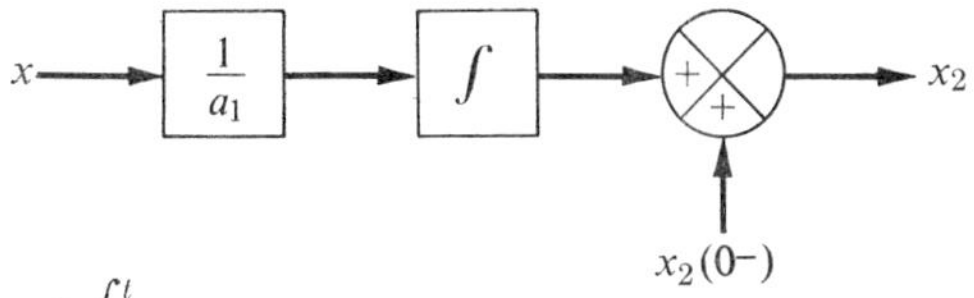

Equation: $\frac{1}{a_1}\int_0^t x\,dt + x_2(0-) = x_2$

FIG. 7–10. Operational block diagram of Eq. (7–1) in integral form.

Diagram:

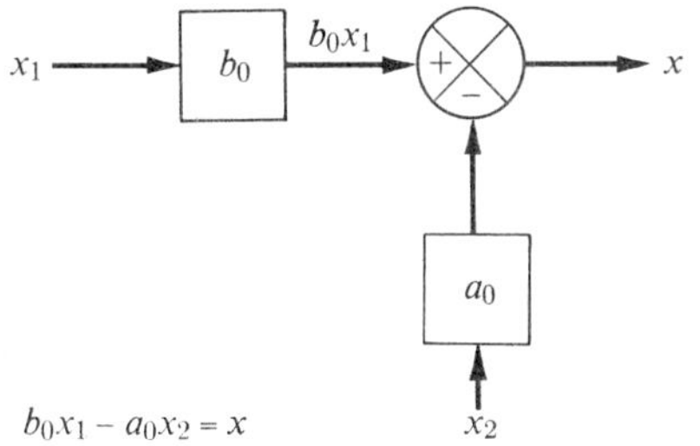

Equation: $b_0x_1 - a_0x_2 = x$

FIG. 7–11. Operational block diagram of Eq. (7–2).

7–3. OPERATIONAL BLOCK DIAGRAMS FOR FIRST-ORDER DIFFERENTIAL EQUATIONS

Block diagrams of very complex systems are developed most readily by working out the block diagrams for small parts of the system and then combining these diagrams later into a single over-all diagram.

Frequently a small part of a large system is described by one or two elementary equations such as the following:

$$x = a_1(dx_2/dt), \tag{7–1}$$

$$x = b_0x_1 - a_0x_2. \tag{7–2}$$

To proceed with drawing an operational block diagram for this subsystem, we must decide which variables are to be considered outputs. In this case, since integration will be much easier to accomplish than differentiation if the system is to be put on a computer, Eq. (7–1) will be integrated with respect to time, and x_2 will then be considered the output (dependent variable), as shown in Fig. 7–10. Then, in Eq. (7–2) the variable x will be the output (dependent variable), as shown in Fig. 7–11. The two block diagrams shown in Figs. 7–10 and 7–11 may now be combined (Fig. 7–12). The result is similar to the block diagram shown for a first-order system in Chapter 5. It is interesting to note in a qualitative way how the different parts of this system respond to a sudden (step) change in x_1. The variable x changes immediately with no lag or delay, but the variable x_2 can have no instantaneous response because time must elapse before the integrator can develop a change in its output even though its input changes suddenly. If the negative feedback through a_0 did not exist, the variable x_2 would continue to increase so long as the step in x_1 was maintained. The presence of the nega-

* The basic reason for this is that signal noise and instability make it difficult to perform differentiation in computing machines, and since block diagrams are often used to program computers, they do not call for operations which would be difficult to carry out on a computer.

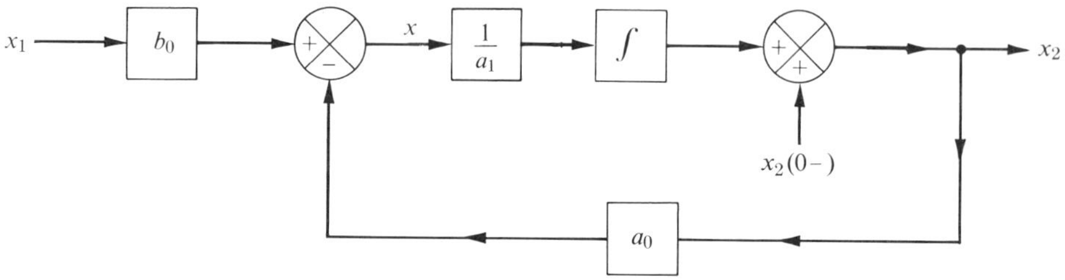

FIG. 7–12. Operational block diagram of Eqs. (7–1) and (7–2).

tive feedback through a_0 diminishes the rate at which x_2 increases with time, because as x_2 increases, more feedback through a_0 enters the summer, counteracting the effects of the input x_1 acting through b_0, and thus it reduces the input to be integrated with respect to time in the integrator.

It is interesting to note that if the feedback through a_0 were positive, the amounts to be integrated by the integrator would increase, x_2 would increase faster, and the system would be unstable. All the dynamic systems which survive in nature are stable and, when portrayed by operational block diagrams, normally exhibit negative feedback around their integrators. First-order systems with positive feedback around an integrator bring about their own complete destruction, go through a transfiguration, or go into a set of circumstances in which the positive-feedback effect is eliminated so that they become stable. An integrator by itself is an example of a marginally stable system. An infinitesimal amount of negative feedback will make it stable, and an infinitesimal amount of positive feedback will make it unstable.

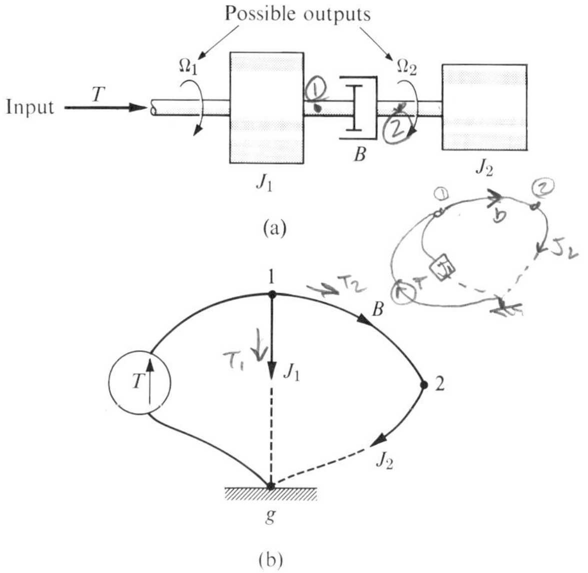

FIG. 7–13. Schematic and circuit diagrams of rotational system. (a) Schematic diagram. (b) Linear graph.

7–4. BLOCK DIAGRAM OF A COMPLETE HIGHER-ORDER SYSTEM

To demonstrate the evolution of operational block diagrams for higher-order systems, let us consider the rotational system in Fig. 7–1. The linear graph for this system is shown in Fig. 7–13 along with the schematic diagram. The describing equations for the individual elements are given by

$$T_1 = J_1(d\Omega_1/dt), \tag{7–3}$$

$$T_2 = J_2(d\Omega_2/dt), \tag{7–4}$$

$$T_2 = B(\Omega_1 - \Omega_2) = B\Omega_{12}. \tag{7–5}$$

The continuity condition requires that

$$T = T_1 + T_2$$

or

$$T_1 = T - T_2. \tag{7–6}$$

The compatibility condition is

$$\Omega_1 = \Omega_{12} + \Omega_2. \tag{7–7}$$

It is very important to note at this point that one should not combine the fundamental equations mathematically before starting the construction of an operational block diagram for the system. The operational block diagram is most readily constructed from these equations as they stand, and doing it in this way usually leads to a clearer understanding of the system. The equations then are systematically combined as the diagram is completed.

Since Eqs. (7–3) and (7–4) contain time derivatives, they should be integrated. The first step in preparing the block diagram can then be to set up these two integrations:

$$\Omega_1 = \frac{1}{J_1}\int_0^t T_1\,dt + \Omega_1(0-), \tag{7–8}$$

$$\Omega_2 = \frac{1}{J_2}\int_0^t T_2\,dt + \Omega_1(0-). \tag{7–9}$$

Since it is evident that J_1 will respond first to the input torque T_1, and since we have established a general pattern of proceeding from left to right (just as in writing a sentence) it is logical to have the integration associated with J_1 appear to the left of that for J_2 (Fig. 7–14). The next step is to add the summation operations required to satisfy Eqs. (7–6) and (7–7) (Fig. 7–15). The final step is to add the feedback paths for T_2 and Ω_2 (also Fig. 7–15).

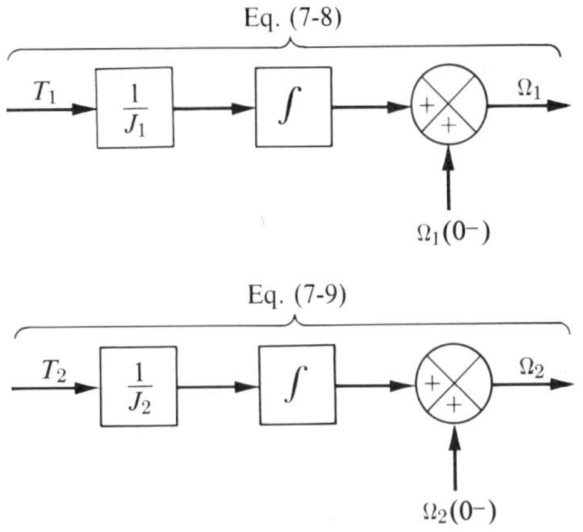

FIG. 7–14. Operational block diagrams of Eqs. (7–8) and (7–9).

The fact that two integrators are required in the block diagram of this system means that it is a second-order system. It is also evident that the variable T_1 is the first to experience a change in response to a change in T. Because Ω_1 and T_2 follow the first integrator, they experience changes only after time has elapsed for integration to take place in the first integrator, and their response to the change in input torque T is not so immediate as the response of T_1. Similarly the response of Ω_2 is delayed even more because it follows the second integrator. Note that both feedback effects around the individual integrators are negative. This discussion of system response in terms of the operations in the block diagram should be compared with the qualitative discussion given earlier in Section 7–1.

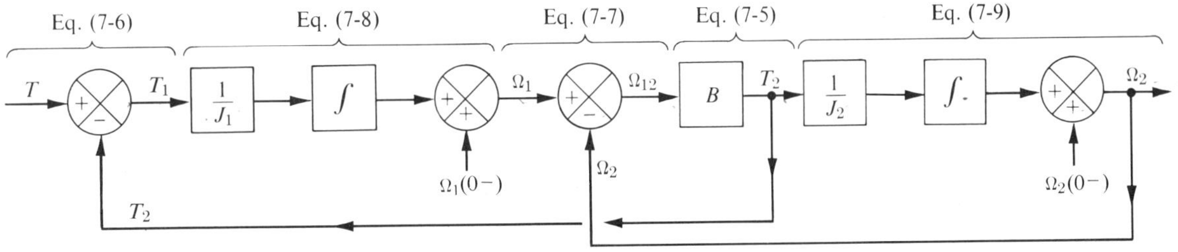

FIG. 7–15. Complete operational block diagram of rotational system.

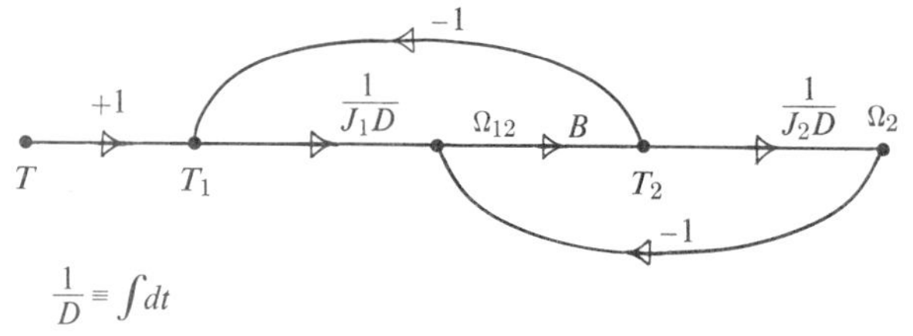

FIG. 7–16. Signal-flow graph of rotational system.

7–5. SIGNAL-FLOW GRAPHS

We can considerably simplify the operational block diagrams which have been employed in the previous section by omitting many of their details. When this is done as shown in Fig. 7–16 for the block diagram in Fig. 7–15, the resulting diagram is called a *signal-flow graph*, not to be confused with the linear graphs which have been already introduced in earlier chapters.

In addition to providing a much quicker means of sketching out the structure of a dynamic system from the signal-flow point of view, the signal-flow graph has been effectively used as an aid in carrying out the algebraic manipulations required to combine system equations. Some people who work extensively with systems analysis find signal-flow graphs very useful, and they are widely used in control-system texts.*

7–6. ELECTRONIC IMPLEMENTATION OF BASIC LINEAR OPERATIONS

There are obviously many ways to implement physically the various mathematical operations depicted by the operational block diagrams which have been used in this chapter. After all, the block diagrams are simply models of, or symbols representing, mathematical models of physical systems which we have been studying. As a matter of fact, several different means have been employed to implement these operations physically in attempts to develop differential analyzers and analog computers for studying the behavior of dynamic systems. One of the earliest of these developments was the electromechanical differential analyzer developed and used at the Massachusetts Institute of Technology between 1937 and 1948, in which mechanical devices were employed to amplify, sum, and integrate the rotational motions of shafts. Hydraulic and pneumatic computers have been used similarly in some instances for computation or control purposes.

The most successful means of simulating continuously the performance of dynamic systems has been the electronic analog computer, in which the basic operations are carried out electronically. There are several reasons why electronic implementation proved to be so successful. Electronic components have been highly developed for many applications such as radio, television, telephone, etc.; they are made in large quantities at low cost; they are interchangeable; and they are readily connected together to form large flexible computing systems. The required electronic operations are carried out reliably at low energy levels, with capability for fast response and corresponding high speeds of operation. The voltages which are used to represent system variables are easily generated, measured, and displayed with conventional electronic instrumentation.

Nearly all the important operations which are performed electronically in an electronic analog computer employ one or more high-gain direct-current (dc) voltage amplifiers for each operation. Thus it is necessary to have a plentiful supply of these "operational amplifiers" when undertaking the simulation of a large dynamic system.

An operational amplifier is a dc-voltage amplifier, an electronic device containing vacuum tubes or transistors which instantaneously* delivers, within its operating range, an output voltage proportional to an input voltage, even when the input is constant. The current required at the input is usually small enough to be negligible, and the amplifier can maintain its output voltage even when it must produce demanded load currents at its output (within prescribed limits).

* Reference 5 contains a thorough treatment of signal-flow graphs and their uses.

* "Instantaneous response" here implies that there is negligible lag or delay between the output and the input over the range of frequencies for which it is to be operated.

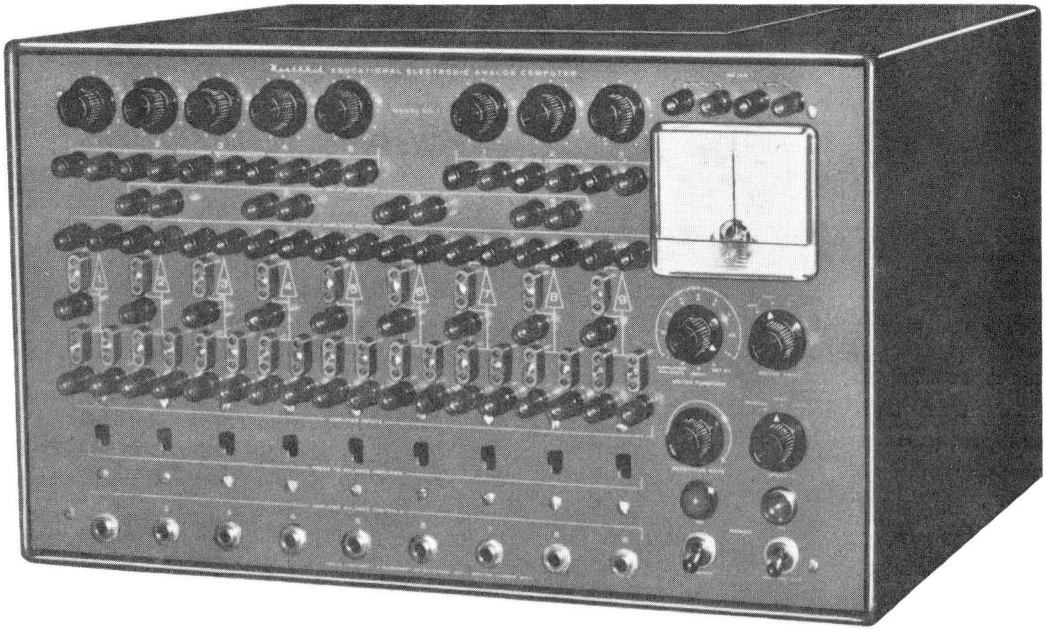

FIG. 7–17. A relatively inexpensive analog computer. (Courtesy Heath Company, Benton Harbor, Michigan.)

These amplifiers, of which there are now many highly developed varieties, have a voltage gain of from 100,000 to higher than 10,000,000, and they have very small voltage drift, constant gain, low noise, high input impedance (resistance), and low output impedance (nearly pure resistance). Earlier versions employ vacuum tubes and more recently developed units are completely transistorized.

It is possible for the novice to literally build his own analog computer by assembling commercially available operational amplifiers and standard electronic components such as resistors, capacitors, diodes, etc. At the other extreme, one may purchase or rent complete computer systems containing hundreds of built-in operational amplifiers, built-in connecting networks, carefully planned programming systems, and specially developed display or read-out equipment. In some cases hybrid computers, which combine analog and digital methods, are also available.

Photographs of two different low-cost analog computers are shown in Figs. 7–17 and 7–18. These

FIG. 7–18. Another relatively inexpensive analog computer. (Courtesy Pastoriza Electronics, Inc., Newton, Massachusetts.)

units are particularly useful for instruction and for the study of small systems or subsystems. At the other end of the scale of cost and complexity are large, integrated analog computer systems which are built around elaborate programming and control consoles. The table-top computers shown in Figs. 7–19 and 7–20, containing 48 operational amplifiers and 64 operational amplifiers, respectively, are medium-sized computers which are provided with removable panels for patch-programming (wiring). All necessary controls are provided for setting coefficients and initial conditions and for operating the units as integrated systems.

Only a brief treatment and discussion of the analog computing art will be presented in this chapter—enough, it is hoped, to enable the reader to grasp the most important factors and to gain valuable insight into the help which this form of computation can lend to the analysis of dynamic systems. The typical operational units of analog computers will be presented and used, but their design, construction, accuracy, etc., will not be discussed in detail.

Some analog computers are designed and built so that the various basic operations are accomplished by individual units corresponding to the blocks employed in Figs. 7–2 through 7–15. They are not intended to simply provide high-gain operational amplifiers for the user to patch up with resistors and capacitors. Both types of diagrams will be employed in this chapter. The computers shown in the photographs of Figs. 7–17 through 7–20 are of the operational amplifier type in which the user must make all the connections for the input and feedback elements when programming a problem on the computer. Figure 7–21 shows a photograph of an electronic analog computer which employs operational units (with input and feedback elements permanently installed within each unit) to perform the various operations, so that the user merely sets calibrated knobs, positions switches, and provides signal connections when programming a problem.

When an analog computer is to be used with a digital computer for hybrid computation, additional features are provided in the analog computer to expe-

FIG. 7–19. Medium-sized analog computer system. (Courtesy Electronic Associates, Inc., West Long Branch, New Jersey.)

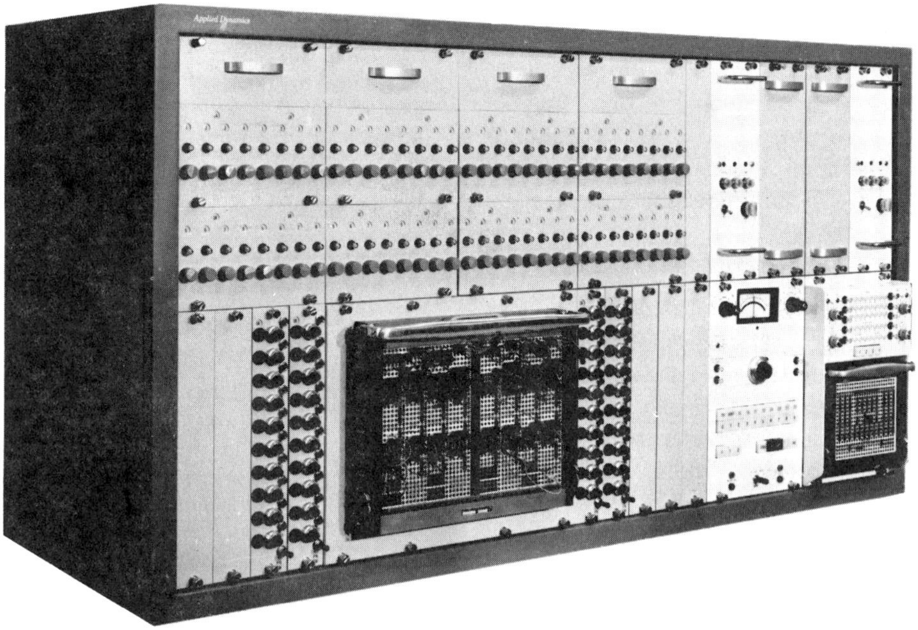

FIG. 7-20. Another medium-sized analog computer system. (Courtesy Applied Dynamics, Inc., Ann Arbor, Michigan.)

FIG. 7-21. Photograph of an analog computer employing operational units, i.e., with input and feedback resistors and capacitors permanently connected within each unit). (Courtesy of G. A. Philbrick Researchers, Inc., Boston, Mass.)

dite the conversion back and forth between the analog and digital signals, and to facilitate the programming of the analog computer in unison with the digital computer. More elaborate control logic (mainly high-speed switching) is then needed in the analog to start, stop, and hold the analog computation processes as required by the discontinuous nature of most digital computer routines.

Operational units to be described in this chapter are:

1) *Potentiometer*—used to multiply a variable input voltage by a constant less than unity. This is simply a voltage divider (Table 7-1) and it delivers an output voltage which is a fraction (determined by the wiper-arm position on the potentiometer) of the input voltage. The potentiometer (pot) is usually set to obtain the desired attenuation with load resistance already connected. In a computer diagram it is represented by the circular symbol.

2) *Operational amplifier*—a dc-amplifier having negative gain of large magnitude which is used in creating the operational units described below. The computer diagram for this operation is a modified triangle.

3) *Constant coefficient*—an operational amplifier connected with an input resistance R_1 and a feedback resistance R_0 to provide the input-output relationship shown in Table 7-1. Best accuracy of this unit is attained when a very large gain is present in the operational amplifier (see Problem 7-3). This operation of multiplying by a constant coefficient is represented by the triangular symbol.

4) *Constant-coefficient summer*—the operation of multiplying each of several variables by a constant coefficient, followed by summation. This operation is achieved by an operational amplifier with input resistors $R_1, R_2, \ldots$, together with a feedback resistor R_0, as shown in Table 7-1. Each input voltage $e_1, e_2, \ldots$ is multiplied by a constant coefficient $C_1, C_2, \ldots$ and then added negatively to obtain the output e_0. In a computer diagram this operation is represented by the modified triangular symbol. Problem 7-5 deals with the derivation of the input-output relationship for this electronic unit.

5) *Integrator*—achieved by replacing the feedback resistor in a constant-coefficient unit with a capacitor (Table 7-1). Several inputs may be accommodated, as in the constant-coefficient summer, and then the unit is called a *summing integrator.* The accuracy of the integration also depends on the duration of the computation and will be degraded over long periods of time by the presence of amplifier drift and capacitor leakage.

Although the problem of setting initial conditions into the outputs of the integrators will not be discussed in detail here, it should be pointed out in this abbreviated treatment of the topic of analog computation that they are established by using relays and voltage sources before the computation is started, as shown in Table 7-1. These relays connect the feedback capacitors to appropriately set voltage sources during the *reset mode of operation* of the computer, thereby establishing the correct initial charge on each capacitor. Then the *run mode of operation* can be initiated by opening the relays simultaneously, thus disconnecting the voltage sources from the operational amplifiers.

If very good integrators are used which can hold their output voltages when their input signals are disconnected, it is possible to stop and hold the computation at a given point in time. This is accomplished by another set of relays which simultaneously disconnect the inputs of all integrators in response to a hold command signal. The integrator outputs which are held in this way are then the proper initial conditions for starting up the computation again when a start command signal is sent to the relays, causing the input signals to be reconnected to the integrators.

Table 7-1 provides a summary of the block diagrams, computer symbols, and operational circuits for the most common analog-computer operations. The references at the end of this chapter have been selected as good sources of additional information for the serious reader who is interested in acquiring more detailed knowledge of the application of these techniques to system analysis.

TABLE 7–1

Summary of Block Diagrams, Symbols, and Computer Circuits for Analog Computer Operations

Operation	Operational block diagram	Computer symbol	Computer circuit
Constant coefficient less than unity	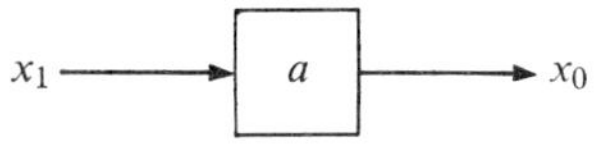	$e_0 = C_a e_1$ 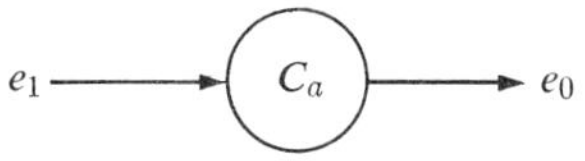 $0 < C_a < 1$	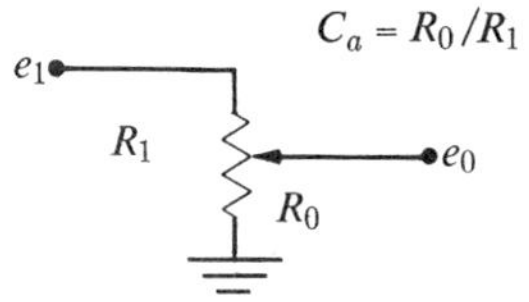
Operational amplification	(None)	$e_0 = -Ae_1$ 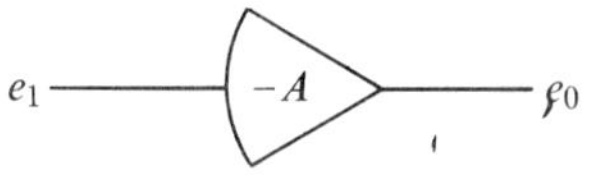 $A \ggg 1$	Complex electronic circuit
Constant coefficient	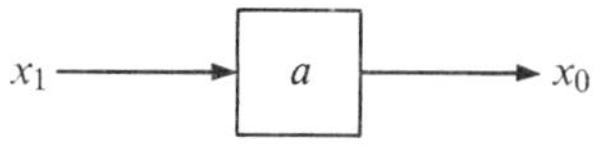	$e_0 = -C_a e_1$ 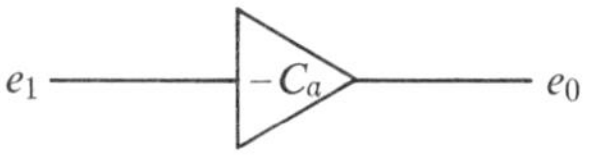 $C_a > 0$	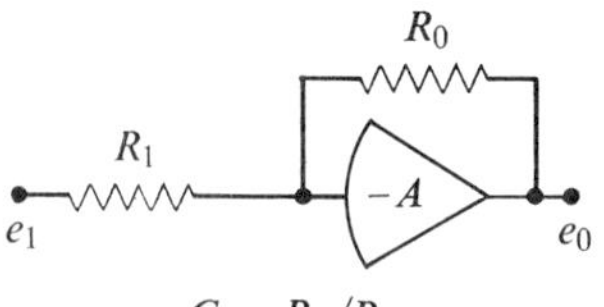 $C_a = R_0/R_1$
Constant coefficients and summation	$x_0 = \Sigma a_n x_n$ 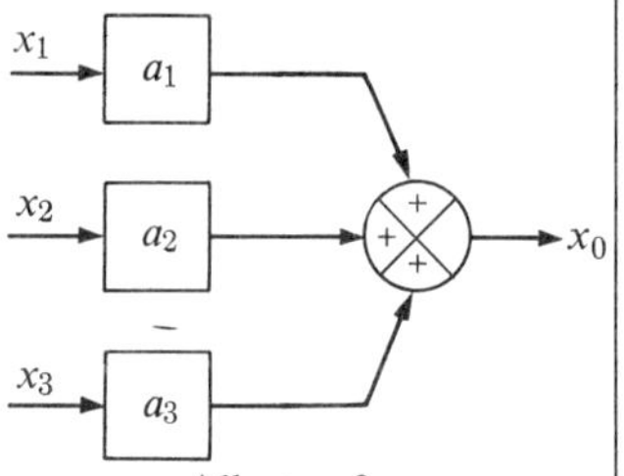 All a's > 0	$e_0 = -\Sigma C_n e_n$ 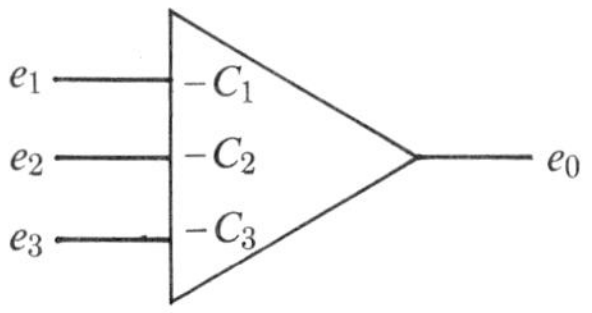 All C's > 0	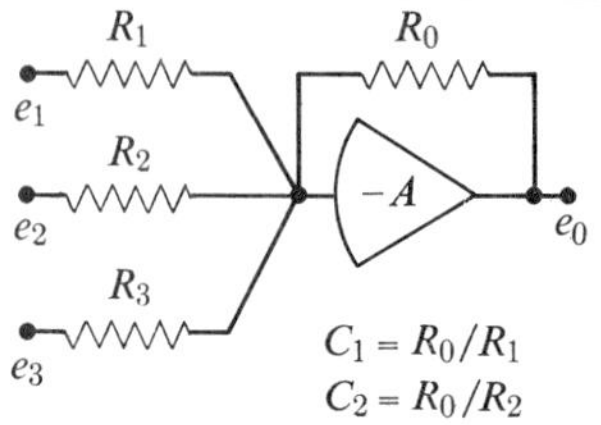 $C_1 = R_0/R_1$ $C_2 = R_0/R_2$ $C_3 = R_0/R_3$
Constant coefficients, summation, and integration	$x_0 = \int_{0-}^{t} (\Sigma a_n x_n)\,dt + x_0(0-)$ 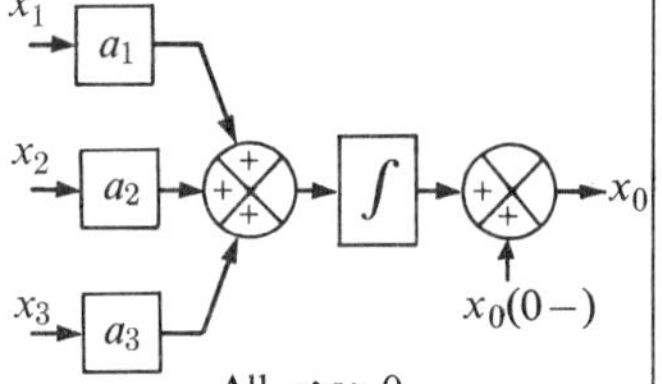 All a's > 0	$e_0 = -\int_{0-}^{t} (\Sigma C_n e_n)\,dt + e_0(0-)$ 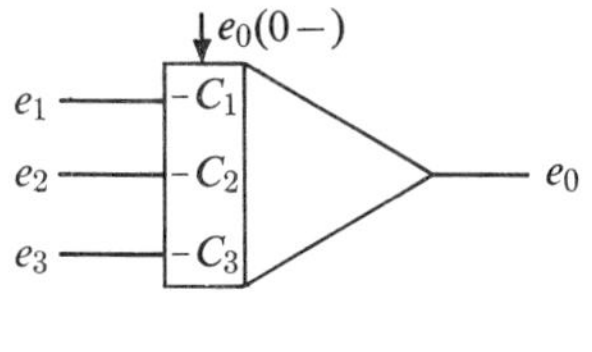 All C's > 0	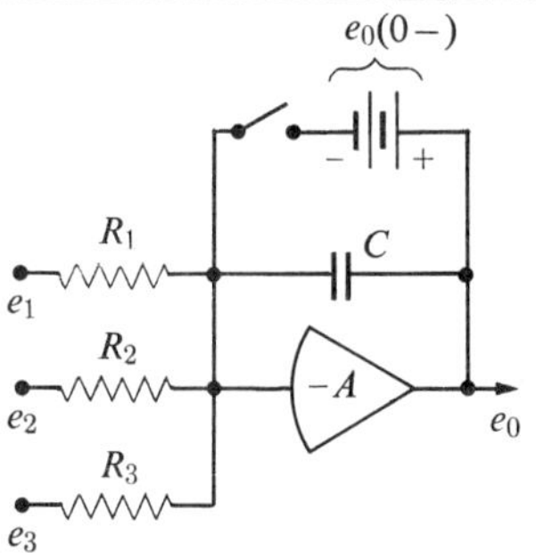

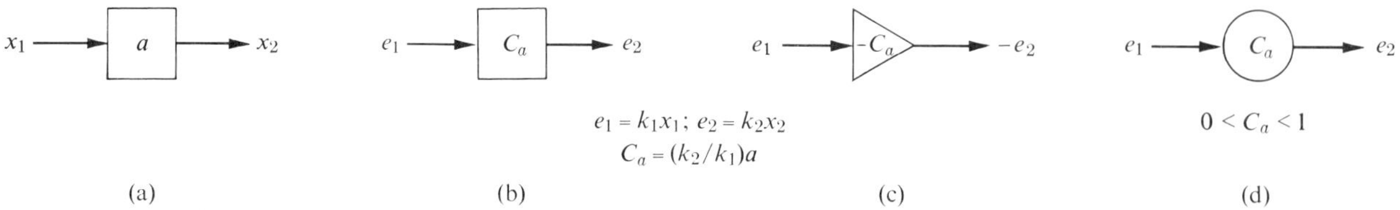

FIG. 7-22. Scale factors for simulation of the constant-coefficient operation.

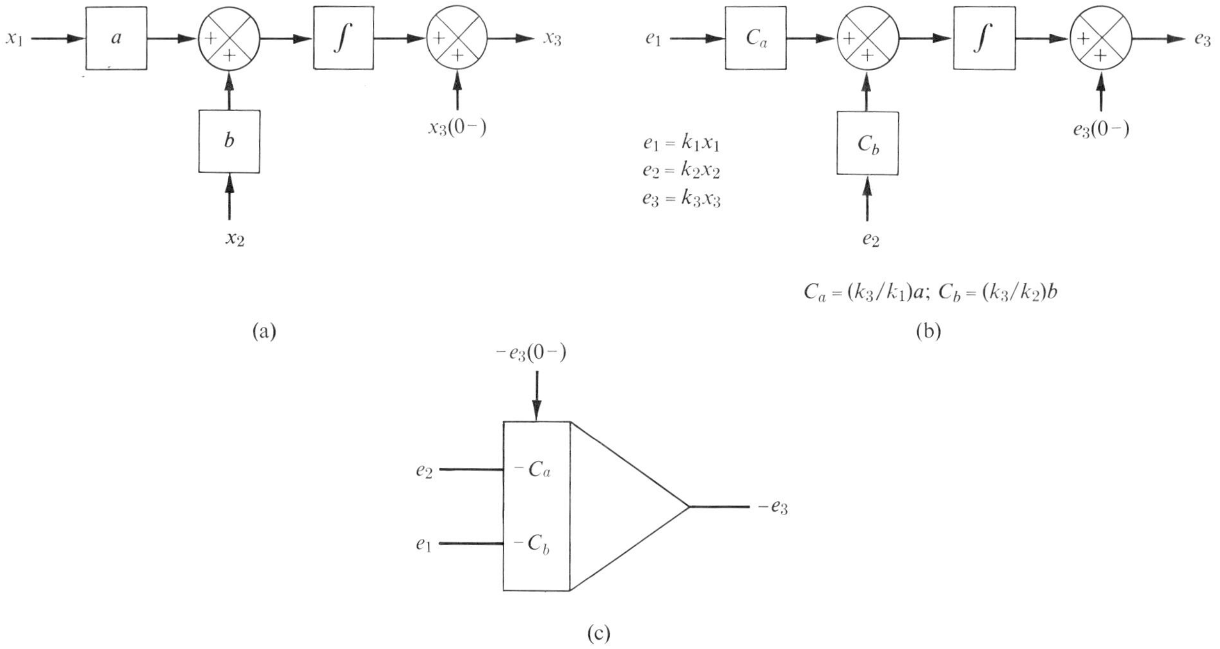

FIG. 7-23. Scale factors for simulation of combined coefficients, summation, and integration operation.

7-7. AMPLITUDE SCALING AND TIME SCALING

In an electronic analog computer, each variable is a voltage which must be scaled to represent properly the physical variable to be simulated. The range through which the analog signal voltages can vary is limited by the characteristics of the operational amplifiers and other electronic components in the computer. Typical ranges are ± 10 v, ± 50 v, and ± 100 v. The ranges of values for the physical system variables may be quite different, depending on the units chosen in the physical system equations. Therefore amplitude scale factors are required for programming a physical system on a computer. It is interesting to note that if special units of measure were employed in setting up the differential equations of the physical system, it might be possible to program the system on a computer with all scale factors equal to unity! In fact this is one way to approach the problem of programming a system on a computer.

To illustrate what is involved in working with scale factors, let us consider the simple problem of simulating a constant-coefficient operation between two physical system variables x_1 and x_2 such that $x_1 a = x_2$ [represented by the block diagram in Fig. 7–22(a)]. The computer simulation of this operation is represented by the analog block diagram in Fig. 7–22(b), where the analog variable e_1 represents the system variable x_1 by the use of the scale factor k_1: $e_1 = k_1 x_1$. Similarly, the scale factor k_2 is employed to relate e_2 to x_2: $e_2 = k_2 x_2$. It is then evident that since $e_1 C_a = e_2$, then $C_a = (k_2/k_1)a$, and the fraction k_2/k_1 is a scale factor relating C_a to a. The computer block diagram shown in Fig. 7–22(c) covers the special conditions which are involved when the constant-coefficient operation is simulated with an operational amplifier having input and feedback resistors (see Table 7–1), and the block diagram in Fig. 7–22(d) covers the case where a simple attenuator is used to simulate the constant-coefficient operation.

Another illustration is the combined operation of multiplying by coefficients, summing, and integrating as shown in Fig. 7–23(a). The computer simulation in real time is shown in Figs. 7–23(b) and 7–23(c). In addition to the amplitude scaling which was described in the previous illustration and which also applies in this case, time scaling also can be illustrated because of the presence of an integrator.

Let us assume that the amplitude scale factors have all been established for real-time computation (Fig. 7–23). Now we can investigate what would be required to change the size of the unit of time used in the integration process, i.e., to change the time scale of the computation. This can be done by measuring time in terms of τ instead of t, where C_i units of τ are equivalent to one unit of t (that is, $\tau = C_i t$), as shown in Fig. 7–24(a). Integration with respect to τ is represented symbolically by the upper block diagram in Fig. 7–24(b), where, strictly speaking, the input $e_1(t)$ should be also described as a function of τ, $e_a(\tau)$, to expedite the following analysis.

The integral of $e_a(\tau)$ with respect to τ may be expressed as follows:

$$e_b(\tau) = \int_{\tau_1}^{\tau_2} e_a(\tau)\, d\tau. \tag{7–10}$$

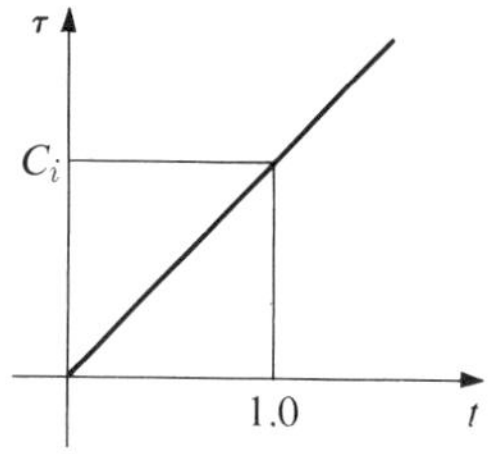

(a)

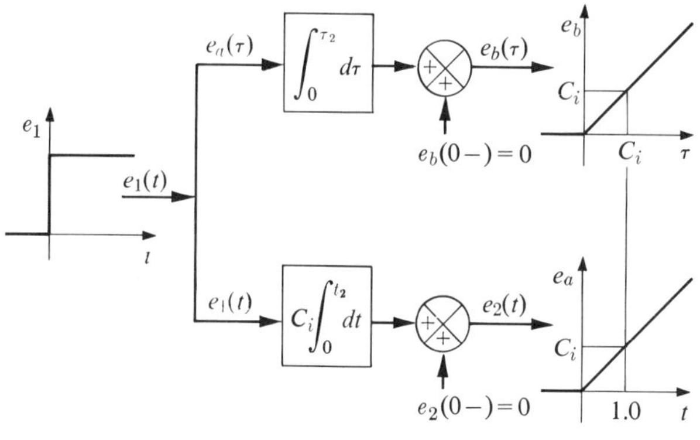

(b)

FIG. 7–24. Integration with respect to different units of time.

Since $\tau = C_i t$, $d\tau = C_i\, dt$, and we may write

$$e_b(\tau) = \int_{t_1}^{t_2} e_a(\tau) C_i\, dt = C_i \int_{t_1}^{t_2} e_1(t)\, dt. \tag{7–11}$$

Now it is evident that $e_b(\tau)$ is equal to C_i times the integral of $e_1(t)$ with respect to t. Therefore the output $e_2(t)$ of the lower block diagram is the same as the output $e_b(\tau)$ of the upper block diagram in Fig. 7–24(b), and it is seen that increasing the gain (i.e., multiplying by a coefficient greater than unity) in an integrator is equivalent to integrating with respect to a proportionately smaller unit of time. Moreover, this is equivalent to computing at a faster rate than the real-time rate of computation. It follows that if the computation rate in a computer simulation is to be speeded up, all integrators in the analog must be speeded up by the same factor.

When a complete system is to be scaled onto an analog computer, usually there will be closed loops in

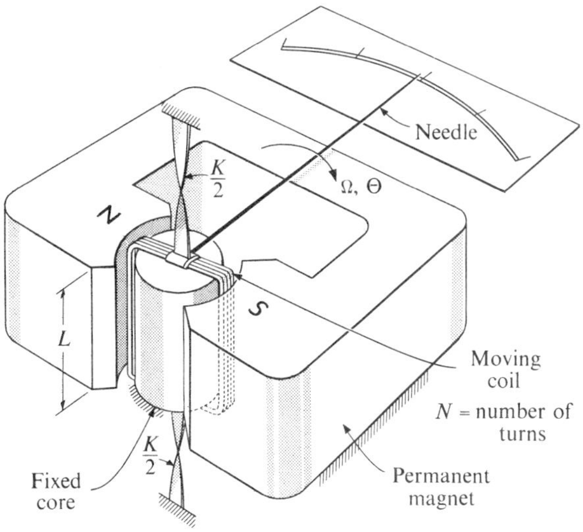

FIG. 7-25. Schematic diagram of a D'Arsonval galvanometer used to measure a voltage.

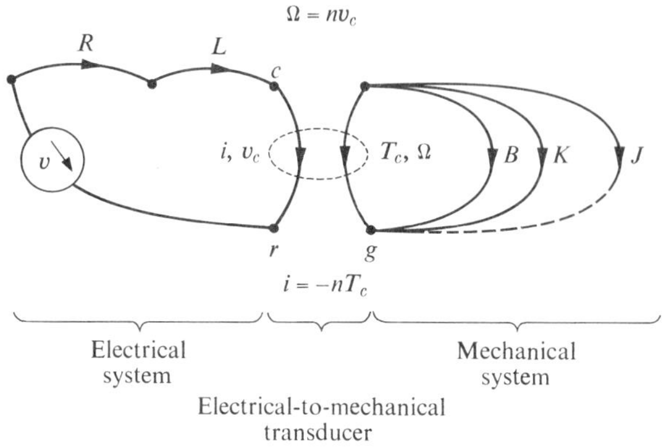

FIG. 7-26. Linear graph of D'Arsonval meter system.

the block diagram. The following general rules apply to the scaling of closed loops:

RULE 1. For each closed loop in the system, the loop gain (the product of all the coefficients or "gains" of the individual blocks in the loop) must remain the same as in the original operational block diagram if the loop is to be properly scaled for real-time computation. This is intuitively evident, but it can be proved rigorously by carefully accounting for all the scale factors discussed in the beginning of this section for each operation in the closed loop.

RULE 2. For the case when time scaling is carried out, the gain of each closed loop is changed by the nth power of the time-scale factor, where n is the number of integrators in the loop. This is essentially a corollary of Rule 1, and it can be proved in the same way.

RULE 3. Pure time scaling does not change the amplitude scaling of a problem, and pure amplitude scaling does not change the time scaling of the problem.

7-8. SIMULATION OF A TYPICAL SYSTEM ON AN ANALOG COMPUTER

To illustrate the most important factors involved in using an analog computer to solve a dynamic system problem, let us undertake the simulation of a D'Arsonval galvanometer (Fig. 7-25).

The interaction between the steady magnetic field (provided by a permanent magnet) and the current passing through a spring-restrained rotating coil provides the torque necessary to oppose the spring, overcome damping, and accelerate the inertia of the coil and needle. Under steady-state conditions the current in the coil is proportional to the applied voltage, and the deflection of the needle is proportional to the current passing through the coil. Under unsteady conditions of operation, however, coil inductance may cause the current to lag the voltage, and needle inertia may cause the deflection of the needle to lag the current. Moreover, a voltage, called *back-emf*, is induced in the coil due to its rate of motion through the magnetic field, and this voltage may significantly affect the current in the coil. This galvanometer is also discussed in Section 2-6.5.

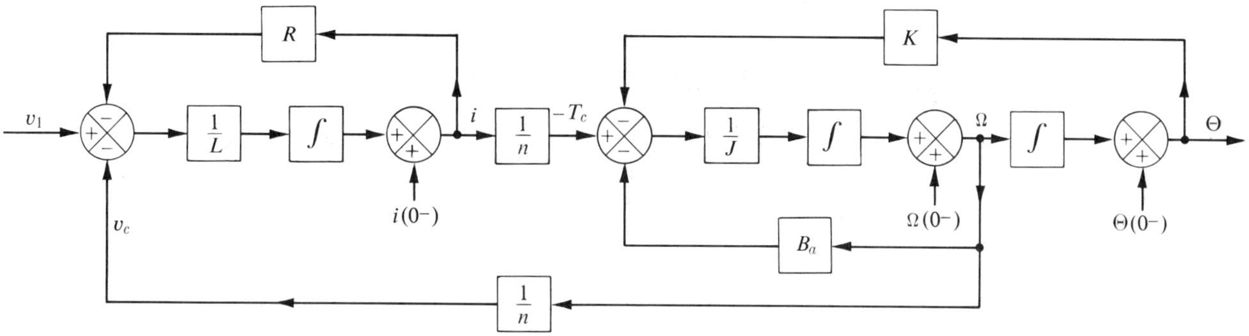

FIG. 7–27. Complete operational block diagram of D'Arsonval meter system.

The problem which we shall pose here is to determine the dynamic response of this meter to a step change in voltage v_1 applied to the coil at time $t = 0$. The system is initially relaxed so that all variables are zero before the application of the step input.

7–8.1 Modeling of System and Preparation of Operational Block Diagram

The first step is to achieve a suitable model of the system. The electrical circuit may be modeled as a combination of a resistance and inductance in series with one side of an electromagnetic transducer, and the mechanical part of the system may be modeled as a parallel-mechanical network in series with the other side of the transducer (Fig. 7–26).

The constitutive relations for the transforming transducer, as derived in Section 2–6.5, are:

$$v_c = (1/n)\Omega, \tag{7–12}$$

$$T_c = -(1/n)i, \tag{7–13}$$

where

v_c = back-emf induced by motion of coil in field,

T_c = torque produced by current in coil, and

n = the transducer constant determined by the geometry of the system and by the units employed for the system variables and parameters (Eqs. 2–96 and 2–97).

Employing the continuity and compatibility equations for the electrical circuit, we obtain

$$v = Ri + L(di/dt) + v_c, \tag{7–14}$$

where

R = resistance of the coil,

L = self inductance of the coil,

v = voltage applied to coil, and

i = current in coil.

Similarly, the vertex equation for the mechanical system is,

$$-T_c = J(d\Omega/dt) + B\Omega + K\Theta, \tag{7–15}$$

where

J = rotational inertia of coil and needle,

B = rotational damping on coil and needle (mostly due to air friction on the needle),

Ω = angular velocity of the coil and needle, and

Θ = angular position of coil and needle $\Theta_0\Omega\, dt$ measured relative to zero position.

A complete operational block diagram for the system is shown in Fig. 7–27. To minimize both the number of operational amplifiers required and the initial scaling of the problem on the analog computer, it is helpful to rearrange the block diagram by incorporating the $1/L$ and $1/J$ coefficients into the coeffi-

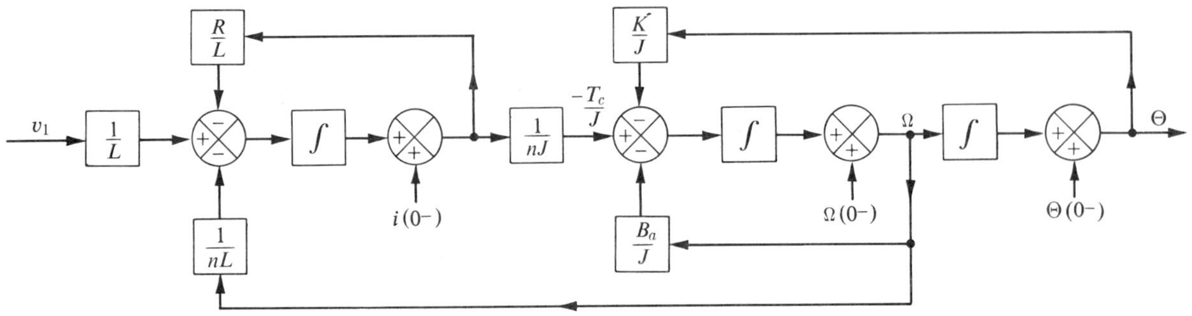

FIG. 7-28. Operational block diagram resulting when $1/L$ and $1/J$ coefficients are incorporated into preceding coefficients.

cients ahead of their respective preceding summers (Fig. 7-28).

The next step is to employ the values of the physical parameters of the system to compute the values of the coefficients shown in Fig. 7-28. The following data are available on the system parameters:

$R = 10$ ohms, $K = 0.025$ n·m/rad,

$L = 1.0$ h, $J = 3 \times 10^{-4}$ n·m·sec^2/rad,

$B = 9 \times 10^{-4}$ n·m·sec/rad.

Full-scale output, Θ_m, is 1.0 rad when the maximum input voltage v_m is applied.

For the maximum voltage v_m we find the steady-state current i_m from Eq. (7-14) by noting that in the steady state, $di/dt = v_c = 0$, so that

$$i_m = v_m/R, \tag{7-16}$$

which gives

$$i_m = 1.0/10 = 0.1 \text{ amp}.$$

Similarly, we may combine Eq. (7-13) with Eq. (7-15), noting that in the steady state $d\Omega/dt = \Omega = 0$, to obtain

$$i_m/n = K\Theta_m, \tag{7-17}$$

which gives

$$n = \frac{1}{(0.025)(1.0)/(0.1)} = 4.0 \text{ amp/n·m}.$$

The values of the required coefficients are now computed to give

$R/L = 10.00,$ $K/J = 83.3,$

$1/L = 1.000,$ $1/nJ = 833,$

$1/nL = 0.250,$ $B/J = 3.00.$

7-8.2 Scaling on the Computer

To program the system on a computer, we must scale the problem so that the computer can accept the system data which have just been prepared. For some computers, such as the Philbrick Analog Computer (Fig. 7-21), the problem may be scaled directly on a block diagram (Fig. 7-28). However, a majority of the commercially available computers, such as those shown in Figs. 7-17 through 7-20, employ the operational-amplifier concept and it is necessary to draw a computer block diagram such as the one shown in Fig. 7-29.

Each operational amplifier produces a sign change. Therefore, each loop having negative gain (and most of them do) must contain an odd number of operational amplifiers. An inverting amplifier was added to illustrate changing the sign of e_{-i} to e_i. Because the coefficient potentiometers can only attenuate, each coefficient must be divided by a power of 10 so that it can be set on its corresponding pot. Then a corresponding power of 10 may be inserted in the appropriate place on the operational block diagram to make up for this change in pot setting. The scale factors

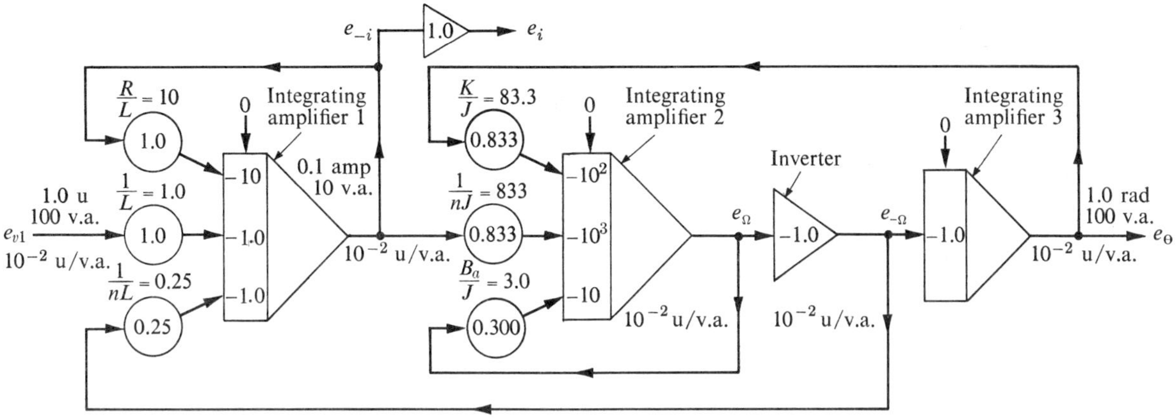

FIG. 7–29. Analog computer block diagram employing conventional operational amplifiers. First-trial, real-time scaling is illustrated.

for initial conditions are the same as the scale factors at the outputs of the corresponding integrators. In this problem all initial conditions are zero. The procedure described so far results in a tentative, or first-trial, program of the system on the computer (Fig. 7–29). To discuss the merits or demerits of this program, we would do well to apply the general rules given in Section 7–7 for the scaling of closed loops.

There are at least one or two reasons why the first trial program usually is not satisfactory:

1) Not all the variables are represented by suitable voltages. The analog voltage (v.a., meaning volts analog) representing a given variable may be so small that it is cluttered with noise, or it may be so large that it imposes limiting conditions of one or more of the operational amplifiers. This aspect of scaling is often called *amplitude scaling*.

2) The computer may not be capable of operating effectively in real time. The choice of input resistors and feedback capacitors available is usually limited to two or three decades of values which give computing capability commensurate with the drift rate, impedance, and speed-of-response characteristics of the operational amplifiers. This aspect of scaling is often called *time scaling*.

In this problem it is evident that the first trial program would result in the same numerical scale factor for all variables, as shown in Fig. 7–29. The symbol u is used in the computer diagram to denote units of the physical system variable, and the symbol v.a. denotes volts analog for the corresponding analog voltage variable in the system. The scale factor which relates physical system units to analog units, then, is called units per volt analog (u/v.a.). Thus the maximum steady current $i_m = 0.1$ amp would be represented by only 10 v on the analog computer if the maximum steady voltage $v_m = 1.0$ v were represented by 100 v.a. An attempt should be made also to represent i_m and all other variables by 100 v.a. (or by whatever voltage the computer is designed to work with most effectively).* Furthermore, there may be problems associated with the amplitude scaling for the variable Ω. However, the scaling for the variable Θ is correct and it should be preserved because $\Theta_m = 1.0$ rad is represented by 100 v.

To improve the scaling for the variable i, we must increase the steady-state gain of the portion of the analog from e_v to e_{-i} by a factor of 10. This can be done by using a gain of 10 instead of 1 at the second input position of integrating amplifier 1, as shown in

* This scaling will be performed under the assumption that no variables will exceed their steady-state value. This may not always occur, and the programmer should be prepared to revise scale factors accordingly.

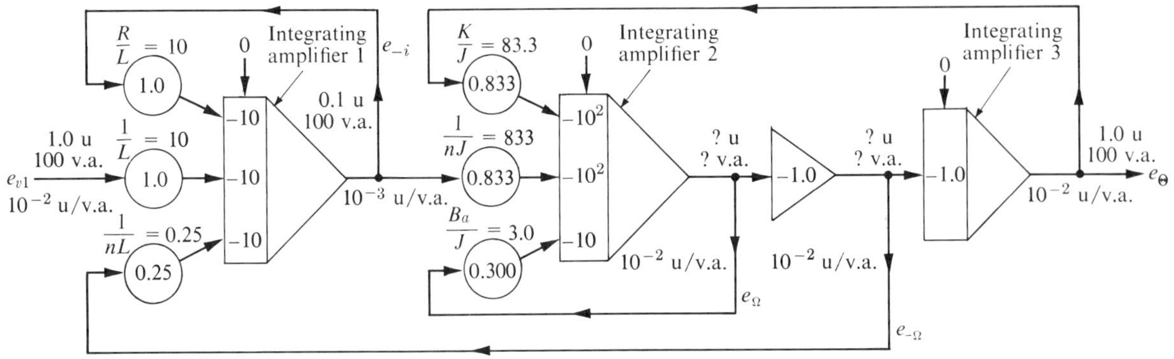

FIG. 7–30. Analog computer diagram rescaled to improve amplitide scaling for e_{-i}.

Fig. 7–30. Now the gain at the second input position of integrating amplifier 2 should be decreased from 10^3 to 10^2 to compensate for the factor of 10 increase in the steady-state gain from e_v to e_{-i}. Also the gain at the third input to integrating amplifier 1 must now be increased by a factor of 10 to preserve the gain of the loop which contains integrating amplifiers 1 and 2. The system is still scaled to compute in real time.

There is still a question about the amplitude scaling of e_Ω. We know that the steady-state value of Ω is always zero, but we have no knowledge yet of the maximum value which may occur during dynamic operation of the system. The computer is ready to tell us, however. In this case the problem can be tried on a conventional computer because the real-time response of the system is readily handled by the computer without time scaling. *If time scaling had been required, it could have been accomplished readily by changing all integrating amplifier gains by the same factor* (usually some power of 10).*

This factor would be chosen so that all integrations were speeded up to avoid excessively long computing time, or slowed down enough for the rate of computation to stay within the capabilities of the computer. If the computation rate is too fast, parasitic lags and delays in the operational amplifiers may no longer be negligible. When this occurs, it can be readily detected by comparing high-speed computation to lower-speed computation. They should, of course, agree with each other.

* See Section 7–7 on amplitude scaling and time scaling.

When the scaling shown in Fig. 7–30 was tried, it was found that when e_v was suddenly changed from 0 to 100 v.a., e_Ω very quickly tended to exceed its maximum allowable value of 100 v.a. Therefore it was decided to change the scaling of e_Ω downward by a factor of 10, as shown in Fig. 7–31. This was accomplished by decreasing each of the gains at the first and second inputs of integrating amplifier 2 by a factor of 10, increasing the gain at the input of integrating amplifier 3 by a factor of 10, and increasing the gain of the third input of integrating amplifier 1 by a factor of 10. Note that all loop gains have been preserved! To verify this, compare all the loop gains in Fig. 7–31 with the loop gains which existed in Fig. 7–29 before scaling was started. This verification is an excellent means of rapidly checking for errors in scaling.

Most programmers prefer to carry out amplitude scaling and time scaling separately; some time scale first and others prefer to amplitude scale first. To carry out time scaling first, one must be able to quickly estimate system speed of response by inspection of the system equations, a skill which is developed only with considerable experience. Hence, the approach used here was to start with amplitude scaling. The system variables are known. It is usually

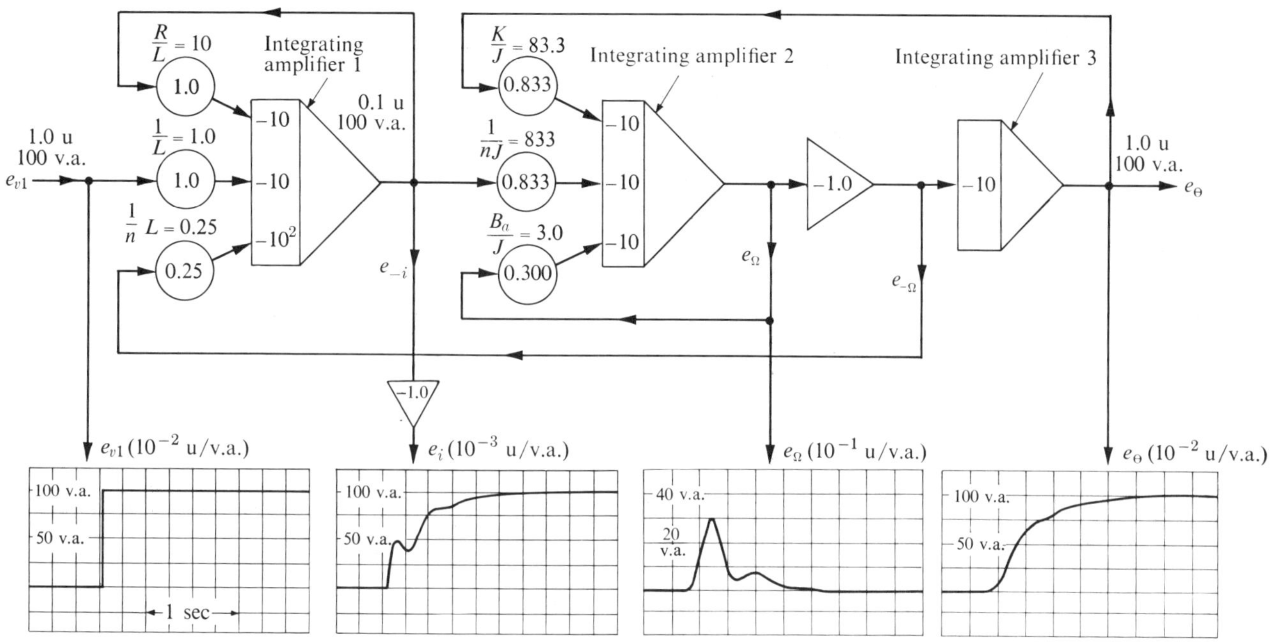

FIG. 7–31. Analog computer diagram with final scale factors and step-response solutions for voltage. All initial conditions are zero.

possible to know the maximum steady-state values of at least some of the system variables. This provides a starting point. Then the computer can be employed to ferret out the rest and enable the programmer to completely amplitude scale the problem in a short time.

7–8.3 Computed Solutions

The responses of all the available system variables to a step change in v from an initial value of zero, computed on the Applied Dynamics Computer in the Dynamic Simulation Laboratory at The Pennsylvania State University, are shown in the individual strip-chart recordings which have been inserted in Fig. 7–31. Note that no values exceeded their steady-state value, as anticipated, except for e_Ω treated separately in the last section.

Sometimes, during initial formulation of system equations, certain parameters are included which later may turn out to be negligible. When this happens, it may be rather difficult to scale the system on the computer until the negligible parameter has been dropped from the analysis. It is often obvious that a parameter should be dropped when this occurs. In some cases, however, it is necessary to get the system scaled onto the computer and have the computer solution demonstrate whether certain parameters are small enough (and have a small enough effect on the solution) to be dropped from the analysis. Unfortunately, the operational block diagram may have to be appreciably altered when a parameter is dropped from the analysis, and a great deal of the job must then be done over again.

It is very interesting to explore this system a little further, now that it has been scaled for the computer. If the meter is to be driven by a current source instead of a voltage source, it is necessary only to apply an input analog voltage directly to the pot at the second input of integrating amplifier 2 instead of using the output of integrating amplifier 1 (i.e., the output of integrating amplifier 1 is disconnected). The analog

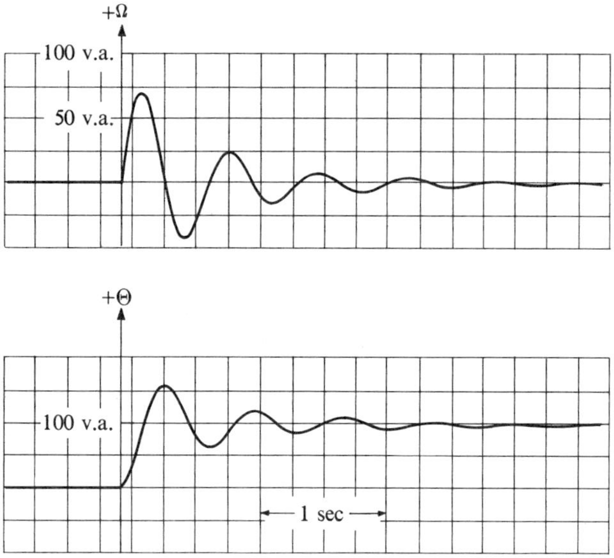

FIG. 7-32. Results of analog computer solution for response of galvanometer to a 0.1-amp step change in current from an initially zero value.

simulation of the response of the system to 0.1-amp step change in i is shown in Fig. 7-32. Now that the feedback loop due to the back-emf induced in the coil is broken, the only damping on the inertia-spring characteristics of the coil-needle assembly is the air damping B, which has quite a small effect on the tendency of the system to oscillate (i.e., the oscillation dies out very slowly). This is now a simple second-order rotary mechanical system which is subjected to a step change in input torque (the torque induced by the step change in current i). Note that in this case Θ overshoots its steady-state value. The operator must carefully check for amplifier overload when this occurs. In this example the computer had sufficient linear range to cover the overshoot.

Comparison of the responses obtained for voltage and current inputs reveals very dramatically how the dynamic response of a system may depend on the manner in which it is driven.

REFERENCES

1. G. A. KORN and T. M. KORN, *Electronic Analogue Computers*, 2nd Ed. New York: McGraw-Hill Book Company, Inc., 1956.
2. H. M. PAYNTER, *A Palimpsest on the Electronic Analog Art*. Boston: George A. Philbrick Researches, Inc., 1955.
3. C. L. JOHNSON, *Analog Computer Techniques*, 2nd Ed. New York: McGraw-Hill Book Company, Inc., 1963.
4. R. J. ASHLEY, *Introduction to Analog Computation*. New York: John Wiley and Sons, Inc., 1963.
5. S. J. MASON and H. J. ZIMMERMAN, *Electronic Circuits, Signals and Systems*. New York: John Wiley and Sons, Inc., 1960.

PROBLEMS

7-1. In each of the systems shown in Fig. 7-33 there are one or two inputs (directly forced variables) and two outputs (dependent variables of primary interest). First, write the governing relationships for the elements of a system and draw the operational block diagram (no differentiations, please), showing the inputs and outputs as well as the other variables and the characteristics of the various elements. Second, derive the differential equation for each output of the system showing how the output (and its derivatives) is related to the inputs (and their derivatives).

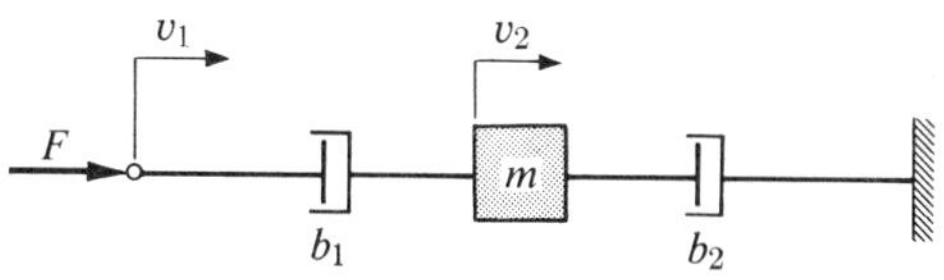

Input: v_1
Outputs: v_2 and F
(a)

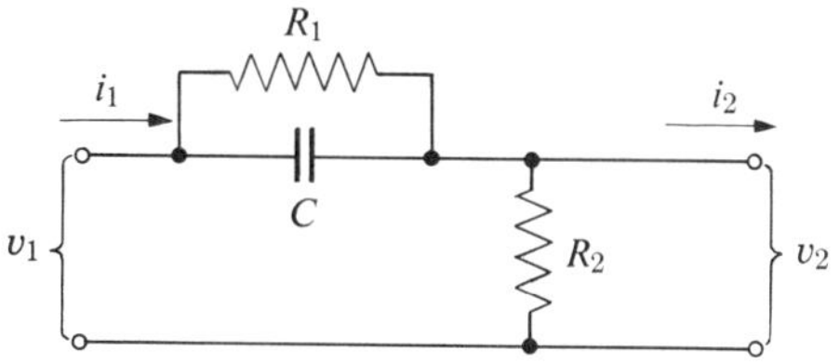

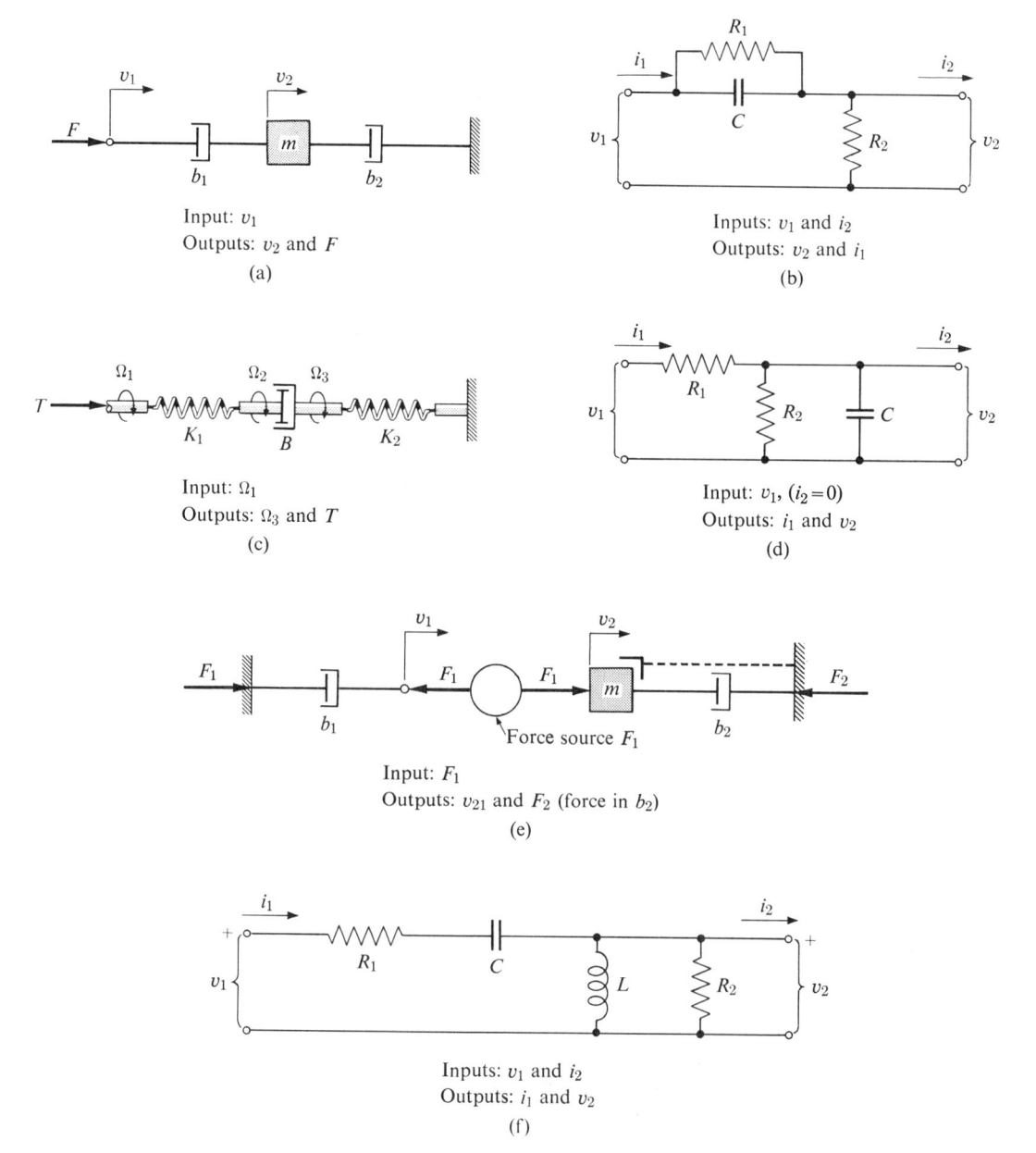

Inputs: v_1 and i_2
Outputs: v_2 and i_1
(b)

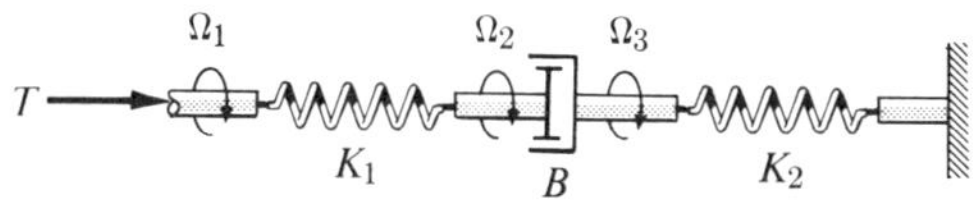

Input: Ω_1
Outputs: Ω_3 and T
(c)

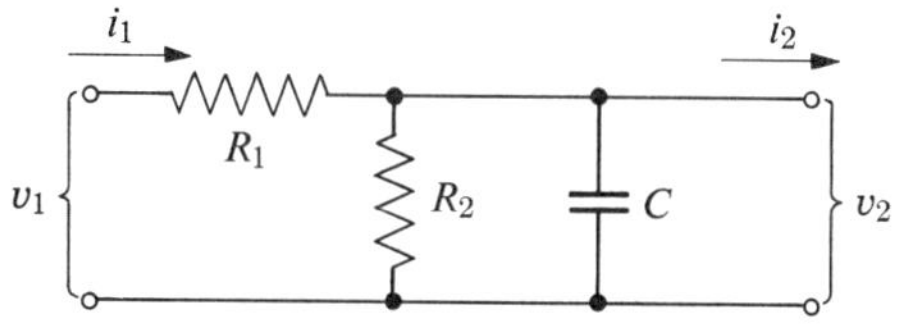

Input: v_1, $(i_2 = 0)$
Outputs: i_1 and v_2
(d)

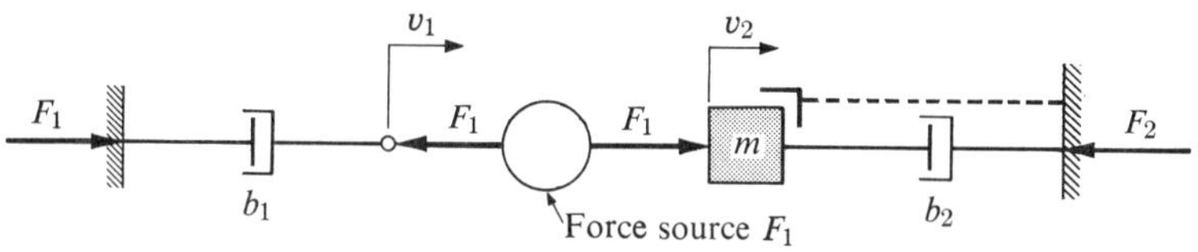

Input: F_1
Outputs: v_{21} and F_2 (force in b_2)
(e)

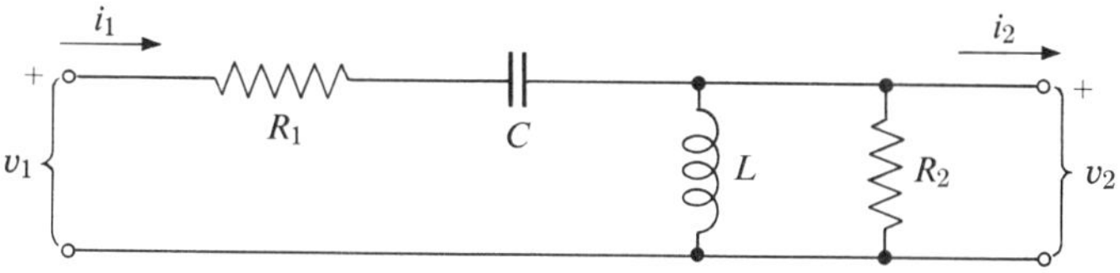

Inputs: v_1 and i_2
Outputs: i_1 and v_2
(f)

FIGURE 7–33

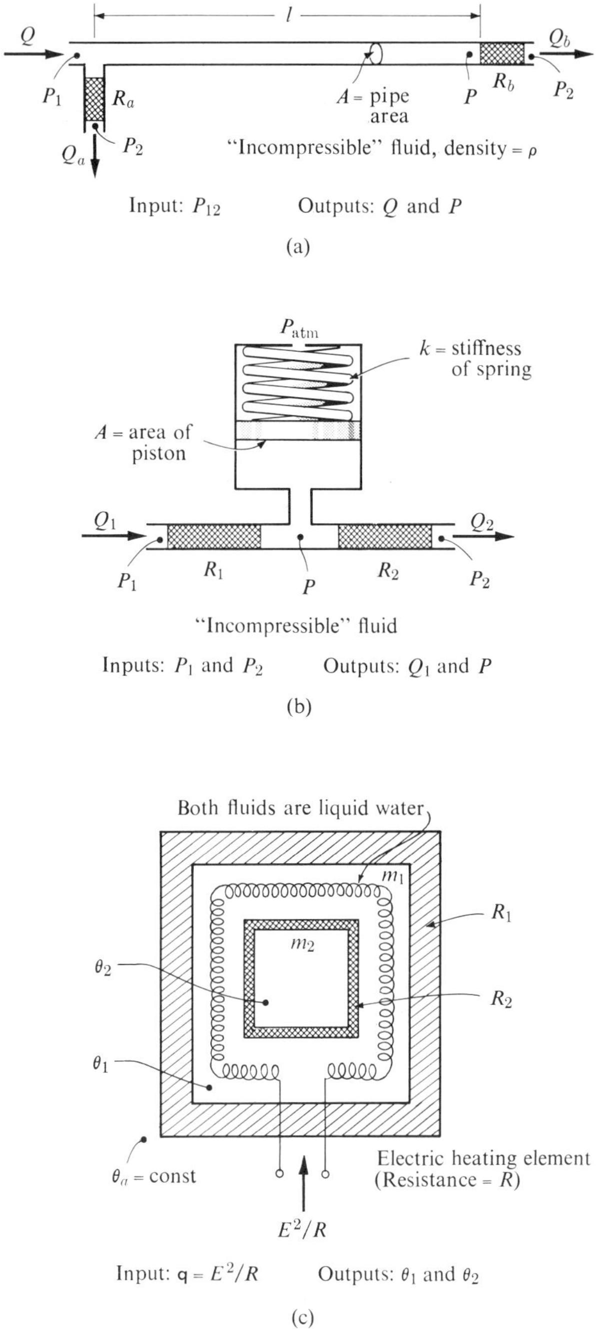

FIGURE 7–34

7–2. Carry out the same analysis as in Problem 7–1 for the fluid and thermal systems shown in Fig. 7–34.

7–3. The circuit in Fig. 7–35 shows how a high-gain operational amplifier may be connected with an input resistor and a feedback resistor to perform the function of multiplying an input voltage by a constant so that $e_o = e_i \times \text{const.}$

a) In this problem you are expected to determine analytically how the output voltage e_o is related to the input voltage e_i and the bias adjustment voltage e_b. (In this case the operational amplifier is a difference amplifier, amplifying the voltage difference, $e_b - e_a$, by a very large gain. The second input e_b is useful in certain special computing circuits as well as in providing a means of making a bias adjustment in the output to overcome drift in the operational amplifier.) As a first approximation, you may assume that the amplifier input current i_a is zero.

b) When you have determined the analytical expression showing how e_o is related to e_i and e_b, proceed to show that as k_a approaches infinity the relationship between e_o and e_i, when $e_b = 0$, approaches $e_o = e_i(R_2/R_1)$.

c) Use the following numerical data to determine the percentage departure (error) of e_0 from the ideal value of $e_i(R_2/R_1)$ for the following two cases:

CASE I	CASE II
R_1 = 10,000 ohms (10 K)	R_1 = 10 K
R_2 = 100,000 ohms (100 K)	R_2 = 100 K
$k_a = 10^5$	$k_a = 10^6$

Does the value of k_a have much effect on this error?

d) Next determine the percentage error in the output voltage due to an amplifier input current i_a of 10^{-10} amp when $e_i = 1.0$ v for the two cases analyzed in part (c). Does the value of k_a have much effect on this error?

7–4. For the summation circuit shown in Fig. 7–36 develop the analytical expression relating e_o to e_1, e_2, e_3, and e_4. Then show that as $k_a \to \infty$, $e_o \to -(R_2/R_1)(e_1 + e_2 + e_3 + e_4)$.

7–5. For the "weighted input" summation circuit shown in Fig. 7–37, relate e_o to e_1 and e_2 and show that as $k_a \to \infty$, $e_o \to -(R_2/R_1)e_1 - (R_2/5R_1)e_2$.

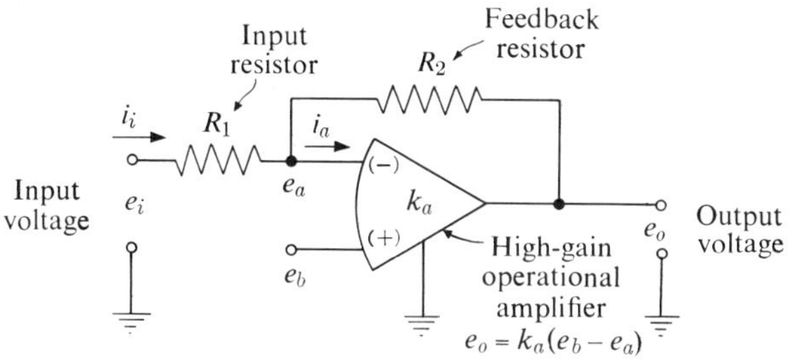

FIGURE 7–35

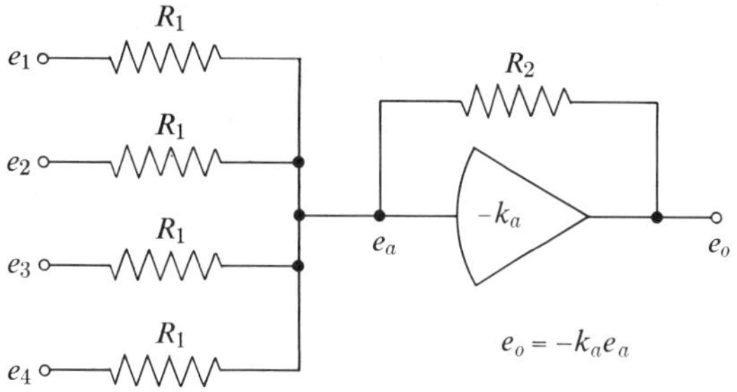

FIGURE 7–36

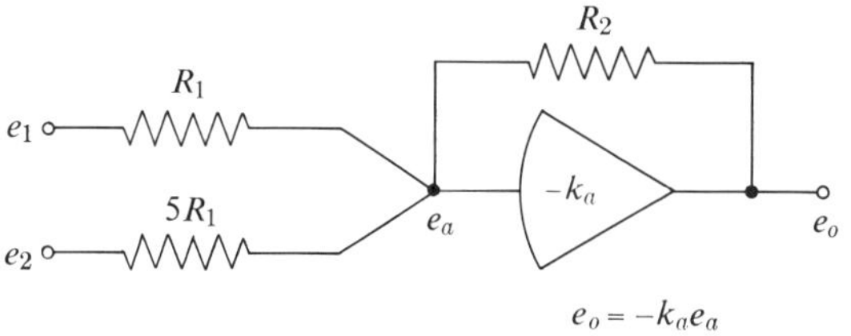

FIGURE 7–37

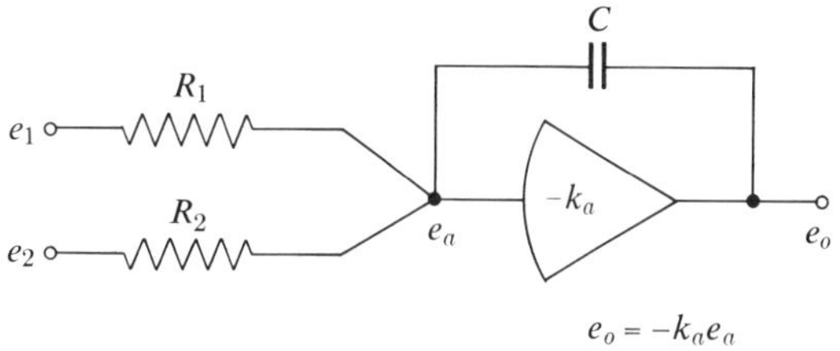

FIGURE 7–38

7–6. For the "weighted input" integrator circuit shown in Fig. 7–38, relate e_o to e_1 and e_2 and show that as $k_a \to \infty$,

$$e_o \to -\int_0^t [(1/R_1C)\, e_1 + (1/R_2C)\, e_2]\, dt + e_0(0-).$$

7–7. For the systems having the following differential equations with constant coefficients, draw the complete analog-computer diagram showing all R- and C-values and all initial condition circuits to find x_r as a function of time when the excitation function x_f is a unit step at $t = 0$. Scale each variable for 1.0 units per v.a. Time is to be measured in seconds.

a) $\dfrac{dx_r}{dt} + 2x_r = x_f;\qquad (x_r)_i = x_r(0-) = 0,$

b) $\dfrac{d^2x_r}{dt^2} + 5\dfrac{dx_r}{dt} + 3x_r = x_f;\qquad \begin{cases}(x_r)_i = x_r(0-) = 0,\\ (\dot{x}_r)_i = \dot{x}_r(0-) = 0,\end{cases}$

c) $\dfrac{dx_r}{dt} + 3x_r = \dfrac{dx_f}{dt} + x_f;\qquad (x_r)_i = x_r(0-) = 0.$

[Use no differentiator please in either c) or d).]

d) $\dfrac{d^2x_r}{dt^2} + 4\dfrac{dx_r}{dt} + 2x_r = 2\dfrac{dx_f}{dt} + x_f;$

$(x_r)_i = x_r(0-) = 0,\ (\dot{x}_r)_i = \dot{x}_r(0-) = 0.$

ANALOG COMPUTER LABORATORY PROBLEMS

LP–1.

Given the differential equation with constant coefficients

$$\frac{d^2x}{dt^2} + b\frac{dx}{dt} + x = \mathsf{f}(t).$$

The initial conditions are

$$x(0-) = \dot{x}(0-) = 0.$$

a) Draw a complete analog computer diagram (including initial-condition circuit and R- and C-values).

b) Let $\mathsf{f}(t) = 20$ for the Heath-kit* computer and let $\mathsf{f}(t) = 2$ for the TR-10* computer. Record (1) the solution of the equation, $x(t)$, and (2) its derivative, $(dx/dt)(t)$, for

$$b = 0, 0.5, 1.0, 2.0, 5.0, 10.0.$$

c) Find the value of b corresponding to critical damping.

* The terms "Heath-kit" and "TR-10" imply computers with $\pm$100-v and $\pm$10-v working-signal ranges, respectively.

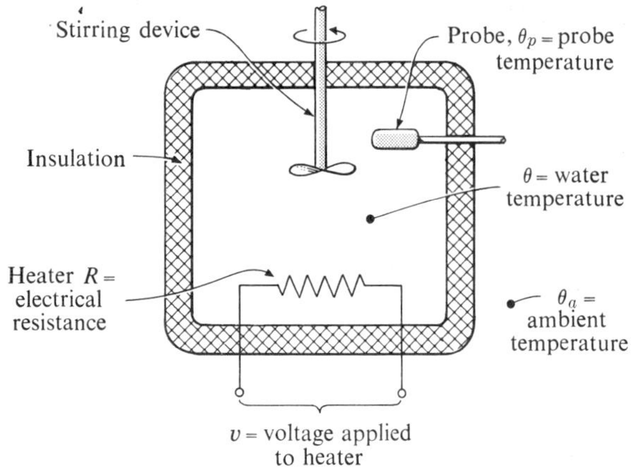

FIGURE 7–39

LP–2.

Given the set of simultaneous differential equations

$$\frac{d^2x}{dt^2} + 0.2\frac{d^2y}{dt^2} + x(t) = 0,$$

$$0.2\frac{d^2x}{dt^2} + \frac{d^2y}{dt^2} + y(t) = 0.$$

The given initial conditions are

$$\dot{x}(0-) = 50, \qquad x(0-) = y(0-) = \dot{y}(0-) = 0.$$

a) Solve the given set of differential equations on the analog computer and record the solutions for x and y as functions of time.

b) Record the solutions of the same set of equations for the following initial conditions:

$$\dot{x}(0-) = \dot{y}(0-) = 50, \qquad x(0-) = y(0-) = 0.$$

LP–3.

Given

$$\frac{d^2x}{dt^2} + 3\frac{d^2y}{dt^2} + 4\frac{dx}{dt} + 14x = 2,$$

$$\frac{d^2y}{dt^2} + 5\frac{d^2x}{dt^2} + 3\frac{dy}{dt} + 7\frac{dx}{dt} + 12y = 0,$$

formulate the analog computer diagram to get solutions for x, y, assuming all initial conditions to be zero for $t < 0$.

LP–4.

With initial conditions $x(0-) = \dot{x}(0-) = 0$, use time and amplitude scale factors to solve

$$\frac{d^2x}{dt^2} + 0.7\frac{dx}{dt} + 10^5x = 10$$

on the analog computer. (Try to use different time scales and see the effect on the solution, the problem workability, etc.)

LP–5.

A stirred tank is heated by an electric element as shown in Fig. 7–39. The tank is insulated from its surroundings by a $\frac{1}{2}$-in. layer of insulation material having a thermal conductivity of 2×10^{-4} (gm·cal)/(sec·cm^2·°C/cm). The temperature of the water is to be measured by a probe weighing 2.0 lb and having a specific heat of 1.5 (gm·cal)/(gm·°C). The coefficient of heat transfer from the water to the probe is 1.0 (gm·cal)/(gm·°C). The tank volume is 1 ft^3 and its surface area is 6 ft^2. The heating element has a resistance of 10 ohms. The ambient temperature $\theta_a = 20$°C.

a) Review the discussion for the described system. It can be seen that an approximation for the probe temperature may be obtained by neglecting the heat flow to the probe to determine the water temperature and using this result to determine the heat flow to the probe and an approximate probe temperature. Program the analog computer for this approximate solution and observe how well the probe and water temperatures compare for a step change in heat input from the electrical heater. Note that the times involved are very long and that time scaling is required.

b) Although the above approximate solution is easier for mathematical computations than the "exact" solution, the exact solution can be handled just as easily on the analog computer. Modify the computer set-up used above so that the exact solution is obtained (i.e., do not neglect the effect of the heat flow to the probe on the water temperature). Compare the results with those obtained in (a) and thus determine the validity of the approximate solution.

Note: The use of the term "exact solution" above should not be interpreted to mean that we have an exact solution to the physical problem posed. We have an exact solution to the mathematical model of the real physical system;

however, this model is only an approximation in the first place. The assumptions made in modeling thermal systems by lumped elements have been discussed earlier.

c) If the thermal capacitance of the probe is four times that given originally, compare the exact and approximate solutions obtained on the analog computer.

LP–6. Design of a dynamic absorber using the analog computer

There are numerous applications of rotating equipment where vibratory excitations inherent in the operation of the equipment may cause large vibratory forces to be transmitted to the attached foundation. Devices such as pumps, compressors, and turbines have a rotor with blades attached, a stator with vanes, and a working fluid. Vibratory forces may be caused by an unbalance of the rotor and/or by pulsing effects of the fluid acting on the blades and/or stator vanes. In either case the vibratory force is a function of the rotor speed, equal in frequency to the rotor speed in the case of unbalance and an integral multiple of the speed in the case of pulsating-flow effects. There are cases where a turbine rotor cannot be completely balanced (heating of the parts can produce unbalance in a cold-balanced rotor), so that an unbalance force exists. This force has a magnitude $F = (m_1 r)\omega^2$, Fig. 7–40(a), where m_1 is the mass unbalance and ω is the rotor speed in rad/sec.

If the exciting frequency, ω, is equal to the natural frequency of the turbine on supports ($\omega_n = \sqrt{k/m_2}$), then a condition of resonance exists and large forces are transmitted to the foundation. Where the exciting frequency is near the natural frequency of the turbine on its supports, large forces also can be transmitted to the foundation. If the machine operates at nearly constant speed, then a device called a *dynamic absorber* may be attached to the machine to eliminate the transmitted force. The dynamic absorber is a mass-spring system tuned to have a natural frequency equal to the operating frequency ω.

Since turbines are used for generating 60-cps electricity and since they are operated at 1800 rpm or 3600 rpm with small variation in speed, a dynamic absorber may be used to attenuate the vertical force transmitted into the foundation. The original system may be idealized as shown in Fig. 7–40(a).

The system with a dynamic absorber is idealized as shown in Fig. 7–40(b). Note that only vertical motion is considered here.

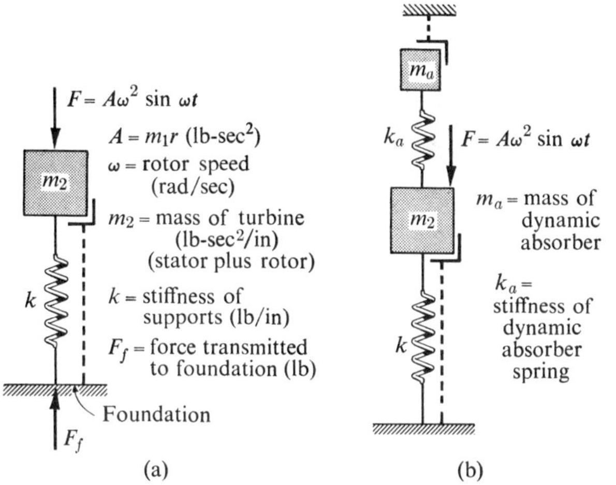

FIGURE 7–40

Let us consider a dynamic absorber designed with the help of the analog computer. The turbine operating speed ω is 1800 rpm, and, due to unbalance, a large vibratory force is transmitted into the foundation. The turbine weighs 2000 lb. Calculations show that space limitations prescribe a maximum of 400 lb for the weight of the dynamic absorber and that the vertical stiffness of the turbine structure supporting this weight may vary from 2500 lb/in. to 25,000 lb/in.

To design this system, we adopt the following procedure.

a) Draw the linear graph of the system with the dynamic absorber added.

b) Write the necessary path and vertex equations and simplify to two simultaneous equations.

c) Draw the analog-computer flow diagram.

d) Scale the problem, modify the computer diagram.

e) Using $m_a/m_2 = 0.1$ (m_a weighs 200 lb), plot a curve for F_f vs. ω for the two extreme values of k as ω is varied from 0 to 3600 rpm (60 rps). (Be sure that all starting transients have died out before taking each measurement. If you have difficulty with this, ask your instructor.)

Note what would happen if the speed ω were varied considerably to either side of the operating rotor speed ω of 1800 rpm.

f) The best design for a dynamic absorber would be one that is least sensitive to small variations in exciting fre-

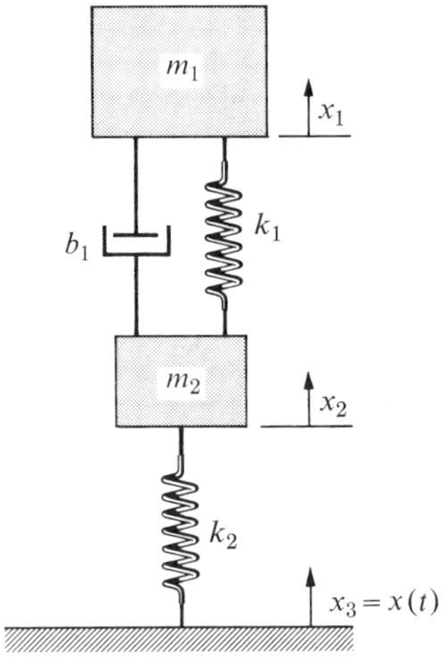

FIGURE 7–41

quency. As we have seen, it is possible to obtain zero transmitted force at the natural frequency of the added mass-spring system, but the exciting frequency may vary slightly. Determine the effect on F_f of absorbers with various masses over a $\pm 5\%$ range of exciting frequency (around the nominal center frequency of 1800 rpm). Consider $m_a/m_2 = 0.05$, 0.15 and 0.20. Recall that the absorber must always be tuned to the center frequency [$\omega_0 = \sqrt{k_a/m_a}$] so that when m_a is changed, k_a must be changed accordingly.

g) For the over-all design conditions stated, select the best absorber and discuss the reasons for your selection. It would be desirable to plot on one graph the F_f vs. ω variation (over the $\pm 5\%$ frequency range) for the four mass ratios given above.

LP–7. Consecutive reactions of chemical kinetics

Consider the chemical reaction

$$A \xrightarrow{k_1} B.$$

The "kinetics" or rate equation is

$$\frac{-dC_A}{dt} = k_1 C_A^N,$$

where

C_A = molal concentration of A,

k_1 = absolute reaction-rate constant,

N = "order" of reaction (for chemical systems reactions up to third order are known).

The radioactive decay of elements can be described by simple first-order chemical-kinetics equations. The problem is treated like consecutive reactions (chemical). Consider $A \xrightarrow{k_1} B \xrightarrow{k_2} C$. The kinetic equations are

$$\begin{aligned} -dC_A/dt &= k_1 C_A, \\ dC_B/dt &= +k_1 C_A - k_2 C_B, \\ dC_C/dt &= +k_2 C_B. \end{aligned}$$

It is known by experiment, in the case of radioactive decay, that these are first-order reactions. If a is the initial amount of A present (B and C are not present initially), then for any time $C_A + C_B + C_C = a$.

PROBLEM: Find the changes in the concentrations of A, B, and C with respect to reaction time for the case of $k_1 = 1$ and $k_2/k_1 < 1.0$ (use two cases: $k_2 = 0.08$ and $k_2 = 0.23$). Also solve the differential equations analytically for C_A, C_B, and C_C as functions of reaction time. Finally, determine analytically: (a) the time and (b) the concentration, where B is a maximum. You may assume that a (the initial concentration of A) is unity.

LP–8. Design of an automobile suspension system

Problem description. To investigate the response of an automobile suspension system for selected disturbances (Fig. 7–41). The system response to these disturbances for various values of the system design parameters are obtained. The most suitable values of the system parameters are determined by selecting the desirable response from the computer solution.

m_1 = one-quarter mass of automobile,

m_2 = mass of the wheel and axle,

k_1 = spring constant of main auto spring,

k_2 = spring constant of tire (assumed linear),

b_1 = shock absorber damping constant,

x_1 = displacement of auto body,

x_2 = displacement of wheel,

x_3 = roadway profile displacement.

The physical constants are:

$m_1 = 25$ slugs, $m_2 = 2$ slugs,

$k_1 = 1000$ lb/ft, $k_2 = 4500$ lb/ft,

b_1 = variable,

(1 slug = 1 lb·sec^2/ft).

The initial conditions and forcing functions are:

$$x_1 = x_2 = \frac{dx_1}{dt} = \frac{dx_2}{dt} = 0 \quad \text{at} \quad t = 0-,$$

$$x_3 = x(t).$$

System equations. The differential equations of motion of the system are derived by equating the forces acting on the mass involved in the system. They are:

$$\frac{d^2x_1}{dt^2} = -\frac{b_1}{m_1}\left(\frac{dx_1}{dt} - \frac{dx_2}{dt}\right) - \frac{k_1}{m_1}(x_1 - x_2),$$

$$\frac{d^2x_2}{dt^2} = -\frac{b_1}{m_2}\left(\frac{dx_2}{dt} - \frac{dx_1}{dt}\right) - \frac{k_1}{m_2}(x_2 - x_1) - \frac{k_2}{m_2}(x_2 - x_3).$$

Simulate the automobile suspension systems on the analog computer for the given data for b_1 of

(a) 100 lb-sec/ft, (b) 50 lb-sec/ft, (c) 20 lb-sec/ft.

In each case, plot the normalized displacements of the auto body (x_1/x_3) and the wheel (x_2/x_3) for a deflection in road profile (let x_3 be a square pulse whose duration is 35 msec).

LP–9. Investigation of the process parameters affecting the control of a stirred-tank reactor

Discussion. The continuous-stirred-tank reactor is an essential part of many chemical processes. This problem illustrates, in simplified form, how a model of a CSTR may be set up and studied.

Problem description. To investigate the effect of parameter changes on the time response of the output concentration of a stirred-tank reactor (Fig. 7–42).

Equations. The reaction taking place is

$$A \xrightarrow{k} B$$

where the specific reaction rate is defined as

$$k = k_{gm} + b\theta_o. \tag{1}$$

Material balance (per unit time):

material in: QC_i, material reacted: kV_tC_o,

material out: QC_o, material accumulated: $V_t(dC_o/dt)$.

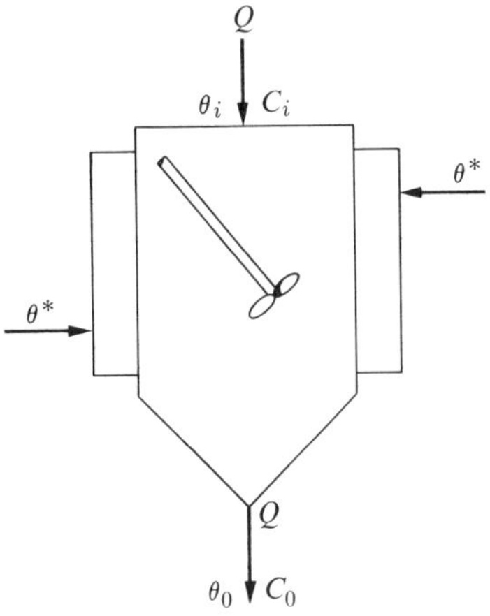

FIGURE 7–42

Summing, we have

$$\frac{dC_o}{dt} = \frac{Q}{V_t}C_i - \frac{Q}{V_t}C_o - kC_o. \tag{2}$$

Heat balance (per unit time):

heat in: $Q_\rho C_p\theta_i$, heat transferred: $hA^*(\theta_o - \theta^*)$,

heat out: $Q_\rho C_p\theta_o$, heat due to reaction: $-HV_tkC_o$,

heat accumulated: $\rho V_tC_p(d\theta_o/dt)$.

Summing, we obtain

$$\frac{d\theta_o}{dt} = \frac{Q}{V_t}\theta_i - \frac{hA}{\rho V_tC_p}(\theta_o - \theta^*) - \frac{HkC_o}{\rho C_p} - \frac{Q}{V_t}\theta_o, \tag{3}$$

where

θ_o, θ_i, θ^*	=	temperature,
Q	=	flow rate of material,
V_t	=	reactor volume,
t	=	time,
h	=	heat transfer coefficient,
A^*	=	effective area of heat transfer,
ρ	=	average density of material,
H	=	heat due to reaction,
C_p	=	average specific heat,
C_i	=	input concentration,
C_o	=	output concentration.

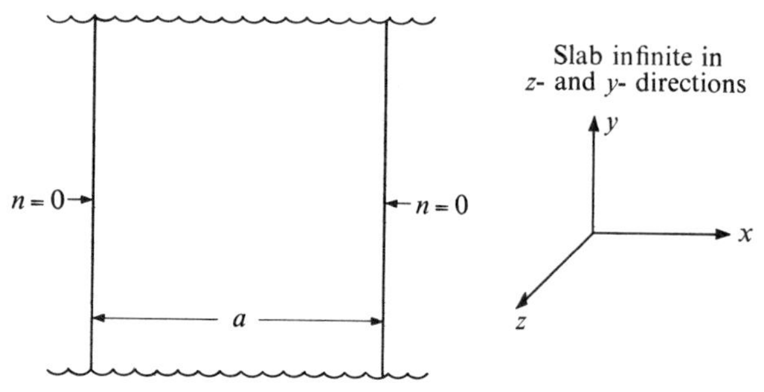

FIGURE 7–43

By simulation of the process (Eqs. 1, 2, and 3) on the analog computer determine:

a) the response of output concentration for a step disturbance to input concentration ($Q/V_t = 0.5\ \text{sec}^{-1}$),

b) the response as in (a) except $Q/V_t = 0.167\ \text{sec}^{-1}$,

c) response of output temperature of reaction for a step disturbance to temperature of material entering reactor $Q/V_t = 0.167\ \text{sec}^{-1}$.

In the above, Q is in ft^3/sec, V_t is in ft^3; $(\text{ft}^3/\text{sec})/\text{ft}^3 = \text{sec}^{-1}$.

Given:

$k = (-31.6 \times 10^{-2} + 5.74 \times 10^{-3}\, T_o)\ \text{sec}^{-1}$,

$Q = -500$ Btu/lb, $\quad V_t = 300\ \text{ft}^3$,

$h = 100\ (\text{Btu/hr})/(\text{ft}^2\text{-°F/ft})$ $\quad \theta_o(0-) = 65.0$°F,

$C_p = 4.0$ Btu/lb-°F, $\quad C_i = 0.45$,

$\rho = 70\ \text{lb/ft}^3$, $\quad \theta_i = 50$°F,

$A = 100\ \text{ft}^2$, $\quad \theta^* = 0$,

$C_o(0-) = 0.4$.

LP–10. Critical size of a nuclear reactor

It is widely known that a minimum mass of fissionable material must be used to sustain a "chain nuclear reaction." The "critical" mass of the atomic bomb is a well-known example. As we will see, this is really more a problem of geometry and size than of total amount of material.

In this laboratory problem, we will use the analog computer to determine the critical size of a nuclear reactor and thus

a) obtain further practice in the use of analog computers;

b) gain an appreciation of an important design problem in nuclear engineering;

c) gain an appreciation of how to handle a distributed system by approximate methods so that the analog computer can be used.

Nuclear power reactors are designed so that a specific quantity of fissionable material exists within a certain geometry. To demonstrate the principles involved, we will consider a reactor which is very long in two dimensions. The problem will be to determine the minimum length of the third dimension to have a "critical" reactor. In a supercritical reactor, any preestablished neutron density would rise without bound as time increases.

Once this condition is established, the reaction is controlled to usable limits by the insertion of control rods which absorb neutrons. The power output of the reactor is controlled by the degree of insertion of these rods. The control problem will not be considered here; we only wish to determine the critical length.

The reactor we are considering is shown schematically in Fig. 7–43.

We assume that the neutron density may be zero at $x = 0$ and $x = a$. The neutron density will vary with x

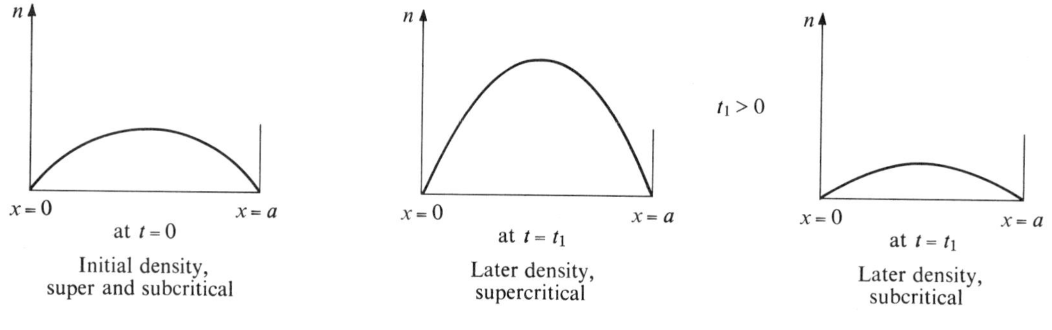

FIGURE 7–44

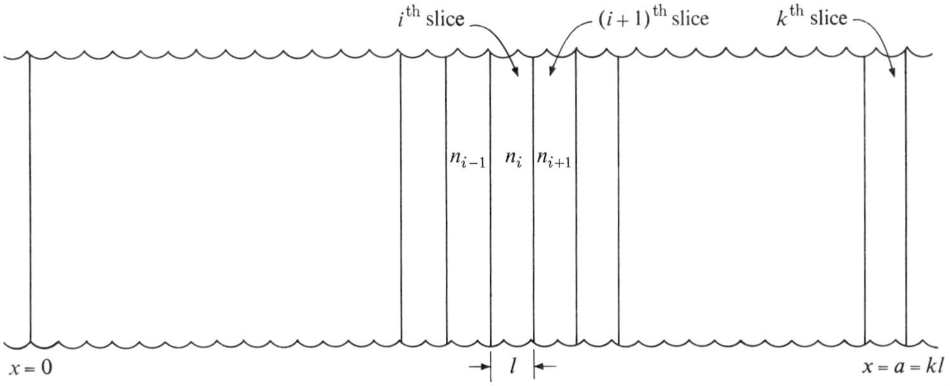

FIGURE 7–45

and time. That is, at any particular time, the neutron density may be as shown in Fig. 7–44. If the reactor is subcritical, the density will decrease; otherwise it will increase as time passes.

The reactor is really a distributed system since the fissionable material and neutron density are continuously distributed between $x = 0$ and $x = a$. The "exact" describing equation would be a *partial* differential equation involving derivatives taken with respect to both distance and time. Such a problem could not be handled on the analog computer because it can handle only a finite number of variables. However, we can approximate the situation quite accurately by considering the reactor to be divided into slices (hopefully a small number of slices). Then, by applying the conservation-of-mass principle to neutrons, we can derive a number of simultaneous ordinary differential equations which can be simulated on the computer.

Consider the following assumptions:

1) Neutrons diffuse from regions of high concentration to those of lower concentration, and the time rate of diffusion across an element of surface will be proportional to the area of that surface element and to the space rate of change of neutron density normal to the surface element (i.e., the neutron concentration gradient). The constant of proportionality is called the *diffusion coefficient*, D, and is assumed constant.

2) The time rate of absorption of neutrons by the material is proportional to neutron density, neutron speed, and inversely proportional to the mean free path for absorption.

3) The rate of neutron production by fission is proportional to the rate of absorption.

Figure 7–45 shows the slab broken into slices. Then, taking a unit cross-sectional area, we obtain

(diffusion into section i from $i - 1$) − (diffusion out of section i to $i + 1$) + (production of neutrons in section i) − (absorption of neutrons in section i) = (time rate of increase of neutrons in section i),

$$\frac{D(n_{i-1} - n_i)}{l} - \frac{D(n_i - n_{i+1})}{l} + l\frac{k_c v}{\lambda_c} n_i - l\frac{v}{\lambda_c} n_i = l\frac{dn_i}{dt},$$

where

$$\text{dimensions of each term} = \frac{\text{neutrons}}{\text{time} \times \text{area}}$$

D = neutron diffusion coefficient, cm^2/sec

n_i = neutron density in section i, neutrons/cm^3

v = neutron speed, cm/sec

λ_c = neutron mean free path, cm

k_c = average number of neutrons produced per capture by fission.

Dividing through by D and multiplying by l, we have

$$(n_{i-1} - 2n_i + n_{i+1}) + l^2 K n_i = \frac{l^2}{D}\frac{dn_i}{dt},$$

where

$$K = \frac{(k_c - 1)}{D\lambda_c} v = \text{const } (1/\text{cm}^2).$$

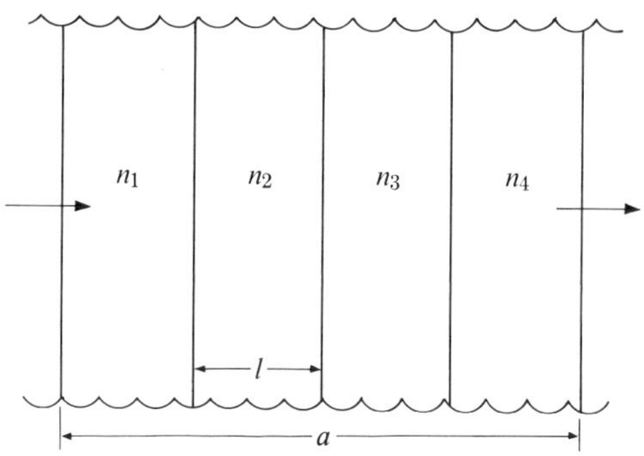

FIGURE 7–46

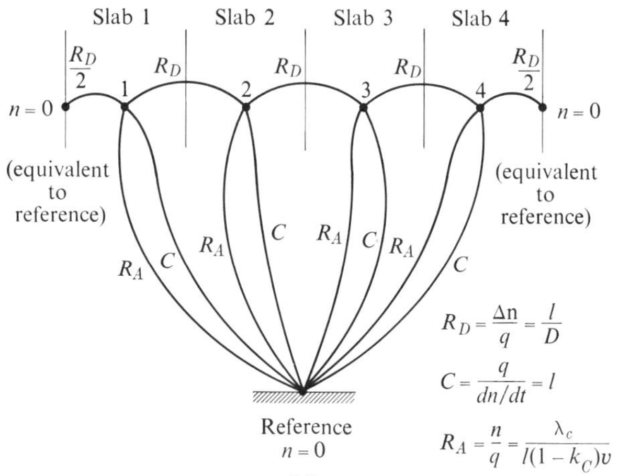

(a)

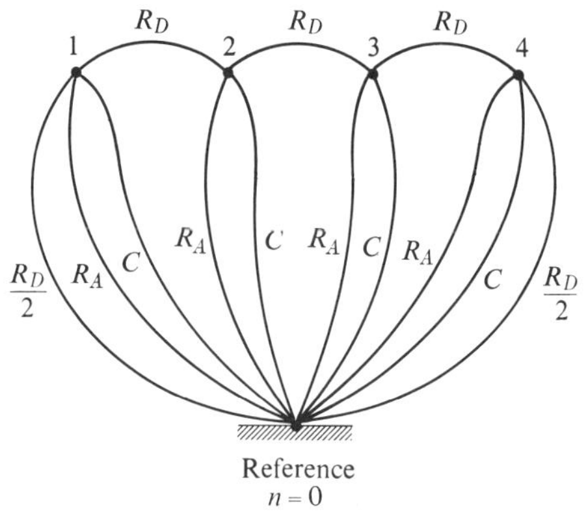

(b)

FIGURE 7–47

An equation of this form can be written for each of the k slices. These represent (when divided by l^2) an approximation to the following partial differential equation which may be derived by more advanced methods:

$$\frac{\partial^2 n}{\partial x^2} + Kn = \frac{1}{D}\frac{\partial n}{\partial t}.$$

If we consider the slab broken into four slices (Fig. 7–46), we have $l = a/4$. For the two outside slices (n_1 and n_4 have one unconfined side), we need slightly different equations. Thus for n_1:

$$\text{diffusion in} = D\frac{(0 - n_1)}{l/2},$$

for n_4:

$$\text{diffusion out} = D\frac{(n_4 - 0)}{l/2};$$

in fact, if we consider

$$q = \text{through-variable} = \text{neutron flow rate} = \frac{\text{neutrons}}{\text{time} \times \text{area}},$$

$$n = \text{across-variable} = \text{neutron density} = \frac{\text{neutrons}}{\text{volume}},$$

we can draw a linear graph to represent this system, using a diffusion resistance between sections, a "resistance" (which may be negative) at each section representing neutron absorption and generation, and a "capacitance" at each section representing storage of neutrons. The system graph is presented in Fig. 7–47.

The two end diffusion resistances are only half as large as those between sections. Thus our four equations are

$$(-3n_1 + n_2) + l^2Kn_1 = \frac{l^2}{D}\frac{dn_1}{dt},$$

$$(n_1 - 2n_2 + n_3) + l^2Kn_2 = \frac{l^2}{D}\frac{dn_2}{dt},$$

$$(n_2 - 2n_3 + n_4) + l^2Kn_3 = \frac{l^2}{D}\frac{dn_3}{dt},$$

$$(n_3 - 3n_4) + l^2Kn_4 = \frac{l^2}{D}\frac{dn_4}{dt}.$$

Consider a uranium-235 reaction for which

$$K = 0.25\ \text{cm}^{-2}, \qquad D = 33 \times 10^4\ \text{cm}^2/\text{sec}, \qquad \text{and}$$

$n_1 = n_2 = n_3 = n_4$ initially, and take the following steps to determine the critical size:

a) Determine a time-scale factor. To get some idea of time variation, take the equation for n_i and assume n_{i+1} and n_{i-1} are zero. This is then a simple first-order equation which you can solve to determine the time constant.

b) Scale the equations.

c) Program the scaled equations for the computer.

d) Observe the solution for various values of a and adjust a until you have a just-critical reactor. Note that when a is changed, l changes if we still keep four slices.*

e) For this case, vary separately both D and K and determine which of these affects the critical size and in what way. Thus derive an empirical formula for the critical size of a reactor of this shape. [Note: By simple dimensional reasoning it can be seen that the critical a must be inversely proportional to $\sqrt{K}$. You should check this and determine the proper constant of proportionality.]

---

* We can also vary a by keeping l fixed and adding or subtracting slices.

8 SYSTEM GRAPHS AND EQUATION FORMULATION

8–1. INTRODUCTION

In the preceding chapters of this book, five types of physical systems have been considered: mechanical translational, mechanical rotational, electrical, fluid, and thermal. The primary reason for a joint study of these various systems is the striking similarity of their dynamic behavior. Thus an understanding of the behavior of the forces in a mechanical system aids in the understanding of the behavior of currents in an electrical circuit, of fluid flows in a hydraulic system, or of the flow of heat in a thermal system.

The similarity between the characteristics of these different physical systems is due to two factors which seem to be nearly universal in lumped-element representations of systems.

First is the similarity between the elemental equations describing the ideal lumped elements, which has been completely summarized in Chapter 4. This similarity alone is not sufficient to ensure similar performance of different systems, however. A second requirement is the similarity of the physical laws governing the relationships between the variables of the system. That is, there must exist similar compatibility and continuity relationships which are demanded by the manner in which the ideal elements are interconnected. In general, these relations will determine the different behaviors of the same collection of elements when they are connected together in different fashions.

To take the utmost advantage of these similarities and to form a basis for dealing with such diverse systems in a coherent manner, it is desirable to develop a generalized approach to the analysis of lumped-element systems [1, 3]. Such a general basis is possible through the concepts of generalized lumped elements, linear graphs, and generalized compatibility and continuity relationships. The stage has been set for this generalization by the previous consideration of system elements in terms of through- and across-variables and by the classification of single-port elements as *A*-type (energy stored because of the across-variable), *T*-type (energy stored because of the through-variable), and *D*-type (energy dissipators) elements. Multiport elements also can be treated by linear graphs; their representation was given in Chapter 4 and will be used here. The development will now be continued by means of the concepts of linear graphs which will allow a general statement of compatibility and continuity.

The continuity and compatibility requirements are statements of physical relationships which must exist between the variables of a connected set of lumped elements because of the manner in which they are interconnected. These relations deal with the structure of the system rather than with the nature of its constituent parts. To portray the compatibility and continuity conditions for a system of elements, linear graphs are useful. The concept of linear graphs which was introduced in Chapters 2 and 4 and used slightly in Chapter 5 will be developed in more detail in this chapter [1].

8–2. LINEAR GRAPH REPRESENTATION OF ELEMENTS

Graphical symbols and diagrams are frequently used as aids for the visualization and understanding of physical systems. For example, the circuit or network diagram, which has been used extensively, is of great value because it allows visualization of the performance of a system and the determination of important effects without a complete formulation and solution of the system equations. In addition, the network diagram is a useful aid in the formulation of the system equations when these equations are required. Operational block diagrams, discussed previously, are also useful for understanding physical system behavior.

The linear graph provides a general method of diagramming lumped physical systems of all kinds in a way which is both a visualization and formulation aid. Not only does it allow any particular physical problem to be treated, but it also allows us to see and use the similarities of different physical systems which have similar structure.

A *linear graph* is a set of interconnected lines. The study of linear graphs involves the branch of mathematics called *topology*, which deals with the properties of figures which are related to the way in which their various parts are interconnected. Thus the properties which are unchanged under distortions such as stretching, bending, and squeezing are topological properties, as are those of any deformation which does not involve tearing or joining. Linear graphs will be used here first as an aid in visualizing the *structure* of systems and second as a basis for a general technique for formulating system equations. Linear graphs may also be used to allow the expedient formulation of system equations in matrix form. References 1 and 2 consider this approach.

The word "linear" in linear graph connotes "line" in the geometrical sense and should not be confused with the use of linear in the algebraic sense in reference to the elemental equations of lumped elements. There is no reason why the systems considered should not be allowed to contain nonlinear elements. *The linear graph technique can be applied to any lumped system, whether or not it is composed of ideal elements.*

The linear graph will bear a close resemblance to the geometry of the electrical network, but this geometrical resemblance will not always hold for mechanical systems. When the technique is extended to multiterminal elements, any direct geometrical similarity is lost even with an electrical network. The basis for drawing the linear graph of a system will be the manner in which the elements are interconnected, and thus the relationships between the various through- and across-variables will determine the appropriate graph of any system.

The linear graph will be composed of lines called *branches.* A branch will indicate that there is a specific relationship between a through-variable and an across-variable in a system. Thus, for systems containing only two-terminal elements, the branches can be considered to represent the presence of an element in the system.

f, v means that $f = \mathsf{f}(v)$,

such as $f = C\dfrac{dv}{dt}$,

or $v = Rf$,

or $v = L\dfrac{df}{dt}$,

or even $f = \dfrac{20v^2}{v+5}$.

An additional objective is accomplished in drawing the graphs. Reference conditions for all variables will be selected, and these references will be shown on the graph by means of arrows associated with each line. A graph formulated in this fashion will be called an *oriented* linear graph. As we have seen, elements can be shown with separate reference conditions for each through- and across-variable. *In the linear-graph approach, one arrow is used to show both references.* In mechanical, electrical, and fluid systems, this definition states that power, the product of the through- and across-variable, is taken to be positive when it flows into the element. In thermal systems, of course, power is the through-variable, **q**, alone.

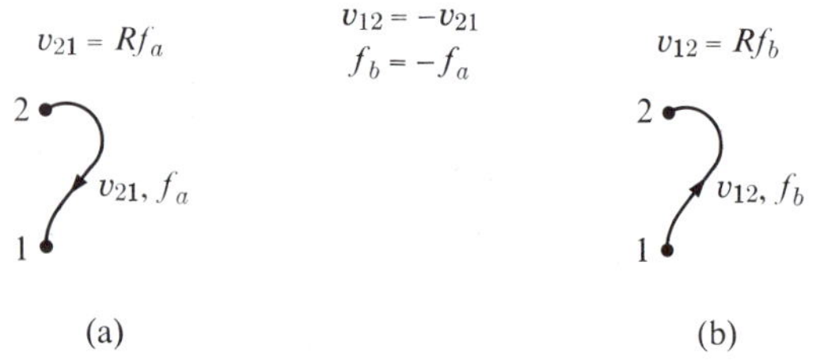

FIG. 8–1. Effect of reversal of orientation of linear graph branch.

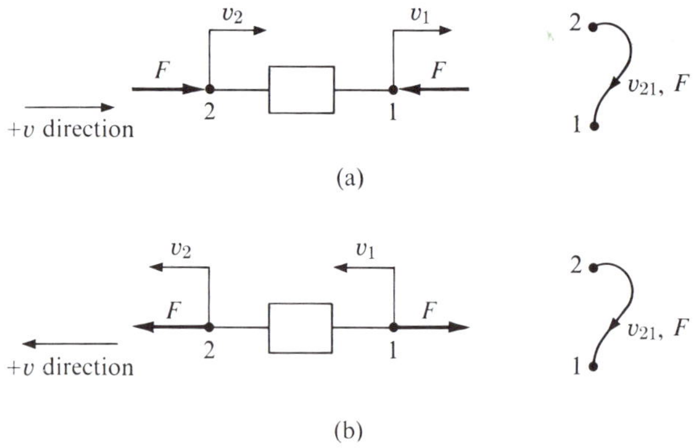

FIG. 8–2. Nonrelation of direction to linear graph orientation. (a) Velocity positive to right. (b) Velocity positive to left.

8–2.1 Single-Port Passive Elements

Table 4–2 shows the linear graph representation of the pure system elements (see inside back cover). The orientation of the graph branches shown in the table is arbitrary. Any or all arrows shown in Table 4–2 could be reversed. The significance of a reversed arrow is that the variables then used with the graph are the negative of those which would have been used with the original orientation. Figure 8–1 shows this relationship. Thus, the chosen orientation is associated with the selection of a particular reference condition in the given system. Note that when the arrow is reversed, the references for both the across- and through-variable are reversed and the sign in the elemental equation stays the same, as shown at the top of Fig. 8–1.

The *only* significance of the arrow is that (a) the across-variable *difference* which will be employed to formulate the equations is the across-variable at the tail of the arrow minus the across-variable at the head and (b) the through-variable reference which will be used in the formulation of equations is the one which will give a positive sign in the elemental equation when the across-variable is as given in (a). (One must refer to the physics of a mechanical situation to determine which direction the through-variable reference has in the actual system. Refer to Table 2–1.)

Special note should be made of mechanical (rotational as well as translational) systems for which an additional factor is present and must be considered. This is the notion of *direction. The arrows on the linear graphs do not convey any information concerning direction* but rather indicate only which across-variable difference will be used and the associated through-variable. Thus we should pre-establish a convention for direction for any given system we wish to consider.

Figure 8–2 shows two situations which have the same linear-graph representation. It will be noted that for both cases the arrow on the linear graph indicates the reference condition for v_{21} and means that if v_{21} is positive, the velocity of 2 in the previously assumed positive direction is algebraically greater than that of 1 in the same positive direction.

Note that the arrows on the linear graph representation of both Fig. 8–2(a, b) are from 2 to 1; however, when v_{21} is positive, the element in (a) will be tending to go into compression (point 2 is getting closer to point 1), whereas when v_{21} is positive, the element in (b) will be tending to go into tension (point 2 is getting farther from point 1). *Thus the linear-graph representation does not indicate whether an element is in tension or compression.* This is because the reference for direction is set independently and one must also use that reference to determine the state of stress. Note that the same symbol F is used in both cases (and will be positive when v_{21} is positive if this is a damper) since the through-variable reference is conveniently changed to go with the across-variable reference.

In summary, for mechanical systems the orientation on the linear graph is independent of the chosen reference direction. However, once this direction is chosen (and sources then given proper algebraic signs) a positive value for v_{21} indicates that the relative velocity $(v_2 - v_1)$ is in the chosen positive direction.

Comments about the mass, inertia, fluid and thermal capacitance elements are in order. In a complicated system in which several masses or inertias may appear, care must be taken to realize that all mass or inertia velocities are taken with respect to the same reference. That is, their second terminals must all be connected to the same point. [In general, it would be possible for the velocities of various masses to be taken with respect to references whose velocity differed by a constant. This would be equivalent to having a constant-velocity source applied between the second terminals of two masses so described. This source, however, must be constant for all time (from minus infinite to plus infinite time) and therefore could not participate in the dynamic behavior of the system but only enter as an additive constant. In systems work there is ordinarily no advantage in such a representation and we will not consider it further.]

The fluid capacitance is similar to the mass element in that the second terminal must be connected to the reference across-variable, which is pressure in this case (Figs. 3–5 and 3–8). Atmospheric pressure is usually taken as reference. However, it should be noted that when this is done, the potential energy calculated by Eq. (3–14) is that stored in excess of the energy stored at the reference pressure (P_1 in Fig. 3–6b).

For the thermal capacitance, again, the second terminal of the capacitance is the reference temperature. Any convenient reference may be used rather than absolute zero since the difference between references will be a constant temperature (i.e., room temperature, zero Fahrenheit or Centigrade). However, if one uses an arbitrary reference, when computing stored energy he must be careful to realize that the energy is that in excess of the amount stored at the reference temperature. The energy in the thermal system is simply the integral of the heat flow.

8–2.2 Source Elements

In addition to the various ideal elements associated with the several physical systems which we have considered, we must also deal with sources which act as input or forcing functions. Sources supply to systems of ideal elements the power and energy which is stored and dissipated by the systems, and consequently are to be distinguished from the ideal elements themselves. Associated with any source is a through-variable and an across-variable. In an *ideal* source, one of these variables is assumed to be independent of the other, as discussed in Chapter 4. For example, a current source supplies a specified current independent of the voltage demanded by the elements to which the source is attached. The representation of real sources by lumped elements and ideal sources has been discussed in Chapter 6, "Modeling of Physical Systems."

Ideal electrical sources of current or voltage clearly have two terminals, and their linear graphs are a line segment whose two ends correspond to these terminals. To designate a source and to indicate the type of source being represented, the appropriate source variable will be enclosed in a circle, with an arrow which indicates the positive reference for the source variable, as shown for the ideal sources in Table 4–2. Figure 8–3(a) shows an ideal voltage source of magnitude v driving a resistance R and other interconnected elements which are not shown but are designated the remainder of the system. Figure 8–3(b) shows a similar system driven by a current source of intensity i. Figure 8–3(c, d) shows the linear graph representation of these sources and resistances and illustrates the sign conventions which will be used.

A common basis is retained for the meaning of the arrow direction in both sources and passive elements (all of the ideal elements considered except sources have been passive; that is, they cannot supply energy continuously to a system.) This arrow was taken in the direction of voltage *drop*. Since voltage *rises* when going from g through the voltage source in Fig. 8–3(a), the arrow within the source is drawn toward g. In other words, the tail of the arrow corresponds to the positive terminal of the voltage source.

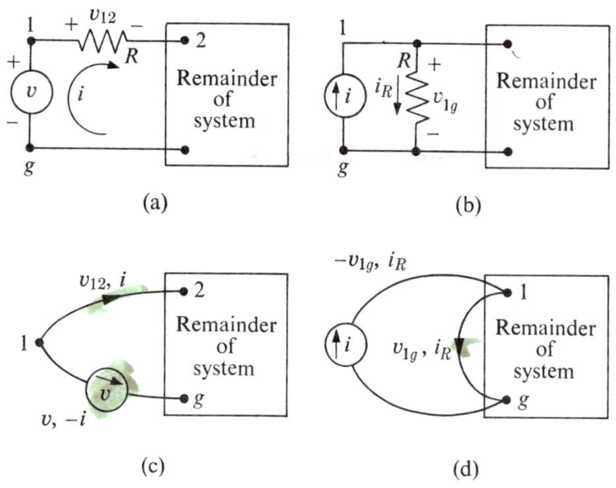

FIG. 8-3. Electrical sources and their linear graphs.

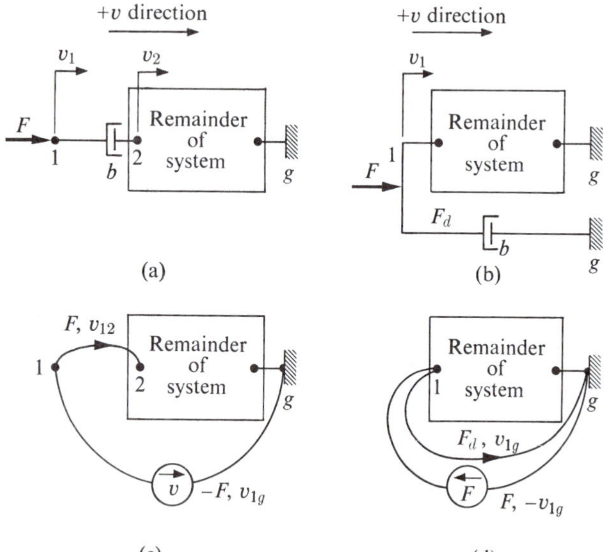

FIG. 8-4. Mechanical sources and their linear graphs.

In Fig. 8-3(b) the current source puts out current in the direction of the arrowhead, which is consistent with the current convention for elements. In both sources shown in Fig. 8-3, the dependent variable which is associated with the source [current i, in (a) and voltage, v_{1g}, in (b)] must be negative according to the conventions established. For instance, in (a) the source current is opposite from the resistance current; if the resistance current is called i, then the source current must be $-i$, since continuity must hold at vertex 1. Similar considerations hold for v_{1g} in (b); the voltage represented by the arrow in d is v_{g1}, which is $-v_{1g}$. Since the current in the voltage source and the voltage across the current source need seldom be considered in formulation of system equations, this negative sign will not be troublesome.

Ideal mechanical sources must also be two-terminal in character. Newton's third law (action equals reaction) demands that any force exerted on one place in a system causes an equal and opposite force to be exerted on some other point. It is quite possible to have a force or torque exerted between two points in a system, both of which are in relative motion. A motor in which both the rotor and stator are allowed to turn is an example of such a system. Normally an indicated force on a system implies that the reaction force acts on ground. The reader may imagine that he is the force generator and that the force is exerted by his hand. The second terminal is the place where he is standing. Similarly, velocity sources have two terminals and the velocity source prescribes the relative velocity between these two points. Figure 8-4 shows examples of force and velocity sources for mechanical translational systems. Note that the graph orientation of the sources is consistent with the chosen positive reference direction if we assume that the source values are algebraically positive.

Fluid and thermal system sources may be represented in a similar fashion. In some cases the two-terminal nature of these sources may not be clear; for example, if water from a dam is used to drive an electric generator, the return of the efflux water to the upstream side of the dam is devious to say the least. In such cases, the source would be taken as connected to ground. Then the return flow will not appear in any of the system equations. Many fluid and thermal sources will clearly have two terminals. For example, a positive-displacement fluid pump has continuous

If ideal:

$v_b = nv_a$

$f_b = -\frac{1}{n} f_a$

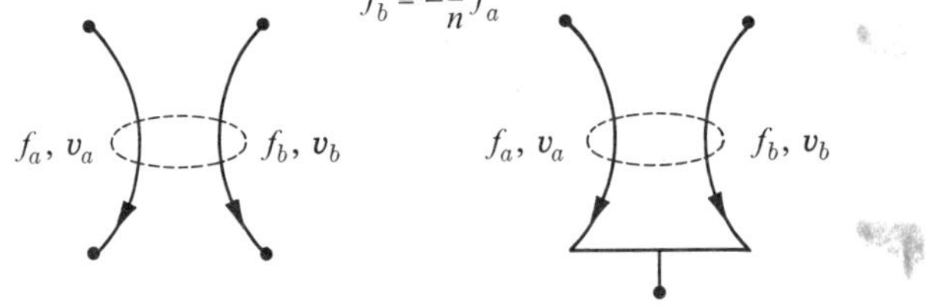

(a) Four-terminal (b) Three-terminal

FIG. 8–5. Linear graph of transformer and transforming transducer.

If ideal:

$v_b = rf_a$

$f_b = -\frac{1}{r} v_a$

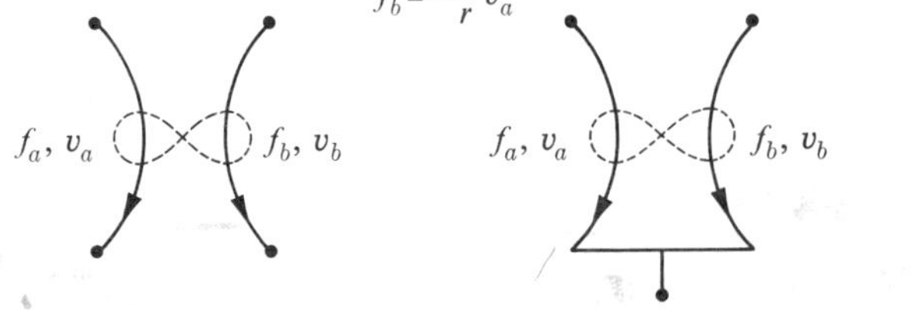

(a) Four-terminal (b) Three-terminal

FIG. 8–6. Linear graph of gyrator or gyrating transducer.

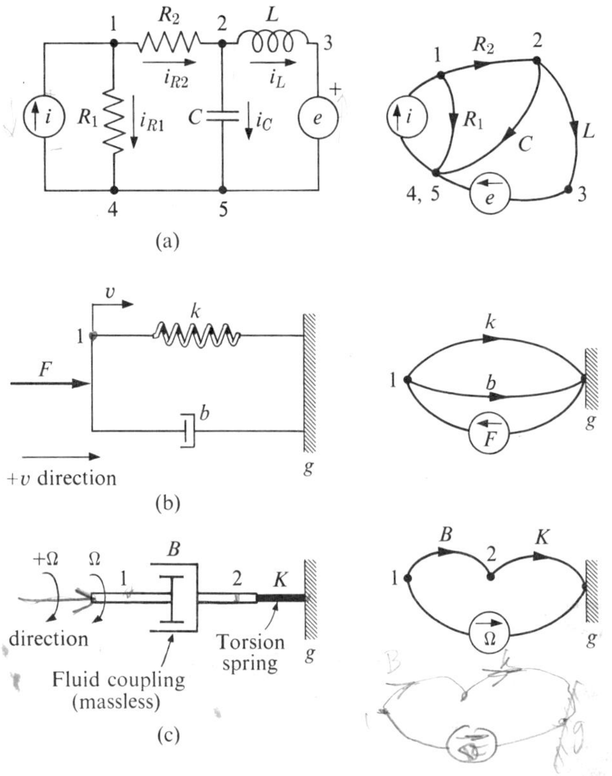

FIG. 8–7. System graphs of some simple systems.

flow from inlet to discharge and behaves approximately like a flow source.

8–2.3 Two-Port Transforming Elements

Transformers, gyrators, and transducers were discussed in general in Chapter 4 as two energy-port elements. The representation used there was essentially the linear graph representation in anticipation of this method of depicting systems. Figures 8–5 and 8–6 show the graphs of these elements. Since the references are chosen so that positive power is flowing into the element on both sides, we see that one of the transformation equations must carry a negative sign, and when the element is in operation at least one of the four variables must be negative at any specific time.

8–3. THE SYSTEM GRAPH

The process of constructing linear graphs for various physical systems will now be illustrated by means of several examples. A linear graph of a complete system is called the *system graph.*

8–3.1 Simple Examples

Figure 8–7 shows the graphs of some simple systems. There should be little difficulty in verifying that they are correct. It should be noted that the geometrical position of the branches in a linear graph, as well as their length and curvature, is of no significance. Only the way in which the branches are interconnected is of importance. Thus the two graphs in Fig. 8–8 are topologically equivalent and could represent the same physical system.

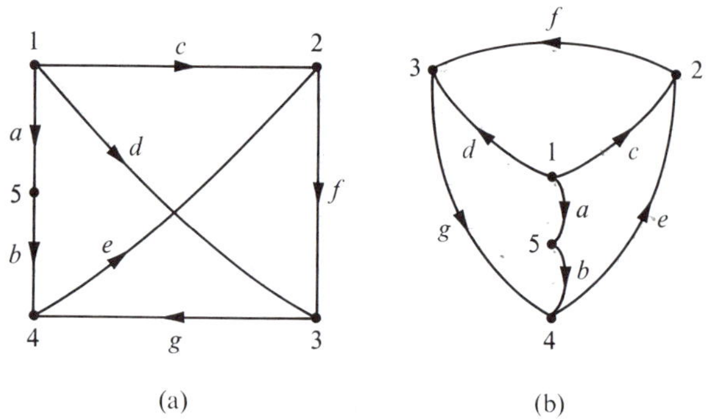

FIG. 8–8. Topologically equivalent graphs.

8–3.2 General Procedure for Drawing System Graphs

A general procedure for drawing the system graph of any system of lumped two-terminal elements may now be given:

a) Identify the lumped elements (this will involve some modeling). The number of lumped elements (including sources) will be equal to the number of branches in the final system graph.

b) Establish a reference point within the system. This point serves as a reference for measurement of all across-variables. A reference vertex in the system graph will correspond to this reference point. (If a mechanical-translational, mechanical-rotational, or fluid-flow system is being considered, the reference must be an inertial reference if any inertial effects are involved.)

c) At each terminal of every two-terminal element, a distinct across-variable will exist. Each across-variable will be measured between the terminal in question and the reference point. A vertex dot for each of these across-variables may be placed in the developing system graph.

d) For each two-terminal element of the system, a branch is entered in the developing system graph. These branches must be between the various vertices established in the previous step; however, every pair of vertices need not have an associated branch in any particular system. All inertial branches must be inserted between the reference vertex and one of the remaining vertices.

e) For each source, a branch is entered to complete the system graph. The two vertices for any source depend on the stipulated method of application of that source.

f) Orient the system graph, if desired, by placing arrows on each branch. This has the effect of establishing reference conditions for the through-variable in each element and the across-variable difference across each element. Note (at least mental, if not formal) should be made of the corresponding references in the physical system. (If the system is mechanical, a positive reference for velocity should be chosen at this time.) The arrows (and thus reference conditions) may be chosen arbitrarily; however, in most cases they will be chosen to show a general flow away from sources.

This procedure may be easily extended to include multiterminal elements such as transformers, gyrators, and transducers. Now, however, the graphs shown in Section 8–2.3 will be used for these elements, and thus two branches will appear for each element of this type. There will be two equations associated with each of these elements, however, and the idea that each branch in the graph accounts for one equation still holds. Consideration of the across-variable will indicate the proper points between which the multiterminal element should be connected.

8–3.3 More Examples

In mechanical systems with masses and in systems for which it is not obvious where the two terminals of elements are connected, a procedure which examines in detail the number of points which have different velocity and the elements which act between those points will ensure that the proper graph is obtained. The points in the system which have different velocities will correspond to the vertices of the linear graph.

The mechanical translational system of Fig. 8–9(a) will be used as an example. The following steps will yield the graph shown in Fig. 8–9(b).

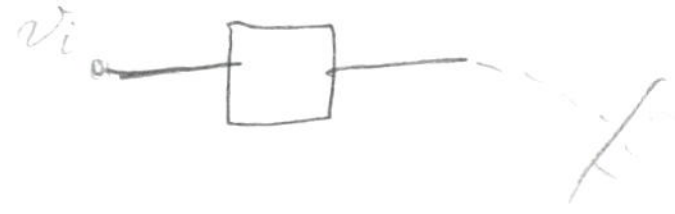

a) There are six elements in the system, (one mass, two dampers, one spring, one velocity source, and one force source), and thus six branches will appear in the graph.

b) A reference vertex is necessary; this corresponds to the reference for all velocity measurements. If any masses are in the system this reference must be an inertial reference frame.

c) The number of distinct velocities is determined to be three, corresponding to points 1, 3, and 7 in the system. These establish three additional vertices which will appear in the graph. Note that rigid connections require that points 2, 3, 4, and 5 have the same velocity and hence will be represented by one vertex. Likewise for points 6 and 7.

d) Damper b_1 is obviously connected between points 1 and 2, and a branch is inserted to represent it. Physically, this means that the relative velocity of this dashpot is the difference between the velocities of points 1 and 2. In like manner, damper b_2 and spring k are connected between the 2, 3, 4, 5 vertex and the 6, 7 vertex and their branches are entered.

e) As indicated earlier, mass elements must have their second terminal connected to ground. The velocity of mass m is that of point 3, so the branch for the mass element is entered between 2, 3, 4, 5, and the reference vertex.

f) The velocity source is applied at 1, and this velocity is measured with respect to ground, so the source branch is inserted as shown. The force source is applied at 6, 7 and reacts on ground and is appropriately inserted.

g) The graph can be oriented by the addition of arrows for each of the elements. This orientation is arbitrary (except for the relative orientation of the two sources). It is to be noted particularly that the orientation has nothing to do with the selection of the positive velocity reference to the right as shown on the figure. (This was clearly indicated in Fig. 8–2.) The positive velocity direction could be changed without changing the graph orientation, but v (the

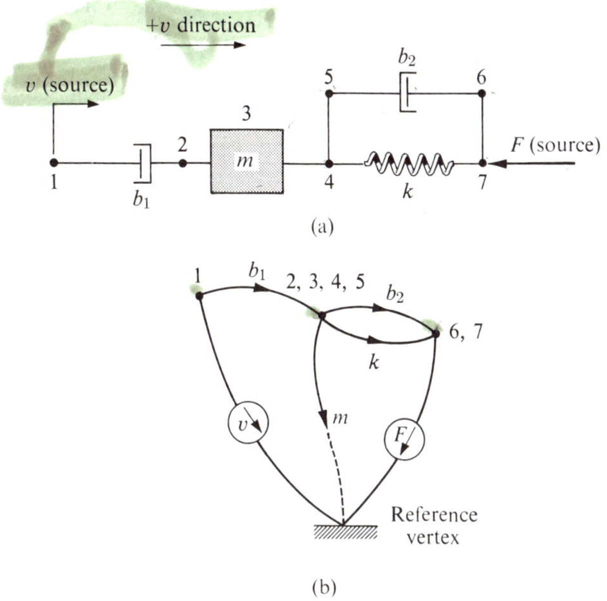

FIG. 8–9. Mechanical system and its graph.

source) would become a negative number if it was previously positive (and similarly with the force source). It should be clearly realized that *the assignment of the orientation to the branches of the graph (when related to the chosen positive direction for velocity) implies the choice of reference conditions for the various forces and velocities in the system.*

For the chosen direction of positive velocity, since v is in that direction, its source arrow will be from 1 to g. Note that the source F is acting so as to make velocity opposite to the reference direction; therefore this source is shown toward the reference vertex (in keeping with the conventions established for across- and through-sources). This can be seen to be correct since that source would tend to make velocity in the mass (m) branch, for instance, opposite to the effect of the velocity source which agrees with conditions in the system. The other arrows are completely arbitrary and only relate to which velocities will be used as variables. Choosing them as shown would be consistent with a positive v, tending to make all the forces and velocities positive.

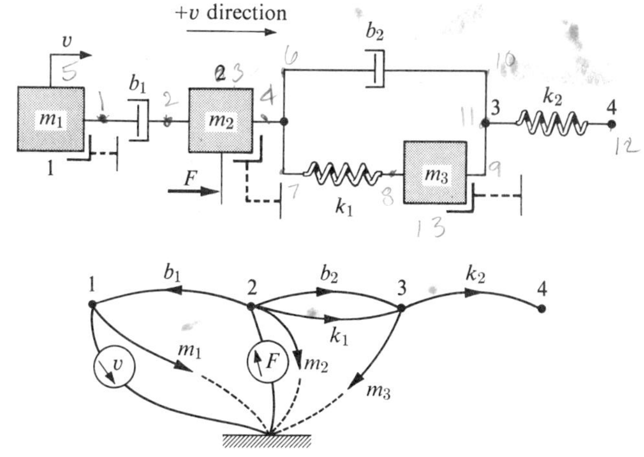

FIG. 8–10. Mechanical translational system and graph.

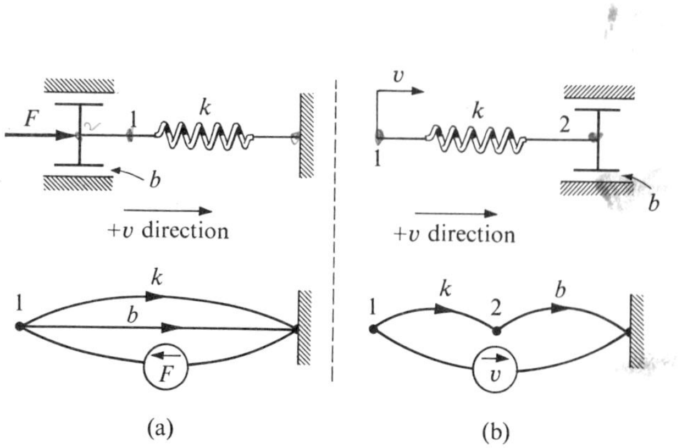

FIG. 8–11. Illustration of different structure of seemingly similar systems.

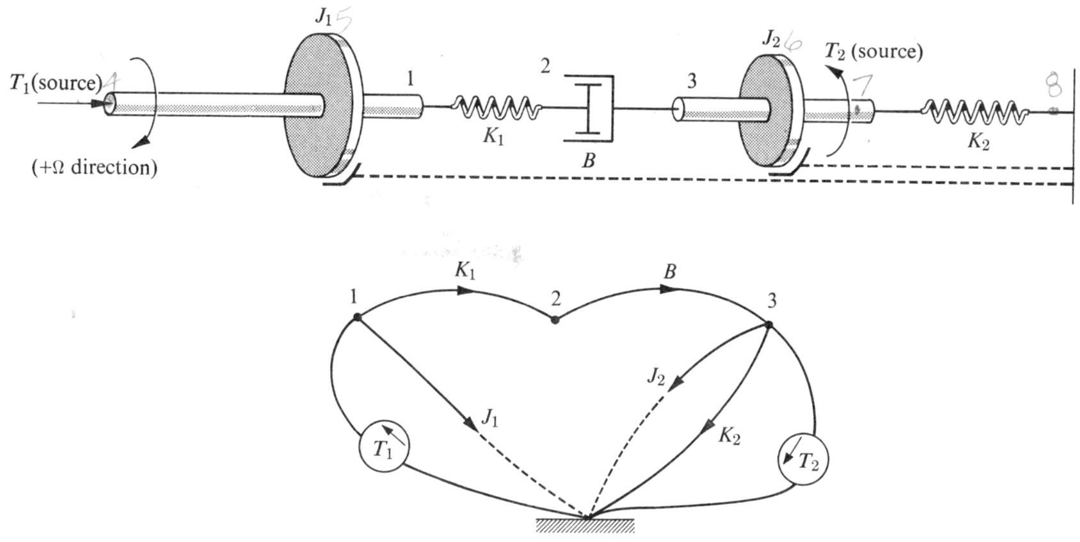

FIG. 8–12. Mechanical rotational system and graph.

Additional examples of graph construction are given in Figs. 8–10, 8–11, 8–12, 8–13, 8–14, 8–15, 8–16, and 8–17. The reader should, by a procedure similar to the previous example, determine the graphs of the systems and check with the result shown in the figures.

Note that element k_2 in Fig. 8–10 does not really affect the system performance since there is no way for a force to exist in the spring with one end free. This clearly shows in the graph since element k_2 is not in any closed path of the graph. Also the velocity source is applied directly to the mass, m_1, and thus

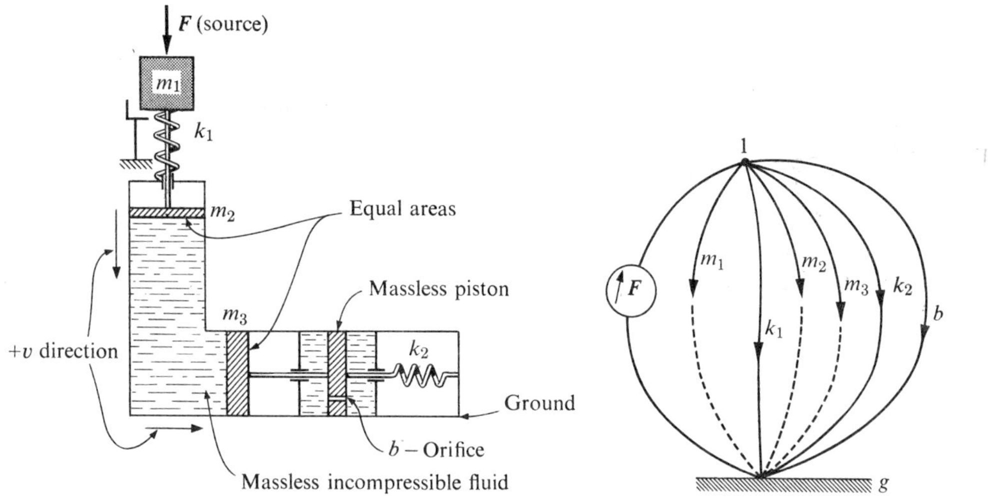

FIG. 8–13. System with motion in more than one direction.

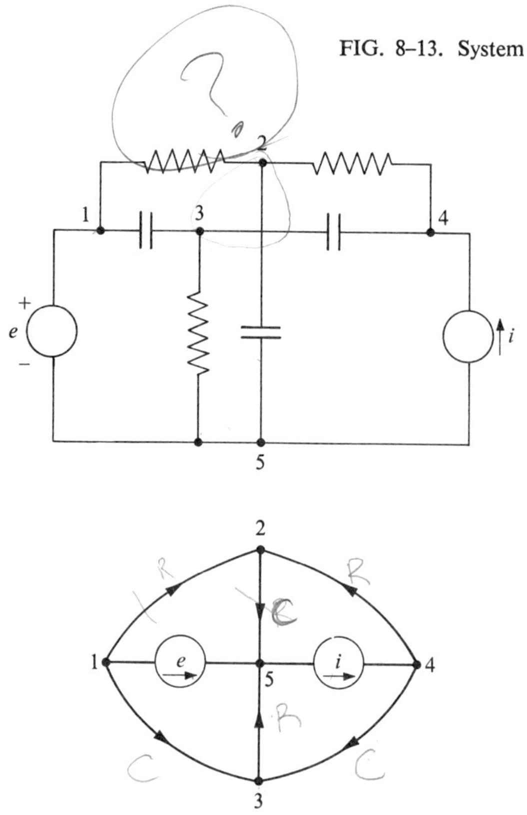

FIG. 8–14. Bridged-T circuit redrawn.

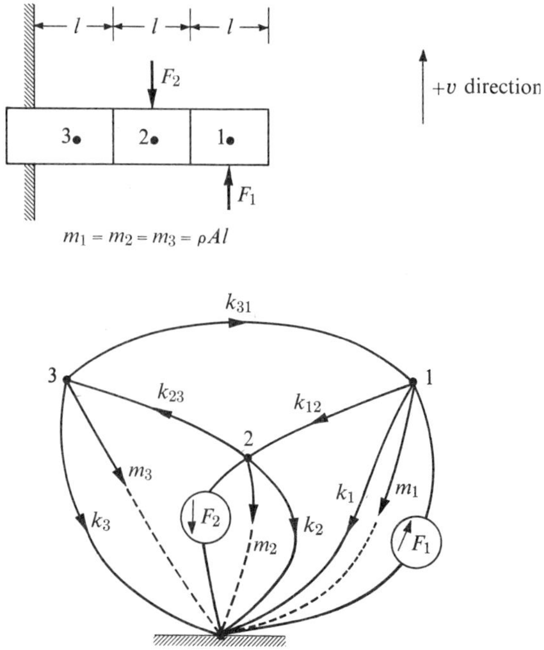

FIG. 8–15. Cantilever beam approximated by 3 lumped masses with stiffness coefficients between each point, including ground.

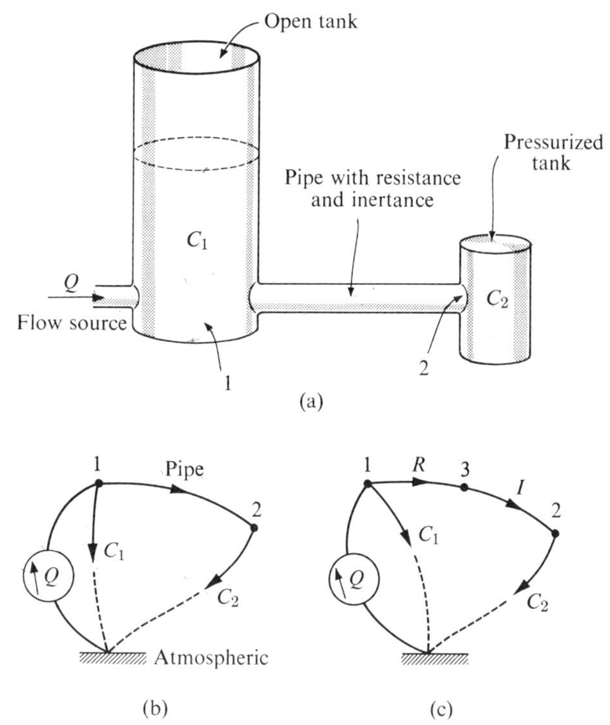

FIG. 8–16. Fluid system. (a) System. (b) System graph for pipe shown in single element. (c) Introduce fictitious vertex 3, for pipe shown as equivalent resistance and inertance.

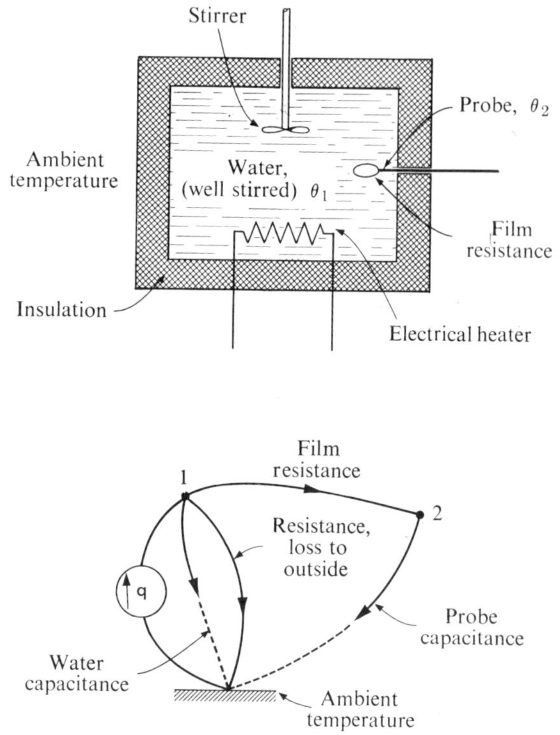

FIG. 8–17. Thermal system.

its velocity is determined and does not depend on the rest of the system. Whether point 1 is considered to have mass or just be a rigid connection to the damper b_1 is inconsequential. This also shows clearly in the graph since m_1 is directly across the fixed velocity source and whether it were present or removed would not change the velocity between point 1 and ground. The force associated with the velocity source (force necessary to give point 1 a velocity v) would of course be different in the two cases.

Some work has been done on extension of through- and across-variable concepts to more complicated systems involving human behavior, such as traffic flow in cities and the economic system as discussed in Section 4–7. The across- and through-concepts are qualitatively thought-provoking, but limited progress has been made in developing these fields quantitatively in this fashion.

System graphs may be drawn for mixed systems such as electromechanical, electrohydraulic, etc. This is an area of powerful application of the graph technique since the graph reveals the interrelated structure of a system involving different types of physical across- and through-variables. This structure is frequently difficult to picture without such an aid. Mixed systems will necessarily involve transducing elements with more than two terminals to represent the transduction from one form of energy to another. An example of a mixed system is the D'Arsonval meter system of Chapter 7 (Fig. 7–24).

8–4. GENERALIZED CONTINUITY AND COMPATIBILITY

The generalized through- and across-variables as defined in Section 4–2 will be used. Table 4–1 shows the relationship of these variables to the variables of the various physical systems:

f = generalized through-variable

v = generalized across-variable

8–4.1 Continuity—The Vertex Law

Continuity and its meaning in the various physical systems has been discussed in Section 5–3.2. Briefly, continuity is a statement of the following physical laws:

electrical—conservation of charge, or Kirchhoff's current law

mechanical—conservation of momentum, or Newton's first and third laws

fluid—conservation of matter

thermal—conservation of energy.

The vertex law is a statement of the continuity condition. It states that, for an oriented linear graph of a system, the algebraic sum of the through-variables entering any vertex must be zero.

$$\sum_i f_{ik} = 0,$$

$$k = 1, 2, \ldots, n, \qquad i = 1, 2, \ldots, l, \tag{8–1}$$

where k indicates one of the n vertices in the linear graph, and l is the number of branches incident to the kth vertex. Figure 8–18 shows a typical vertex (a) of a graph. For this vertex Eq. (8–1) gives

$$f_1 + f_2 + f_3 - f_4 = 0.$$

Obviously the continuity law could also be stated that the sum of the through-variables entering a vertex must equal the sum of the through-variables leaving the vertex or alternatively that the algebraic sum of the through-variables leaving the vertex is zero.

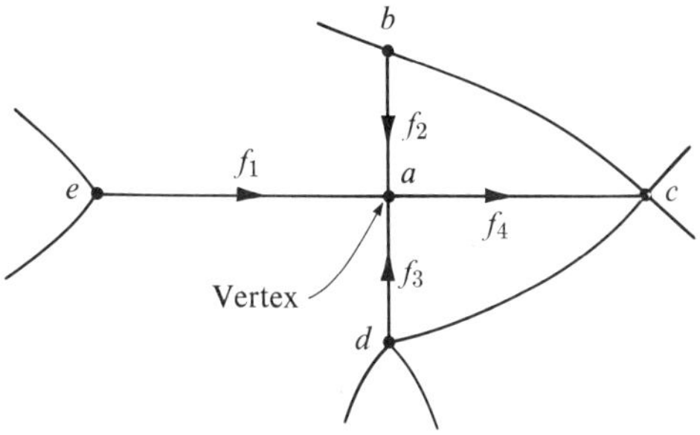

FIG. 8–18. Typical vertex showing through-variables.

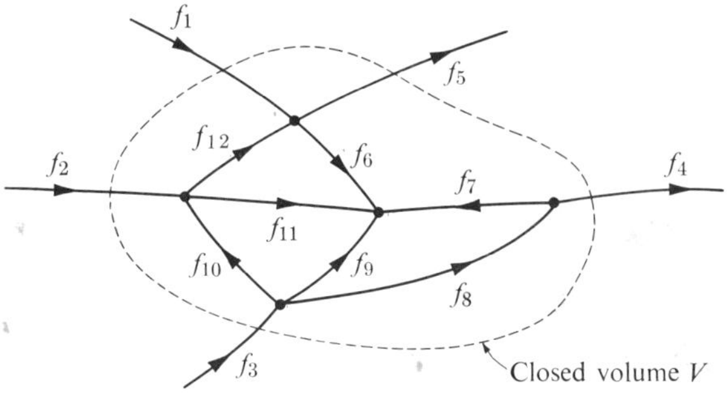

FIG. 8–19. Continuity applied to closed volume.

Using the stated vertex law, we can easily prove that a similar relation applies to *any closed volume* which cuts through a system graph. Figure 8–19 shows such a situation. The vertex law may be applied at each vertex, internal to the closed volume V. In these equations, each f associated with a branch which has its vertices internal to V will enter twice, once in each of the equations for its vertex pair. It will appear with opposite signs in each of these equations since it must be entering one vertex and leaving the other. Thus a simple addition of all the equations for internal vertices will result in a canceling of all the f's associated with branches whose vertices are internal to V (which could be called internal branches, but the statement as given also applies to branches which intersect V an even number of times). Thus we will be left with an equation involving only those f's associated with branches which have one vertex internal to V and the other vertex external to V,

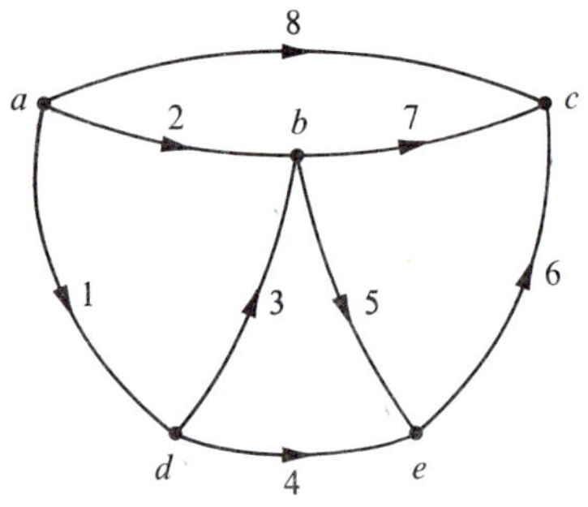

FIG. 8-20. Illustration of generalized compatibility law.

which states that the algebraic sum of these through-variables is zero. Thus, for the example of Fig. 8-19 we have, for volume V,

$$f_1 + f_2 + f_3 - f_4 - f_5 = 0.$$

A direct corollary of the above generalization is: *given a system graph of two-terminal elements with n vertices; only (n − 1) of the vertex equations are linearly independent.* The equation obtained at the nth vertex is the negative sum of the equations at the other $(n - 1)$ vertices. We can easily see this by drawing a closed volume of $(n - 1)$ internal vertices.

If multiterminal elements are included in a graph, then the graph may have separate parts which are not connected (for instance, the electrical transformer). If there are p separate parts, then it can easily be seen that there will be $(n - p)$ linearly independent vertex equations since each separate part will obey the earlier statement.

8-4.2 Compatibility—The Path Law

Compatibility and its meaning in the various physical systems have been discussed in Section 5-3.2. Briefly, compatibility is a statement of the following truths:

electrical—potential is a scalar or Kirchhoff's voltage law,
mechanical—distance is a scalar or geometrical constraints,
fluid—pressure is a scalar,
thermal—temperature is a scalar.

The path law is a statement of the compatibility condition. It states that, for an oriented graph, the algebraic sum of the across-variables around any closed path is zero. The across-variable differences are considered positive if their orientation is in the direction of traverse around the closed path. Thus,

$$\sum_{q} v_{qp} = 0,$$

$$q = 1, 2, \ldots, m \quad \text{when } p = 1, 2, \ldots, \tag{8-2}$$

where the symbol $\sum$ means summation around a closed path p. It is evident that this sum will consist of m terms when a path containing m elements is traversed. The maximum number of different across-variable terms v_{qp} is equal to b, the number of branches in the system graph. However, since the same element could be traversed more than once, the number of terms in (8-2) may have any value (in fact infinity, since devious routes with subpaths which are repeatedly traversed may be selected). Also an infinite number of different paths, p, may be chosen.

Figure 8-20 shows an example in which the following path equation may be obtained from Eq. (8-2). For path *a-b-d-a*,

$$v_2 - v_3 - v_1 = 0.$$

Note that the v's are negative when one progresses opposite to the arrow direction on the branch being traversed. For path *b-e-d-b*,

$$v_5 - v_4 + v_3 = 0.$$

For path *a-b-e-d-a*,

$$v_2 + v_5 - v_4 - v_1 = 0.$$

Note that the equation for path *a-b-e-d-a* can be obtained by adding the equations for paths *a-b-d-a* and *b-e-d-b*, since v_3, which is common to the latter two paths, cancels out in the sum. For path *a-c-b-a-c-e-d-a*,

$$2v_8 - v_7 - v_2 - v_6 - v_4 - v_1 = 0.$$

In this path branch 8 is traversed twice.

It is obvious that not all possible path equations are linearly independent. In fact: *given a system graph of two-terminal elements with n vertices and b branches; only b − (n − 1) of the path equations are linearly independent.* (If there are p separate parts, then $b - (n - p)$ path equations will be independent. The following discussion will assume only one separate part, but the reader can easily extend the conclusions to the more general case.)

To prove this, we introduce the concept of a *tree*: A tree of a graph is a subgraph which contains all the vertices and the maximum number of branches which can be included without forming any closed paths.

By constructing a tree we can see that it must contain $(n - 1)$ branches if there are n vertices. This is so because the first branch included is incident on two vertices and each additional branch added includes one new vertex. There are many trees which can be drawn for any graph. Figure 8–21 shows some possibilities.

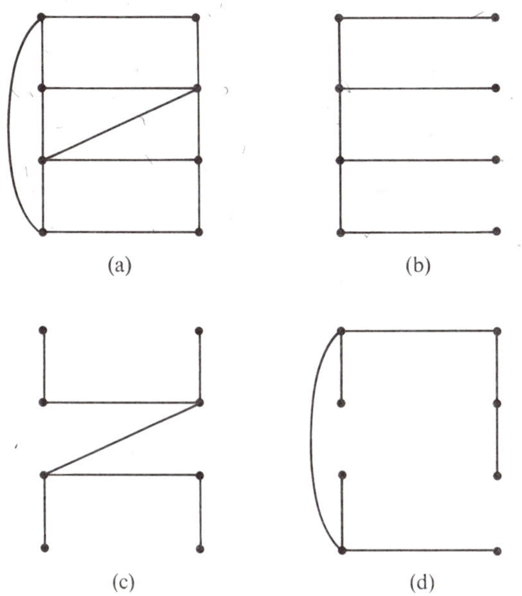

FIG. 8–21. A graph (a) and some of its trees (b), (c), and (d).

Now we define a *tree-link*: A tree-link is any one of the branches of the original graph which is not included in the tree under consideration.

If there were b branches in the original graph, then there must be $[b - (n - 1)]$ tree-links since each tree must contain $(n - 1)$ branches.

If now the insertion of any single tree-link is visualized, it will be seen that this addition will cause a closed path to be formed since by the definition of a tree, the omission of the tree-links was necessary for no closed paths to be present. Also each closed path formed by separate addition to the links, one at a time, will be a new path since it will contain a branch which was not in any previous subgraph. If the path law is applied to each of the paths formed in this way, then each of the $[b - (n - 1)]$ equations will be independent of the others since each equation will have a variable which does not appear in *any* of the other equations. Additionally, the application of the path law to any path other than the ones formed by the addition of tree-links will produce an equation which can also be obtained by a linear combination of the previously obtained $[b - (n - 1)]$ equations, since any additional closed path can be constructed by the combination of the paths previously formed by the tree-links. *This proves that there are only* $(b - n + 1)$ *linearly independent path or compatibility equations.*

a) *General choice of paths.* The foregoing discussion indicates that any arbitrary choice of $(b - n + 1)$ closed paths of the original system graph will not necessarily produce independent path equations. The closed paths formed by the addition of tree-links to a particular tree will produce one set of independent path equations, but it is obvious that there are many other sets also, since there are many trees for any given graph. Any of these sets of independent paths is suitable for equation formulation.

b) *Planar graphs—meshes.* It is expedient to make a simple choice of paths which will be clearly obvious for the great majority of graphs we will encounter. Let us consider the category called *planar* graphs.

A planar graph is one which may be topologically deformed so that all branches will lie on the surface of a plane in such a way that no branches cross each

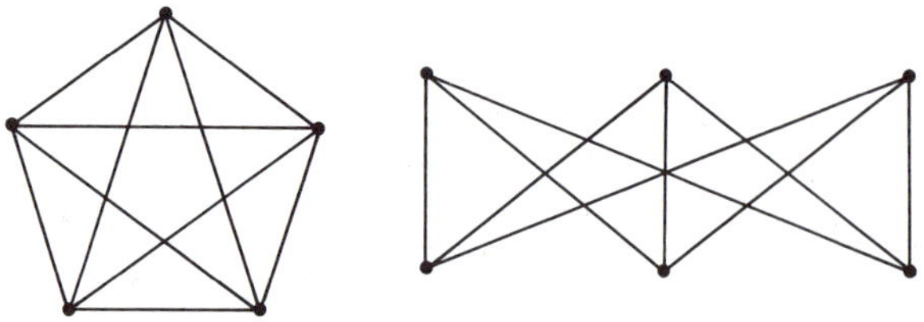

(a) (b)

FIG. 8–22. Fundamental nonplanar graphs.

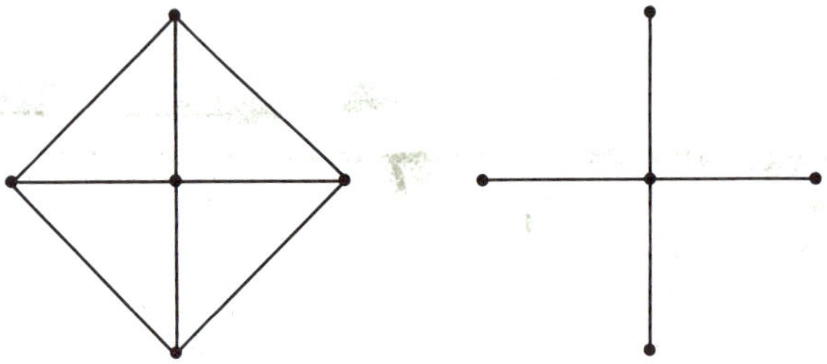

(a) Graph (b) Tree

FIG. 8–23. Graph with no internal meshes.

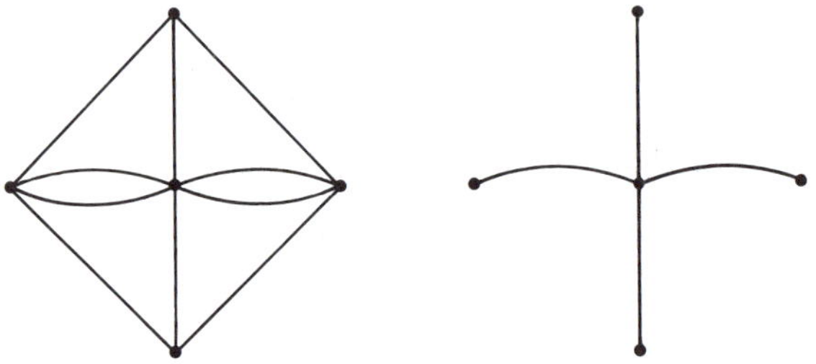

(a) Graph (b) Tree

FIG. 8–24. Graph with internal meshes.

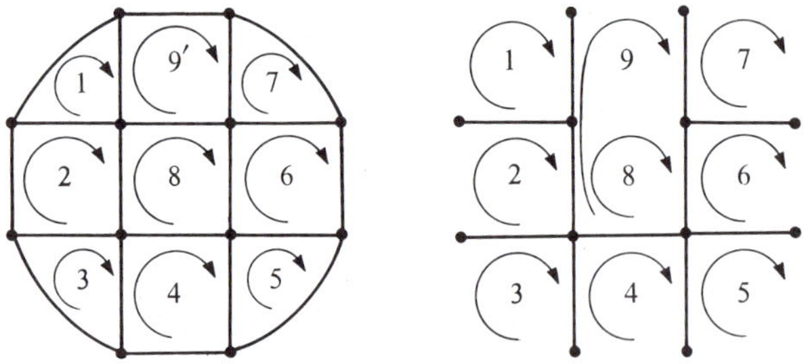

$b = 20$

$n = 12$

$b - n + 1 = 9$

(a) (b)

FIG. 8–25. Example of path addition. (a) Graph with nine independent paths. (b) Tree from which nine independent paths can be formed.

other. The two graphs in Fig. 8–22 are fundamental nonplanar graphs.

A *mesh* of a planar graph will be defined as a closed path for which any branch in the path may be reached from an interior point without crossing any other branch. Thus in a drawing of a planar graph, the meshes are the "windows" of the graph. A planar graph may be mapped on the surface of a sphere without crossing branches. If the graph is imagined as drawn on a sphere and then sufficient branches are removed so that a tree is formed, there will be n vertices, $(n - 1)$ branches in the tree, and $(b - n + 1)$ branches removed. The addition of these $(b - n + 1)$ branches will cause the formation of $(b - n + 2)$ meshes on the sphere since the last branch added will cause two meshes to be formed; one to the right of the last branch and one to the left. If the planar graph mapped on the sphere is now deformed so that it lies in a plane without crossing branches, it will have $(b - n + 1)$ meshes because the periphery will have been a mesh on the sphere. Thus the *number* of meshes of the planar graph when mapped on a plane is equal to the number of independent path equations which can be written. Is it true that the mesh paths will yield independent equations?

For a simple graph which contains no "internal" meshes, this may easily be shown to be true. Consider the graph of Fig. 8–23(a) [with a tree shown in Fig. 8–23(b)]. The tree-links will form paths which are the mesh paths and thus these must be independent. However, for a "complex" planar graph with "internal" meshes, it is impossible to select a tree which has links that will form the mesh paths. This is seen in Fig. 8–24. The reader is invited to attempt to find a tree whose links will form the mesh paths. It cannot be done!

However, for these "complex" planar graphs with "internal" meshes, it can be seen that the path equations for the internal meshes can *always* be formed by selective addition of the independent path equations which are obtained from the tree. *Thus, the set of path equations obtained from the meshes is also an independent set.* Figure 8–25 shows a particular example. Path 9′ in (a) can be obtained by 9 minus 8. The set of equations obtained from the 9 meshes shown in (a) is independent.

For planar graphs, an independent set of paths is immediately evident in the meshes of the graph. It will be expedient to use this set of paths and this will usually be done in succeeding work.

8–5. FORMULATION OF SYSTEM EQUATIONS

The vertex and path laws along with the elemental equations allow the system equation to be formulated. From the results of Section 8–4 it now may be shown that: *A sufficient set of equations for determining the system equation for any output of any system (linear or nonlinear) is obtained by using a set of $(n - 1)$ linearly independent vertex equations, a set of $(b - n + 1)$ linearly independent path equations, and $(b - s)$ elemental equations.* For a linear system, this set of equations is necessary to obtain a solution.

For a graph with a total of b branches of which s branches are sources, there are $[2(b - s) + s] = (2b - s)$ unknowns in the aforementioned set of equations since each nonsource branch has two unknowns (one through- and one across-variable) and each source has one unknown (the complementary variable for that source branch which has not been prescribed). The foregoing set of equations is linearly independent and contains exactly $(2b - s)$ equations, since we have

$$\underset{\text{vertex}}{(n-1)} + \underset{\text{path}}{(b-n+1)} + \underset{\text{elemental}}{(b-s)} = \underset{\text{total}}{(2b-s)}. \tag{8–3}$$

For a linear system this forms the necessary and sufficient set of linear equations for it to be possible to eliminate from this set all but any one particular variable.

For a nonlinear system, this set of equations is sufficient to determine the system performance, but it is not always possible to eliminate all but one variable; the set may have to be treated simultaneously since it may involve transcendental relations between some or all variables. The entire set may not be necessary for a nonlinear system, however, since in some cases, it may be possible to determine the response of a portion of the system using less than the full foregoing complement of equations (e.g., if it contained a two-way relay which coupled two parts of the system, then it would be possible to have conditions such that the relay did not close and only one part of the system need be considered for a solution).

Various examples of the formulation of system equations will now be considered which will illustrate the application of continuity and compatibility by use of the vertex and path laws applied to system graphs. It is assumed that the reader has already developed equations for three-element systems such as were introduced in Chapter 5.

The first example will develop the equations in detail and discuss possible troublesome points. The relative simplicity of the procedure will be illustrated by straightforward application of the procedure by two alternative approaches in the later examples.

8–5.1 General Procedure

We have seen that for a graph which has b branches, of which s branches are sources, $(2b - s)$ independent equations can be written. This set of equations in $(2b - s)$ unknowns (a through- and across-variable unknown for each branch; for source branches there is *either* a through- or across-variable unknown) may be solved for any selected unknown, which will be called the "output" of the system, by elimination of the other unknowns. The resulting differential equation is called the *system equation* for the selected output. To illustrate the number of equations with the least extra complexity, a circuit with only resistive branches will be used.

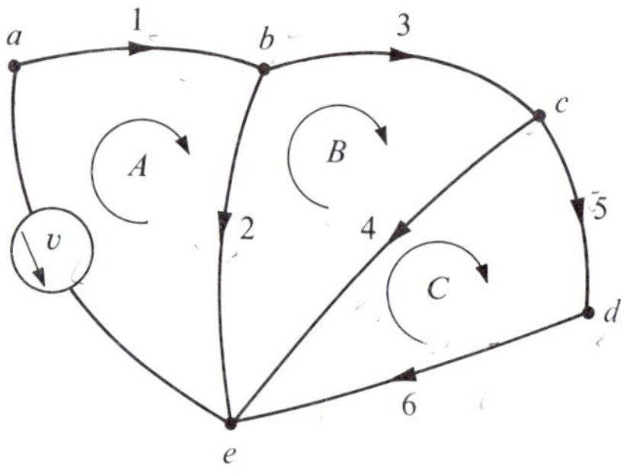

FIG. 8-26. System graph for Example 1.

Example 1. For the graph shown in Fig. 8-26, the following is noted: input, v; output, v_6; $b = 7$, $s = 1$, $n = 5$.

Unknowns:

$2b - s = 13.$ $v_1, f_1,$ $v_4, f_4,$
$v_2, f_2,$ $v_5, f_5,$
$v_3, f_3,$ $v_6, f_6,$

f (through-variable of v source).

Vertex equations:

$n - 1 = 4.$

At vertex a, $$-f - f_1 = 0, \quad (8\text{-}4)$$

at vertex b, $$f_1 - f_2 - f_3 = 0, \quad (8\text{-}5)$$

at vertex c, $$f_3 - f_4 - f_5 = 0, \quad (8\text{-}6)$$

at vertex d, $$f_5 - f_6 = 0. \quad (8\text{-}7)$$

Path equations:

$b - n + 1 = 3.$

Path A, $$v_1 + v_2 - v = 0, \quad (8\text{-}8)$$

path B, $$v_3 + v_4 - v_2 = 0, \quad (8\text{-}9)$$

path C, $$v_5 + v_6 - v_4 = 0. \quad (8\text{-}10)$$

Elemental equations:

$b - s = 6.$

$$v_1 = R_1 f_1, \quad (8\text{-}11)$$
$$v_2 = R_2 f_2, \quad (8\text{-}12)$$
$$v_3 = R_3 f_3, \quad (8\text{-}13)$$
$$v_4 = R_4 f_4, \quad (8\text{-}14)$$
$$v_5 = R_5 f_5, \quad (8\text{-}15)$$
$$v_6 = R_6 f_6. \quad (8\text{-}16)$$

Equations (8-4) through (8-16) form a set of 13 linear algebraic equations in the 13 unknowns indicated. These may be solved for the desired output v_6 in terms of the input v. This could be done by a straightforward formal technique such as the use of determinants and Cramer's rule. However, most of these equations (the elemental equations particularly) contain only two unknowns, and thus it is ordinarily simpler to make some substitutions before resorting to formal methods. Some observations on the stated equations will first be noted.

Equation (8-4) is the only equation which contains the unknown through-variable of the source, f. If we are not immediately interested in this quantity, we may exclude both f and Eq. (8-4) from consideration. This reduces our set to 12 equations in 12 unknowns. If we were interested in f, Eq. (8-4) tells us that $f = -f_1$, and we could still exclude this equation, solve the remaining set for f_1, and then use $f = -f_1$. (Whether the through-variable out of the source is considered to be f or f_1 is only a matter of reference conditions as discussed earlier when source references were established.)

8-5.2 Reduction of Simultaneous Equations

The first problem of the statement of the proper set of equations has been resolved (how many equations should be written? which equations should be written?) and now the next problem of how to solve this set in an orderly fashion will be treated. It is appropriate to consider the ways in which substitutions may be made in these equations to obtain the solution in a straightforward manner. These considerations are important since the large number of equations which arises even in relatively simple graphs sometimes results in confusion for the neophyte, who may end up going in circles and not getting anywhere if an orderly procedure is not followed.

There are two basically different approaches which will readily lead to a reduction in the number of equations. These methods are more easily presented in terms of the variables which are chosen as primary unknowns, and are called the *node* and *loop methods*. It is to be noted that these methods may be used to formulate equations for either a linear or nonlinear system; however, with the nonlinear system there is no assurance that the equations so formulated may be solved.

a) *Node method.* In the node method, the across-variables from each vertex to some reference vertex are chosen as the unknowns in terms of which the final set of equations will be formulated. These variables are called the *node variables*. These variables automatically satisfy the path laws since the across-variable between nodes is simply the difference between the appropriate node variables. The vertex equation is then written at each node, and the through-variables are then expressed directly in terms of the node variables as related by the elemental equations. We then have eliminated all variables except the node variables and have a number of equations equal to the number of these unknowns, which in general will be $(n-1)$.

This procedure will be demonstrated with the previous example of Fig. 8–26.

i) Choose the node variables to be from vertices a, b, c and d, respectively, to e. Since $v_{ae} = v$, which is the source and is known, one unknown is eliminated immediately. Thus $v_2 = v_{be}$, $v_4 = v_{ce}$, and $v_6 = v_{de}$ are the node variables.

ii) Apply the vertex law at b, c, and d. The preferred form for the vertex law statement in this method is that the net flow *out of* the node is zero. Thus

$$\text{node } b, \quad -f_1 + f_2 + f_3 = 0,$$
$$\text{node } c, \quad -f_3 + f_4 + f_5 = 0,$$
$$\text{node } d, \quad -f_5 + f_6 = 0.$$

These equations are equivalent to Eqs. (8–5), (8–6), and (8–7). Equation (8–4) is not necessary since v is known at node a.

iii) Substitute for the f's in terms of the node variables by using the elemental equations (8–11) through (8–16). Express the across-variables between nodes in terms of the difference between the node variables, which is equivalent to a statement of the path laws of Eqs. (8–8), (8–9), and (8–10); that is, $v_1 = v - v_2$, $v_3 = v_2 - v_4$, $v_5 = v_4 - v_6$. Then

$$\text{node } b, \quad -\frac{(v - v_2)}{R_1} + \frac{v_2}{R_2} + \frac{(v_2 - v_4)}{R_3} = 0, \qquad (8\text{–}17)$$

$$\text{node } c, \quad -\frac{(v_2 - v_4)}{R_3} + \frac{v_4}{R_4} + \frac{(v_4 - v_6)}{R_5} = 0, \qquad (8\text{–}18)$$

$$\text{node } d, \quad -\frac{(v_4 - v_6)}{R_5} + \frac{v_6}{R_6} = 0. \qquad (8\text{–}19)$$

unknowns
v_2, v_4
v_2, v_4, v_6
v_4, v_6
$= v_2, v_4, v_6$

These equations may be rewritten by grouping terms as

$$v_2\left(\frac{1}{R_1} + \frac{1}{R_2} + \frac{1}{R_3}\right) - v_4\frac{1}{R_3} = \frac{v}{R_1}, \qquad (8\text{–}20)$$

$$-v_2\frac{1}{R_3} + v_4\left(\frac{1}{R_3} + \frac{1}{R_4} + \frac{1}{R_5}\right) - v_6\frac{1}{R_5} = 0, \qquad (8\text{–}21)$$

$$-v_4\left(\frac{1}{R_5}\right) + v_6\left(\frac{1}{R_5} + \frac{1}{R_6}\right) = 0. \qquad (8\text{–}22)$$

These equations may now be solved by determinants and the use of Cramer's rule.

Note that with the node method one quickly reduces the number of equations to that equal to the number of node variables (which is one less than the number of vertices less any nodes at which the across-variable is known such as at a in this example).

b) *Loop method.* In the loop method, the variables are chosen so that the vertex law is automatically satisfied. We will consider only planar graphs. The variables will then be taken as "circulating" through-variables in each of the meshes. These variables are called the *loop variables.* Thus each branch through-variable will be the difference between the loop variables on each side of the branch (if it is an "outside" branch then its through-variable is the loop variable alone since the one on the other side is zero). In this way, the vertex equations are automatically satisfied and need not be considered further. If the loop variables are chosen with the same sense in each mesh, then the equations will be symmetrical, as we will see.

The path law for each mesh is then written and substitutions are made for the across-variables in terms of the loop variables using the elemental equations. In this way the system is quickly reduced to a number of equations equal to the number of meshes (which is $b - n + 1$, as found earlier).

This procedure will be demonstrated with the same example as used previously. It is repeated as Fig. 8–27.

i) Choose the meshes shown previously as A, B, and C for applying the path laws. Define f_A, f_B, and f_C (Fig. 8–27) as loop variables.

ii) Applying the path law around A, B, and C, we have

loop A, $\quad -v + v_1 + v_2 = 0,$

loop B, $\quad -v_2 + v_3 + v_4 = 0,$

loop C, $\quad -v_4 + v_5 + v_6 = 0.$

These equations are equivalent to Eqs. (8–8), (8–9), and (8–10).

iii) Substitute for the v's in terms of the *loop variables* by using the elemental equations (8–11) through (8–16). Express the branch across-variables for internal branches in terms of the difference between the loop variables, which is equivalent to a statement of the vertex laws of Eq. (8–5), and (8–6), that is, $f_2 = f_A - f_B, f_4 = f_B - f_C$. Then

loop A, $$-v + R_1 f_A + R_2(f_A - f_B) = 0, \tag{8–23}$$

loop B, $$-R_2(f_A - f_B) + R_3 f_B + R_4(f_B - f_C) = 0, \tag{8–24}$$

loop C, $$-R_4(f_B - f_C) + R_5 f_C + R_6 f_C = 0. \tag{8–25}$$

These equations obtained by the loop method should be compared to Eqs. (8–17), (8–18), and (8–19) obtained by the node method. These equations may be rewritten by grouping terms as

$$f_A(R_1 + R_2) - f_B R_2 = v, \tag{8–26}$$

$$-f_A R_2 + f_B(R_2 + R_3 + R_4) - f_C R_4 = 0, \tag{8–27}$$

$$- f_B R_4 + f_C(R_4 + R_5 + R_6) = 0. \tag{8–28}$$

These equations may now be solved by determinants and the use of Cramer's rule.

Note that with the loop method one quickly reduces the number of equations to that equal to the number of loop variables (which is $b - n + 1$ less the number of meshes which contain a through-variable source; in a manner similar to the node method example, if a through-variable source were present here, one circulating current would be immediately known, thus reducing the number of variables).

b−n+1−Ss

c) *Comparison.* One may compare the two approaches afforded by the node and loop methods.

8-5 | FORMULATION OF SYSTEM EQUATIONS 217

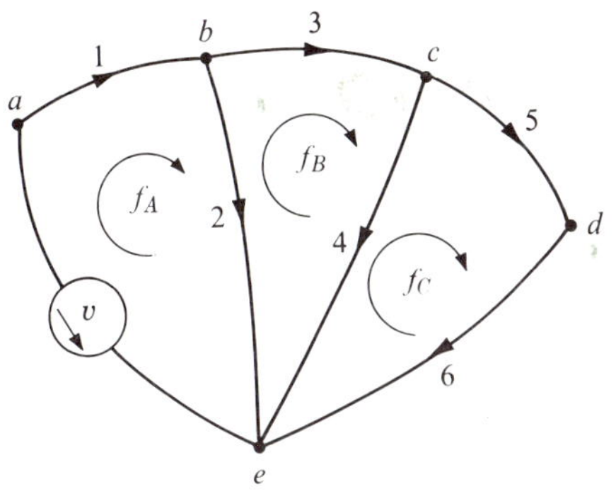

FIG. 8-27. Graph for Fig. 8-26. Arranged for loop method.

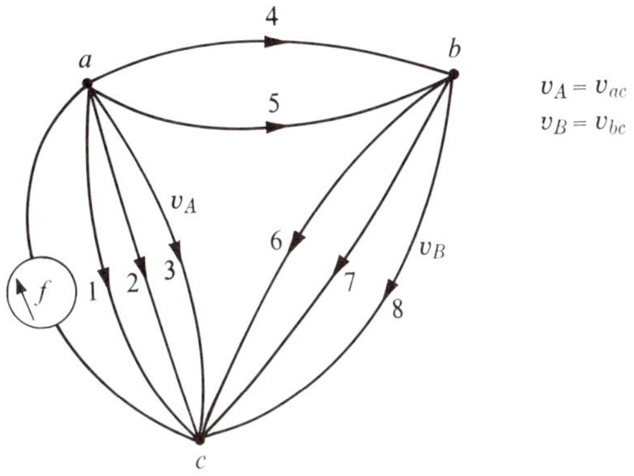

FIG. 8-28. System graph for first example.

$-v + f_A R_1 + f_A R_2 + R_3(f_A - f_B) + R_4(f_A + f_B) \ldots$

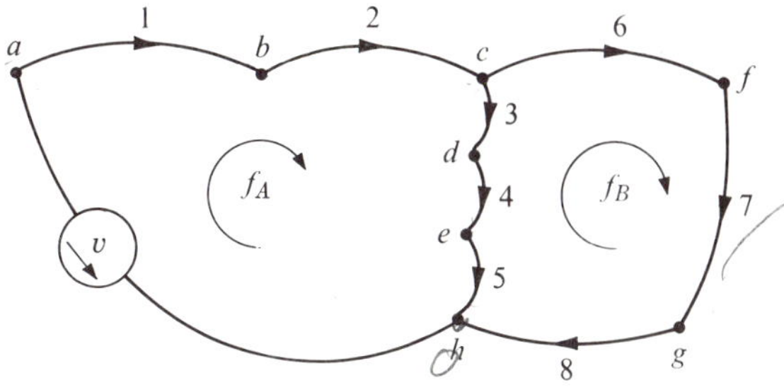

FIG. 8-29. System graph for second example.

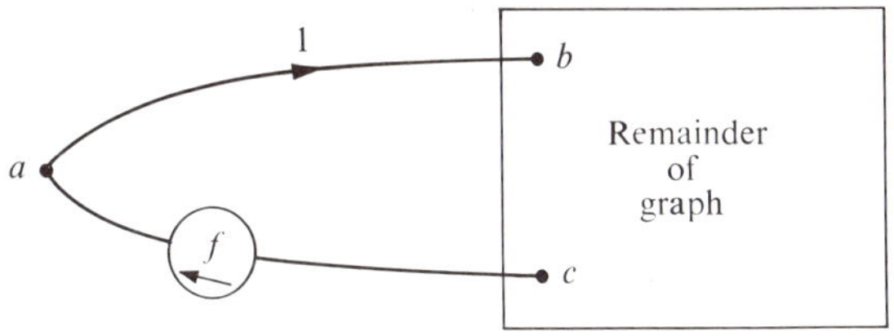

FIG. 8-30. A through-source in series with an element.

In general the node method results in $(n - 1)$ equations to be solved, whereas in general the loop method results in $(b - n + 1)$ equations. Thus the choice of which method to use will in general depend on whether the given system has fewer loops than nodes. In the example used, there was little choice since both methods resulted in three simultaneous equations.

When the node and loop methods are applied as just described, the resulting sets of simultaneous equations will be symmetric, e.g., in Eq. (8-26) the coefficient of f_B is the same as the coefficient of f_A in Eq. (8-27), etc. The form of these equations is discussed further in Chapter 14.

The following examples illustrate other cases.

Example. The example of Fig. 8-28 will not be solved but is given here to point out that, since there are considerably fewer nodes (two) than loops (six, because the source loop variable is known), this situation is handled most simply by the node method, using v_A and v_B as node

variables. If it is desired to solve for some other output (such as f_4), this can be easily done after v_A and v_B are found.

Example. The example of Fig. 8-29 will not be solved but is given to point out that since there are considerably fewer loops (two) than nodes (six, because the source node variable is known), this situation is most simply handled by the loop method, using f_A and f_B as loop variables. If it is desired to solve for some other output (such as v_4), this can easily be done after f_A and f_B are found.

The following two examples discuss situations which sometimes arise and may cause difficulty.

Example. In the situation shown in Fig. 8-30, the vertex equation at a gives $f = f_1$, and since f is specified, f_1 and therefore v_1 (from the elemental equation for branch 1) is immediately determined without the necessity of simultaneous equations. Outputs in other branches may be determined by previous considerations without element 1 entering into the computations. Thus, from the point of view of the remainder of the graph, the f source *alone* may be considered to be applied between b and c directly.

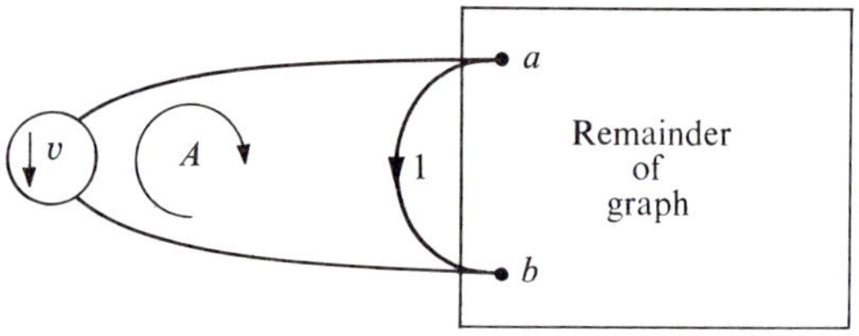

FIG. 8–31. An across-source in parallel with an element.

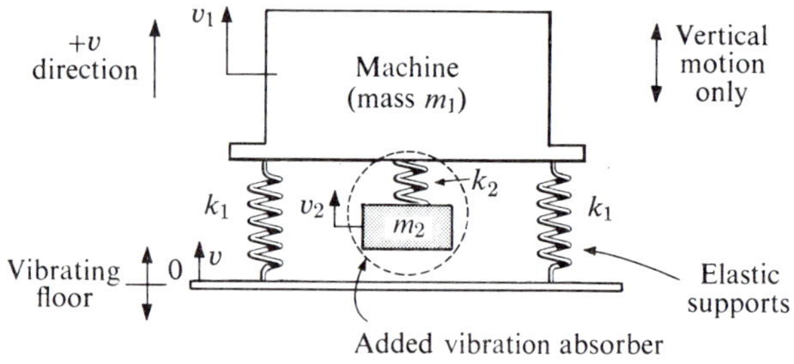

FIG. 8–32. Vibration absorber in pictorial form.

Example. In Fig. 8–31, similarly, the path equation around path A gives $v = v_1$, and since v is specified, v_1 and therefore f_1 (from the elemental equation for branch 1) is immediately determined without the necessity of simultaneous equations. Outputs in other branches may be determined by previous considerations without element 1 entering into the computations. Thus, from the point of view of the remainder of the graph, the v source *alone* may be considered to be applied between a and b.

8–5.3 Application

The preceding section has established the basic principles which allow the formulation of the system equations for any system. Relatively simple examples of systems of idealized elements have been used to illustrate the application of these principles. We will now consider some more complex systems which are realistic examples of the types of dynamic system problems of practical interest.

In accordance with the statement in Section 5–1 (as simplified by the methods of this chapter) of the steps involved in attacking a dynamic system problem, the procedure to be used here will be:

1) *System definition and modeling:* (a) identify the *lumped* elements in the system, and (b) identify the *system variables* and decide on inputs and outputs.

2) *Formulation:* (a) draw the system graph and orient it, thus choosing reference conditions, and (b) use either the node or loop method to apply continuity and compatibility and incorporate the elemental equations which describe the system elements.

3) *System equation:* obtain the *system differential equation* for the *output* in terms of the input by using the equations obtained to eliminate all other variables (if the system is nonlinear, it may not be possible to obtain a single equation).

The system definition and modeling involves a combination of modeling and organizing. The reader should try to clearly recognize when the identification is a part of the modeling process. The main disadvantage of the use of the graph method for neophytes is that it is likely to be applied by rote without full consideration of the significance of reference conditions and the over-all relation to the physical problem. Thus, the student will have to guard against the loss of physical reality.

One example of the typical application of this procedure to practical systems follows.

Vibration absorber (node method). Let us consider the application of a vibration absorber to a vibrating machine (Fig. 8–32). The vibrating machine has been conceived of (i.e., modeled) as consisting of a spring-mounted mass m_1 in which the stiffness of each of the two mounting springs is k_1. This mass-spring system is excited by the motion of the vibrating floor at the lower ends of the springs k_1, which may be considered as the system input (the floor will be considered an ideal velocity source). We will not be concerned here with what causes the floor to vibrate (the problem of floor vibration is an example of an even more complex dynamic system problem). It is found that, at the frequency of the input motion, the vibration amplitude of m_1 is excessive. Therefore, the additional spring-mass system, k_2 and m_2, is attached to the mass m_1. It is hypothesized that if this additional spring-mass system can be properly designed, its vibration will produce a force on m_1 which tends to cancel the vibratory force transmitted to m_1 through k_1 by the vibrating floor. If successful then, k_2 and m_2 will act as a "vibration absorber" for m_1.

Formulated in a design problem, k_2 and m_2 must be given values which will reduce the vibration amplitude of m_1 to some satisfactorily low limit. To accomplish this, it is necessary to formulate the describing differential equation (the system equation) for the motion of m_1, solve this equation subject to the appropriate input signal, and then choose values of k_2 and m_2 which will satisfy the specifications. This section will deal with only the formulation of the differential equation for an unspecified type of input signal. The complete solution of this problem is given in Section 15–2.3 for a sinusoidal floor vibration.

1) *System definition and modeling*

a) Identify the lumped elements:

i) the mass m_1 whose motion is of interest,

ii) two springs k_1, each of which has one end attached to the floor and the other end attached to the mass m_1,

iii) the mass m_2,

iv) the spring k_2, which has one end attached to the mass m_1 and the other end attached to the mass m_2,

v) air-resistance damping which, for low velocities, may be assumed to be linear. This damping force will exist primarily on the large areas of the two masses. The magnitude of these damping forces will depend on the mass velocities with respect to the stationary air. This effect is therefore represented by equivalent dampers between the masses and the velocity reference. Note that these dampers are *not*, in effect, connected between the masses and the floor since the forces depend on the absolute velocities of the masses and not on their relative velocity with

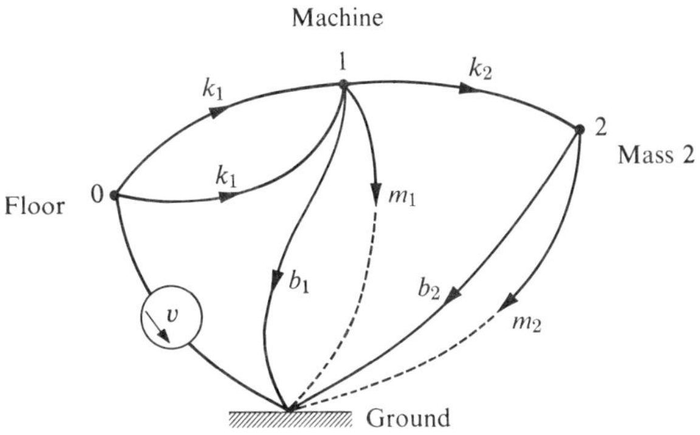

FIG. 8–33. System graph of vibration absorber of Fig. 8–32.

respect to the floor. (*Note:* this is clearly an assumption and not absolute truth; if the floor moves it will tend to set the surrounding air in motion also. By assuming that the damping depends on the absolute velocities, we are in effect assuming that this latter effect is negligible.) We will neglect any damping which may exist in the springs. In some situations, this could be the predominant damping effect. If this effect were to be included, it might be modeled as a damper in parallel with each spring.

b) Identify the system variables, inputs, and outputs: The input to the system is chosen as the velocity of the floor, and the "up" direction is taken as a positive reference direction for all velocities. That is, an across-source v, connected between the floor and ground, drives the system. The vibration of mass m_1 is of vital concern, so that velocity v_1 is the output. Displacement or acceleration of mass m_1 may readily be found from v_1 by integration and differentiation, respectively. Other variables of interest are the velocity of mass m_2 and the forces in the elements.

2) *Formluation*

a) Draw the system graph and orient it. The system graph is drawn by inspection from Fig. 8–32 and the above assumptions concerning damping in the system. The graph is shown in Fig. 8–33 and is oriented arbitrarily (the v source being the only nonarbitrary orientation; if v is to be positive when the floor has an upward velocity, then its orientation must be as shown). When we solve for v_1, with the orientation given v_1 on Fig. 8–33, we should realize that a positive v_1 is an upward velocity.

b) Formulate the equations using either the node or loop methods. Since the graph has only two node variables (v_1 and v_2, since $v_0 = v$ is known), the node method seems to be most suitable here and will be used as follows. The vertex laws give:

$$\textit{node } 1, \quad F_{b1} + F_{m1} + F_{k2} - F_{k1} - F_{k1} = 0, \tag{8–29}$$

$$\textit{node } 2, \quad F_{b2} + F_{m2} - F_{k2} = 0. \tag{8–30}$$

By inspection

$$v_{01} = v - v_1 \qquad \text{and} \qquad v_{12} = v_1 - v_2,$$

which automatically satisfies compatibility. Using the elemental equations for the respective elements, we have by substitution in Eq. (8–29) and (8–30),

$$b_1v_1 + m_1\frac{dv_1}{dt} + k_2\int_0^t (v_1 - v_2)\,dt + (F_{k2})_0 - 2k_1\int_0^t (v - v_1)\,dt - 2(F_{k1})_0 = 0 \tag{8–31}$$

$$b_2v_2 + m_2\frac{dv_2}{dt} - k_2\int_0^t (v_1 - v_2)\,dt - (F_{k2})_0 = 0 \tag{8–32}$$

3) *System equation.* Eliminate all other variables to obtain the system equation for the output. Differentiating (8–31) and solving for v_2, we obtain

$$v_2 = \frac{1}{k_2}\left[(k_2 + 2k_1)v_1 + b_1\frac{dv_1}{dt} + m_1\frac{d^2v_1}{dt^2} - 2k_1v\right]. \tag{8–33}$$

Eq. (8–33) may now be substituted into (8–32) (differentiated) to obtain the final differential equation. The student should work out the steps of substitution himself and verify Eq. (8–34):

$$m_1m_2\frac{d^4v_1}{dt^4} + (b_2m_1 + b_1m_2)\frac{d^3v_1}{dt^3} + [k_2m_1 + b_1b_2 + m_2(k_2 + 2k_1)]\frac{d^2v_1}{dt^2} + [b_1k_2 + b_2(k_2 + 2k_1)]\frac{dv_1}{dt} + 2k_1k_2v_1 = 2k_1k_2v + 2k_1b_2\frac{dv}{dt} + 2k_1m_2\frac{d^2v}{dt^2}. \tag{8–34}$$

In general form, this may be written as

$$a_4\frac{d^4v_1}{dt^4} + a_3\frac{d^3v_1}{dt^3} + a_2\frac{d^2v_1}{dt^2} + a_1\frac{dv_1}{dt} + a_0v_1 = c_0v + c_1\frac{dv}{dt} + c_2\frac{d^2v}{dt^2}. \tag{8–35}$$

This is a fourth-order equation since the system has four independent energy-storing elements (m_1, m_2, k_2, and the two k_1's in parallel).

A second example of the application of this procedure to practical systems follows.

Pulse transformer (loop method). Formulation of the system equation for an electrical circuit will follow the same procedure as in the previous example. It may seem to the reader that the electrical circuit is simpler to handle than the mechanical system since the formulation will be shorter. This is misleading, however, and a more penetrating comparison will show that the difference is almost entirely in the system definition and modeling. In this step in the mechanical example, a considerable part of the discussion was really devoted to the modeling process. This electrical example will start by postulating that a certain circuit is to be analyzed. *When this circuit is drawn, the model* *has already been chosen.* That is, the modeling process is not a part of the discussion here. In actual situations, the difficulty of obtaining a satisfactory model of a real electrical situation can be just as difficult (if not more so, in many cases) than the analogous process for mechanical systems. Such modeling problems with electrical systems are, by choice, not going to be considered here.

1) *System definition and modeling.* The system to be analyzed is the equivalent circuit of a step-up pulse transformer (Fig. 8–34a). The transformer itself is represented by the magnetizing inductance L_1 and the leakage inductance L_2 in this model. The capacitor C is the capacitance of the output winding of the transformer. The capacitance of the input winding is neglected since this is a step-up transformer and the output winding capacitance predominates. R_1 is the resistance of the source e, and R_2 is the load resistance. (The leakage inductance, capacitance, and load-resistance values shown are those referred to the primary side of the transformer.)

a) Identify the lumped elements. The system is composed of five elements, as is obvious from the circuit diagram. They are all assumed linear.

b) Identify system variables, input, and output. The system variables are the currents in each branch and the corresponding voltages. There are five distinct currents and five distinct voltages including the source e. The input is taken to be the voltage source e and the output is selected as the source current (current through R_1).

2) *Formulation.* The system graph (Fig. 8–34b) looks identical to the circuit itself and is primarily useful for establishing and displaying reference conditions. The graph is oriented arbitrarily as usual. This graph has two node-variables (since e at node 1 is given) and three loops. This would seem to indicate that the node method should be used. However, since we wish to solve for i_1, it would be necessary to solve first for a voltage and then to relate this to i_1. There is therefore little choice here, and we will use the loop method to illustrate this approach. The reader is invited to do this problem by the node method and check the result.

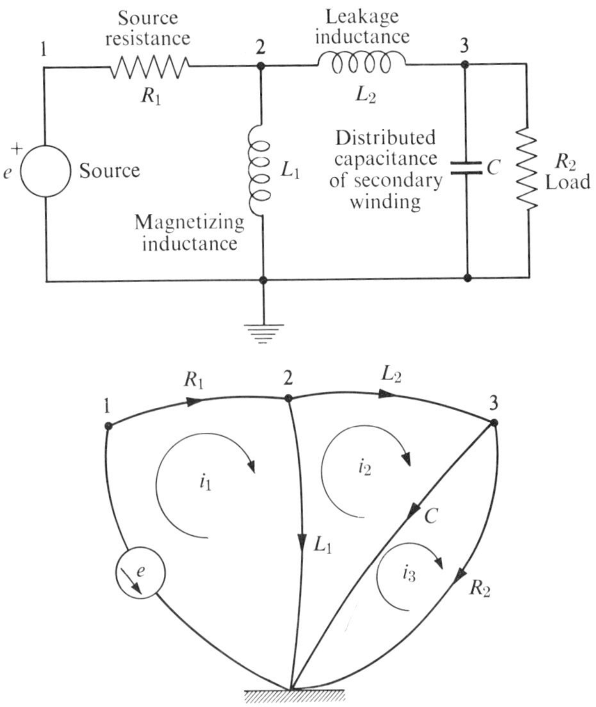

FIG. 8–34. Equivalent circuit of step-up pulse transformer. (a) Circuit schematic. (b) Oriented system graph.

The loop-variables will be i_1, i_2, and i_3 (Fig. 8–34b). The path laws give

$$v_{R1} + v_{L1} = e, \tag{8–36}$$

$$-v_{L1} + v_{L2} + v_C = 0, \tag{8–37}$$

$$-v_C + v_{R2} = 0. \tag{8–38}$$

By inspection,

$$i_{L1} = i_1 - i_2, \; i_C = i_2 - i_3,$$

automatically satisfying continuity. Using the elemental equations for the respective elements, we have by substitution in Eqs. (8–36), (8–37), and (8–38),

$$i_1R_1 + L_1 \frac{d}{dt}(i_1 - i_2) = e, \tag{8–39}$$

$$-L_1 \frac{d}{dt}(i_1 - i_2) + L_2 \frac{di_2}{dt} + \frac{1}{C}\int_0^t (i_2 - i_3)\,dt + (v_3)_0 = 0, \tag{8–40}$$

$$-\frac{1}{C}\int_0^t (i_2 - i_3)\,dt - (v_3)_0 + i_3R_2 = 0. \tag{8–41}$$

We may differentiate the last two equations to remove the integrals. Then we may express i_3 in terms of i_1 and i_2, using (8–40) (differentiated):

$$i_3 = C\left[\frac{i_2}{C} + (L_1 + L_2)\frac{d^2i_2}{dt^2} - L_1\frac{d^2i_1}{dt^2}\right]. \tag{8–42}$$

This may now be substituted into (8–41) (differentiated) to give, with Eq. (8–39), the following two equations in two unknowns:

$$i_1R_1 + L_1\frac{di_1}{dt} - L_1\frac{di_2}{dt} = e, \tag{8–39}$$

$$-L_1\frac{d^2i_1}{dt^2} - R_2CL_1\frac{d^3i_1}{dt^3} + R_2\frac{di_2}{dt} + (L_1 + L_2)\frac{d^2i_2}{dt^2} + R_2C(L_1 + L_2)\frac{d^3i_2}{dt^3} = 0. \tag{8–43}$$

Now Eq. (8–39) can be solved for i_2 and this result substituted into Eq. (8–43). All unknowns have been eliminated except i_1, and thus the desired system differential equation has been formulated. The reader should follow through these steps and verify Eq. (8–44). It is to be noted that Eq. (8–39) does not yield i_2 directly, but rather gives di_2/dt. Since in Eq. (8–43) the lowest-order term in i_2 is di_2/dt, this is quite satisfactory. If this were not the case, di_2/dt from Eq. (8–39) would have to be integrated to make the substitution. We see that the system equation is

$$R_2CL_1L_2\frac{d^3i_1}{dt^3} + [R_1R_2C(L_1 + L_2) + L_1L_2]\frac{d^2i_1}{dt^2} + [R_1(L_1 + L_2) + L_1R_2]\frac{di_1}{dt} + R_2R_1i_1 = R_2e + (L_1 + L_2)\frac{de}{dt} + R_2C(L_1 + L_2)\frac{d^2e}{dt^2}. \tag{8–44}$$

The general form of this equation is

$$a_3\frac{d^3i_1}{dt^3} + a_2\frac{d^2i_1}{dt^2} + a_1\frac{di_1}{dt} + a_0i_1 = b_0e + b_1\frac{de}{dt} + b_2\frac{d^2e}{dt^2}. \tag{8–45}$$

This is a third-order equation since the system has three independent energy-storage elements (L_1, L_2, and C).

8–5.4 Combination of Simultaneous Linear Differential Equations

In the two examples considered in Section 8–5.3, it was rather easy to eliminate the desired unknowns to get the system equation in one unknown. There was considerable algebra involved, but in each case one could choose a single equation in which one unknown could be solved for directly in terms of the other unknowns. This is not always the case, as will be illustrated here.

Let us consider the two equations

$$f = a_1x_1 + a_2\frac{dx_1}{dt} - a_3x_2,$$

$$0 = -a_4x_1 + a_5x_2 + a_6\frac{dx_2}{dt}. \tag{8–46}$$

One may easily eliminate either x_1 or x_2 to obtain an equation in a single unknown. For instance, solving the first equation for x_2 yields

$$x_2 = \frac{1}{a_3}\left(a_2\frac{dx_1}{dt} + a_1x_1 - f\right), \tag{8–47}$$

which can be substituted into the second equation to obtain one equation in the single unknown x_1. The two examples treated were of the form of (8–46).

Let us consider, however, the two equations

$$f = a_1 x_1 + a_2 \frac{dx_1}{dt} - a_3 x_2 - a_4 \frac{dx_2}{dt},$$

$$0 = -a_5 x_1 - a_6 \frac{dx_1}{dt} + a_7 x_2 + a_8 \frac{dx_2}{dt}. \qquad (8\text{–}48)$$

For this situation, one cannot directly solve either equation for either x_1 or x_2 as was done in (8–47). The following method may be used to obviate this difficulty.

If these were algebraic (rather than differential) equations, one could then eliminate a variable by appropriate multiplications and addition. For example,

$$f = a_1 x_1 - a_2 x_2, \qquad 0 = -a_3 x_1 + a_4 x_2.$$

Multiply the first equation by a_4, multiply the second equation by a_2, and add:

$$\begin{aligned} a_4 f &= a_1 a_4 x_1 - a_2 a_4 x_2 \\ 0 &= -a_3 a_2 x_1 + a_2 a_4 x_2 \\ \hline a_4 f &= (a_1 a_4 - a_3 a_2) x_1 \end{aligned}.$$

A similar technique can be used on the differential equations, except that now some of the operations must include differentiation as well as multiplication.

Returning to Eq. (8–48), we see that if one operates on the first equation with $a_7 + a_8(d/dt)$ and on the second equation with $a_3 + a_4(d/dt)$, the resulting equations are

$$a_7 f + a_8 \frac{df}{dt} = a_1 a_7 x_1 + (a_2 a_7 + a_1 a_8)\frac{dx_1}{dt} + a_2 a_8 \frac{d^2 x_1}{dt^2} - a_3 a_7 x_2 - (a_4 a_7 + a_3 a_8)\frac{dx_2}{dt} - a_4 a_8 \frac{d^2 x_2}{dt^2},$$

$$0 = -a_5 a_3 x_1 - (a_6 a_3 + a_5 a_4)\frac{dx_1}{dt} - a_6 a_4 \frac{d^2 x_1}{dt^2} + a_7 a_3 x_2 + (a_8 a_3 + a_7 a_4)\frac{dx_2}{dt} + a_8 a_4 \frac{d^2 x_2}{dt^2}.$$

If the two equations are now added, the terms in x_2 cancel and an equation only in x_1 is obtained.

An example of a system in which this difficulty occurs will be given. A slight modification in the vibration-absorber example of Section 8–5.3 results in this type of situation. If a damper, say b_3, is placed in parallel with spring k_2 (as was a suggested model if the spring exhibited any energy dissipation), then an extra force will appear in Eqs. (8–29) and (8–30). This will result in the derivatives of Eqs. (8–31) and (8–32) being modified as follows:

$$-2k_1(v - v_1) + k_2(v_1 - v_2) + b_3\left(\frac{dv_1}{dt} - \frac{dv_2}{dt}\right) + m_1 \frac{d^2 v_1}{dt^2} + b_1 \frac{dv_1}{dt} = 0 \qquad (8\text{–}49)$$

and

$$k_2(v_1 - v_2) + b_3\left(\frac{dv_1}{dt} - \frac{dv_2}{dt}\right) - b_2 \frac{dv_2}{dt} - m_2 \frac{d^2 v_2}{dt^2} = 0. \qquad (8\text{–}50)$$

It is now not possible to solve for v_1 or v_2 directly using either equation since each variable appears also with derivative(s). Rearranging (8–49) and (8–50), we have

$$2k_1 v = (2k_1 + k_2)v_1 + (b_1 + b_3)\frac{dv_1}{dt} + m_1 \frac{d^2 v_1}{dt^2} - k_2 v_2 - b_3 \frac{dv_2}{dt},$$

$$0 = -k_2 v_1 - b_3 \frac{dv_1}{dt} + k_2 v_2 + (b_2 + b_3)\frac{dv_2}{dt} + m_2 \frac{d^2 v_2}{dt^2}.$$

If the first equation is operated on with

$$k_2 + (b_2 + b_3)\frac{d}{dt} + m_2 \frac{d^2}{dt^2}$$

and the second equation is operated on with

$$k_2 + b_3 \frac{d}{dt},$$

when the equations are added, all terms in v_2 will cancel and leave the following system equation for v_1 which should be compared to (8–34), the equation for the system without b_3. The student should carry out the above operations and verify (8–51):

$$m_1 m_2 \frac{d^4 v_1}{dt^4} + [b_2 m_1 + b_1 m_2 + b_3(m_1 + m_2)] \frac{d^3 v_1}{dt^3}$$
$$+ [k_2 m_1 + b_1 b_2 + m_2(k_2 + 2k_1) + b_3(b_1 + b_2)] \frac{d^2 v_1}{dt^2}$$
$$+ [b_1 k_2 + b_2(k_2 + 2k_1) + 2k_1 b_3] \frac{dv_1}{dt} + 2k_1 k_2 v_1$$
$$= 2k_1 k_2 v + 2k_1(b_2 + b_3)\frac{dv}{dt} + 2k_1 m_2 \frac{d^2 v}{dt^2}. \tag{8–51}$$

The difficulty of this procedure indicates the desirability of being able to work with algebraic equations rather than differential equations. Such a technique will be developed in Chapter 13 by use of generalized impedance. One major reason for the use of generalized impedance and the Laplace transform (Chapter 16) is to avoid this difficulty. These methods also make it easier to deal with systems such as those given in Section 8–5.3. The student should learn the basic formulation approach given in this chapter first, however, since it is applicable also to nonlinear systems, whereas the simplifications which will be introduced later only apply to linear systems.

It should be noted here that, in general, for nonlinear systems it may not be possible to obtain a single system equation for the output since the elemental equations may be transcendental.

8–6. GENERALIZATIONS ABOUT SYSTEM EQUATIONS

At this point in the development, we have done sufficient work with systems to be able to make some generalizations. These generalizations deal with the general properties of the system equations and will be helpful for checking and for appreciation of possible performance of systems.

8–6.1 General Form of System Equations

Inspection of the particular results given by Eqs. (8–34), (8–44), and (8–51) should convince the reader that, in general, system equations for lumped-parameter linear systems can be expressed in the form

$$a_n \frac{d^n x_r}{dt^n} + a_{n-1} \frac{d^{n-1} x_r}{dt^{n-1}} + \cdots + a_1 \frac{dx_r}{dt} + a_0 x_r$$
$$= b_0 x_f + b_1 \frac{dx_f}{dt} + \cdots + b_m \frac{d^m x_f}{dt^m}. \tag{8–52}$$

The notation x_r indicates a responding variable (an output) while x_f indicates a forcing variable (an input).

In general, there are no restrictions on the relative magnitudes of m and n. That is, it might be possible to have $n > m$, $\quad n = m$, $\quad n < m$.

However, for real physical systems with realistic sources, it is true in general that $n \geq m$. If this were not true, that is, if the order of the highest derivative of the input were larger than the order of the highest derivative of the output, then the output would increase without bound with increased rate of variation of the input (for instance, at high frequency). This can be seen by choosing a sinusoidal input and output.

We have seen in Chapter 5 that for first-order systems the steady-state response to a sinusoidal input is also a sinusoid of the same frequency. This is also true for an nth order system. Thus if the input to Eq. (8–52) is a sinusoid, $x_f = X_f \sin \omega t$, its steady-state response is of the form

$$x_r = X_r \sin(\omega t + \phi).$$

These functions substituted into (8–52) give

$$a_n \omega^n X_r \frac{d^n[\sin(\omega t + \phi)]}{d(\omega t)^n}$$
$$+ a_{n-1}\omega^{n-1} X_r \frac{d^{n-1}[\sin(\omega t + \phi)]}{d(\omega t)^{n-1}} + \cdots$$
$$= b_m \omega^m X_f \frac{d^m[\sin \omega t]}{d(\omega t)^m} + \cdots.$$

The sinusoids in this equation and their derivatives vary only between ± 1. Comparing the amplitudes of the various terms, we see that the nth term on the left and the mth term on the right predominate as $\omega \to \infty$. Therefore, as $\omega \to \infty$, we may write

$$a_n \omega^n X_r \cong b_m \omega^m X_f, \qquad X_r \cong \frac{b_m}{a_n} \omega^{m-n} X_f.$$

Thus if $m > n$, then $X_r \to \infty$ as $\omega \to \infty$; if $m = n$, then $X_r \to (b_m/a_n)X_f$ as $\omega \to \infty$; and if $m < n$, then $X_r \to 0$ as $\omega \to \infty$.

Ordinarily, a real physical system is not capable of following very fast variations in input (that is, $X_r \to 0$ as $\omega \to \infty$). In any analysis, it must be remembered that the analysis is being performed on an idealized model of the real system and for sufficiently high frequency, some parasitic elements which may have been neglected will become important. When these other elements are included they will, in general, cause the output to be incapable of following very fast variations of the input. Thus, when good enough models are used so that they describe the very-high-frequency behavior of real systems, then as a rule, $n > m$.

An illustration of this point is afforded by a simple electronic amplifier. If the interelectrode capacitances in the vacuum tube are not included (a justifiable model at low frequencies), one would conclude that the amplifier output would have a constant value as the input frequency was increased (that is, $m = n$). However, the proper model for high frequencies must include these capacitances, and when this is done, one finds that the output will go to zero as the input frequency increases (that is, $n > m$).

Similarly, the main reason for inclusion of the winding capacitance in the model of the pulse transformer example in Section 8–5.3 is to obtain an accurate representation of performance for high-frequency variations of the input.

8–6.2 Order of System

It is useful to be able to make some general statements about the order of differential equations which result from systems of the types analyzed here. For example, why did the order of Eq. (8–44) come out to be three for the i_1 (output) terms and two for the e (input) terms? Could this be predicted by inspection? We will consider the order of the output terms (called the order of the system) in this section. The order of the input terms is considered in Problem 8–24. We would like to be able to predict the order of the system equation without going through all the previous steps of setting up and solving the equations.

It should be rather obvious that the order of the system cannot exceed the number of energy-storing elements in the system. It is well known that n initial conditions are needed to solve an nth order differential equation (see Chapter 11 if this is not well known to *you*). These n initial conditions describe the state of the system at the instant of application of an input. Physically, the state of a system at any instant is specified by the values of the across-variables for A-type elements and through-variables for T-type elements (since these respective variables fix the energy storage in each element). From this reasoning, one would say offhand that the order of the system is the number of A-type elements plus the number of T-type elements.

However, we have already seen that for some systems the resulting differential equation is of lower order than the number of energy-storage elements. An obvious case is the vibration absorber, where the two springs k_1 were in parallel and thus entered as one effective energy storage element rather than two. Other cases can be cited in which the order is lower than the number of energy-storage elements but in which there are no obvious series or parallel connections of similar elements. There can be topological constraints which restrain the critical variables from being independent. There also can be degenerate situations in which the element values are numerically such that the order is lower than expected. For all these reasons the determination of the system order is a complicated business. Although the theory is not beyond the scope of this text, space limitations preclude an extended discussion here. We will only indicate the type of results which occur.

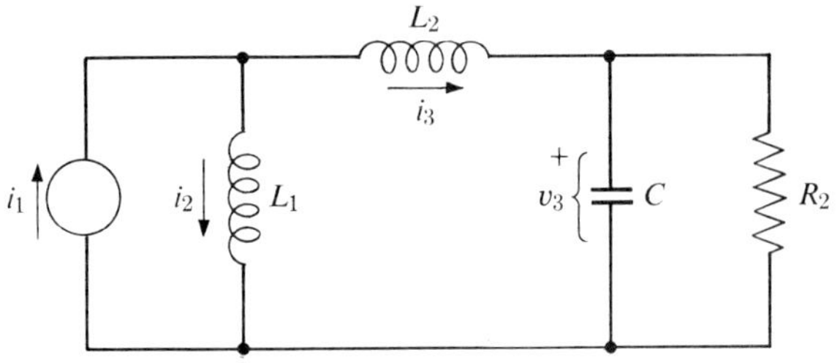

FIG. 8–35. Pulse transformer circuit with applied current source i_2 (second-order system).

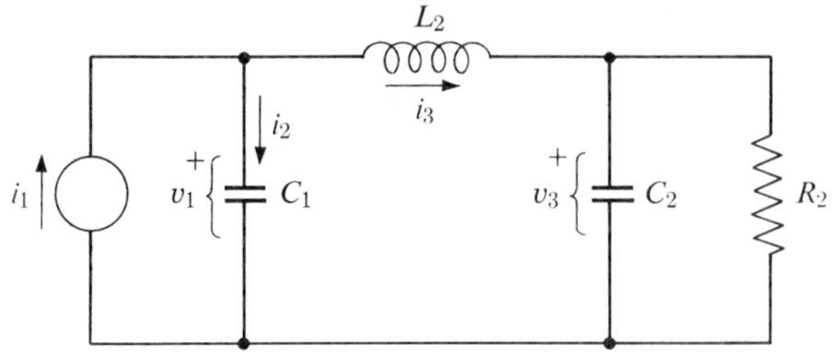

FIG. 8–36. Third-order system with applied current source i_1.

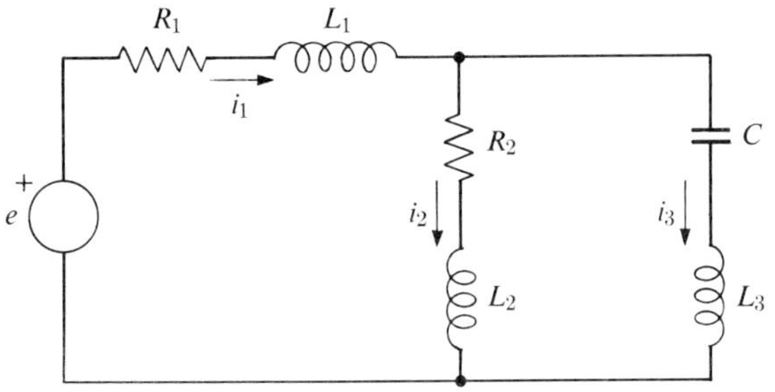

FIG. 8–37. Circuit of order three.

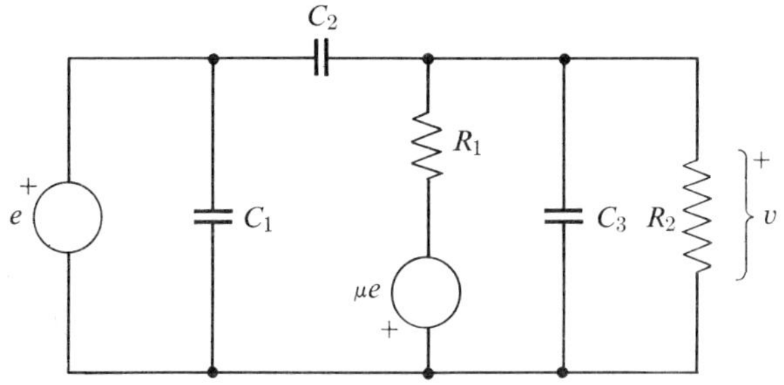

FIG. 8–38. Equivalent circuit of vacuum-tube amplifier of order one.

The "normal" situation occurs when the order of the system is equal to the number of energy-storing elements. All of the previous examples in this text were of this nature (except for some obvious exceptions of series or parallel energy storers which act as one). A more complete statement of the criterion is that the order is equal to the number of *independent* energy storers. The theory of determination of the independence of interconnected energy storers in a system is extensive. It will suffice here to give a few examples. The reader who is interested in this subject is referred to Guillemin [8], who discusses the problem from the viewpoint of network topology. (This reference is recommended for interesting reading in many areas related to dynamic systems.)

Let us consider the vibration absorber of Section 8–5.3 (Fig. 8–33). The order of this system (for v input) may be determined by investigating the number of independent initial energy storages which may be imposed on the system. That is, we will try to store energy in all ways possible such that at $t = 0$, the system is "released" and then allowed to "go its own way," which will be determined by the elements, inputs, and initially stored energies. The energy storage in a given element is independent if it may be given any arbitrary value without changing any other previously established energy storage.

For this vibration-absorber system, energy may be stored independently in four places, i.e., in each mass, spring k_2, and the pair of springs k_1, because the velocities of the masses and the forces in the springs may be set at any desired values without the establishment of any one value restricting the establishment of any other value. One may make a similar argument about energy storages for the pulse transformer of Section 8–5.3 and conclude that the order should be three, which agrees with Eq. (8–44).

It is worthy of note that the types of sources and their manner of application to a system can affect the order by imposing new constraints. For instance

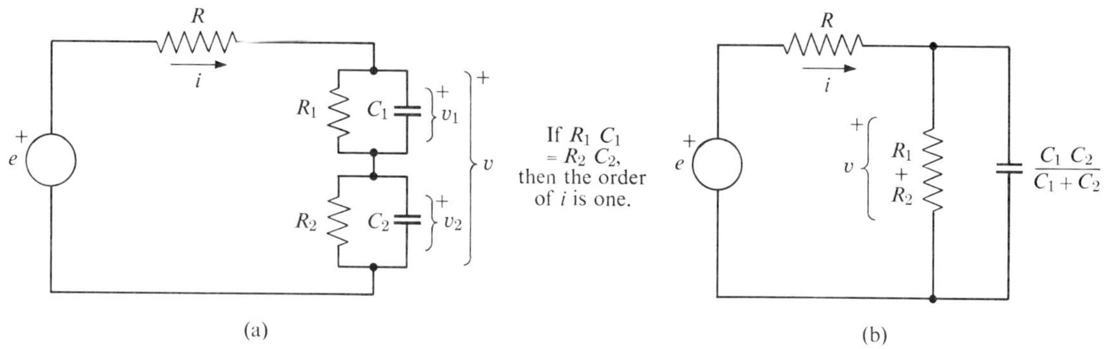

FIG. 8-39. (a) Second-order system with element-value degeneracy which reduces order to one. (b) Equivalent circuit for $R_1C_1 = R_2C_2$.

if, in Fig. 8-34, the circuit is excited by a current source i_1 rather than the voltage source e shown, the order of the system will be changed. As discussed in Section 8-5.2, a current source i_1 gives a situation like that shown in Fig. 8-30, and R_1 has no effect on the rest of the system since its current is determined. The network thus reduces to that shown in Fig. 8-35. The initial energy of C can be specified arbitrarily by setting v_3. The initial energy of L_2 can be specified arbitrarily by setting i_3 independent of v_3. With the current source specified, however, one cannot now set the current i_2 arbitrarily and independent of i_3 since $i_2 = i_1 - i_3$. Both i_1 and i_3 have been previously established, and i_2 is therefore fixed. The inductance L_1 is not therefore an independent storage location and the system with current forcing is of order two rather than three, as it was with the original voltage forcing.

It is thus seen that the order depends on the types of elements and their topology as well as on the types of sources and their application in relation to this topology. Figure 8-36 shows a situation similar to Fig. 8-35 except that L_1 is replaced by a capacitance. The order of this system is three since the fixed current i_2 does not specify the energy storage in the capacitance C_1 and the voltage v_1 may be specified independent of i_1, i_3, and v_3.

Figures 8-37 and 8-38 show two situations in which the order is less than the number of energy-storage elements due to topological constraints. In Fig. 8-37, the capacitance and only two of the three inductances can be independent energy storers since $i_1 = i_2 + i_3$ and if two of the inductance currents are specified, the third is then no longer arbitrary. Thus, the order of the system is three. However, if one of the inductances were shunted by a resistance, the inductance currents would all then be arbitrary and the order would be four.

Figure 8-38 is an equivalent circuit of a vacuum-tube amplifier which has a dependent source (μe). The capacitances are the interelectrode capacitances. The order of this system is one, not three. The voltage across C_1 is specified by the source, so it is not independent (in fact C_1 cannot affect the response of the rest of the circuit since its voltage is fixed). If the energy storage in C_3 is specified by fixing v, then the voltage across C_2 is not independent since $v_{C_2} = e - v$, and thus only one of C_2 and C_3 is independent.

The case of element-value degeneracy will be illustrated by an example. The circuit shown in Fig. 8-39(a) will, in general, be second-order. However, if the element values are such that $R_1C_1 = R_2C_2$, then the equation for i is only first-order and the circuit acts as shown in 8-39(b). It is worthy of note that the two initial energies of C_1 and C_2 may be set independently even when this condition is imposed. Problems 8-20 and 8-21 investigate this type of degeneracy further.

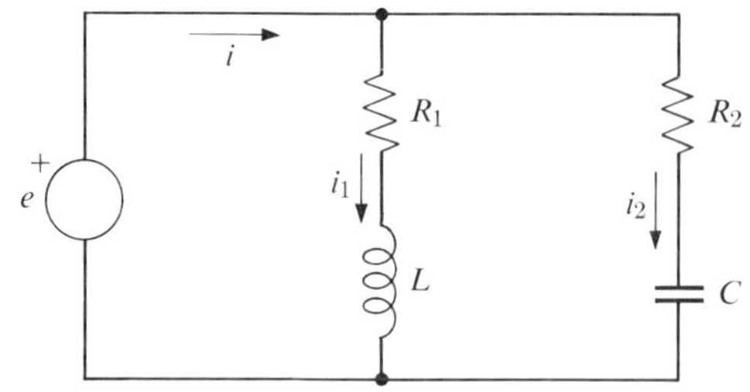

FIG. 8-40. Second-order system with element-value degeneracy which reduces order to zero.

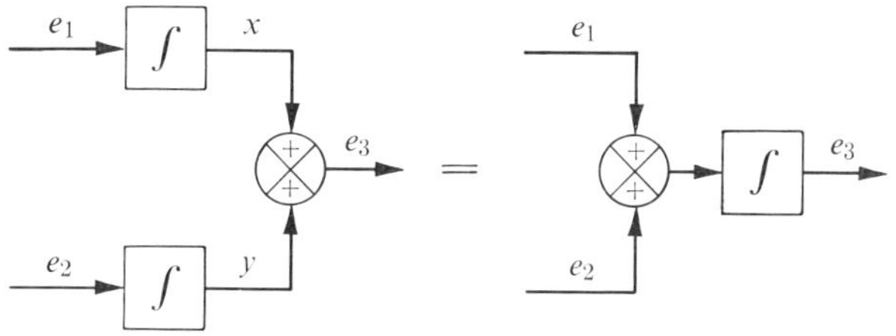

FIG. 8-41. Superfluous integrators in a block diagram.

A final classic example of element-value degeneracy is given in Fig. 8-40. Here the two currents are independent so that the order of i_1 and i_2 is one. The order of i in general would be two, without restrictions on the values of elements and energy storage. However, if there is zero initial energy storage and

$$R = R_1 = R_2 = \sqrt{L/C},$$

the order of i is zero and we obtain $iR = e$ for the system equation! This means that the circuit appears to the source e as a pure resistance R even though it contains inductance and capacitance. It is left as an exercise for the student to derive this result. Note that if the initial energy storage were not specially chosen so that $i_0 = e/R$, then some transient would have to occur. However, it is true that for the steady-state response to a sinusoid, $i = e/R$ if

$$R = R_2 = R_1 = \sqrt{L/C}$$

for all frequencies, and this combination yields a frequency-independent network.

The order of the system can also be determined from a block (or analog computer) diagram. Since each integrator must have an initial condition specified, and n initial conditions are needed for an nth order system, the order of the system is equal to the *minimum* possible number of integrators in a system block diagram. Notice, however, the qualification of the minimum number of integrators possible. Normally one will draw the system block diagram with this minimum number, but there is evidently no way to prove whether one has successfully obtained the minimum number. Thus again, this is not a conclusive method. Also, in general, element-value degeneracies are not discovered by this procedure. Figure 8-41 shows an example of a block diagram which has more than the minimum number of integrators.

8-6.3 Signs of the Coefficients

It is to be noted that every term in Eqs. (8-34) and (8-44) has a positive sign. This is not coincidental, but a general property of passive systems, and this fact should be used as a check on the correctness of the system equation of any passive system. A passive system is one which does not contain any elements with energy sources. All the ideal elements are passive, but elements such as modulators (vacuum tubes, transistors, hydraulic valves, etc.) usually involve active elements which include a power source connected to the third port (see Section 4-5). A system containing a dependent source which can supply more power at its output than it receives at its input is not a passive system.

It has been seen that the coefficients of the system equation, a_k, in general are composed of a number of terms which involve the system element parameters. The a_k must be positive for all possible values of the element parameters, so one may additionally conclude that every term in any a_k must also be positive. This fact is most useful as a check on the correctness of the equation.

In Section 13–4.3 this property of the positiveness of the signs of the a_k will be proved. Such a general statement cannot be made concerning the signs of the b_k, however; if the output is a variable other than the complementary variable of the source, the b_k may be of different signs. Problem 8–22 shows such a situation.

8–6.4 System Equation for Various Outputs

Let us suppose that v_3 had been selected as the output for the system of Fig. 8–34, rather than i_1. The reader may verify that the system equation for this case, retaining e as the source, is

$$R_2CL_1L_2\frac{d^3v_3}{dt^3} + [R_1R_2C(L_1 + L_2) + L_1L_2]\frac{d^2v_3}{dt^2} + [R_1(L_1 + L_2) + L_1R_2]\frac{dv_3}{dt} + R_2R_1v_3 = R_2L_1\frac{de}{dt}. \tag{8–53}$$

It should be noted that the left-hand side of this equation is exactly the same as the left-hand side of Eq. (8–44) with v_3 substituted for i_1. That is, the characteristic equation is the same regardless of whether i_1 or v_3 is the output. This is a general property of *linear* systems. A general proof of the truth of this statement is given in Section 14–3. However, the following should convince the reader of the validity of this important aspect of system behavior.

The characteristic equation (the left-hand side) determines the exponents which appear in the exponentials in the solution. The system variables must obey the continuity and compatibility conditions, which are simultaneous linear differential equations when the system is composed entirely of linear elements. If every system variable did not have the same characteristic equation, then the solution for the various variables would be composed of exponentials which had different time constants. The impossibility of a number of terms adding to zero when they did not have common exponential factors should be apparent. Therefore, it is quite reasonable that all system variables will have the same characteristic equation. This fact means that the characteristic equation is characteristic of the *system* and not only of the particular output selected.

It should be noted, however, that the right-hand sides of Eqs. (8–44) and (8–53) are different. This must of necessity be true since the right-hand sides will determine the steady-state value of the various outputs. For instance, if the source is a constant (dc), $e = E$, then the steady-state solution for i_1 is $i_1 = E/R$, whereas the steady-state for v_3 is zero (because the inductance L_1 is a short circuit to dc). Equation (8–44) gives $i_1 = E/R$ for the case of all derivatives equal to zero, whereas Eq. (8–53) gives $v_3 = 0$ for this case. Problems 8–23 and 8–24 discuss the determination of the order of the right-hand side of the system equation.

8–6.5 Checking

Equations with the complexity of Eqs. (8–34) and (8–44) are very likely to contain an error due to incorrect algebraic manipulation in the earlier steps. Before such equations are used for engineering work, they should be thoroughly checked. A certain amount of checking is easy to accomplish and will frequently save many hours of subsequent worthless effort. There are three especially useful means of checking system equations. They are by (a) dimensions, (b) signs and order, and (c) limiting cases.

a) *Dimensions.* Every equation must be dimensionally homogeneous if it is to be valid. That is, each term to be added into a sum must have the same dimensions. A statement of the dimensions of all the quantities in (8–34) is easily determined from the elemental relations. We have, dimensionally,

F = force,
L = length,
T = time,
b_1, b_2 = force/velocity = FT/L,
k_2, k_1 = force/length = F/L,
m_2, m_1 = force/acceleration = FT^2/L,
v_1, v = velocity = L/T,
t = time = T.

This gives, if we choose terms from (8–34),

$$m_1 m_2 \frac{d^4 v_1}{dt^4} = \left(\frac{FT^2}{L}\right)^2 \left(\frac{L}{T}\right)\left(\frac{1}{T^4}\right) = \frac{F^2}{LT},$$

and therefore all terms must have dimensions of F^2/LT:

$$b_2 m_1 \frac{d^3 v_1}{dt^3} = \left(\frac{FT}{L}\right)\left(\frac{FT^2}{L}\right)\left(\frac{L}{T}\right)\left(\frac{1}{T^3}\right) = \frac{F^2}{LT},$$

$$k_2 m_1 \frac{d^2 v_1}{dt^2} = \left(\frac{F}{L}\right)\left(\frac{FT^2}{L}\right)\left(\frac{L}{T}\right)\left(\frac{1}{T^2}\right) = \frac{F^2}{LT},$$

$$b_1 b_2 \frac{d^2 v_1}{dt^2} = \left(\frac{FT}{L}\right)^2 \left(\frac{L}{T}\right)\left(\frac{1}{T^2}\right) = \frac{F^2}{LT},$$

and so on for the rest of the terms. The reader should check that the remaining terms are dimensionally correct in Eq. (8–34).

The final result of Eq. (8–44) will be checked by dimensions also. Dimensionally, we have

$$R = \frac{V}{A}, \qquad e = V \qquad \text{(volts)},$$

$$L = \frac{VT}{A}, \qquad i = A \qquad \text{(amperes)},$$

$$C = \frac{AT}{V}, \qquad t = T \qquad \text{(time)},$$

$$R_2 C L_1 L_2 \frac{d^3 i_1}{dt^3} = \left(\frac{V}{A}\right)\left(\frac{AT}{V}\right)\left(\frac{VT}{A}\right)^2\left(\frac{A}{T^3}\right) = \frac{V^2}{A}.$$

It will help to note that

$$L/R = T, \qquad RC = T, \qquad LC = T^2,$$

and the reader can verify that the equation is dimensionally homogeneous.

b) *Signs and order.* As noted in Section 8–6.3, the signs of all the coefficients of the left-hand side of the system equation of a passive system must be the same for all values of the parameters. This check should always be made on a final system equation. Algebraic sign errors made in the manipulation of the simultaneous equations are most likely to be found by this check.

The order of the equation should be checked by inspection of the system, using the ideas of Section 8–6.2 (insofar as the system allows a simple determination of order). This can easily pick up an error in differentiation made during manipulation of simultaneous differential equations. Incorrect statement of elemental equations may also be found in this way.

c) *Limiting cases.* Another valuable check is obtained by selecting certain parameters equal to either zero or infinity (or 1.0 in some more general problems). A proper choice of parameters to have limiting values will cause the physical situation to be so simplified that the correctness (or incorrectness) of the simplified equation can be determined by inspection or a simplified analysis. For the vibration-absorber system, the following checks are made:

1) $k_1 \to \infty$. If the two springs connecting v and v_1 are infinitely stiff, then a rigid connection exists and the equation should reduce to $v_1 = v$. To let $k_1 \to \infty$ and get useful results, we must divide Eq. (8–34) by k_1, and then k_1 may be allowed to approach infinity. We see that such an operation yields

$$2m_2 \frac{d^2 v_1}{dt^2} + 2b_2 \frac{dv_1}{dt} + 2k_2 v_1 = 2k_2 v + 2b_2 \frac{dv}{dt} + 2m_2 \frac{d^2 v}{dt^2},$$

and the terms on both sides have precisely the same coefficients and derivatives, so that $v_1 = v$, as required.

2) $k_1 \to 0$. In this case, the input is uncoupled from mass m_1 and an equation should result in which v does not appear. Inspection of (8–34) for $k_1 = 0$ shows this to be true. In addition, we note that one term in v_1 disappears also. This is expected because the order of the system should be reduced by one when the springs k_1 are removed.

3) $k_2 \to \infty$. In this case m_1 and m_2 are rigidly connected together and should only appear lumped together as $(m_2 + m_1)$. In addition, the dampers b_1 and b_2 have the same velocity, so these quantities also should only appear lumped together as $(b_1 + b_2)$.

Letting $k_2 \to \infty$ in Eq. (8–34) yields

$$(m_1 + m_2)\frac{d^2 v_1}{dt^2} + (b_1 + b_2)\frac{dv_1}{dt} + 2k_1 v_1 = 2k_1 v.$$

Note also that if $k_2 \to \infty$, the system has effectively one mass and one spring (the two springs k_1 are in parallel and equivalent to one spring of $2k_1$), thereby giving the above second-order equation.

4) $m_2 \to \infty$. In this case $v_2 = 0$ since m_2 is so massive; point 2 cannot move. Therefore k_2 is essentially connected to ground, and thus it is in parallel with $2k_1$ and only the combination $(k_2 + 2k_1)$ should appear with v_1, whereas $2k_1$ without k_2 should appear with v. Also b_2 should disappear. The system should be second order. Letting $m_2 \to \infty$ yields

$$m_1 \frac{d^4 v_1}{dt^4} + b_1 \frac{d^3 v_1}{dt^3} + (k_2 + 2k_1)\frac{d^2 v_1}{dt^2} = 2k_1 \frac{d^2 v}{dt^2}.$$

Since this equation has second derivatives as lowest-order terms, it can be simply integrated twice and the expected result is obtained. This process may be continued with all the seven elements until the engineer feels confident of having the correct equation.

In general, the parameters allowed to take on limiting values should be selected so that each term in the equation is checked at least once. Occasionally, verifying the resulting equation for a limiting case is a major task. For instance, the reader is invited to make a check for $k_2 \to 0$. In this case, inspection of the system tells us that mass m_2 is uncoupled from m_1 and therefore that m_2 and b_2 should not appear in the final result. Also the system should reduce to second-order. The result given by (8–34) for $k_2 = 0$ does not, on first inspection, seem to bear this out, but both the expected performance and the limiting form of (8–34) for $k_2 = 0$ are correct. The reason for this is that when $k_2 = 0$ there are essentially two independent systems: (1) b_2 and m_2 with no input and (2) b_1, m_1, and $2k_1$ with input v. Since these systems were coupled before k_2 was allowed to approach zero, the equations of two systems are mixed together. However, it can be shown that they can be separated by factoring m_2 out of the terms in which it appears and b_2 out of the terms in which it appears. Then equating terms which only have m_2 or only b_2 yields the correct result. The reader should work through this case on his own for experience with this type of situation.

The reader is invited to make the following checks on the pulse-transformer equation (8–44).

1) Either $C \to \infty$ or $R_2 \to 0$. There is a short circuit from point 3 to ground and the circuit reduces to a simple first-order R_1-L system with an equivalent inductance equal to the parallel combination of L_1 and L_2 [that is, $L_1L_2/(L_1 + L_2)$].

2) $C \to 0$ and $R_2 \to \infty$. The circuit is open between point 3 to ground and $i_3 = 0$ so that the system is a simple series R_1-L_1 circuit.

3) $L_1 \to 0$. This should reduce to $i_1 = e/R_1$. Watch coupling effect between R_1 and L_2, C and R_2, which are really independent circuits if $L_1 = 0$.

4) $L_2 \to 0$, $L_1 \to \infty$, $R_2 \to \infty$. This is seen to result in a simple R_1-C series circuit.

An additional check may also be employed. It is useful for checking the correctness of the highest- and lowest-order terms on each side of the equation. This idea can also be used for the vibration-absorber (or any other) system, of course.

1) Let us consider what happens to the circuit for very slow variations of the input (i.e., low frequency). In the absence of sudden changes, the output will also vary slowly. Therefore the derivative terms will be negligible compared to the terms without derivatives. We then get from Eq. (8–44)

$$R_2 R_1 i_1 = R_2 e$$

or

$$i_1 = e/R_1.$$

Inspection of the circuit shows this result to be correct (using the well-known facts that were established in Chapter 6, that for dc conditions capacitances are open circuits and inductances are short circuits).

2) Let us consider very fast variations of the input (i.e., high frequency). The output will also vary rapidly. In this case, the predominant terms will be the highest derivatives present (see Section 8–6.1).

Thus, for this limit from Eq. (8–44), canceling the common factor (R_2C_2), one obtains

$$L_1L_2\frac{d^3i_1}{dt^3} = (L_1 + L_2)\frac{d^2e}{dt^2}.$$

Consider that

$$e = E_0 \sin \omega t, \qquad \frac{d^2e}{dt^2} = -\omega^2 E \sin \omega t,$$

and the solution for i_1 will be of the form

$$i_1 = I_1 \cos \omega t,$$

so that

$$\frac{d^3i_1}{dt^3} = +\omega^3 I_1 \sin \omega t,$$

and

$$L_1L_2\omega^3 I_1 = -(L_1 + L_2)\omega^2 E_0$$

or

$$I_1 = \frac{(L_1 + L_2)}{L_1L_2}\frac{E_0}{\omega} = -\frac{E_0}{\omega L_{eq}},$$

where

$$L = \frac{L_1L_2}{L_1 + L_2} = \text{parallel combination of } L_1 \text{ and } L_2.$$

From this, we conclude that $i_1 \to 0$ as $\omega \to \infty$, and we can verify this from the circuit by using the well-known facts that for high frequencies, capacitances are short circuits and inductances are open circuits (if the student is unsure of this, he should examine the elemental equations of *A*- and *T*-type storage elements for sinusoidal variation and observe what occurs as $\omega \to \infty$). This conclusion would be reached for any differential equation for which the order of the highest derivative of the input is less than the order of the highest derivative of the output. Thus it has been verified for Eq. (8–44) that there should be no terms in e higher than d^2e/dt^2. This check for high frequencies is of additional value, however, for actually checking the correctness of the coefficients of the highest derivatives. It is seen from the circuit (Fig. 8–34) that, for high frequencies, it reduces to simply the parallel combination of L_1 and L_2 as the elements which limit the current i_1. The result given above verifies this.

8–7. ANALOGS, DUALS, AND DUALOGS

In this section we shall consider a more complete formulation of the concepts of analogs and duals which were introduced in Chapter 5. The basis for this discussion will be the system graphs and the ideal elements.

Although the ideas presented here make it possible to build and test analogous (and/or dual) systems, the reader should not be misled into feeling that this is the reason for these considerations. In the past, considerable effort was devoted to this type of activity. (Reference 4 is the fundamental work in this direction. Reference 5 surveys the ideas and has an extensive bibliography.) In general this activity centered around building the electrical analog of mechanical or other systems and testing it to determine the performance of the original system. The reason for doing this was that electrical components were readily available and easy to assemble compared to mechanical construction, and electrical instrumentation was simpler. The advent of the electronic analog computer has essentially caused a complete cessation of this kind of simulation since it is now easier and more accurate to use the computer.

Our interest in analogs and duals is to increase our understanding of system behavior. It will be seen that if we know the behavior of any one system we also know the behavior of at least seven other systems (the analog systems and the dual system and the duals of the analogs; it may be greater than seven systems if the particular problem allows a thermal realization). Thus, one should get an appreciation of the fact that the study of any one particular system is not so limited as it might appear at first, since the results will find wide application to many other situations which may be similar (as well as identical) to the analog and dual systems.

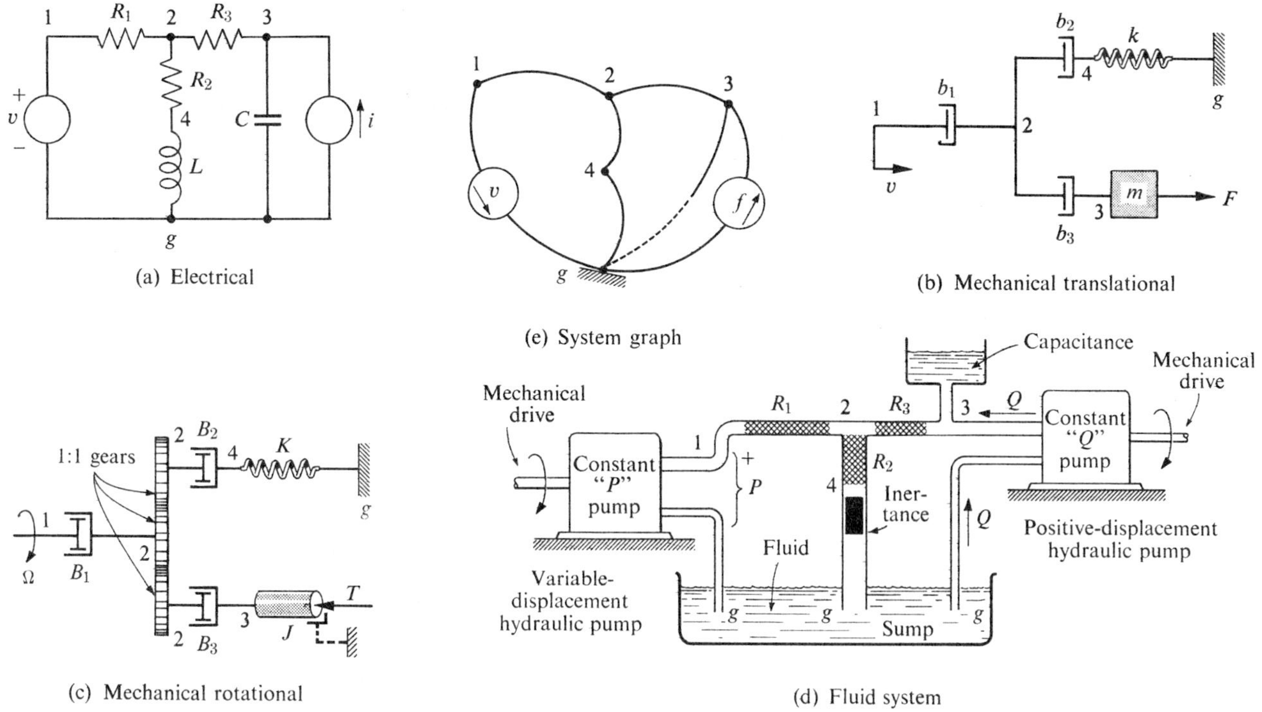

FIG. 8–42. Analogous systems.

8–7.1 Analogs

If two different physical systems have the same system graph they will be called *structurally analogous*. This means that the same path and vertex equations apply to the two systems (when the generalized variables f and v are used, the form of the equations for the two systems will be identical). This identity of structure must also apply to sources as well as element branches for the systems to be considered analogous. That is, if an across-variable source is applied in one type of physical system, the other physical system must also have an across-variable source applied at the same points in its graph for the structural analogy to be complete.

If two physical systems of the same type (i.e., both electrical) have the same graph, they will be called *structurally identical*. If, in addition, corresponding branches have the same type of elements then, of course, the two systems are completely identical.

If two different types of physical systems are structurally analogous and, in addition, if corresponding branches have elements for which the through- and across-variable relation is the same, they will be called *completely analogous* or simply *analogous*. If the systems are composed of ideal elements, then to be completely analogous, corresponding branches must contain elements listed under the same type of energy-storage element in Table 4–2. However, the analogous property is not restricted to ideal elements but also might apply to systems with nonlinear elements if the f-v relations are the same. Considerable use has been made of analogies of many types in engineering analysis [5].

Figure 8–42 shows four systems which are completely analogous. The general system graph is also shown.

It is readily recognized that the thermal analog of any system which contains a T-type (generalized

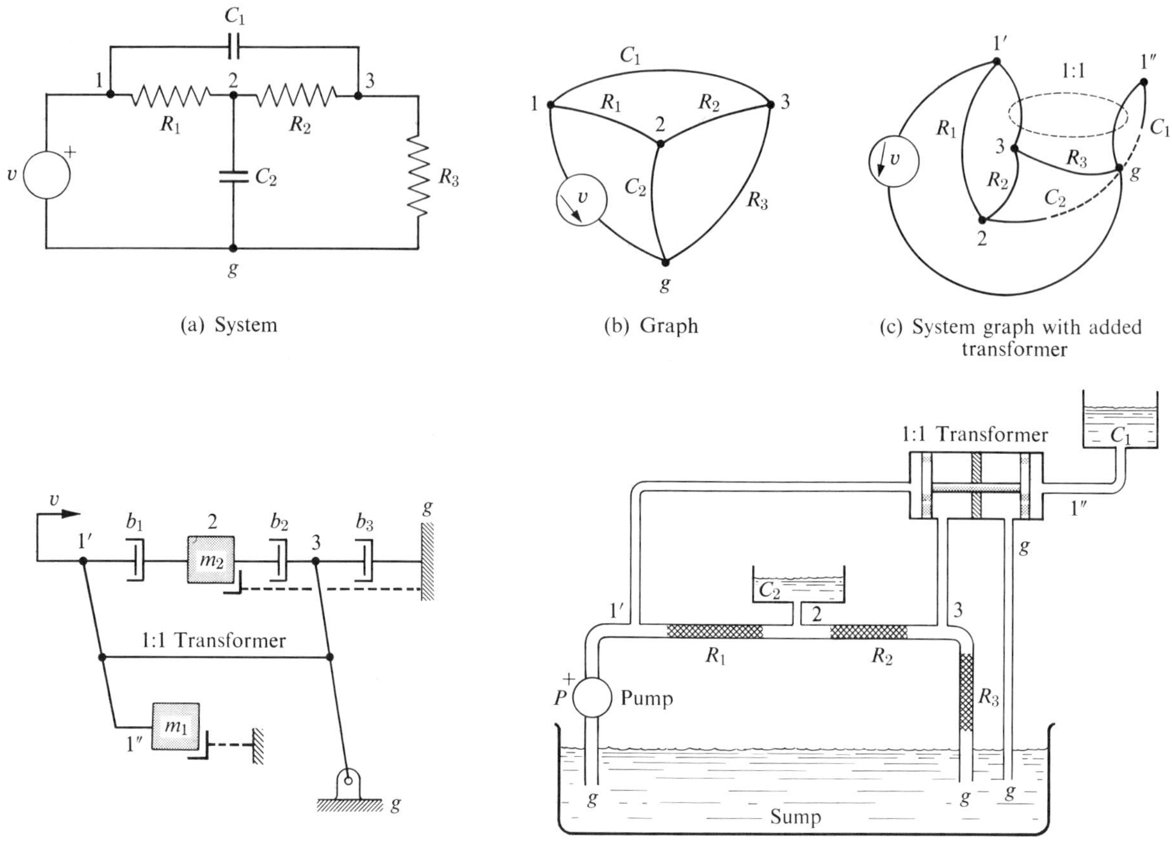

(a) System (b) Graph (c) System graph with added transformer

(d) Mechanical analog (e) Fluid analog

FIG. 8–43. An electric circuit without a directly physically realizable analog.

inductance) element will not be physically realizable since the analogous thermal element does not exist. Additionally, not all electrical systems will possess physically realizable analogs in the various other systems. In an electrical circuit, all capacitors are not restricted to having one terminal in common, whereas this condition must be met with the capacitance elements in the other systems. Figure 8–43(a) and (b) shows an electrical circuit which does not possess a directly physically realizable mechanical, fluid, or thermal analog.

There is a scheme whereby a mechanical, or fluid, analog of any electrical system may be constructed regardless of the location of capacitors in the electrical system. This involves inserting ideal 1 : 1 transformers in the electrical system to isolate the various capacitors from one another. Then one terminal of each capacitor may be connected to a common point without changing the circuit performance. The ideal electrical transformers are analogous to four-terminal mechanical or fluid transformers. In this way, "floating" or ungrounded masses or fluid capacitances may be realized. Figure 8–43(c), (d), and (e) shows such a realization for the analogs of Fig. 8–43(a). A gyrator connected to a spring can also be used in place of a mass, but this is stretching the analogy.

8–7.2 Duals

In dual systems the roles played by across-variables in one system are taken by through-variables in the other system, and vice-versa. Duality is a property displayed in the same type of physical system (i.e., two electrical systems may be duals of each other).

Given a particular type of physical system, if another system of the same physical type has vertex equations which are identical with the path equations of the original system (when v is replaced by f) and has path equations which are identical to the vertex equations of the original system (when f is replaced by v), then the two systems are said to be *structurally dual.*

We desire to determine a procedure for drawing the dual graph of a system graph. To do this, we note that for the above properties to exist, there must be a vertex in the dual graph corresponding to every closed path in the original graph. (We need consider only the most direct closed paths here, since if the dual has vertex laws which correspond to the path laws for these simple paths, then the more complex path-law duals may be obtained by linear addition of the vertex laws.) To completely discuss this problem we will consider a planar graph mapped on the sphere. There will be $(b - n + 2)$ meshes as shown previously in Section 8–4.2.

For the dual graph to obey the earlier definition, it must have a vertex equation for each of the meshes in the original graph as mapped on the sphere. Only $(b - n + 1)$ of these are independent, and the equation written for the last mesh corresponds to the extra vertex equation of the dual, which is also not independent. The vertices of the dual may be imagined to be placed interior to each of the meshes on the sphere. If we now connect the vertices of the dual graph by b branches in such a way that every vertex of the original graph finds itself interior to a mesh of the dual graph, we will easily see that the path equations of the dual graph will correspond to the vertex equations of the original graph. When the dual diagram is completed there will be no way of determining which graph was constructed first; they are dual in the true sense since there is no preferred sequence of passage from one to the other.

Let us consider now a nonplanar graph which can be constructed from a planar graph mapped on a sphere by adding a single branch between two vertices so that it must cross another branch. The dual of the original graph, before addition of this extra element, can be imagined with a vertex interior to each mesh. This additional branch would require that we be able to write another independent path equation for the original graph. A proposed dual of the new nonplanar graph would thus require another vertex in addition to those already present. This additional vertex may be imagined as being placed above the surface of the sphere. One additional branch would need to be added to the proposed dual. Since all other branches and vertices in the proposed dual are already established, the extra vertex would have only one extra branch attached to it (with a decision yet to be made concerning where the other terminal of this branch should be connected). Thus the vertex equation at this new vertex could not possibly be the dual of any path in the original graph. Therefore, it can be seen by this heuristic reasoning that a *nonplanar graph does not possess a dual.*

Figure 8–44(b) shows a pair of dual graphs and indicates the method of construction of a dual when graphs are drawn on a plane rather than on a sphere. In addition to a vertex of the dual placed interior to each mesh of the original, another vertex is placed outside the original graph. This vertex corresponds to the one interior to the mesh formed by the periphery of the plane graph when it is mapped on a sphere. Branches are then drawn between vertices so that every original branch has one and only one branch of the dual crossing it.

If two graphs are structurally dual and if, in addition, in corresponding branches the roles played by the through- and across-variables in the elemental relations for one system are interchanged in the other system, then the two systems are said to be *completely dual* or simply *dual.* If the systems are composed of ideal elements, then this implies a replacement of capacitance by inductance, inductance by capacitance,

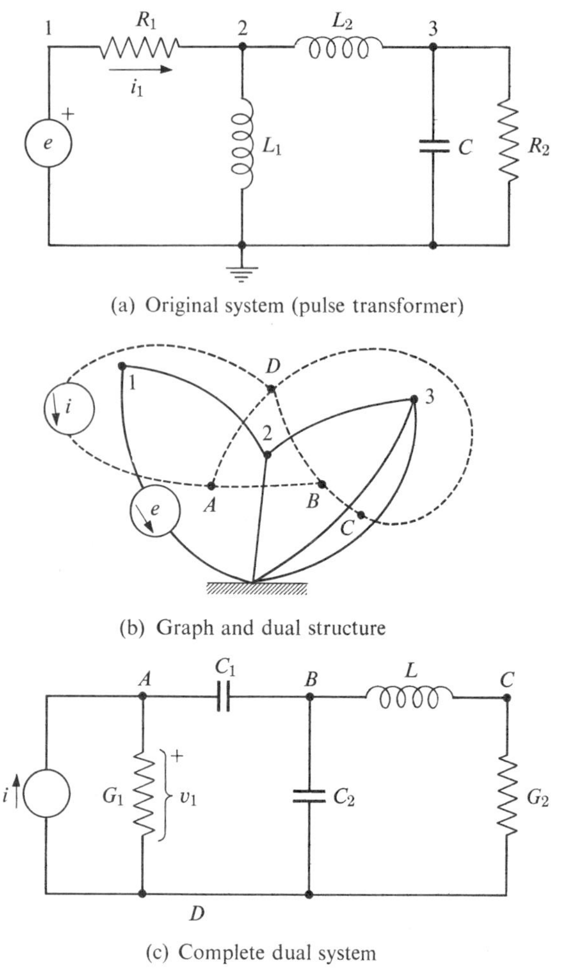

FIG. 8–44. Dual electrical systems.

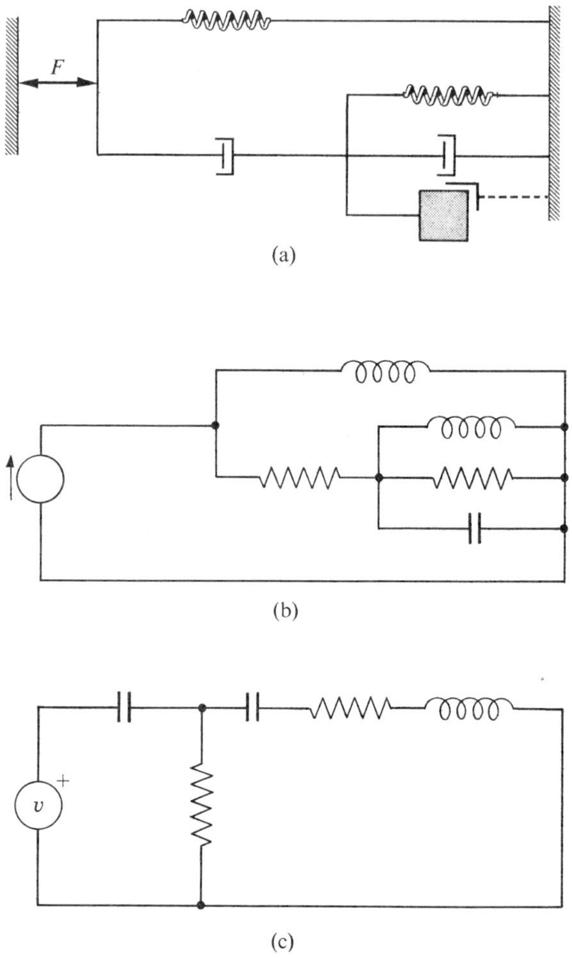

FIG. 8–45. Analogous systems.

and resistance by reciprocal resistance (or conductance). The dual concept is not limited to linear systems, however.

For the dual electrical systems in Fig. 8–44, if, in the equation for i in the system shown in Fig. 8–44(a), the substitutions $i_1 \to v_1$, $e \to i$, $R_1 \to G_1 = 1/R_1$, $R_2 \to G_2 = 1/R_2$, $L_1 \to C_1$, $L_2 \to C_2$, $C \to L$ are made, the equation for v_1 in the system shown in Fig. 8-44(c) will be obtained. [The system equation for Fig. 8–44(a) is Eq. (8–44).] We therefore see that the performance of many systems (the analogs and the dual and the duals of the analogs) can be determined if one has knowledge of the performance of a single system.

8–7.3 Dualogs

Using the previously developed concepts of analogs and duals, we see that another basis for analogs may be established. If, for instance, one draws the dual of an electrical system and then the mechanical analog of the dual system, the resulting system will be related

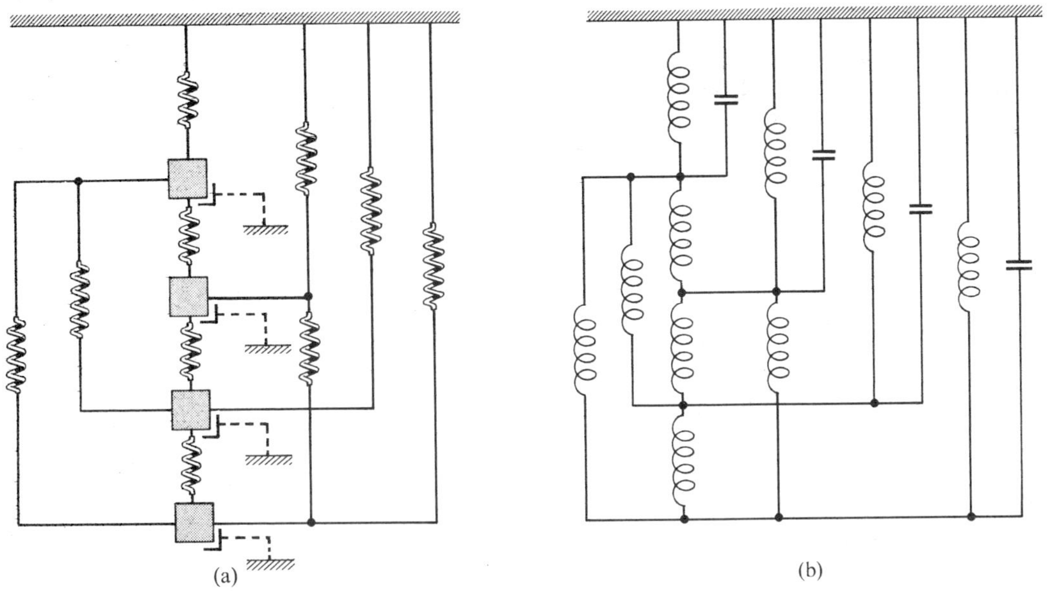

FIG. 8-46. Mechanical system for which mass-inductance analog does not exist.

to the original system such that

$$\begin{aligned}
\text{force} &\longrightarrow \text{voltage,} \\
\text{velocity} &\longrightarrow \text{current,} \\
m &\longrightarrow L, \\
1/k &\longrightarrow C, \\
b &\longrightarrow R.
\end{aligned}$$

This analogy is called the *mass-inductance analog* as opposed to the mass-capacitance analog used in previous discussions. The analog of a dual will be called *the dualog*. Using electrical and mechanical systems as examples, Table 8-1 shows the relations between duals, analogs, and dualogs. In a similar manner, the dualog systems of fluid and thermal systems can be formulated.

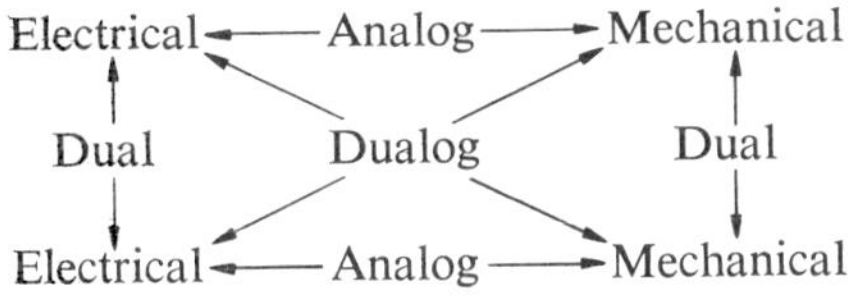

The mass-inductance analogy historically preceded the mass-capacitance analogy [4, 6]. However, in this book, the preference for the mass-capacitance analog is based on the concept of through- and across-variables which results in a structural similarity between systems. There is no point to a discussion concerning which analogy is correct; on a purely mathematical basis there is no preference; the approaches are mathematically analogous. However, the mass-capacitance choice allows one to draw system graphs by inspection, a virtue which is sufficient to justify that approach. Thus the most significant basis for choosing the mass-capacitance analogy is the structural similarity of the various physical systems and not only the mathematical similarity of the elemental equations. Figure 8-45 shows an example of the two analogies; we can see that the structural similarity of the mass-capacitance systems is strong, whereas there is no apparent structural similarity for the mass-

TABLE 8-1

Relations Between Duals, Analogs, and Dualogs

Electrical ←— Analog —→ Mechanical
↕ ↘ ↙ ↕
Dual   Dualog   Dual
↕ ↙ ↘ ↕
Electrical ←— Analog —→ Mechanical

inductance systems (actually the similarity exists through the dual-forming structure, a relation which is not apparent by inspection). Reference 7 gives an abstract mathematical basis for analogies.

Another interesting point concerning the two types of analogs is brought out by considering a nonplanar system. Since a nonplanar graph does not have a dual, the dualog system does not exist. Therefore the mass-inductance analog of a nonplanar mechanical system does not exist. There is never any difficulty in obtaining the mass-capacitance analog, however. Figure 8–46 shows a nonplanar mechanical system. Any attempt to draw the mass-inductance analog of this system will result in short circuits [5].

It is worth pointing out here that this occurrence has nothing to do with the fact that all mass velocities must be referred to ground; it is purely based on the fact that in the mass-inductance analogy, force summations become voltage summations and velocity summations become current summations. In a nonplanar system, the number of independent velocity summations is greater than the number of independent vertices which can be constructed with the given number of elements; i.e., a nonplanar graph does not have a dual graph.

8–8. SUMMARY

The main points developed in this chapter are briefly summarized as follows:

a) The structure of any system is representable by a system graph having b branches, n vertices, and s sources. It may also have separate parts.

b) A complete set of system equations consists of $2b - s$ simultaneous equations for all unknowns.

c) A sufficient set of equations for the primary loop or node variables can be formulated easily by defining the variables so that they satisfy continuity or compatibility automatically. The node method gives $(n - 1)$ simultaneous equations, and the loop method gives $(b - n + 1)$ simultaneous equations.

d) For linear systems, a single differential equation can be derived for any output. The order of this equation equals the number of energy storers which can independently store energy.

e) Once a solution to one dynamic system is obtained, the performance of many others is also known by means of the analog and dual relationships. The dualog affords another type of analogy which is not so general as the direct analog.

REFERENCES

1. H. E. Koenig and W. A. Blackwell, *Electromechanical System Theory*. New York: McGraw-Hill Book Company, Inc., 1961.
2. S. Seshu and M. B. Reed, *Linear Graphs and Electrical Networks*. Reading, Mass.: Addison-Wesley Publishing Co., Inc., 1961.
3. H. M. Trent, "Isomorphisms between Oriented Linear Graphs and Lumped Physical Systems," *J. Acoust. Soc. Am.*, 500–527 (1955).
4. C. A. Nickle, "Oscillographic Solution of Electromechanical Systems," *Trans. AIEE* **44,** 844–856 (1925).
5. W. J. Karplus and W. W. Soroka, *Analog Methods*. New York: McGraw-Hill Book Company, Inc., 1959.
6. F. A. Firestone, "A New Analogy between Mechanical and Electrical Systems," *J. Acoust. Soc. Am.* **4,** 249–267 (1932–33).
7. F. H. Branin, Jr., "An Abstract Mathematical Basis for Network Analogies and Its Significance in Physics and Engineering." IBM Technical Publication, TR 00.781, Poughkeepsie, New York (February 1961).
8. E. A. Guillemin, *Theory of Linear Physical Systems*. New York: John Wiley and Sons, Inc., 1963. See pp. 181–189 for discussion of the order of a system.

PROBLEMS

8–1. Draw linear graphs for the systems in Fig. 8–47(a) through (j). Label all vertices. Orient each graph completely. Formulate the path and vertex equations for each system. Discuss any similarities or differences among the systems grouped in each of Fig. 8–47(a), (b), (e), (f), (g), (i), and (j).

8–2. Draw the linear graphs of the systems in Fig. 8–48(a) and (b).

8–3. For each system in Problem 8–2:

a) draw the linear graph,

b) orient the graph,

c) specify the necessary and sufficient *number* of path and vertex equations required to describe the system, and

d) write a necessary and sufficient set of equations.

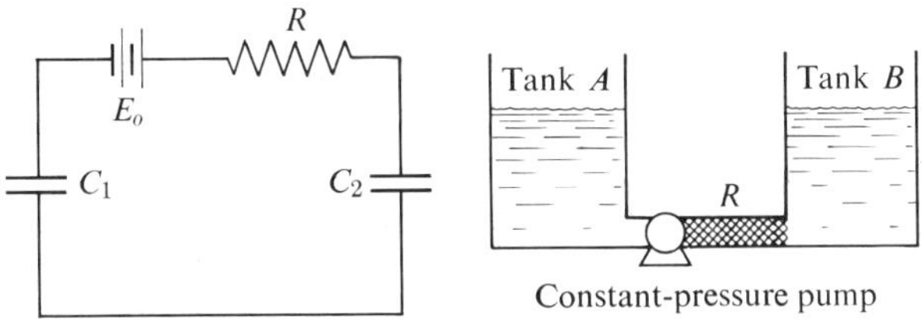

(a)

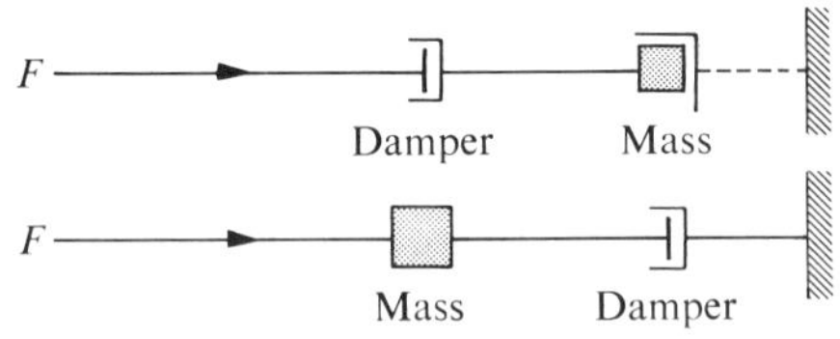

(b)

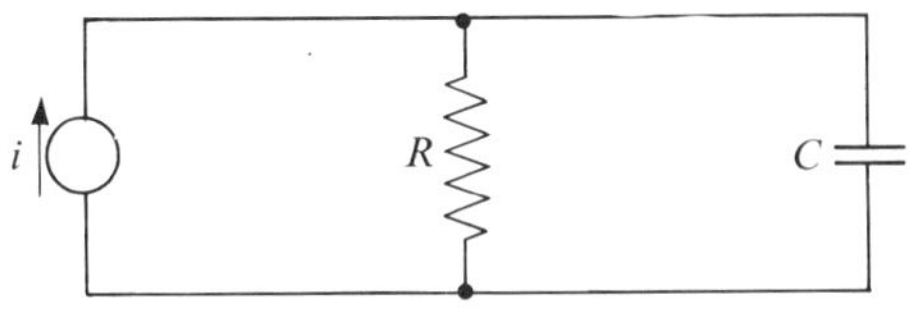

(c)

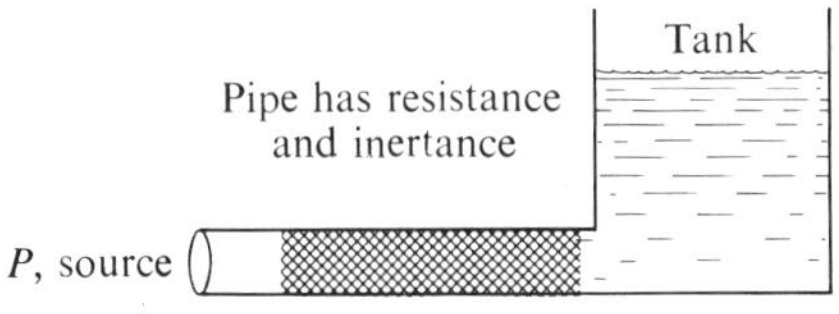

(d)

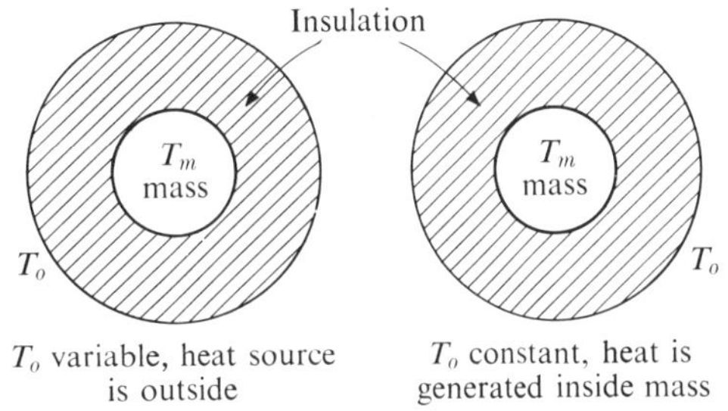

(e)

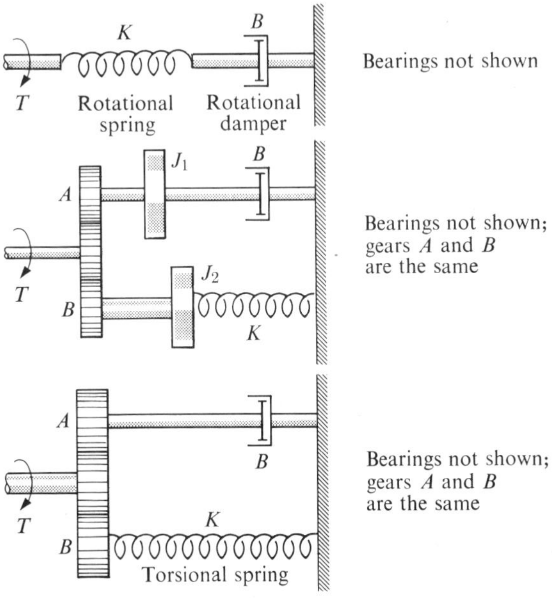

(f)

FIGURE 8–47 (cont.)

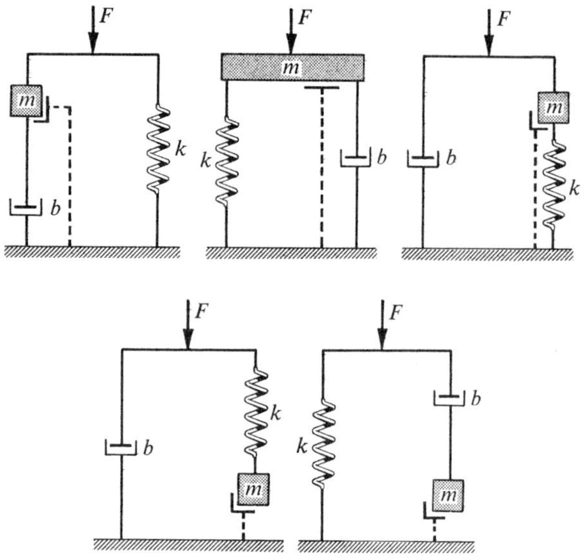

(g)

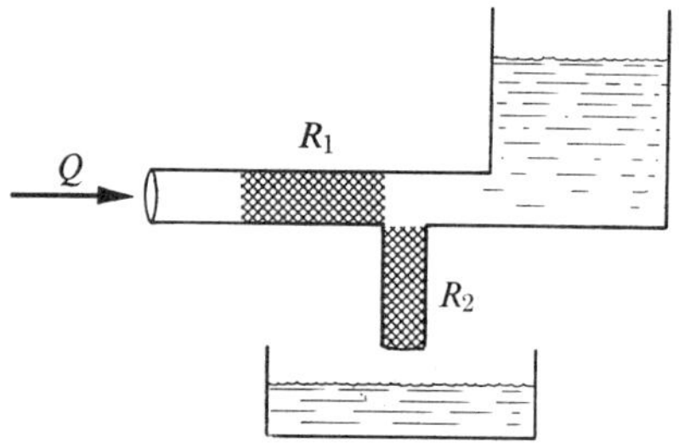

(h)

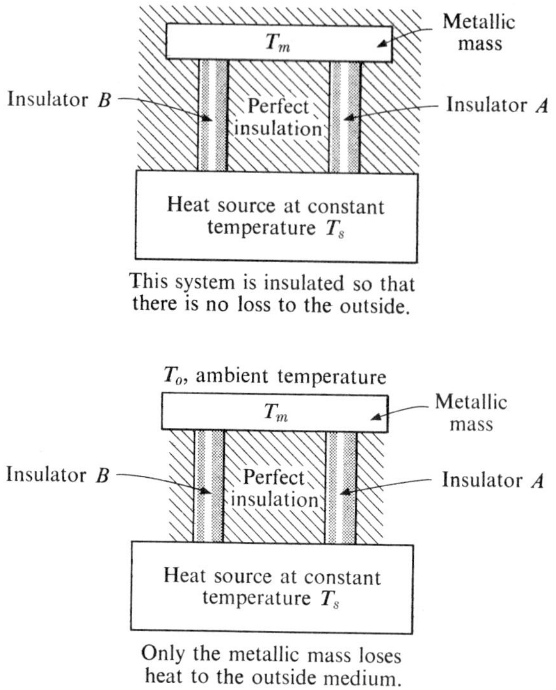

(i)

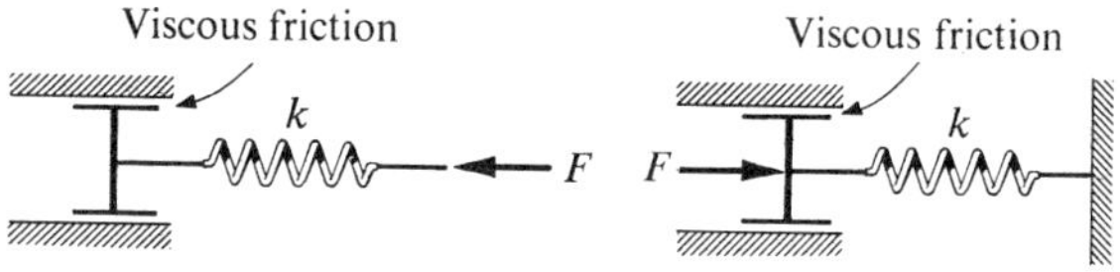

(j)

FIGURE 8–47

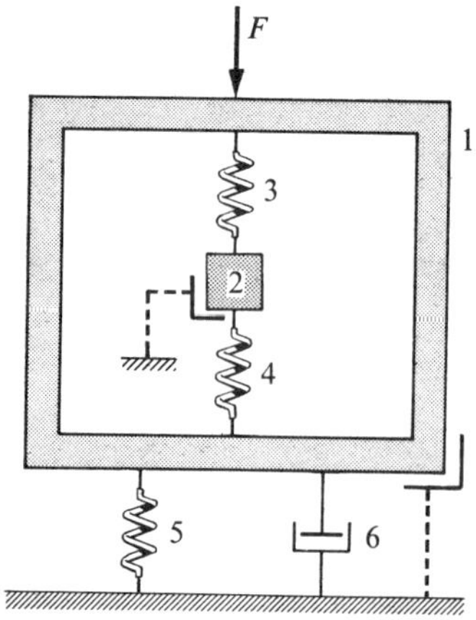

Element 1 = m_1
2 = m_2
3 = k_1
4 = k_2
5 = k_3
6 = b

(a)

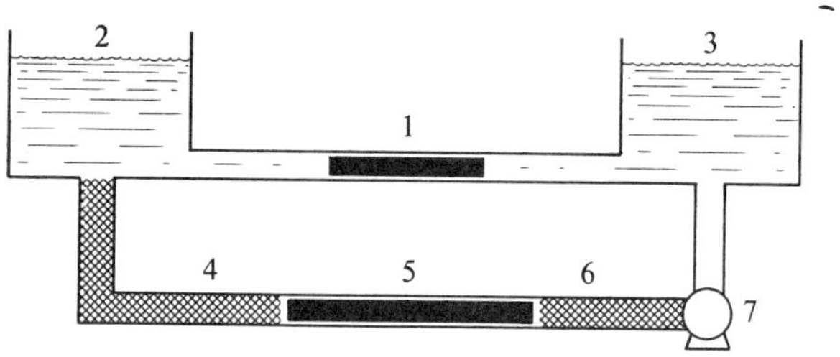

Element 1 = inertance A
2 = tank A
3 = tank B
4 = resistance A
5 = inertance B
6 = resistance B
7 = constant-flow pump

(b)

FIGURE 8–48

8–4. For the system and linear graph in Fig. 8–49:

a) Write the necessary and sufficient set of vertex equations.

b) For the tree in Fig. 8–50 successively add branches to write the necessary and sufficient path equations.

c) Write the necessary and sufficient path equations by successively adding branches to the tree in Fig. 8–51.

d) By selective addition and substitution, show that the path equations in (b) are equivalent to those in (c).

8–5. Given the linear graph in Fig. 8–52.

a) Write the path and vertex equations for this linear graph.

b) Sketch the five systems for which this is the linear graph and label each element:

1) mechanical-translational,
2) mechanical-rotational,
3) electrical,
4) fluid,
5) thermal.

8–6. Using the path and vertex equations developed for the systems in Problem 8–1, set up the system equations for *all* the systems for the variables listed below from Fig. 8–47.

a) Voltage across C_2; pressure at tank B.
b) Velocity of mass in both cases.
c) Current through C.
d) Flow.
e) T_m in both cases.
f) Velocity of B in each case.
g) Velocity of m in each case (except the last one, in which you are to find the velocity of k).
h) Pressure at the tank.
i) T_m in both cases.
j) Velocity of damper.

8–7. Solve for v_6 as a function of v using both Figs. 8–26 and 8–27 of Example 1 in the text. This may be done by solving the final node and loop equations, given in Example 1, by using Cramer's rule. Compare both answers by the loop and node method, noting that $v_6 = R_6 f_6 = R_6 f_C$.

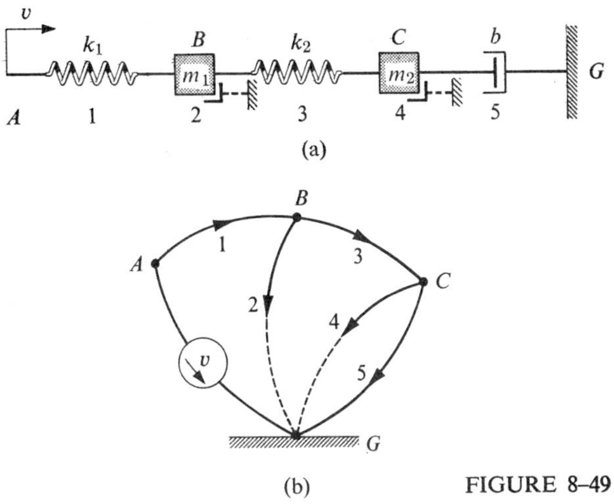

FIGURE 8–49

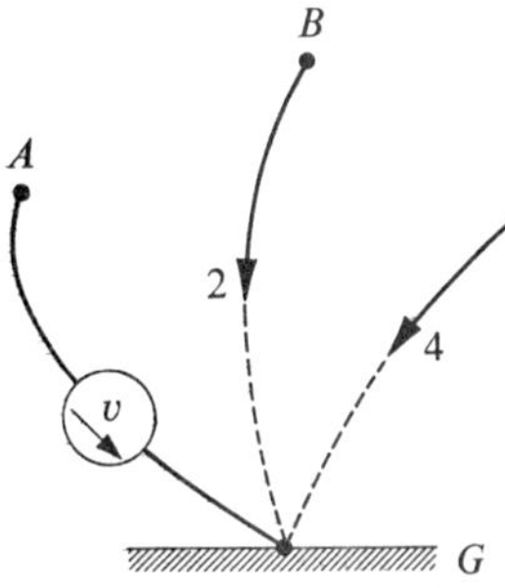

FIGURE 8–50

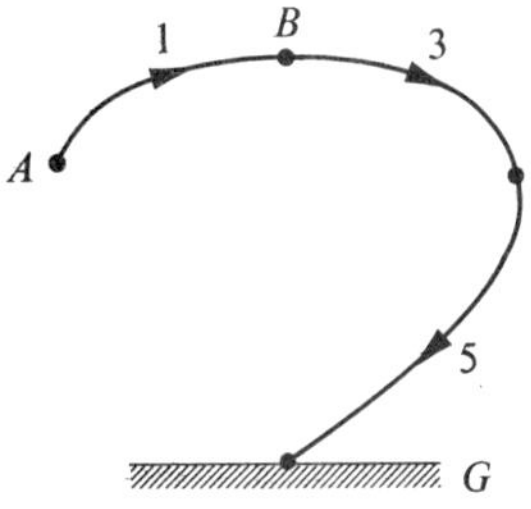

FIGURE 8–51

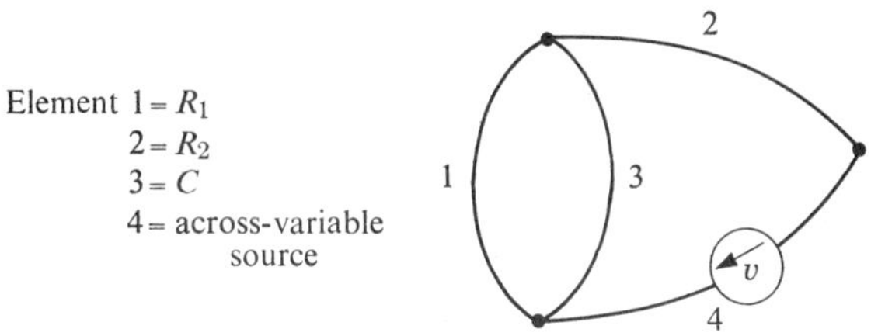

FIGURE 8–52

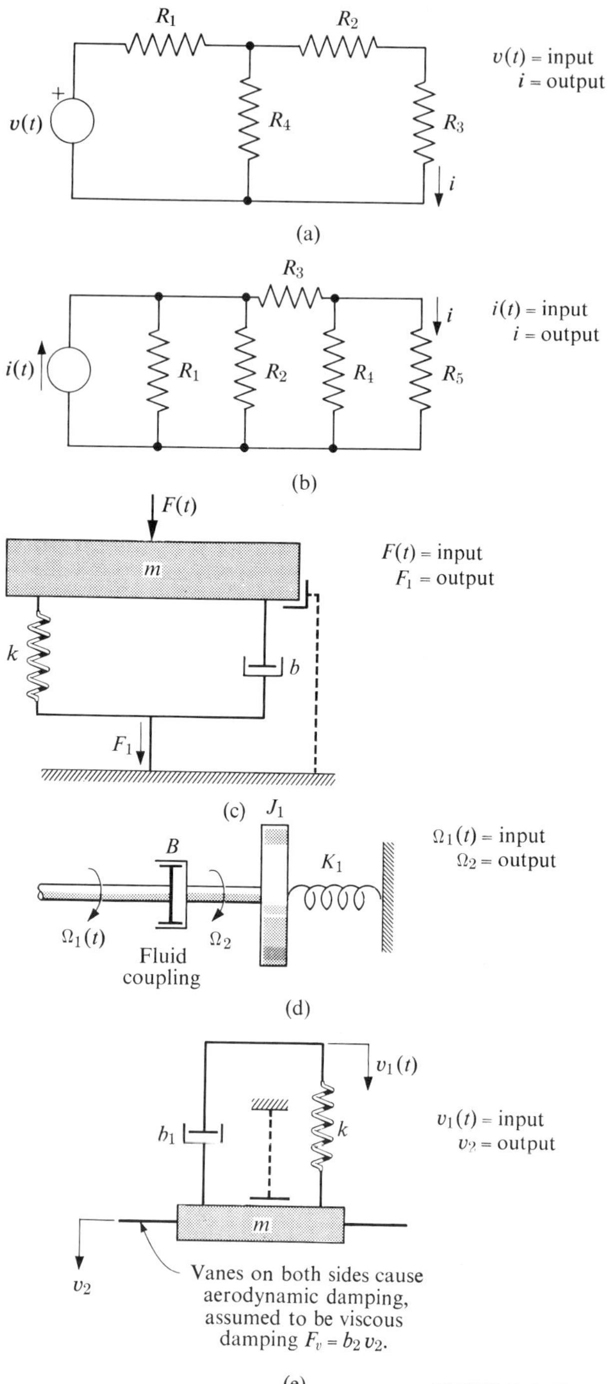

FIGURE 8–53

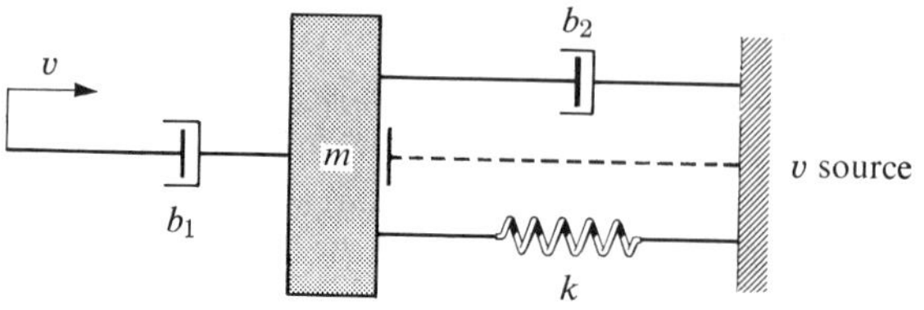

FIGURE 8–54

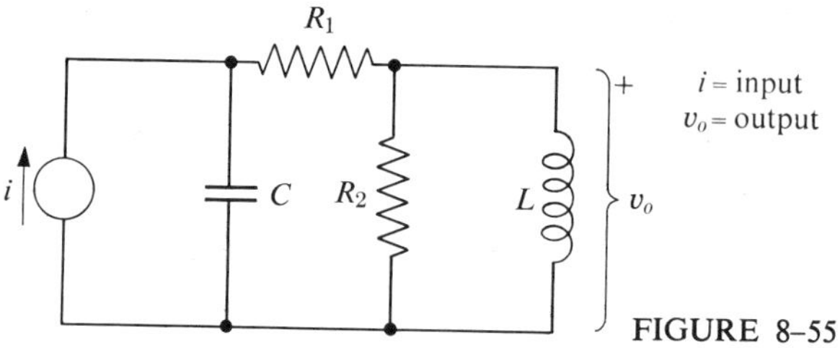

FIGURE 8–55

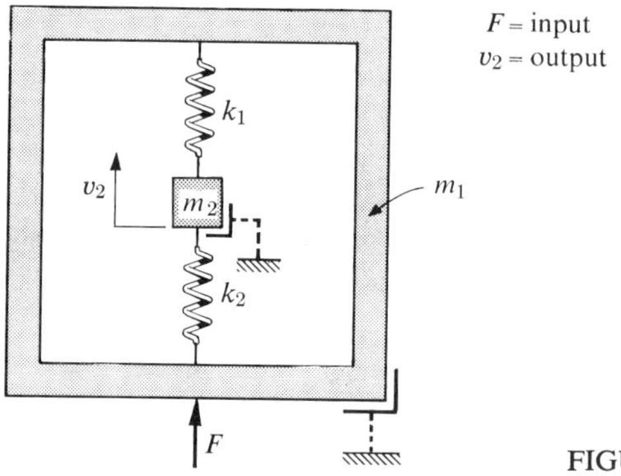

FIGURE 8–56

8–8. For the systems shown in Fig. 8–53(a) through (e) solve for the system equation relating the source input and the variable shown.

8–9. Determine the system equation a) for the mass velocity in the idealized mechanical-translational system in Fig. 8–54; b) for the idealized electrical circuit in Fig. 8–55.

8–10. Develop the system equation for the mechanical translational system in Fig. 8–56.

8–11. Identify the variable which may be considered as an input perturbation at $t = 0$ in the systems in Fig. 8–57(a) and (b), determine the value of the output at $t = 0+$, and solve for the system response.

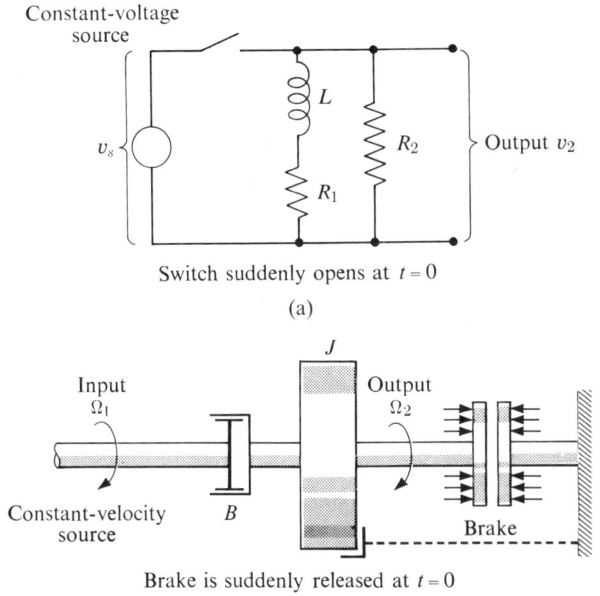

(b) FIGURE 8–57

8–12. A thermal system consisting of a thin plate whose thickness b is small compared to its length and having one side at temperature θ_A and the other side at temperature θ_B (Fig. 8–58a) can be considered to be composed of both thermal resistive and capacitive elements. If one assumes a straight-line variation of temperature across the thickness of the plate, then a first approximation to the actual system is a capacitive element C having the midpoint temperature (under static conditions) $(\theta_A - \theta_B)/2$ and two resistive elements $R/2$. This can be depicted by Fig. 8–58(c). A second approximation of the real system is to consider the plate as composed of two capacitive and three resistive elements (Fig. 8–58b), in which each capacitive element is at the temperature of the one-third thickness and the two-thirds thickness, respectively. Figure 8–58(d) shows the system graph. The second approximation more nearly represents the real (distributed) system, and higher approximations may be obtained by considering the system as composed of more capacitive and resistive elements. For the second approximation system, solve for the system equation relating $q_A(t)$ the input, and $q_o(t) = q_{GB}(t)$, the output. What are the values of R and C in terms of the material properties and dimensions?

8–13. Find how the spring force is related to the driving force F and gravitational field g for the system in Fig. 8–59.

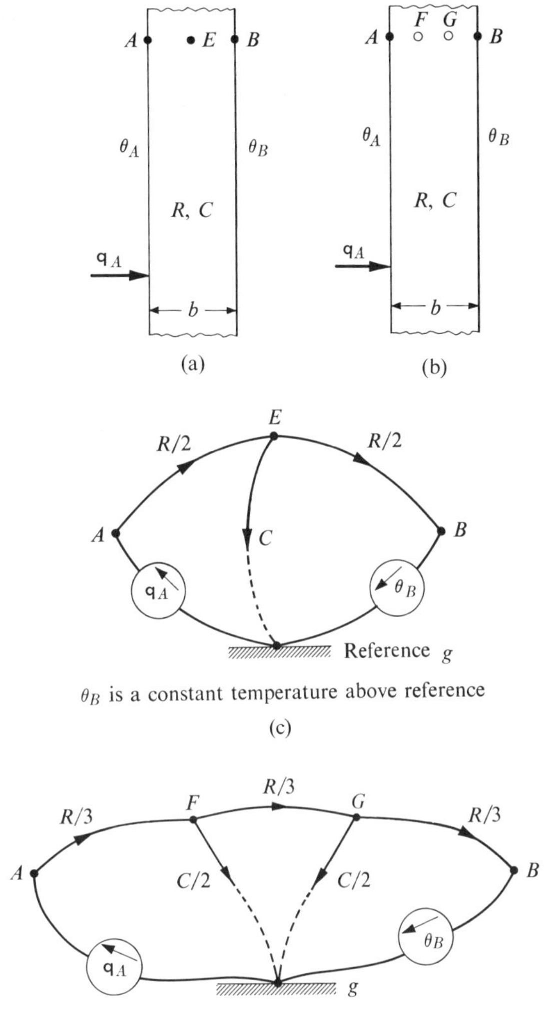

(d) FIGURE 8–58

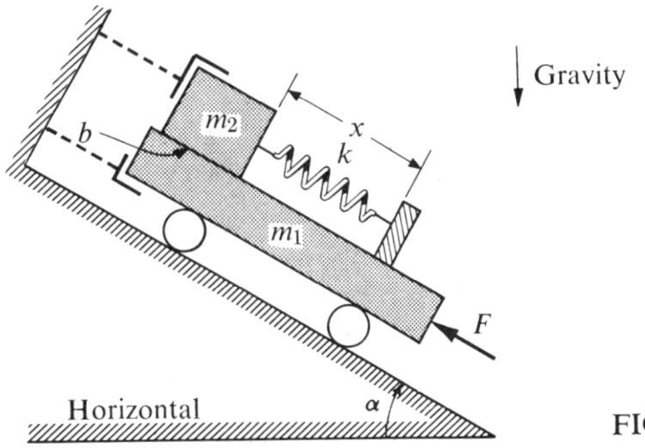

FIGURE 8–59

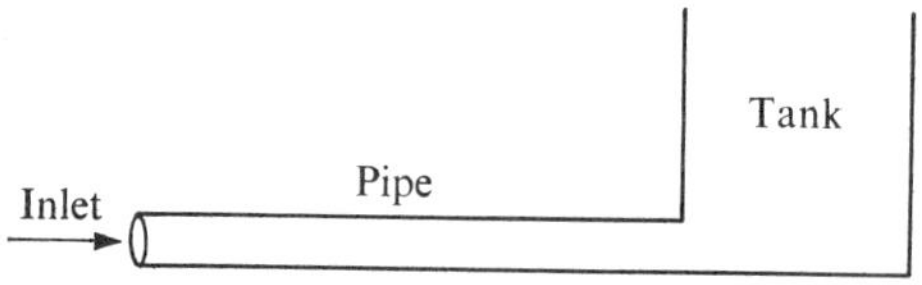

FIGURE 8–60

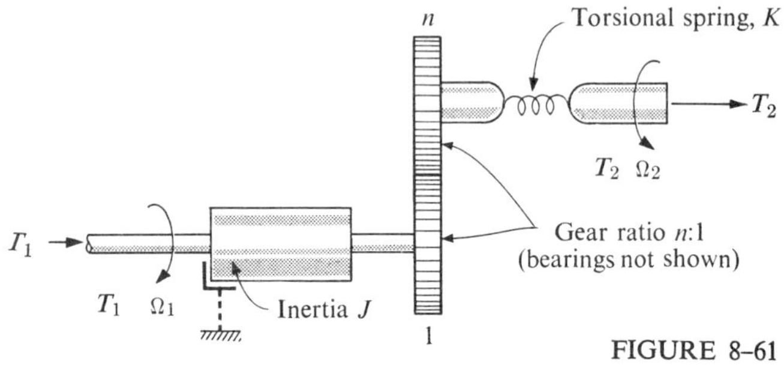

FIGURE 8–61

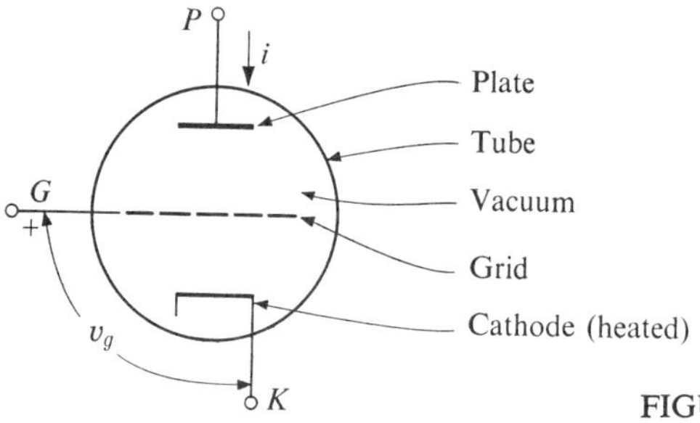

FIGURE 8–62

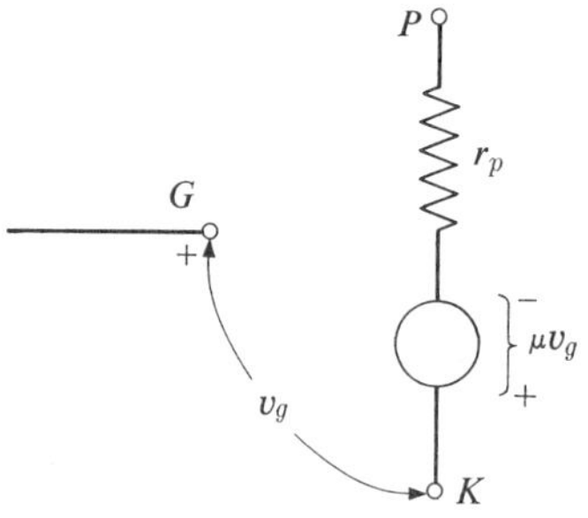

FIGURE 8–63

8–14. Consider the fluid system in Fig. 8–60, where laminar flow of incompressible fluid exists in a pipe which has both resistance and inertance.

The following test is performed. With the system at rest and the tank empty, the inlet pressure is suddenly changed from zero (atmospheric) to some value. The following measurements are made:

inlet pressure change,	10 lb/in^2,
initial flow rate,	zero,
initial rate of increase of flow rate,	25(ft^3/sec)/sec,
total fluid in tank after system settles.	40 ft^3.

After the above steady state is reached, it is found that when the inlet pressure is increased to 25 psi (lb/in^2), the tank overflows after taking 20 ft^3 more fluid, and the steady flow rate then observed is 10 ft^3/sec.

a) Find the fluid resistance of the pipe section.

b) Find the fluid inertance of the pipe section.

c) Find the fluid capacitance of the tank.

8–15. The rotational mechanical system of Fig. 8–61 has two inputs, T_1 and T_2.

a) Derive an integro-differential equation relating the angular velocity Ω_1 to T_1 and T_2 (the velocity Ω_2 should not appear).

b) Derive an integro-differential equation relating the angular velocity Ω_2 to T_1 and T_2 (the velocity Ω_1 should not appear).

8–16. A triode vacuum tube is constructed as shown in Fig. 8–62. Electrons emitted from a heated cathode K move past a screen or grid G to a receiving electrode or plate P. A voltage v_g applied between the cathode and the grid can control the flow of current through the tube much as a faucet or valve controls the flow of water through a pipe. Although the vacuum tube is inherently nonlinear, for sufficiently small variations in currents and voltages, the tube may be represented by a *dependent voltage source* and a resistance called the *plate resistance* r_p (Fig. 8–63). The quantity μ is a constant for small variations of current and voltage. In this circuit no steady current is assumed to enter the tube through the grid terminal.

Any triode will have capacitance between the three electrodes. Figure 8–64 is an equivalent lumped-parameter circuit representation of a single-triode vacuum-tube amplifier. The *load* connected to the tube is resistance R_L, which could represent a loudspeaker, for example.

a) Find the system equation relating the load voltage v_L to the input voltage v_g as a function of time.

b) Find the voltage v_L at time $t = 0+$. *(Continued)*

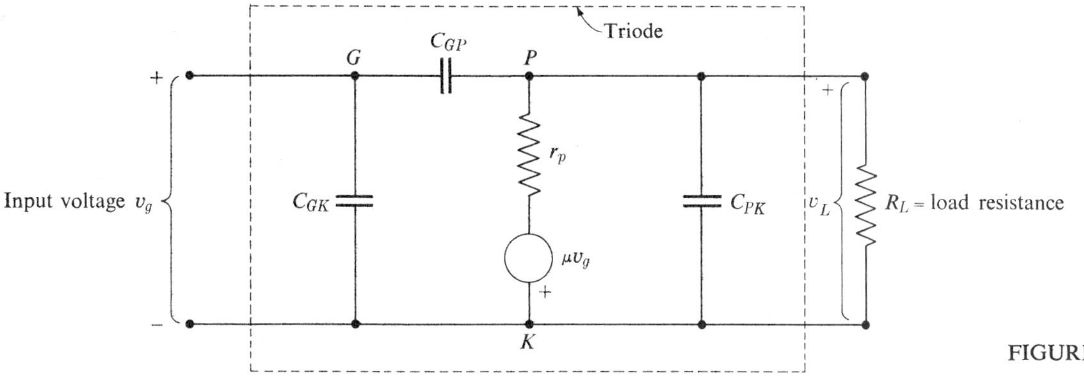

FIGURE 8–64

c) Given that

$$
\begin{aligned}
C_{GK} &= 4\,\mu\mu\text{f}, & \mu &= 15,\\
C_{PK} &= 16\,\mu\mu\text{f}, & r_p &= 70{,}000 \text{ ohms},\\
C_{GP} &= 2\,\mu\mu\text{f}, & R_L &= 140{,}000 \text{ ohms},
\end{aligned}
$$

find the response of the load voltage v_L to a step change of input voltage v_g and sketch.

8–17 For the vibration-absorber example discussed in Section 8–5.3 and shown in Fig. 8–32, derive the system equation (8–34) using the free-body method from mechanics, and compare this method to the formulation by the linear-graph method.

8–18. Consider the circuit shown in Fig. 8–39(a) under the conditions $R_1C_1 = R_2C_2$. Prove the following relative to the element-value degeneracy statements given in the discussion in Section 8–6.2.

a) Derive the system equation for i in terms of the input e, thus demonstrating that this equation is first order when $R_1C_1 = R_2C_2$.

b) Show that the circuit in Fig. 8–39(b) is an equivalent circuit (from the point of view of the current i) by investigating the equation relating v and i.

c) Note that the two initial energies of C_1 and C_2 may be set independently even when $R_1C_1 = R_2C_2$. If the initial conditions in the original network are v_{10} and v_{20}, then $v_0 = v_{10} + v_{20}$ is the initial condition in the equivalent network. Is the initial energy stored in the equivalent network [$v_0 = v_{10} + v_{20}$ applied to $C_1C_2/(C_1 + C_2)$] equal to the initial energy stored in the original network (v_{10} applied to C_1 and v_{20} applied to C_2)? (*Answer:* No, not unless the two initial charges are equal.)

8–19. Prove the statements given in Section 8–6.2 relative to the element-value degeneracy displayed by the circuit shown in Fig. 8–40. Consider:

a) the transient case with zero and nonzero initial conditions (solve for the response of i_1, i_2, and i);

b) the sinusoidal steady-state case.

8–20. Consider the electrical circuit in Fig. 8–65, with the initial conditions $i_1 = i_2 = I_o$ and the parameter symmetry $L_1/R_1 = L_2/R_2$ but $R_1 \neq R_2$. The initial currents are established by an auxiliary set of switches and sources.

a) Determine the system equation for i. Note its order.

b) Solve for the response of i. How many exponentials appear in the solution?

c) Determine the system equation for i_1. Note its order.

d) Solve for the response of i_1. How many exponentials appear in the solution? (Note that the solution for i_2 can also be obtained from this by symmetry. Let $i_1 \to i_2$, $R_1 \to R_2$, $R_2 \to R_1$, etc.) (*Continued*)

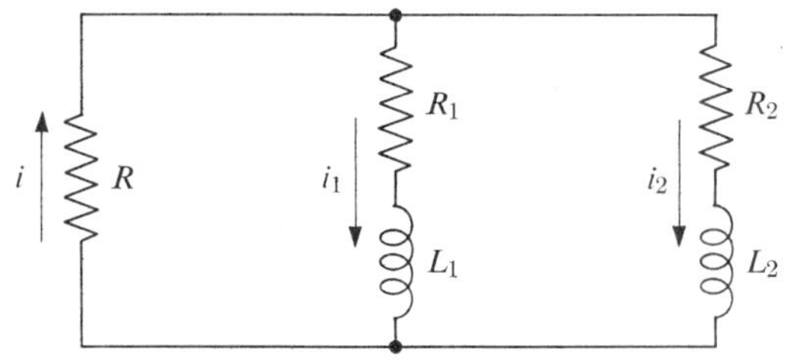

FIGURE 8–65

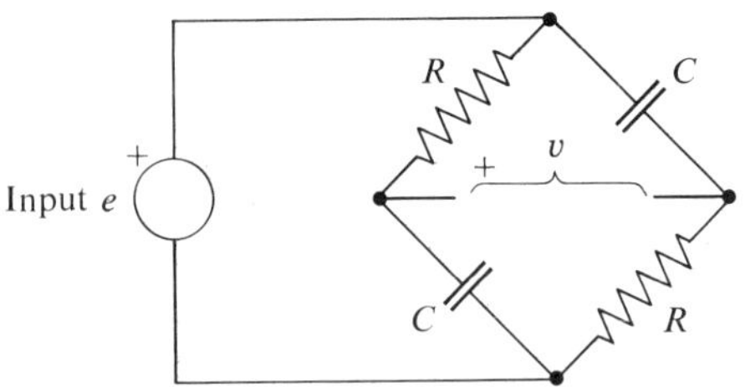

FIGURE 8–66

e) Sketch response curves for i, i_1, and i_2 and note how the second exponential which appears in i_1 and i_2 cancels out in i.

f) From the point of view of the resistance R, the L-R combinations look equivalent to a single R and L in series. What values should these equivalent components have, and what is the initial current in the equivalent L?

g) What is the energy stored in the original circuit initially? Compare to the initial energy in the equivalent L. Are they equal? Why or why not?

h) What initial conditions will cause i_1 and i_2 to have only a first-order response? For this initial condition compare the initial energy stored in the original circuit and the equivalent circuit.

8–21. In the previous problem, it was seen that the initial energy stored in the original circuit and the initial energy stored in the equivalent circuit would not be the same for arbitrary initial conditions. The purpose of this problem is to investigate this discrepancy and account for it. The figure for Problem 8–20 applies.

The final values of the above currents are all zero. The energy dissipated in all the resistances must therefore be equal to the initially stored energy.

a) Calculate the energy dissipated in the three resistances and show that this equals the initially stored energy.

b) Calculate the energy dissipated in R and the equivalent resistance and show that this equals the energy stored initially in the equivalent inductance.

c) If $R_1 = R_2 = 0$, then find the response of i. What is the equivalent inductance, and how much energy should be initially stored in it? Compare to the actual initial energy storage.

d) For $R_1 = R_2 = 0$, all the energy is dissipated in R only. Since the final current in the equivalent inductance must be zero, this energy dissipation must equal the energy stored initially in the equivalent inductance. Since this energy is not equal to the actual initially stored energy in the two inductances, where did all the energy go? Note that now there are no other resistances to take up the slack. [*Hint:* Are the final values of i_1 and i_2 zero?]

e) Could the situation considered in (d) ever occur in the real world? Consider each of mechanical-translational, mechanical-rotational, and fluid systems as well as the electrical system. Discuss completely.

8–22. The bridge circuit shown in Fig. 8–66 is given.

a) Find the system equation for v.

b) Note that the signs of the coefficients of the terms in v (left-hand side) are all positive, whereas this is not so for the signs of the terms in e (right-hand side). This relates to the statements in Section 8–6.3.

c) Consider the sinusoidal steady-state when $e = E \sin \omega t$. What is the amplitude of v when $\omega = 1/RC$?

8–23. Using the general linear system equation given in Eq. (8–52), prove the following theorem by integrating this equation m times (m is the order of the right-hand side) between $0-$ and t (and finally letting $t = 0+$) and then investigating which values of x_r and its derivatives are zero, finite, or infinite at $t = 0$ when x_f is a step function at $t = 0$.

> "Given a linear system equation whose output variable x_r and all its derivatives are zero at $t = 0-$ and a step function input, the only finite, nonzero derivative of x_r between $t = 0-$ and $t = 0+$ is the $(n - m)$th derivative."

This theorem can be used to determine the order of the right-hand side of a system equation by inspection of the block diagram (see Problem 8–24).

8–24. The theorem given in Problem 8–23 can be used to determine the order m of the right-hand side, if the order n of the system is known, with the use of the system block diagram (recall that n is equal to the minimum total number of integrators in the system block diagram). Assume that a step function is applied to the input of a block diagram (if there is more than one input, treat each separately since the order of the input derivatives will be different for each input) and that all the integrators have zero initial conditions. The only nonzero quantity at $t = 0+$ is then the input (and anything not separated from it by integrators). Therefore, according to our theorem, the input must be proportional to the $(n - m)$th derivative of the output being considered. Consider the integrator whose output variable is under consideration. Its input

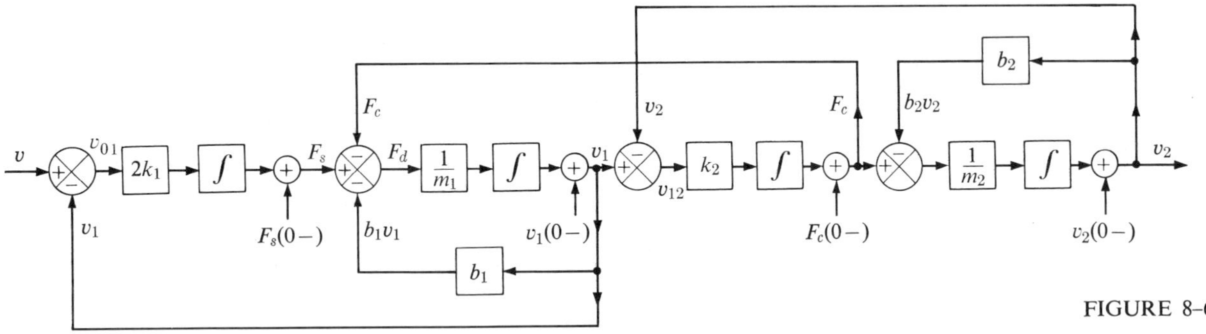

FIGURE 8–67

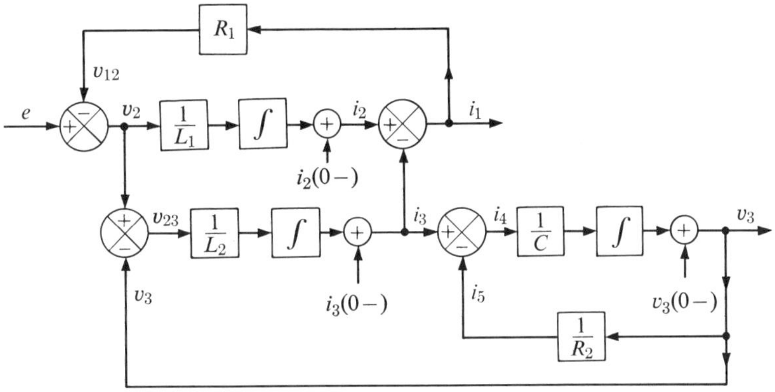

FIGURE 8–68

must be the derivative of this output. Likewise each integrator between the output and input must successively have an input which is proportional to one-order-higher derivative of the output. Thus the minimum number of integrators between (on a path which is opposite to the signal flow arrows) the selected output and the input must equal $(n - m)$. One can also see that the greater the "separation" (in terms of integrators) between the input and output, the lower the order of the right-hand side for that output and the slower this output will respond to changes in the input.

a) The block diagram of the vibration absorber considered in Section 8–5.3 is shown in Fig. 8–67. Use the above method to verify that the order of the right-hand side is 2 when v_1 is the output and compare to Eq. (8–34), which is the derived system equation for v_1. Use the block diagram to determine the order of the right-hand side when v_2, F_s, and F_c are respectively the outputs.

b) The block diagram of the pulse-transformer system considered in Section 8–5.3 is shown in Fig. 8–68. Use the above method to verify that the order of the right-hand side is 2 when i_1 is the output and 1 when v_3 is the output and compare to Eqs. (8–44) and (8–53), which are the respective system equations. Use the block diagram to determine the order of the right-hand side when i_2, i_3, v_2, and v_{23} are respectively the outputs. Is there any output of this circuit for which the order of the right-hand side is zero?

8–25. Derive the equations needed to relate the designated output(s) to the given input(s) for the systems in Fig. 8–69, and draw operational block diagrams. Check the order of both the left- and right-hand sides by using the block-diagram method (see Problem 8–24).

8–26. The system in Fig. 8–70 consists of ideal elements. The input is x, and there are three outputs, x_1, x_2, and F (an external force applied at x).

a) Draw the operational block diagram for this system, using adders, coefficients, and integrators only. Show each of the outputs. *(Continued)*

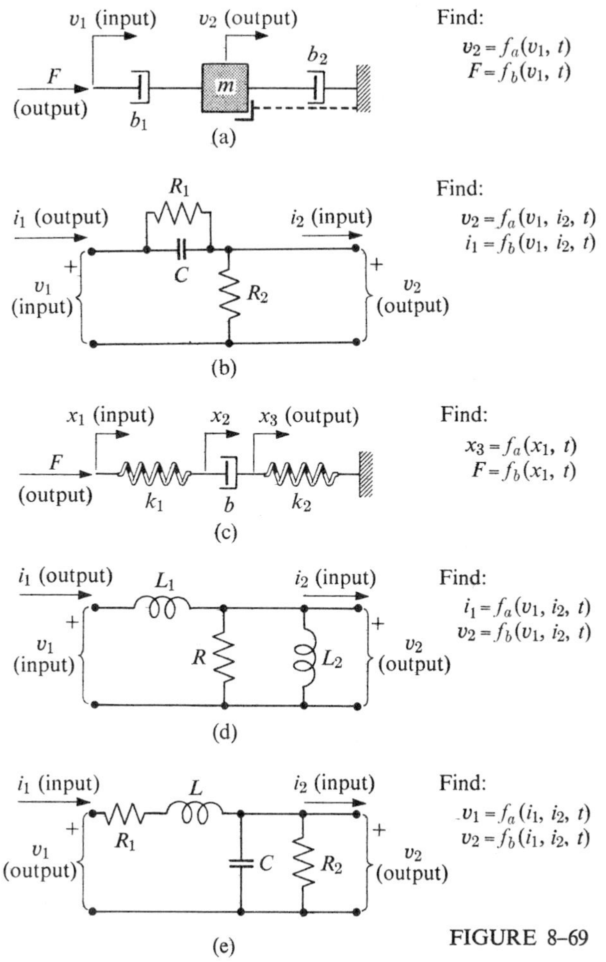

FIGURE 8–69

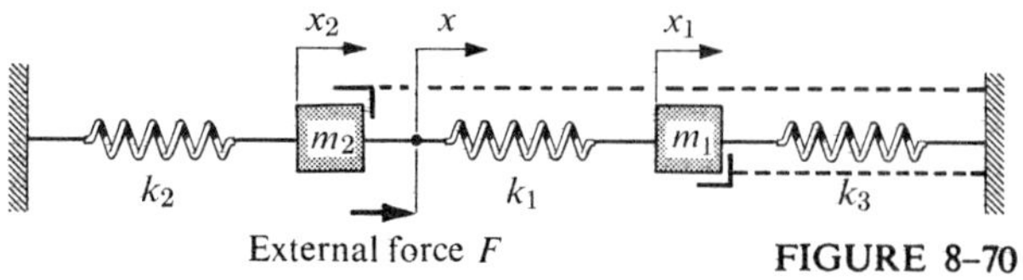

FIGURE 8–70

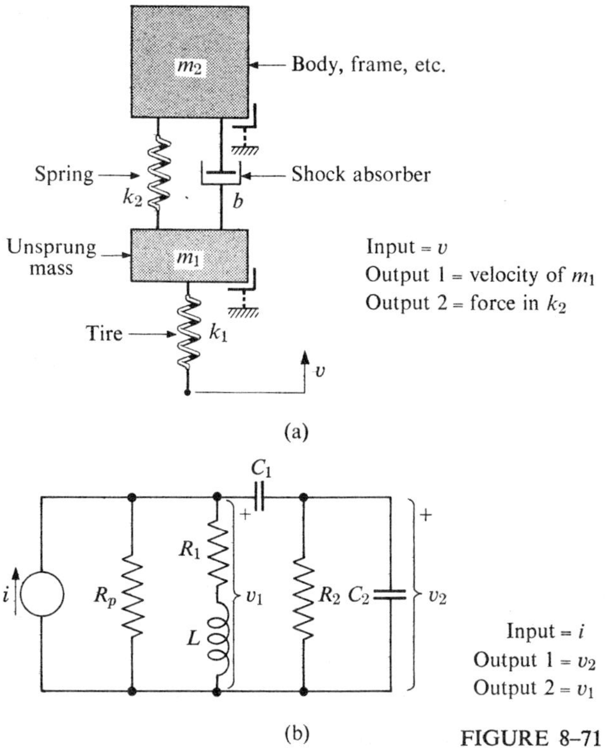

FIGURE 8–71

b) If x is the input, what is the order of the governing equation for the dependent variable?

i) If x_1 is the dependent variable?

ii) If x_2 is the dependent variable?

iii) If F is the dependent variable?

c) What is the order of the independent variable x, in the governing equation for each of the cases in part (b)? (i.e., what is the order of the *right-hand* side of the differential equations?) Explain how you arrived at your answers. (See Problem 8–24.)

8–27. Figure 8–71(a) depicts a simplified model of an automobile suspension in which the whole car has the same vertical motion at all points. Figure 8–71(b) depicts an equivalent circuit of the video amplifier stage of a television receiver.

For each system shown in Fig. 8–71(a) and (b), perform the following steps.

a) Predict the order of the system.

b) Determine the system equation for output 1.

c) Draw an operational block diagram of the system. (No differentiators allowed.)

d) Check the order of the output terms found in (b) with the results of (a) and (c). *(Continued)*

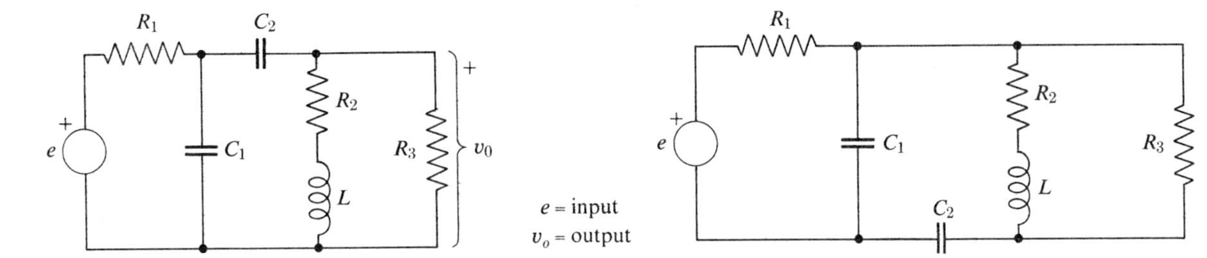

(a) FIGURE 8–72 (b)

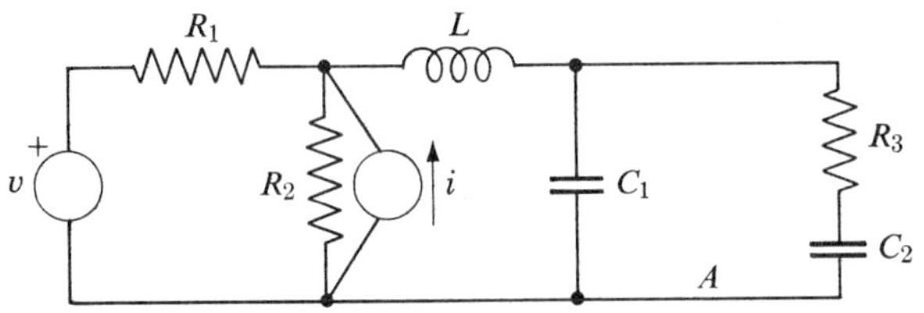

FIGURE 8–73

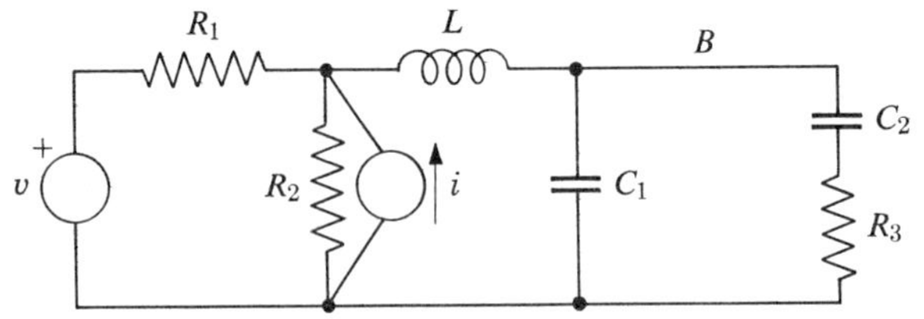

FIGURE 8–74

e) Check the order of the input terms (right-hand side) found in (b) with the results of (c). (See Problem 8–24.)

f) Check the dimensions of all terms.

g) Determine the system equation for output 2. Check that the left-hand side (characteristic equation) is the same as that in (b).

h) Check your result in part (b) by examining the validity of your equation for the following limiting cases. In other words, does the limiting case of your result in (b) agree with the equation which you would get directly from the simplified system?

For the system of Fig. 8–71(a):

CASE 1: $k_2 \to \infty$,

CASE 2: $k_2 \to 0$ and $m_1 \to 0$ simultaneously,

CASE 3: let one other parameter of your choice go to zero or infinity as a limiting case.

For the system of Fig. 8–71(b):

CASE 1: $L \to 0$ and $C_1 \to \infty$ simultaneously,

CASE 2: $C_1 \to 0$,

CASE 3: One other of your choice.

i) Check the low-frequency performance with the result of (b) at low frequencies.

j) Check the high-frequency performance with the result of (b) at high frequencies.

8–28. Develop the system equation for the electrical system in Fig. 8–72(a). Also find the following:

a) The analog of the above electrical system in the (1) rotational mechanical system, and (2) translational mechanical system.

b) The system equation for the analogous rotational system.

c) The analogous quantities to the voltages across the capacitors.

d) Find the translational and rotational mechanical analogs for Fig. 8–72(b): and compare with the original electrical system and its analogs.

8–29. For the electrical system in Fig. 8–73:

a) Draw the system linear graph.

b) Draw a schematic of the mechanical-translational analog.

c) Draw a schematic of the mechanical-rotational analog.

d) Draw a schematic of the lumped-fluid-system analog. (*Continued*)

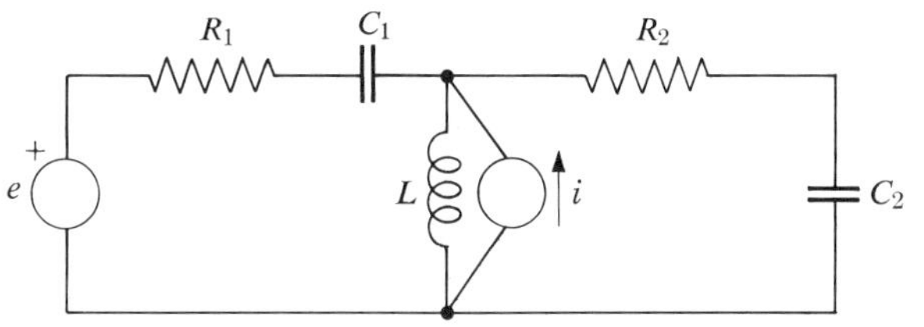

FIGURE 8–75

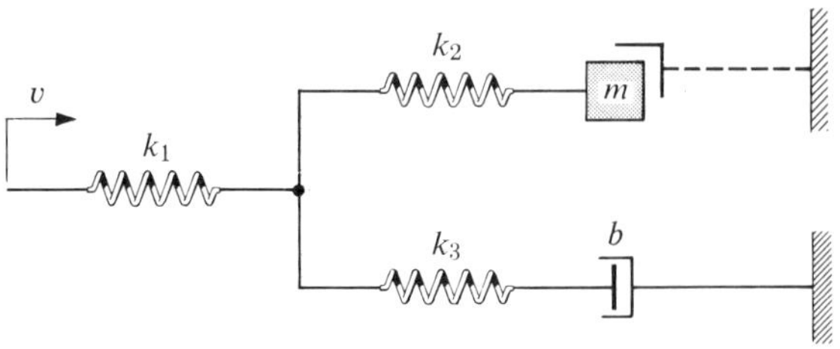

FIGURE 8–76

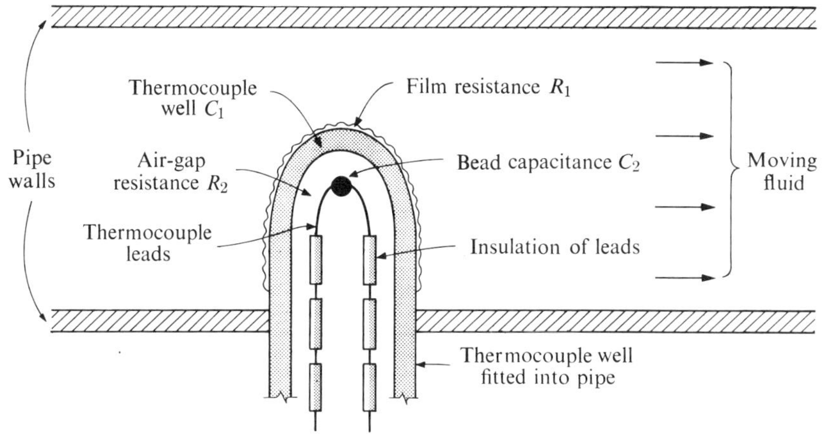

FIGURE 8–77

e) Determine the dual mechanical system of (b) (which is the analog of the original electrical system). The reference vertex in the dual graph may be taken at either end of the single mass element. Show the two systems obtained by these separate choices and discuss their differences.

8–30. In the previous problem the choice of a reference vertex for the graph of the mechanical system was dictated to be point A, since this was the common terminal of the two capacitors C_1 and C_2. The electrical system will perform identically, however, if R_3 and C_2 are interchanged. In this case a different vertex B in the graph would be chosen as the reference vertex (Fig. 8–74). Draw the mechanical analog for this case and compare it to that obtained in Problem 8–29 and discuss the differences.

8–31. Write the system equation, as discussed in the text, for Fig. 8–44(c) by making the duality substitution into the system equation for Fig. 8–44(a).

8–32. Find a physically realizable mechanical-translational analog of the electrical system in Fig. 8–75. [*Hint:* Consider changing the above to another equivalent electrical system and then find the mechanical analog of the modified electrical system.]

8–33. Determine the dual mechanical-translational system of Problem 8–9(a) and write the corresponding system equation by inspection.

8–34. Determine the electrical *analog* to the system in Problem 8–9(a) and write the corresponding system equation by inspection.

8–35. Determine the electrical *dualog* of the system in Problem 8–9(a) and write the corresponding system equation by inspection.

8–36. Find the electrical analog for the mechanical system of Problem 8–10 and write the system equation for the analogous system.

8–37. (a) Find the dual of the mechanical system presented in Problem 8–10. Write the system equation for the dual system.

b) Find the electrical analog (i.e., the dualog of the original system) of the dual found in (a). Write the system equation for this system. Compare to the analog of Problem 8–36.

8–38. (a) Find the dual of the system represented in Problem 8–28. Write the system equation for the dual system.

b) Find the mechanical-translational analog of the dual found in (a) (i.e., the dualog). Write the system equation for this system. Compare to the analog of Problem 8–28.

8–39. Show that the four-terminal transformers shown in Fig. 8–43 actually perform as 1:1 transformers (both mechanical and fluid). What would n:1 four-terminal mechanical and fluid transformers look like?

8–40. Since a mass is the dual of a spring, only those mechanical systems in which all springs are within a single closed path possess a straightforward physically realizable dual (since all second-mass terminals must be common). The system in Fig. 8–76 does not have a directly physically realizable dual. Consider making a dual by means of inserting a four-terminal 1:1 transformer in the dual graph which will allow the realization of an "ungrounded" mass (see Fig. 8–43).

8–41. Figure 8–77 shows a thermocouple in a protective well fitted into a pipe to measure the temperature of a moving fluid. The purpose of the well is to protect the bead of the thermocouple from corrosion by the fluid. The thermocouple is to be used as the error-sensing element in a feedback control loop. It is, therefore, pertinent to know the dynamics of the situation to fully design the device.

To model this system, one may follow the through-variable, heat-flow, as it passes through discrete elements toward the thermocouple bead, which is assumed to be at a lower temperature. Temperature, of course, is the across-variable. One may first assume that a film resistance, R_1, exists on the surface of the thermocouple well. From the well surface, inside the film resistance element, the well itself may be assumed to be of such size and material that it has a finite heat capacitance, C_1. No resistance element will be assumed here for the heat passing through the well, i.e., the outside and inside wall temperatures of the well are assumed to have the same value. From the inside wall of the well the heat must pass through a finite air gap of resistance R_2. The heat then goes into the thermocouple bead, which, although of small mass, does have a finite heat capacity, C_2. It is assumed that the thermocouple leads do not carry away any significant amount of heat.

Draw the system graph and find the system equation for the bead temperature for a change in fluid temperature input.

9 COMPLEX NUMBERS

9–1. INTRODUCTION

All the variables in real systems are real quantities, and to be measured they must be observed physically in terms of real-number scales on measuring instruments and meters. Nevertheless, it is often convenient and sometimes necessary to use *complex numbers* in the analysis of dynamic systems. Complex numbers arise in the determination of the factors of certain polynomials (historically this was the first use of complex numbers), and they are very useful in describing certain types of dynamic phenomena.

The roots of the equation $x^2 + 1 = 0$ are $x = \pm\sqrt{-1}$, for example, so that the factors of $(x^2 + 1)$ are $(x + \sqrt{-1})$ and $(x - \sqrt{-1})$. A number whose square is a negative real number is not encountered in ordinary measurements of physical phenomena. Hence it is called an *imaginary number*. It is perhaps unfortunate that the name "imaginary number" was chosen and is still widely used for such numbers, because imaginary numbers are as important as real numbers in the mathematical analysis of real systems.

The use of complex numbers, which consist of combinations of real and imaginary numbers, often leads to a great deal of simplification in working with harmonic and exponential functions in the analysis of dynamic systems. A study or review of the properties of these numbers can be very helpful in preparing for work which involves finding the solutions of differential equations.

The symbol j will be used here to represent the square root of minus one:

$$j = \sqrt{-1}. \tag{9–1}$$

Thus a real number may be represented by a letter such as a, and an imaginary number will then be represented by a letter preceded by or followed by j, for example, jb or bj. A complex number, which is the sum of a real number and an imaginary number, may then be written

$$\mathbf{Z}_1 = a + jb. \tag{9–2}$$

Just as real numbers may be represented by points along a real scale, imaginary numbers may be represented by points along an imaginary scale. It is customary to employ the x- and y-axes of the conventional cartesian coordinates for the real and imaginary axes, respectively. When this is done (Fig. 9–1), it is evident that a complex number can be thought of as a vector having the properties of magnitude and direction. Because complex numbers will be used here as vectors, they will be designated in vector fashion (bold-faced capital letters). The real and imaginary parts will be denoted by particular values of x and y, respectively:

$$\mathbf{Z}_1 = x_1 + jy_1. \tag{9–3}$$

The *magnitude* or *absolute value* of $\mathbf{Z}_1$ is defined as the length of the hypotenuse of the triangle in

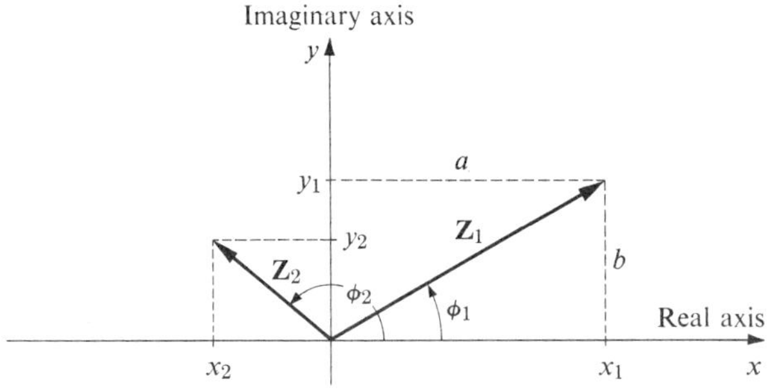

FIG. 9–1. Cartesian-coordinate representation of complex numbers.

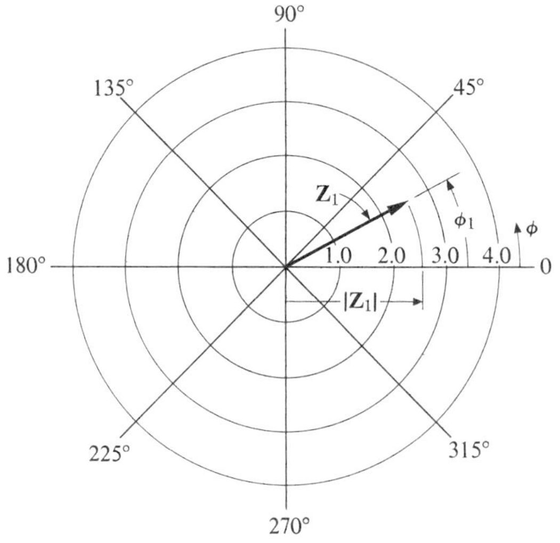

FIG. 9–2. Polar-coordinate representation of a complex number $\mathbf{Z}_1$.

Fig. 9–1, and the *angle* of $\mathbf{Z}_1$ is the angle between the hypotenuse and the real axis:

$$|\mathbf{Z}_1| = \sqrt{x_1^2 + y_1^2} = \text{magnitude of } \mathbf{Z}_1, \tag{9-4}$$

$$\phi_1 = \tan^{-1}(y_1/x_1) = \text{angle of } \mathbf{Z}_1. \tag{9-5}$$

This naturally leads to the possibility of the polar-coordinate representation (Fig. 9–2), which can be plotted readily on polar graph paper. The cartesian coordinates are readily recovered from the polar coordinates by the use of the trigonometric identities:

$$x_1 \equiv |\mathbf{Z}_1| \cos \phi_1, \tag{9-6}$$

$$y_1 \equiv |\mathbf{Z}_1| \sin \phi_1. \tag{9-7}$$

Two complex numbers are *equal* only if (a) their real parts are equal and their imaginary parts are equal or (b) their magnitudes are equal and their angles are equal. Thus if

$$\mathbf{Z}_1 = x_1 + jy_1 \quad \text{and} \quad \mathbf{Z}_2 = x_2 + jy_2,$$

then

$$\mathbf{Z}_1 = \mathbf{Z}_2$$

only if

$$x_1 = x_2 \quad \text{and} \quad y_1 = y_2,$$

or if

$$|\mathbf{Z}_1| = |\mathbf{Z}_2| \quad \text{and} \quad \phi_1 = \phi_2.$$

Purely real or purely imaginary numbers may be considered as special cases of complex numbers having appropriate parts equal to zero. For example,

$$a = a + j0; \qquad jb = 0 + jb.$$

Two complex numbers are *conjugate* if (a) their real parts are equal and their imaginary parts are equal but of opposite sign or if (b) their magnitudes are equal and their angles are of opposite sign. The vector $\mathbf{Z}_1^* = x_1 - jy$ is the conjugate of the vector $\mathbf{Z}_1 = x_1 + jy_1$ (Fig. 9–3).

9–2. BASIC OPERATIONS WITH COMPLEX NUMBERS

9–2.1 Addition

Complex numbers are added by simply collecting terms of the same type in cartesian form:

$$\mathbf{Z}_1 + \mathbf{Z}_2 = (x_1 + jy_1) + (x_2 + jy_2)$$
$$= (x_1 + x_2) + j(y_1 + y_2). \tag{9-8}$$

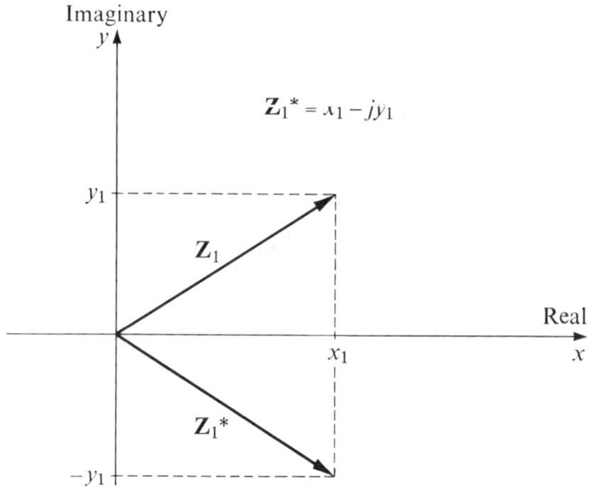

(a)

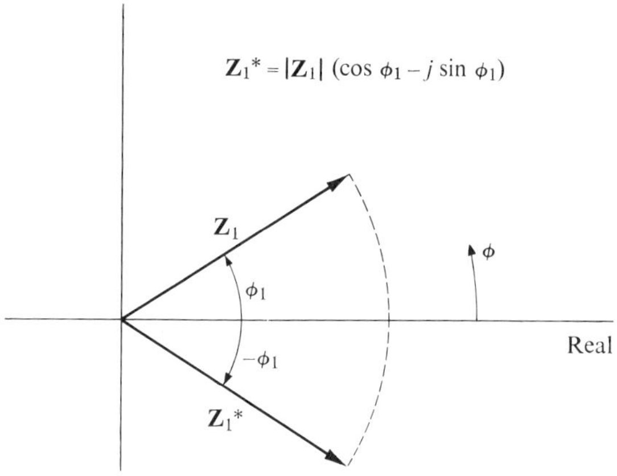

(b)

FIG. 9–3. Illustration of conjugate complex numbers. (a) In cartesian coordinates. (b) In polar coordinates.

This is illustrated graphically with cartesian coordinates in Fig. 9–4. If the complex numbers to be added are given in polar coordinates, they should be converted into cartesian coordinates and then added. As evident from Fig. 9–4, the sum of a pair of complex numbers is readily attained graphically by constructing a parallelogram on the vectors representing the two numbers to be added.

When a pair of conjugate complex numbers are added, the sum is a real number:

$$Z_1 + Z_1^* = 2x_1. \tag{9–9}$$

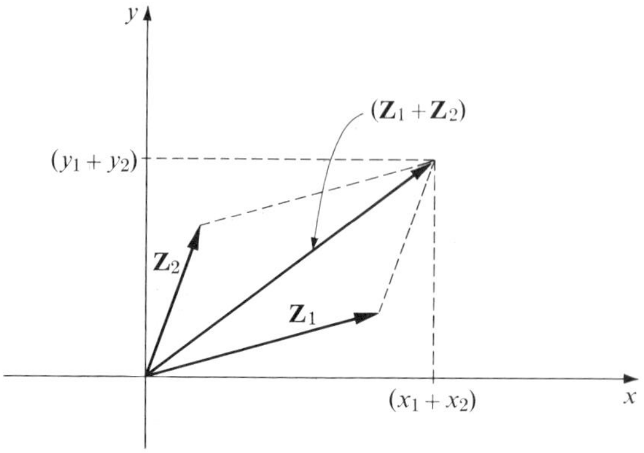

FIG. 9–4. Graphical addition of two complex numbers.

When one of a pair of conjugate complex numbers is subtracted from the other, the difference is an imaginary number:

$$\mathbf{Z}_1 - \mathbf{Z}_1^* = j2y_1 \tag{9–10}$$

or

$$-\mathbf{Z}_1 + \mathbf{Z}_1^* = -j2y_1. \tag{9–11}$$

9–2.2 Multiplication

The multiplication of two complex numbers in cartesian coordinates may be accomplished by using the ordinary rules of algebra for the product of two binomials:

$$\begin{aligned}\mathbf{Z}_1\mathbf{Z}_2 &= (x_1 + jy_1)(x_2 + jy_2) \\ &= x_1x_2 + jx_1y_2 + jx_2y_1 + j^2y_1y_2\end{aligned}$$

and making use of the identity $j^2 \equiv -1$,

$$\mathbf{Z}_1\mathbf{Z}_2 = (x_1x_2 - y_1y_2) + j(x_1y_2 + x_2y_1). \tag{9–12}$$

As will be shown next, this is rather cumbersome in comparison to the multiplication of two complex numbers in polar coordinates. However, it is readily seen from Eq. (9–12) that the product of a pair of conjugate complex numbers is a real number:

$$\mathbf{Z}_1\mathbf{Z}_1^* = x_1^2 + y_1^2. \tag{9–13}$$

Making use of the trigonometric identities (9–6) and (9–7), we see that the product of two complex numbers may be given by:

$$\begin{aligned}\mathbf{Z}_1\mathbf{Z}_2 &= |\mathbf{Z}_1|\,(\cos\phi_1 + j\sin\phi_1)\,|\mathbf{Z}_2|\,(\cos\phi_2 + j\sin\phi_2)\\ &= |\mathbf{Z}_1|\,|\mathbf{Z}_2|\,[(\cos\phi_1\cos\phi_2 - \sin\phi_1\sin\phi_2)\\ &\qquad + j(\cos\phi_1\sin\phi_2 + \sin\phi_1\cos\phi_2)].\end{aligned}$$

Making use of the identities

$$\begin{aligned}\cos(\phi_1+\phi_2) &= \cos\phi_1\cos\phi_2 - \sin\phi_1\sin\phi_2,\\ \sin(\phi_1+\phi_2) &= \cos\phi_1\sin\phi_2 + \sin\phi_1\cos\phi_2,\end{aligned}$$

we obtain

$$\mathbf{Z}_1\mathbf{Z}_2 = |\mathbf{Z}_1|\,|\mathbf{Z}_2|\,[\cos(\phi_1+\phi_2) + j\sin(\phi_1+\phi_2)] \tag{9–14}$$

$$\mathbf{Z}_1\mathbf{Z}_2 = \mathbf{Z}_3, \tag{9–15}$$

where

$$|\mathbf{Z}_3| = |\mathbf{Z}_1|\,|\mathbf{Z}_2|, \tag{9–16}$$

$$\angle\mathbf{Z}_3 = \phi_3 = \phi_1 + \phi_2.† \tag{9–17}$$

Thus the product of two complex numbers in polar coordinates is a complex number whose magnitude equals the product of the magnitudes of the two numbers and whose angle equals the sum of the angles of the two numbers. Thus multiplication of two complex numbers is more easily accomplished in polar coordinates than in cartesian coordinates. (Fig. 9–5).

The product of a pair of conjugate complex numbers is a real number equal to the square of the magnitude of one of the complex numbers:

$$\mathbf{Z}_1\mathbf{Z}_1^* = \left(\sqrt{x_1^2 + y_1^2}\right)^2 = x_1^2 + y_1^2. \tag{9–18}$$

9–2.3 Division

The quotient of two complex numbers is found by applying the ordinary rules of algebra to the product of two complex numbers:

$$\mathbf{Z}_4\mathbf{Z}_2 = \mathbf{Z}_1,\quad |\mathbf{Z}_4|\,|\mathbf{Z}_2| = |\mathbf{Z}_1|,\quad \phi_4 + \phi_2 = \phi_1,$$

$$\mathbf{Z}_4 = \mathbf{Z}_1/\mathbf{Z}_2, \tag{9–19}$$

† The symbol $\angle$ means "the angle of" the vector.

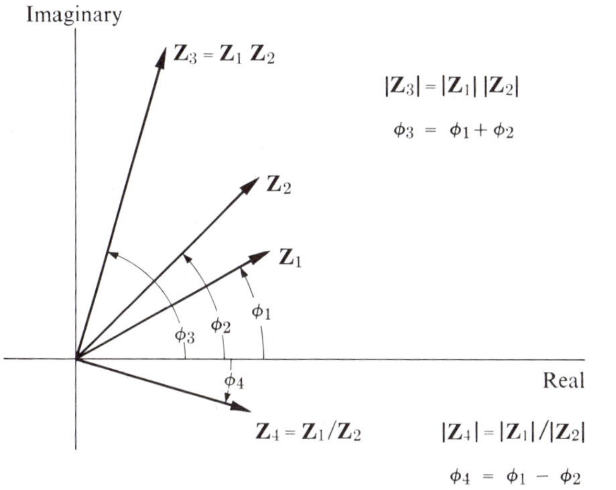

FIG. 9–5. Multiplying and dividing complex numbers in polar coordinates.

where

$$|\mathbf{Z}_4| = |\mathbf{Z}_1|/|\mathbf{Z}_2|, \tag{9–20}$$

$$\phi_4 = \phi_1 - \phi_2. \tag{9–21}$$

Thus the quotient of two complex numbers is a complex number whose magnitude is the quotient of the magnitudes of the two numbers and whose angle is the difference of the angles of the two numbers.

Occasionally it is necessary to find the quotient of two complex numbers in cartesian coordinates. This operation can be expedited considerably by multiplying the numerator and the denominator by the conjugate of the denominator, thereby converting the denominator into a real number:

$$\frac{\mathbf{Z}_1}{\mathbf{Z}_2} = \frac{x_1 + jy_1}{x_2 + jy_2}, \tag{9–22}$$

$$\frac{\mathbf{Z}_1\mathbf{Z}_2^*}{\mathbf{Z}_2\mathbf{Z}_2^*} = \frac{(x_1 + jy_1)(x_2 - jy_2)}{x_2^2 + y_2^2}, \tag{9–23}$$

or

$$\frac{\mathbf{Z}_1}{\mathbf{Z}_2} = \frac{(x_1x_2 + y_1y_2) + j(y_1x_2 - x_1y_2)}{x_2^2 + y_2^2}. \tag{9–24}$$

The operations of multiplying and dividing complex numbers are illustrated in Fig. 9–5.

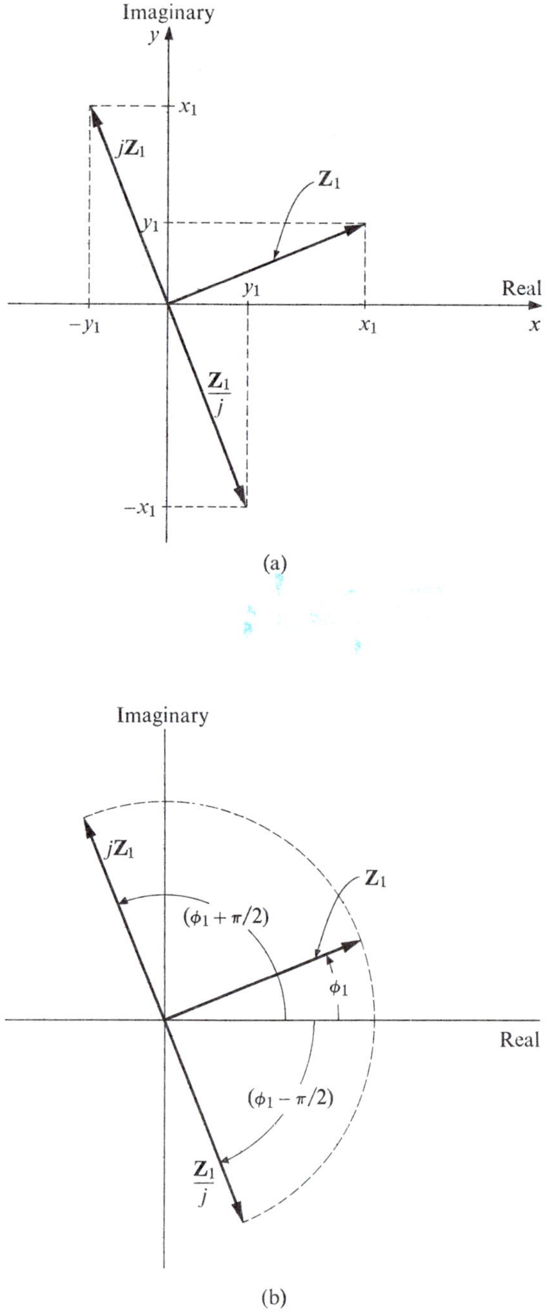

FIG. 9–6. Multiplying and dividing a complex number by j. (a) In cartesian coordinates. (b) In polar coordinates.

The quotient of a pair of conjugate complex numbers is a complex number of unit magnitude and an angle twice the angle of the complex number in the numerator of the quotient:

$$\mathbf{Z}_1/\mathbf{Z}_1^* = \mathbf{Z}_3, \tag{9–25}$$

where

$$|\mathbf{Z}_3| = 1 \tag{9–26}$$

and

$$\angle \mathbf{Z}_3 = \phi_3 = 2\phi_1. \tag{9–27}$$

Another special case of importance is multiplication of a complex number by j, which is shown as follows to be equivalent to counterclockwise rotation of the complex number vector by an angle of $\pi/2$ (i.e., 90°):

$$\mathbf{Z}_1 = x_1 + jy_1,$$
$$j\mathbf{Z}_1 = jx_1 + j^2y_1 = -y_1 + jx_1, \tag{9–28}$$

or

$$j\mathbf{Z}_1 = \mathbf{Z}_3, \tag{9–29}$$

where

$$|\mathbf{Z}_3| = |j|\,|\mathbf{Z}_1| = |\mathbf{Z}_1| \tag{9–30}$$

and

$$\angle \mathbf{Z}_3 = \angle j + \angle \mathbf{Z}_1 = \pi/2 + \phi_1. \tag{9–31}$$

Similarly, it can be shown that dividing a complex number by j is equivalent to clockwise rotation of the vector representing the complex number by $\pi/2$:

$$\mathbf{Z}_1 \div j = \mathbf{Z}_3, \tag{9–32}$$

where

$$|\mathbf{Z}_3| = |\mathbf{Z}_1| \tag{9–33}$$

and

$$\angle \mathbf{Z}_3 = \phi_1 - \pi/2. \tag{9–34}$$

The operations of multiplying and dividing by j are illustrated in Fig. 9–6.

9–3. EXPONENTIAL FORM OF COMPLEX NUMBERS

The rules which have been developed for adding angles when multiplying one complex number by another in polar coordinates are reminiscent of the rules for addition of exponents when multiplying two exponentials. For instance, $u^a \cdot u^b = u^{a+b}$. This leads one to suspect that complex numbers may be capable of some form of exponential representation. This turns out to be the case if the base e of the Naperian logarithm is used for the exponential function. To show this, one may employ Eq. (9–35) to represent a complex number $\mathbf{Z}$,

$$\mathbf{Z} = |\mathbf{Z}|\,(\cos\phi + j\sin\phi), \tag{9–35}$$

and then use the power series expansions for $\cos\phi$ and $\sin\phi$ to obtain a power series expansion of $\cos\phi + j\sin\phi$:

$$\cos\phi = 1 - \frac{\phi^2}{2!} + \frac{\phi^4}{4!} - \frac{\phi^6}{6!} + \cdots, \tag{9–36}$$

$$\sin\phi = \phi - \frac{\phi^3}{3!} + \frac{\phi^5}{5!} - \frac{\phi^7}{7!} + \cdots, \tag{9–37}$$

$$\cos\phi + j\sin\phi = 1 + (j\phi) + \frac{(j\phi)^2}{2!} + \frac{(j\phi)^3}{3!} + \frac{(j\phi)^4}{4!} + \cdots \tag{9–38}$$

The right-hand side of Eq. (9–38) has the form of a power series expansion of an exponential function of e:

$$e^x = 1 + x + \frac{x^2}{2!} + \frac{x^3}{3!} + \cdots \tag{9–39}$$

or

$$e^{j\phi} = 1 + (j\phi) + \frac{(j\phi)^2}{2!} + \frac{(j\phi)^3}{3!} + \cdots \tag{9–40}$$

Therefore we obtain *Euler's formula*

$$e^{j\phi} = \cos\phi + j\sin\phi, \tag{9–41}$$

so that we can express a complex number in polar coordinates by the exponential function

$$\mathbf{Z} = |\mathbf{Z}|e^{j\phi} \tag{9–42}$$

rather than having to use words to say $\mathbf{Z} = |\mathbf{Z}|$ at an angle ϕ, or having to write Eq. (9–35) in terms of $\sin\phi$ and $\cos\phi$. The operation of multiplying one complex number by another can now be carried out in polar coordinates by using the ordinary rules of algebra for exponentials:

$$\mathbf{Z}_1\mathbf{Z}_2 = |\mathbf{Z}_1|e^{j\phi_1}|\mathbf{Z}_2|e^{j\phi_2} = |\mathbf{Z}_1|\,|\mathbf{Z}_2|e^{j(\phi_1+\phi_2)}. \tag{9–43}$$

Similarly, for division

$$\frac{\mathbf{Z}_1}{\mathbf{Z}_2} = \left|\frac{\mathbf{Z}_1}{\mathbf{Z}_2}\right| \frac{e^{j\phi_1}}{e^{j\phi_2}} = \left|\frac{\mathbf{Z}_1}{\mathbf{Z}_2}\right| e^{j(\phi_1-\phi_2)}. \tag{9–44}$$

The relations derived earlier for product and quotient of two complex numbers follow directly from Eqs. (9–43) and (9–44).

9–4. EXPONENTIAL REPRESENTATION OF SINE AND COSINE FUNCTIONS

We may now use Euler's formula given in Eq. (9–41) to express the sine and cosine functions in terms of exponential functions of e. First, we write

$$e^{+j\phi} = \cos\phi + j\sin\phi, \tag{9–41}$$

$$e^{-j\phi} = \cos\phi - j\sin\phi. \tag{9–45}$$

Note that $e^{-j\phi}$ is the complex conjugate of $e^{+j\phi}$. Solving Eqs. (9–41) and (9–45) by addition and then subtraction, we obtain

$$\cos\phi = (e^{+j\phi} + e^{-j\phi})/2, \tag{9–46}$$

$$\sin\phi = (e^{+j\phi} - e^{-j\phi})/2j, \tag{9–47}$$

which are also known as Euler's formulas. Although Eqs. (9–46) and (9–47) indicate that both $e^{+j\phi}$ and $e^{-j\phi}$ are needed to describe $\cos\phi$ and $\sin\phi$, it is evident from Eq. (9–41) that one may use only the real part of the exponential $e^{+j\phi}$ to describe $\cos\phi$ and that one may use only the imaginary part of $e^{+j\phi}$ to describe $\sin\phi$. Thus we may write

$$\cos\phi = \mathrm{Re}[e^{+j\phi}], \tag{9–48}$$

$$\sin\phi = \mathrm{Im}[e^{+j\phi}] \tag{9–49}$$

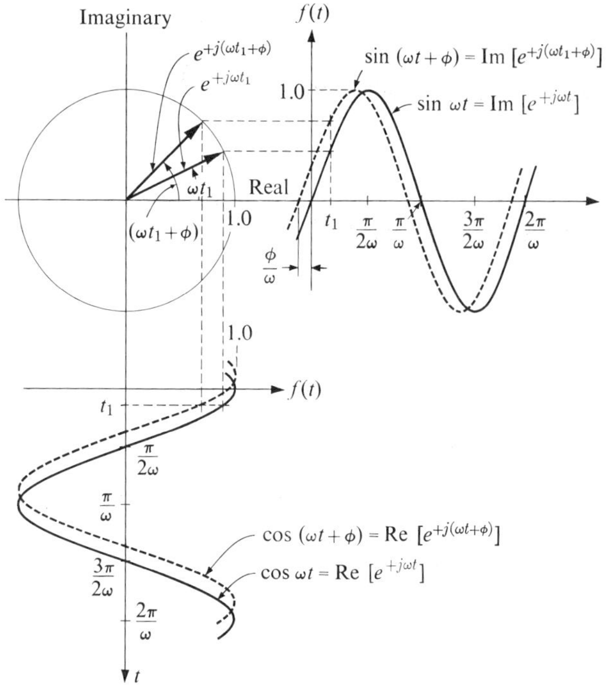

FIG. 9–7. Graphical illustration of generation of sine waves and cosine waves from rotating vectors.

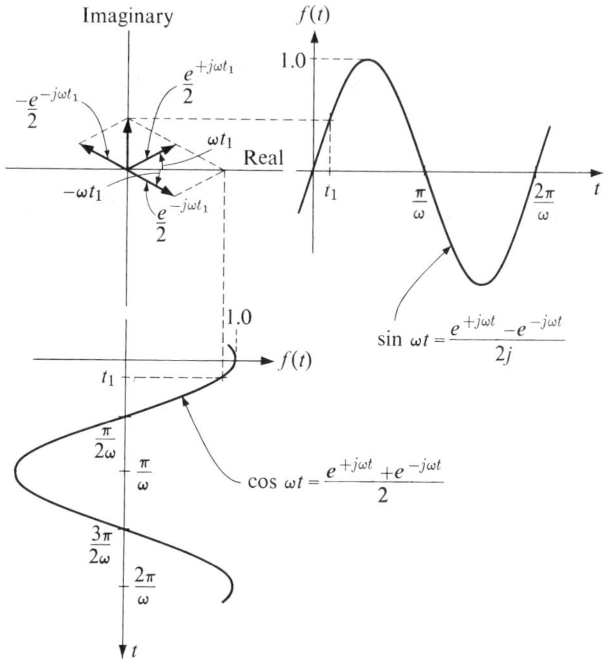

FIG. 9–8. Combination of $e^{+j\omega t}/2$ and $e^{-j\omega t}/2$ to form $\sin \omega t$ and $\cos \omega t$.

to describe the operations of taking the real and imaginary part of $e^{+j\phi}$. (It should be noted that since j has been dropped from the imaginary part in this process, this part is not, technically speaking, imaginary any more. We have simply used the real-number aspect of the imaginary part.) The operations Re[ ] and Im[ ] are linear operations,† and they will be used frequently in work with exponential representation of sinusoidal functions. In this way we may use only $e^{+j\phi}$ to describe $\sin \phi$ or $\cos \phi$.

As a corollary, it is interesting to note again (see Section 9–2) that the sum of a complex number and its conjugate is equal to twice the real part of the complex number, Eq. (9–9), and that the difference between a complex number and its conjugate is equal to twice the imaginary part of the complex number, Eqs. (9–10) and (9–11). Thus combining a complex number with its conjugate by addition or subtraction corresponds to taking twice its real part or twice its imaginary part, respectively. Note that in the second instance, the difference must be divided by j if one is to end up with only the real-number aspect of the imaginary part (corresponding to the operation Im[ ]):

$$\frac{\mathbf{Z}_1 - \mathbf{Z}_1^*}{j} = 2y_1. \tag{9–50}$$

The following relations are often useful:

$$\text{Re}[Z] = (\mathbf{Z} + \mathbf{Z}^*)/2, \tag{9–51}$$

$$\text{Im}[Z] = (\mathbf{Z} - \mathbf{Z}^*)/2j. \tag{9–52}$$

† $\text{Re}[\mathbf{Z}_1 + \mathbf{Z}_2] = \text{Re}[\mathbf{Z}_1] + \text{Re}[\mathbf{Z}_2]$, for example (see Section 9–2).

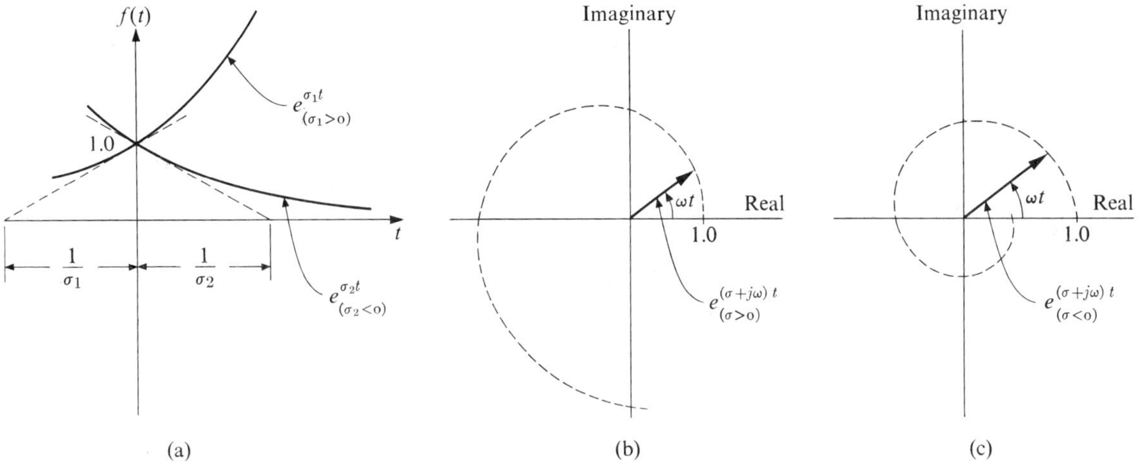

FIG. 9–9. Exponential decay and growth. (a) Real exponentials. (b) Complex exponential growth. (c) Complex exponential decay.

9–5. EXPONENTIAL REPRESENTATION OF FUNCTIONS OF TIME (OR SOME RUNNING VARIABLE)

The exponential $e^{+j\omega t}$ is widely used to represent steady periodic time functions. For example,

$$\text{Re}[e^{+j\omega t}] = \cos \omega t \tag{9–53}$$

and

$$\text{Im}[e^{+j\omega t}] = \sin \omega t. \tag{9–54}$$

A phase shift of a sine or cosine function along the time axis is readily accomplished by multiplying $e^{+j\omega t}$ by $e^{+j\phi}$,

$$e^{+j\omega t}e^{+j\phi} = e^{+j(\omega t+\phi)},$$

so that

$$\text{Re}[e^{+j(\omega t+\phi)}] = \cos (\omega t + \phi), \tag{9–55}$$

$$\text{Im}[e^{+j(\omega t+\phi)}] = \sin (\omega t + \phi). \tag{9–56}$$

The sine and cosine waves generated by the rotating unit vectors $e^{+j\omega t}$ and $e^{+j(\omega t+\phi)}$ used in Eqs. (9–53) through (9–56) are illustrated graphically in Fig. 9–7.

Equations similar to Eqs. (9–46) and (9–47) (Euler's formulas) may also be used to describe sine and cosine waves as follows:

$$\cos \omega t = \frac{e^{+j\omega t} + e^{-j\omega t}}{2}, \tag{9–57}$$

$$\sin \omega t = \frac{e^{+j\omega t} - e^{-j\omega t}}{2j}. \tag{9–58}$$

The combination of the conjugate contrarotating vectors $e^{+j\omega t}/2$ and $e^{-j\omega t}/2$, representing the terms in Euler's formulas, to form sine and cosine waves is shown graphically in Fig. 9–8. Here it is again evident that all the information necessary to describe sin ωt or cos ωt is provided by the rotating vector $e^{+j\omega t}$. The use of the vector $e^{-j\omega t}$ together with the vector $e^{j\omega t}$ simply makes it unnecessary to use the operations Re[ ] and Im[ ].

When the rotating unit vector $e^{+j\omega t}$ is multiplied by a constant such as k or e^c, where k and c are real numbers, it then becomes a rotating vector of constant magnitude k or constant magnitude e^c, respectively.

When the rotating unit vector $e^{+j\omega t}$ is multiplied by the exponential $e^{\sigma t}$, where σ is a real number, the resulting product is a rotating vector of amplitude $e^{\sigma t}$. The decay or growth (corresponding to whether the sign of σ is negative or positive) of the exponential $e^{\sigma t}$ is illustrated in Fig. 9–9(a). The growth or decay

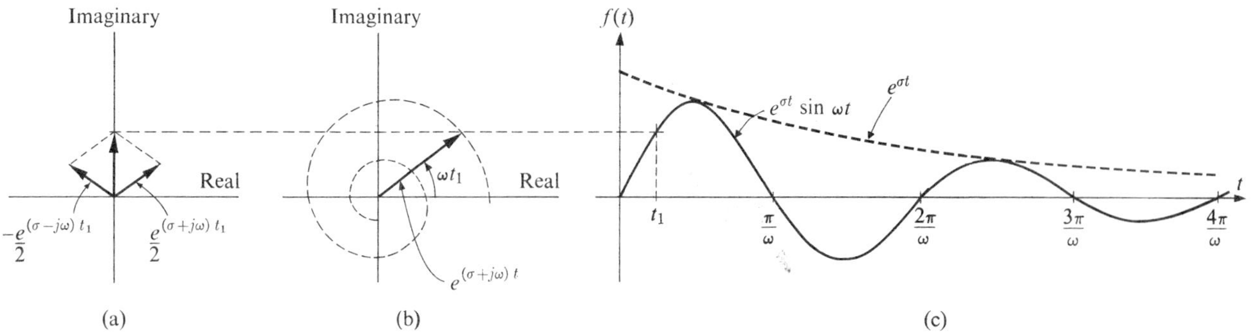

FIG. 9–10. Formation of a decaying sinusoid ($\sigma < 0$). (a) From Euler's formulation using contrarotating vectors. (b) Using the imaginary part of a single rotating vector.

of the product $e^{\sigma t}e^{+j\omega t} = e^{(\sigma+j\omega)t}$ is illustrated in Figs. 9–9(b) and (c), where time is now indicated by the angular position of the rotating vector.

At this point we are able to describe decaying or growing sinusoidal waves in terms of complex numbers. For instance, the decaying sinusoid $e^{\sigma t}\sin\omega t$ can be represented by

$$f(t) = e^{\sigma t}\sin\omega t = \text{Im}(e^{(\sigma+j\omega)t}) \tag{9–59}$$

or

$$f(t) = e^{\sigma t}\sin\omega t = (e^{(\sigma+j\omega)t} - e^{(\sigma-j\omega)t})/2j, \tag{9–60}$$

where σ is a negative real number. Graphical illustrations of Eqs. (9–59) and (9–60) are given in Fig. 9–10. It is left for the reader to formulate expressions similar to (9–59) and (9–60) for $f(t) = e^{\sigma t}\cos\omega t$ and to visualize graphically the two ways of representing such a function of time with complex exponentials.

9–6. DIFFERENTIATION OF EXPONENTIALS AND SINUSOIDS

One reason that exponential functions are so useful in the analysis of dynamic systems is because all their derivatives and integrals are of the same form. The general form of complex exponential is e^{st}, where s is a complex number given by

$$s = \sigma + j\omega. \tag{9–61}$$

By applying the ordinary rules for differentiation of such an exponential with respect to time, we obtain

$$f(t) = e^{st},\quad f'(t) = se^{st},\quad f''(t) = s^2e^{st},\quad \text{etc.} \tag{9–62}$$

For one example, let us consider the special case when $\sigma = 0$:

$$\begin{aligned} f(t) &= e^{+j\omega t},\\ f'(t) &= +j\omega e^{+j\omega t},\\ f''(t) &= (+j\omega)^2e^{+j\omega t} = -\omega^2e^{+j\omega t},\\ f'''(t) &= (+j\omega)^3e^{+j\omega t} = -j\omega^3e^{+j\omega t}. \end{aligned} \tag{9–63}$$

We now employ Eqs. (9–63) to determine the derivatives of a cosine wave:

$$\begin{aligned} \cos\omega t &= \text{Re}[e^{+j\omega t}],\\ \frac{d}{dt}(\cos\omega t) &= \text{Re}[+j\omega e^{+j\omega t}]\\ &= \text{Re}[j\omega(\cos\omega t + j\sin\omega t)]\\ &= -\omega\sin\omega t,\\ \frac{d^2}{dt^2}(\cos\omega t) &= \text{Re}[(+j\omega)^2e^{+j\omega t}]\\ &= \text{Re}[-\omega^2(\cos\omega t + j\sin\omega t)]\\ &= -\omega^2\cos\omega t,\\ \frac{d^3}{dt^3}(\cos\omega t) &= \text{Re}[(+j\omega)^3e^{+j\omega t}]\\ &= \text{Re}[= j\omega^3(\cos\omega t + j\sin\omega t)]\\ &= +\omega^3\sin\omega t, \qquad \text{etc.} \end{aligned}$$

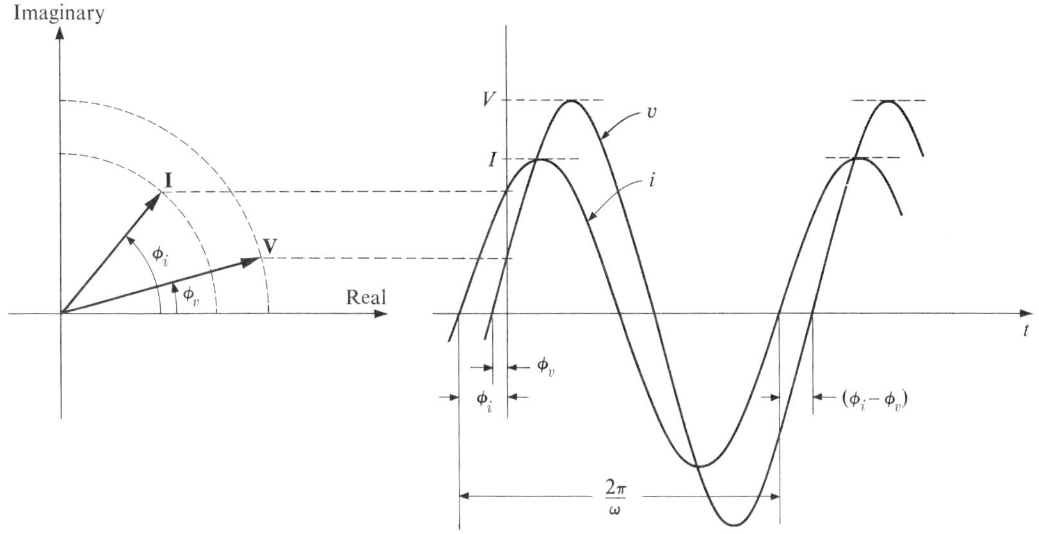

FIG. 9–11. Use of phasors to describe sinusoids.

A similar procedure applied to a sine wave yields

$$
\begin{aligned}
\sin \omega t &= \text{Im}[e^{+j\omega t}], \\
\frac{d}{dt}(\sin \omega t) &= \text{Im}[+j\omega e^{+j\omega t}] \\
&= \omega \cos \omega t, \\
\frac{d^2}{dt^2}(\sin \omega t) &= \text{Im}[(+j\omega)^2 e^{+j\omega t}] \\
&= -\omega^2 \sin \omega t, \\
\frac{d^3}{dt^3}(\sin \omega t) &= \text{Im}[(+j\omega)^3 e^{+j\omega t}] \\
&= -\omega^3 \cos \omega t.
\end{aligned}
$$

Thus it is seen that one can represent a sine wave or cosine wave time function by $e^{+j\omega t}$, carry out as many differentiations as necessary on $e^{+j\omega t}$, and then recover the required derivatives of $\sin \omega t$ or $\cos \omega t$ by taking the imaginary or real parts respectively of the differentiated exponential. This procedure will be especially useful in the study of linear dynamic systems in which all variables will be undergoing steady sinusoidal variations with time and thus can be represented by exponentials of the form $e^{+j\omega t}$. Then differentiation of all variables can be accomplished in a very simple manner. The reason why differentiation of sinusoids is so easily accomplished by the use of exponentials is that the imaginary part of the derivative of a complex number is equal to the derivative of the imaginary part of the same complex number. The validity of this statement is demonstrated as follows:

$$\frac{d}{dt}\text{Im}(\mathbf{Z}) = \frac{dy}{dt},$$

$$\text{Im}\left(\frac{d}{dt}\mathbf{Z}\right) = \text{Im}\left(\frac{dx}{dt} + j\frac{dy}{dt}\right) = \frac{dy}{dt}.$$

Therefore

$$\frac{d}{dt}\text{Im}(\mathbf{Z}) = \text{Im}\left(\frac{d}{dt}\mathbf{Z}\right).$$

9–7. PHASORS, ADDITION OF SINUSOIDS, AND MULTIPLICATION OF SINUSOIDS

It was shown in Section 9–5 that a sine wave or a cosine wave could be expressed in terms of the exponential $e^{j\omega t}$ and that it could be shifted along in time by multiplying $e^{j\omega t}$ by $e^{j\phi}$. Two or more sinusoids of the same frequency ω (or the same period $2\pi/\omega$) are thus described by vectors rotating at the same angular speed ω. It is then convenient to employ a

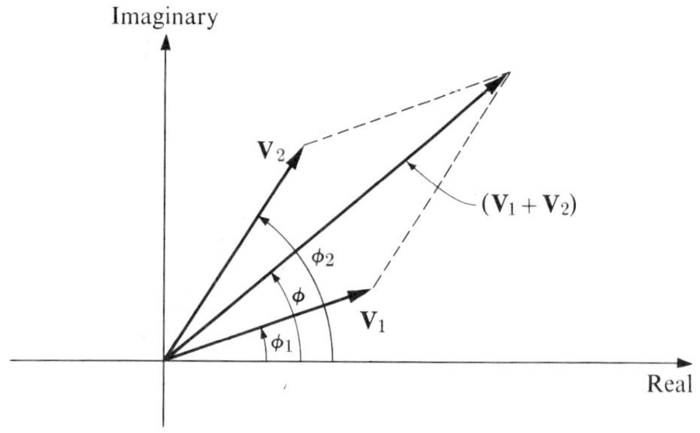

FIG. 9–12. Vector addition of phasors to sum sinusoids of the same frequency.

special kind of vector called a *phasor* to depict the amplitude and phase of each of the sinusoids. For example, the voltage v and the current i at some point in an electrical circuit can be described as follows:

$$v = \text{Im}[\mathbf{V}e^{j\omega t}] = V \sin(\omega t + \phi_v),$$
$$i = \text{Im}[\mathbf{I}e^{j\omega t}] = I \sin(\omega t + \phi_i),$$

where $V = |\mathbf{V}|$ and $I = |\mathbf{I}|$ are the amplitudes of v and i, respectively. The voltage phasor $\mathbf{V}$ and the current phasor $\mathbf{I}$, illustrated in Fig. 9–11, may be visualized as the rotating vectors used to generate v and i "frozen" in the positions which they occupy when $t = 2n\pi$.

When it is necessary to add two sinusoidal waves, this is readily accomplished by simply summing their phasors vectorially. For example, suppose we wish to add the velocities v_1 and v_2, which are given by

$$v_1 = V_1 \sin(\omega t + \phi_1) = \text{Im}[\mathbf{V}_1 e^{j\omega t}],$$
$$v_2 = V_2 \sin(\omega t + \phi_2) = \text{Im}[\mathbf{V}_2 e^{j\omega t}].$$

We may now sum v_1 and v_2 algebraically to obtain

$$v_1 + v_2 = \text{Im}[\mathbf{V}_1 e^{j\omega t} + \mathbf{V}_2 e^{j\omega t}]$$
$$v_1 + v_2 = \text{Im}[(\mathbf{V}_1 + \mathbf{V}_2)e^{j\omega t}]$$
$$= |\mathbf{V}_1 + \mathbf{V}_2| \sin(\omega t + \phi),$$

so that it is seen that we simply need the vector sum of $\mathbf{V}_1$ and $\mathbf{V}_2$ (Fig. 9–12). For example, let us suppose that

$$v_1 = 10 \sin(\omega t + 37.5°),$$
$$v_2 = 15 \sin(\omega t + 60°),$$

then

$$\mathbf{V}_1 = 10e^{j37.5°} = 8 + j6,$$
$$\mathbf{V}_2 = 15e^{j60°} = 7.5 + j12.98,$$
$$\mathbf{V}_1 + \mathbf{V}_2 = 15.5 + j18.98 = Ae^{j\phi},$$

where

$$A = \sqrt{(15.5^2 + (18.98)^2} = \sqrt{600} = 24.5,$$
$$\phi = \tan^{-1}(18.98/15.5) = 50.8°.$$

Thus,

$$v_1 + v_2 = 24.5 \sin(\omega t + 50.8°).$$

It *is very important to note*, however, that when it is desired to multiply one sinusoid by another, it is necessary to use both terms in the Euler formula for each sinusoid to obtain the correct result. For instance, the product $\sin \omega t \sin(\omega t + \phi)$ cannot be found by taking the imaginary part of $e^{j\omega t}e^{(\omega t + \phi)}$ The use of Euler's formulas gives

$$\sin \omega t \sin(\omega t + \phi)$$
$$= \left[\frac{e^{+j\omega t} - e^{-j\omega t}}{2j}\right]\left[\frac{e^{+j(\omega t+\phi)} - e^{-j(\omega t+\phi)}}{2j}\right]$$
$$= \frac{e^{+j(2\omega t+\phi)} + e^{-j(2\omega t+\phi)} - e^{j\phi} - e^{-j\phi}}{-4}$$
$$= -\tfrac{1}{2}\cos(2\omega t + \phi) + \tfrac{1}{2}\cos\phi,$$

which can be verified by referring to a table of trigonometric identities. It is left to the reader to show in a similar manner that

$$\sin(\omega_1 t + \phi_1)\cos(\omega_2 t + \phi_2)$$
$$= \tfrac{1}{2}\{\sin[(\omega_1 + \omega_2)t + \phi_1 + \phi_2] + \sin[(\omega_1 - \omega_2)t + \phi_1 - \phi_2]\}. \tag{9–64}$$

As an example of situations in which this type of problem arises, let us consider the case of an electrical system or network with alternating currents and voltages, for which we wish to determine the power being transmitted through a given point. The power is equal to the product vi, where v and i are steady sinusoids:

$$\mathcal{P} = vi,$$

and v and i can be expressed as the real parts of exponentials,

$$\begin{aligned} v &= V\cos(\omega t + \phi_v) \\ &= \mathrm{Re}[Ve^{+j(\omega t+\phi_v)}] = \mathrm{Re}[\mathbf{V}e^{+j\omega t}], \\ i &= I\cos(\omega t + \phi_i) \\ &= \mathrm{Re}[Ie^{+j(\omega t+\phi_i)}] = \mathrm{Re}[\mathbf{I}e^{+j\omega t}], \end{aligned}$$

where

$$\begin{aligned} \mathbf{V} &= Ve^{+j\phi_v}, & \mathbf{I} &= Ie^{+j\phi_i}, \\ V &= |\mathbf{V}|, & I &= |\mathbf{I}|, \quad \text{(amplitudes of } v \text{ and } i\text{, respectively)} \\ \phi_v &= \angle\mathbf{V}, & \phi_i &= \angle\mathbf{I}. \end{aligned}$$

The power is then determined as follows:

$$\mathcal{P} = \mathrm{Re}[\mathbf{V}e^{+j\omega t}]\,\mathrm{Re}[\mathbf{I}e^{+j\omega t}],$$

or

$$\mathcal{P} = \mathrm{Re}[Ve^{+j(\omega t+\phi_v)}]\,\mathrm{Re}[Ie^{+j(\omega t+\phi_i)}].$$

The amplitudes, being real numbers, may be brought outside the brackets, and conjugates of the exponentials may now be employed as in Eq. (9–5) to give

$$\mathcal{P} = VI\left[\frac{e^{+j(\omega t+\phi_v)} + e^{-j(\omega t+\phi_v)}}{2}\right] \times \left[\frac{e^{+j(\omega t+\phi_i)} + e^{-j(\omega t+\phi_i)}}{2}\right],$$

$$\mathcal{P} = VI\left[\frac{e^{+j(2\omega t+\phi_v+\phi_i)} + e^{-j(2\omega t+\phi_v+\phi_i)} + e^{+j(\phi_v-\phi_i)} + e^{-j(\phi_v-\phi_i)}}{4}\right],$$

$$\mathcal{P} = \frac{VI}{2}\{\mathrm{Re}[e^{+j(2\omega t+\phi_v+\phi_i)}] + \mathrm{Re}[e^{j(\phi_v-\phi_i)}]\},$$

$$\mathcal{P} = \frac{VI}{2}[\cos(2\omega t + \phi_v + \phi_i) + \cos(\phi_v - \phi_i)]. \tag{9–65}$$

It is very important to note here that

$$\mathrm{Re}[\mathbf{V}e^{j\omega t}\mathbf{I}e^{j\omega t}] \neq \mathrm{Re}[\mathbf{V}e^{j\omega t}]\,\mathrm{Re}[\mathbf{I}e^{j\omega t}],$$

and therefore that multiplying the complete exponentials first and then taking the real part will not yield the correct result.

9–8. SUMMARY

A thorough knowledge of the structure of complex numbers, the ways in which they can be represented, and the basic operations which can be carried out with them are all needed in the analysis of linear dynamic systems. The use of complex numbers offers a mathematical means of dealing with transient and steady sinusoidal phenomena. This means capitalizes on the use of the simple operations of addition and subtraction with real numbers in the mathematical manipulation and graphical representation of exponential functions of time. The use of complex exponential representation of sinusoidal functions of time also simplifies the process of differentiation with respect to time, a process which is encountered at every turn in solving and working with differential equations.

REFERENCES

1. M. R. SPIEGEL, *Theory and Problems of Complex Variables.* New York: Schaum Publishing Co. (Outline Series) (pp. 1–32) 1964.

2. R. E. JOHNSON, L. L. LENDSEY, W. E. SLESNICK, and G. E. BATES, *Modern Algebra-Second Course.* Reading, Mass., Addison-Wesley Publishing Co., Inc., 1962.

PROBLEMS

9–1. Find numerically the magnitude and phase of the following complex numbers and sketch in the complex plane:

a) $\mathbf{Z} = 10e^{j3\pi/4}$, b) $\mathbf{Z} = 2.5e^{j\pi/3}$,

c) $\mathbf{Z} = -1 - j3$, d) $\mathbf{Z} = \dfrac{1}{2 + j5}$,

e) $\mathbf{Z} = \dfrac{2 + j3}{1 + j5}$.

9–2. Find numerically the real and imaginary parts of the following complex numbers and sketch in the complex plane:

a) $\mathbf{Z} = 10e^{j3\pi/4}$, b) $\mathbf{Z} = 2.5e^{j\pi/3}$,

c) $\mathbf{Z} = Ae^{j\phi}$, where $A = 3.65$ in. and $\phi = 26.7°$,

d) $\mathbf{Z} = Ae^{j\phi}$, where $A = 65$ v and $\phi = 135°$,

e) $\mathbf{Z} = Ae^{j\phi}$, where $A = 16°$F and $\phi = 2.6\pi$ rad.

9–3. Perform the summation $\mathbf{Z}_1 + \mathbf{Z}_2 = \mathbf{Z}$ for the following complex numbers by using the cartesian coordinates of each number:

a) $\mathbf{Z}_1 = 3.5e^{(\pi/4)j}$; $\mathbf{Z}_2 = 1.0 + j2.0$,

b) $|\mathbf{Z}_1| = 28$, $\angle\mathbf{Z}_1 = 45°$; $|\mathbf{Z}_2| = 16$, $\angle\mathbf{Z}_2 = 135°$,

c) Check (a) by performing the summation graphically (parallelogram).

d) Determine the magnitude and the angle (direction) of the result found in (b); check your answer by means of graphical summation.

9–4. Obtain the difference $\mathbf{Z}_1 - \mathbf{Z}_2$ for the complex numbers given in Problem 9–3.

9–5. Carry out the following operations. Express your answer first in polar and then in cartesian coordinates:

a) $(1.5 + j0.5)(-2.0 + j1.5)$,

b) $(-1.0 - j2.0)/(2.0 - j1.5)$,

c) $3.0/(1.0 + j1.5)$, d) $(6.5 + j3.0)/j2.0$.

9–6. Express the following functions as the imaginary part of a complex exponential function:

a) $y = A \sin(\omega t + \phi)$, b) $y = A \cos \omega t$,

c) $y = Ae^{-\alpha t} \cos(\omega t + \pi/4)$,

d) $y = \sin \omega t - 2 \cos \omega t$.

9–7. Perform the summation $\mathbf{Z}_1 + \mathbf{Z}_2 = \mathbf{Z}$ for the following sinusoids using complex numbers (Section 9–7):

a) $\mathbf{Z}_1 = 10 \sin 5t$; $\mathbf{Z}_2 = 2 \cos 5t$,

b) $\mathbf{Z}_1 = -2 \sin 2t$; $\mathbf{Z}_2 = 3 \sin(2t + \pi/2)$,

c) $\mathbf{Z}_1 = 3 \sin 10t$; $\mathbf{Z}_2 = 4 \sin(10t - 30°)$.

Express your answer in the form $\mathbf{Z} = |\mathbf{Z}| \sin(\omega t + \phi)$ and determine the values of $|\mathbf{Z}|$ and ϕ.

9–8. Using the polar form, show the following. Also indicate in a sketch how the complex numbers on the left-hand side are combined to give the result on the right-hand side.

a) $(-1 + j)^7 = -8(1 + j)$,

b) $\dfrac{5j}{2 + j} = 1 + 2j$,

c) $j(1 - j\sqrt{3})(\sqrt{3} + j) = 2 + 2j\sqrt{3}$.

9–9. a) Prove the following theorem, which is attributed to de Moivre:
$(\cos\theta + j \sin\theta)^n = \cos n\theta + j \sin n\theta$.
Using this, show that

$$\mathbf{Z}^n = r^n(\cos n\theta + j \sin n\theta),$$

where $\mathbf{Z} = re^{j\theta}$.

b) Using Euler's formula, Eq. (9–41), show that

$$\cos x = \frac{e^{jx} + e^{-jx}}{2}, \qquad \sin x = \frac{e^{jx} - e^{-jx}}{2j},$$

where x is real.

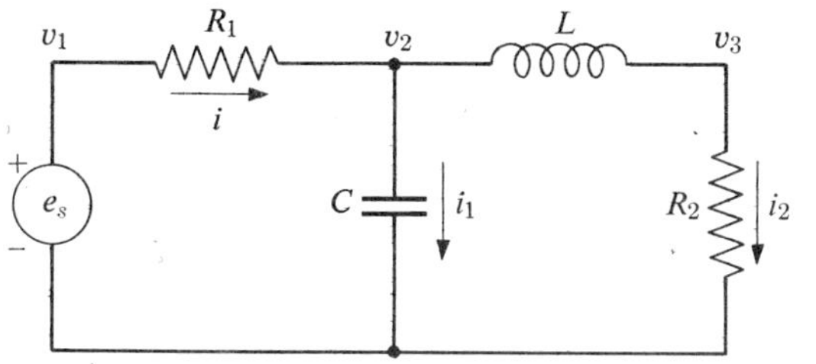

FIGURE 9–13

9–10. The voltage source applied to the electrical circuit in Fig. 9–13 is a steady sinusoid so that all the voltages and currents throughout the circuit are also steady sinusoids. When the amplitude of the steady sinusoidal current i_2 is 1.0 amp (e.g., when $I_2 = 1.0$ amp in $i_2 = I_2 \operatorname{Re}[e^{j\omega t}]$), determine the amplitude of the voltage source e_s. Take $R_1 = R_2 = 1$ ohm, $L = 1$ h, $C = 1$ f, and $\omega = 1$ rad/sec, use Kirchhoff's laws, the elemental equations, and combine sinusoids using complex numbers. [*Hint:* Start with i_2 and work back to e_s.]

10 SYSTEM EXCITATION

10–1. INTRODUCTION

The form of the stimulus or input that is applied to a system is an important factor in studies of system dynamic behavior. The dynamic behavior, as measured or described by other system variables, is intimately related to the inputs acting to excite the system. Therefore we must be able to describe each input as a function of time.

For real physical systems, the inputs may be of two types; a special signal generator (often called a *function generator*) devised to produce a particular form of input, or the natural output response of another system or device. The inputs to real physical systems are physically realizable phenomena, and hence they are measurable in terms of real numbers.

In addition to the physically realizable inputs that may be applied to real systems, there are other inputs which can be described mathematically in terms of complex numbers and which can be used mathematically as inputs in solving the system differential equation or equations. Solving for the system output to one of these physically unrealizable inputs is equivalent to carrying out a thought experiment. Although these inputs may be very difficult or even impossible to generate physically, they play an important role in helping to reveal the nature of the dynamic-response characteristics of physical systems.

Most inputs used for analysis of elementary dynamic systems are simple functions of time expressed mathematically. As such, they are often good approximations to the real inputs that appear in many physical systems. There is a close analogy between mathematical time functions used to represent real physical inputs and idealized models used to represent real systems. Suddenly pressing down on the brake pedal of a moving vehicle and suddenly turning on the switch of an electric stove are examples of *step* functions. The action of ocean waves on a ship and the effects of rotor unbalance in a high-speed motor are examples of *steady sinusoidal* excitations. The blow struck by a hammer on a bell and a sudden gust of wind blowing on an antenna tower are examples of pulse functions.

We can often consider these system excitations to be generated by sources of the types discussed in Chapter 4. In many cases, the device or system producing the excitation is not ideal, and important dynamic interactions then may occur between the source and the system being studied. For instance, if the network being excited by a sine-wave voltage generator requires too much current, the output of the generator deteriorates, perhaps even becoming distorted from a true sinusoidal; and the manner in which a hammer rebounds after striking a bell depends on the manner in which the bell deforms during the period of impact, thereby influencing the shape of the pulse generated by the hammer striking the bell.

In this chapter we shall assume that the functions being discussed have been generated by ideal sources.

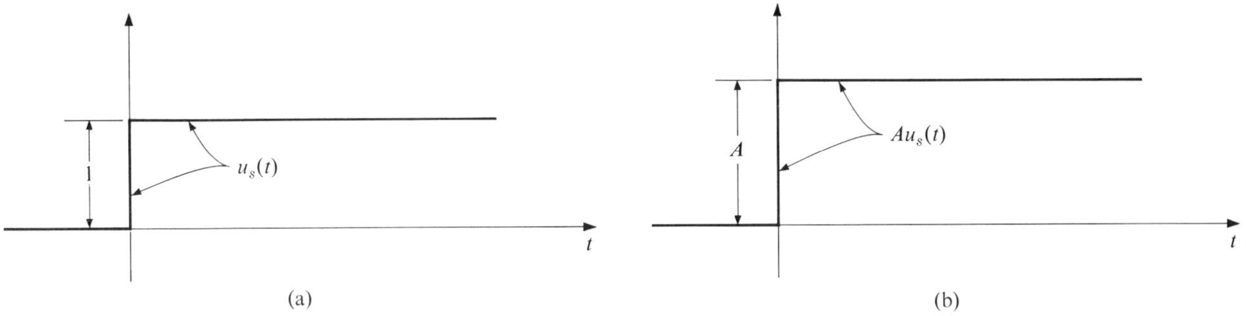

FIG. 10–1. Step function occurring at $t = 0$.

10–2. SINGULARITY FUNCTIONS

10–2.1 Step Functions

One of the simplest and most widely used inputs is the *step* change or step function. Although creating a sudden change of a physical variable is impossible to accomplish perfectly because it would require an infinite power source, sufficiently rapid changes of many variables are realizable physically for us to speak of them as step changes. A step input is one of the most drastic inputs that can be applied to a system, and it usually excites the system in such a way that it reveals all the natural modes of its behavior in its response.

A *unit step* function, designated $u_s(t)$, is shown in Fig. 10–1(a). A step function of amplitude A, designated $Au_s(t)$, is shown in Fig. 10–1(b). In addition to serving as a useful input function itself, a unit step function is also useful mathematically in deleting another time function up to the point in time where the step occurs. For instance, $u_s(t) \cdot \mathsf{f}(t)$ denotes the time function shown in Fig. 10–2. When the step change occurs at some time other than at $t = 0$, it is denoted by $u_s(t - T)$, as shown in Fig. 10–3(a). The time function $u_s(t - T) \cdot \mathsf{f}(t)$ is shown in Fig. 10–3(b).

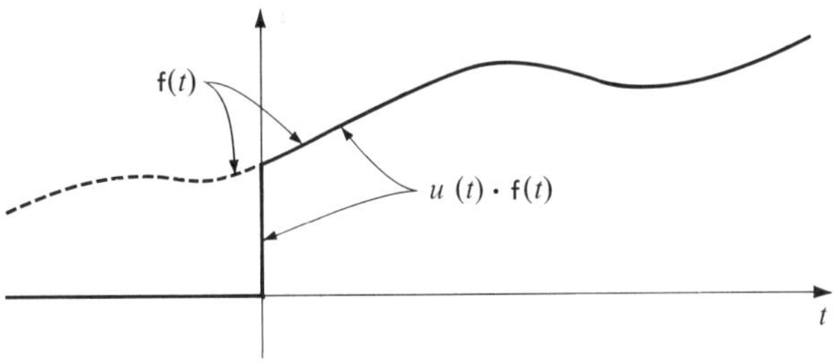

FIG. 10–2. A step function used to delete another time function up to time $t = 0$.

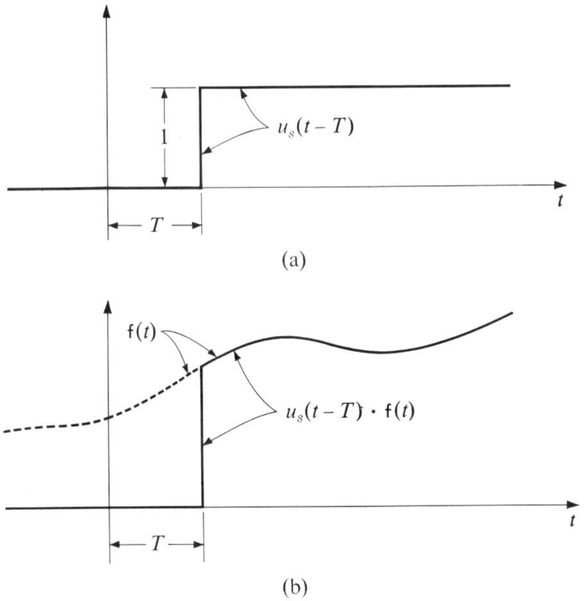

FIG. 10–3. Delayed step functions.

10–2.2 Ramp Functions

When a unit step function is integrated with respect to time, the result is a *unit ramp*, $u_r(t)$, whose slope is

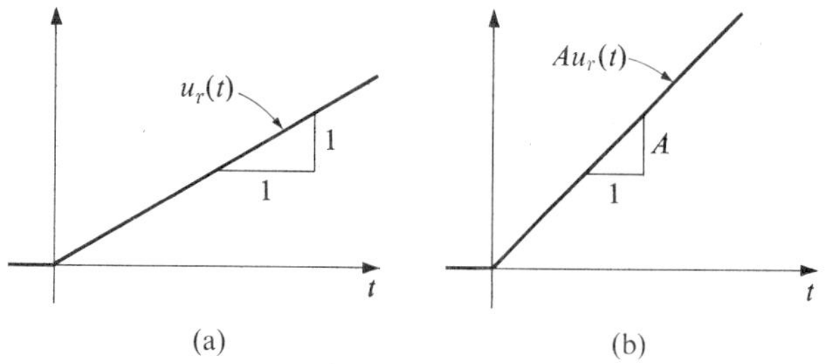

FIG. 10–4. Ramp functions starting at $t = 0$.

unity (Fig. 10–4a). A ramp of amplitude A is the integral of a step of amplitude A and has slope A (Fig. 10–4b). Starting the ramp at time T rather than at $t = 0$ is accomplished in the same way as in the case of the step, namely by writing $u_r(t - T)$. When a unit step is applied as the input to an integrator whose output is zero at $t = 0$ (Fig. 10–5), the output is a unit ramp.

A ramp function is not so drastic an input to a system as a step function, and for some systems it is preferred because it is not so likely as a step input to damage the system. Also, ramp functions are frequently encountered in the natural working of systems, such as the steadily increasing angular position of a motor after it is suddenly started or the steadily rising level of water in a tank after a constant flow rate of water is suddenly diverted into it. When a unit ramp function is integrated, a unit parabola function, $u_p(t)$, is generated (Fig. 10–6). Parabolic functions are sometimes used as cam profiles in automatic machinery, and they sometimes occur in the natural working of systems, such as the height of a free-falling body as a function of time.

10–2.3 Pulse Functions

Another important class of functions is that of the *pulse* functions. A pulse is usually thought of as a single excursion of a time function from zero or from some constant value (Fig. 10–7). Its duration is T, and its strength A_p is the time integral of the amplitude of its variation from its initial value (shaded area) over the duration of the pulse. Pulses of many kinds are generated for specific purposes, especially in electronic systems. Much of the information that is processed and transmitted in computer systems is in the form of pulses. Pulses also occur in the natural working of systems, as in the case of a bat striking a ball or in the case of an electron striking the phosphorescent material in the face of a cathode-ray tube.

A *rectangular pulse* can be thought of as resulting from one step change followed a short time later by another step change of opposite sign (Fig. 10–8). Therefore the pulse can be described mathematically in terms of unit steps as in Eq.(10–1). The strength of such a pulse is AT, the shaded area shown in Fig. 10–8:

$$\mathrm{f}(t) = A[u_s(t) - u_s(t - T)] \qquad (10\text{–}1)$$

A unit rectangular pulse is a pulse of unit strength and may be designated by $u_T(t)$. A unit rectangular pulse can have any width (Fig. 10–9). *Therefore it is necessary to specify the width of a rectangular pulse as well as its strength* to define it completely. Thus the subscript T in $u_T(t)$ may be used to specify the pulse width in the mathematical symbol for this function.

When the width of a unit rectangular pulse is decreased so that it approaches zero, the result is a function that approaches a unit impulse, $u_i(t)$. The limiting process is shown in Eq. (10–2):

$$u_i(t) = \lim_{T \to 0} u_T(t). \qquad (10\text{–}2)$$

A unit impulse may be thought of as a unit pulse of infinitesimal width having unit area under it. An impulse of strength A is a pulse of infinitesimal width having area A under it. The amplitude of an impulse $Au_i(t)$ at t precisely equal to zero is undefined (however, it is very, very large). At all other times it is zero. Because of this, it is impossible to draw the function itself on a graph. Instead it is represented by a vertical arrow of height A (Fig. 10–10).

A unit impulse is the time derivative of a unit step. This can be demonstrated by applying the limiting process to the time function $\mathrm{f}_1(t)$ which is composed of two ramp functions (Fig. 10–11). The func-

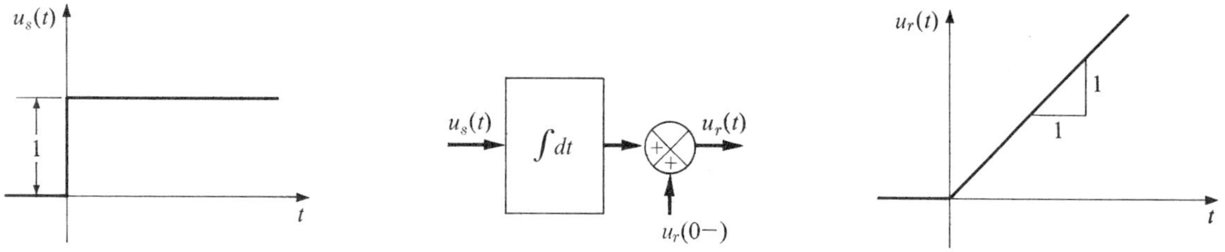

FIG. 10–5. Unit ramp obtained by integrating a unit step.

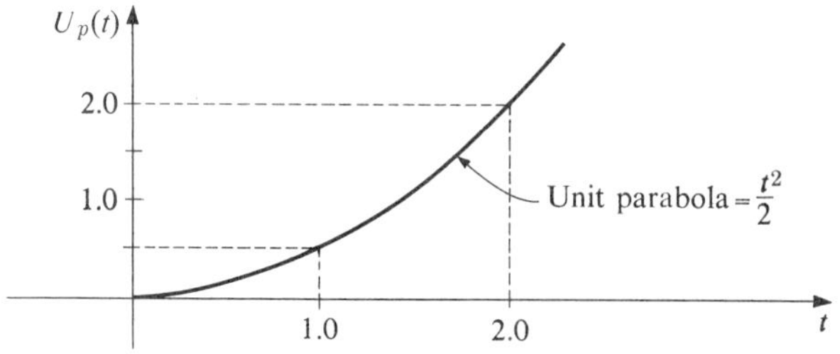

FIG. 10–6. Unit parabola function.

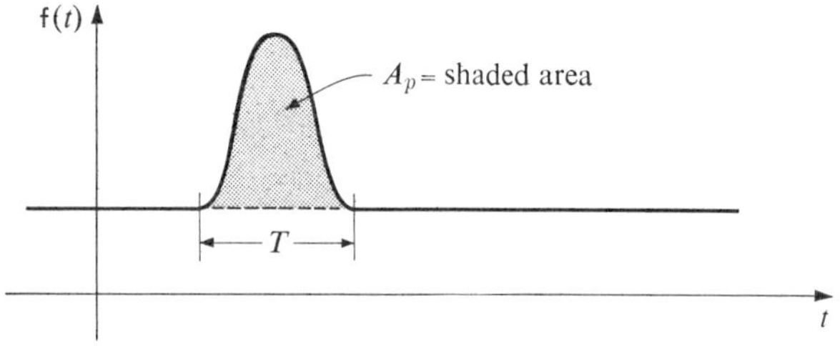

FIG. 10–7. A pulse of strength A_p.

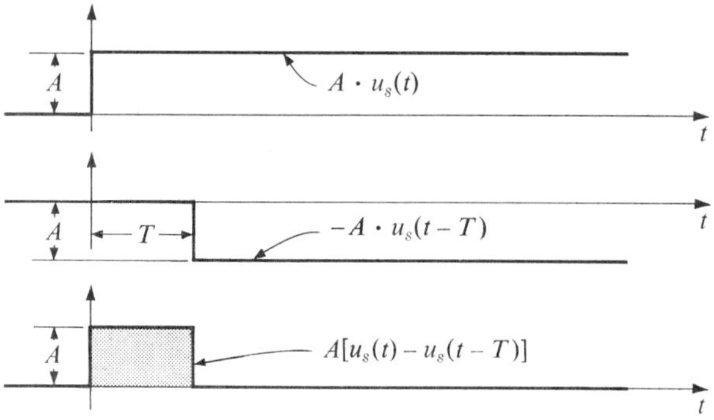

FIG. 10–8. Formation of a rectangular pulse of strength AT from two step functions.

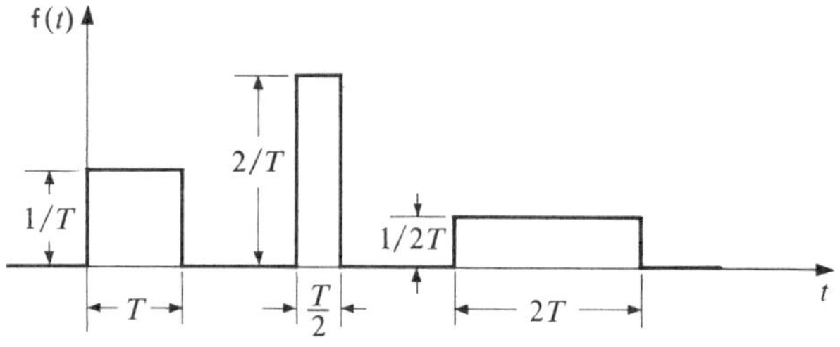

FIG. 10–9. Unit rectangular pulses of different widths.

tion $f_2(t)$ is the derivative of $f_1(t)$ with respect to time. As $T \to 0$, $f_1(t)$ approaches a unit step, and $f_2(t)$ approaches a unit impulse. It should be noted that there are many other functions besides ramp functions which may be employed to describe a function like $f_1(t)$ which approaches a unit step as $T \to 0$, while at the same time $f_2(t) = f_1'(t)$ approaches a unit impulse [1].

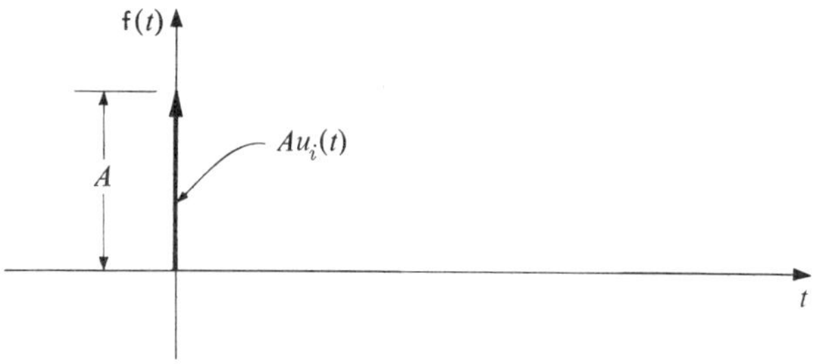

FIG. 10–10. Graphical representation of an impulse occurring at time $t = 0$.

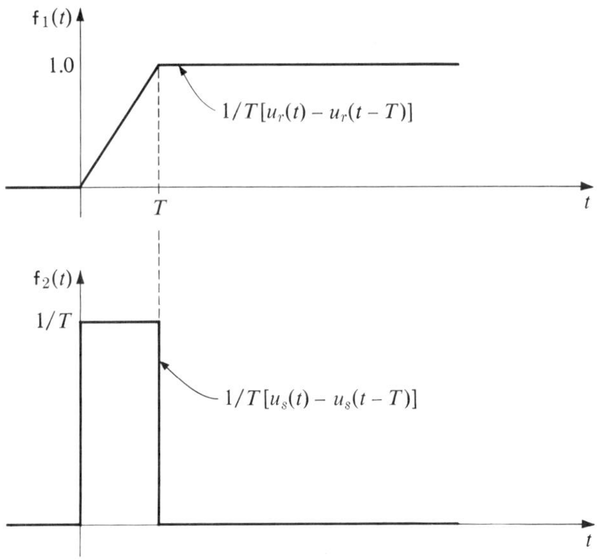

FIG. 10–11. Simple functions which may be used to demonstrate that $du_s(t)/dt = u_i(t)$. As $T \to 0$, $f_1(t) \to u_s(t)$ and $f_2(t) \to u_i(t)$.

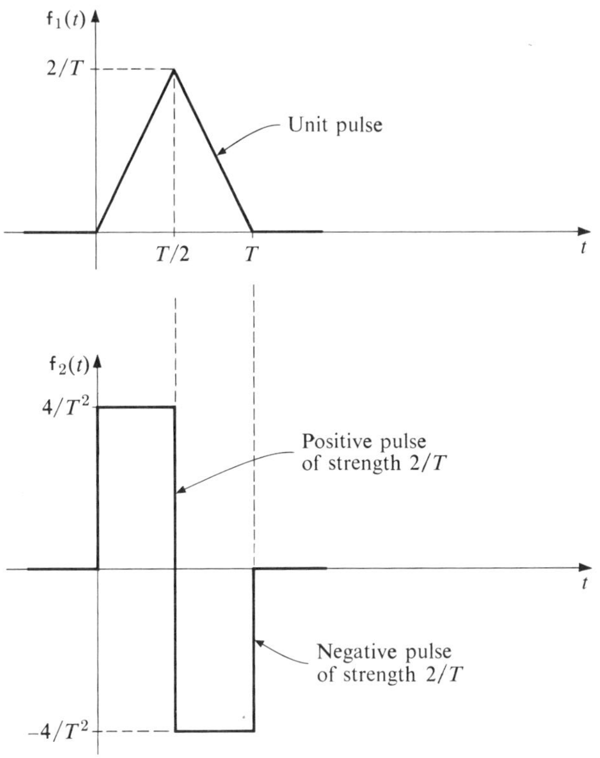

FIG. 10–12. Simple functions which may be used to demonstrate that $du_i(t)/dt = u_d(t)$. As $T - 0$, $f_1(t) - u_i(t)$ and $f_2(t) - u_d(t)$.

A similar line of reasoning applied to the functions $f_1(t)$ and $f_2(t)$ shown in Fig. 10–12 leads to a demonstration that the derivative of a unit impulse is a unit doublet $u_d(t)$, that is, a positive impulse of infinite strength immediately followed by a negative impulse of infinite strength.

Whereas a perfect step change is impossible to obtain physically because of power limitations of a real source, an impulse is impossible to achieve on two counts, power limitations and limitations on the variable itself. Real materials in dynamic system elements are incapable of withstanding infinite forces, velocities, currents, or voltages! Nevertheless, it is often possible to use short-duration pulses of finite amplitudes which occur in short enough time that their effect in producing a dynamic-system response is very nearly the same as that predicted mathematically from a true impulse input. Thus the vibration of a mechanical structure when struck by a hammer (which produces a short-duration finite-amplitude pulse in force on the structure) is the same as that which a mathematical analysis would predict based on an impulse input of force having the same strength as the strength of the pulse produced by the hammer.

An impulse or a very short pulse is an even more drastic input to a system than a step change, and it tends to strongly excite the natural modes of the response of a system. A short pulse can be destructive to sensitive parts of a system if it is not used carefully.

10–2.4 The Family of Singularity Functions

The family of functions which includes parabolas, ramps, steps, and impulses has been given the name *singularity functions* because of the singularity which exists at the point of discontinuity in each of the func-

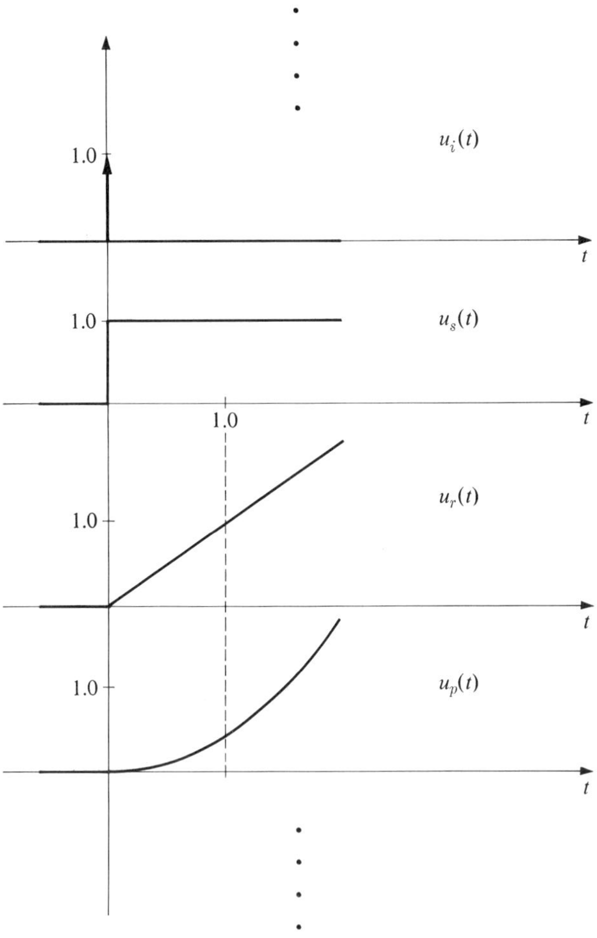

FIG. 10–13. The family of singularity functions.

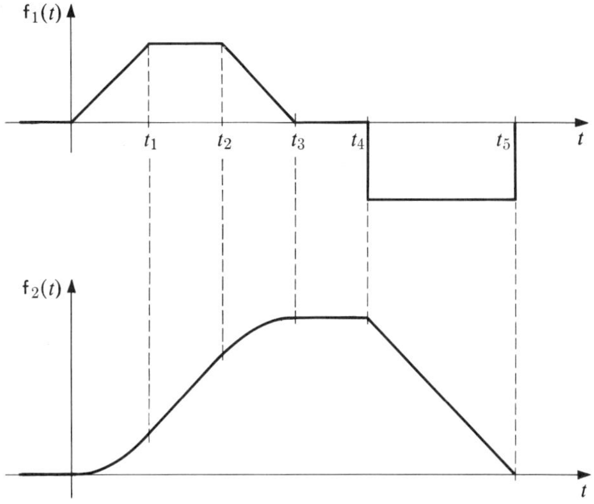

FIG. 10–14. Formation of complicated time functions from singularity functions.

tions in the family. The close relationship between the members of this family is evident when it is realized that all the rest of the members are obtained by successive differentiation or integration of one member of the family. Several members of the family of singularity functions are illustrated in Fig. 10–13.

Often singularity functions can be combined to form more complicated time functions. The formation of a rectangular pulse from two steps in Fig. 10–8 and the formation of a pulse from three ramps in Fig. 10–12 are simple examples of this. Formation of more complicated functions from singularity functions is illustrated in Fig. 10–14. The function $f_1(t)$ may be expressed mathematically by

$$f_1(t) = u_r(t) - u_r(t - t_1) - u_r(t - t_2) + u_r(t - t_3) - u_s(t - t_4) + u_s(t - t_5), \quad (10\text{–}3)$$

and the function $f_2(t)$ may be expressed by

$$f_2(t) = u_p(t) - u_p(t - t_1) - u_p(t - t_2) + u_p(t - t_3) - u_r(t - t_4) + u_r(t - t_5). \quad (10\text{–}4)$$

In this case $f_2(t)$ also happens to be the time integral of $f_1(t)$. This combination was chosen to illustrate the ease with which integration or differentiation of time functions can be accomplished if they are formed from combinations of singularity functions.

10–3. EXPONENTIAL FUNCTIONS

An exceedingly useful and versatile time function in working with dynamic systems is the exponential function, which was also discussed briefly in Chapter 9:

$$f(t) = e^{st}, \quad (10\text{–}5)$$

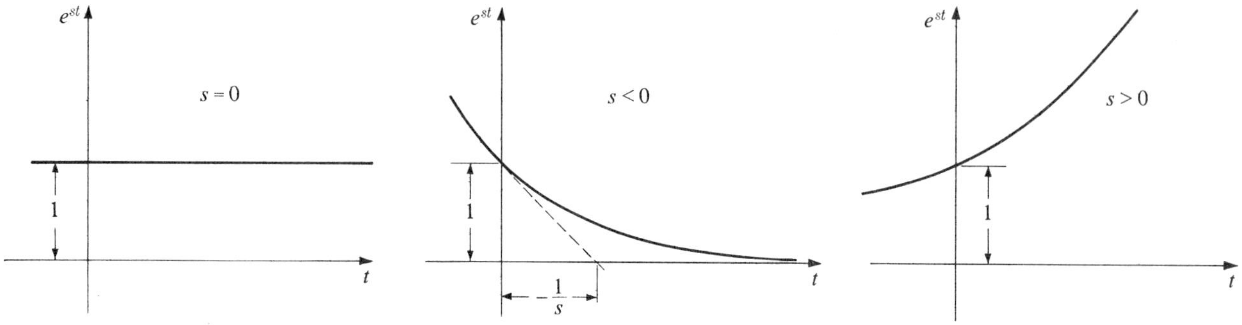

FIG. 10–15. Graphical representation of e^{st} for various cases when s is real.

where s is a constant which may be real, imaginary, or complex. When the constant s is real, the exponential time function is also real, and it may be displayed as a function of time (Fig. 10–15).

When the constant s is imaginary, the exponential time function is complex so that real and imaginary axes are needed to describe it (Fig. 10–16). It is described in a two-dimensional manner in Fig. 10–16(a) in terms of a vector of unit amplitude rotating at a constant angular velocity ω, so that the locus of $f(t)$ is a circle of unit radius. Also, it may be described analytically by Euler's formula (Eq. 10–6). The real part of this function is a cosine function and the imaginary part is a sine function:

$$f(t) = e^{j\omega t} = \cos \omega t + j \sin \omega t. \quad (10\text{–}6)$$

Thus one may write

$$\text{Re}\,[e^{j\omega t}] = \cos \omega t \quad (10\text{–}7)$$

and

$$\text{Im}\,[e^{j\omega t}] = \sin \omega t. \quad (10\text{–}8)$$

The time function $f(t) = e^{j\omega t}$ may also be depicted three-dimensionally as a helix generated by a unit vector rotating at a constant angular velocity ω and moving at a constant velocity along the time axis (Fig. 10–16b).

When the constant s is complex, the time function $f(t) = e^{st}$ is also complex. By expressing s in terms

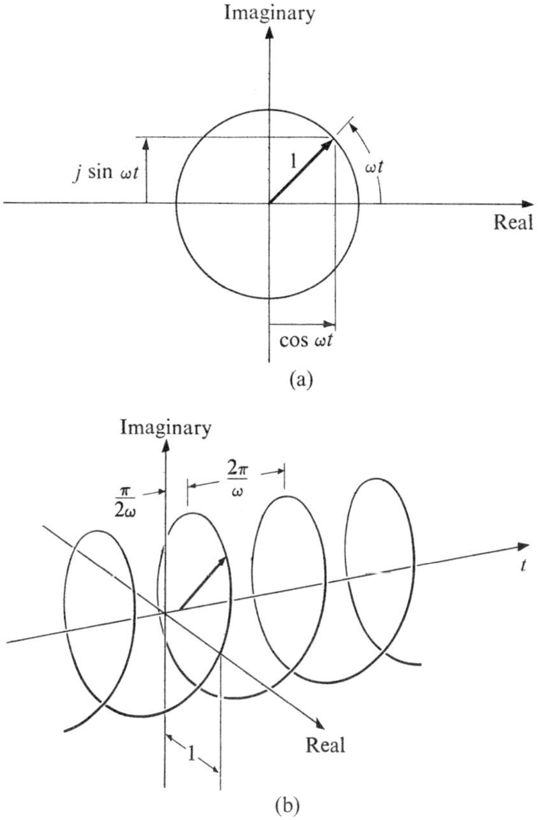

FIG. 10–16. Graphical representation of e^{st} when s is imaginary.

of σ and $j\omega$, Eq. (10–9),

$$s = \sigma + j\omega, \tag{10–9}$$

one may express this exponential time function as the product of two exponentials (Eq. 10–10). The first exponential $e^{\sigma t}$ is a real time function,

$$e^{st} = e^{(\sigma+j\omega)t} = e^{\sigma t} \cdot e^{j\omega t}, \tag{10–10}$$

and the second exponential $e^{j\omega t}$ is a complex time function. The product of these two time functions may be portrayed graphically, as in Fig. 10–17. Making use of Euler's formula, one also may express this complex time function in terms of its real and imaginary parts, Eqs. (10–11), (10–12), and (10–13):

$$\begin{aligned} e^{st} &= e^{\sigma t}(\cos \omega t + j \sin \omega t) \\ &= e^{\sigma t} \cos \omega t + je^{\sigma t} \sin \omega t, \end{aligned} \tag{10–11}$$

$$\text{Re}[e^{st}] = e^{\sigma t} \cos \omega t, \tag{10–12}$$

$$\text{Im}[e^{st}] = e^{\sigma t} \sin \omega t. \tag{10–13}$$

Thus we have a way of generating the convergent and divergent oscillations shown in Fig. 10–18 from an exponential function of time.

The exponential function is not only useful as an input and for generating inputs (two or more exponentials can be combined to form more complicated time functions). It also happens to be the form of the terms in the homogeneous solution of a linear constant-coefficient differential equation. Also, it is a function whose derivatives are all of the same form, which makes it easy to use when several successive differentiations are required.

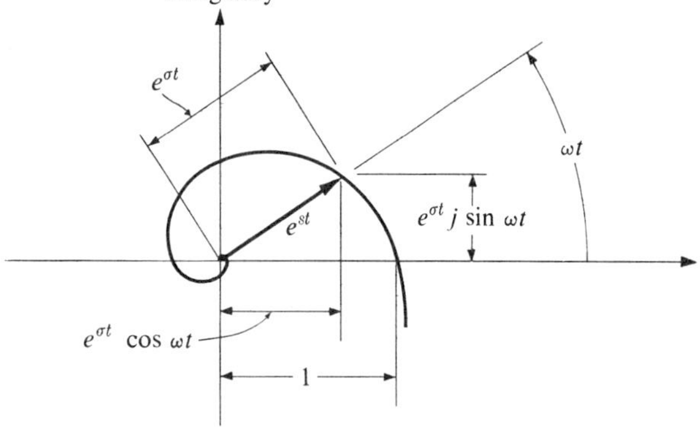

FIG. 10–17. Graphical representation of e^{st} when s is complex ($\sigma + j\omega$) for the case when $\sigma < 0$.

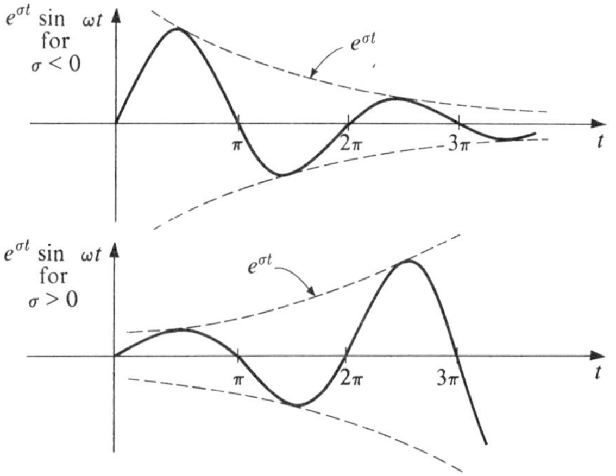

FIG. 10–18. Illustration of convergent and divergent oscillations.

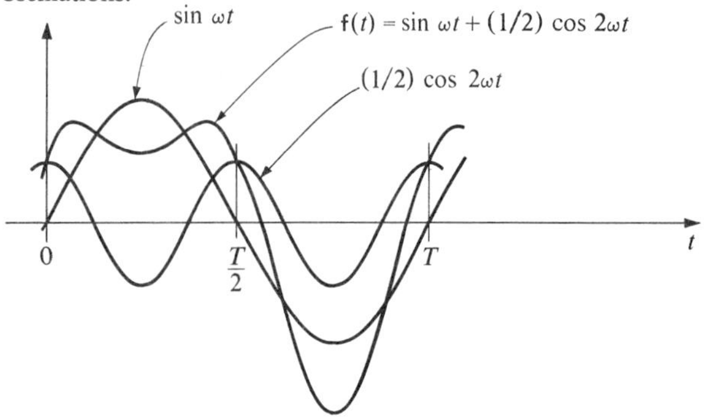

FIG. 10–19. Summation of $\sin \omega t$ and $\frac{1}{2} \cos 2\omega t$ to form a function $f(t)$ having period $T = 2\pi/\omega$.

10–4. PERIODIC FUNCTIONS AND FOURIER SERIES

When two or more sinusoidal functions whose periods are integral multiples of each other are summed, a periodic function results whose period equals that of the sinusoid with the longest period. The summation of $\sin \omega t$ and $\frac{1}{2} \cos 2\omega t$ to form a function $f(t)$ of period $T = 2\pi/\omega$ is illustrated in Fig. 10–19. As one would expect, it is possible to express any periodic function in terms of a series of sine and cosine com-

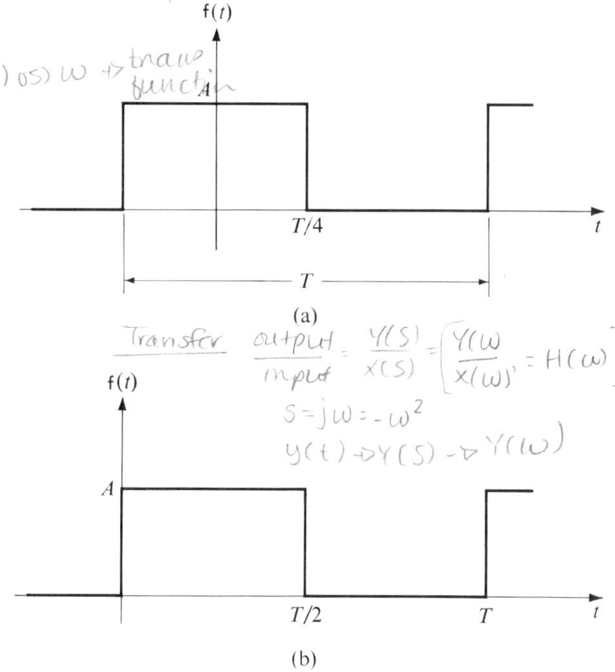

FIG. 10–20. Square waves having period T. (a) Symmetric about axis $t = 0$. (b) Antisymmetric about axis $t = 0$.

ponents. In some cases, many sinusoidal components of different frequencies, called *harmonics*, may be required to describe the given periodic function with sufficient accuracy. In other cases only a few harmonics may be needed.

The infinite series of sine and cosine terms, whose periods are integral multiples of each other, which is needed to precisely describe an arbitrary periodic function, is called a *Fourier series* [2] after the famed French mathematician J. B. J. Fourier, who was responsible for the application of trigonometric series in the solution of problems involving the unsteady flow of heat in solids. The Fourier series for a periodic function of time is given by

$$\mathrm{f}(t) = a_0 + \sum_{n=1}^{\infty} a_n \cos n\omega t + \sum_{m=1}^{\infty} b_m \sin m\omega t, \qquad (10\text{–}14)$$

where $\omega = 2\pi/T$ is the angular frequency (rad/sec) of the function having period T, and

$$a_0 = \frac{\omega}{2\pi}\int_{t_1}^{(t_1+2\pi/\omega)} \mathrm{f}(t)\,dt, \qquad (10\text{–}15)$$

$$a_n = \frac{\omega}{\pi}\int_{t_1}^{(t_1+2\pi/\omega)} \mathrm{f}(t)\cos(\omega nt)\,dt, \qquad (10\text{–}16)$$

$$b_m = \frac{\omega}{\pi}\int_{t_1}^{(t_1+2\pi/\omega)} \mathrm{f}(t)\sin(\omega mt)\,dt. \qquad (10\text{–}17)$$

It is readily seen that a_0 is simply the time average value of the periodic function. The sine and cosine terms in Eq. (10–14) then describe the variations of $\mathrm{f}(t)$ from its average value.

For an example of the application of Fourier series, let us consider the square wave $\mathrm{f}(t)$ (Fig. 10–20a), which is to be the input to a linear system. If we can determine the amplitudes and the frequencies of the Fourier components of the square wave, we can then find the response of the linear system to $\mathrm{f}(t)$ by finding the responses to each of the Fourier components and, using superposition, summing them. Here we shall find only the Fourier components, leaving until a later chapter the problem of finding the responses to the components.

Applying Eqs. (10–15), (10–16), and (10–17), and noting that $\omega = 2\pi/T$, we find that

$$a_0 = A/2, \qquad (10\text{–}18)$$

$$\begin{aligned} a_n &= (2/T)\int_{-T/2}^{+T/2} \mathrm{f}(t)\cos(2\pi nt/T)\,dt \\ &= (2/T)\int_{-T/4}^{+T/4} A\cos(2\pi nt/T)\,dt \\ &= (A/\pi n)\sin(2\pi nt/T)\Big|_{-T/4}^{+T/4}, \end{aligned}$$

$$a_n = (2A/n\pi)\sin(n\pi/2), \qquad (10\text{–}19)$$

or

$$\begin{aligned} a_n &= 0 &&\text{if } n = 2, 4, 6, \ldots, \\ a_n &= 2A/n\pi &&\text{if } n = 1, 5, 9, \ldots, \\ a_n &= -2A/n\pi &&\text{if } n = 3, 7, 11, \ldots. \end{aligned}$$

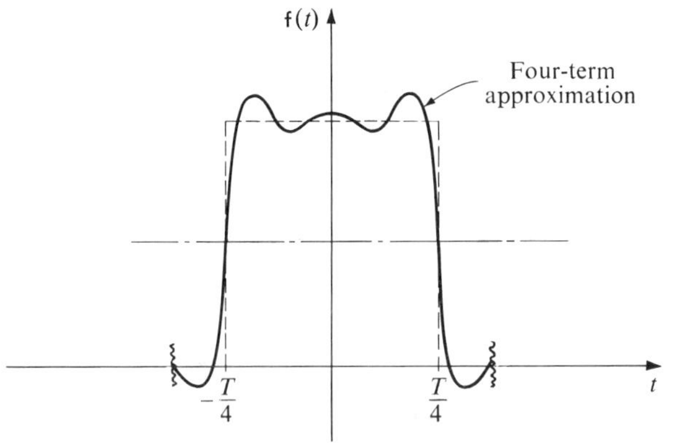

FIG. 10–21. Four-term Fourier-series approximation for the square wave shown in Fig. 11–20(a).

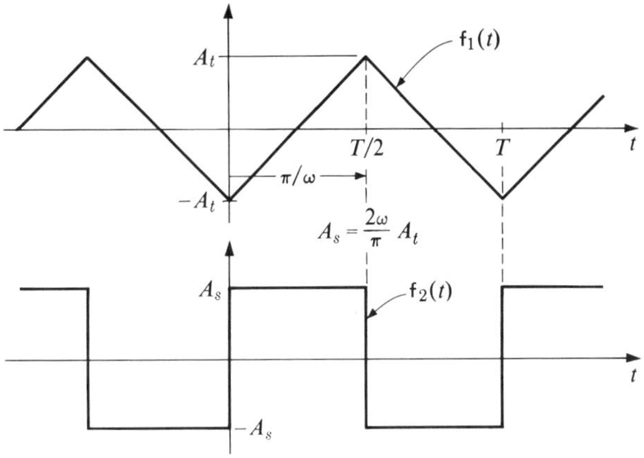

FIG. 10–22. Sawtooth wave and its time derivative, a square wave.

Because this function is symmetrical about the axis $t = 0$, the coefficient b_m is equal to zero for all values of m. It is interesting to note that the amplitude of each harmonic component in the Fourier series for a square wave is inversely proportional to n. Thus the contributions of the higher-frequency terms diminish as the frequency increases.

Summarizing the results so far, we see that the symmetrical square wave shown in Fig. 10–20(a) may now be expressed as follows:

$$\mathsf{f}(t) = A/2 + (2A/\pi)(\cos \omega t - \tfrac{1}{3}\cos 3\omega t + \tfrac{1}{5}\cos 5\omega t - \cdots). \tag{10–20}$$

When only four of the terms indicated in Eq. (10–20) are used to approximate $\mathsf{f}(t)$, the result is as shown in Fig. 10–21.

When the position of the square wave is shifted so that it is antisymmetric with respect to the axis $t = 0$, the resulting Fourier series contains only positive sine terms instead of only cosine terms having alternating signs, but the resulting series is like that given in Eq. (10–20) in every other respect.

Another interesting example of the determination of a Fourier series is the case when $\mathsf{f}(t)$ is a sawtooth or triangular wave $\mathsf{f}_1(t)$ (see Fig. 10–22). Rather than starting from the beginning and using Eqs. (10–15),

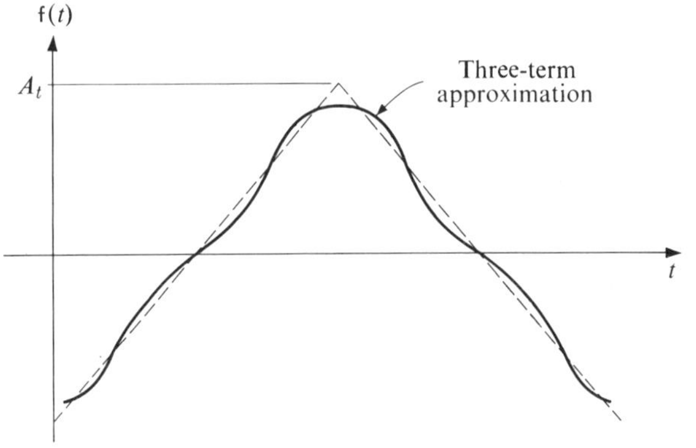

FIG. 10–23. Three-term Fourier series approximation for sawtooth wave shown in Fig. 11–22.

(10–16), and (10–17) to find the terms in the Fourier series for this function of time, we can make use of the fact that this function happens to be the time integral of the antisymmetrical square wave $\mathsf{f}_2(t)$ shown directly below it. Therefore we may employ what we have already learned about the Fourier series of square waves to find the Fourier series of $\mathsf{f}_2(t)$, and then integrate $\mathsf{f}_2(t)$ with respect to time to find $\mathsf{f}_1(t)$. The Fourier series for $\mathsf{f}_2(t)$ is given by

$$\mathsf{f}_2(t) = \frac{4A_s}{\pi}(\sin \omega t + \tfrac{1}{3}\sin 3\omega t + \tfrac{1}{5}\sin 5\omega t + \cdots), \tag{10–21}$$

so that the Fourier series for $f_1(t)$ is given by

$$f_1(t) = \int_0^t f_2(t)\,dt$$
$$= \frac{8A_t}{\pi^2}\left(-\cos\omega t - \frac{1}{3^2}\cos 3\omega t - \frac{1}{5^2}\cos 5\omega t - \cdots\right). \quad (10\text{–}22)$$

As shown in Fig. 10–22, $A_t = A_s\pi/2\omega$.

It is interesting to note that the amplitude of the high-frequency terms of the Fourier series of a sawtooth is inversely proportional to the square of the frequency and that they thus die out much more rapidly with increasing frequency than in the case of a square wave.

When only three terms of the Fourier series given by Eq. (10–22) are used to approximate a sawtooth wave, the result is as shown in Fig. 10–23.

The abbreviated treatment of Fourier series given here should be supplemented by more detailed study [2, 3, 4] to gain facility with a number of techniques which have been developed for their use.

REFERENCES

1. B. van der Pol and H. Bremmer, *Operational Calculus*, 2nd Ed. (Reprinted). Cambridge, England: Cambridge University Press (pp. 56–84) 1959.
2. W. T. Martin and E. Reissner, *Elementary Differential Equations*, 2nd Ed. Reading, Mass.: Addison-Wesley Publishing Company, Inc. (pp. 280–281) 1961.
3. D. K. Cheng, *Analysis of Linear Systems*. Reading, Mass.: Addison-Wesley Publishing Company, Inc. (pp. 121–135) 1959.
4. R. L. Halfman, *Dynamics, Vol. II*. Reading, Mass.: Addison-Wesley Publishing Company, Inc., 1962.

PROBLEMS

10–1. Express the time functions in Fig. 10–24 (possible input excitations) in terms of the simple excitation functions discussed in this chapter.

10–2. Find the maximum instantaneous difference between the following pairs of time functions:

a) parts (c) and (f) of Problem 10–1 (make $t_1 = \pi/\omega$),

b) parts (e) and (g) of Problem 10–1 (make $t_1 = \pi/\omega$).

10–3. Determine the strength of each of the following pulses and then find an expression for and sketch the corresponding pulse having the same strength but only one-tenth the duration of the original pulse:

a) part (a) of Problem 10–1,

b) part (b) of Problem 10–1,

c) part (c) of Problem 10–1,

d) part (f) of Problem 10–1.

10–4. Express the following sinusoids in exponential form:

a) $f(t) = F\cos(\omega t + \phi)$,

b) $f(t) = Fe^{-\zeta t}\sin(\omega t - \pi)$,

c) $f(t) = Fe^{\sigma t}\cos(\omega t + \pi/2)$.

10–5. a) Show that the exponential series

$$f(t) = \sum_{k=-\infty}^{+\infty} C_k e^{jk\omega t}$$
$$= \cdots + C_{-2}e^{-j2\omega t} + C_{-1}e^{-j\omega t} + C_0 + C_1 e^{j\omega t} + C_2 e^{j2\omega t} + \cdots$$

is equivalent to the Fourier series given by Eq. (10–14) in this chapter. (Note that n and m may only be positive, whereas k may be positive or negative.) [*Hint:* Let

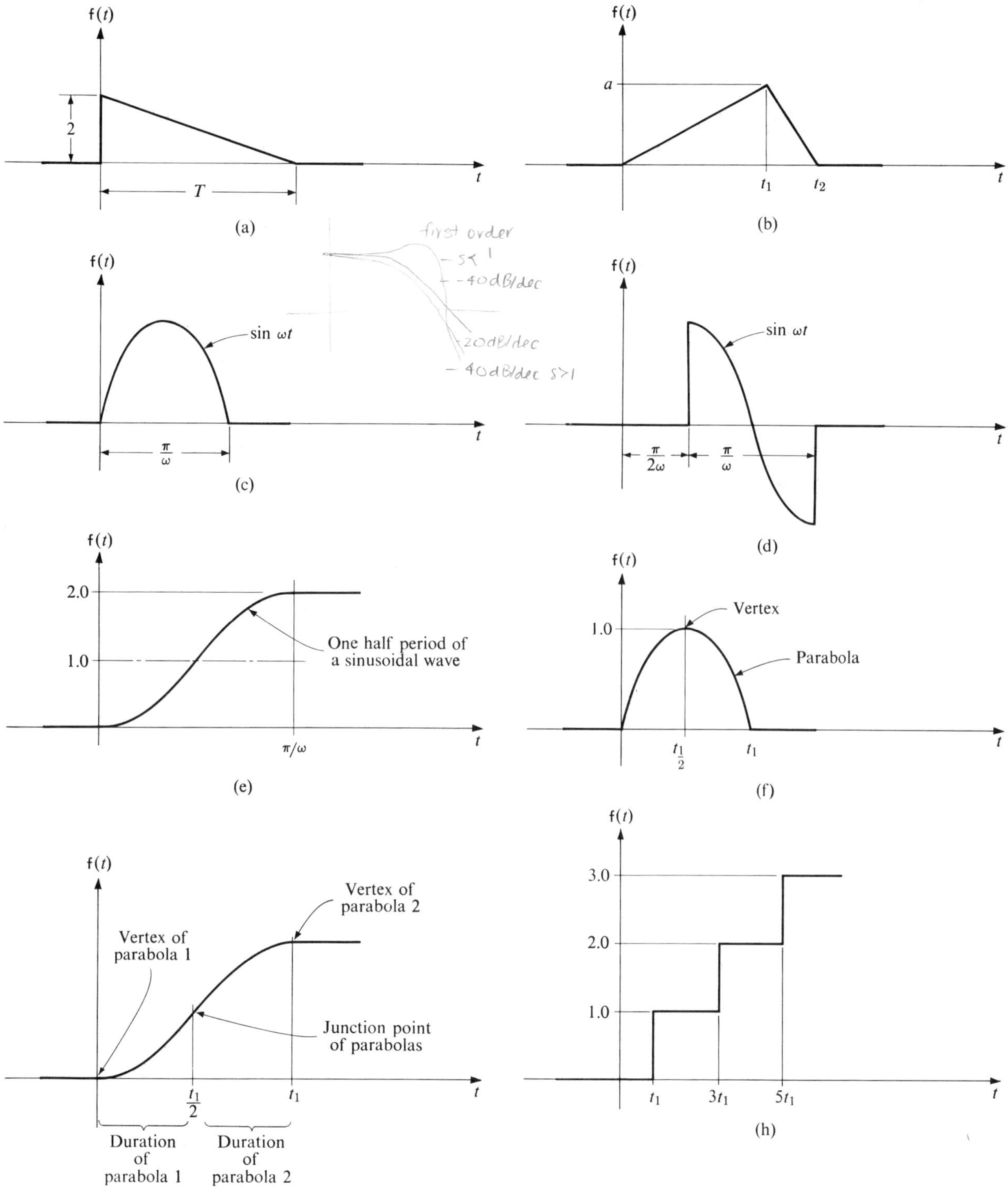

f(t)
2
T
t
(a)
f(t)
a
t1
t2
t
(b)
f(t)
sin ωt
π/ω
t
(c)
f(t)
sin ωt
π/2ω
π/ω
t
(d)
f(t)
2.0
1.0
One half period of a sinusoidal wave
π/ω
t
(e)
f(t)
Vertex
1.0
Parabola
t1/2
t1
t
(f)
f(t)
Vertex of parabola 2
Vertex of parabola 1
Junction point of parabolas
t1/2
t1
t
Duration of parabola 1
Duration of parabola 2
(g)
f(t)
3.0
2.0
1.0
t1
3t1
5t1
t
(h)

FIGURE 10–24

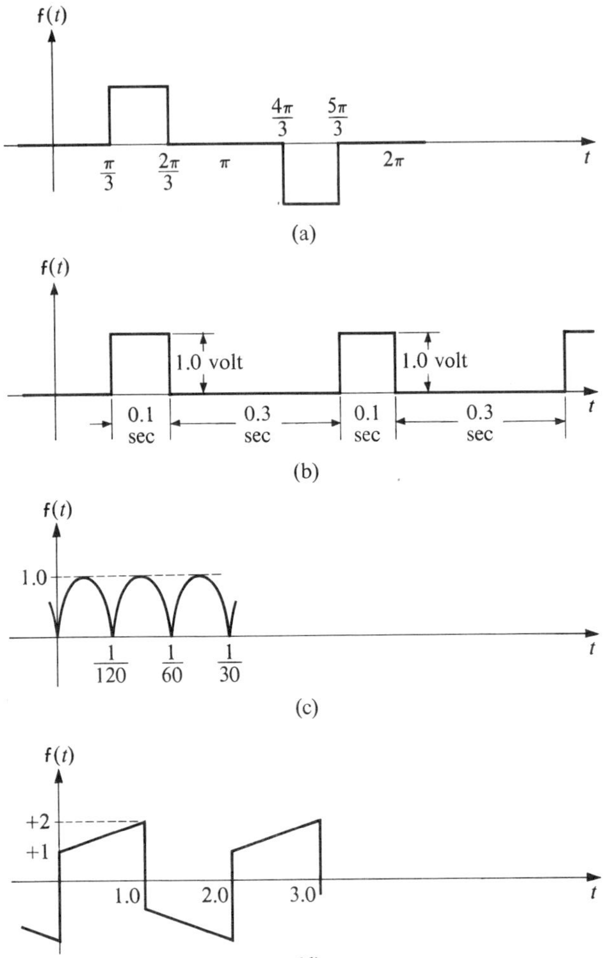

FIGURE 10–25

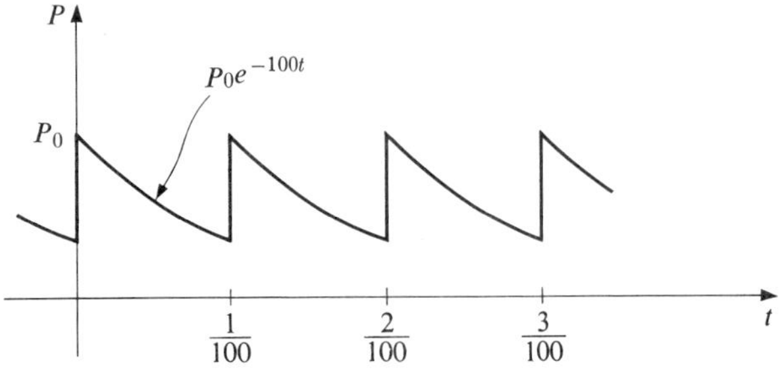

FIGURE 10–26

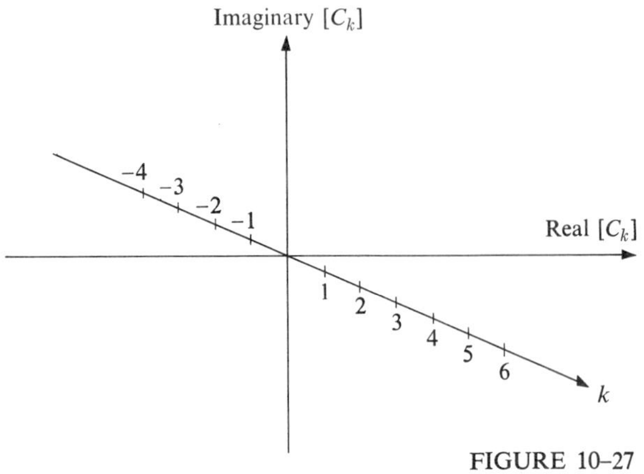

FIGURE 10–27

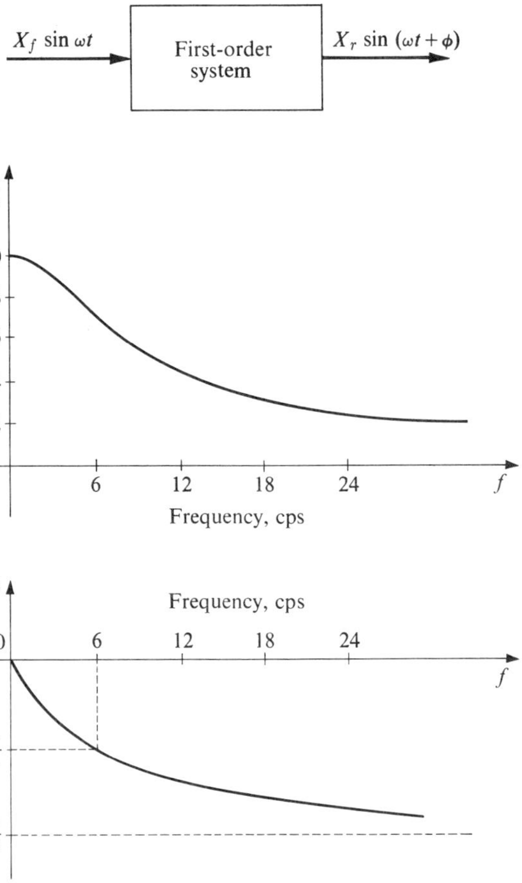

FIGURE 10–28

$n = m = p$ where $p > 1$ and then show that

$$C_p = (a_p - jb_p)/2, \quad \text{and} \quad C_{-p} = (a_p + jb_p)/2.]$$

b) Develop the integral for the Fourier coefficients (in exponential form) for C_k. (This would correspond to the integrals given in Eqs. (10–15), (10–16), and (10–17) for the Fourier coefficients in real form.)

10–6. Find the Fourier series for the periodic waves shown in Fig. 10–25. Sketch an approximation for each case based on the first two terms of the series.

10–7. Plot the Fourier coefficients for the sawtooth and square waves given in Fig. 10–22 as functions of the integer p (i.e., values of n and m).

10–8. The pressure in a fluid passage has been found to vary as shown in Fig. 10–26; in other words, regular pulses consisting of a sudden surge followed by an exponential decay occurring at the rate of 100/sec. Find the Fourier series for this periodic wave of pressure vs. time.

10–9. Plot the complex Fourier coefficients C_k (see Problem 10–5) for the periodic wave shown in Fig. 10–26 as a function of the integer k. Use a three-dimensional presentation, as indicated in Fig. 10–27.

10–10. Find the Fourier series expansions for the following periodic functions with period 2π, using $\cos n\omega t$ and $\sin m\omega t$ terms:

a) $f(x) = x, \qquad -\pi < x < +\pi,$

b) $f(x) = 0, \qquad -\pi < x < 0,$
$f(x) = \sin x, \qquad 0 < x < +\pi.$

10–11. Find the complex Fourier coefficients C_k (see Problem 10–5) for parts a) and b) of Problem 10–10, and plot as in Problem 10–9.

10–12. Suppose that a first-order system has the steady-state amplitude and phase characteristics for the response to unit-amplitude sine wave inputs at various values of input frequency f (Fig. 10–28). If a square wave, such as that given in Fig. 10–22, of unit amplitude and a frequency of three cps is applied to this system, show how you would obtain the output, and make a rough sketch of the output. (A complete, formal mathematical treatment is not required here, but you should try to explain your reasoning and the steps which need to be taken in obtaining an answer.)

11 CLASSICAL SOLUTION OF DIFFERENTIAL EQUATIONS

11–1. INTRODUCTION

The equations which describe the behavior of lumped-element models of physical systems are usually *differential* equations. When the models are composed of ideal (linear) elements, these differential equations are linear and have constant coefficients. If nonlinear or time-varying elements are involved in the system model, nonlinear equations or equations with nonconstant coefficients will result from the system analysis. To determine the behavior of the system model, we must solve these equations. As discussed in Chapter 5, techniques of mathematical solution, analog computation, numerical methods and digital computation, or graphical techniques can be used for this purpose. In this chapter the classical methods for obtaining explicit mathematical solutions of differential equations are discussed. We shall see that linear differential equations with constant coefficients can be solved readily, whereas for other classes of equations explicit solutions are possible only in restricted cases if at all.

The treatment of differential equations in this chapter gives a complete methodology for the solution of ordinary differential equations with constant coefficients and forms the basis for subsequent development of more efficient analytical methods for determining the dynamic behavior of linear systems.

11–2. DEFINITIONS AND NOMENCLATURE

In the differential equations arising from the analysis of a physical system, there are one or more *dependent variables* and one or more *independent variables.* These equations contain the variables and their derivatives. The dependent variables are the *unknown* or *output* quantities of the system, and the independent variables are the *assigned* or *input* quantities of the system. The solutions of the equations are mathematical expressions (or computer outputs or graphical representations) for the dependent variables as functions of the independent variables, these functions being free from derivatives of the dependent variables.

In the equations

$$\frac{\partial^2 V}{\partial x^2} + 3x\frac{\partial T}{\partial t} = \sin \omega t,$$

$$\frac{\partial V}{\partial t} + t\frac{\partial^2 T}{\partial t\,\partial x} = \cos kx,$$

there are two independent variables, x and t, and two dependent variables, V and T. It is possible to obtain solutions for the dependent variables only if there are as many individual equations that are independent of each other as there are dependent variables (two in this example). Such sets of equations are called *simultaneous differential equations.* Equations in which the dependent variables are functions of more than one independent variable are called *partial differential equations.* Hence the above equations are partial differential equations in V and T. If the dependent variables are functions only of one independent variable, the equations are called *ordinary differential equations.* We shall deal in this chapter only with ordinary differential equations.

The most general ordinary differential equation may be written in the functional form

$$\mathsf{g}\left(\frac{d^n x}{dt^n}, \frac{d^{n-1}x}{dt^{n-1}}, \ldots, x, t\right) = \mathsf{f}(t), \tag{11-1}$$

where x is the output or dependent variable, t is the input or independent variable, and f and g indicate functional relationships. The *order* of the equation is the order of the highest derivative (n in 11–1). If the equation contains any powers or products of the *dependent* variable or its derivatives, it is said to be *nonlinear*. If it consists of a linear combination of x and its derivatives, it is a *linear* equation. Thus the following equation is nonlinear and of third order:

$$\frac{d^3x}{dt^3} + x\left(\frac{dx}{dt}\right)^2 + x^4 = \sin t.$$

The most general *linear* ordinary differential equation is of the form

$$a_n \frac{d^n x}{dt^n} + a_{n-1}\frac{d^{n-1}x}{dt^{n-1}} + \cdots + a_0 x = \mathsf{f}(t). \tag{11-2}$$

The a's are called the *coefficients* of the equation, and they may be constant or functions of the independent variable t.

The function $\mathsf{f}(t)$ in Eqs. (11–1) and (11–2) is called the *forcing function*. If $\mathsf{f}(t)$ is zero, the differential equation is said to be *homogeneous*. In many instances, $\mathsf{f}(t)$ will consist of two or more individual functions of time. For instance, $\mathsf{f}(t) = \sin t + e^t$ or $\mathsf{f}(t) = y + dy/dt$.

11–3. EXISTENCE AND UNIQUENESS THEOREMS

Two of the most basic theorems in the theory of differential equations are the existence and uniqueness theorems. These theorems will be stated here without proof; the proofs may be found in Reference 1 or other comprehensive texts on differential equations.

11–3.1 Existence Theorem

Given a differential equation of order n and of the form (11–1) *in which* g *is defined and continuous in a region R specified by t, x, and the first n − 1 derivatives of x, and in which the first n − 1 derivatives of* g *are continuous. A solution x(t) then exists which satisfies the differential equation and the n initial conditions at* $t = t_0$:

$$x(t_0) = C_0, \quad \frac{dx}{dt}(t_0) = C_1, \ldots, \frac{d^{n-1}x}{dt^n}(t_0) = C_{n-1}. \tag{11-3}$$

11–3.2 Uniqueness Theorem

A solution function x(t) meeting the conditions of the existence theorem is the one and only solution which will satisfy the differential equation and the n initial conditions (11–3).

These two theorems tell us that under appropriate conditions of continuity a solution to a differential equation of interest will exist and that if a solution is found by formal procedure, by guess, or by accident, which satisfies the equation and the n given initial conditions, there is no need to look further.

The *general solution* to a differential equation is the solution in which the n initial conditions have not been assigned specific values. Hence the general solution contains n *arbitrary independent* constants whose values may be adjusted so that the solution satisfies the n independent initial conditions (11–3). A constant in the general solution is said to be arbitrary if when it is assigned any value the resulting function still satisfies the differential equation. The constants in the solution are independent if the equation cannot be rearranged to form a solution with fewer constants. The *specific solution* to a differential equation is a general solution in which the n arbitrary constants have been assigned values such that the n initial conditions, particular to the problem at hand, are satisfied. Clearly, the general solution of a differential equation encompasses all possible specific solutions.

11–4. LINEAR ORDINARY DIFFERENTIAL EQUATIONS

We have seen that the analysis of system models based on ideal lumped elements leads to linear differential equations having constant coefficients. Thus the solutions of linear differential equations are of great engineering importance.

In this section we will discuss the nature of solutions of linear differential equations in general (which may have nonconstant coefficients) and will then develop in Section 11-5 a specific method of solution for linear equations having constant coefficients.

11-4.1 General and Specific Solutions

Let us consider first the most general *homogeneous* linear equation

$$a_n \frac{d^n x}{dt^n} + a_{n-1} \frac{d^{n-1} x}{dt^{n-1}} + \cdots + a_0 x = 0, \tag{11-4}$$

where the a's may be functions of t. Let x_1 be any function of time which happens to satisfy the homogeneous equation (11-4). Then the function Ax_1, where A is any constant, is also a solution since, by substitution into (11-4),

$$A\left(a_n \frac{d^n x_1}{dt^n} + a_{n-1} \frac{d^{n-1} x_1}{dt^{n-1}} + \cdots + a_0 x_1\right) = 0.$$

This is the property of *proportionality*.

Now let us suppose that x_2 is a second function which satisfies Eq. (11-4). It is easy to show by direct substitution into (11-4) that the linear combination

$$x = A_1 x_1 + A_2 x_2$$

is also a solution where A_1 and A_2 are arbitrary. This is the property of *superposition*.

This procedure can be extended to any number of functions which satisfy (11-4). If n linearly independent functions $x_1, \ldots, x_n$ can be found, then

$$x_c = \sum_{k=1}^{n} A_k x_k \tag{11-5}$$

is a solution. The n functions are linearly independent if none of these functions can be expressed as a linear combination of the others. Mathematically, it can be shown that the functions are linearly independent providing that their *Wronskian*

$$W = \begin{vmatrix} x_1 & x_2 & \cdots & x_n \\ dx_1/dt & \cdots & \cdots & dx_n/dt \\ \vdots & & & \vdots \\ \dfrac{d^{n-1}x_1}{dt^{n-1}} & \cdots & \cdots & \dfrac{d^{n-1}x_n}{dt^{n-1}} \end{vmatrix} \tag{11-6}$$

is not zero.* If $W \neq 0$, then x_c contains n arbitrary independent constants A_k, satisfies the homogeneous equation (11-4), and thus fulfills the conditions for a general solution to the homogeneous equation. Equation (11-5) therefore *is* the general solution of (11-4).

Now let us consider the general *nonhomogeneous* linear equation (11-2) and let x_p be any function whatsoever, having no arbitrary constants, which satisfies this complete equation. We shall call x_p a *particular integral.* Now if we add x_c, the general solution to the homogeneous equation corresponding to (11-2), to the particular integral, we obtain

$$x(t) = x_c + x_p. \tag{11-7}$$

We shall call x_c the *complementary function*. By substituting (11-7) into (11-2) we find

$$\left(a_n \frac{d^n x_c}{dt^n} + \cdots + a_0 x_c\right) + \left(a_n \frac{d^n x_p}{dt^n} + \cdots + a_0 x_p\right) = \mathsf{f}(t). \tag{11-8}$$

The first enclosed term is zero by definition of the complementary function, and the second enclosed term equals $\mathsf{f}(t)$ by definition of a particular integral. Therefore $x = x_c + x_p$ satisfies (11-2). Furthermore, since x contains n independent arbitrary constants A_k, it must be the unique general solution of (11-2).

In dealing with physical systems we may think of the complementary function x_c as the *natural part* of

* For a proof of this statement, see Reference 1, Section 4-3.

the system response to an input or forcing function and of the particular integral x_p as the *forced part* of the response. The general form of x_c is clearly independent of the nature of the forcing function $\mathsf{f}(t)$, whereas x_p depends directly on $\mathsf{f}(t)$, hence the above terminology. In some cases the terms *transient response* and *steady-state response* have been erroneously applied to x_c and x_p, respectively. This nomenclature is improper since x_c may not be transient and x_p may not be steady. We will use the term transient response to describe a solution $x(t)$ or a part of $x(t)$ which dies away to zero or a constant as time increases. The term steady-state response will refer to a part of $x(t)$ which has either a constant or a periodically varying value.

The specific solution of a given linear differential equation subject to a set of initial conditions can be found in three steps.

STEP 1. Determine n independent solutions of the homogeneous equation and form their linear combination with n arbitrary constants to obtain the *complementary function.*

STEP 2. Find the *particular integral,* which contains no arbitrary constants and satisfies the complete equation. Add this to the complementary function to obtain the *general solution*

$$x = \sum_{k=1}^{n} A_k x_k + x_p. \tag{11–9}$$

STEP 3. Determine the n arbitrary constants A_k by setting $x(t)$ and its first $n - 1$ derivatives equal to the specified initial conditions, Eq. (11–3), at $t = t_0$. The resulting $x(t)$ is the specific solution for the given forcing function and initial conditions.

11–4.2 The Principle of Superposition

Let us consider any linear equation which has one or more distinct forcing functions and is subject to a nonzero set of initial conditions at $t = 0$:

$$a_n \frac{d^n x}{dt^n} + a_{n-1} \frac{d^{n-1} x}{dt^{n-1}} + \cdots + a_0 x = \mathsf{f}_a(t) + \mathsf{f}_b(t) + \cdots \tag{11–10}$$

We see that from the preceding section the solution is

$$x(t) = \sum_{k=1}^{n} A_k x_k + x_{p_a} + x_{p_b} + \cdots, \tag{11–11}$$

where $x_{pa}, x_{pb}, \ldots$ are the particular integrals for $\mathsf{f}_a(t), \mathsf{f}_b(t), \ldots$, respectively. The A_k are found by setting x and its first $n - 1$ derivatives equal to the given initial conditions at $t = 0$. Thus at $t = 0$,

$$\begin{aligned} A_1 x_1 + A_2 x_2 + \cdots + A_n x_n &= x(0) - x_{p_a}(0) + x_{p_b}(0) + \cdots \\ \vdots \qquad\qquad &\qquad \vdots \\ A_1 \frac{d^{n-1} x_1}{dt^{n-1}} + \cdots + A_n \frac{d^{n-1} x_n}{dt^{n-1}} &= \frac{d^{n-1} x(0)}{dt^{n-1}} - \frac{d^{n-1} x_{p_a}(0)}{dt^{n-1}} + \cdots \end{aligned} \tag{11–12}$$

Now let A_{ki} be the values of the constants given by Eqs. (11–12) when the forcing functions $\mathsf{f}_a(t), \mathsf{f}_b(t), \ldots$ are zero.

The solution $x = x_i$ is then the response due only to the initial conditions

$$x_i(t) = \sum_{k=1}^{n} A_{ki}x_k. \tag{11–13}$$

If now the initial conditions are zero and all forcing functions except $f_a(t)$ are zero, Eqs. (11–12) may be solved for a second set of constants A_{k_a} which determine the response to $f_a(t)$ only:

$$x_a(t) = \sum_{k=1}^{n} A_{k_a}x_k + x_{p_a}. \tag{11–14}$$

For each additional $f(t)$ there will correspond a set of A's and a time solution of the form (11–14).

If the initial conditions are nonzero and several forcing functions act simultaneously, it is obvious that the sum of the A's determined by allowing the initial conditions and the individual forcing functions to act separately will satisfy Eqs. (11–12), and this sum will comprise the appropriate A's to be used in the solution, Eq. (11–9):

$$x(t) = \sum_{k=1}^{n} [A_{ki} + A_{k_a} + A_{k_b} + \cdots]x_k + x_{p_a} + x_{p_b} + \cdots \tag{11–15}$$

or, breaking this sum into separate terms, we have

$$x(t) = \underbrace{\left\{\sum_{k=1}^{n} A_{ki}x_k\right\}}_{\substack{\text{Response}\\ \text{due to initial}\\ \text{conditions}}} + \underbrace{\left\{\sum_{k=1}^{n} A_{k_a}x_k + x_{p_a}\right\}}_{\substack{\text{Response due to}\\ f_a(t)}} + \underbrace{\left\{\sum_{k=1}^{n} A_{k_b}x_k + x_{p_b}\right\}}_{\substack{\text{Response due to}\\ f_b(t)}} + \cdots \tag{11–16}$$

Therefore, we may draw the following very important conclusion: *The specific response or solution function for a linear differential equation having given initial conditions and one or more forcing functions is equal to the sum of the specific solution due to the initial conditions (with all forcing functions zero) and the specific solutions for each forcing function considered to act separately (with zero initial conditions).*

In effect the initial conditions constitute a type of forcing function to the differential equation.

The property described by Eq. (11–16), called the *superposition principle*, of the solutions of linear differential equations, forms the basis of much of the theory of linear dynamic systems.

A corollary to the superposition principle is the proportionality property. *If any forcing function (including the initial conditions) of a linear differential equation is multiplied by a constant factor γ, the specific solution due to that forcing function is also multiplied by γ.*

11–4.3 Further Consequences of Linearity

The superposition principle can be used to show that if any linear operation (e.g., one which does not involve powers or products of the variables) is applied to a given forcing function, the same operation applied to the specific solution for that forcing function will yield the specific solution for the modified forcing function.

Three important cases are worthy of mention here. Suppose that the specific solution of a linear equation having only a forcing function $f(t)$ (zero initial conditions) is

$$x = \sum_{k=1}^{n} A_k x_k + x_p.$$

CASE 1. If the forcing function is differentiated, the specific solution is differentiated.

CASE 2. If the forcing function is integrated from $t = 0$ to t, the specific solution is also integrated from $t = 0$ to t.

CASE 3. If the forcing function is complex, i.e., consists of a real and imaginary part,

$$\mathsf{f}(t) = \mathsf{f}_a(t) + j\mathsf{f}_b(t),$$

then the real part of the specific solution for $\mathsf{f}(t)$ will be the specific solution for the real part of $\mathsf{f}(t)$ and the imaginary part of the specific solution will be the specific solution for the imaginary part of $\mathsf{f}(t)$.

We will make extensive use of Case 3 in Chapter 15 to study the response of systems to sinusoidal forcing functions.

The proofs of the above three cases will be considered in Problems 11–4 and 11–5.

In general it is very difficult to find either the complementary function or the particular integral for a linear equation which has nonconstant coefficients. However, when the coefficients are constant, which is often the case in practice, a straightforward method of finding these functions is usually possible.

11–5. SOLUTION OF LINEAR EQUATIONS WITH CONSTANT COEFFICIENTS

11–5.1 Homogeneous Equations and Complementary Functions

In Chapter 5 we saw that the solution of linear first-order differential equations with constant coefficients was exponential in form. Let us consider as a possible solution of the general homogeneous equation (11–4), where the a's are now assumed constant, the function

$$x_1 = Ae^{st}, \tag{11–17}$$

where A and s are constants. Direct substitution of (11–7) into (11–4) gives

$$Ae^{st}(a_n s^n + a_{n-1} - s^{n-1} + \cdots + a_0) = 0.$$

Therefore the assumed exponential is a solution providing

$$a_n s^n + a_{n-1}s^{n-1} + \cdots + a_0 = 0. \tag{11–18}$$

Equation (11–18) is called the *characteristic equation* of the differential equation. It is known that there will be n values of s which will satisfy this algebraic equation. These values are called the *roots* of the characteristic equation and they depend on the values of the coefficients $a_0, \ldots, a_n$. For large n it is difficult to factor the polynomial (11–18) and hence to find the roots. For the present, however, we will assume that this can be done and will designate these n roots by $s = r_1, r_2, \ldots, r_n$. Then, assuming all roots are distinct, we see that there are n independent exponential functions of the form of Eq. (11–17) which may be combined by Eq. (11–5) to obtain the complementary function:

$$x_c = A_1e^{r_1t} + A_2e^{r_2t} + \cdots + A_ne^{r_nt} = \sum_{k=1}^{n} A_ke^{r_kt}. \tag{11–19}$$

The characteristic equation (11–18) which determines the roots can be written from inspection of (11–4) by substituting s^m for d^mx/dt^m. This substitution is justified by the fact that the derivative of an exponential e^{st} is simply se^{st}, so that one s appears for each differentiation.

If all n roots are not distinct, that is, if some roots are repeated, then Eq. (11–19) apparently does not yield n independent functions. Consider the case when $r_1$and r_2 are nearly equal, so that $r_2 = r_1 + \epsilon$, where ϵ is small. The complementary function corresponding to r_1 and r_2 is

$$x_c = A_ae^{r_1t} + A_be^{(r_1+\epsilon)t} = A_ae^{r_1t} + A_be^{r_1t}e^{\epsilon t}.$$

For small ϵ, $e^{\epsilon t}$ can be expanded in a power series, giving

$$x_c = A_ar^{r_1t} + A_be^{r_1t}\left(1 + \epsilon t + \frac{\epsilon^2t^2}{2!} + \cdots\right)$$

or

$$x_c = (A_a + A_b)e^{r_1t} + A_ce^{r_1t}\left(t + \frac{\epsilon t^2}{2!} + \cdots\right),$$

where $A_c = \epsilon A_b$ is a new arbitrary constant. If we now let ϵ approach zero, the limit of x_c corresponds to $r_1 = r_2$. Let $A_1 = A_a + A_b$ and $A_c = A_2$. Then

$$x_c = A_1 e^{r_1 t} + A_2 t e^{r_1 t}. \tag{11-20}$$

This reasoning is readily extended to show that if a root r_i is repeated m times, the m terms in the complementary function due to the m equal roots will be

$$(x_c)_{r_i} = A_1 e^{r_i t} + A_2 t e^{r_i t} + \cdots + A_m t^{m-1} e^{r_i t}, \tag{11-21}$$

where the A's are arbitrary independent constants. The remaining $(n - m)$ distinct roots give complementary functions of the form (11–19).

The nature of the roots r_k of the characteristic equation (11–18) are of interest since they determine the behavior of the exponentials $e^{r_k t}$ as functions of time. A polynomial of degree n can be written as the product of n factors. By the definition of the r_k as roots of Eq. (11–18), we have

$$a_n s^n + a_{n-1} s^{n-1} + \cdots + a_0 = a_n[(s - r_1)(s - r_2), \ldots, (s - r_n)] = 0. \tag{11-22}$$

To study the properties of these roots and their corresponding exponentials, we may consider the polynomial of any nth order characteristic equation to be composed of the product of lower-order polynomials.

If n is even, the product of $n/2$ second-order polynomials will give an nth order equation. If n is odd, then the product of $(n - 1)/2$ polynomials of degree two with one polynomial of degree one will give an nth order polynomial. Thus we may conclude that the properties of the n roots can be no more complicated than those of first- and second-order characteristic equations. Therefore, let us consider the homogeneous solutions of first- and second-order differential equations.

First-order. The homogeneous first-order equation with constant coefficients is

$$a_1 \frac{dx}{dt} + a_0 = 0, \tag{11-23}$$

which has a characteristic equation

$$a_1 s + a_0 = 0 \tag{11-24}$$

and one root given by

$$r_1 = -a_0/a_1. \tag{11-25}$$

Therefore the solution of Eq. (11–23) is

$$x_c = A_1 e^{r_1 t} = A_1 e^{(-a_0/a_1)t}. \tag{11-26}$$

The quantity $a_1/a_0 = -1/r_1 = \tau$, where τ is the time constant introduced in Chapter 5. Since a_0 and a_1 are real, r_1 is real. If a_0 and a_1 have the same sign, r_1 is negative and x_c tends exponentially to zero for large time. If a_0 and a_2 are of opposite sign, r_1 is positive and x_c tends to infinity for large time. These are the only possible types of behavior of x_c for a first-order system.

Second-order. The most general homogeneous second order equation is:

$$a_2 \frac{d^2x}{dt^2} + a_1 \frac{dx}{dt} + a_0 x = 0. \tag{11-27}$$

The characteristic equation

$$a_2 s^2 + a_1 s + a_0 = 0 \tag{11-28}$$

has two roots, $s = r_1$ and $s = r_2$, given by

$$r_1 = -\frac{a_1}{2a_2} + \sqrt{(a_1/2a_2)^2 - a_0/a_2}, \tag{11-29}$$

$$r_2 = -\frac{a_1}{2a_2} - \sqrt{(a_1/2a_2)^2 - a_0/a_2}. \tag{11-30}$$

Assuming that the coefficients a_0, a_1, and a_2 are real constants, we see that the roots r_1 and r_2 may take several forms, depending on the signs and relative magnitudes of the coefficients. Altogether there are seven cases of interest. In most practical systems the values of a_0, a_1, and a_2 will all be positive. There are then three possible cases to consider.

CASE 1. $a_1^2 > 4a_0a_2$. In this case, from Eqs. (11–29) and (11–30), we see that r_1 and r_2 are *negative, real,* and *unequal.* The complementary solution is therefore

$$x_c = A_1e^{r_1t} + A_2e^{r_2t}. \tag{11-31}$$

The two components of x_c are decaying exponentials whose values tend to zero for large time. This is thus equivalent to two first-order solutions.

CASE 2. $a_1^2 = 4a_0a_2$. Here $r_1 = r_2 = -a_1/2a_2$ since the radicals in Eqs. (11–29) and (11–30) vanish. Equation (11–20) gives the solution to be

$$x_c = (A_1 + A_2t)e-^{(a_1/2a_2)t}. \tag{11-32}$$

This solution also tends to zero for large time.

CASE 3. $a_1^2 < 4a_0a_2$. The roots are now *complex conjugates* of the form

$$r_1 = -\alpha + j\omega_d, \qquad r_2 = -\alpha - j\omega_d, \tag{11-33}$$

where

$$\alpha = a_1/2a_2, \tag{11-34}$$

$$\omega_d = \sqrt{(a_0/a_2) - (a_1/2a_2)^2}. \tag{11-35}$$

The complementary function is

$$\begin{aligned} x_c &= A_1e^{(-\alpha+j\omega_d)t} + A_2e^{(-\alpha-j\omega_d)t} \\ &= e^{-\alpha t}(A_1e^{j\omega_dt} + A_2e^{-j\omega_dt}). \end{aligned} \tag{11-36}$$

By using Euler's formulas, Eqs. (9–41) and (9–42), we can eliminate $e^{j\omega_dt}$ and $e^{-j\omega_dt}$ from (11–36) to get

$$x_c = e^{-\alpha t}(B_1 \sin \omega_dt + B_2 \cos \omega_dt), \tag{11-37}$$

where B_1 and B_2 are real, arbitrary constants. The complementary function in this case is a damped sinusoid whose amplitude decays to zero as time increases.

Occasionally a second-order linear differential equation is encountered in which one or two of the coefficients are negative (if all three are negative, the equation should be multiplied by -1 to give all positive signs). These cases cause responses which contain exponentials with positive real parts and which therefore approach infinity as time increases. These *unstable* cases are considered in the problems at the end of this chapter. The physical significance and examples of stable second-order systems will be discussed in detail in Chapter 12.

As we have stated, the behavior of the complementary functions of any nth order differential equation is always some combination of the first- and second-order behavior discussed in the preceding paragraphs. In general the complementary function will consist of terms of the form (11–26), (11–37), and of limiting forms corresponding to repeated roots.

11–5.2 Particular Integral or Forced Part of the Solution to Nonhomogeneous Equations

We shall now consider the problem of determining a particular integral x_p of a nonhomogeneous equation; i.e., an equation with a nonzero forcing function $\mathfrak{f}(t)$.

Although several methods exist for finding the particular integral of a linear differential with constant coefficients [1, 2], only the method of "undetermined coefficients" will be presented here. This technique is applicable to any equation whose forcing function $\mathfrak{f}(t)$ has only a finite number of linearly independent derivatives. Most forcing functions encountered in physical systems satisfy this restriction.

In the method of undetermined coefficients, the particular integral x_p for Eq. (11–2), where the a's are constant, consists of the sum of $\mathfrak{f}(t)$ and all its derivatives, each multiplied by an undetermined constant. These constants may be found by substituting x_p into the differential equation and determining the values which the constants must have to satisfy the equation.

For illustration, we will consider several examples in which the differential equation is second order.

1) Constant forcing function, $\mathfrak{f}(t) = K$:

$$a_2\frac{d^2x}{dt^2} + a_1\frac{dx}{dt} + a_0x = K. \tag{11-38}$$

Since K has no derivatives, we assume that $x_p = C$, where C is constant. Then Eq. (11–38) is satisfied if

$$C = K/a_0. \tag{11-39}$$

2) Power function in time, $f(t) = Kt^2$:

$$a_2 \frac{d^2x}{dt^2} + a_1 \frac{dx}{dt} + a_0 x = Kt^2. \tag{11-40}$$

Assume a linear combination of the derivatives of Kt^2:

$$x_p = C_1 t^2 + C_2 t + C_3.$$

Substitution of x_p into (11-40) gives

$$2a_2C_1 + 2a_1C_1t + a_1C_2 + a_0C_1t^2 + a_0C_2t + a_0C_3 = Kt^2.$$

When like powers of t are equated, we find

$$C_1 = \frac{K}{a_0},$$

$$C_2 = -\frac{2a_1K}{a_0^2},$$

$$C_3 = \frac{2K}{a_0^3}(a_1^2 - a_0a_2),$$

and therefore

$$x_p = \frac{K}{a_0}\left[t^2 - 2\frac{a_1}{a_0}t + 2\frac{(a_1^2 - a_0a_2)}{a_0^2}\right] \tag{11-41}$$

is a particular integral of (11-40). Clearly, for any other power of t a combination of that power of t and all its derivatives would be assumed.

3) Sinusoidal forcing function, $f(t) = K \sin \omega t$:

$$a_2 \frac{d^2x}{dt^2} + a_1 \frac{dx}{dt} + a_0 x = K \sin \omega t. \tag{11-42}$$

Assume

$$x_p = C_1 \sin \omega t + C_2 \cos \omega t.$$

Substitute x_p into (11-42) and collect terms in $\sin \omega t$ and $\cos \omega t$:

$$(-a_2\omega^2C_1 - a_1\omega C_2 + a_0C_1)\sin \omega t + (-a_2\omega^2C_2 + a_1\omega C_1 + a_0C_2)\cos \omega t = K \sin \omega t.$$

Thus x_p is a solution if

$$(a_0 - a_2\omega^2)C_2 + a_1\omega C_1 = 0,$$
$$-a_1\omega C_2 + (a_0 - a_2\omega^2)C_1 = K,$$

or, solving for C_1 and C_2, we obtain

$$x_p = \frac{K[(a_0 - a_2\omega^2)\sin \omega t - a_1\omega \cos \omega t]}{[(a_0 - a_2\omega^2)^2 + (a_1\omega)^2]}. \tag{11-43}$$

It is left as an exercise for the reader to determine the x_p for $f(t) = K \cos \omega t$

The principle of superposition is often useful in determining particular integrals. For instance, a particular integral of the following equation is obtained simply by adding the particular integrals for each term in the forcing function taken separately:

$$a_2 \frac{d^2x}{dt^2} + a_1 \frac{dx}{dt} + a_0 x = K(1 + t^2 + \sin \omega t).$$

The particular solution is thus the sum of Eqs. (11-39), (11-41), and (11-43).

4) Exponential function, $f(t) = Ke^{st}$:

$$a_2 \frac{d^2x}{dt^2} + a_1 \frac{dx}{dt} + a_0 x = Ke^{st}. \tag{11-44}$$

Assume

$$x_p = Ae^{st}. \tag{11-45}$$

Substitute into (11-44) and solve for A to obtain

$$x_p = \frac{Ke^{st}}{(a_2s^2 + a_1s + a_0)}. \tag{11-46}$$

5) Exception: if any term or derivative of any term on the right-hand side of the differential equation to be solved occurs also in the complementary function, the previous method does not apply. For example if s in Eq. (11-44) is a root of the characteristic equation (11-27), a term Ae^{st} will occur both in x_p and in the complementary function. The denominator of (11-46) then vanishes and x_p cannot be determined from (11-46).

When a term in the assumed particular integral, formed by the method of undetermined coefficients, occurs in the complementary function it is necessary to multiply each such term, its derivatives, and the terms from which it derives, by the lowest integral power of t for which no term in the complementary function appears in the assumed particular integral.

This special situation is illustrated by the following equation:

$$\frac{d^2x}{dt^2} + 3\frac{dx}{dt} + 2x = e^{-t}.$$

The characteristic equation

$$s^2 + 3s + 2 = 0$$

has two roots, $s = -1$ and $s = -2$, and the complementary function is

$$x_c = A_1e^{-t} + A_2e^{-2t}.$$

The assumed particular integral e^{-t} coincides with one term of x_c, so we assume

$$x_p = Ate^{-t}.$$

Substitution into the differential equation gives

$$x_p = te^{-t}.$$

11–5.3 General Solution

The general solution to a linear equation consists of the complementary function of the homogeneous equation (which may be determined as in Section 11–5.1), plus a particular integral (which may usually be found by the method of Section 11–5.2). This general solution of an nth order equation has n independent arbitrary constants and includes all possible solutions which satisfy the differential equation.

11–5.4 Specific Solution, Summary

A specific solution to a differential equation is a general solution in which the n arbitrary constants have been assigned values such that the solution satisfies n specified initial conditions, Eq. (11–3), on x and its first $n - 1$ derivatives.

The procedure for determining the specific solution to a linear nonhomogeneous differential equation with *constant coefficients* may now be summarized.

Given an equation of the form

$$a_n\frac{d^nx}{dt^n} + a_{n-1}\frac{d^{n-1}x}{dt^{n-1}} + \cdots + a_0x = \mathsf{f}(t),$$

where the coefficients a_k are real constants; proceed as follows:

STEP 1. Find the complementary function or natural part of the solution x_c. This is done by solving for the n roots, $r_1, \ldots, r_n$, of the characteristic equation

$$a_ns^n + a_{n-1}s^{n-1} + \cdots + a_0 = 0. \tag{11–47}$$

The complementary function x_c is then

$$x_c = A_1e^{r_1t} + A_2e^{r_2t} + \cdots = \sum_{k=1}^{n} A_ke^{r_kt}. \tag{11–48}$$

If there are repeated roots, refer to Section 11–5.1.

STEP 2. Determine the particular integral or forced part of the solution x_p. If the forcing function $\mathsf{f}(t)$ has a finite number of linearly independent derivatives, a linear combination of $\mathsf{f}(t)$ and its derivatives may be assumed. The constants in the assumed solution are determined by substituting the function into the differential equation.

If the assumed particular integral contains terms which also appear in the complementary function, see Section 11–5.2.

STEP 3. Add the complementary function and the particular integral to obtain the general solution $x = x_c + x_p$.

STEP 4. Evaluate the n arbitrary constants in x by causing x and its first $n - 1$ derivatives to satisfy known initial conditions at a given value of t (usually at $t = 0$), thus forming the desired specific solution.

An example of the above procedure is now given. Let

$$0.25\frac{d^2x}{dt^2} + 0.1\frac{dx}{dt} + x = 10\cos t.$$

The roots of the characteristic equation

$$0.25s^2 + 0.1s + 1 = 0$$

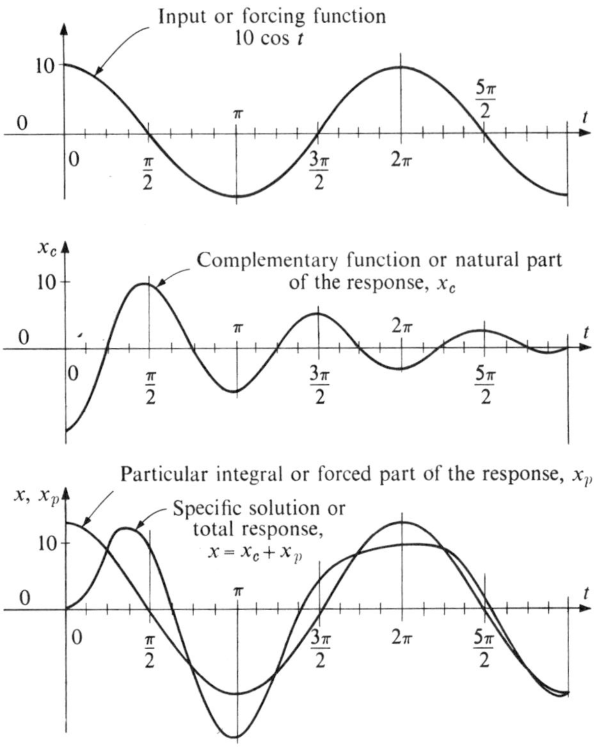

FIG. 11-1. Complementary function, particular integral and specific solution for the equation $0.25\, d^2x/dt^2 + 0.1\,(dx/dt) + x = 10 \cos t$ with $x = 0$ and $dx/dt = 0$ at $t = 0$.

are $r_1 = -0.2 + j1.98$, $r_2 = -0.2 - j1.98$, approximately, and the complementary function is

$$x_c = e^{-0.2t}(A_1 e^{j1.98t} + A_2 e^{-j1.98t}).$$

This may be written in terms of sin $1.98t$ and cos $1.98t$ by the use of Euler's formula, Eq. (11-37):

$$x_c = e^{-0.2t}(B_1 \sin 1.98t + B_2 \cos 1.98t).$$

The particular integral is found by assuming a solution

$$x_p = C_1 \sin t + C_2 \cos t.$$

Substituting this into the differential equation, as in Example 3, Section 11-5.2, gives

$$x_p = 1.75 \sin t + 13.11 \cos t.$$

The general solution is then

$$x = e^{-0.2t}(B_1 \sin 1.98t + B_2 \cos 1.98t) + 1.75 \sin t + 13.11 \cos t.$$

Now determine B_1 and B_2 by setting $x(0+) = \dot{x}(0+) = 0$:

$$\dot{x}(0+) = B_2 + 13.11 = 0,$$

$$\dot{x}(0+) = 1.98B_1 - 0.2B_2 + 1.75 = 0,$$

or $B_2 = -13.11$, $B_1 = -2.2$. The specific solution is therefore

$$x = e^{-0.2t}(-2.2 \sin 1.98t - 13.11 \cos 1.98t) + 1.75 \sin t + 13.11 \cos t.$$

Figure 11-1 shows this solution, together with its components x_c and x_p. This procedure will be used frequently in our later work in this book.

11-6. LINEAR EQUATIONS WITH NONCONSTANT COEFFICIENTS

Linear differential equations whose coefficients are not constant but depend on the independent variable t obey the superposition principle of Section 11-4.2. Their solutions also consist of a complementary part containing n arbitrary constants and a particular integral which depends on the forcing function $f(t)$. In spite of these properties, however, the determination of analytical solutions to this type of equation is very difficult, and solutions have been found for only a few special types of equations. When solutions must be obtained, graphical, numerical, and analog computer methods are usually employed because direct analytical techniques either do not exist or are excessively complicated.

11-7. NONLINEAR EQUATIONS

When nonlinear differential equations are encountered, the problems of obtaining analytical solutions are extremely formidable. Superposition does not hold, so the solution cannot be constructed by adding together a natural part and a forced part. Proportionality also does not hold, and the solution for a particular forcing function may change character completely as the amplitude of the forcing function is changed. In the case of linear equations, we have seen that for stable situations the forced part of the response is similar to the forcing function. For instance, if the forcing function is sinusoidal at frequency ω, the forced part of the solution is also sinusoidal at frequency ω. This is not so with a nonlinear equation. In fact, a nonlinear system forced by a sinusoid of frequency ω may respond with nonsinusoidal periodic oscillations at frequency ω/α, where α is an integer greater than one (subharmonic resonance). The steady-state response of a nonlinear system may also depend on the initial conditions of the system when the forcing function is applied, whereas in linear systems the steady-state response is independent of these initial conditions.

Very few nonlinear equations have explicit analytical solutions, but various techniques exist which permit nonlinear solutions to be found by combining mathematical and numerical analysis. Reference 3 describes some of the more useful methods of treating nonlinear equations. Analog or digital computation is almost always required to solve nonlinear equations of higher than second order.

Sometimes the solution of a nonlinear differential equation is required only for small values of the dependent variable. Then it is often possible to *linearize* the equation and obtain an approximate linear solution. For example, the reader may verify that the differential equation for the motion of the simple pendulum in Fig. 11–2 is

$$\frac{d^2\theta}{dt^2} + \frac{g}{l}\sin\theta = \frac{T}{ml^2}. \tag{11–49}$$

This equation is nonlinear because of the $\sin\theta$ term. If we are interested only in small variations $\Delta\theta$ from an initial equilibrium point $\theta = \theta_0$, we can write

$$\theta = \theta_0 + \Delta\theta,$$

$$\frac{d^2(\theta_0 + \Delta\theta)}{dt^2} + \frac{g}{l}\sin(\theta_0 + \Delta\theta) = \frac{T_0 + \Delta T}{ml^2},$$

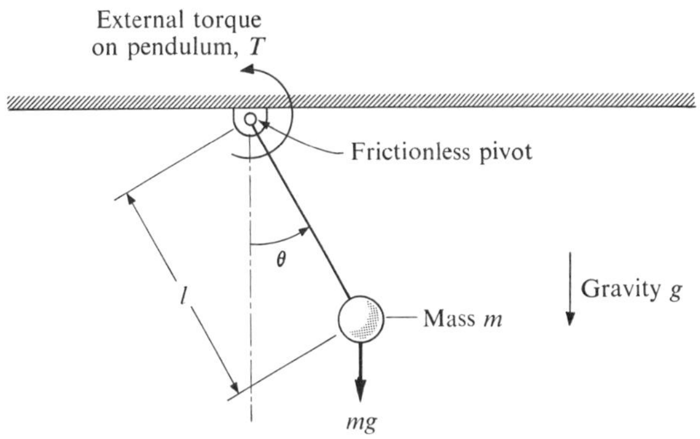

FIG. 11–2. Simple pendulum in a gravity field.

where T_0 is the equilibrium torque corresponding to θ_0 and ΔT is a small change in T. Expanding $\sin(\theta_0 + \Delta\theta)$, we obtain

$$\frac{d^2\Delta\theta}{dt^2} + \frac{g}{l}(\sin\theta_0\cos\Delta\theta + \cos\theta_0\sin\Delta\theta) = \frac{T_0 + \Delta T}{ml^2}. \tag{11–50}$$

For equilibrium, $\Delta\theta = 0$, and (11–50) gives

$$\frac{g}{l}\sin\theta_0 = \frac{T_0}{ml^2}. \tag{11–51}$$

If $\Delta\theta$ is small so that $\cos\Delta\theta \approx 1$ and $\sin\Delta\theta \approx \Delta\theta$, Eq. (11–50) combined with (11–51) yields

$$\frac{d^2\Delta\theta}{dt^2} + \frac{g\cos\theta_0}{l}\Delta\theta = \frac{\Delta T}{ml^2}. \tag{11–52}$$

This equation is now linear in $\Delta\theta$ with constant coefficients and is readily solved for $\Delta\theta$ given the forcing function ΔT. If the variations $\Delta\theta$ become so large that $\sin\Delta\theta$ differs considerably from $\Delta\theta$, then the linear solution is invalid and the original nonlinear equation must be solved to determine θ. Problem 11–27 at the end of this chapter illustrates a useful method, called *phase-plane analysis*, which permits Eq. (11–49) to be solved.

In the remainder of this book our primary concern will be with the analysis of linear systems, to which Section 11–5 of this chapter is applicable.

REFERENCES

1. Wilfred Kaplan, *Ordinary Differential Equations.* Reading, Mass.: Addison-Wesley Publishing Company, 1961.
2. W. T. Martin and E. Reissner, *Elementary Differential Equations.* Reading, Mass.: Addison-Wesley Publishing Company, Inc., 1961.
3. R. A. Struble, *Nonlinear Differential Equations.* New York: McGraw-Hill Book Company, Inc., 1962.

PROBLEMS

11–1. Derive the governing equations for the systems in Figs. 11–3, 11–4, 11–5, and 11–6. Find the order of each equation and state whether it is linear or nonlinear, homogeneous or nonhomogeneous, and whether it has constant or nonconstant coefficients. Obtain solutions for those equations that can be solved with the technique given in this chapter.

a) The displacement of m is the dependent variable in Fig. 11–3. The friction coefficient (friction force/normal force) between the inclined plane and the mass is proportional to the square of the velocity. The spring is an ideal spring.

b) The vertical displacement of m is the dependent variable in Fig. 11–4. V is constant and all elements are ideal. Note that this is a model of an automobile suspension system.

c) The dependent variable is the current in Fig. 11–5. The capacitor is charged and then at $t = 0$ the switch is closed. The resistance is proportional to the sinh of the current; $R = R_0 \sinh i$.

d) The flywheel in Fig. 11–6 is given an initial angular displacement and released. All elements are ideal.

11–2. Solve the differential equations

a) $(x^2 + x)\dfrac{dv}{dx} = (1 + v^2)$,

b) $(x^2 + 4xy + 6)\dfrac{dy}{dx} + (4 + 2xy + 2y^2) = 0$.

11–3. Given the equation

$$\frac{dx}{dt} + \frac{2}{t}x = 3 \sin \omega t + \omega t \cos \omega t.$$

a) What type of differential equation is this?

b) Verify that $x_p = t \sin \omega t$ is a particular integral.

c) Find the complementary function and the general solution.

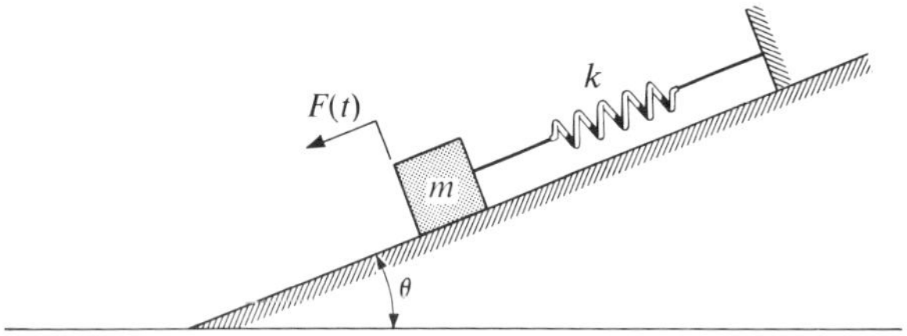

FIGURE 11–3

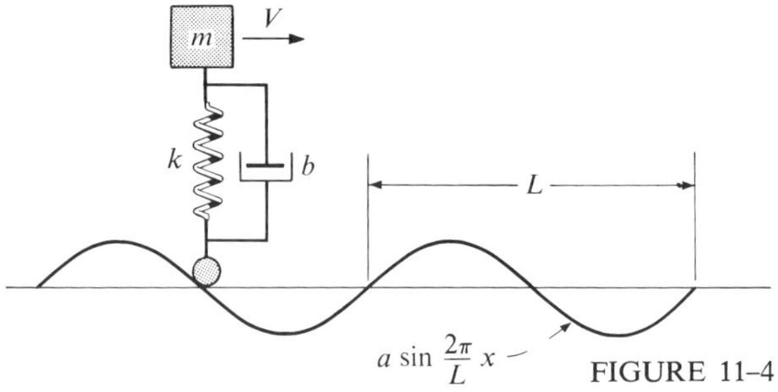

FIGURE 11–4

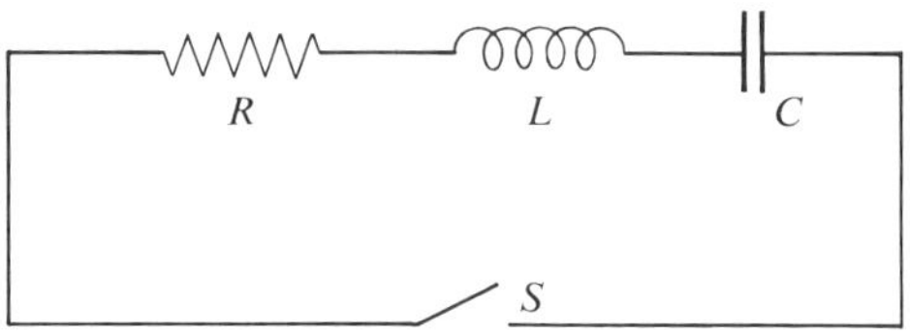

FIGURE 11–5

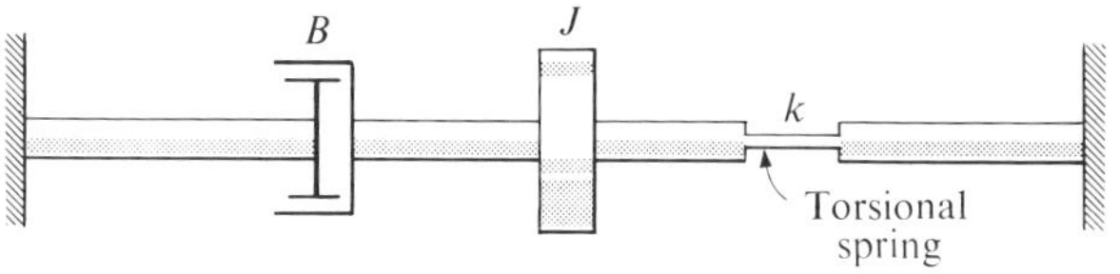

FIGURE 11–6

11–4. In the equation

$$a_2 \frac{d^2x}{dt^2} + a_1 \frac{dx}{dt} + a_0 x = y(t),$$

$$x(0+) = \dot{x}(0+) = 0,$$

Show that if $x(t)$ is the specific solution,

a) dx/dt is the specific solution when the input is dy/dt,

b) $\int_0^t x\,dt$ is the specific solution when $\int_0^t y\,dt$ is the input.

11–5. In the equation

$$a_n \frac{d^n x}{dt^n} + \cdots + a_1 \frac{dx}{dt} + a_0 x = \mathsf{f}(t),$$

all n initial conditions are zero at $t = 0$. Let $\mathsf{f}(t) = y_1 + jy_2$, where y_1 and y_2 are real functions of time.

a) Show that the solution $x(t)$ must be expressible as $x = x_1 + jx_2$ where x_1 and x_2 are real functions.

b) Show that x_1 is the response due to y_1 and x_2 is the response due to y_2, thus proving Case 3 of Section 11–4.3.

11–6. For the equation

$$\frac{d^2x}{dt^2} + 4\frac{dx}{dt} + 5x = 0, \qquad x(0+) = 10 \text{ and } \dot{x}(0+) = 0.$$

a) Determine the response; i.e., solve the differential equation.

b) Roughly plot the response.

11–7. Solve the following differential equations:

$$3\frac{dv}{dt} + 2v = 0 \qquad \text{at } t = 0, \quad v = 2.$$

$$5\frac{di}{dt} - 3i = 0 \qquad \text{at } t = 0, \quad i = 3.$$

$$\frac{d^2x}{dt^2} + 25x = 0 \qquad \text{at } t = 0, \quad \begin{cases} dx/dt = 0, \\ x = 2. \end{cases}$$

$$3\frac{d^2i}{dt^2} + 9i = 0 \qquad \text{at } t = 0, \quad \begin{cases} di/dt = 3, \\ i = 1. \end{cases}$$

$$3\frac{d^2i}{dt^2} + 12\frac{di}{dt} + 9i = 0; \quad \text{at } t = 0+, \quad i = 1, \quad di/dt = 1.$$

$$3\frac{d^2i}{dt^2} + 12\frac{di}{dt} + 9i = 0; \quad \text{at } t = 0, \quad i = 0, \quad di/dt = 1.$$

$$\frac{dq}{dt} + 3q = F_0; \quad F_0 = 5; \quad \text{at } t = 0, \quad q = 1.$$

$$\frac{dq}{dt} + 3q = 5e^t - 4t; \qquad \text{at } t = 0, \quad q = 1.$$

11–8. Solve the differential equations

$$\frac{dq}{dt} + 3q = 5e^{-3t}; \qquad \text{at } t = 0+, \quad q = 1.$$

$$\frac{d^2x}{dt^2} + 25x = e^t + 5\cos 2t + t^2;$$

find the particular integral only.

$$\frac{dq}{dt} + 3q = 2\sin t; \qquad \text{at } t = 0+, \quad q = 1.$$

$$\frac{d^2q}{dt^2} + 9q = 9\sin t; \quad \text{at } t = 0+, \quad q = 0, \quad \frac{dq}{dt} = 0.$$

11–9. Obtain the complete solutions of all the following equations for which the particular integrals can be obtained by the method of undetermined coefficients. Do not solve the equations to which this method does not apply.

a) $\dfrac{d^2y}{dx^2} - 3\dfrac{dy}{dx} + 2y = x^2 + 2e^{-x} - \cos x.$

b) $\dfrac{d^2y}{dx^2} + 5y = 2e^x + 2e^{-x} + x.$

c) $\dfrac{d^3y}{dx^3} + 3\dfrac{d^2y}{dx^2} - \dfrac{dy}{dx} - 3y = a\cosh x + b\sinh x.$

d) $\dfrac{dy}{dx} + 4y = x^2 + \ln x.$

e) $\dfrac{d^3y}{dx^3} + \dfrac{dy}{dx} = x^2 + \cosh x + \sinh x.$

11–10. Graphically construct the rotating phasors which represent the complementary function and the particular integral of the following equation assuming that $x_r(0+) = 1$, $\dot{x}_r(0+) = 0$. Sketch the two components of the solution and the complete solution and check that the initial conditions are satisfied.

$$3\frac{d^2x_r}{dt^2} + 5\frac{dx_r}{dt} + 27x_r = 5\frac{dx_f}{dt} + 2x_f,$$

$$x_f = A\sin \omega t.$$

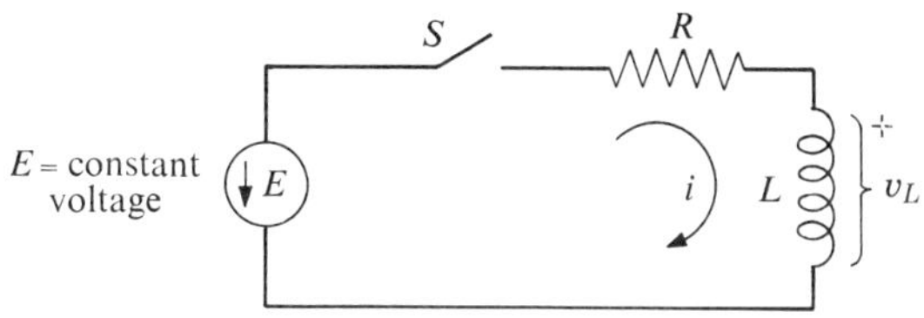

FIGURE 11–7

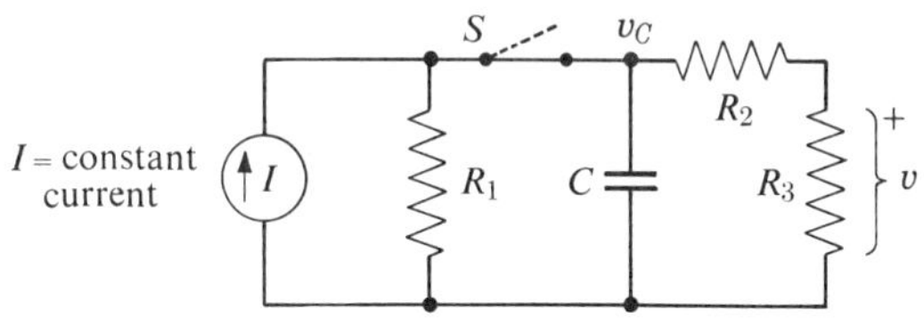

FIGURE 11–8

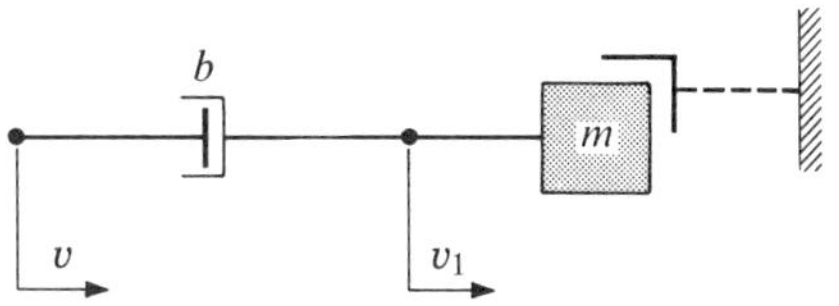

FIGURE 11–9

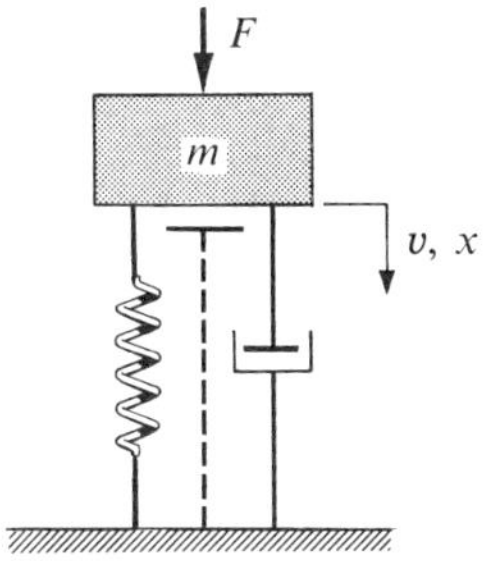

FIGURE 11–10

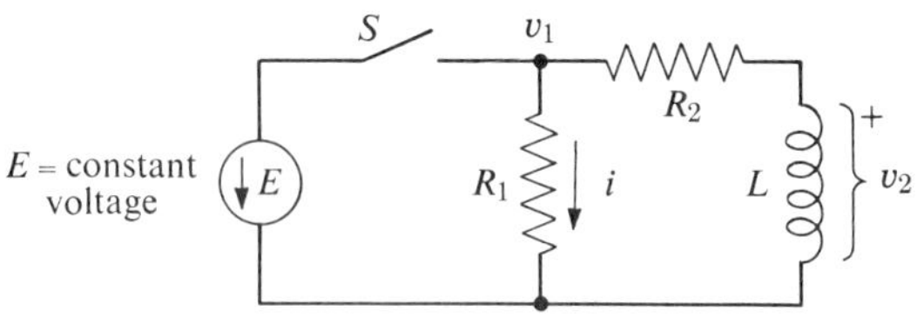

FIGURE 11–11

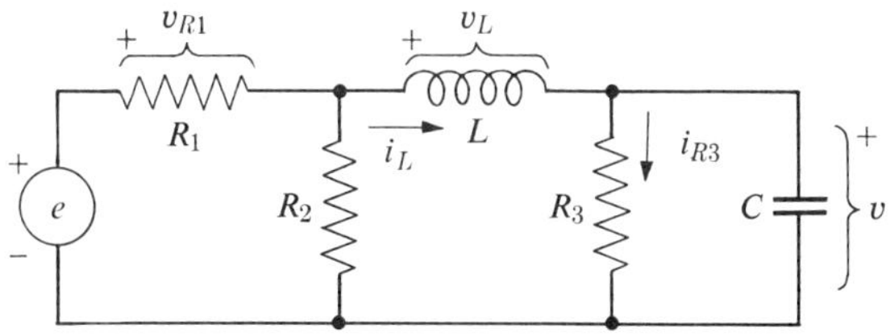

FIGURE 11–12

11–11. In the circuit in Fig. 11–7 S is suddenly closed at $t = 0$. Find $i(0+)$ and $v_L(0+)$.

11–12. At $t = 0$, S is suddenly opened (Fig. 11–8). Find $v(0+)$ and $v_C(0+)$. Prior to $t = 0$, S has been closed for a long time.

11–13. In the mechanical system in Fig. 11–9, v has been constant equal to V_0 for a long time and suddenly becomes zero at $t = 0$. Find $v_1(0+)$ and $\dot{v}_1(0+)$.

11–14. The force F acting on the mass-spring-damper system (Fig. 11–10) has been zero for a long time, so $x = v = 0$ and is then changed at $t = 0$ to the function $F = F_0 \cos \omega t$. Find $x(0+)$, $v(0+)$ and solve for the motion $v(t)$ for $t > 0$.

11–15. In the circuit in Fig. 11–11 switch S has been closed for a long time and is suddenly opened at $t = 0$. Find $v_1(0-)$, $v_2(0-)$, $i(0-)$, $v_1(0+)$, $v_2(0+)$, $i(0+)$.

11–16. At $t = 0$ when i_L and v are both zero in Fig. 11–12, e is suddenly increased from zero to E and held constant. Find $v_{R1}(0+)$, $v_L(0+)$, $i_{R3}(0+)$, $v(0+)$, and $\dot{v}(0+)$.

11–17. A differential equation for a physical system is

$$a_2 \frac{d^2x}{dt^2} + a_1 \frac{dx}{dt} + a_0 x = b_2 \frac{d^2y}{dt^2} + b_1 \frac{dy}{dt} + b_0 y,$$

where x is the output, y is the input, and the a's and b's are constant. For $t < 0$, $x \equiv 0$. The input is $y(t) = Yu_s(t)$, a step function of magnitude Y.

a) By integrating the differential equation between $t = 0^-$ and $t = 0+$, find the initial conditions $x(0+)$ and $\dot{x}(0+)$.

b) Show that the output acceleration d^2x/dt^2 suddenly jumps at $t = 0$ from zero to $b_2 Y/a_2$.

11–18. Solve the following differential equation and roughly plot the result:

$$2\frac{d^2x}{dt^2} + 4\frac{dx}{dt} + 3x = 9 \sin 2t,$$

$$x(0+) = \dot{x}(0+) = 0.$$

11–19. Solve the following differential equation and plot each case:

$$\frac{d^2x}{dt^2} + \beta^2 x = 10 \sin \omega t, \qquad x(0+) = \dot{x}(0+) = 0.$$

a) for $\beta = 2$ and $\omega = 4$, b) for $\beta = 3$ and $\omega = 3$.

11–20. Solve the following differential equation and plot x_c, x_p, and $x(t)$:

$$M\frac{d^2x}{dt^2} + B\frac{dx}{dt} + Kx = 10, \qquad x(0+) = \dot{x}(0+) = 0,$$

for

a) $B = 4, M = 2, K = 8,$
b) $B = 4, M = 2, K = 2,$
c) $B = 4, M = 2, K = 1.$

11–21. Given the equation

$$4\frac{d^2v}{dt^2} + 2\frac{dv}{dt} + v = 2t + e^t,$$

$t = 0, \quad v = 10, \quad$ and $\quad dv/dt = 0.$

a) Find v_1, the response due to the initial conditions.
b) Find v_2, the response to $2t$.
c) Find v_3, the response to e^t.
d) Separately find v the specific solution when $2t$, e^t, and the initial conditions are present.
e) Verify that $v = v_1 + v_2 + v_3$, as stated by Eq. 11–16.

11–22. Assume that three roots of a system characteristic equation are r, $r + \epsilon_1$, and $r + \epsilon_2$, where ϵ_1 and ϵ_2 are small. By expanding $e^{\epsilon_1 t}$ and $e^{\epsilon_2 t}$ in power series about $t = 0$ and letting ϵ_1 and ϵ_2 approach zero, show that the complementary function for three equal roots can be written $x_c = (A_1 + A_2 t + A_3 t^2)e^{rt}$.

11–23. Prove that Eq. (11–37) is equivalent to (11–36) and solve for B_1 and B_2 in terms of A_1 and A_2.

11–24. In the following equation suppose that $a_1 < 0$ and a_2 and a_0 are positive. Solve for $x(t)$, and sketch the result qualitatively.

$$a_2\frac{d^2x}{dt^2} + a_1\frac{dx}{dt} + a_0 x = 0,$$

$$x(0+) = 0, \qquad \dot{x}(0+) = V_0,$$

11–25. Repeat Problem 24, assuming that $a_0 < 0$, a_1 and a_2 are positive, and that $a_1^2 < 4|a_0|a_2$.

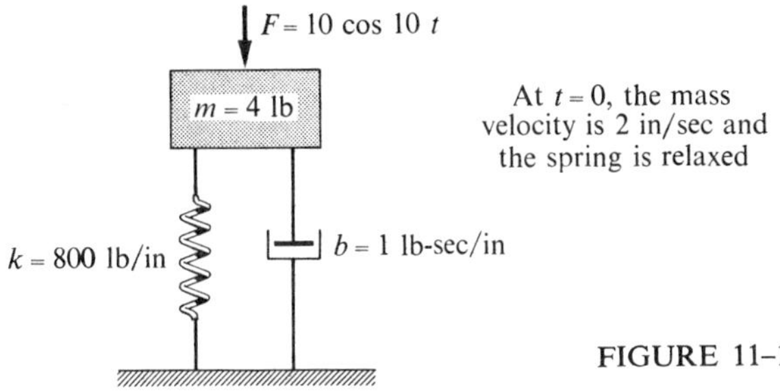

FIGURE 11–13

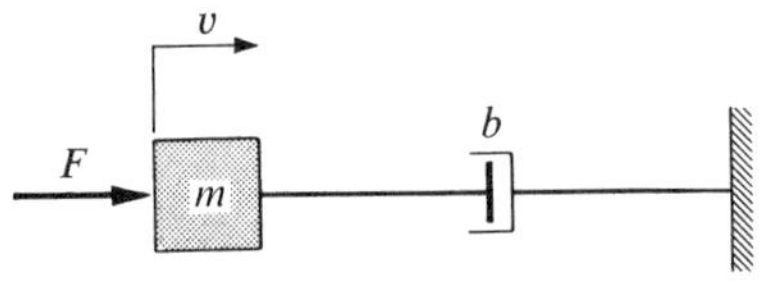

FIGURE 11–14

11–26. Repeat Problem 24, assuming that $a_2 < 0$, a_1 and a_0 are positive and that $a_1^2 < 4a_0|a_2|$.

11–27. Consider Eq. (11–49) for $T = 0$:

$$\frac{d^2\theta}{dt^2} + \frac{g}{l}\sin\theta = 0.$$

Let $d_2\theta/dt^2 = (d\dot{\theta}/d\theta)\dot{\theta}$ and eliminate time as an explicit variable in the above equation. Solve for $\dot{\theta}$ as a function of θ for given initial conditions $\dot{\theta}_0$ and θ_0. The locus of $\dot{\theta}$, θ in cartesian coordinates is called a *phase-plane* trajectory. Plot loci of $\dot{\theta}$ vs. θ for the following cases:

a) $\theta_0 = 0, \quad \dot{\theta}_0 = \sqrt{g/l},$
b) $\theta_0 = 0, \quad \dot{\theta}_0 = \sqrt{2g/l},$
c) $\theta_0 = 0, \quad \dot{\theta}_0 = 2\sqrt{g/l}.$

What conclusions can you draw from these trajectories regarding the types of possible motions of the pendulum?

11–28. Find and sketch the velocity of the mass as a function of time (Fig. 11–13).

11–29. A sinusoidal force $F = F_0 \sin\omega t$ is *suddenly* applied at $t = 0$ to the mass-dashpot system in Fig. 11–14. Find the motion v of the mass as a function of time. Sketch the velocity vs. time for each of the following cases:

a) $\frac{m}{b}\omega = 0.1, \quad$ b) $\frac{m}{b}\omega = 1.0, \quad$ c) $\frac{m}{b}\omega = 10.0.$

For each case, about how many cycles of the input must elapse before the natural part of the output velocity falls to 5% of the amplitude of the steady sinusoidal part of the response?

12 TRANSIENT RESPONSE OF SIMPLE LINEAR SYSTEMS

12–1. INTRODUCTION

In Chapter 11 it was shown that the solution of a linear differential equation consists of two parts: the *complementary function* and the *particular integral.* When analyzing the response of physical systems to specified inputs, the more descriptive terms *natural part* and *forced part* are most often used for the two parts of the response, corresponding respectively to the *complementary function* and the *particular integral* of the complete solution of the differential equation.

The natural part, closely related to dependent variable terms of the equation (left-hand side of the differential equation), contains n arbitrary constants; and the forced part, closely related to the forcing function (right-hand side of the differential equation), is of the same form as the input forcing function. The n constants are determined by requiring the complete response to satisfy n known initial conditions.

In this chapter, the methods of Chapter 11 will be used to find the response of a system to a single input forcing function (Fig. 12–1) which is a simple singularity function (see Chapter 10). The real system, described by an approximate idealized mathematical model, will have an actual response which is approximated by the response determined from this idealized mathematical model. The degree of approximation involved in the mathematical modeling is often discernible when the computed response can be compared with an actual response. Thus determination of system response is important from an experimental as well as an analytical point of view.

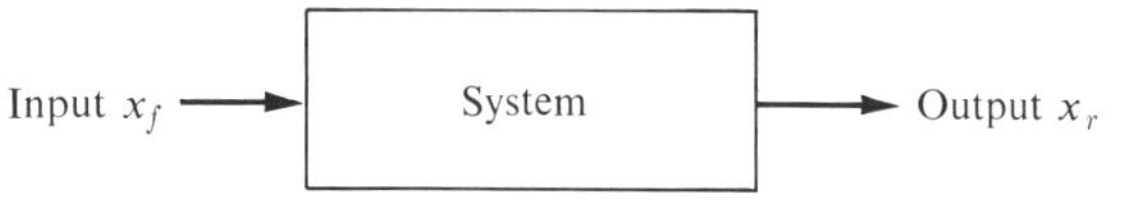

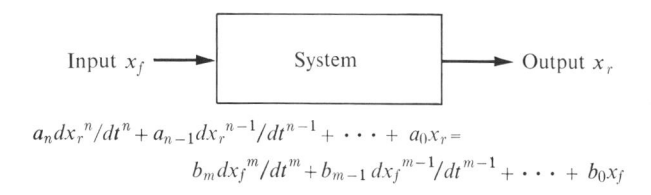

$$a_n dx_r^{\,n}/dt^n + a_{n-1}dx_r^{\,n-1}/dt^{n-1} + \cdots + a_0 x_r = b_m dx_f^{\,m}/dt^m + b_{m-1}\,dx_f^{\,m-1}/dt^{m-1} + \cdots + b_0 x_f$$

FIG. 12–1. Block diagram of system showing input or forcing function and output or response.

When a linear dynamic system is excited or forced by one of the singularity functions discussed in Chapter 10, the natural part of the response is clearly discernible from the forced part of the response. When the natural part of the response decays with time and approaches zero for large values of time, sometimes it is called the *transient* part of the response. The forced part of the response, being of the same family as the input singularity function, has at least some steady character and sometimes it is called the *steady-state* part of the response. Although only the natural part of the response is really a transient, it is common to find that the complete response is called a transient response.

In this chapter we shall deal only with systems described by linear differential equations with constant coefficients such as those composed of the ideal linear elements discussed earlier in this book. Moreover, we shall concentrate on systems in which only one input variable is present, although the principle of superposition makes it possible to extend these results to situations where several inputs may occur simultaneously. Also, we shall insist that no changes in the structure or composition of the system shall be allowed during the interval in which a system response is to be analyzed. Most of the material will deal with responses to step changes of the input variable, although a brief discussion of responses to ramp, impulse, and pulse inputs is included in the last section.

12–2. RESPONSE OF FIRST-ORDER SYSTEMS TO STEP INPUTS

In Chapter 5 the transient response of a simple first-order system was actually determined by four different methods: analytical, graphical, analog computer, and digital computer. Thus there is little need here to be concerned with these methods of solution. The main purpose of this section is to summarize the step-response characteristics of first-order systems in terms of the classical methods discussed in the previous chapter, and to demonstrate by example the characteristics of the natural and forced parts of these responses. This summary will provide an effective base on which to build a discussion of the transient responses of second-order systems.

12–2.1 Response of Primary Output Variable

Usually there is one particular variable in the system which is of primary interest as an output. Since all the other dependent variables of the system also can be considered outputs, we consider them outputs of secondary interest.

The system consisting of a spring and two dampers shown in Fig. 12–2 is a first-order system which can serve very effectively to illustrate and summarize all the important aspects of the transient response of first-order systems. We shall determine the response of the velocity v_2 to a step change in v_1, $v_1 = V_1 u_s(t)$, for a time when the system has been in a motionless state before the occurrence of the step input. The forces F_{b1}, F_{k1}, and F are possible outputs of secondary interest.

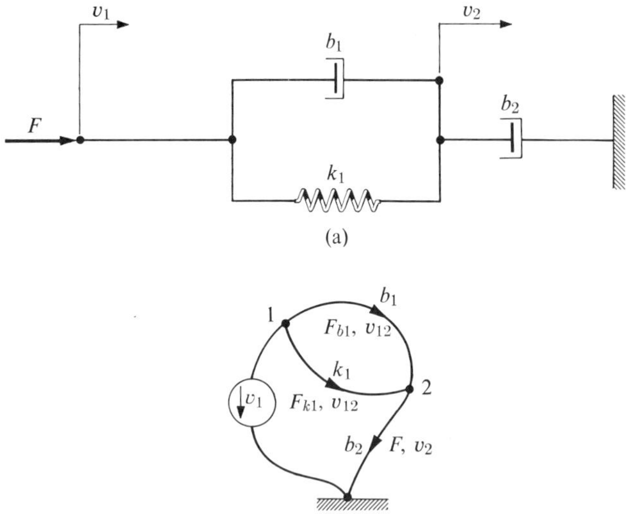

FIG. 12–2. System consisting of a spring and two dampers analyzed in Chapter 5.

The differential equation relating the force F to the velocity v_1 was derived in Chapter 5 and given in Eq. (5–72):

$$(b_1 + b_2)(dF/dt) + k_1 F = b_2(b_1\, dv_1/dt + k_1 v_1). \tag{5–72}$$

Noting that $F = b_2 v_2$, we may substitute $b_2 v_2$ for F in Eq. (5–72),

$$(b_1 + b_2)\, dv_2/dt + k_1 v_2 = b_1\, dv_1/dt + k_1 v_1, \tag{12–1}$$

to obtain the system differential equation relating v_2 to v_1. Because Eq. (12–1) contains both the input and its first derivative on the right-hand side, it is a more general form of first-order system differential equation than the ones which were solved in Chapter 5.

The *natural part* of the response is the natural part of the solution of the system differential equation discussed in Chapter 11, and it is of the form

$$(v_2)_n = Ae^{rt}, \tag{12–2}$$

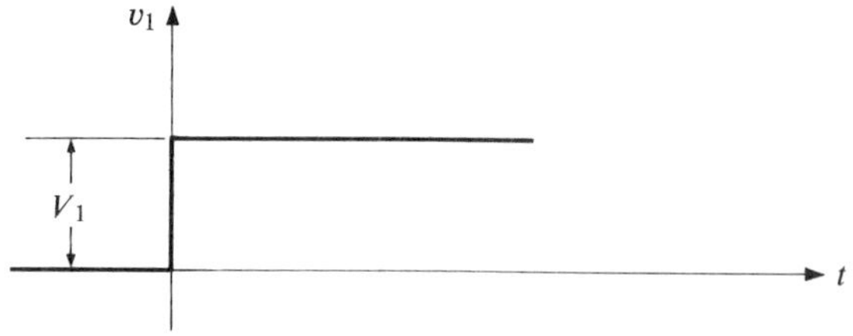

FIG. 12–3. Graphical representation of step change in input v_1.

where r is the root of the characteristic equation

$$(b_1 + b_2)s + k_1 = (b_1 + b_2)(s - r), = 0$$

$$r = -\frac{k_1}{b_1 + b_2} = -\frac{1}{\tau_1}. \tag{12–3}$$

Thus the natural part of the response is a transient consisting of a decaying exponential having a time constant τ_1.

The *forced part* of the response is the forced part of the solution of the system differential equation which, being of the same form as the input, is a constant satisfying the system equation for $t > 0$,

$$k_1(v_2)_f = k_1(v_1) = k_1V_1,$$

$$(v_2)_f = V_1. \tag{12–4}$$

The *complete response* is then given by

$$v_2 = (v_2)_n + (v_2)_f = Ae^{-t/\tau_1} + V_1. \tag{12–5}$$

To determine A, we need to know the value of v_2 at $t = 0+$. In this case it is necessary to determine if v_2 suddenly changes at $t = 0$, and if so, how much. Because of continuity, the force transmitted to b_2 must be F. Also the force F_{k1} at $t = 0+$ must be zero unless $(v_1 - v_2)$ goes to infinity at $t = 0$. Here it is evident that neither v_1 nor v_2 becomes infinite at $t = 0$ because v_1 is finite as given (Fig. 12–3) and there is no source of infinite force to make v_2 infinite. Thus F_{b1} at $t = 0+$ is equal to F, so that

$$b_1(v_1 - v_2) = b_2v_2,$$

$$(b_1 + b_2)v_2 = b_1v_1,$$

$$v_2(0+) = b_1v_1/(b_1 + b_2) = b_1V_1/(b_1 + b_2). \tag{12–6}$$

Thus v_2 suddenly changes from zero to $b_1V_1/(b_1 + b_2)$ at $t = 0$, so that

$$v_2(0+) = b_1V_1/(b_1 + b_2) = A + V_1$$

and

$$A = \left(\frac{b_1}{b_1 + b_2} - 1\right)V_1 = \left(\frac{-b_2}{b_1 + b_2}\right)V_1. \tag{12–7}$$

The complete response thus is given by

$$v_2 = \underbrace{\left(\frac{-b_2}{b_1 + b_2}\right)V_1e^{-t/\tau_1}}_{\text{Natural part}} + \underbrace{V_1}_{\text{Forced part}}. \tag{12–8}$$

The natural and forced parts of the response, together with the complete response, are illustrated in Fig. 12–4.

12–2.2 Responses of Secondary Output Variables

Having determined the transient response of v_2 to a step change in v_1, we may now use this result to determine the transient responses of F_{b1}, F_{k1}, and F to the same input. To find F_{b1} it is necessary only to find $(v_1 - v_2)$ and multiply it by b_1. For $t > 0$,

$$v_1 - v_2 = \left(\frac{b_2}{b_1 + b_2}\right)V_1e^{-t/\tau_1},$$

so that

$$F_{b1} = \left(\frac{b_1b_2}{b_1 + b_2}\right)V_1e^{-t/\tau_1}. \tag{12–9}$$

Thus the steady-state part of the response of F_{b1} to a step change in v_1 is zero. This response of F_{b1}, together with the differential equation relating F_{b1} to v_1, is shown in Fig. 12–4.

To find F_{k1}, we may integrate v_{12} with respect to time and multiply the integral by k_1,

$$F_{k1} = k_1\int_{0+}^{t} v_{12}\,dt + F_{k1}(0+).$$

It was shown earlier that $F_{k1}(0+) = 0$, so that

$$F_{k1} = \left(\frac{k_1b_2}{b_1 + b_2}\right)(-\tau_1V_1)[e^{-t/\tau_1} - 1]. \tag{12–10}$$

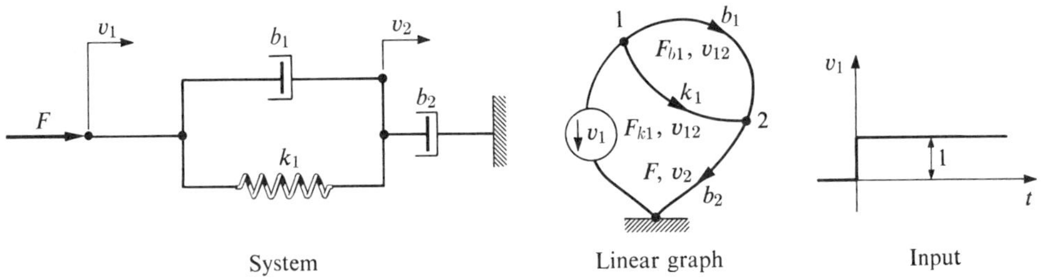

Output variable	v_2	F_{b1}	F_{k1}	F
Equation	$(b_1+b_2)\,dv_2/dt+k_1v_2=$ $b_1dv_1/dt+k_1v_1$	$(b_1+b_2)\,dF_{b1}/dt+k_1F_{b1}=$ $b_1b_2\,dv_1/dt$	$(b_1+b_2)\,dF_{k1}/dt+k_1F_{k1}=$ $b_2k_1v_1$	$(b_1+b_2)\,dF/dt+k_1F=$ $b_2b_1dv_1/dt+b_2k_1v_1$

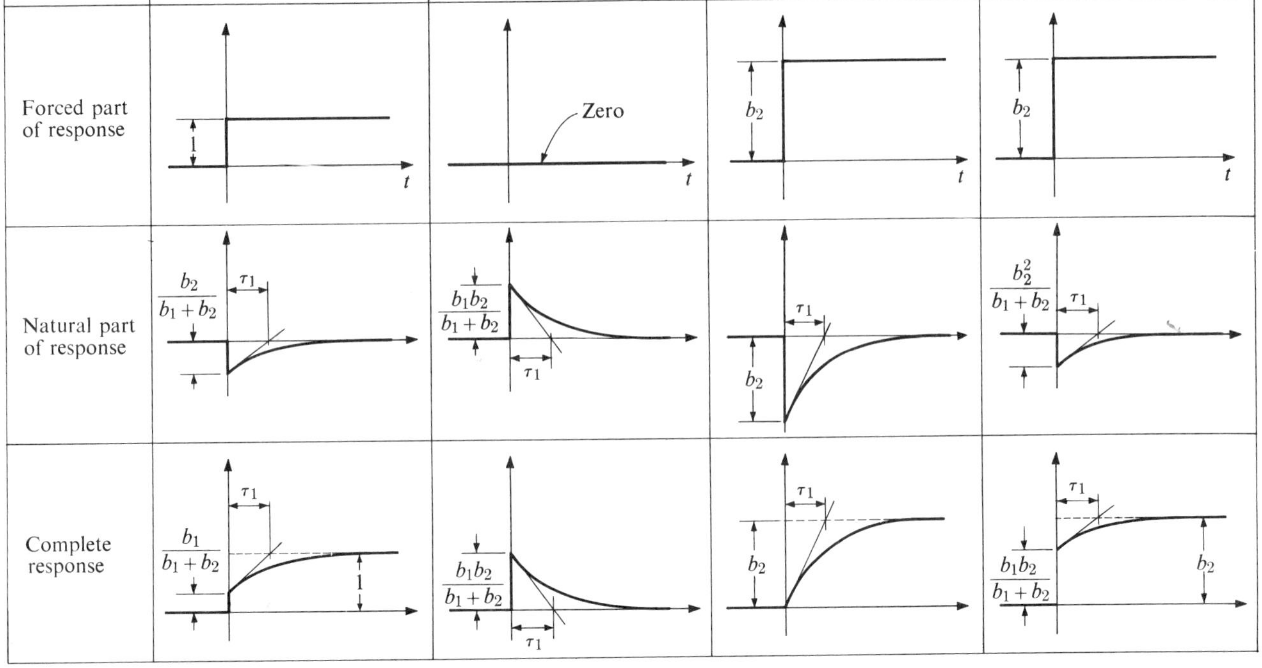

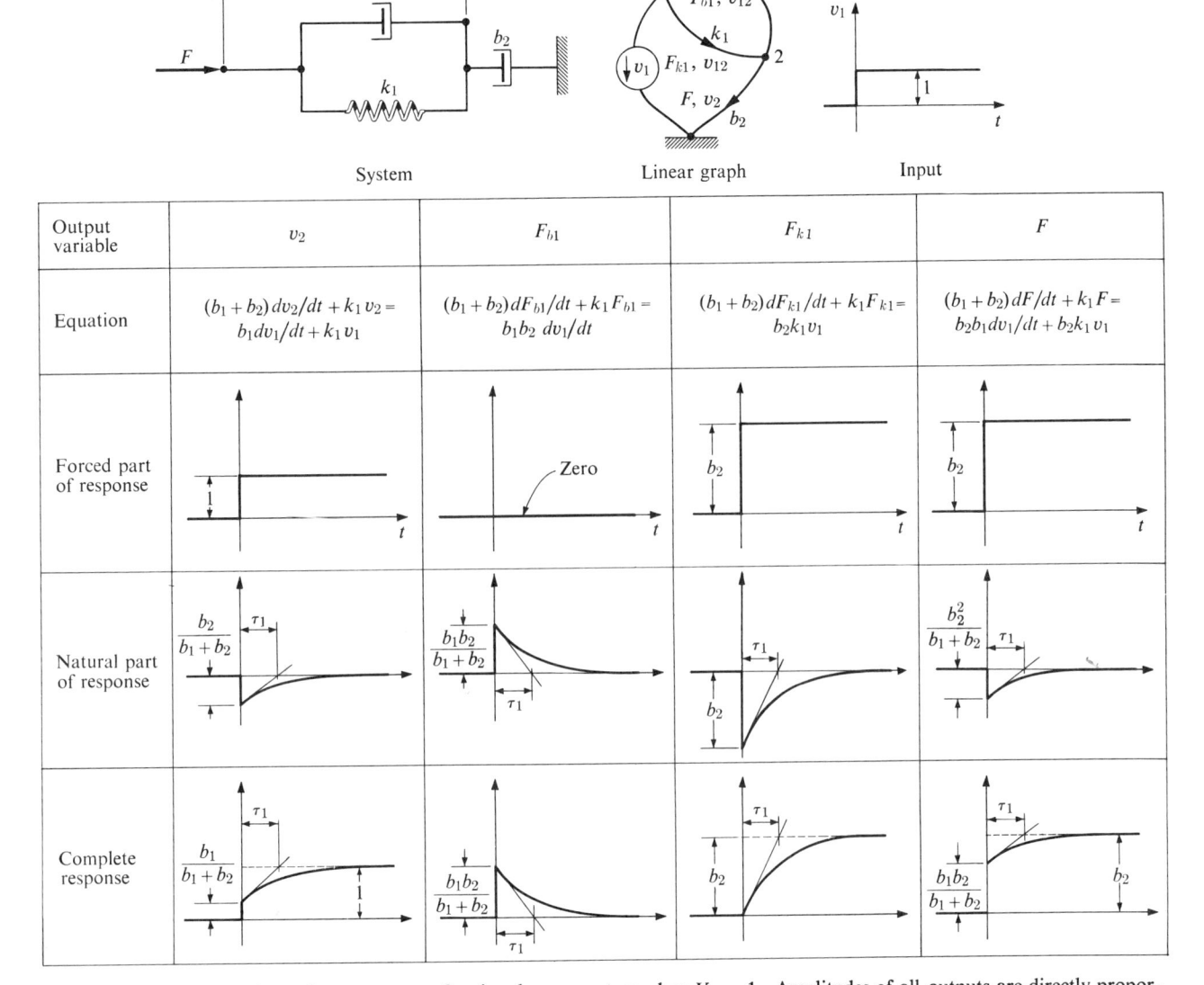

FIG. 12–4. Summary of transient responses of spring-damper system when $V_1 = 1$. Amplitudes of all outputs are directly proportional to V_1, and $\tau_1 = (b_1 + b_2)/k_1$.

And since

$$\tau_1 = \frac{b_1 + b_2}{k_1},$$

$$F_{k1} = \underbrace{-b_2 V_1 e^{-t/\tau_1}}_{\text{Natural part}} + \underbrace{b_2 V_1}_{\text{Forced part}}. \tag{12–11}$$

The natural part, forced part, and complete response of F_{k1}, together with the differential equation relating F_{k1} to v_1, are shown in Fig. 12–4.

Finally, the force F is determined by summing F_{b1} and F_{k1}:

$$F = \left(\frac{b_1 b_2}{b_1 + b_2} - b_2\right) V_1 e^{-t/\tau_1} + b_2 V_1$$

or

$$F = \underbrace{\left(\frac{-b_2^2}{b_1 + b_2}\right) V_1 e^{-t/\tau_1}}_{\text{Natural part}} + \underbrace{b_2 V_1}_{\text{Forced part}}. \tag{12–12}$$

This result checks with that obtained by multiplying the v_2 response given by Eq. (12–8) by b_2. The responses and the differential equation relating F to v_1 also are shown in Fig. 12–4.

Now we have a complete picture, summarized in Fig. 12–4, of the transient-response characteristics of a first-order system to a step input. It should be noted that each differential equation relating an output to the input v_1 has the same set of terms on the left-hand side. Of particular interest also is the fact that the terms on the right-hand side of the equation relating F to v_1 are the sums of the terms appearing on the right-hand sides of the equations relating F_{b1} and F_{k1} to v_1. Our ability to sum the responses to individual terms on the right-hand side of the differential equation stems from the principle of superposition, and it affords an independent means of obtaining a solution if it is needed.

Also, the response of F to a step change in v_1 is shown to be equal to $b_2 v_2$, and the terms on the right-hand side of the d.e. (differential equation) for F are the terms on the right-hand side of the d.e. for v_2 multiplied by b_2. This is really another demonstration of the principle of superposition in that it shows the amplitude of the response to be proportional to the amplitude of the terms on the right-hand side of the differential equation.

In each case, the decaying exponential of the transient part of the response has the same time constant τ_1, and the slope of the complete response at any time is such that the output would reach its final value τ_1 units of time later if it were to keep going at its initial rate. This property of a first-order exponential response is a very important characteristic not possessed by many responses which may seem to be simple first-order exponentials; thus a response which appears to have different time constants at different points in time based on local slope measurements is not a simple first-order exponential.

It is interesting to note that the output does not change suddenly at $t = 0$ unless the order of the highest derivative of the input variable is the same as the order of the highest derivative of the output variable.

Although a detailed discussion of unstable systems is not intended here, we may at least note that if the sign of one of the coefficients on the left-hand side of the differential equation had been negative, then the system would have been unstable. This is because the natural part of the response then would be, instead of a decaying exponential, a growing exponential because of a positive root in the characteristic equation and the complete response would not have reached an ultimate steady state.

12–3. RESPONSE OF SECOND-ORDER SYSTEMS TO STEP INPUTS

The classical method of finding the response of a second-order system to a step input is mainly an extension of the procedure which was used for a first-order system. For a second-order system, however, the transient part of the response consists of a pair of exponentials. Moreover, each exponential will have an imaginary part if the roots are not real. We shall begin with the case where the roots are real.

12–3.1 Step Response of a Second-Order System Having Real Roots

The series-connected *RLC* network shown in Fig. 12–5 will be used to illustrate the determination of transient responses of second-order systems.

To begin with, we shall consider that the voltage v_3 across the capacitance C is the output of primary interest and we shall determine its response to a step change in the input voltage v_1 (Fig. 12–3) when the system is initially relaxed (all voltages and currents zero before the step is applied).

The system differential equation is obtained as follows:

$$v_1 - v_3 = Ri + L\,di/dt, \qquad (12\text{–}13)$$

$$i = C\,dv_3/dt; \qquad di/dt = C\,d^2v_3/dt^2,$$

$$v_1 - v_3 = RC\,dv_3/dt + LC\,d^2v_3/dt^2. \qquad (12\text{–}14)$$

Rearranging Eq. (12–14), we have

$$LC\,d^2v_3/dt^2 + RC\,dv_3/dt + v_3 = v_1. \qquad (12\text{–}15)$$

The *natural part* of the response is of the form

$$(v_3)_n = A_1e^{r_1t} + A_2e^{r_2t}, \qquad (12\text{–}16)$$

where r_1 and r_2 are the roots of the characteristic equation,

$$LCs^2 + RCs + 1 = LC(s - r_1)(s - r_2) = 0$$

$$r_1, r_2 = \frac{-RC \pm \sqrt{R^2C^2 - 4LC}}{2LC}. \qquad (12\text{–}17)$$

When $R^2C^2 > 4LC$, the roots are real and we may think of the system as having two time constants,

$$\tau_1 = -1/r_1 \qquad \text{and} \qquad \tau_2 = -1/r_2,$$

which can be expressed in binomial factors of the characteristic equation as follows:

$$LCs^2 + RCs + 1 = (\tau_1s + 1)(\tau_2s + 1) = 0. \qquad (12\text{–}18)$$

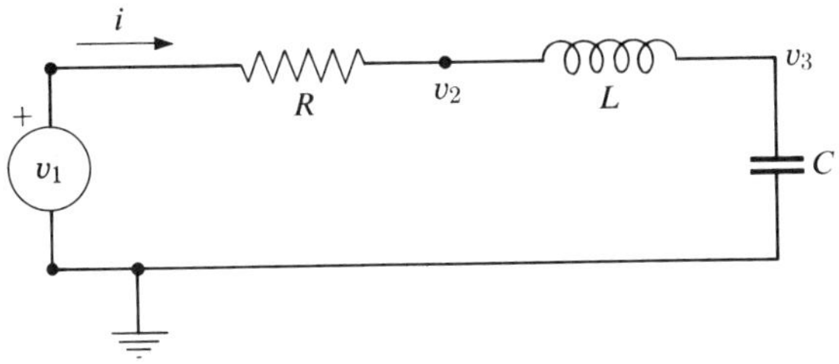

FIG. 12–5. Series-connected resistance-inductance-capacitance network.

Also, the time constants can be expressed directly in terms of the system parameters,

$$\tau_1, \tau_2 = \frac{-2LC}{-RC \pm \sqrt{R^2C^2 - 4LC}}$$

$$= \frac{-2LC(-RC \pm \sqrt{R^2C^2 - 4LC})}{R^2C^2 - R^2C^2 + 4LC},$$

$$= \frac{RC \pm \sqrt{R^2C^2 - 4LC}}{2}. \qquad (12\text{–}19)$$

The *forced part* of the response is a constant which satisfies Eq. (12–15) for $t > 0$,

$$(v_3)_f = (v_1) = V_1. \qquad (12\text{–}20)$$

The *complete response* is then given by

$$v_3 = (v_3)_n + (v_3)_f = A_1e^{-t/\tau_1} + A_2e^{-t/\tau_2} + V_1. \qquad (12\text{–}21)$$

To determine A_1 and A_2 we must know v_3 and $\dot{v}_3$ at $t = 0+$. Here we can readily see that the current i cannot change suddenly because there is no source to provide an infinite voltage across the inductance L at $t = 0$. Therefore $i(0+) = 0$. The only way that the voltage v_3 across C could change in zero time would be for a finite charge to flow into it in zero time, i.e., for i to be infinite at $t = 0$. Therefore $v_3(0+) = 0$. The rate of change of v_3 at $t = 0+$ is equal to $i(0+)/C$, so that $\dot{v}_3(0+) = 0$ also. Substituting $v_3(0+) = 0$ in Eq. (12–21) at $t = 0+$ gives

$$0 = A_1 + A_2 + V_1. \qquad (12\text{–}22)$$

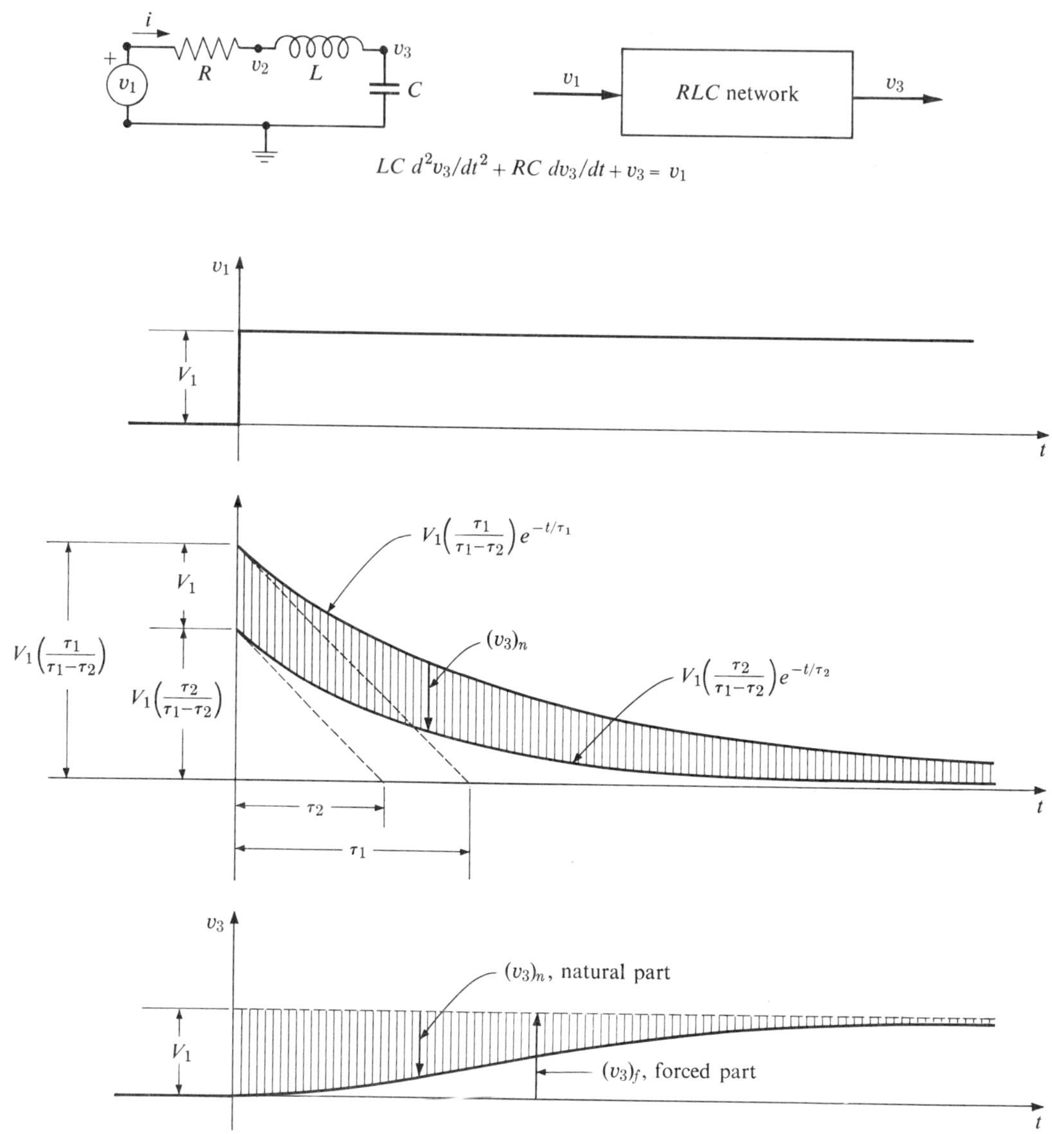

FIG. 12–6. Response of series *RLC* network to a step change in v_1; $\tau_1, \tau_2 = (RC \pm \sqrt{R^2C^2 - 4LC})/2$ (overdamped).

Differentiating Eq. (12–21) to obtain an expression for $\dot{v}_3$, we find

$$\dot{v}_3 = (-A_1/\tau_1)e^{-t/\tau_1} + (-A_2/\tau_2)e^{-t/\tau_2}. \tag{12–23}$$

Substituting $\dot{v}_3(0+) = 0$ in Eq. (12–23) at $t = 0+$ gives

$$0 = A_1/\tau_1 + A_2/\tau_2. \tag{12–24}$$

Solving Eqs. (12–22) and (12–24) for A_1 and A_2, we see that

$$A_1 = [-\tau_1/(\tau_1 - \tau_2)]V_1, \tag{12–25}$$

$$A_2 = [\tau_2/(\tau_1 - \tau_2)]V_1, \tag{12–26}$$

so that

$$v_3 = \frac{-\tau_1 V_1}{\tau_1 - \tau_2} e^{-t/\tau_1} + \frac{\tau_2 V_1}{\tau_1 - \tau_2} e^{-t/\tau_2} + V_1. \tag{12–27}$$

A typical step response for this case is shown in Fig. 12–6, where the natural part of the response is represented by the shaded region between the exponential curves. The individual components of the transient part, A_1e^{-t/τ_1} and A_2e^{-t/τ_2}, are of opposite sign. As R^2C^2 approaches $4LC$, τ_1 and τ_2 (or r_1 and r_2) approach the same value, and the value of each component of the transient part at $t = 0+$ tends toward infinity. However, their sum at $t = 0+$ remains constant and equal to $-V_1$. When $R = 2\sqrt{L/C}$, the roots are equal and the method for the special case of equal roots discussed in Chapter 11 should be employed. When this critical condition is reached, any further decrease in R relative to $2\sqrt{L/C}$ results in complex roots of the characteristic equation.

Rather than looking next at the response when R is less than $2\sqrt{L/C}$, let us first examine the step responses of the variables of secondary interest in this system.

The current i is related to v_3 by the elemental equation of the capacitance C,

$$i = C\,dv_3/dt, \tag{12–28}$$

so that the response of i to a step change in v_1 is

$$i = \frac{CV_1}{\tau_1 - \tau_2} e^{-t/\tau_1} - \frac{CV_1}{\tau_1 - \tau_2} e^{-t/\tau_2}. \tag{12–29}$$

The steady-state part of this response is zero, as it should be, because the flow of charge into C must be zero when v_3 reaches a steady state.

The voltage v_2 is the sum of the voltages across C and L,

$$v_2 = v_3 + L\,di/dt = v_3 + LC\,d^2v_3/dt^2, \tag{12–30}$$

so that the response of v_2 to a step change in v_1 is

$$v_2 = A_1(1 + LC/\tau_1^2)e^{-t/\tau_1} + A_2(1 + LC/\tau_2^2)e^{-t/\tau_2} + V_1. \tag{12–31}$$

It should be noted that $LC = \tau_1\tau_2$, so that using (12–25) and (12–26) we may write

$$v_2 = -\frac{(\tau_1 + \tau_2)V_1}{\tau_1 - \tau_2} e^{-t/\tau_1} + \frac{(\tau_1 + \tau_2)V_1}{\tau_1 - \tau_2} e^{-t/\tau_2} + V_1. \tag{12–32}$$

The responses of v_3, i, and v_2 to a unit step change in v_1 are summarized in Fig. 12–7. The role of derivative terms on the right-hand side of the differential equation is well portrayed in this illustration. When the order of the highest derivative term on the right-hand side of the differential equation is the same as the order of the highest derivative term on the left-hand side, the output suddenly changes at $t = 0$. When the order of the right-hand side is one lower than the left-hand side, only the derivative of the output suddenly changes at $t = 0$. When it is two lower than the left-hand side, only the second derivative suddenly changes, and so on.

This is another example of a stable system, because the natural part of the step response is composed of decaying exponentials. It would be necessary for one of the terms on the left-hand side of

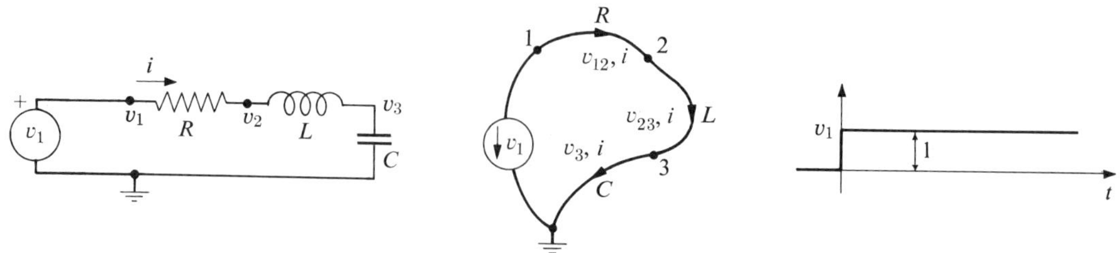

Output variable	v_3	i	v_2
Equation	$LC\frac{d^2v_3}{dt^2}+RC\frac{dv_3}{dt}+v_3=v_1$	$LC\frac{d^2i}{dt^2}+RC\frac{di}{dt}+i=C\frac{dv_1}{dt}$	$LC\frac{d^2v_2}{dt^2}+RC\frac{dv_2}{dt}+v_2=LC\frac{d^2v_1}{dt^2}+v_1$
Forced part of response	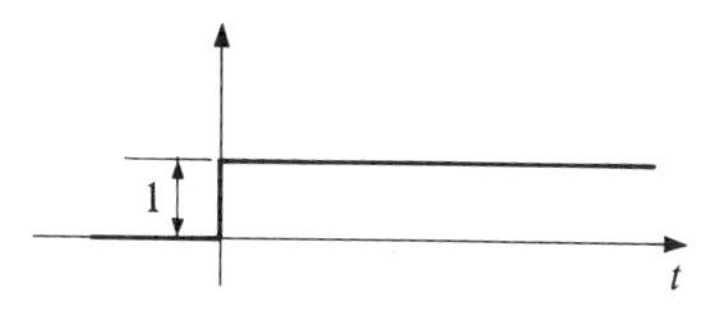	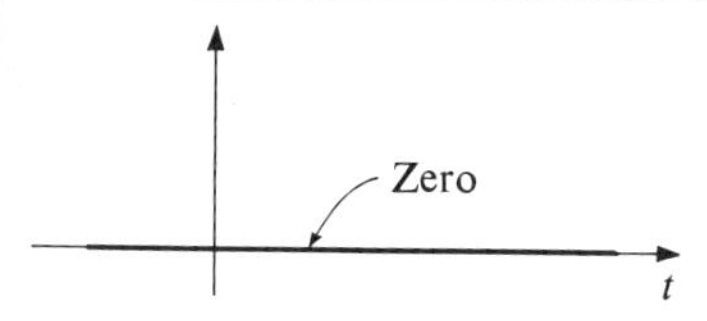 Zero	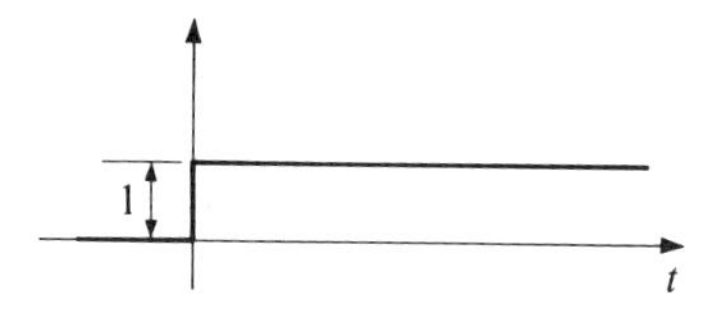
Natural part of response	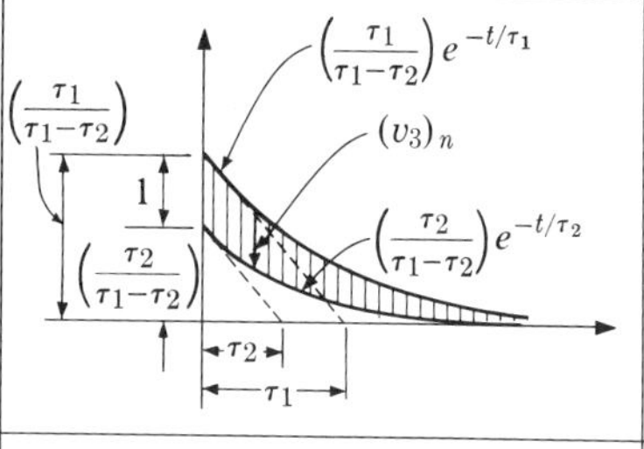 $(v_3)_n$	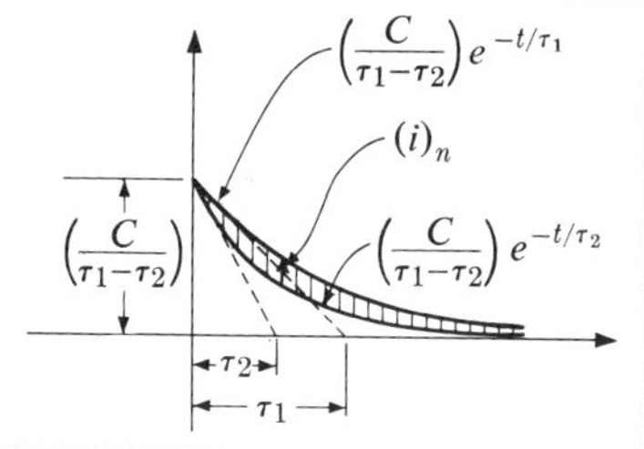 $(i)_n$	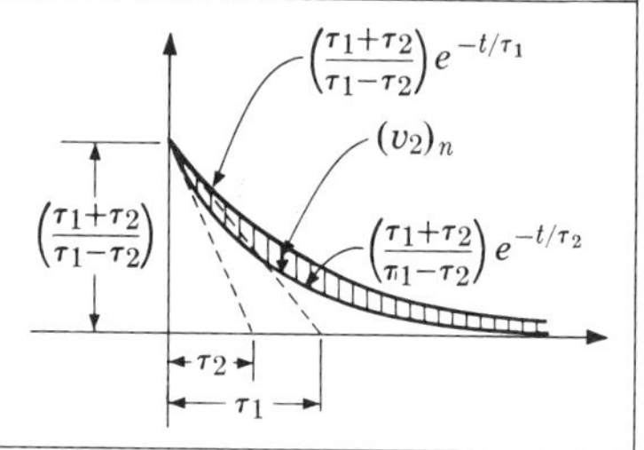 $(v_2)_n$
Complete response	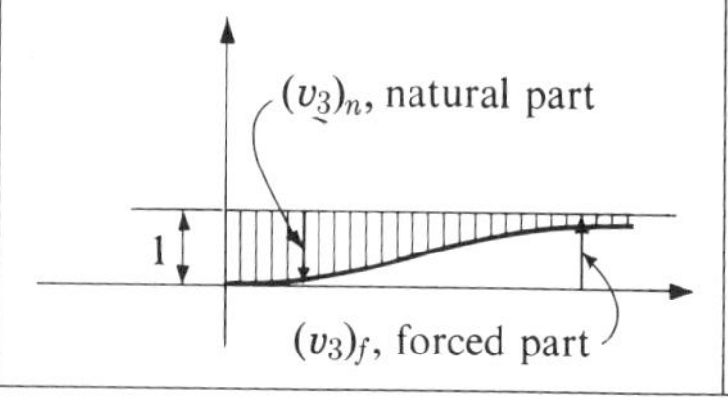 $(v_3)_n$, natural part; $(v_3)_f$, forced part	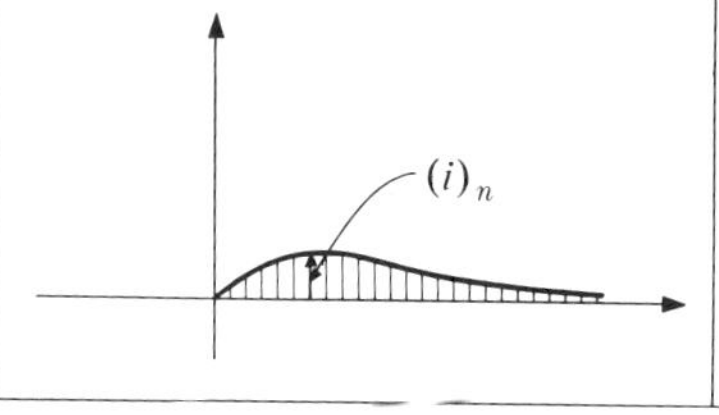 $(i)_n$	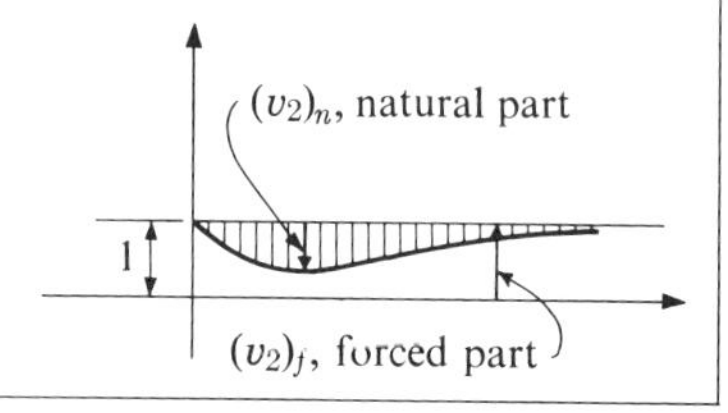 $(v_2)_n$, natural part; $(v_2)_f$, forced part

FIG. 12–7. Summary of responses of *RLC* network to a step change in v_1 when $V_1 = 1$. Amplitudes of all outputs are directly proportional to V_1.

Eq. (12–15) to be negative for one or both of the roots of the characteristic equation to be positive and for the transient part of the output to contain a growing instead of a decaying exponential. Thus this system, being composed of ideal elements and having positive values of *R*, *C*, and *L*, will always be a stable system, and it will reach a steady state when subjected to a step input.

12–3.2 Step Response of a Second-Order System Having Complex Roots

When the resistance *R* is zero (or so small that it is negligible) the roots of the characteristic equation are found from Eq. (12–17) to be

$$r_1, r_2 = \pm j\sqrt{1/LC}. \tag{12–33}$$

Now that the roots are complex (in this case imaginary), there is no need to determine the time constants because they would be found to be complex and would not have any meaning in the domain of the real variable, time. However, this does not prevent us from using imaginary roots to determine the response of v_3 to a step change in v_1. The forced part of the response to a step change in v_1 of magnitude V_1 is still V_1, and the equation for the complete response is

$$v_3 = A_1 e^{r_1 t} + A_2 e^{r_2 t} + V_1. \tag{12-34}$$

The arbitrary constants A_1 and A_2 may be found by using the initial conditions for v_3 and $\dot{v}_3$ at $t = 0+$ (they are the same as in the previous discussion of Section 12-3.1 when R was not zero),

$$v_3(0+) = 0, \qquad \dot{v}_3(0+) = 0.$$

We may now solve for A_1 and A_2 as for real roots, and obtain

$$A_1 = [r_2/(r_1 - r_2)]V_1, \tag{12-35}$$

$$A_2 = [-r_1/(r_1 - r_2)]V_1. \tag{12-36}$$

Thus the solution for v_3, in terms of r_1 and r_2, is

$$v_3 = \frac{r_2 V_1}{r_1 - r_2} e^{r_1 t} - \frac{r_1 V_1}{r_1 - r_2} e^{r_2 t} + V_1. \tag{12-37}$$

We may now employ the expressions for r_1 and r_2 given by Eq. (12-33) to obtain

$$v_3 = \left(\frac{-j/\sqrt{LC}}{2j/\sqrt{LC}} e^{jt/\sqrt{LC}} - \frac{j/\sqrt{LC}}{2j/\sqrt{LC}} e^{-jt/\sqrt{LC}} + 1 \right) V_1,$$

or

$$v_3 = \left[-\left(\frac{e^{jt/\sqrt{LC}} + e^{-jt/\sqrt{LC}}}{2} \right) + 1 \right] V_1. \tag{12-38}$$

Using Euler's equation, $(e^{j\alpha} + e^{-j\alpha})/2 = \cos \alpha$, we find

$$v_3 = \underbrace{-V_1 \cos (t/\sqrt{LC})}_{\text{Natural part}} + \underbrace{V_1}_{\text{Forced part}}. \tag{12-39}$$

The forced part and the natural part of the response are shown along with the complete response in Fig. 12-8. Because the natural part of the response is an undamped sinusoid, it is not strictly correct to refer to it as the transient part of the response—it does not decay with time and thus it has a steady character. It is evident that the forced part of the response represents the steady value which the output would reach if the sinusoid were damped so that it would decay with time. Having $R = 0$ is a limiting case of course, because it is impossible to construct a real inductance which does not have some resistance in it (real capacitors also have a very small amount of leakage due to conductivity of the dielectric material, but this is modeled as a resistance in parallel with the capacitance, a resistance which is not provided for in the network model used in this example).

Thus we are vitally interested in how this *RLC* network responds when R is small. Before developing the solution of the differential equation for this case, we should point out that the period of the undamped response which results when $R = 0$ is given by

$$T = 2\pi\sqrt{LC} \text{ sec} \tag{12-40}$$

when L is measured in henries and C is measured in farads and the angular frequency ω_n (often called the *undamped natural frequency*) is given by

$$\omega_n = 1/\sqrt{LC} \text{ rad/sec.} \tag{12-41}$$

When a small value exists for R, the roots of the characteristic equation which were given in Eq. (12-17) may be expressed as follows:

$$r_1, r_2 = \frac{1}{\sqrt{LC}} \times \left(-\frac{RC}{2\sqrt{LC}} \pm \sqrt{R^2C^2/4LC - 4LC/4LC} \right),$$

or

$$r_1, r_2 = \omega_n(-\zeta \pm j\sqrt{1 - \zeta^2}), \tag{12-42}$$

where ω_n is defined in Eq. (12-41), and the damping ratio ζ is

$$\zeta \equiv (R/2)\sqrt{C/L}.$$

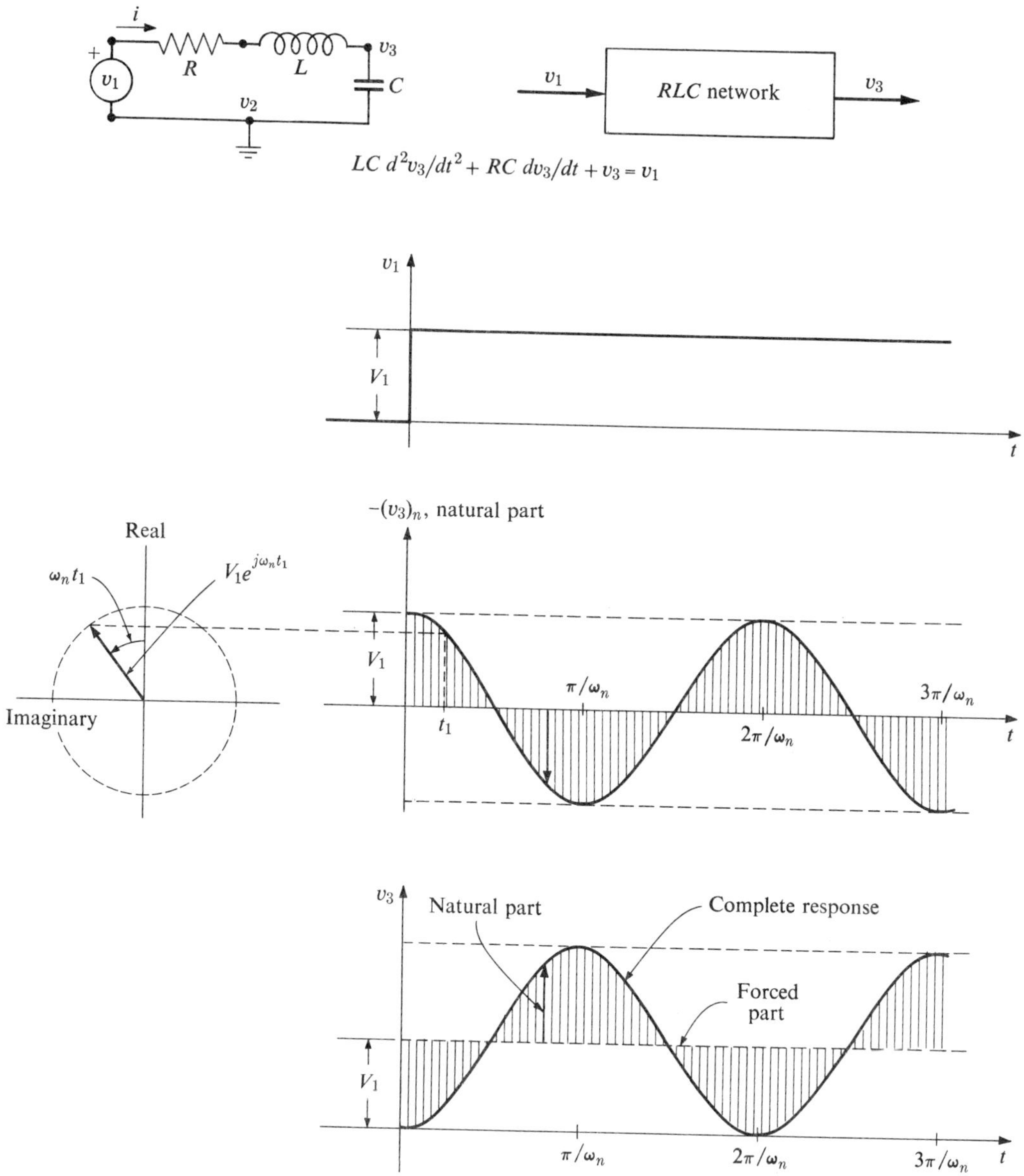

FIG. 12-8. Response of series *RLC* network to a step change in v_1, when $R = 0$ (undamped); $\omega_n = 1/\sqrt{LC}$.

The response of v_3 to a step change v_1 of magnitude V_1 is now determined from Eq. (12–37) by substituting for r_1 and r_2,

$$v_3 = \frac{(-\zeta - j\sqrt{1-\zeta^2})V_1}{2j\sqrt{1-\zeta^2}} e^{\omega_n(-\zeta + j\sqrt{1-\zeta^2})t} - \frac{(-\zeta + j\sqrt{1-\zeta^2})V_1}{2j\sqrt{1-\zeta^2}} e^{\omega_n(-\zeta - j\sqrt{1-\zeta^2})t} + V_1.$$

Collecting terms, we have

$$v_3 = \frac{V_1 \zeta e^{-\zeta\omega_n t}}{\sqrt{1-\zeta^2}} \times \left(\frac{-[1 + j\sqrt{(1/\zeta^2) - 1}]e^{j\omega_d t} + [1 - j\sqrt{(1/\zeta^2) - 1}]e^{-j\omega_d t}}{2j} \right) + V_1, \tag{12–43}$$

where $\omega_d = \omega_n\sqrt{1-\zeta^2}$. The terms

$$[1 + j\sqrt{(1/\zeta^2) - 1}]$$

and

$$[1 - j\sqrt{(1/\zeta^2) - 1}]$$

may be expressed by exponentials as shown in the geometric construction of Fig. 12–9,

$$[1 \pm j\sqrt{(1/\zeta^2) - 1}] = (1/\zeta)e^{\pm j\psi}, \tag{12–44}$$

where $\psi = \cos^{-1}\zeta$. Use of these exponentials in Eq. (12–43) gives

$$v_3 = \frac{V_1 e^{-\zeta\omega_n t}}{\sqrt{1-\zeta^2}} \left[\frac{-e^{j(\omega_d t + \psi)} + e^{-j(\omega_d t + \psi)}}{2j} \right] + V_1.$$

Now we may employ Euler's equation to obtain

$$v_3 = \underbrace{-V_1 \frac{1}{\sqrt{1-\zeta^2}} e^{-\zeta\omega_n t} \sin(\omega_d t + \psi)}_{\text{Natural part}} + \underbrace{V_1}_{\text{Forced part}}. \tag{12–45}$$

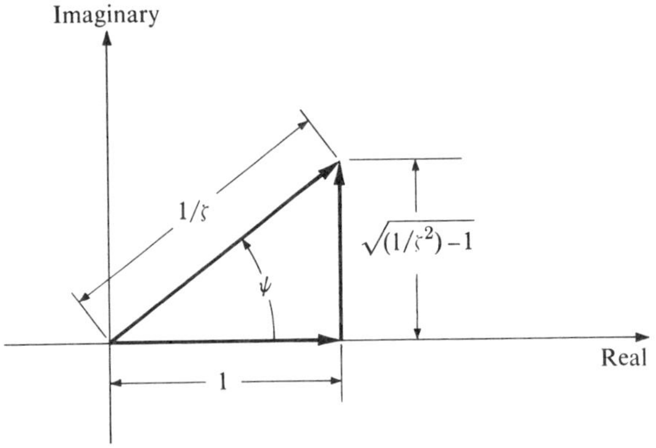

FIG. 12–9. Geometrical construction to show that $\psi = \cos^{-1}\zeta$ and that $(1 \pm j\sqrt{1/\zeta^2 - 1}) = (1/\zeta)e^{\pm j\psi}$.

The natural and forced parts of the response and the complete solution are shown in Fig. 12–10. It is interesting to note that the damped sinusoid which constitutes the transient part of the response may be visualized as the imaginary part of the rotating vector,

$$\frac{V_1 e^{-\zeta\omega_n t} e^{j(\omega_d t + \psi)}}{\sqrt{1-\zeta^2}}, \tag{12–46}$$

with an angular frequency $\omega_d = \omega_n\sqrt{1-\zeta^2}$.

At this point it should be evident that the resistance R causes the natural part of the response to decay with time, so that it truly may be called a transient. The rate of decay of this damped sinusoid is governed by the exponential $e^{-\zeta\omega_n t}$, and the per-cycle decay ratio DR is given by

$$DR = \frac{e^{-\zeta\omega_n[t + (2\pi/\omega_d)]}}{e^{-\zeta\omega_n t}} = e^{-\zeta 2\pi\omega_n/\omega_d} = e^{-2\pi\zeta/\sqrt{1-\zeta^2}}. \tag{12–47}$$

The parameter ζ which we have been using since Eq. (12–42) is thus a direct measure of the degree of damping caused by the resistance R, and here it is easy to see why it is given the name *damping ratio*. From its definition,

$$\zeta \equiv (R/2)\sqrt{C/L}, \tag{12–48}$$

308 TRANSIENT RESPONSE OF SIMPLE LINEAR SYSTEMS | 12-3

$$LC\, d^2v_3/dt^2 + RC\, dv_3/dt + v_3 = v_1$$

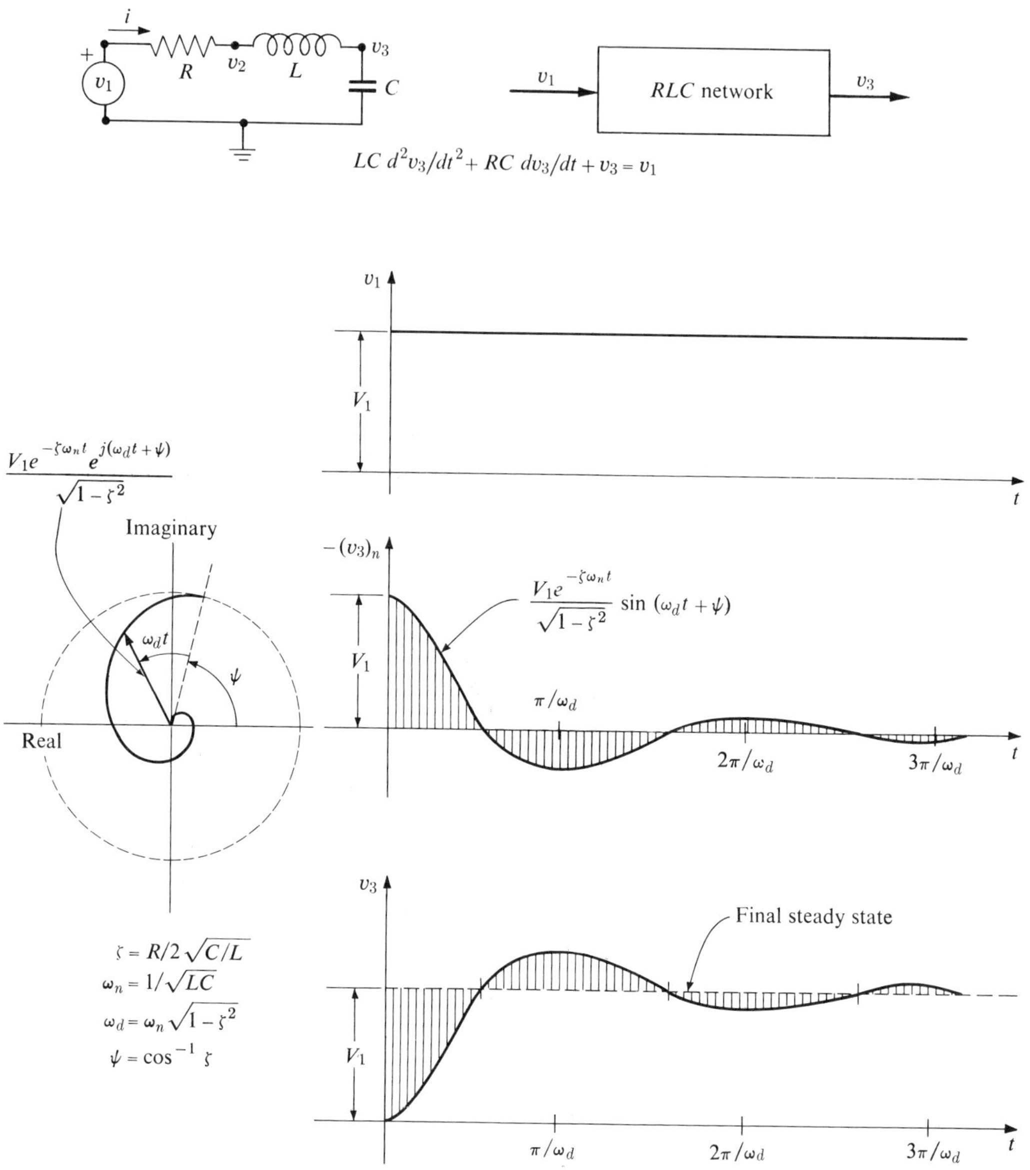

FIG. 12-10. Response of a series *RLC* network to a step change in v_1 (underdamped).

$\zeta = R/2\sqrt{C/L}$
$\omega_n = 1/\sqrt{LC}$
$\omega_d = \omega_n\sqrt{1-\zeta^2}$
$\psi = \cos^{-1}\zeta$

we can visualize it as the ratio of the resistance R to the critical value $R_{\text{crit}} = 2\sqrt{L/C}$ which would be needed to give a critically damped response (real, equal roots in the characteristic equation, Eq. 12-17):

$$\zeta = R/R_{\text{crit}} = R/2\sqrt{L/C} = (R/2)\sqrt{C/L}.$$

A system will have a critically damped response when its damping ratio is equal to unity. This is a special case which divides the overdamped class of systems ($\zeta > 1$) from the underdamped class of systems ($0 < \zeta < 1$). When a second-order system is overdamped, the roots of its characteristic equation are real and unequal and the natural part of its response consists of decaying real exponentials. When a second-order system is underdamped, the roots of its characteristic equation are complex and the natural part of its response is a decaying sinusoid.

12–4. HIGHER-ORDER SYSTEMS

When the system differential equation is higher than second order, the same procedure can be used for finding the response to a step input as for first- and second-order systems. However, it becomes more difficult to find the roots of the characteristic equation when the order of the system equation is higher; and evaluation of the undetermined coefficients, A_n, of the exponentials in the transient part of the response also becomes more tedious.

For an illustration, let us consider the case of a third-order system differential equation,

$$a_3\, d^3x_r/dt^3 + a_2\, d^2x_r/dt^2 + a_1\, dx_r/dt + a_0x_r = b_0x_f. \tag{12–49}$$

Systems having this order of differential equation have been discussed in Chapters 7 and 8, and the analog-computer simulation of such a system (D'Arsonval galvanometer) was described in Chapter 7. The pulse transformer of Section 8–5.3 also results in a third-order differential equation.

The characteristic equation of this system is

$$a_3s^3 + a_2s^2 + a_1s + a_0 = a_3(s - r_1)(s - r_2)(s - r_3) = 0,$$

where the roots r_1, r_2, and r_3 are values of s for which the polynomial in s is equal to zero. The methods employed to find these roots, often tedious when used without the aid of a computer, are discussed in several references [2, 3, 4, 5].

The response to a step change in x_f of magnitude X_f will be of the form

$$x_r = A_1e^{r_1t} + A_2e^{r_2t} + A_3e^{r_3t} + X_fb_0/a_0, \tag{12–50}$$

where the coefficients A_1, A_2, and A_3 are determined with the help of three initial conditions such as $x_r(0+)$, $\dot{x}_r(0+)$, and $\ddot{x}_r(0+)$, by solving the three algebraic equations

$$x_r(0+) = A_1 + A_2 + A_3 + X_fb_0/a_0 \tag{12–51}$$

$$\dot{x}_r(0+) = r_1A_1 + r_2A_2 + r_3A_3, \tag{12–52}$$

$$\ddot{x}_r(0+) = r_1^2A_1 + r_2^2A_2 + r_3^2A_3, \tag{12–53}$$

for A_1, A_2, and A_3.

Because of all the tedious effort often required in employing these classical methods for solving higher-order system equations, it is frequently simpler to employ Laplace transforms (see Chapter 16) to obtain the desired solutions. It is interesting to note here, however, that it is not often necessary to carry out the complete solution for a transient response. The most useful information to an engineer is contained in the roots of the characteristic equation, which reveal the time constants, natural frequencies, and damping ratios of the system. Also, if any of the roots contain a positive real part, the system will be unstable because the corresponding exponential term in the natural part of the response will grow exponentially with time.

The response of Θ to a step change in v determined by the analog computer study of D'Arsonval's galvanometer in Chapter 7 is typical of a third-order system having two roots which are complex, thereby describing a damped sinusoidal component of the transient response which is evident in the measured response shown in Fig. 7–38.

12–5. SUMMARY OF STEP-RESPONSE CHARACTERISTICS

The most general form of linear first-order system differential equation with constant coefficients is given in Eq. (12–54), where x_f is the input or forcing function and x_r is the output or response function:

$$a_1\, dx_r/dt + a_0x_r = b_1\, dx_f/dt + b_0x_f. \tag{12–54}$$

When the coefficient b_1 is zero, the output x_r will not change suddenly at $t = 0$ when a step change in x_f occurs, and the response will be as shown in

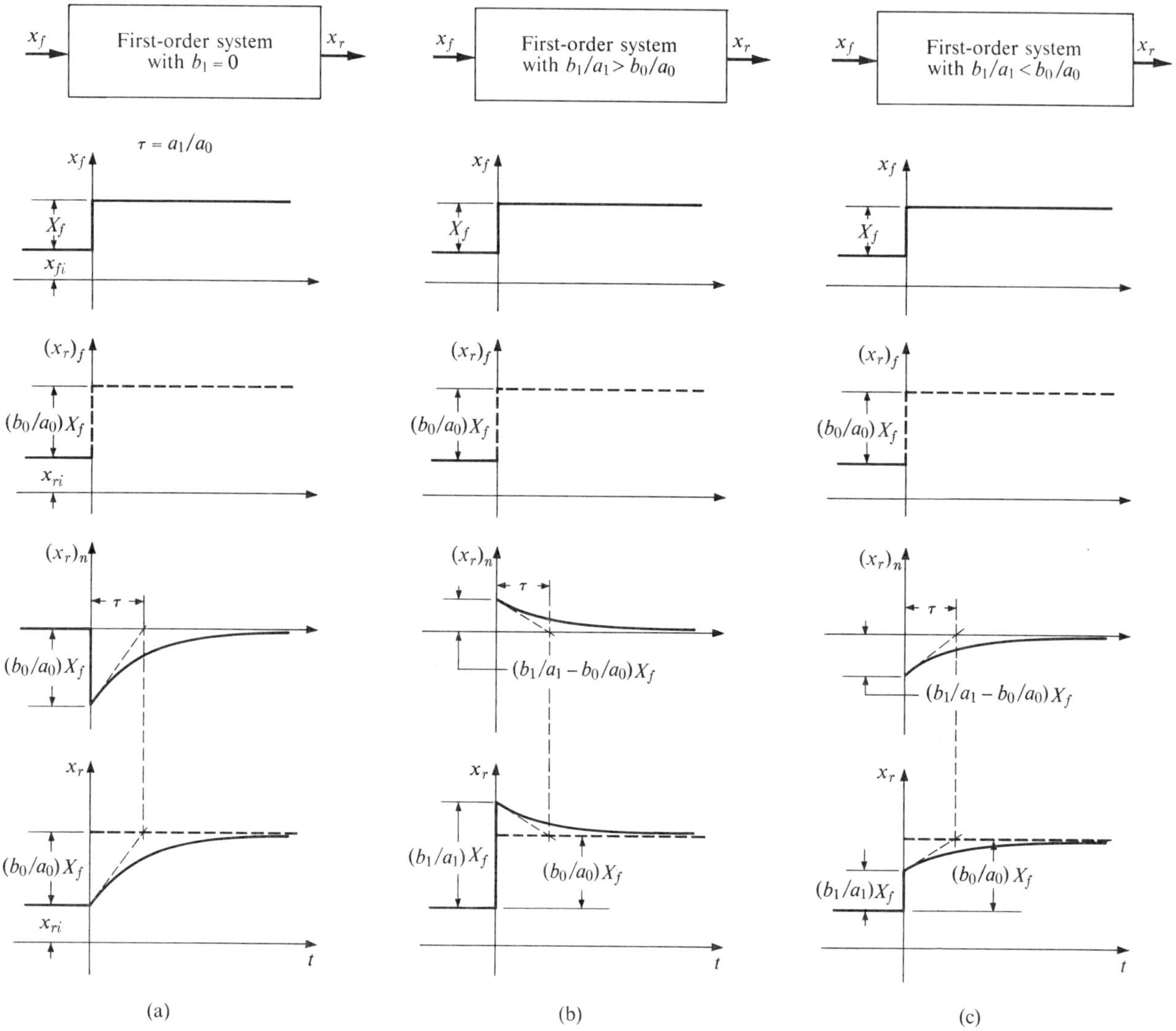

FIG. 12–11. Summary of the transient response of first-order systems to a step change in input. The system equation is $a_1\, dx_r/dt + a_0 x_r = b_1\, dx_f/dt + b_0 x_f$.

Fig. 12–11(a). Note that initial steady values of x_f and x_r are also included for the sake of generality. The changes which occur after $t = 0$ from the initially steady values correspond to the variations from zero which were discussed for a particular mechanical system in Section 12–2.

When b_1 is not zero, there are two cases of particular interest: Fig. 12–11(b), where $b_1/a_1 > b_0/a_0$, and Fig. 12–11(c), where $b_1/a_1 < b_0/a_0$. When $b_1/a_1 > b_0/a_0$, the output x_r changes suddenly at $t = 0$ by an amount which is greater than the final steady-state change, and it then approaches the final steady state with a decaying exponential having time constant $\tau = a_1/a_0$. When $b_1/a_1 < b_0/a_0$, the output x_r changes suddenly at $t = 0$ by an amount which is less than the final steady-state change, and it then approaches the final steady state with a decaying exponential having the same time constant τ.

The general form of linear second-order differential equations with constant coefficients and no derivative terms on the right-hand side is given by

$$a_2\, d^2x_r/dt^2 + a_1\, dx_r/dt + a_0 x_r = x_f. \tag{12–55}$$

The response of a system having this differential equation is summarized in Fig. 12–12 for various values of the damping ratio ζ. When the damping ratio is greater than unity, there is no damped oscillation in the response and the two time constants τ_1 and τ_2 are used to describe the exponential components of the transient response.

When a first derivative term is present on the right-hand side of the differential equation, there will be a sudden change in the derivative of the output x_r at $t = 0$, and when a second derivative term is present on the right-hand side of the differential equation, there will be a sudden change in the output x_r at $t = 0$. The characteristics of the responses for these cases with various values of the damping ratio ζ will be similar in many respects to those shown in Fig. 12–12. The major differences appear shortly after $t = 0$ and are closely associated with the different initial conditions at $t = 0+$ which depend on the derivative terms on the right-hand side of the differential equation.

12–6. RESPONSES TO OTHER SINGULARITY FUNCTIONS

It was shown in Chapter 10 that the impulse, step, ramp, etc., functions were all related to each other by simple differentiation or integration with respect to time. As might be expected, the responses of a given linear system to this family of singularity functions are related to each other in a similar manner. This means that once the response to one member of this family of inputs is known, it is relatively simple to find the response to any other member. In conducting experiments on real linear systems, we frequently find it more feasible to apply one singularity function than the others to the system. Then it is a simple matter to use the measured response to this input in determining what the response to the other inputs would be.

12–6.1 Differentiation and Integration of the Input and Response Functions

We have already seen that as a result of superposition, doubling the amplitude of the input to a linear system doubles the amplitude of the response. This also applies when the linear operations of integration and differentiation are applied to the input.

For example, if the response x_r of a linear system is $\mathsf{f}_r(t)$ when the input x_f is $\mathsf{f}(t)$, then

$$x_r = 2\mathsf{f}_r(t) \quad \text{when} \quad x_f = 2\mathsf{f}(t), \tag{12–56}$$

$$x_r = d\mathsf{f}_r(t)/dt \quad \text{when} \quad x_f = d\mathsf{f}(t)/dt, \tag{12–57}$$

and

$$x_r = \int \mathsf{f}_r(t)\, dt \quad \text{when} \quad x_f = \int \mathsf{f}(t)\, dt. \tag{12–58}$$

This is easily demonstrated, for instance, when a function $\int \mathsf{f}(t)\, dt$ is applied instead of $\mathsf{f}(t)$ at the input x_f in a system described by the equation

$$a_n\, d^n x_r/dt^n + a_{n-1}\, d^{n-1}x_r/dt^{n-1} + \cdots + a_0 x_r = x_f. \tag{12–59}$$

This corresponds to using $\mathsf{f}(t)$ as the forcing function when the system input is $\dot{x}_f$. The equation relating x_r to x_f is then given by

$$a_n\, d^n \dot{x}_r/dt^n + a_{n-1}\, d^{n-1}\dot{x}_r/dt^{n-1} + \cdots + a_0 \dot{x}_r = \dot{x}_f, \tag{12–60}$$

and it is evident that the response of $\dot{x}_r$ will be equal to $\mathsf{f}_r(t)$ and that the response of x_r to this input will be equal to $\int \mathsf{f}_r(t)/dt$.

This means that the response of a linear system to a unit-ramp input is the time integral of the response

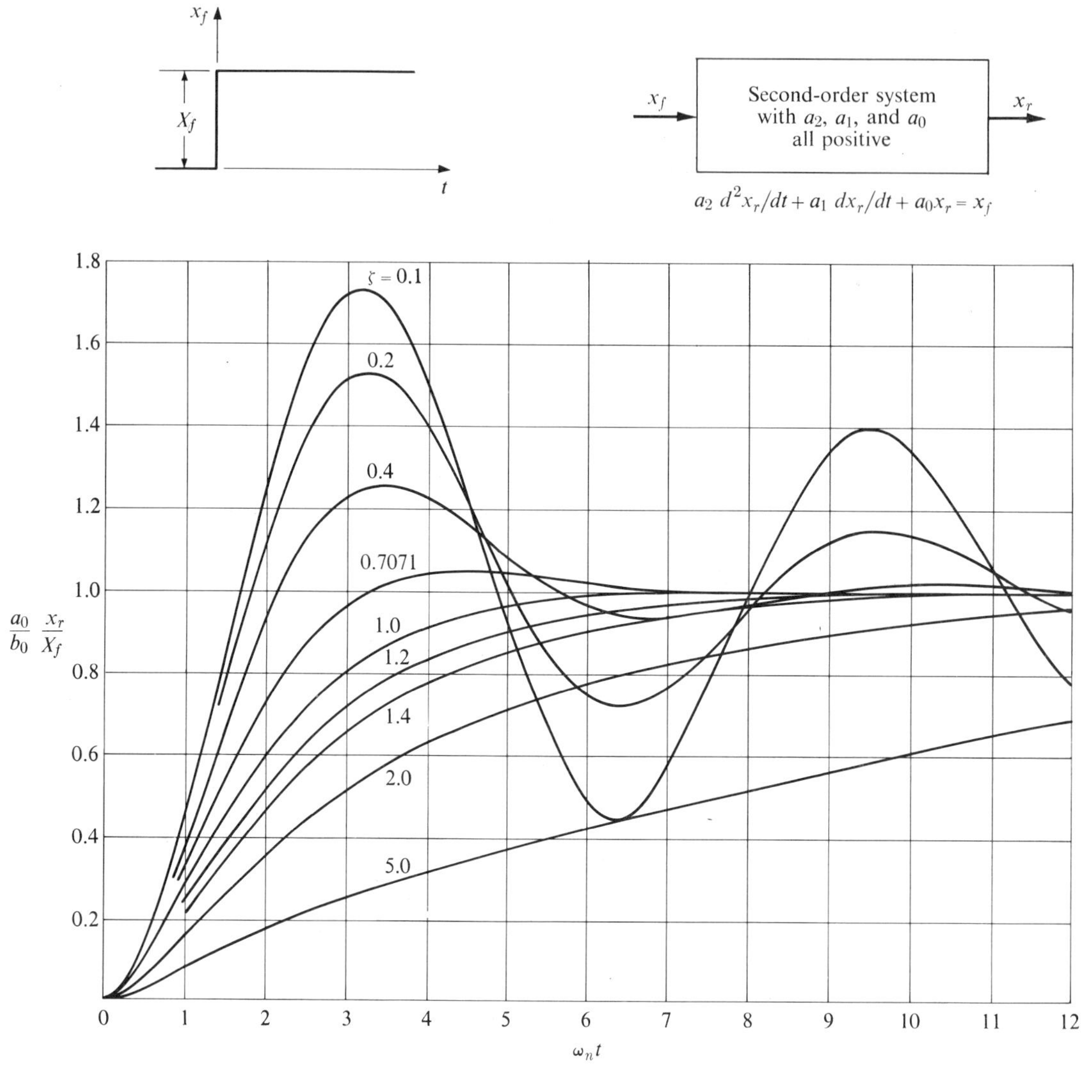

FIG. 12–12. Complete family of transient response curves for a second-order system having no derivative terms on the right-hand side of the differential equation.

$$\omega_n^2 = a_0/a_2, \qquad \tau_1, \tau_2 = (a_1/2a_0)(1 \pm \sqrt{1 - 4a_0a_2/a_1}),$$

or

$$\zeta = a_1/2\sqrt{a_2a_0}, \qquad \tau_1, \tau_2 = (\zeta \pm \sqrt{\zeta^2 - 1})/\omega_n, \qquad \tau_1/\tau_2 = 2\zeta^2 + 2\zeta\sqrt{\zeta^2 - 1} - 1.$$

to a unit-step input because a unit ramp is the time integral of a unit step. Similarly, we see that the response of a linear system to a unit impulse input is the time derivative of the response to a unit-step input, because a unit impulse is the time derivative of a unit step.

12–6.2 Pulse Response—Approximation to Impulse Response

It was shown in Chapter 10 that a unit impulse can be expressed as the sum of two step changes. This means that the response of a linear system to a pulse is readily computed by finding the response to each step change and summing the two step responses.

If the pulse width is small compared to the time constants and the inverse natural frequencies of the system, the response to the pulse will be very nearly the same as the response to an impulse of the same strength.

This means that if the pulse is carefully chosen, pulse testing of a real system can be employed as a means of experimentally determining its impulse response. In addition to having small enough width, the pulse must have a height which is not too great. True impulse testing is not physically possible because no system could stand the infinite excursion of the impulse which occurs at $t = 0$. The trick is to employ a pulse which satisfactorily approximates an impulse for the system being studied.

Pulse testing is necessary with some systems because it is either not possible or not allowable to expose them to a step input. It is interesting to note that if the pulse meets the width and height requirements mentioned above, it can have an infinite variety of shapes, e.g., it does not have to have sharp corners.

12–7. COMMENTS ABOUT LINEARITY

It is very important to note that this chapter has been concerned only with linear systems, whereas the earlier chapters in general were not limited to discussions of linear systems.

The characteristics such as time constant, natural frequency, and damping ratio have real meaning only when working with linear systems or with systems operating in linear ranges of their variables. Thus a truly nonlinear system does not have a time constant, natural frequency, or damping ratio as discussed in this chapter.

Neither does the principle of superposition apply when the system is nonlinear. As a matter of fact, one good way to experimentally determine that a system is nonlinear is to double the amplitude of a step input and find that the output amplitude is not doubled.

Also, a system which responds to a step input (or a ramp or impulse) with an oscillation having a natural period which varies with time is obviously nonlinear. The same is true if the decay per cycle varies with time.

REFERENCES

1. C. S. Draper, W. McKay, and S. Lees, *Instrument Engineering*, Vol 2. New York: McGraw-Hill Book Company, Inc. (pp. 101–286) 1953.
2. E. V. Bohn, *The Transform Analysis of Linear Systems*. Reading, Mass.: Addison-Wesley Publishing Company, Inc. (pp. 145–155) 1963.
3. P. Calingaert, *Principles of Computation*. Reading, Mass.: Addison-Wesley Publishing Company, Inc. (pp. 133–139) 1965.
4. H. L. Harrison and J. G. Bollinger, *Introduction to Automatic Controls*. Scranton, Pa.: International Textbook Company, Inc. (pp. 319–324) 1963.
5. T. V. Karman and M. A. Biot, *Mathematical Methods in Engineering*. New York: McGraw-Hill Book Company, Inc., 1940.

PROBLEMS

12–1. For the circuit shown in Fig. 12–13, the switch is closed at $t = 0$.

a) What is the time constant?

b) What is the initial value of each current at $t = 0+$?

c) What is the final value of each current (at $t \to \infty$)?

12–2. The circuit shown in Fig. 12–14 has a step function input $e = Eu_s(t)$ applied. The time constant may be determined to be

$$\tau = [R_1R_2/(R_1 + R_2) + R_3]C.$$

a) Sketch curves of the variation of i_1, i_2, i_3, and v_C as functions of time. Compute and show on these curves the initial and final values of each variable (paying particular attention to whether the final values are greater or less than the initial values).

b) The system differential equation must be of the form

$$a(dy/dt) + by = ce + d(de/dt),$$

where y is any of i_1, i_2, i_3, or v_c. For each variable indicate which of the coefficients a, b, c, or d are zero. Clearly state also which coefficients are not zero. You probably will not have time to derive each system equation. You are expected to answer this by means of physical and mathematical reasoning and knowledge of the effects of each type of term.

12–3. Consider the system shown in Fig. 12–15. The electric motor is connected to the inertia by means of a rigid shaft. The load torque, T_l, is an arbitrary input just as is the voltage, e_m. The damping torque is proportional to Ω. The proportionality constant is B. The motor characteristics are given in Fig. 12–16.

a) Derive the relationship between the output, Ω, and the two inputs e_m and T_l. Put the relationship into the form

$$a_n\, d^n\Omega/dt^n + \cdots + a_1\, d\Omega/dt + a_0\Omega = f_2(e_m, t) + f_3(T_l, t)$$

Draw the detailed block diagram.

b) What is the time constant of the system?

c) Assuming that e and T_l are constant after a long time, what is the expression for Ω (steady state)?

d) What is the effect of the slope of the characteristics, K_m, on the time constant? What about B, the damping constant? What can you say about K_m and B?

12–4. Consider the system shown in Fig. 12–17.

a) Repeat questions a), b), and c) of Problem 12–3 for this problem.

b) Assume that K_a is increased to a very large value so that $K_aK_DK_e \gg (b + K_m)$. What is the time constant under these conditions? Is the system dynamic response better or worse? Why? Is Ω (steady state) sensitive to *parameter* changes and, if so, which ones? What does your answer imply about the *required quality* (cost) of the various components if it is required that Ω be proportional to e_i in the steady state (with $T_l = 0$)? Is this system as sensitive to T_l as the previous system (without feedback)? (What do you think of "feedback" now?)

c) The motor is now connected to the inertia by means of a springy shaft (of spring constant K). Repeat question a) of Problem 12–3 for a system with no feedback. What is the *order* of the differential equation?

12–5. For the equation

$$dx^2/dt^2 + 4\, dx/dt + 5x = 0,$$

where

$$x(0) = 10 \quad \text{and} \quad \dot{x}(0) = 0,$$

a) determine the response, i.e., solve the differential equation;

b) roughly plot the response;

c) what is the damping ratio?

12–6. The differential equation relating x_r and x_f (for *all* time t) for a certain dynamic system is

$$d^2x_r/dt^2 + 4\, dx_r/dt + 13x_r = 5x_f,$$

where

for $t < 0$, $\quad x_r = -1.0 = \text{const}$

and

for $t > 0$, $\quad x_f = 0 = \text{const.}$

a) Sketch the time response of x_r from $t = 0-$ to t very large.

b) Find the maximum or peak value of x_r and find the time at which this maximum occurs.

12–7. In the systems in Fig. 12–18(a), (b), (c), (d), and (f), find the frequencies at which the indicated variables oscillate when the system is disturbed.

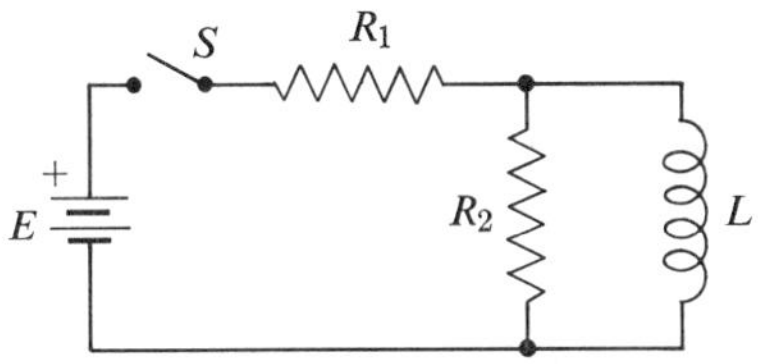

Circuit relaxed at $t = 0-$

FIGURE 12–13

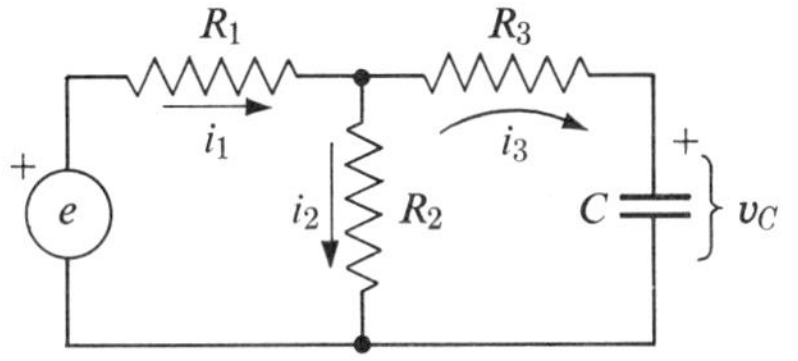

No energy stored at $t = 0-$

FIGURE 12–14

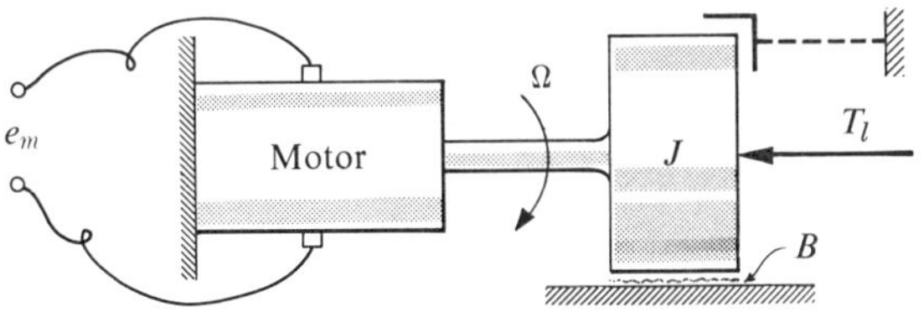

FIGURE 12–15

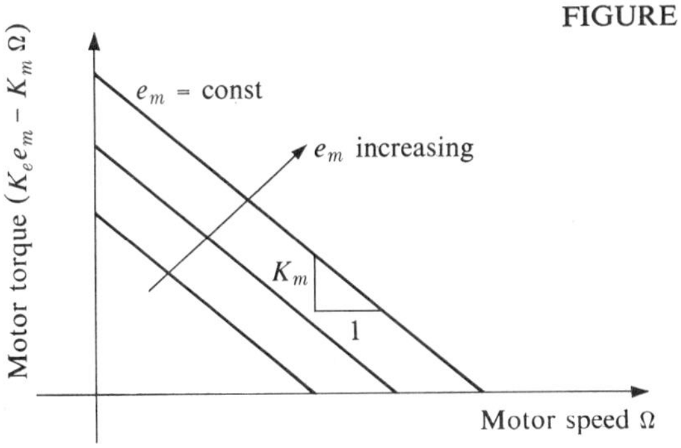

FIGURE 12–16

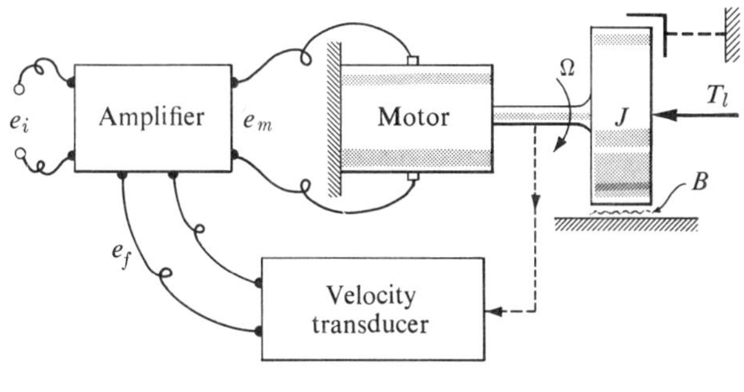

Velocity transducer: $e_f = K_d \, \Omega$

Amplifier: $e_m = K_a(e_i - e_f)$

FIGURE 12–17

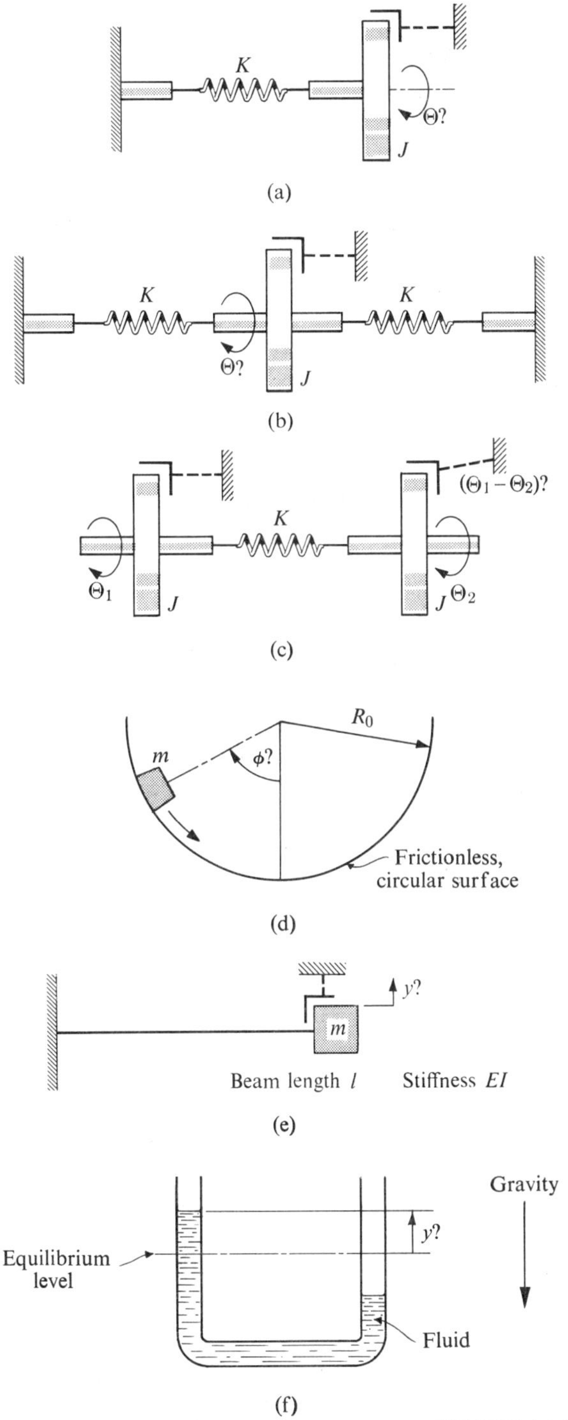

FIGURE 12–18

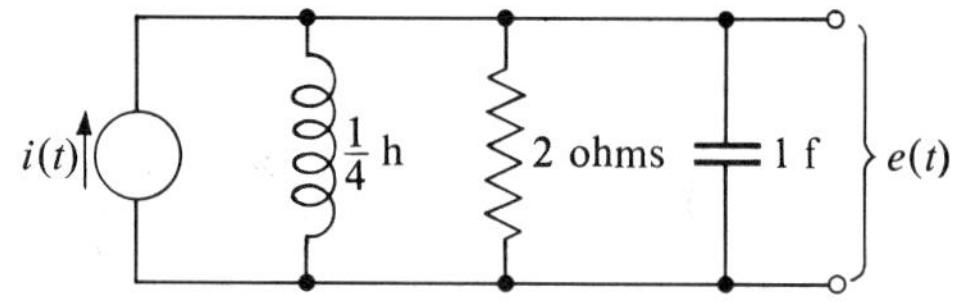

FIGURE 12–19

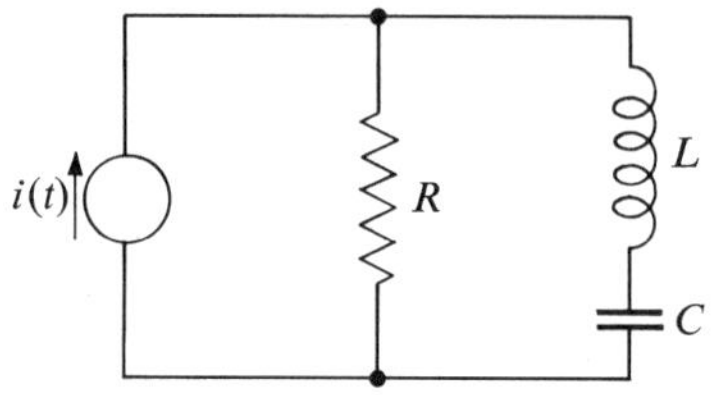

Circuit relaxed at $t = 0-$.

FIGURE 12–20

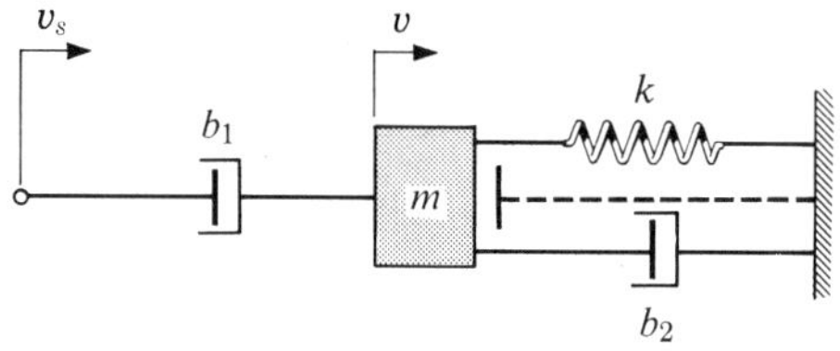

FIGURE 12–21

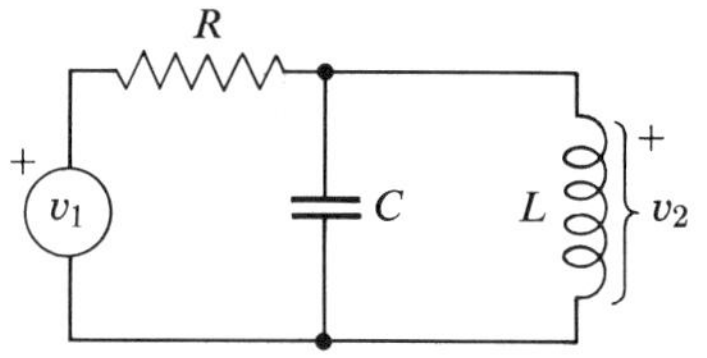

FIGURE 12–22

12–8. In the circuit shown in Fig. 12–19, the input is $i(t)$ and the output is $e(t)$. Using the solutions developed in Chapter 12, find the response of $e(t)$ to a step input in i. Use either the time constant(s) or damping ratio and undamped natural frequency, as appropriate.

12–9. For the system shown in Fig. 12–20,

a) develop the system differential equation for the voltage across R;

b) assuming that i is a step function, $i = Iu_s(t)$ and that $R = 1$, $C = 1$, $L = 1$, determine analytically and sketch the response for the voltage across R;

c) draw the analogous mechanical translational system and the dual electrical system and show the correspondence between parameters.

12–10. The system in Fig. 12–21 is at rest prior to $t = 0$. The system input for $t > 0$ is v_s.

a) Draw the linear graph for this system.

b) Formulate the system differential equation for v.

c) If v_s is a *step change* $v_s = Vu_s(t)$, find $v(0+)$ and $\dot{v}(0+)$.

d) When v_s is the above step change, solve for v. The parameter values are $b_1 = b_2 = 1$ lb-sec/in, $m = 2$ lb-sec^2/in, $k = 2$ lb/in, $V = 1$ in/sec. Sketch the solution and estimate the maximum force in damper b_2.

e) From d) determine the force in the spring as a function of time and sketch. Estimate the maximum force in the spring.

12–11. The circuit shown in Fig. 12–22 is composed of ideal elements as indicated schematically. The differential equation for the output v_2 in terms of the input v_1 is

$$RC\frac{d^2v_2}{dt^2} + \frac{dv_2}{dt} + \frac{R}{L}v_2 = \frac{dv_1}{dt}.$$

For a step input of v_1 from zero to v_{10}, the initial conditions for v_2 are

$$v_2(0+) = 0, \qquad \dot{v}_2(0+) = \frac{v_{10}}{RC}.$$

The parameter values are $R = 0.5$ ohm, $L = 1$ h, $C = 2$ f.

a) Determine the steady-state value of v_2.

b) Sketch the response v_2 to a step input in v_1 properly showing the main qualitative features of this response. It is neither necessary nor even desirable to obtain an equation for the curve for this part of the problem.

c) Determine the mathematical solution for v_2 as a function of time with a step input for v_1.

12–12. The system shown in Fig. 12–23 is a highly simplified model of the power transmission system of a ship.

a) Derive the equation(s) needed to show how the propeller speed Ω_3 is related to Ω_1 and T.

b) When the system is running steadily with $(\Omega_3)_i = 0.95(\Omega_1)_i$, $T_i = k_1(\Omega_3)_i$ and $B_2 = B_1/20$, find B_1 and B_2 in terms of k_1 and $(\Omega_1)_i$.

c) After the engine has been running steadily under the above conditions, its speed is suddenly dropped at $t = 0$ to one-half the initial value. Find the response of Ω_3 to this change in Ω_1 assuming that the shaft stiffness K is such that the system damping ratio is 0.5 ($T = k_1\Omega_3$ during the transient).

12–13. An elevator in a 20-story building (Fig. 12–24) is supported by one multiple-strand steel cable which is attached to a motor-driven pulley located on the roof. When the elevator is fully loaded with passengers and located at the ground floor, the cable is stretched 1 ft from its unstressed length.

We desire to investigate the operation of the elevator under various conditions of acceleration and deceleration. To get an appreciation of the factors involved in this problem, we assume that the system has no damping or other drag forces acting on it. As a first approximation, it may also be assumed that the upper end of the cable has either zero or constant speed.

a) If the upper end of the cable is suddenly stopped when the elevator is moving upward at its maximum velocity, there is a possibility that the elevator will continue moving upward sufficiently to cause the cable to go slack. If this happens, the elevator will be "snapped" when its subsequent downward motion loads the cable again, which would make the passengers uncomfortable. Determine the maximum upward velocity that the elevator may have without allowing the cable to become slack. Assume that the elevator is traveling up from the sub-subbasement and is traveling at constant velocity when the upper end of the cable is suddenly brought to rest as the elevator reaches the ground floor.

b) Consider that the fully loaded elevator is at rest at the ground floor when the upper end of the cable is given a constant speed upward to lift the elevator. Find the maximum deflection (from unstressed) which the cable must withstand if the speed of the upper end is that found in (a). If you were unable to get an answer for (a), use v_0 as the speed.

12–14. A fishing boat weighing 32,200 lb is to be towed by a much larger ship. The tow cable is linearly elastic and elongates 0.40 ft for each 1,000 lb of tension in it. The wave and viscous drag on the fishing boat can be assumed to be linearly proportional to its velocity, and equal to 3,500 lb-sec/ft. At time $t = 0$, the large tow ship starts moving with constant velocity $v_0 = 5$ ft/sec. There is no initial slack in the cable.

a) Propose a model of this system consisting of ideal lumped elements.

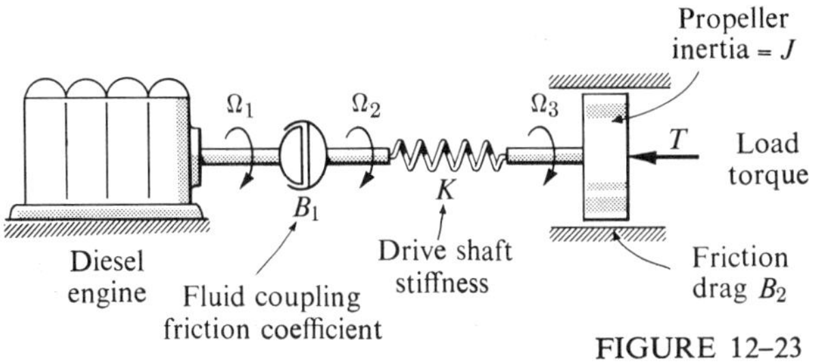

FIGURE 12–23

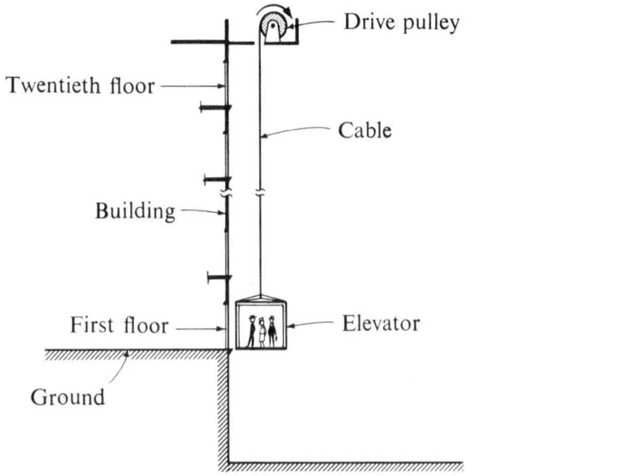

FIGURE 12–24

b) Find an expression for the fishing boat *displacement*, x_2, as a function of time.

c) What is the maximum force in the cable and at what time does it occur?

d) What is the elongation of the tow cable due to the drag of the fishing boat at $t = \infty$?

e) It is desired to change the stiffness of the cable so the fishing boat will approach the *velocity* of the tow ship as fast as possible without oscillating. What should the cable stiffness be in this case?

f) If the tow cable were 0.15 the length of the cable whose stiffness is given above, what would be the peak *velocity* obtained by the fishing boat, and at what time would this velocity occur?

12–15. The following questions deal with the use of a galvanometer (see Chapter 2) in various ways. A typical unit is shown in Fig. 12–25.

a) Find an expression for the galvanometer sensitivity in terms of the meter parameters. Torsional stiffness of the

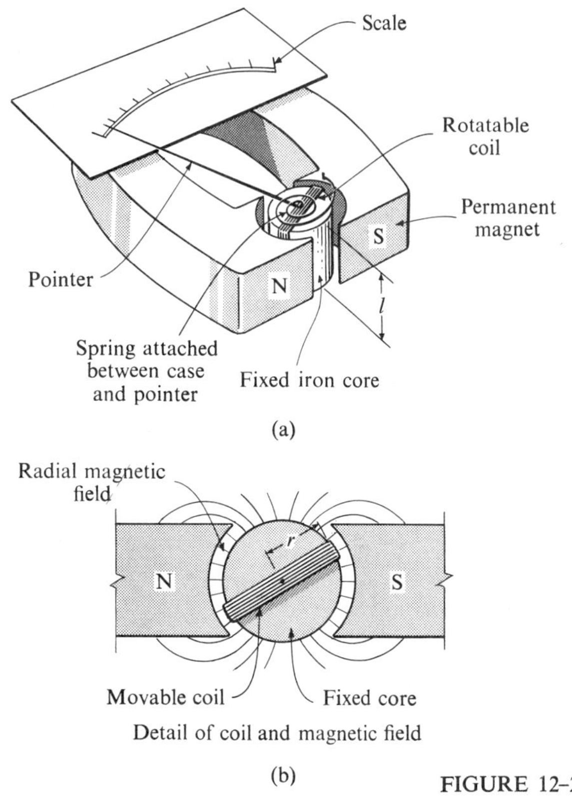

FIGURE 12–25

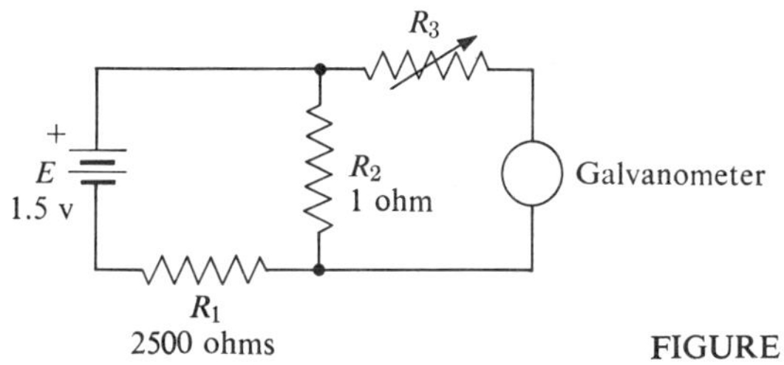

FIGURE 12–26

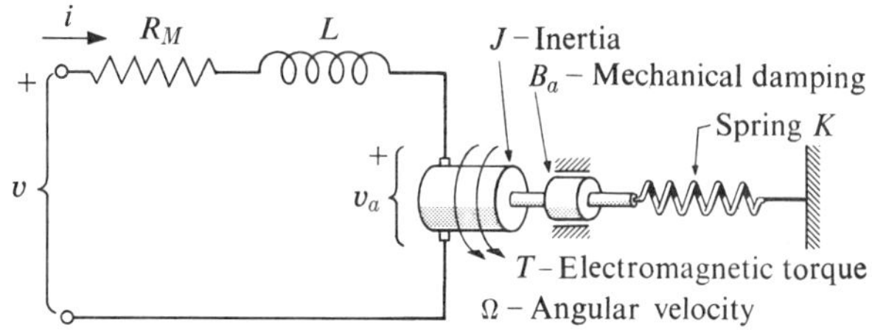

FIGURE 12–27

spring is K. The sensitivity k_s is defined as the angular deflection of the pointer per unit of steady current supplied to the coil. Discuss the problem of designing a sensitive galvanometer.

b) A galvanometer is tested using the circuit shown in Fig. 12–26. When $R_3 = 450$ ohms, the deflection is 0.15 rad. When $R_3 = 950$ ohms, the deflection is 0.075 rad. Find the meter resistance and sensitivity.

c) If full-scale deflection is 75 deg, find the current required for full-scale deflection.

d) The galvanometer may be used to construct a voltmeter or an ammeter. Find the series resistance required to produce a 2-v full-scale voltmeter. Find the shunt resistance required to make a 1-amp full-scale ammeter.

e) Galvanometers must exhibit acceptable dynamic characteristics. The movement has inertia and is restrained by a torsional spring. Air damping is present to some extent and is often augmented by adding a vane to the movement. Other instruments achieve damping by an electrical method which involves the flow of "eddy currents" in an aluminum frame around which the coil is wound. This frame is a closed circuit (in a magnetic field) around which a voltage is induced when the coil moves. This voltage causes a current to flow through the frame resistance. If the frame inductance is negligible, the current is proportional to the voltage. The presence of this current in a magnetic field causes a proportional torque to act on the coil movement. This torque is therefore proportional to the velocity and constitutes a linear damping. An equivalent model of the galvanometer is shown in Fig. 12–27.

Determine the differential equation relating galvanometer deflection to voltage input to the galvanometer coil. What is the order of this dynamic system? If the inductance of the galvanometer coil is negligible, what is the order of the system? Find an expression for the damping ratio of the system for this case. How can the damping be varied electrically? Compared with the damping which is present when the meter terminals are open-circuited ($i = 0$), what is the damping in the two cases when the meter terminals are (1) short circuited ($v = 0$) and (2) driven by a current source? (*Note:* An analog computer study of this system is described in Chapter 7.)

12–16. For the system shown in Fig. 12–28,

a) derive the system equation in $v_C(t)$,

b) determine the initial conditions assuming that there is no energy stored in L and C before the switch is closed.

12–17. The system shown in Fig. 12–29 is an armature-controlled dc motor driving an inertia load (actually its own inertia). The field excitation is fixed, so that the torque and armature-voltage equations for the motor are

$$T = k_t i_a, \qquad v_a = k_g \Omega.$$

a) Develop the differential equation relating the motor shaft speed Ω to the input voltage v_1.

b) Using the following data, find the response of the motor speed Ω after the voltage v_1 is suddenly increased from zero to 110 v.

$L_a = 10.0$ h, $\quad k_t = 0.5$ ft-lb/amp,

$R_a = 1.0$ ohm, $\quad k_g = 0.071$ v/rpm,

J = a rotor which weighs 16 lb and has a 3-in. radius of gyration.

12–18. To make a remote measurement of the level of fuel in a large tank in a refinery "tank farm," the scheme shown in Fig. 12–30 is to be employed to avoid installing electrical lines in the neighborhood of the large tank.

a) Determine the resistance of the 300-ft line (rate of flow in this line is always very small because it only has to supply flow to pressurize the 10-ft^3 chamber). Viscosity μ of the fuel is 2.0×10^{-6} lb-sec/in^2.

b) Determine the inertance of the 300-ft line. The mass density of the fuel is 8×10^{-5} lb-sec^2/in^4.

c) Determine the capacitance of the two tanks. The bulk modulus of the fuel is $\beta = 3 \times 10^5$ lb/in^2.

d) Draw an electrical analog for this fluid system based on pressure being like voltage and flow rate being like current.

e) If the input flow Q_1 is suddenly changed, comment on whether it is valid to assume that the signal line is not present in order to compute the pressure in the storage tank, P_1, and then use this (computed) pressure as an input to the signal line to find the pressure P_2. Discuss your assumption in terms of the values of components and the electrical analog.

f) Determine the capacitance of the 300-ft line and express your opinion as to whether or not it would be wise to neglect this line capacitance in an analysis of this system. The bulk modulus of the fuel is $\beta = 3 \times 10^5$ lb/in^2.

g) Make some estimates of rate of change of flow Q_2 to determine whether the inertance of the line is important. (*Hint:* consider that P_1 changed instantly; assume that inertance equals zero; find dQ_2/dt initially; compare $I(dQ_2/dt)$ to P_1.)

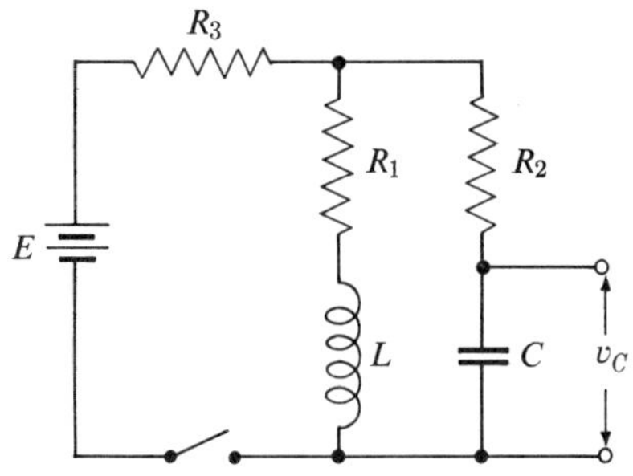

FIGURE 12–28

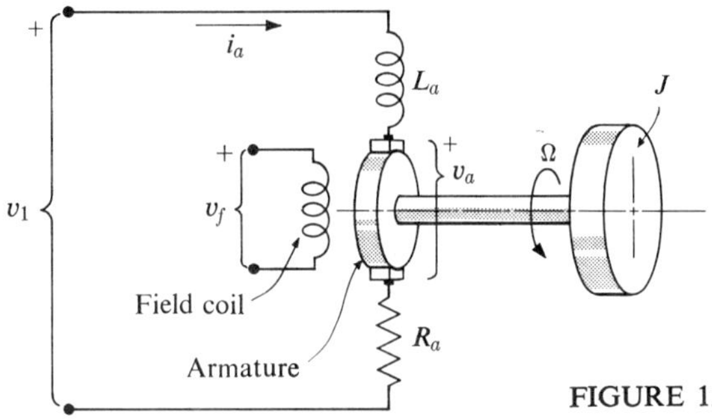

FIGURE 12–29

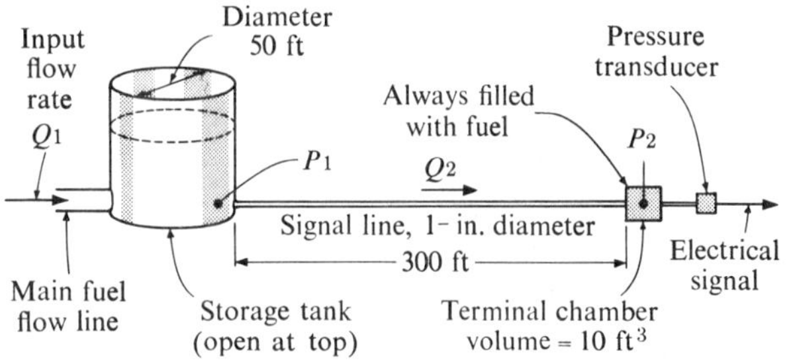

FIGURE 12–30

12–19. To design various kinds of seagoing equipment, we must have some basic information on wave motion in the open sea. (Though people have gone to sea for thousands of years, the belated invention of calculus delayed observations until times which are quite recent in geological terms. For this reason we are now interested in good measurements of wave motion.) It is proposed to make these measurements from a "stationary" platform which is actually a very long spar weighted on one end so that

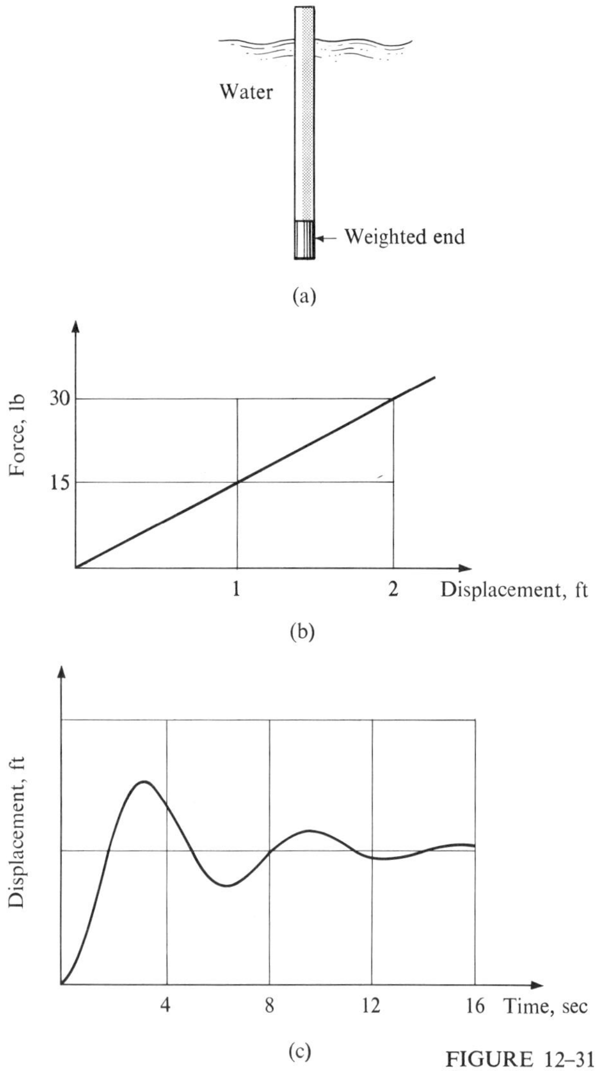

FIGURE 12–31

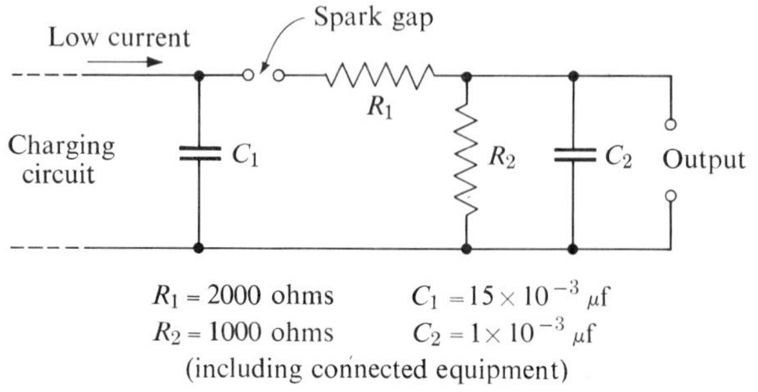

FIGURE 12–32

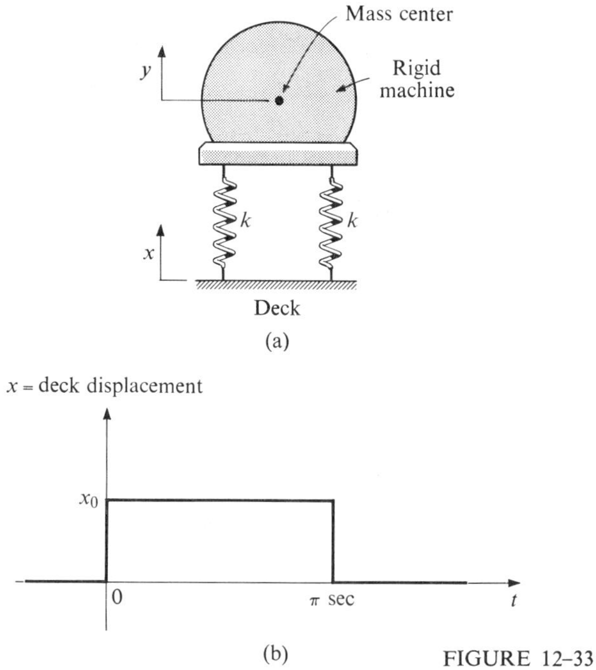

FIGURE 12–33

about $\frac{1}{10}$ of its length projects out of the water. (A spar is much like a telephone pole.) Ultimately it is desired to learn how stable this platform is. To begin with, however, our concern is simply to model the spar in preparation to making a dynamic analysis of the problem.

A physical model of a spar has been made and two tests have been run. In the first test the spar was pushed down from its free-floating position so that its top was under water, and the force needed to do this was measured. In the second test the spar was released after having been pushed down so that its top was under water, and its displacement from the free-floating position was noted as a function of time. The two curves resulting from these measurements are given in Fig. 12–31:

a) Write a differential equation which will describe the vertical motion of the spar, putting an arbitrary forcing function on the right-hand side of the equation.

b) Evaluate the physical constants on the left-hand side of this equation from the graphical data in Fig. 12–31.

c) After three years the spar will have lost half its buoyancy and the seaweed which has grown on it will have increased its drag in the water by a factor of 3. Sketch the new curves for force vs. displacement and displacement vs. time.

12–20. In this chapter, the general expression for the response of a second-order system to a step input is derived for the case in which the input is *not* an impulse. Obtain the expression for the response of such a system if the input is an impulse:

$$x_f = X_f u_i(t),$$

where $u_i(t)$ is a unit impulse.

12–21. The circuit shown in Fig. 12-32 is proposed for impulse testing of electrical power equipment (transformer insulation, etc.). The parameter values given are proposed to produce an output voltage which rises very quickly to a large value and then falls off rather slowly. The spark gap breaks down when the voltage across it reaches a preset value, and the gap may be considered to have negligible voltage drop across it when conducting. A high-resistance source slowly charges C_1.

The resistor R_1 is introduced so that the output voltage will not rise immediately to its peak value, and the resistor R_2 is included to cause the voltage to fall off after peaking.

a) Determine the time lapse between the spark gap breakdown and the peak output voltage.

b) Determine the time for the output voltage to fall from the peak to half the peak value.

c) What spark-gap setting (i.e., what breakdown voltage) should be used so that the peak output voltage will be 25 kvs?

12–22. A circulating pump in a submarine weighs 644 lb and is mounted on shock mounts to protect it against damage from underwater blast. A simplified dynamic model of this system is shown in Fig. 12–33(a). Gravity is neglected. Prior to $t = 0$ the system is at rest. The displacement x of the deck caused by a distant nuclear explosion is idealized as the *rectangular pulse* shown in Fig. 12–33(b).

a) Write the differential equation for vertical mass motion y vs. deck displacement x as an input.

b) Find y vs. time for $0 < t < \pi$ and for $t > \pi$ and sketch a plot of the result. $k = 90$ lb/ft, $m = 644$ lb, and $x_0 = 1$ ft.

c) Sketch the *force* on the mass as a function of time $0 < t < \infty$ and compute its maximum value.

d) Your junior officer, upon inspecting the shock mounts (k's), recommends that a damper of $b = 60$ lb-sec/ft be installed between the deck and the pump. He says this will reduce oscillations of m and thus lower the forces on the pump. Show qualitatively on your sketches of b) and c) how the damper will change the force-time curve. Will you accept his suggestion?

12–23. A man in a rubber life raft wishes to put a nail part-way through a piece of wood in order to make a float for a shark-fishing line. For fear of damaging the boat he holds the piece of wood in his hand while hammering. Tests he had the foresight to make before the ship went down showed that the nail could be made to penetrate the wood with a steady force of 100 lb. The wood is a block weighing 1 lb. The hammer weighs 3 lb and strikes the block with a velocity of 14 ft/sec. The nail weighs 0.005 lb, is 3 in. long, and is made of steel. The man hammers three times in 12 sec.

a) Assuming the man holds the block very closely while hammering, sketch the net force-vs.-time curve for the block. Estimate the maximum force exerted on the block by the hammering.

b) Estimate how long the hammering force from each blow is exerted on the block.

c) Using the ideal elements we have used so far, draw a model for the block and support. Evaluate, very roughly, the magnitude of the constants for these elements.

d) Write the differential equation describing the motion of the block appropriate for the block and support, and put down the initial conditions needed to complete the formulation of the problem.

12–24. For the system shown in Fig. 12–28.

a) Show why the current through R_1, at $t = 0+$ must be zero.

b) Show why the currents through R_3 and R_2 are both equal to $E/(R_1 + R_2)$ at $t = 0+$.

c) Determine the voltage across L at $t = 0+$.

d) Determine the current i_C at $t = 0+$.

e) Determine dv_C/dt at $t = 0+$.

f) Determine the time rate of change of i_L at $t = 0+$.

g) Determine the time rate of change of i_C at $t = 0+$.

This method of determining initial conditions at $t = 0+$ involves systematic inspection of elemental equations, path equations, and vertex equations at $t = 0+$.

12–25. For the system shown in Fig. 12–28, use the method of successive integration of the system differential equation in $i_C(t)$ from $t = 0-$ to $t = 0+$ (see Problem 11–17) to determine the initial values of i_C and di_C/dt.

12–26. Consider the problem of finding the response of a wheel suspended mass (a unicycle, or the simple model of an automobile suspension system shown in Fig. 6–10c) to a step change in road surface height.

a) Draw the linear graph for this system.

b) Show that the differential equation for this system in $x_c(t)$ (that is, x_c the output) is: $m\,d^2x_c/dt + b\,dx_c/dt + kx_c = b\,dx/dt + kx - mg$.

c) Use the method of successive integration from $t = 0-$ to $t = 0+$ to determine $x_c(0+)$ and $\dot{x}_c(0+)$. Assume $x_c(0-) = 0$ and $x(0-) = 0$.

d) What is the force in the shock absorber at $t = 0-$? At $t = 0$? At $t = 0+$?

13 RESPONSE OF LINEAR SYSTEMS— SYSTEM FUNCTION; POLES AND ZEROS

13-1. INTRODUCTION

At this point in our development, it is instructive to review our progress. This review will set into proper context the application of the principles which we have developed and will indicate the limitations of the approaches presented in the following chapters.

After a general introduction (Chapter 1), the various dynamic system elements were considered (Chapters 2, 3, and 4). Distinction was made among a physical element (such as a spring as it exists in the real world), a pure element (such as an element which has spring-like characteristics but free of the presence of other effects such as mass and damping), and an ideal element (such as a pure linear spring). The simple interconnection of dynamic system elements and subsequent behavior was considered in Chapter 5 with examples limited to ideal systems. Chapter 6 discussed the way in which a real system could be approximated (modeled) by means of the elements and interconnections introduced earlier. There was considerable emphasis on the fact that not all systems could be satisfactorily modeled by ideal elements.

In Chapter 7, the investigation of system performance was considered using the approach offered by the analog computer. Nonlinear effects can be included in an analog computer solution with relative ease. Although linear examples were used, this method is perfectly general and not restricted to linear systems. The mathematical formulation of system equations, using the system graph, was then considered in Chapter 8. These methods were not restricted to linear systems, and the statement of continuity and compatibility conditions (vertex and path laws) and elemental equations in differential equation form may be applied directly to nonlinear systems.

Chapter 9 reviewed some mathematical techniques of complex numbers in preparation for (a) Chapter 10, where the types of inputs used for system analysis were considered and (b) Chapter 11, where the solution of differential equations was reviewed. Chapter 12 then applied the methods of Chapters 9, 10, and 11 for the determination of the transient response of simple linear systems. The concepts of time constant, natural frequency, and damping ratio which were developed there are, strictly speaking, only applicable to linear systems.

We thus have developed, to this point, all the background required for the modeling, formulation, and solution of lumped dynamic systems, whether linear or nonlinear. None of the methods have been restricted to linear systems, although primarily linear systems have been used for illustration because of their relative simplicity and because the results for linear systems are important. Specifically the results of Chapter 12 are limited to linear systems.

We have seen that there is considerable effort involved in the development of the system equations for complex systems when the differential equation formulation is used. Chapter 8 mentioned methods which would be used for simplification. This chapter introduces methods which not only simplify the equation formulation for *linear* systems but also offer a compact way of presenting results from which the complete solution may be obtained or inferred. The next chapter will present additional simplifications which can be used on *linear* systems. Chapter 15 will present the sinusoidal response of *linear* systems. These three chapters (13, 14, and 15) also offer general results which may be applied to any linear system and thus these results are called the backbone of *linear system theory*. Chapter 16 introduces the Laplace transform which is also primarily useful for *linear* systems.

13–2. FORCED RESPONSE TO EXPONENTIAL INPUTS

In Chapter 8 we saw that the use of the system graph simplifies the formulation of system equations for all types of systems by making evident the vertices and paths to which continuity and compatibility should be applied. Use of the vertex and path laws along with the appropriate elemental equations will allow determination of the system equation, as we have seen. However, we are still faced with the difficulty of solving simultaneous differential equations which may involve cumbersome manipulation of differential operators. This was clearly illustrated in Chapter 8 when some complex systems were treated. We will now consider a method which will simplify this process for *linear systems* by enabling one to work with algebraic rather than differential equations.

13–2.1 Generalized Impedance

Let us consider a system composed of ideal elements, i.e., a linear system, which may have various sources. If we consider all sources to be exponential in nature (proportional to e^{st} with the same s if more than one source exists), then the forced part of the solution for any variable will also be exponential in nature (see Sections 10–6 and 11–5.2). Therefore, *considering only the forced part of the solution*, we have for every through- and across-variable:

$$f = \mathbf{F}e^{st} \quad \text{and} \quad v = \mathbf{V}e^{st}, \tag{13–1}$$

where **F** and **V** are, in general, functions of the complex variable s and will be called *complex amplitudes*. The variable s is called the *complex frequency*, is the same for all variables, and is determined by the source variation.

The generalized ideal two-terminal elements obey the following equations (see Section 4–3):

$$f = C\frac{dv}{dt} \quad \text{(generalized capacitance)}, \tag{13–2}$$

$$v = Rf \quad \text{(generalized resistance)}, \tag{13–3}$$

$$v = L\frac{df}{dt} \quad \text{(generalized inductance)}. \tag{13–4}$$

Substituting the exponential forms, we have

$$\mathbf{F}e^{st} = sC\mathbf{V}e^{st},$$
$$\mathbf{V}e^{st} = R\mathbf{F}e^{st},$$
$$\mathbf{V}e^{st} = sL\mathbf{F}e^{st},$$

and it is seen that the exponential factors may be canceled, resulting in the following relations between the complex amplitudes:

$$\mathbf{F} = sC\mathbf{V}, \tag{13–5}$$

$$\mathbf{V} = R\mathbf{F}, \tag{13–6}$$

$$\mathbf{V} = sL\mathbf{F}. \tag{13–7}$$

Thus we may think of the system as having a complex amplitude across-variable existing across each element and a complex amplitude through-variable through each element. The elements themselves are described by the algebraic quantities sC, R, and sL. In these terms, time does not appear explicitly anywhere; the complex amplitudes and element descriptions are constants and not functions of time. The actual through- and across-variables are time-varying in accordance with the exponential factor (13–1), but the complex amplitudes **F** and **V** are functions of s and independent of time. Therefore it is possible to

TABLE 13–1

Impedances for Lumped Linear Elements

Domain	A-type generalized capacitance	D-type generalized resistance	T-type generalized inductance
General	$1/sC$	R	sL
Mechanical translational	mass, $1/sm$	damping, $1/b$	spring, s/k
Mechanical rotational	inertia, $1/sJ$	damping, $1/B$	spring, s/K
Electrical	capacitance, $1/sC$	resistance, R	inductance, sL
Fluid	fluid capacitance, $1/sC_f$	fluid resistance, R_f	inertance, sI
Thermal	thermal capacitance, $1/sC_t$	thermal resistance, R_t	does not exist so far as is known.

work with algebraic relations rather than with differential equations in studying interconnected systems of linear elements. This will result in considerable simplification in the determination of the system equations for complex systems such as those treated in Chapter 8. The fact that only the forced part of the solution is used is not a limitation, since at any point in the calculations it is possible to return to the general differential form of any equation by recognizing that each s was the result of a time differentiation. In this manner both the natural and forced parts of the solution may be obtained. The latter procedure will be illustrated in the next section.

This procedure is definitely limited to linear systems since, in general, an exponential input to a nonlinear system will not result in an exponential response. (In fact, for a nonlinear system it is not in general possible to separate the solution into forced and natural parts.)

The ratio of the complex amplitudes of the across-variable to the through-variable for any element is defined as the *generalized impedance* of that element:*

$$\mathbf{Z} = \mathbf{V}/\mathbf{F}. \tag{13–8}$$

From (13–5, –6, and –7), we see that the impedances for the three types of elements are

$$\mathbf{Z} = 1/sC \quad \text{(generalized capacitance)}, \tag{13–9}$$

$$\mathbf{Z} = R \quad \text{(generalized resistance)}, \tag{13–10}$$

$$\mathbf{Z} = sL \quad \text{(generalized inductance)}. \tag{13–11}$$

Table 13–1 summarizes the impedances of physical two-terminal elements.

* Various definitions of impedance for mechanical systems will be found in the literature. The ratio of complex force amplitude to complex velocity amplitude is frequently used, but one can also find the ratio of complex force amplitude to complex displacement amplitude. Reference 1 uses $\mathbf{Z} = \mathbf{F}/\mathbf{V}$, which is the reciprocal of the definition of Eq. (13–8). The reason for this definition is that the analogy we have called the "dualog" was most popular in the past. This led to use of the reciprocal of our definition for mechanical-system impedance since it was then, using the dualog analogy, similar to the definition for electrical systems. The reciprocal impedance for mechanical systems is sometimes called "mobility." As was discussed in Chapter 8, the authors prefer to use the analogy based on through- and across-variables, which results in analogous systems having the same linear graph.

It should be clearly understood that the way of defining impedance is arbitrary—one can do as he pleases so long as he is consistent thereafter. However, the student who becomes familiar with the system used here should, in reading elsewhere, be wary and check which definition of impedance is used.

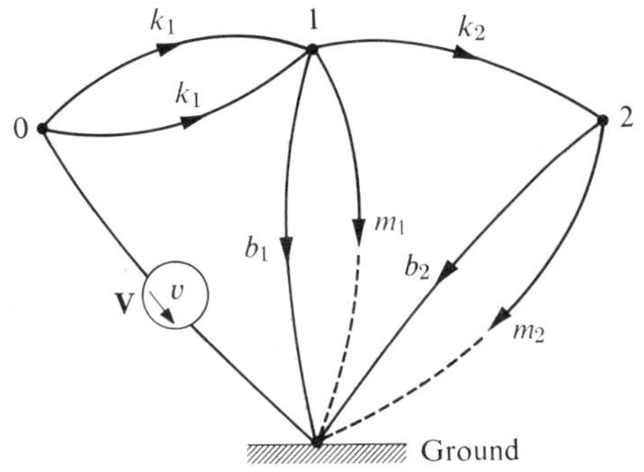

FIG. 13–1. System graph of vibration absorber of Fig. 8–33.

It is sometimes convenient to work with *admittance*, the reciprocal of impedance:*

$\mathbf{Y} = 1/\mathbf{Z}.$

13–2.2 Solution for Output

A complete procedure for the formulation of system equations using impedance will now be illustrated by the vibration-absorber example of Chapter 8. Figure 13–1 shows the system graph of the vibration absorber as shown in Fig. 8–33 and as modeled in the discussion of Section 8–5.3.

The source is considered to be exponential, $v = \mathbf{V}e^{st}$. We may then select variables, using the node method as in Section 8–5.3. Our variables will be $\mathbf{V}_1$ and $\mathbf{V}_2$, with $\mathbf{V}$ as input. The vertex law applied at nodes 1 and 2 then gives directly:

$$\mathbf{F}_{b1} + \mathbf{F}_{m1} + \mathbf{F}_{k2} - \mathbf{F}_{k1} - \mathbf{F}_{k1} = 0, \tag{13–12}$$

$$\mathbf{F}_{b2} + \mathbf{F}_{m2} - \mathbf{F}_{k2} = 0. \tag{13–13}$$

By inspection, we see that $\mathbf{V}_{01} = \mathbf{V} - \mathbf{V}_1$ and $\mathbf{V}_{12} = \mathbf{V}_1 - \mathbf{V}_2$. The elemental equations are written, using generalized impedances, as

$$\mathbf{V}_{01} = \frac{s}{k_1}\mathbf{F}_{k1}, \tag{13–14}$$

$$\mathbf{V}_{12} = \frac{s}{k_2}\mathbf{F}_{k2}, \tag{13–15}$$

$$\mathbf{V}_1 = \frac{1}{sm_1}\mathbf{F}_{m1}, \tag{13–16}$$

$$\mathbf{V}_1 = \frac{1}{b_1}\mathbf{F}_{b1}, \tag{13–17}$$

$$\mathbf{V}_2 = \frac{1}{b_2}\mathbf{F}_{b2}, \tag{13–18}$$

$$\mathbf{V}_2 = \frac{1}{sm_2}\mathbf{F}_{m2}. \tag{13–19}$$

Substituting these elemental equations into (13–12) and (13–13), we obtain

$$b_1\mathbf{V}_1 + sm_1\mathbf{V}_1 + \frac{k_2}{s}(\mathbf{V}_1 - \mathbf{V}_2) - \frac{2k_1}{s}(\mathbf{V} - \mathbf{V}_1) = 0, \tag{13–20}$$

$$b_2\mathbf{V}_2 + sm_2\mathbf{V}_2 - \frac{k_2}{s}(\mathbf{V}_1 - \mathbf{V}_2) = 0. \tag{13–21}$$

Equations (13–20) and (13–21) may be rewritten in a form for solution:

$$\left(\frac{2k_1 + k_2}{s} + b_1 + sm_1\right)\mathbf{V}_1 - \frac{k_2}{s}\mathbf{V}_2 = \frac{2k_1}{s}\mathbf{V}, \tag{13–22}$$

$$-\frac{k_2}{s}\mathbf{V}_1 + \left(\frac{k_2}{s} + b_2 + sm_2\right)\mathbf{V}_2 = 0. \tag{13–23}$$

Solving (13–23) for $\mathbf{V}_2$ and substituting into (13–22) gives one equation in one unknown which, when cleared of fractions, gives the solution for $\mathbf{V}_1$ in terms of $\mathbf{V}$:

$$\{[(2k_1 + k_2) + b_1s + s^2m_1] \times [k_2 + b_2s + s^2m_2] - k_2^2\}\mathbf{V}_1 = 2k_1(k_2 + b_2s + s^2m_1)\mathbf{V}$$

or

$$\{m_1m_2s^4 + (b_2m_1 + b_1m_2)s^3 + [k_2m_1 + b_1b_2 + m_2(2k_1 + k_2)]s^2 + [b_1k_2 + b_2(2k_1 + k_2)]s + 2k_1k_2\}\mathbf{V}_1 = 2k_1(k_2 + b_2s + m_2s^2)\mathbf{V}. \tag{13–24}$$

* In mechanical systems, our admittance is sometimes called "mobility" [1].

Now the system differential equation for v_1 output can be obtained from (13–24) by replacing terms with s multiplied by $\mathbf{V}$ or $\mathbf{V}_1$ by time derivatives of v or v_1; thus

$$s^k\mathbf{V} \rightarrow \frac{d^k v}{dt^k}, \qquad s^k\mathbf{V}_1 \rightarrow \frac{d^k v_1}{dt^k}.$$

In this manner, the student can check that (13–24) is equivalent to Eq. (8–34). With practice, this method can be carried out with considerable ease compared to the use of simultaneous differential equations. It is suggested that the student use this method on the vibration absorber with an extra damper, as described in Section 8–5.4. The advantage of working with algebraic equations will be clearly seen for that example, and the student should be able to obtain Eq. (8–51) easier and without the necessity for the cumbersome method described in Section 8–5.4.

13–3. THE SYSTEM FUNCTION

In a manner similar to that in which impedance was defined for an individual element as the ratio of complex amplitudes of the element variables, a quantity can be defined which relates a system output to a given input for an over-all system excited by an exponential input. The quantity is called the *system function*.

13–3.1 Definition of System Function

The general form of a system equation is given by Eq. (8–52), which is repeated here:

$$a_n \frac{d^n x_r}{dt^n} + a_{n-1}\frac{d^{n-1} x_r}{dt^{n-1}} + \cdots + a_1 \frac{dx_r}{dt} + a_0 x_r = b_m \frac{d^m x_f}{dt^m} + \cdots + b_1 \frac{dx_f}{dt} + b_0 x_f. \tag{8–52}$$

The notation x_r indicates a responding variable (an output), whereas x_f indicates a forcing variable (an input). Using the concept of an exponential input,

$$x_f = \mathbf{X}_f e^{st}, \tag{13–25}$$

we note that the *forced part* of the output will also be exponential:

$$x_r = \mathbf{X}_r e^{st}. \tag{13–26}$$

If (13–25) and (13–26) are substituted into (8–52) and the common exponential factors are canceled, the equation becomes

$$(a_n s^n + a_{n-1}s^{n-1} + \cdots + a_0)\mathbf{X}_r = (b_m s^m + b_{m-1}s^{m-1} + \cdots + b_0)\mathbf{X}_f. \tag{13–27}$$

If the impedance method described in the previous section is used to solve the system, Eq. (13–27) is obtained as the final solution directly without first obtaining Eq. (8–52). For example, Eq. (13–24) for the vibration absorber is of the form of Eq. (13–27). In any event, this equation takes the form of two polynomials in s multiplied by the input and output complex amplitude. The ratio of the complex amplitude of the output, $\mathbf{X}_r$, to the complex amplitude of the input, $\mathbf{X}_f$,

$$\mathbf{T}(s) = \frac{\mathbf{X}_r}{\mathbf{X}_f} = \frac{(b_m s^m + b_{m-1}s^{m-1} + \cdots + b_0)}{(a_n s^n + a_{n-1}s^{n-1} + \cdots + a_0)}, \tag{13–28}$$

is defined as the *system function*. As an example, the system function of the vibration-absorber system is obtained directly from Eq. (13–24) and is

$$\mathbf{T}(s) = \frac{\mathbf{V}_1}{\mathbf{V}} = \frac{2k_1(m_2s^2 + b_2s + k_2)}{m_1m_2s^4 + (b_2m_1 + b_1m_2)s^3 + [k_2m_1 + b_1b_2 + m_2(2k_1 + k_2)]s^2 + [b_1k_2 + b_2(2k_1 + k_2)]s + 2k_1k_2}. \tag{13–29}$$

It should always be kept in mind that the system function is the ratio of the complex amplitude of the *forced part* of the output to the complex amplitude of the input for an exponential input function.

If $\mathbf{X}_r$ and $\mathbf{X}_f$ are both across- or through-variables, $\mathbf{T}(s)$ will be called the *transfer function*.* If the ratio $\mathbf{T}(s)$ is that of an across-variable to a through-variable, the system function will be an *impedance*, as defined previously. A ratio of a through-variable to an across-variable will be an *admittance*.

It is worth while to note here that in other literature the system function will be found using the operators $p = d/dt$ or $D = d/dt$ in place of s. The development here can be considered to be a justification of the use of operators in this manner.

The complete (natural plus forced) response to an exponential input may be determined by using the system function. The denominator of the system function should be recognized as the characteristic equation of the system. Since the roots of the characteristic equation determine the natural part of the response, the system function contains the information necessary to determine the complete response. The initial conditions must, of course, be specified in addition.

If we have

$$\mathbf{T}(s) = \mathbf{N}(s)/\mathbf{D}(s), \tag{13-30}$$

then the n roots of the characteristic equation $(r_1, r_2, \ldots, r_n)$ are given by

$$\mathbf{D}(r_k) = 0, \qquad k = 1, 2, \ldots, n. \tag{13-31}$$

If $\mathbf{D}(s)$ is high order, the problem of finding numerical values for the roots becomes formidable. There are many books which deal with the numerical evaluation of the roots. The reader is invited to consult Reference 2 for an appreciation of the methods used. The widespread availability and use of digital computers has made this problem less formidable.

* Many workers use "transfer function" in the same sense we use system functions: as the ratio of $\mathbf{X}_r$ to $\mathbf{X}_f$ regardless of what variables are involved.

The complete response to a suddenly applied exponential input $x_f = \mathbf{X}_f e^{st}$ is then (if there are no repeated roots)

$$x_r = \mathbf{T}(s)\mathbf{X}_f e^{st} + \sum_{k=1}^{n} A_k e^{r_k t}, \tag{13-32}$$

where the initial conditions must be applied as before (see Chapters 11 and 12) to determine the values of the A_k.

13-3.2 Use for Sinusoidal Input, $s = j\omega$

The system function is particularly useful for finding the *steady-state* response to a sinusoidal input. If the input is sinusoidal, then

$$x_f = X_f \sin \omega t = \mathrm{Im}[X_f e^{j\omega t}], \tag{13-33}$$

or

$$x_f = X_f \cos \omega t = \mathrm{Re}[X_f e^{j\omega t}], \tag{13-34}$$

where X_f is the amplitude of the sinusoidal input.

According to the results of Section 11-4.3, Case 3, we may find the response of a linear system to the inputs in the equation (13-33) or (13-34) by using only $X_f e^{j\omega t}$ as the input and then taking either the real or imaginary part of the output.

Considering the input

$$x_f = X_f e^{j\omega t} \tag{13-35}$$

applied to our general system, we see that the output is

$$x_r = \mathbf{X}_r e^{j\omega t}. \tag{13-36}$$

We may thus use the system function with $s = j\omega$. So $\mathbf{X}_r = \mathbf{T}(j\omega)X_f$ and $x_r = \mathbf{T}(j\omega)X_f e^{j\omega t}$.

The actual time function of the forced part of the output for a $\sin \omega t$ input is

$$x_r = \mathrm{Im}[\mathbf{T}(j\omega)X_f e^{j\omega t}].$$

Now

$$\mathbf{T}(j\omega) = |\mathbf{T}(j\omega)|e^{j\phi}, \tag{13-37}$$

so

$$x_r = \mathrm{Im}[|\mathbf{T}(j\omega)|X_f e^{j(\omega t+\phi)}]$$

or

$$x_r = |\mathbf{T}(j\omega)|X_f \sin(\omega t + \phi) \tag{13–38}$$

is the forced part of the system response to a sinusoid. If we are only interested in the steady-state response, this is the desired solution.

Summary. To find the steady-state response of a linear system to a sinusoid of angular frequency ω:

a) let $s = j\omega$ in the system function,

b) determine the magnitude, $|\mathbf{T}(j\omega)|$, and phase ϕ of $\mathbf{T}(j\omega)$ as in Eq. (13–37),

c) the amplitude of the output sinusoid is $|\mathbf{T}(j\omega)|$ times the amplitude of the input sinusoid, as shown in Eq. (13–38),

d) the output sinusoid leads the input sinusoid by the phase angle ϕ, as shown in Eq. (13–38).

Example 1. Let

$$x_f = X_f \sin \omega t \quad \text{and} \quad \mathbf{T}(s) = \frac{2s+1}{3s^2+s+1}.$$

Then

$$\mathbf{T}(j\omega) = \frac{2j\omega + 1}{-3\omega^2 + j\omega + 1} = \frac{1 + j2\omega}{(1-3\omega^2) + j\omega},$$

$$|\mathbf{T}(j\omega)| = \frac{\sqrt{1+4\omega^2}}{\sqrt{(1-3\omega^2)^2+\omega^2}}.$$

$$\phi = \tan^{-1} 2\omega - \tan^{-1}\left(\frac{\omega}{1-3\omega^2}\right),$$

$$x_r = |\mathbf{T}(j\omega)|X_f \sin(\omega t + \phi)$$

$$= \frac{\sqrt{1+4\omega^2}\,X_f}{\sqrt{(1-3\omega^2)^2+\omega^2}} \times \sin\left[\omega t + \tan^{-1} 2\omega - \tan^{-1}\left(\frac{\omega}{1-3\omega^2}\right)\right].$$

Example 2. Let

$$x_f = 10\cos(3t) \quad \text{and} \quad \mathbf{T}(s) = \frac{s^2+2s+3}{s^3+2s^2+2s+1}.$$

Now $\omega = 3$ and

$$\mathbf{T}(j3) = \frac{-9+2j3+3}{-j27-2\cdot 9+2j3+1} = \frac{-6+j6}{-17-j21}$$

$$= \frac{6-j6}{17+j21} = \frac{8.46\angle -45^\circ}{27.0\angle +51^\circ} = 0.313\angle -96^\circ,$$

$$x_r = 3.13\cos(3t - 96^\circ).$$

13–3.3 Use for Step Input, $s = 0$

For a step input, $x_f = X_f u_s(t)$, which corresponds to $s = 0$ in our general exponential form. Thus the steady-state or forced part of the response to a constant input is found by letting $s = 0$ in $\mathbf{T}(s)$:*

$$x_r = \mathbf{T}(0)X_f. \tag{13–39}$$

From Eq. (13–28), we see that this is

$$x_r = \frac{b_0}{a_0} X_f, \tag{13–40}$$

a result which is also easily found from the system equation and by simple inspection of the system for the steady-state performance. Thus the use of $s = 0$ is a convenient check on the correctness of $\mathbf{T}(s)$.

The system function contains complete information on the system response. It was noted that the complete response can be found by determining the roots of the characteristic equation which is the denominator of the system function, as discussed earlier and indicated in Eq. (13–31).

Summary. To find the complete response of a linear system to a step input $x_f = X_f u_s(t)$:

a) let $s = 0$ in $\mathbf{T}(s)$. The forced part of the response is then $\mathbf{T}(0)X_f$;*

---

* If $a_0 = 0$ or, more generally, if any power of s is factorable out of the denominator of $\mathbf{T}(s)$, this does not give the correct forced response since the input is then of the same form as a term in the complementary function; see the exception in Section 11–5.2, pp. 288 and 289. The result for $s = 0$ does give the correct "steady state" since $x_r \to \infty$ as $t \to \infty$ for this case.

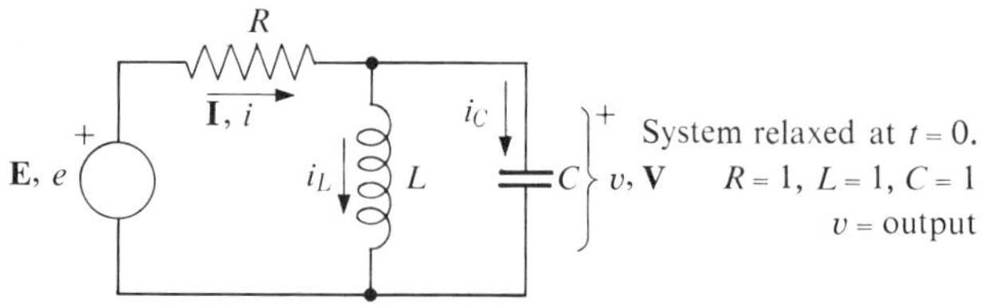

FIG. 13–2. Circuit for Example 2.

b) determine the n roots of the denominator of $\mathbf{T}(s)$, $(r_1, r_2, \ldots, r_n)$ from $\mathbf{D}(r_k) = 0; k = 1, 2, \ldots, n$. The natural part of the response is then (if no roots are repeated)

$$\sum_{k=1}^{n} A_k e^{r_k t},$$

where A_k are the arbitrary constants and the r_k are the roots of $\mathbf{D}(s)$.

c) The complete solution is then

$$x_r = \mathbf{T}(0) X_f + \sum_{k=1}^{n} A_k e^{r_k t}.$$

d) By applying the n initial conditions, determine the A_k.

Example 1. Let $x_f = X_f u_s(t)$, $x_r = 0$ at $t = 0$ and $\mathbf{T}(s) = 10/(s + 2)$. Then $x_r = 5X_f + Ae^{-2t}$ and $x_r = 5X_f(1 - e^{-2t})$ when the zero initial condition is applied.

Example 2. Determine the response of the system shown in Fig. 13–2 to a step input of 10 v:

$$\mathbf{E} - \mathbf{V} = \mathbf{I}R, \qquad \mathbf{I} = sC\mathbf{V} + \frac{\mathbf{V}}{sL},$$

$$\mathbf{E} - \mathbf{V} = R\left(sC + \frac{1}{sL}\right)\mathbf{V} = \left(\frac{s^2RLC + R}{sL}\right)\mathbf{V},$$

or

$$(s^2RLC + sL + R)\mathbf{V} = sL\mathbf{E},$$

$$\mathbf{T}(s) = \frac{\mathbf{V}}{\mathbf{E}} = \frac{sL}{s^2RLC + sL + R} = \frac{s}{s^2 + s + 1}.$$

For $e = 10u_s(t)$, we note that from $\mathbf{T}(0)$, $v_{ss} = 0$ (which checks since L is a short circuit for dc conditions), the roots of the denominator are $r_{1,2} = -\frac{1}{2} \pm j\sqrt{3}/2$. Since the roots are complex, the form using sines and cosines should be employed:

$$v = e^{-t/2}[A \sin(\sqrt{3}/2)t + B\cos(\sqrt{3}/2)t].$$

Since $v = 0$ at $t = 0$, $B = 0$, then

$$v = Ae^{-t/2} \sin(\sqrt{3}/2)t.$$

From the circuit, $i_L = 0$ at $t = 0$, so $i_C = i$, and since $v = 0$ at $t = 0$, $i = i_C = C(dv/dt) = e/R$ at $t = 0$ or $dv/dt = 10$, so

$$10 = A[-\tfrac{1}{2}e^{-t/2}\sin(\sqrt{3}/2)t + (\sqrt{3}/2)e^{-t/2}\cos(\sqrt{3}/2)t]_{(t=0)} = A(\sqrt{3}/2).$$

Finally, $v = (20/\sqrt{3})e^{-t/2}\sin(\sqrt{3}/2)t$.

13–4. POLES AND ZEROS

The system function can be described in terms of its poles and zeros, which may be used to discuss system response.

A *zero* of the system function is a value of s which makes $\mathbf{T}(s) = 0$, i.e., *a root of the numerator of* $\mathbf{T}(s)$. A *pole* of the system function is a value of s which makes $\mathbf{T}(s) = \infty$, i.e., *a root of the denominator.*

The poles and zeros are properties of the system, and their values depend on the element parameters and on the way in which the elements are interconnected (i.e., on the structure of the system). Conversely, a knowledge of the poles and zeros is sufficient to describe completely the response of a system (within a constant factor) to any specified input function.

In general, for lumped systems,

$$\mathbf{T}(s) = \frac{b_m s^m + b_{m-1}s^{m-1} + \cdots + b_0}{a_n s^n + a_{n-1}s^{n-1} + \cdots + a_0}$$

or

$$\mathbf{T}(s) = \frac{b_m}{a_n}\frac{(s - r_{01})\cdots(s - r_{0m})}{(s - r_{p1})\cdots(s - r_{pn})}, \tag{13–41}$$

where $r_{01}, \ldots, r_{0m}$ = zeros of $\mathbf{T}(s)$, $r_{p1}, \ldots, r_{pn}$ = poles of $\mathbf{T}(s)$.

In general the poles and zeros may be complex and some may be repeated. The poles and zeros are frequently shown in an s-plane plot, with poles indicated by $\times$ and zeros by $\bigcirc$.

Figure 13–3 shows the s-plane plot for

$$\mathbf{T}(s) = \frac{b_m}{a_n} \frac{s(s+b)}{(s+a)(s+\alpha+j\beta)(s+\alpha-j\beta)} = \frac{b_m}{a_n} \frac{s(s+b)}{(s+a)(s^2+2\zeta\omega_n s+\omega_n^2)},$$

where $\alpha = \zeta\omega_n$, $\beta = \omega_n\sqrt{1-\zeta^2}$.

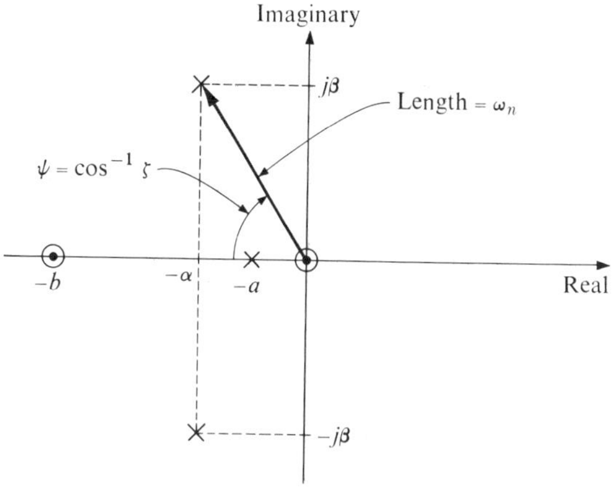

FIG. 13–3. Poles and zeros of system functions.

13–4.1 Transient Response

The s-plane plot of the *poles* of a transfer function is very useful for visualizing the character of the natural or transient response of the system. For example, the presence of a pair of complex conjugate poles (Fig. 13–3) indicates oscillatory response. The speed of response associated with this pair of poles, as measured by the undamped natural frequency, is given by the phasor connecting the origin with the pole. As the poles become closer to the imaginary axis, the damping ratio for the particular pair of poles decreases and the response due to these poles becomes more oscillatory.

A pole lying on the real axis is indicative of a single-order exponential term in the system response. In Fig. 13–3, for example, the pole at $-a$ indicates an exponential with time constant $\tau = 1/a$.

Figure 13–4 shows the s-plane plot of a system function and the transient response expected from the system. It consists of a real exponential response combined with an underdamped second-order response.

The discussion of Section 8–6.2. concerning the order of systems, is, of course, equally applicable to the system function as to the system equation. In fact, the degeneracy which occurs due to element-value symmetries can be significantly interpreted in the light of poles and zeros. This reduction of order is due to the coincidence of a pole and a zero. When this occurs a common factor may be canceled from the numerator and denominator of the system function, and, as

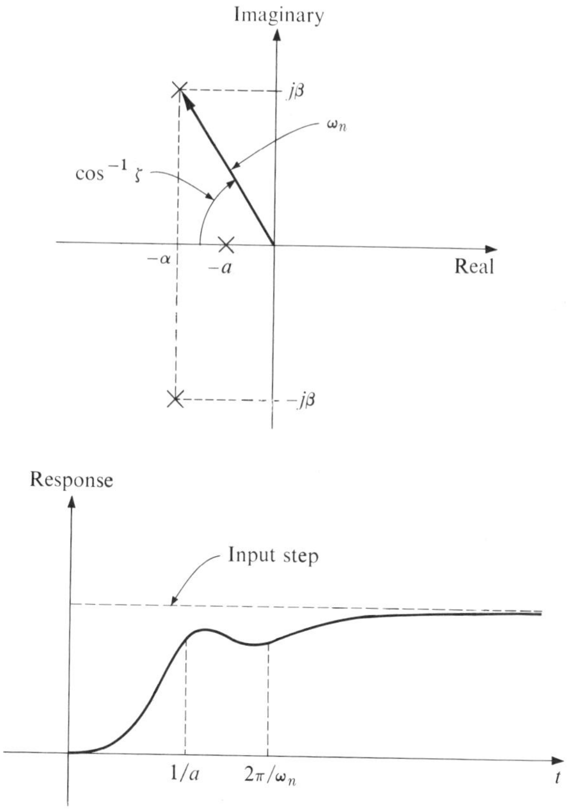

FIG. 13–4. Qualitative transient response related to pole configuration.

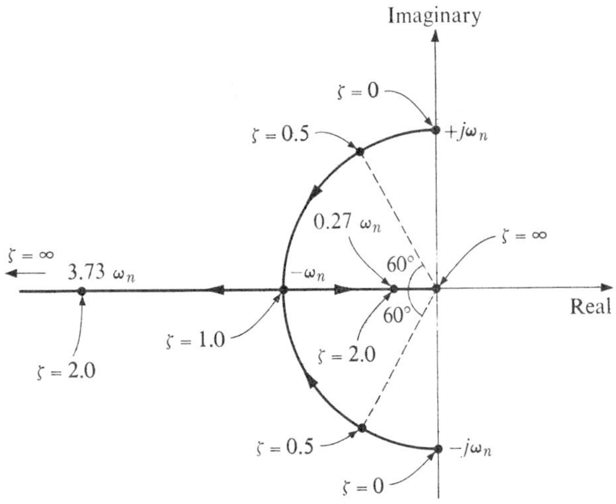

FIG. 13–5. Root-locus of second-order system.

was seen in Section 8–6.2. one must be wary of whether he will get the complete and correct answer for a transient response. If the system is initially relaxed, then it is always allowable to cancel factors. If there are arbitrary nonzero initial conditions, then it may be allowable to cancel factors for some outputs but not for others, as was seen earlier. If one always *uses* n initial conditions for a system with n energy-storage elements, then any difficulty will be evident. If the nth initial condition is not automatically satisfied by a response with $(n-1)$ poles when a common factor was canceled, then the natural response due to the canceled poles should be included.

13–4.2 Locus of the Poles in the s-plane

It is of interest to observe the locus of the roots on the s-plane as a system parameter is varied. This gives some insight into the factors affecting the speed and nature of the system response. We will consider only first- and second-order systems here.

The single pole of a first-order system lies on the real axis and is equal to $r_1 = -a_0/a_1 = -1/\tau$. The larger r_1 is, the faster the response (small τ). Therefore the distance of the pole from the origin indicates the speed of response. As a_0 and a_1 change, the locus of r_1 is on the negative real axis.

The locus of a second-order system is more interesting since it can be both over- and underdamped. Let us consider the locus as the damping ratio, ζ, is varied from 0 to ∞.

For $a_2 s^2 + a_1 s + a_0 = 0$ or

$$s^2 + 2\zeta\omega_n s + \omega_n^2 = 0, \tag{13–42}$$

where $\omega_n^2 = a_0/a_2$ and $\zeta = a_1/2\sqrt{(a_0 a_2)}$, we have the poles (for the underdamped case, $\zeta < 1$):

$$\begin{aligned} r_{p1} &= \omega_n[-\zeta + j\sqrt{1-\zeta^2}] = -\alpha + j\beta, \\ r_{p2} &= \omega_n[-\zeta - j\sqrt{1-\zeta^2}] = -\alpha - j\beta, \end{aligned} \tag{13–43}$$

where $\alpha = \zeta\omega_n$ and $\beta = \omega_n\sqrt{1-\zeta^2}$. As ζ is varied, the locus of r_{p1} and r_{p2} is a circle, as can easily be seen from

$$\alpha^2 + \beta^2 = \zeta^2\omega_n^2 + \omega_n^2(1-\zeta^2) = \omega_n^2, \tag{13–44}$$

which is the equation of a circle. This is shown on Fig. 13–5.

For an overdamped case or $\zeta > 1$, we have

$$\begin{aligned} r_{p1} &= \omega_n[-\zeta + \sqrt{\zeta^2 - 1}] = -\frac{1}{\tau_1}, \\ r_{p2} &= \omega_n[-\zeta - \sqrt{\zeta^2 - 1}] = -\frac{1}{\tau_2}, \end{aligned} \tag{13–45}$$

which lie on the negative real axis; r_{p1} is between the circle and the origin and r_{p2} is outside the circle. These are also shown in Fig. 13–5. The values of the poles for $\zeta = 0, 0.5, 1.0$, and 2.0 are given.

13–4.3 Stability of Linear Systems

If any pole of the system function lies to the right of the imaginary axis, i.e., in the right half of the s-plane, the real part of this pole will be positive. In this case, one exponential factor in the response will contain a positive real exponent, and this factor will tend toward infinity with increasing time. The system is then said to be *dynamically unstable*. On the other hand, if all the poles are in the left half-plane, all the exponentials will die out with increasing time.

It may therefore be concluded that a *necessary and sufficient condition for a linear dynamic system to be dynamically stable is that all its poles lie in the left half of the s-plane.*

It should not be assumed that an externally applied input would be necessary to have instability in a system which has at least one pole in the right half-plane. There are always internal disturbances (electrical noise, vibration, etc.) which will build up exponentially if the system is unstable. This build-up will continue until the system either destroys itself or becomes nonlinear in such a fashion that the amplitudes reach limiting values (often this involves oscillations, since the poles are frequently complex). Unless an oscillator is desired (e.g., a sinusoidal source), this is a most undesirable situation, and great pains are taken to stabilize systems which have the capability of being unstable.

Let us investigate the types of systems for which instability can occur. Consider first a *passive* linear system, i.e., one composed only of the ideal *RLC*-elements and linear energy-transformation elements (transformers, gyrators, and transducers). If this system is given some initial energy storage and released without any input, the response cannot build up indefinitely since there is only a finite amount of energy available and infinite amplitude would imply infinite storage of energy in the energy-storing elements.

The response must die out since the initially stored energy is being dissipated in the resistive elements. At best, if the system has no energy loss (no resistive elements), the response will be indefinitely oscillatory of constant amplitude, but it cannot increase in amplitude. *Thus a passive system must be stable, and hence its poles must be in the left-half plane or on the imaginary axis.* The case of no energy dissipative elements results in poles on the imaginary axis. All real passive systems will have some energy dissipation, however small, and hence the limiting case of poles on the imaginary axis is never attained in practice.

On the other hand, an *active* system is one which has an internal source of energy. Systems which contain vacuum tubes, transistors, hydraulic values, etc., are active. An active system can be unstable. A common form of active system of considerable practical interest in which instability is a common problem is one which contains *feedback*. Feedback occurs when a portion of the output is fed back to the input. Most modern control systems are feedback systems.

The coefficients $a_n, a_{n-1}, \ldots, a_0$ of the characteristic equation determine the poles and hence the stability of systems. What conclusions can be drawn concerning the relationship between these coefficients and the stability of the system? A complete discussion of the determination of the stability of linear systems will not be given in this book, but we will consider here the signs of the coefficients. The reader is referred to the many books which treat this subject [3, 4].

If all the coefficients do not have the same sign (e.g., positive) then there must be at least one pole with a positive real part. This may be seen from an expansion of the characteristic equation polynomial in terms of its roots (see Problem 13–38):

$$(s - r_1)(s - r_2)(s - r_3)\cdots(s - r_n) = 0,$$

$$s^n - \sum_{k=1}^{n} r_k s^{n-1} + \sum_{i\neq j} r_i r_j s^{n-2} - \cdots$$

$$+ (-1)^{n-1} \sum_{k=1}^{n} \frac{r_1 r_2 r_3 \cdots r_n}{r_k} s + (-1)^n r_1 r_2 r_3 \cdots r_n = 0. \tag{13–46}$$

Also, if any of the coefficients are zero, the system will have poles either on the imaginary axis or in the right half-plane (see Problem 13–39). A situation which frequently arises is that of a passive system which is assumed, for purposes of analysis, to have no energy-dissipation elements. In this case, all the coefficients of the odd powers of s will be zero and the poles will be purely imaginary.

We may therefore conclude that, since passive systems cannot have any poles in the right half-plane, all the signs of the coefficients of passive systems must be alike. This proves the statement to this effect made in Section 8–6.3.

On the other hand, if one is considering an active system, it is *not* true that if all the signs are positive the system will be stable since it may have complex roots with positive real parts.

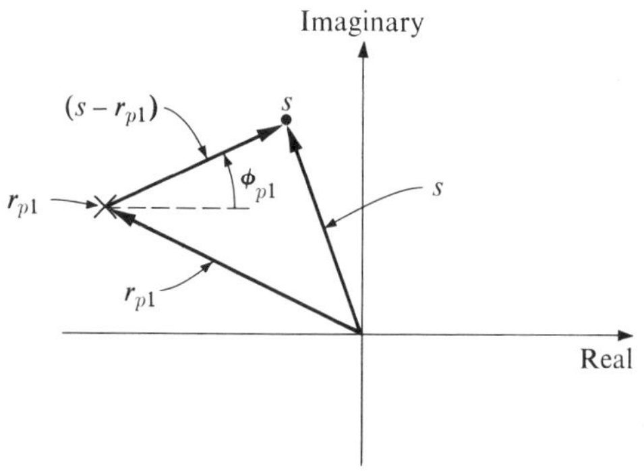

FIG. 13–6. An s-plane plot of s and one pole.

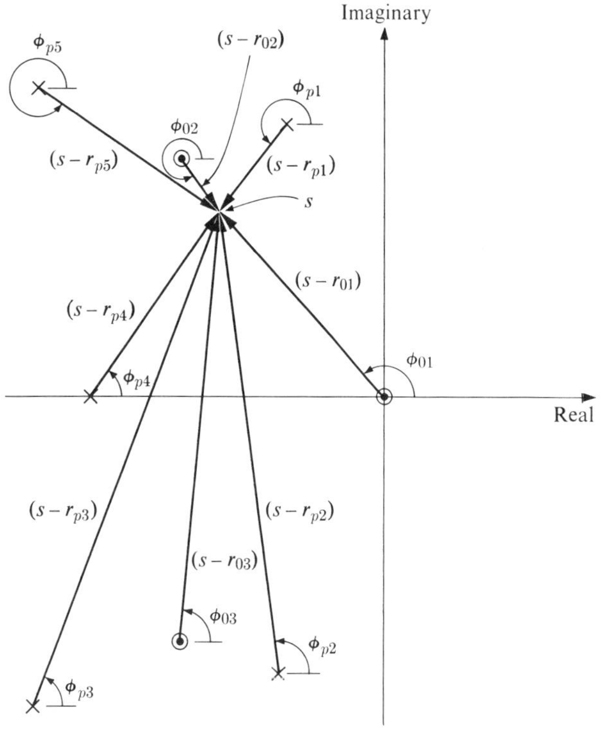

FIG. 13–7. An s-plane plot of s and poles and zeros of $\mathbf{T}(s)$.

Summary of system stability

a) The necessary and sufficient condition for system stability is that the poles of the system function lie in the left half-plane of the s-plane.

b) Passive systems must be stable. Active systems may be unstable.

c) The signs of the coefficients of the denominator of the system function for a passive system must be all positive.

d) If the signs of the coefficients of the denominator of the system function for an active system are not all positive, the system is unstable and has at least one pole with a positive real part. The response will go to infinity either exponentially (without oscillating) if the pole is real, or it will go to infinity in an oscillatory fashion if the root is complex. If any coefficients are zero, the system will have poles on the imaginary axis or in the right half-plane.

e) If the signs of the coefficients of the denominator of the system function of an active system are all positive, it can still be unstable. The instability, if it exists, will be due to a conjugate complex root, and therefore if the system is unstable the response will go to infinity in an oscillatory fashion. The literature on determination of system stability is extensive [3, 4].

13–4.4 Forced Response, T(s), by s-Plane Geometry

A further significance of the poles and zeros relates to the forced response to an exponential input.

The system function is the ratio of some output to the input. If the magnitude and angle of the complex number $\mathbf{T}(s)$ are determined, then the forced part of the output may be found once the input is known. The following discussion will establish the relation of the magnitude and angle of $\mathbf{T}(s)$ to the poles and zeros and the value of s. *It should be clearly recognized that this determines only the forced part of the response.*

If the input is exponential, the determination of the forced part of the output requires a calculation of $\mathbf{T}(s)$ for the known value of s. It is assumed that the system poles and zeros are known. This requires determining the roots of the numerator and denominator polynomials, which may be quite difficult if the order of the system is high.

Consider that $s = \sigma + j\omega$. For this form of s, the exponential is complex. (Real inputs could then be either the real or imaginary part of this complex exponential.) Consider one of the poles or zeros of $\mathbf{T}(s)$ plotted in the s-plane along with s (Fig. 13–6). We see that the factor $(s - r_{p1})$ which appears in the factored form of $\mathbf{T}(s)$ can be interpreted geometrically as a phasor in the s-plane. It has a magnitude, which is equal to the length of the phasor from r_{p1} to s, and an angle, ϕ_{p1}, which is equal to the angle the phasor makes with the horizontal. Thus

$$(s - r_{p1}) = |s - r_{p1}|e^{j\phi_{p1}}. \tag{13–47}$$

A similar construction may be made for every pole and zero of $\mathbf{T}(s)$ (Fig. 13–7). Then

$$|\mathbf{T}(s)| = \frac{b_m}{a_n}\frac{|s - r_{01}|\,|s - r_{02}|\,|s - r_{03}|}{|s - r_{p1}|\,|s - r_{p2}|\,|s - r_{p3}|\,|s - r_{p4}|\,|s - r_{p5}|} = \frac{b_m}{a_n}\frac{\text{product of phasor magnitudes from zeros to } s}{\text{product of phasor magnitudes from poles to } s}. \tag{13–48}$$

Angle of $\mathbf{T}(s)$

$$\begin{aligned}\phi &= +\sum \text{angles of phasors from zeros to } s - \sum \text{angles of phasors from poles to } s\\ &= \phi_{01} + \phi_{02} + \phi_{03} - \phi_{p1} - \phi_{p2} - \phi_{p3} - \phi_{p4} - \phi_{p5}.\end{aligned} \tag{13–49}$$

Care should be taken that the proper angles are used, as shown on Fig. 13–7. The multiplying constant b_m/a_n is determined from the coefficients of s^m in the numerator and s^n in the denominator of the system function before factoring.

It can easily be seen that when s is near a *zero* of $\mathbf{T}(s)$, the forced part of the output will be very small. When s is near a *pole* of $\mathbf{T}(s)$, the output will be very large. These effects are broadly called *resonance*. Resonance is particularly associated with sinusoidal inputs (as described in Chapter 15) but it can be broadened to include any exponential input. There are several definitions, which differ slightly, of resonance for sinusoidal inputs. For the present purpose, *resonance* will be defined as occurring when s is exactly equal to a *pole* of $\mathbf{T}(s)$. *Antiresonance* will be defined as occurring when s is exactly equal to a *zero* of $\mathbf{T}(s)$. Thus the system output at resonance is infinite and the system output at antiresonance is zero.

REFERENCES

1. *Mechanical Impedance Methods for Mechanical Vibrations.* A collection of papers published by the American Society of Mechanical Engineers, New York. Publ. No. G-30, 1958.
2. E. V. Bohn, *The Transform Analysis of Linear Systems.* Reading, Mass.: Addison-Wesley Publishing Company, Inc., 1963.
3. H. Chestnut, *System Engineering Tools.* New York: John Wiley and Sons, Inc., 1965.
4. A. G. J. MacFarlane, *Engineering Systems Analysis.* Reading, Mass.: Addison-Wesley Publishing Company, Inc., 1964.

PROBLEMS

13–1. For the pulse-transformer equivalent circuit shown in Fig. 8–34, derive the system equation Eq. (8–44) using the method of Section 13–2.2.

13–2. Use the methods of Section 13–2.2 to derive the system equation for the vibration absorber with an extra damper, as described in Section 8–5.4, to obtain Eq. (8–51).

13–3. Derive the system function for Fig. 8–35 (v_3 output) to demonstrate that this is a second-order system (the three energy-storage elements are not all independent). $\mathbf{V}_3/\mathbf{I}_1$ is called a *transfer impedance*.

13–4. Derive the system function for Fig. 8–36 (v_3 output) and show that this is a third-order system. The system function here is a transfer impedance, $\mathbf{V}_3/\mathbf{I}_1$.

13–5. Derive the system function for Fig. 8–37 (i_1 output) and show that this is a third-order system (not fourth-order). The system function here is an input admittance, $\mathbf{I}_1/\mathbf{E}$.

13–6. Derive the system equation for Fig. 8–44c (v_1 output) and compare with the dual substitution into Eq. (8–44) as described at the end of Section 8–7.2. The system function here is an input impedance, $\mathbf{V}_1/\mathbf{I}$.

13–7. Find the transfer function $\mathbf{F}_1/\mathbf{F}$ for Problem 8–8c and from this write the system differential equation by replacing s and its powers by appropriate derivatives.

13–8. Find the transfer function $\mathbf{\Omega}_2/\mathbf{\Omega}_1$ for Problem 8–8d and then the system differential equation as indicated in Problem 13–7.

13–9. Find the transfer function $\mathbf{V}_2/\mathbf{V}_1$ for Problem 8–8e and then the system differential equation as indicated in Problem 13–7.

13–10. Solve Problem 8–9a.

13–11. Solve Problem 8–9b for the transfer impedance $\mathbf{V}_0/\mathbf{I}$.

13–12. Solve Problem 8–10 for the transfer impedance $\mathbf{V}_2/\mathbf{F}$.

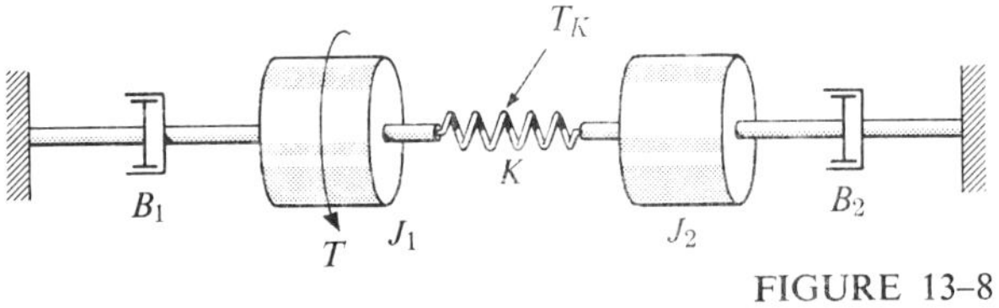

FIGURE 13–8

13–13. Solve for the transfer function $\mathbf{V}_{m1}/\mathbf{V}$ and the transfer admittance $\mathbf{F}_{k2}/\mathbf{V}$ for the system in Problem 8–27a.

13–14. Solve for the transfer impedances $\mathbf{V}_2/\mathbf{I}$ and $\mathbf{V}_1/\mathbf{I}$ for the system in Problem 8–27b.

13–15. Solve Problem 8–28.

13–16. Solve Problem 8–22 and find the sinusoidal response by substituting $s = j\omega$ into the transfer function.

13–17. The differential equation for the torque in the spring T_K for the input T in the system shown in Fig. 13–8 is:

$$\frac{J_1 J_2}{K}\frac{d^3 T_K}{dt^3} + \left(\frac{J_1 B_2 + J_2 B_1}{K}\right)\frac{d^2 T_K}{dt^2} + \left(\frac{B_1 B_2}{K} + J_1 + J_2\right) \times \frac{dT_K}{dt} + (B_1 + B_2)T_K = B_2 T + J_2 \frac{dT}{dt}.$$

a) Write the system function $\mathbf{T}_K/\mathbf{T}$.

b) If $T = T_M \sin \omega t$, then in the steady-state

$T_K = T_{KM} \sin(\omega t + \phi)$.

Determine T_{KM} and ϕ as functions of ω and the system parameters.

c) If $T = T_0$ = constant, then the steady-state value of $T_K = T_{K0}$ = const. Determine T_{K0}.

d) Draw the mechanical dual system and write its differential equation.

e) Draw the electrical analog and write its differential equation.

13–18. Determine the steady-state solution $x_s(t)$ of the differential equation

$$4\frac{dx}{dt} + 3x = \frac{de}{dt} + 2e,$$

where $e = 4 \sin 2t$.

13–19. Determine the steady-state amplitude and phase of y in the system equation

$$m\frac{d^2 y}{dt^2} + b\frac{dy}{dt} + ky = F,$$

where $F = A \sin \omega t$.

13–20. Determine the magnitude of the output v_1 in the steady state to a sinusoidal input v_2 of magnitude 1.0 at angular frequency of 3 rad/sec if the transfer function is

$$\frac{\mathbf{V}_1}{\mathbf{V}_2} = \frac{1 + 10s}{1 + 3s + s^2}.$$

13–21. If two systems have transfer functions

$$\mathbf{T}_1(s) = \frac{1 + 13s}{1 + 3s + 10s^2},$$

$$\mathbf{T}_2(s) = \frac{1 + 130s}{1 + 30s + 1000s^2},$$

which system responds faster? Why?

13–22. Using the system function for the vibration absorber, Eq. (13–29), determine the amplitude for sinusoidal vibration without damping ($b_1 = b_2 = 0$) by letting $s = j\omega$. For what value of ω will $\mathbf{T}(j\omega)$ be zero? This comparison shows that this device will really absorb vibration at this frequency. For what values of ω will the denominator of $\mathbf{T}(j\omega)$ be zero? This shows that the vibration amplitude will still be large at other frequencies. (Read Section 15–2.3 if you are interested in the full story.)

13–23. If $\mathbf{Z} = \mathbf{V}/\mathbf{I}$ is the input impedance of a circuit and

$$\mathbf{Z} = \frac{1 + s}{1 + 2s + 2s^2},$$

what is the average input power when v is a sinusoidal voltage of 10 v and 1 cps?

13–24. Solve Eq. (11–50) by using the system function and $s = j\omega$. Compare the result with Eq. (11–51).

13–25. Assuming that the circuit described in Problem 13–23 is driven with a step-function current, determine the response if there is no initial energy storage.

13–26. Assuming that a system is initially relaxed and has the system function

$$\frac{\mathbf{V}}{\mathbf{F}} = \mathbf{T}(s) = \frac{1}{1 + 2s + 2s^2},$$

determine the magnitude of the first overshoot in v for a step-function F input. At what value of time does this occur?

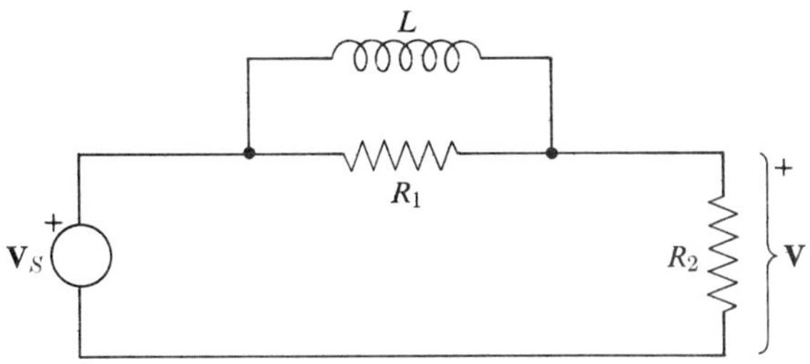

FIGURE 13–9

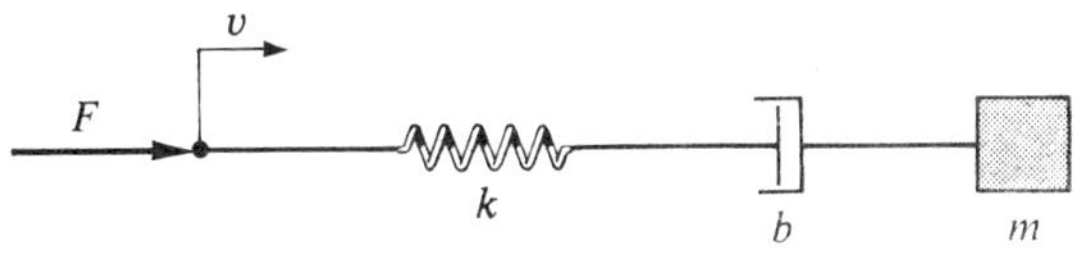

FIGURE 13–10

13–27. Determine the transient response for the mass velocity of the system in Problem 8–9a for a unit step input of v, assuming that $b_1 = b_2 = 1$ lb-sec/ft $k = 2$ lb/ft and $m = 1$ lb-sec^2/ft.

13–28. The electric network in Fig. 13–9 is driven by a voltage source v_s which is sinusoidal.

a) Derive the differential equation relating the output voltage v to the input voltage v_s, and identify the system time constants.

b) Determine the amplitude and phase angle of $\mathbf{V}/\mathbf{V}_s$.

13–29. For the system in Fig. 13–10, perform the following steps:

a) Determine the system equation for v input and F output.

b) Find the forced response if $v = V_0 \sin \omega t$.

c) Express the forced response in the form

$$F = F_0 \sin (\omega t + \phi)$$

and determine F_0 and ϕ.

d) Determine the condition on k, b, and m for the system to have an *underdamped* natural response. Should b be large or small for an oscillatory natural response?

e) Assuming that v is a step function from 0 to V_0 and the system is initially at rest (no stored energy), determine and sketch the complete response if the parameters k, b, and m are such that the system is underdamped. Be careful with initial conditions. (*Continued*)

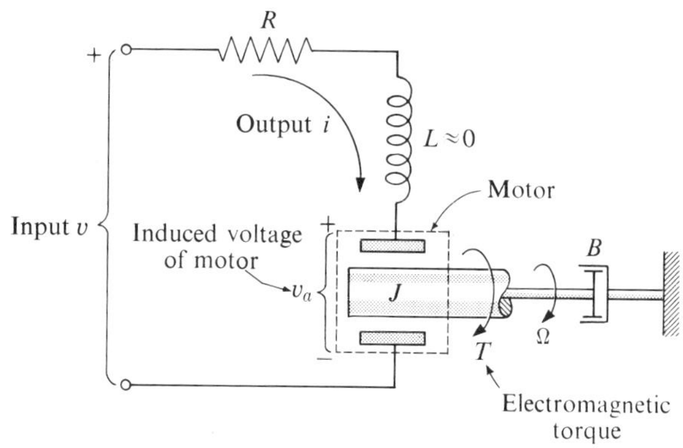

FIGURE 13–11

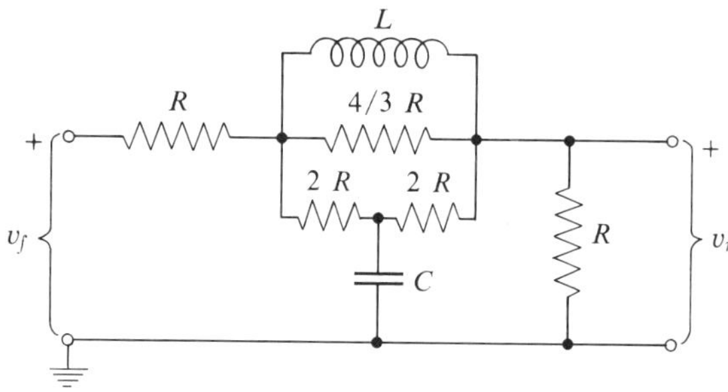

FIGURE 13–12

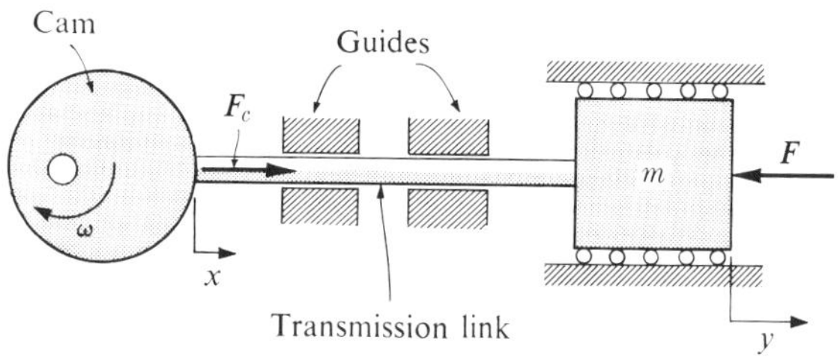

FIGURE 13–13

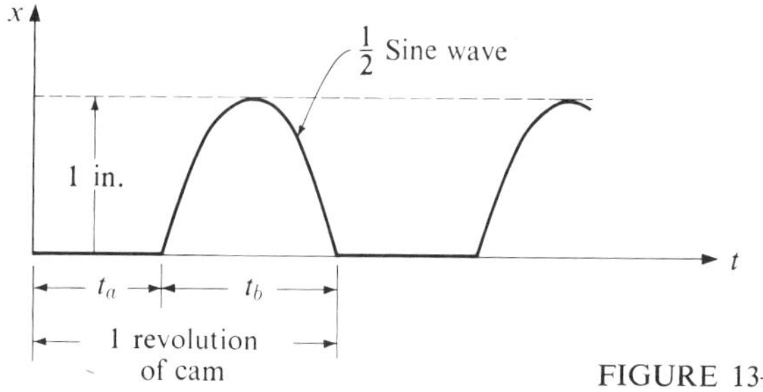

FIGURE 13–14

f) By inspection write the solution for the analogous electric circuit for conditions analogous to those given in (e).

13–30. The motor shown in Fig. 13–11 is to be used in a control system. The input to the electromechanical system shown is the voltage v. As a first approximation, assume the motor *has negligible inductance* $L \approx 0$. Its mechanical elements can be idealized as an inertia J and a damper B. The relations between the electrical and mechanical quantities are given by

$$v_a = k_g \Omega \qquad \text{(back-emf generated per Lenz's Law)},$$

$$T = k_T i \qquad \text{(torque produced per Biot-Savart Law)}.$$

a) Derive the differential equation relating the output i, to the input v.

b) What are the two time constants for this system?

c) Assuming that v is a step function of magnitude V_0, determine the initial condition, $i(0+)$. (The system has $i = 0$, $v = 0$, for $t < 0$.) Solve for $i = f(t)$.

d) Assuming that v is a steady sinusoidal input,

$$v = V \sin \omega t,$$

determine the *steady-state* amplitude and phase shift of the output i with respect to v.

e) If the inductance of the motor (in series with R) were not negligible, what would be the order of the governing equation? Explain how you arrived at your answer.

13–31. The system function relating an output $\mathbf{X}_r$ to an input $\mathbf{X}_f$ is found to be

$$\frac{\mathbf{X}_r}{\mathbf{X}_f} = \frac{s^2 + 2s + 5}{s^2 + 4s + 13}.$$

This system is excited by a steady sinusoid

$$x_f = (13/5) \sin \omega t.$$

a) Determine the poles and zeros of the system function and plot them in the complex (imaginary vs. real) plane. Show the phasors representing the numerator and denominator factors of the system function for $\omega = 2$ rad/sec.

b) *Sketch* the magnitude $|\mathbf{X}_r|$ and the phase ϕ between $\mathbf{X}_r$ and $\mathbf{X}_f$ as a function of frequency ω for $0 < \omega < 6$ rad/sec.

c) When $\omega = 2$ rad/sec, find the steady state sinusoidal time function representing $\mathbf{X}_r$.

13–32. The circuit shown in Fig. 13–12 is called a "bridged-T" filter.

a) Find the system function $\mathbf{V}_r/\mathbf{V}_f$.

b) Assuming that $R^2 = L/C$, $C = 100\,\mu\mu\text{f}$, $R = 10{,}000$ ohms, find the *poles* of the system function which are the *roots* of the characteristic equation. Find also the zeros of the system function which are the values of s which reduce the numerator of the system function to zero. Plot the zeros and poles in the complex plane.

c) For a steady sinusoidal input voltage $v_f = V \sin \omega t$, sketch the maximum amplitude of the output voltage $|v_r(t)|_{\max}$ as a function of frequency ω. Comment on the use of this circuit as a filter.

13–33. In one part of a high-speed pretzel-tying machine (Fig. 13–13), a *cam* is used to index a movable element having mass m. The cam profile produces a *rise* x vs. time which has the shape shown in Fig. 13–14. A constant force F acts on the mass as shown and is large enough to hold the moving parts in contact with the cam surface at all times. The stiffness of the transmission link along the x-direction is $k = 1000$ lb/in, and the mass $m = 5$ lb. The angular speed ω of the cam is 120 rpm. The guides provide a viscous friction force on the link whose magnitude is not known.

a) Draw a lumped-element model of the system.

b) Write the system equations relating displacement y and the cam-to-transmission link force F_c to the input cam displacement x.

c) Assuming that the machine is started from rest at $t = 0$, sketch qualitatively the force at the cam F_c as a function of time for $0 < t < (t_a + t_b)$.

d) Estimate the required value of the constant force F so that the link will remain in contact with the cam ($F_c > 0$ at all times).

e) Sketch the mass motion as a function of time, *qualitatively*, for $0 < t < 2(t_a + t_b)$.

f) What changes could be made in the cam profile $x(t)$ to reduce the maximum cam force F_c and yet obtain approximately the same mass-displacement vs. time function?

13–34. Determine the poles and zeros of the system given in Problem 13–27.

13–35. Determine the transfer function $\mathbf{V}/\mathbf{E}$ for Fig. 8–39 and show that for $R_1C_1 = R_2C_2$ there is cancellation of a pole and a zero.

13–36. Determine the input impedance $\mathbf{E}/\mathbf{I}$ for Fig. 8–40 and show that when $R = R_1 = R_2 = \sqrt{L/C}$ there is complete cancellation of s.

13–37. Given: the system equation

$$a_2 \frac{d^2x}{dt^2} + a_1 \frac{dx}{dt} + a_0 x = \mathsf{f}(t).$$

a) Express this equation in terms of undamped natural frequency ω_n and damping ratio ζ.

b) For constant natural frequency, plot the locus of the system roots or poles as the damping ratio varies from zero to infinity. Label the locations of the roots for $\zeta = 0$, $\sqrt{2}/2$, 1, and 10.

c) How does this plot change if the natural frequency is doubled?

13–38. Prove Eq. (13–46) by expanding the characteristic equation in terms of its roots. By considering each coefficient of s, show that if all coefficients do not have the same sign (e.g., positive) then there must be at least one pole with a positive real part.

13–39. Using Eq. (13–46), show that if any coefficients (other than the highest or lowest) are zero in the characteristic equation, the system will have at least one pole on the imaginary axis or in the right half-plane.

13–40. The characteristic equation for a third-order system (denominator of system function) is

$$a_3 s^3 + a_2 s^2 + a_1 s + a_0 = 0.$$

Show that $a_2 a_1 > a_0 a_3$ for stability by expanding the factors $(s - r)(s - \alpha - j\beta)(s - \alpha + j\beta)$ and considering the cases where $\alpha = 0$ and $\alpha =$ negative.

13–41. Using the result of Problem 13–40, show that the pulse transformer of Fig. 8–34 must be stable because it is a passive system, by applying the stability criterion of Problem 13–40 to the system equation (8–44).

13–42. Using the result of Problem 13–40, determine whether the following system is stable:

$$\mathbf{T}(s) = \frac{1 + s}{1 + 2s + 2s^2 + s^3}.$$

What are the poles? What is the form of the transient response? Does this indicate instability? Why?

13–43. Repeat Problem 13–42 for

$$\mathbf{T}(s) = \frac{1 + s}{2 + s + s^2 + 2s^3}.$$

13–44. A Butterworth filter has a system function

$$|\mathbf{T}(j\omega)| = \frac{1}{\sqrt{1 + \omega^{2n}}}.$$

Show that the poles lie on the unit circle in the complex plane. What is the specific form of $\mathbf{T}(s)$ [not $\mathbf{T}(j\omega)$] for $n = 2$ and $n = 3$?

13–45. Plot the locus in the s-plane of $\mathbf{T}(j\omega)$ as ω goes from zero to infinity for the first-order system

$$\mathbf{T}(s) = \frac{1}{1 + s\tau}.$$

Draw a system that has this system function.

13–46. Plot the locus in the s-plane of $\mathbf{T}(j\omega)$ as ω goes from zero to infinity for the first-order system

$$\mathbf{T}(s) = \frac{s\tau_2}{1 + s\tau_1}.$$

Draw a system that has this system function.

13–47. Plot the poles and zeros of the vibration absorber, Eq. (13–29), without damping ($b_1 = b_2 = 0$), in the complex plane. By letting $s = j\omega$ and considering ω to vary from zero to infinity, sketch the output amplitude vs. frequency by noting the lengths of the respective numerator and denominator factors as in Fig. 13–7 and Eq. (13–48). State the resonances and antiresonances. Assuming that b_1 and b_2 are small but not zero, show how the pole and zero plot would change and describe its effect on the amplitude vs. frequency plot.

14 SIMPLIFICATION OF LINEAR SYSTEM ANALYSIS

14–1. INTRODUCTION

The concept of generalized impedance and the properties of superposition associated with *linear* models of physical systems may be exploited to develop efficient and systematic methods of analysis. Although the methods discussed in this chapter do not permit any problems to be solved which cannot be solved by applying the procedures described earlier in this book, they *will* permit problems to be solved with less difficulty and probability of error.

The reader should bear clearly in mind that the techniques presented in this chapter are applicable only to *linear* systems. If nonlinear elements are employed in the system model to be analyzed, then the approach discussed in earlier chapters (5 to 8) must be retained and the following material is inapplicable.

14–2. FORMULATION OF SYSTEM EQUATIONS

We have seen in Chapter 13 that the system function, and therefore the differential equation relating any chosen output to any chosen input of a linear dynamic system, can be derived *algebraically* by considering only the forced response to an exponential input function. When used in connection with the loop and node methods presented in Chapter 8, this approach permits the process of equation formulation for linear dynamic systems to be simplified and further systematized.

As discussed previously, the application of the loop or node methods is begun by selecting loop variables or node variables in such a way that the vertex or path relations, respectively, are automatically satisfied. It is then necessary only to substitute the expressions for the element impedances into the path or vertex equations, respectively, to formulate a necessary and sufficient set of algebraic equations for the chosen system variables. Solution of this set of simultaneous equations for any desired system variable is readily accomplished by the use of determinants and Cramer's rule. The use of impedance in equation formulation by the loop and node methods is discussed in the following two subsections.

14–2.1 Loop Method of Formulation

The loop method will be discussed in connection with the mechanical translational system of Fig. 14–1(a), whose linear graph is shown in Fig. 14–1(b). We proceed as follows:

a) Select the internal loops or meshes of the system graph for applying the path conditions. We have proved in Chapter 8 that this will yield an independent set of loop equations.

b) Define circulating loop variables in each mesh as indicated in Fig. 14–1(b). Choose the sense of the loop variables to be the same in each loop (e.g., clockwise in Fig. 14–1b). This choice will result in a symmetrical set of equations, as we shall see.

When the through-variables are defined in this way, the vertex equations are automatically satisfied and need not be considered further. For example,

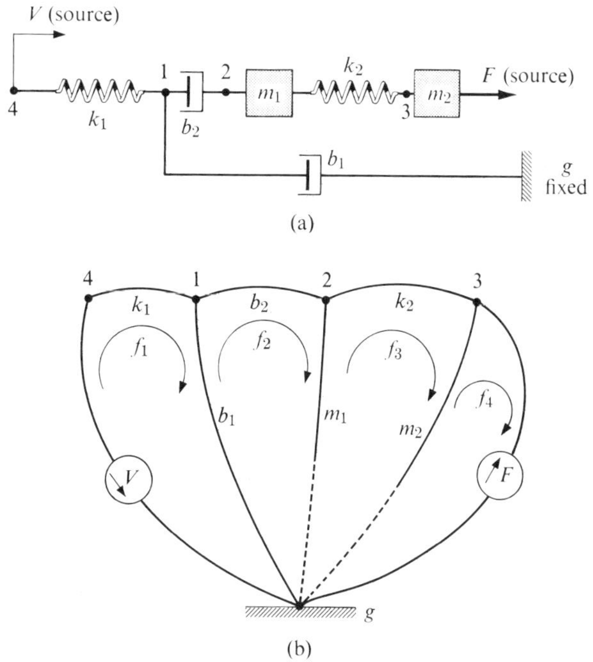

FIG. 14–1. Mechanical translational system and its graph.

at node 2 of Fig. 14–1(b), we have

$$f_2 - (f_2 - f_3) - f_3 = 0$$

regardless of the values of f_2 and f_3.

c) Write the compatibility or path equation for each mesh by expressing the across-variable difference across each branch of the mesh in terms of its net through-variable and its impedance and then summing across-variable differences around the mesh. For a graph containing l meshes this will result in l independent algebraic equations for the l loop through-variables. For the system of Fig. 14–1 this procedure yields the following equations. Let $f = \mathbf{F}e^{st}$, $v = \mathbf{V}e^{st}$, $F = \mathbf{F}e^{st}$, and so forth.

Loop g-4-1-g, taking across-variable decrease as positive,

$$-\mathbf{V} + \frac{s}{k_1}\mathbf{F}_1 + \frac{1}{b_1}(\mathbf{F}_1 - \mathbf{F}_2) = 0. \tag{14–1}$$

Loop g-1-2-g,

$$\frac{1}{b_1}(\mathbf{F}_2 - \mathbf{F}_1) + \frac{1}{b_2}\mathbf{F}_2 + \frac{1}{m_1 s}(\mathbf{F}_2 - \mathbf{F}_3) = 0. \tag{14–2}$$

Loop g-2-3-g,

$$\frac{1}{m_1 s}(\mathbf{F}_3 - \mathbf{F}_2) + \frac{s}{k_2}\mathbf{F}_3 + \frac{1}{m_2 s}(\mathbf{F}_3 - \mathbf{F}_4) = 0. \tag{14–3}$$

Loop g-3-g. Note that this loop contains a through-variable source. Therefore it is not necessary to sum across-variables around this loop to find an equation for $\mathbf{F}_4$; it is specified by the source through continuity:

$$\mathbf{F}_4 = -\mathbf{F}. \tag{14–4}$$

The next step in the loop method is thus

d) Use the continuity relation for each loop which contains a through-variable source to eliminate this through-variable from the other loop equations. If the system contains through-variable sources in q of the l loops, this will reduce the number of unknown variables to $l - q$.

The resulting set of equations may then be written in an orderly form with the corresponding through-variables in each equation aligned in vertical columns. Thus, for our example,

$$\left(\frac{1}{b_1} + \frac{s}{k_1}\right)\mathbf{F}_1 \quad - \left(\frac{1}{b_1}\right)\mathbf{F}_2 \quad - (0)\mathbf{F}_3 = \mathbf{V}, \tag{14–5}$$

$$-\left(\frac{1}{b_1}\right)\mathbf{F}_1 + \left(\frac{1}{b_1} + \frac{1}{b_2} + \frac{1}{m_1 s}\right)\mathbf{F}_2 \quad - \left(\frac{1}{m_1 s}\right)\mathbf{F}_3 = 0, \tag{14–6}$$

$$-(0)\mathbf{F}_1 \quad - \left(\frac{1}{m_1 s}\right)\mathbf{F}_2 + \left(\frac{1}{m_1 s} + \frac{1}{m_2 s} + \frac{s}{k_2}\right)\mathbf{F}_3 = -\left(\frac{1}{m_2 s}\right)\mathbf{F}. \tag{14–7}$$

We can now draw some general conclusions regarding the form of the set of equations written by the loop method. Equation (14–5) is the equation for the loop in which $\mathbf{F}_1$ exists (the $\mathbf{F}_1$ loop). It contains $\mathbf{F}_1$ and the through-variables of all other loops which have branches common to the $\mathbf{F}_1$ loop (in this case only $\mathbf{F}_2$). The coefficient of $\mathbf{F}_1$ in the $\mathbf{F}_1$ loop equation is *positive* and equal to the sum of all the impedances of the branches in the $\mathbf{F}_1$ loop. The coefficient of $\mathbf{F}_2$ is *negative* and equal to the sum of the impedances of all branches in the $\mathbf{F}_1$ loop through which $\mathbf{F}_2$ flows. The right-hand side of (14–5) is the across-variable source in the $\mathbf{F}_1$ loop. The sign of $\mathbf{V}$ is positive because it tends to produce positive $\mathbf{F}_1$ (clockwise). Note that Eq. (14–6) follows the rules just stated. If a loop contains branches common to a loop which has a through-variable source, a source term will appear on the right-hand side of that loop equation equal to the source through-variable times the series impedance of those branches through which the source variable flows. In Eq. (14–7), $\mathbf{F}_4$ has been set equal to $-\mathbf{F}$; hence the term $(1/m_2s)\mathbf{F}_4$ in Eq. (14–3) becomes *known* and is placed on the right-hand side of Eq. (14–7).

The set of loop equations may be completely characterized by the unknown loop variables, $\mathbf{F}_1, \mathbf{F}_2$, and $\mathbf{F}_3$ in the present example, and two arrays of numbers. If the coefficients of $\mathbf{F}_1, \mathbf{F}_2$, and $\mathbf{F}_3$ are written in rows and columns as they appear in the set of equations, we have what is called the *loop-impedance matrix*. Thus, the matrix for Eqs. (14–5), (14–6), and (14–7) is

$$\begin{bmatrix} \left(\frac{1}{b_1}+\frac{s}{k_1}\right) & -\left(\frac{1}{b_1}\right) & 0 \\ -\left(\frac{1}{b_1}\right) & \left(\frac{1}{b_1}+\frac{1}{b_2}+\frac{1}{m_1s}\right) & -\left(\frac{1}{m_1s}\right) \\ 0 & -\left(\frac{1}{m_1s}\right) & \left(\frac{1}{m_1s}+\frac{1}{m_2s}+\frac{s}{k_2}\right) \end{bmatrix}. \tag{14–8}$$

Notice that the loop-impedance matrix is *symmetric* about its diagonal (upper left to lower right). This is a consequence of selecting the loop variables in the same sense in each loop, and also of linearity.

A single-column matrix called the *source matrix* characterizes the source functions for all loops. In the present example, this matrix is

$$\begin{bmatrix} \mathbf{V} \\ 0 \\ \frac{-\mathbf{F}}{m_2s} \end{bmatrix}. \tag{14–9}$$

By analogy, a third column matrix may be defined as the *variable matrix:*

$$\begin{bmatrix} \mathbf{F}_1 \\ \mathbf{F}_2 \\ \mathbf{F}_3 \end{bmatrix}. \tag{14–10}$$

In the loop-impedance matrix, let $\mathbf{Z}_{11}$ be the sum of the impedances in loop 1, $\mathbf{Z}_{12}$ be the sum of the impedances in loop 1 through which f_2 flows (i.e., common impedances between loops 1 and 2), and so on. Then the matrix (14–8) can be written more concisely as

$$\begin{bmatrix} \mathbf{Z}_{11} & -\mathbf{Z}_{12} & -\mathbf{Z}_{13} \\ -\mathbf{Z}_{12} & \mathbf{Z}_{22} & -\mathbf{Z}_{23} \\ -\mathbf{Z}_{13} & -\mathbf{Z}_{23} & \mathbf{Z}_{33} \end{bmatrix}, \tag{14–11}$$

since by the previously mentioned symmetry, $\mathbf{Z}_{pq} = \mathbf{Z}_{qp}$. If we let $\mathbf{V}_{s1}, \mathbf{V}_{s2}$, and $\mathbf{V}_{s3}$ represent the equivalent across-variable sources for each loop, (14–9) becomes

$$\begin{bmatrix} \mathbf{V}_{s1} \\ \mathbf{V}_{s2} \\ \mathbf{V}_{s3} \end{bmatrix}. \tag{14–12}$$

It is possible to define a multiplication process, called *matrix multiplication*, whereby the loop equations are obtained from the impedance, source, and variable matrices.

$$\begin{bmatrix} \mathbf{Z}_{11} & -\mathbf{Z}_{12} & -\mathbf{Z}_{13} \\ -\mathbf{Z}_{12} & \mathbf{Z}_{22} & -\mathbf{Z}_{23} \\ -\mathbf{Z}_{13} & -\mathbf{Z}_{23} & \mathbf{Z}_{33} \end{bmatrix} \cdot \begin{bmatrix} \mathbf{F}_1 \\ \mathbf{F}_2 \\ \mathbf{F}_3 \end{bmatrix} = \begin{bmatrix} \mathbf{V}_{s1} \\ \mathbf{V}_{s2} \\ \mathbf{V}_{s3} \end{bmatrix}. \tag{14–13}$$

Comparison between this expression and Eqs. (14–5), (14–6), and (14–7) shows that the latter equations are obtained if the terms in the rows of the impedance matrix are multiplied by the terms in the variable-column matrix, added, and set equal to the terms in the source matrix. Specifically, (14–13) gives

$$\begin{aligned} \mathbf{Z}_{11}\mathbf{F}_1 - \mathbf{Z}_{12}\mathbf{F}_2 - \mathbf{Z}_{13}\mathbf{F}_3 &= \mathbf{V}_{s1}, \\ -\mathbf{Z}_{12}\mathbf{F}_1 + \mathbf{Z}_{22}\mathbf{F}_2 - \mathbf{Z}_{23}\mathbf{F}_3 &= \mathbf{V}_{s2}, \\ -\mathbf{Z}_{13}\mathbf{F}_1 - \mathbf{Z}_{23}\mathbf{F}_2 + \mathbf{Z}_{33}\mathbf{F}_3 &= \mathbf{V}_{s3}. \end{aligned} \tag{14–14}$$

For our present purposes it will be sufficient to note that the set of equations represented by Eqs. (14–5), (14–6), and (14–7) or Eq. (14–13) may be solved by the use of determinants and Cramer's rule. Thus, if Δ is the *determinant* of the loop-impedance matrix (14–8) or (14–11),

$$\Delta = \begin{vmatrix} \mathbf{Z}_{11} & -\mathbf{Z}_{12} & -\mathbf{Z}_{13} \\ -\mathbf{Z}_{12} & \mathbf{Z}_{22} & -\mathbf{Z}_{23} \\ -\mathbf{Z}_{13} & -\mathbf{Z}_{23} & \mathbf{Z}_{33} \end{vmatrix}, \tag{14–15}$$

the solution for $\mathbf{F}_1$, for example, is

$$\mathbf{F}_1 = \frac{\begin{vmatrix} \mathbf{V}_{s1} & -\mathbf{Z}_{12} & -\mathbf{Z}_{13} \\ \mathbf{V}_{s2} & \mathbf{Z}_{22} & -\mathbf{Z}_{23} \\ \mathbf{V}_{s3} & -\mathbf{Z}_{23} & \mathbf{Z}_{33} \end{vmatrix}}{\Delta}. \tag{14–16}$$

The expansion of (14–16) by introducing the $\mathbf{V}_s$ values and the $\mathbf{Z}$'s may be rather tedious but is straightforward.

Once the phasors of the loop variables, $\mathbf{F}_1, \mathbf{F}_2, \ldots,$ have been found, all other variables may be determined easily by use of simple continuity relations and the elemental equations. For instance, the across-variable $\mathbf{V}_1$ at node 1 in Fig. 14–1(b) is

$$\mathbf{V}_1 = (\mathbf{F}_1 - \mathbf{F}_2)/b_1.$$

After the phasors or complex amplitudes of the unknowns have been found (e.g., $\mathbf{F}_1$, $\mathbf{F}_2$, etc.) as functions of s, the differential equations for these variables when the inputs are any time functions can be recovered by noting that each s is the result of a differentiation. This technique has been outlined in Chapter 13.

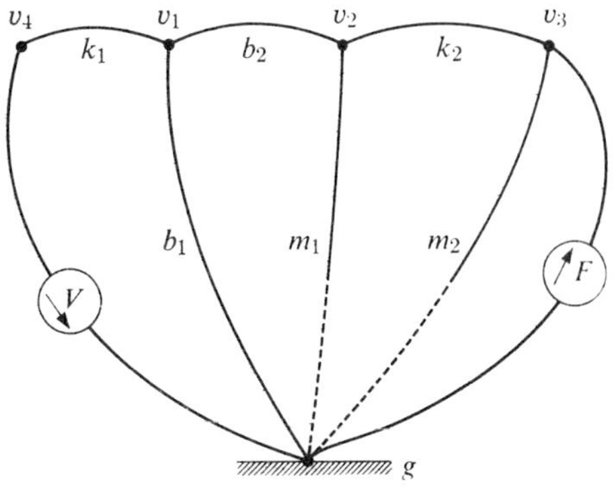

FIG. 14–2. System graph of Fig. 14–1 with node variables defined.

14–2.2 Node Method of Formulation

The node method, which is the dual of the loop method, will now be discussed.

a) Select the *node variables* as the unknown variables. The system graph of Fig. 14–1 is redrawn in Fig. 14–2 with these node variables defined with respect to the reference or ground node g. The loop equations will then automatically be satisfied. For example, in loop g-1-2-g,

$$v_1 + (v_2 - v_1) - v_2 = 0$$

regardless of the values of v_1 and v_2.

b) Write the continuity equations at each node by expressing the amplitude of the through-variable for each incident branch in terms of its across-variable difference and its impedance. This will result in $(n - 1)$ equations for a system having n nodes. For the system of Fig. 14–2, we have the following equations, where $v_1 = \mathbf{V}_1 e^{st}$, etc.

Node 4: Since node 4 is attached to the source, $\mathbf{V}$, $\mathbf{V}_4$ is known, and thus $\mathbf{V}_4$ is eliminated as an unknown:

$$\mathbf{V}_4 = \mathbf{V}. \tag{14–17}$$

Node 1. Take through-variables as positive out of the node:

$$(\mathbf{V}_1 - \mathbf{V}_4)\frac{k_1}{s} + \mathbf{V}_1 b_1 + (\mathbf{V}_1 - \mathbf{V}_2)b_2 = 0. \tag{14–18}$$

Node 2

$$(\mathbf{V}_2 - \mathbf{V}_1)b_2 + \mathbf{V}_2 m_1 s + (\mathbf{V}_2 - \mathbf{V}_3)\frac{k_2}{s} = 0. \tag{14–19}$$

Node 3

$$(\mathbf{V}_3 - \mathbf{V}_2)\frac{k_2}{s} + \mathbf{V}_3 m_2 s - \mathbf{F} = 0. \tag{14–20}$$

These equations may be rearranged as follows:

$$\left(b_1 + b_2 + \frac{k_1}{s}\right)\mathbf{V}_1 \quad - (b_2)\mathbf{V}_2 \quad - (0)\mathbf{V}_3 = \frac{k_1}{s}\mathbf{V}, \tag{14–21}$$

$$-(b_2)\mathbf{V}_1 + \left(m_1 s + b_2 + \frac{k_2}{s}\right)\mathbf{V}_2 \quad - \left(\frac{k_2}{s}\right)\mathbf{V}_3 = 0, \tag{14–22}$$

$$-(0)\mathbf{V}_1 \quad - \left(\frac{k_2}{s}\right)\mathbf{V}_2 + \left(m_2 s + \frac{k_2}{s}\right)\mathbf{V}_3 = \mathbf{F}. \tag{14–23}$$

In Eq. (14–21), the equation for node 1, the coefficient of $\mathbf{V}_1$ is positive and is the sum of the reciprocals of the impedances (i.e., the sum of the *admittances*) of all branches incident on node 1. The coefficient of $\mathbf{V}_2$ is the sum of the admittances of all branches connected between nodes 2 and 1. Since no branches connect $\mathbf{V}_3$ and $\mathbf{V}_1$, the coefficient of $\mathbf{V}_3$ is zero in (14–21). The across-variable source $\mathbf{V}$ produces a known component of through-variable into node 1 equal to $\mathbf{V}$ times the admittance of the branch between $\mathbf{V}$ and $\mathbf{V}_1$. This term appears on the right of Eq. (14–21) and is positive because it tends to produce positive $\mathbf{V}_1$.

In Eq. (14–23) the through-variable source $\mathbf{F}$ incident on node 3 appears directly.

Let $\mathbf{Y}_{pp}$ equal the sum of the admittances of the branches incident on the pth node and $\mathbf{Y}_{pq}$ the sum of the admittances of the branches connected between the pth node and the qth node. Also let $\mathbf{F}_{s1}$, $\mathbf{F}_{s2}$, and $\mathbf{F}_{s3}$ be the equivalent through-variable sources for nodes 1, 2, and 3, respectively. Then Eqs. (14–21), (14–22) and (14–23) can be written

$$\begin{aligned} \mathbf{Y}_{11}\mathbf{V}_1 - \mathbf{Y}_{12}\mathbf{V}_2 - \mathbf{Y}_{13}\mathbf{V}_3 &= \mathbf{F}_{s1}, \\ -\mathbf{Y}_{12}\mathbf{V}_1 + \mathbf{Y}_{22}\mathbf{V}_2 - \mathbf{Y}_{23}\mathbf{V}_3 &= \mathbf{F}_{s2}, \\ -\mathbf{Y}_{13}\mathbf{V}_3 - \mathbf{Y}_{23}\mathbf{V}_2 + \mathbf{Y}_{33}\mathbf{V}_3 &= \mathbf{F}_{s3}, \end{aligned} \tag{14–24}$$

since $\mathbf{Y}_{pq} = \mathbf{Y}_{qp}$. In matrix notation similar to Eq. (14–13),

$$\begin{bmatrix} \mathbf{Y}_{11} & -\mathbf{Y}_{12} & -\mathbf{Y}_{13} \\ -\mathbf{Y}_{12} & \mathbf{Y}_{22} & -\mathbf{Y}_{23} \\ -\mathbf{Y}_{13} & -\mathbf{Y}_{23} & \mathbf{Y}_{33} \end{bmatrix} \cdot \begin{bmatrix} \mathbf{V}_1 \\ \mathbf{V}_2 \\ \mathbf{V}_3 \end{bmatrix} = \begin{bmatrix} \mathbf{F}_{s1} \\ \mathbf{F}_{s2} \\ \mathbf{F}_{s3} \end{bmatrix}. \tag{14–25}$$

Comparison of the through-variable source matrix with Eqs. (14–21) to (14–23) shows that, for our example, $\mathbf{F}_{s1} = (k_1/s)\mathbf{V}$, $\mathbf{F}_{s2} = 0$, and $\mathbf{F}_{s3} = \mathbf{F}$. The matrix of the admittances $\mathbf{Y}$ is called the *node admittance* matrix, and the $\mathbf{V}$ and $\mathbf{F}$ matrices are the variable and source matrices, respectively.

The node equations can be solved by determinants in the manner described in the discussion of Section 14–2.1.

In any particular problem, either the loop method or the node method of solution is possible. In a system which has considerably more nodes than loops, the loop method will usually be simpler since the number of simultaneous equations to be solved will be smaller. Conversely, if there are many more loops than nodes, the node method will ordinarily be simpler.

Elaborate mathematical techniques have been developed which deal with the manipulation of matrices to solve systems of linear equations. It is beyond the scope of this book to deal with these methods. The interested reader is referred to the various texts on matrix algebra or advanced electric circuit theory [1, 2, 3].

14-2.3 Example of a Solution

An example will illustrate the application of the methods just described. In the circuit of Fig. 14-3, v_s is a voltage source and the voltage v_3 is to be found. Assume the variables v_s, v_1, v_2, v_3, i_1, and i_2 to be exponential functions with complex amplitudes $\mathbf{V}_s$, $\mathbf{V}_1$, $\mathbf{V}_2$, $\mathbf{V}_3$, $\mathbf{I}_1$, and $\mathbf{I}_2$, respectively.

a) *Loop method.* From Fig. 14-3, the loop equations are

$$(R_1 + Ls)\mathbf{I}_1 - Ls\mathbf{I}_2 = \mathbf{V}_s,$$

$$-Ls\mathbf{I}_1 + \left(\frac{1}{Cs} + R_2 + Ls\right)\mathbf{I}_2 = 0.$$

The voltage amplitude across R_2 is $\mathbf{I}_2R_2$. The determinant of the loop impedance matrix (14-15) is

$$\Delta = \begin{vmatrix} (R_1 + Ls) & -Ls \\ -Ls & \left(\frac{1}{Cs} + R_2 + Ls\right) \end{vmatrix}$$

$$= \frac{R_1}{Cs} + \frac{L}{C} + R_1R_2 + (R_1 + R_2)Ls.$$

The current $\mathbf{I}_2$ is therefore

$$\mathbf{I}_2 = \frac{\begin{vmatrix} (R_1 + Ls) & \mathbf{V}_s \\ -Ls & 0 \end{vmatrix}}{\frac{R_1}{Cs} + \frac{L}{C} + R_1R_2 + (R_1 + R_2)Ls}$$

$$= \frac{Ls\mathbf{V}_s}{\frac{R_1}{Cs} + \frac{L}{C} + R_1R_2 + (R_1 + R_2)Ls},$$

and $\mathbf{V}_3$ is

$$\mathbf{V}_3 = \frac{R_2LCs^2\mathbf{V}_s}{R_1 + (L + R_1R_2C)s + (R_1 + R_2)LCs^2}. \tag{14-26}$$

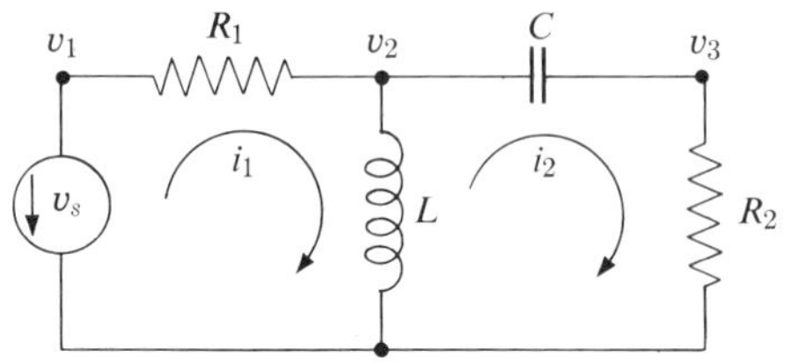

FIG. 14-3. Circuit for application of loop and node methods.

The differential equation for v_3 is obtained by clearing fractions and recognizing that each power of s is the result of a differentiation:

$$(R_1 + R_2)LC\frac{d^2v_3}{dt^2} + (L + R_1R_2C)\frac{dv_3}{dt} + R_1v_3 = R_2LC\frac{d^2v_s}{dt^2}.$$

b) *Node method.* The node equations are

$$\left(\frac{1}{R_1} + \frac{1}{Ls} + Cs\right)\mathbf{V}_2 - Cs\mathbf{V}_3 = \frac{\mathbf{V}_s}{R_1},$$

$$-Cs\mathbf{V}_2 + \left(Cs + \frac{1}{R_2}\right)\mathbf{V}_3 = 0.$$

The determinant of the admittance matrix is

$$\Delta = \begin{vmatrix} \left(\frac{1}{R_1} + \frac{1}{Ls} + Cs\right) & -Cs \\ -Cs & \left(Cs + \frac{1}{R_2}\right) \end{vmatrix}$$

$$= \left(\frac{C}{R_2} + \frac{C}{R_1}\right)s + \left(\frac{C}{L} + \frac{1}{R_1R_2}\right) + \frac{1}{R_2Ls},$$

and the solution for $\mathbf{V}_3$ is

$$\mathbf{V}_3 = \frac{\begin{vmatrix} \left(\frac{1}{R_1} + \frac{1}{Ls} + Cs\right) & \frac{\mathbf{V}_s}{R_1} \\ -Cs & 0 \end{vmatrix}}{\left(\frac{R_1 + R_2}{R_1R_2}\right)Cs + \left(\frac{C}{L} + \frac{1}{R_1R_2}\right) + \frac{1}{R_2Ls}}$$

$$= \frac{R_2LCs^2\mathbf{V}_s}{R_1 + (L + R_1R_2C)s + (R_1 + R_2)LCs^2},$$

which is seen to agree with Eq. (14-26).

14–3. GENERAL FORM OF SYSTEM EQUATIONS; SUPERPOSITION

From the results of the previous section we may conclude that the equations for the amplitudes of the mesh through-variables of any order linear system are of the form

$$\begin{bmatrix} \mathbf{Z}_{11} & -\mathbf{Z}_{12} & -\mathbf{Z}_{13} & \cdots & -\mathbf{Z}_{1n} \\ -\mathbf{Z}_{12} & \mathbf{Z}_{22} & -\mathbf{Z}_{23} & \cdots & -\mathbf{Z}_{2n} \\ -\mathbf{Z}_{13} & -\mathbf{Z}_{23} & \mathbf{Z}_{33} & \cdots & \\ \vdots & \vdots & \vdots & & \vdots \\ -\mathbf{Z}_{1n} & -\mathbf{Z}_{2n} & -\mathbf{Z}_{3n} & \cdots & \mathbf{Z}_{nn} \end{bmatrix} \cdot \begin{bmatrix} \mathbf{F}_1 \\ \mathbf{F}_2 \\ \vdots \\ \mathbf{F}_n \end{bmatrix} = \begin{bmatrix} \mathbf{V}_{s1} \\ \mathbf{V}_{s2} \\ \vdots \\ \mathbf{V}_{sn} \end{bmatrix}, \tag{14–27}$$

where the **Z**'s are impedances, the **F**'s are the mesh variables, and the $\mathbf{V}_s$'s are the equivalent across-variable sources in each loop. The latter are proportional (through an impedance) to the actual sources present and may involve the sum of several actual source variables.

We can solve equations (14–27) for all output through-variables by using Cramer's rule:

$$\mathbf{F}_1 = \frac{\boldsymbol{\Delta}_{11}}{\boldsymbol{\Delta}}\mathbf{V}_{s1} - \frac{\boldsymbol{\Delta}_{12}}{\boldsymbol{\Delta}}\mathbf{V}_{s2} + \cdots + (-1)^{n-1}\frac{\boldsymbol{\Delta}_{1n}}{\boldsymbol{\Delta}}\mathbf{V}_{sn},$$

$$\mathbf{F}_2 = -\left[\frac{\boldsymbol{\Delta}_{21}}{\boldsymbol{\Delta}}\mathbf{V}_{s1} - \frac{\boldsymbol{\Delta}_{22}}{\boldsymbol{\Delta}}\mathbf{V}_{s2} + \cdots + (-1)^{n-1}\frac{\boldsymbol{\Delta}_{2n}}{\boldsymbol{\Delta}}\mathbf{V}_{sn}\right],$$

$$\vdots \qquad \vdots \qquad \vdots \qquad \vdots$$

$$\mathbf{F}_n = (-1)^{n-1} \times \left[\frac{\boldsymbol{\Delta}_{n1}}{\boldsymbol{\Delta}}\mathbf{V}_{s1} - \frac{\boldsymbol{\Delta}_{n2}}{\boldsymbol{\Delta}}\mathbf{V}_{s2} + \cdots + (-1)^{n-1}\frac{\boldsymbol{\Delta}_{nn}}{\boldsymbol{\Delta}}\mathbf{V}_{sn}\right], \tag{14–28}$$

where $\boldsymbol{\Delta}$ is the determinant of the impedance matrix and $\boldsymbol{\Delta}_{pq}$ is the determinant of the submatrix obtained when the pth row and the qth column are crossed out in the impedance matrix. Once all the mesh through-variables are known, any other variable, say $\mathbf{X}_{rm}$, can be found as a linear combination of these mesh variables multiplied by appropriate impedances. Therefore, we may conclude that the equation for any response variable in a system which is forced by n sources $\mathbf{X}_{f1}, \mathbf{X}_{f2}, \ldots, \mathbf{X}_{fn}$ is of the form

$$\mathbf{X}_{rm} = \frac{\mathbf{N}_{m1}(s)\mathbf{X}_{f1}}{\mathbf{D}(s)} + \frac{\mathbf{N}_{m2}(s)\mathbf{X}_{f2}}{\mathbf{D}(s)} + \cdots + \frac{\mathbf{N}_{mn}(s)\mathbf{X}_{fn}}{\mathbf{D}(s)}, \tag{14–29}$$

where $\mathbf{D}(s) = \boldsymbol{\Delta}$ is the impedance matrix determinant and $\mathbf{N}_{mq}$ is the determinant of the appropriate submatrix for $\mathbf{X}_{fq}$ together with the appropriate algebraic sign [e.g., as given in Eqs. (14–28)]. From Eq. (14–29), system functions $\mathbf{T}_{pq}$ may be defined which relate $\mathbf{X}_{rm}$ to the various forcing functions:

$$\mathbf{X}_{rm} = \mathbf{T}_{m1}(s)\mathbf{X}_{f1} + \mathbf{T}_{m2}(s)\mathbf{X}_{f2} + \cdots + \mathbf{T}_{mn}(s)\mathbf{X}_{fn}. \tag{14–30}$$

The differential equation for $x_r(t)$ can be found by multiplying (14–29) by the common factor $\mathbf{D}(s)$ and reverting from s to time differentiation. From the above results, we conclude the following.

a) The characteristic equation $[\mathbf{D}(s) = 0]$ is the same for all response variables regardless of whether one or all forcing functions are present. Therefore, the left-hand side of the differential equation for any response variable is the same (within a constant and one or more derivatives or integrals of the variable).

b) The equation for any response variable can be found by superposition of the forcing function terms ($\mathbf{N}_{m1}\mathbf{X}_{f1}$, $\mathbf{N}_{m2}\mathbf{X}_{f2}$, etc.) considered to act separately while retaining the common left-hand side $[\mathbf{D}(s)\mathbf{X}_{rm}]$. Thus each source in a system will establish its own forcing-function terms on the right-hand side of the differential equation for any variable in the system.

We could also reach the above conclusions by starting with the node across-variables rather than the mesh through-variables.

The reader should recall that we have proved (in Chapter 11) that the solution of a linear differential equation when two or more forcing functions are present is the sum of the solutions for each forcing function taken separately, provided that the system is at rest (all variables and their derivatives zero), prior to the application of the forcing functions.

This property of linear systems is restated by the following *superposition theorem:*

> If a linear model of a physical system which is initially at rest is excited by two or more ideal sources, the total response of the system is the sum of the responses when each source is applied separately with all other sources set equal to zero.

Note that a zero across-variable source between two nodes means that the nodes have the same across-variable (a direct connection or a short circuit). A zero through-variable source between two nodes means no connection (through the source) of the nodes, i.e. an open circuit.

This superposition theorem will be useful in the following section on equivalent networks.

14–4. EQUIVALENT NETWORKS

14–4.1 Impedance Combination

When two elements having impedances $\mathbf{Z}_1$ and $\mathbf{Z}_2$ are connected in series as shown in Fig. 14–4, they may be replaced by a single combined impedance $\mathbf{Z}_s$ given by

$$\mathbf{Z}_s = \mathbf{Z}_{13} = \frac{\mathbf{V}_{31}}{\mathbf{F}} = \frac{\mathbf{V}_{32} + \mathbf{V}_{21}}{\mathbf{F}} = \mathbf{Z}_1 + \mathbf{Z}_2.$$

For n impedances in series we have

$$\mathbf{Z}_s = \mathbf{Z}_1 + \mathbf{Z}_2 + \cdots + \mathbf{Z}_n. \tag{14–31}$$

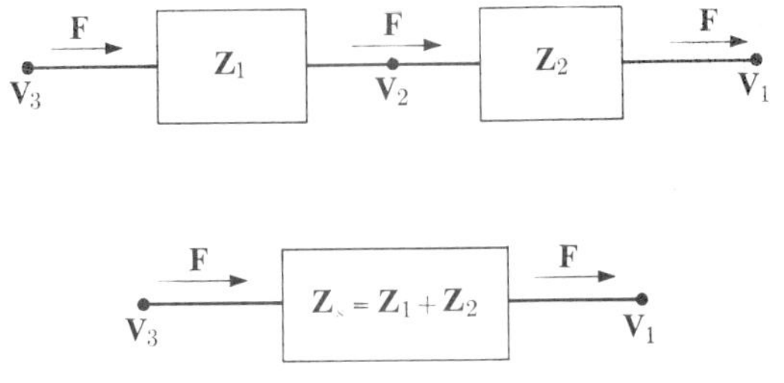

FIG. 14–4. Equivalent impedance of two impedances in series.

This is analogous to the combination of series resistors $R_1, R_2, \ldots, R_n$ into an equivalent single resistance R_s:

$$R_s = R_1 + R_2 + \cdots + R_n.$$

However, the impedances in (14–31) are functions of s.

When two elements are connected in parallel, as in Fig. 14–5, they may be replaced by an equivalent impedance $\mathbf{Z}_p$, given by

$$\mathbf{Z}_p = \frac{\mathbf{V}_{21}}{\mathbf{F}} = \frac{\mathbf{V}_{21}}{\mathbf{F}_1 + \mathbf{F}_2} = \frac{1}{(\mathbf{F}_1/\mathbf{V}_{21}) + (\mathbf{F}_2/\mathbf{V}_{21})}$$

$$= \frac{1}{(1/\mathbf{Z}_1) + (1/\mathbf{Z}_2)}, \tag{14–32}$$

or in terms of the admittances $\mathbf{Y}_1 = 1/\mathbf{Z}_1, \mathbf{Y}_2 = 1/\mathbf{Z}_2$, and $\mathbf{Y}_p = 1/\mathbf{Z}_p$,

$$\mathbf{Y}_p = \mathbf{Y}_1 + \mathbf{Y}_2.$$

Thus admittances add for parallel combinations and impedances add for series combinations. For n impedances in parallel,

$$\frac{1}{\mathbf{Z}_p} = \frac{1}{\mathbf{Z}_1} + \frac{1}{\mathbf{Z}_2} + \cdots + \frac{1}{\mathbf{Z}_n} \tag{14–33}$$

or

$$\mathbf{Y}_p = \mathbf{Y}_1 + \mathbf{Y}_2 + \cdots + \mathbf{Y}_n. \tag{14–34}$$

When only a few of the variables in a network are to be found, the network can often be reduced by

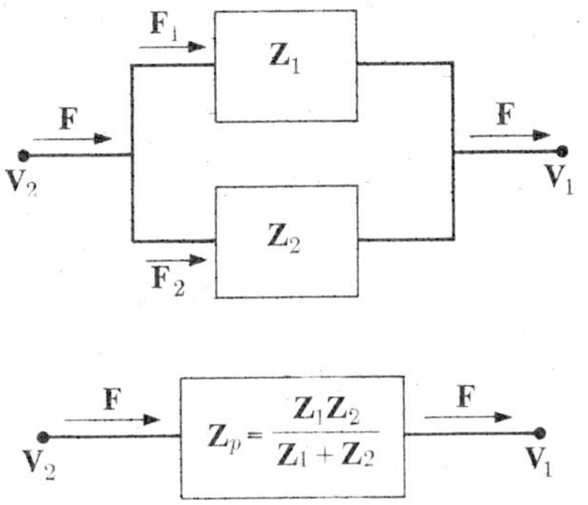

FIG. 14–5. Equivalent impedance of two impedances in parallel.

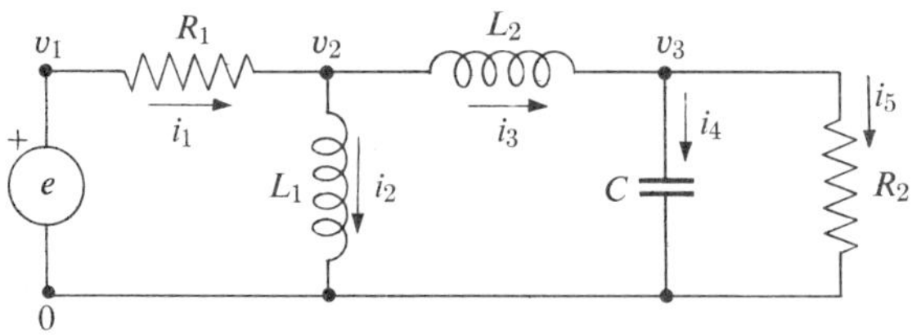

FIG. 14–6. Equivalent circuit of step-up pulse transformer.

combination of impedances to the point where the required variables can be found easily.

For illustration, let us consider the step-up pulse transformer circuit analyzed in Chapter 8. The circuit of Fig. 8–34 is redrawn as Fig. 14–6 for convenience.

Let us suppose that we wish to find the complex amplitude $\mathbf{I}_1$, of the current i_1, given the amplitude $\mathbf{E}$ of the source e. This is the transfer function relating the current from the source to the voltage applied by the source and is called the *driving-point impedance* $\mathbf{Z}_{dp}$ if $\mathbf{Z}_{dp} = \mathbf{E}/\mathbf{I}_1$ or the *driving-point admittance* $\mathbf{Y}_{dp}$ if $\mathbf{Y}_{dp} = \mathbf{I}_1/\mathbf{E}$. To find $\mathbf{I}_1$, we combine impedances by applying the series and parallel rules, successively moving from the right-hand side of the network toward the source. When only a single equivalent impedance $\mathbf{Z}_{dp}$ remains connected directly to the source, we can write the current as $\mathbf{I}_1 = (1/\mathbf{Z}_{dp})\,\mathbf{E}$.

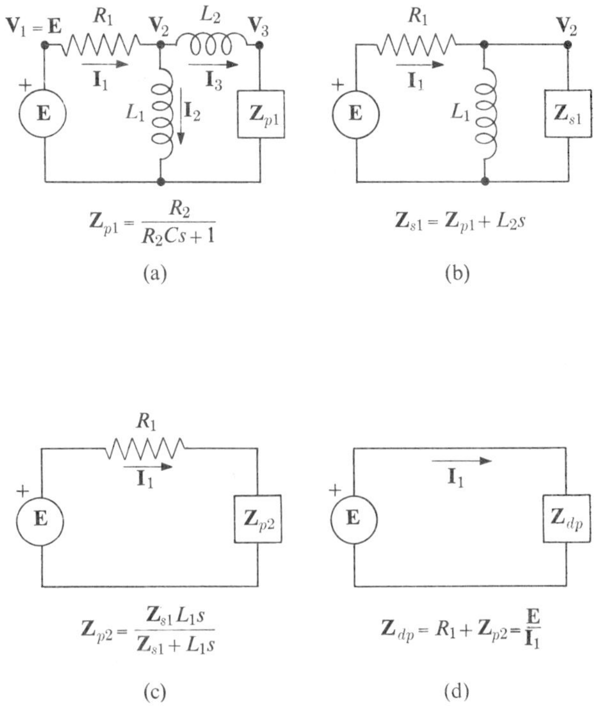

FIG. 14–7. Steps in reduction of the network shown to find the driving point current. (a) Parallel combination of R_2 and C into $\mathbf{Z}_{p1}$. (b) Series combination of $\mathbf{Z}_{p1}$ and L_2 into $\mathbf{Z}_{s1}$. (c) Parallel combination of $\mathbf{Z}_{s1}$ and L_i into $\mathbf{Z}_{p2}$. (d) Series combination of $\mathbf{Z}_{p2}$ and R_1 into driving point impedance $\mathbf{Z}_{dp}$.

First combine the impedances of R_2 and C_2 into a single impedance $\mathbf{Z}_{p1}$ (Fig. 14–7a),

$$\mathbf{Z}_{p1} = \frac{1}{Cs + 1/R_2} = \frac{R_2}{R_2Cs + 1}.$$

Next combine $\mathbf{Z}_{p1}$ and L_2 in series, Fig. 14–7(b), to obtain $\mathbf{Z}_{s1}$,

$$\mathbf{Z}_{s1} = \frac{R_2}{R_2Cs + 1} + L_2s = \frac{R_2L_2Cs^2 + L_2s + R_2}{R_2Cs + 1}.$$

Now combine $\mathbf{Z}_{s1}$ and L_1 in parallel to find $\mathbf{Z}_{p2}$, Fig. 14–7(c),

$$\mathbf{Z}_{p2} = \frac{\mathbf{Z}_{s1}L_1s}{\mathbf{Z}_{s1} + L_1s} = \frac{R_2L_2L_1Cs^3 + L_1L_2s^2 + R_2L_1s}{R_2C(L_1 + L_2)s^2 + (L_1 + L_2)s + R_2}.$$

Finally, combine $\mathbf{Z}_{p2}$ and R_1 in series, Fig. 14-7(d), to yield $\mathbf{Z}_{dp}$,

$$\mathbf{Z}_{dp} = \frac{\mathbf{E}}{\mathbf{I}_1} = \frac{R_2CL_1L_2s^3 + [R_1R_2C(L_1 + L_2) + L_1L_2]s^2 + [R_1(L_1 + L_2) + R_2L_1]s + R_1R_2}{R_2C(L_1 + L_2)s^2 + (L_1 + L_2)s + R_2}.$$

The differential equation for i_1 can be obtained by reverting from s to differentiation:

$$R_2CL_1L_2\frac{d^3i_1}{dt^3} + [R_1R_2C(L_1 + L_2) + L_1L_2]\frac{d^2i_1}{dt^2} + [R_1(L_1 + L_2) + L_1R_2]\frac{di_1}{dt} + R_1R_2i_1$$
$$= R_2C(L_1 + L_2)\frac{d^2e}{dt^2} + (L_1 + L_2)\frac{de}{dt} + R_2e.$$

This result agrees with Eq. (8-44), which was derived by applying the general node, vertex, and elemental equations to the network.

If it is necessary to find a variable which is internal to the network (separated from the source by one or more elements), it is usually necessary to determine the driving-point impedance and then work back through the network to reach the variable of interest. For illustration, let us suppose that v_3 is to be found in the network of Figs. 14-6 and 14-7. Start with Fig. 14-7(a). Since

$$\mathbf{I}_1 = \mathbf{E}/\mathbf{Z}_{dp}$$

has been determined,

$$\mathbf{V}_2 = \mathbf{E} - \mathbf{I}_1R_1 = \mathbf{E}\left(1 - \frac{R_1}{\mathbf{Z}_{dp}}\right) = \mathbf{E}\left(\frac{\mathbf{Z}_{dp} - R_1}{\mathbf{Z}_{dp}}\right).$$

But

$$\mathbf{I}_2 = \frac{\mathbf{V}_2}{L_1s} = \mathbf{E}\frac{(\mathbf{Z}_{dp} - R_1)}{L_1\mathbf{Z}_{dp}s},$$

and by continuity

$$\mathbf{I}_3 = \mathbf{I}_1 - \mathbf{I}_2 = \frac{\mathbf{E}}{\mathbf{Z}_{dp}} - \frac{\mathbf{E}(\mathbf{Z}_{dp} - R_1)}{L_1\mathbf{Z}_{dp}s}$$
$$= \mathbf{E}\left[\frac{L_1s - \mathbf{Z}_{dp} + R_1}{L_1\mathbf{Z}_{dp}s}\right].$$

Finally,

$$\mathbf{V}_3 = \mathbf{V}_2 - \mathbf{I}_3L_2s = \mathbf{E}\left[\frac{(\mathbf{Z}_{dp} - R_1)}{\mathbf{Z}_{dp}} - \frac{(L_2s - (L_2/L_1)\mathbf{Z}_{dp} + R_1L_2/L_1)}{\mathbf{Z}_{dp}}\right]$$
$$= \mathbf{E}\left[\frac{-L_2s + \mathbf{Z}_{dp}(1 + L_2/L_1) - R_1(1 + L_2/L_1)}{\mathbf{Z}_{dp}}\right].$$

The expansion of this expression by substituting the previous result for $\mathbf{Z}_{dp}$ is tedious, as the reader may verify, but results in the following:

$$\frac{\mathbf{V}_3}{\mathbf{E}} = \frac{R_2L_1s}{R_2CL_1L_2s^3 + [R_1R_2C(L_1 + L_2) + L_1L_2]s^2 + [R_1(L_1 + L_2) + R_2L_1]s + R_1R_2}. \qquad (14\text{-}35)$$

The difficulty of finding this result is probably as great as that involved in using the method of Chapter 8 (the equation for v_3 is Eq. 8–53).

The determination of variables internal to a network is considerably simplified by the use of source-impedance equivalents, as described below.

14–4.2 Source-Impedance Equivalents

To investigate what is happening at an internal location in a given network, we find it convenient to divide it, at the section of interest, into two parts, as in Fig. 14–8(a). The across-variable difference v_{mn} and the through-variable f_q exist at this section. We will show that each of these two parts A and B in Fig. 14–8(a) can be represented by a single across-variable source in series with an equivalent impedance *or* by a single through-variable source in parallel with an equivalent impedance. With the two parts thus represented, it is a simple matter to solve for v_{mn} and/or f_q.

a) *Thévenin equivalents.* Assume for the moment that part A of the divided network of Fig. 14–8(a) contains one or more ideal sources (A-type or T-type) and that part B contains no sources. Now imagine that an across-variable source v_e is inserted in series at the dividing section (Fig. 14–8b), which has the exact magnitude and time variation to make the through-variable f_q always zero. Since part B contains no sources, it can be replaced by its equivalent driving-point impedance, and hence if $f_q = 0$, $v_{mn} = 0$. If $f_q = 0$, the across-variable $(v_{mn})_o$ across the terminals of part A must equal the across-variable that would exist if parts A and B were actually *disconnected* in Fig. 14–8(a). In other words, $(v_{mn})_o$ is the open-circuit value of v_{mn} for part A. Application of the loop equation in Fig. 14–8(c) shows $v_e = (v_{mn})_o$. Now we note that the sources in A produced f_q in the original network. If we set all sources equal to zero in parts A *and* B, $f_q = 0$ in Fig. 14–8(a), and by superposition the through-variable in Fig. 14–8(b), would then be $-f_q$. If we reverse the polarity of $v_e = (v_{mn})_o$ as shown in Fig. 14–8(d) then the sign

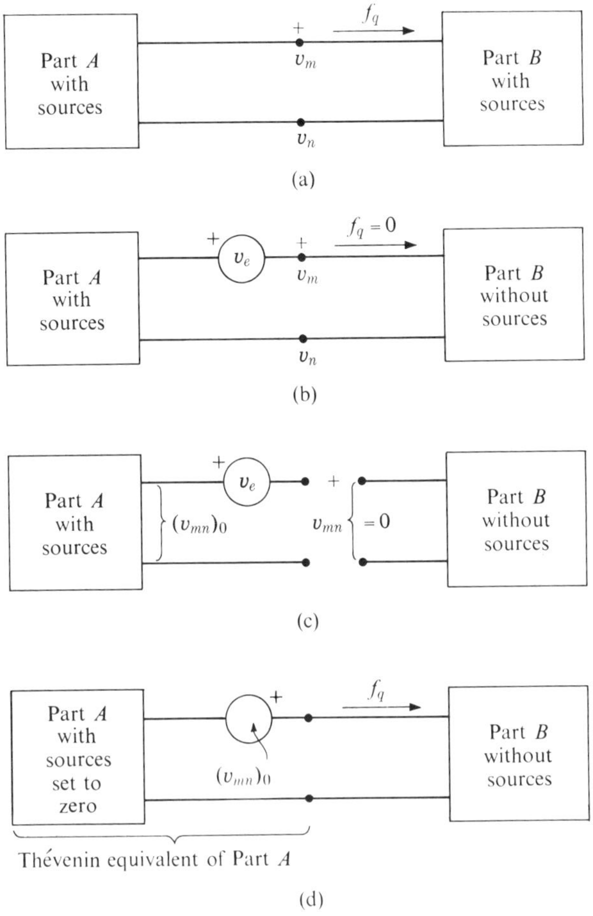

FIG. 14–8. Thévenin equivalent of two-terminal network. (a) Division of network into two parts. (b) Insertion of an across-variable source which results in zero net through-variable at connection. (c) Connection broken after across-variable at connecting points is zero to show that $v_e = (v_{mn})_o$ when $f = 0$ to part B. (d) Removal of sources in part A and reversal of polarity of $v_e = (v_{mn})_o$ to reestablish original through-variable into part B.

of f_q will change. The network of Fig. 14–8(d) will then have f_q flowing into part B and v_{mn} across part B. Therefore, $(v_{mn})_o$ in series with part A, with all sources set equal to zero, is equivalent to the original part A.

We may reason as follows to show that the above equivalence is still true even if part B contains sources. Suppose that the through- and across-variables at the

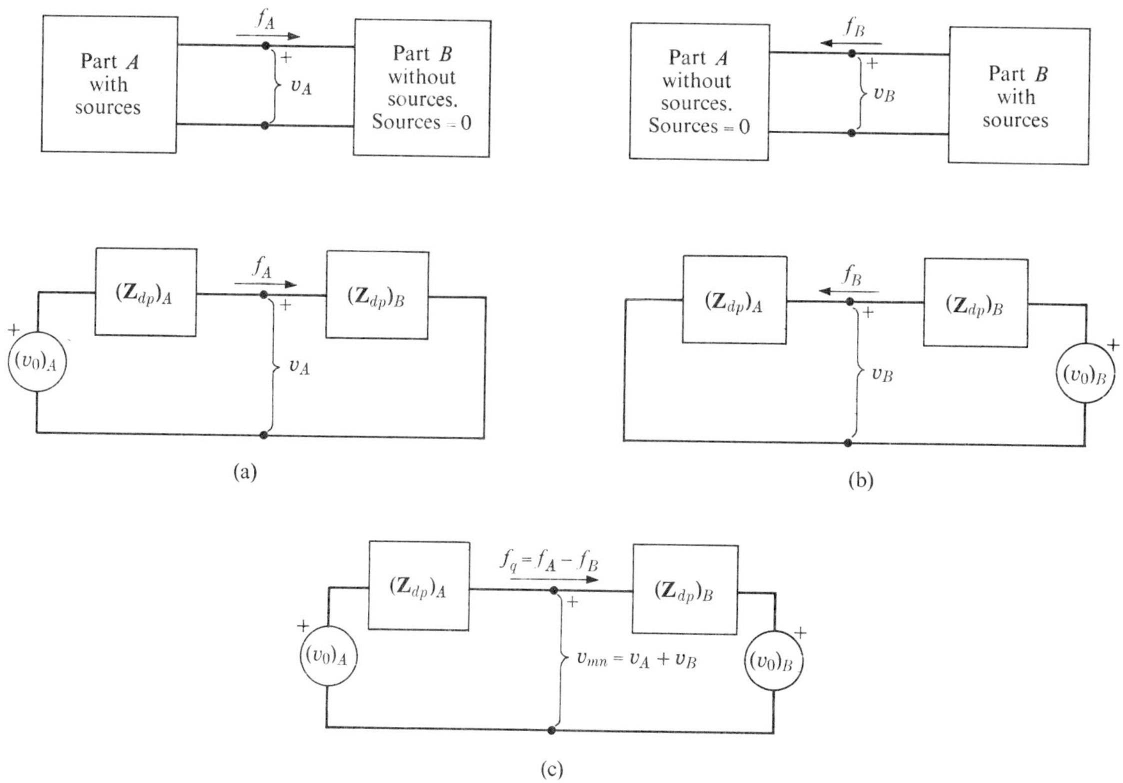

FIG. 14-9. Thévenin equivalents of two parts of a network, both of which contain sources. (a) Thévenin equivalent when sources in part $B = 0$. (b) Thévenin equivalent when sources in part $A = 0$. (c) Thévenin equivalent when parts A and B both contain sources.

section between parts A and B are f_q and v_{mn}, as in Fig. 14-8(a), when A and B both contain sources. These variables are the sum, by superposition, of the variables present when the sources in B are zero and the sources in A are on, and the variables present when the sources in A are zero and those in B are on. In the first case, f_A and v_A can be found by replacing A by its equivalent open-circuit across-variable $v = (v_o)_A$ and series impedance $(\mathbf{Z}_{dp})_A$, and by representing B by its driving-point impedance $(\mathbf{Z}_{dp})_B$ (Fig. 14-9). In the second case, f_B and v_B are found by representing B by its open-circuit across-variable $v = (v_o)_B$ and series impedance $(\mathbf{Z}_{dp})_B$ while A is replaced by its driving-point impedance $(\mathbf{Z}_{dp})_A$. The sums of these two f's and v's, which are f_q and v_{mn} of the original system, are clearly obtained by replacing *both* A and B by their open-circuit v's in series with their driving-point impedances. Therefore the original representation of A is valid even if B contains sources.

The equivalent network just described is called a *Thévenin equivalent* after the French electrical engineer M. L. Thévenin who first discussed it in 1883. *Thévenin's theorem* may be stated as follows:

> Insofar as external or terminal characteristics are concerned, any two-terminal network containing ideal sources and ideal system elements is equivalent to an *across-variable source in series with the network in which all sources are set to zero.* The across-variable source has the instantaneous value of the across-variable appearing at the open-circuit terminals of the original network.

The use of Thévenin's theorem can be illustrated by applying it to the electrical network discussed earlier (Fig. 14–6). Suppose that the differential equation for v_3 is to be found. Assume that all variables are exponential with amplitudes $\mathbf{V}_1$, $\mathbf{V}_2$, etc. Thévenin's theorem says that if the circuit is divided as shown in Fig. 14–10(a), the left-hand portion may be represented by the equivalent network shown in Fig. 14–10(b), where the new source $(v_3)_o$ is the open-circuit voltage which would appear across the terminals 3–0 in Fig. 14–10(b). The equivalent series impedance is found by setting $e = 0$, thus connecting node 1 to ground (node 0). In terms of the amplitudes of exponential variations, the network of Fig. 14–10(a) may be replaced by that of Fig. 14–10(c), where

$$(\mathbf{V}_3)_0 = (\mathbf{V}_3)_{\text{open-circuit}} = \frac{\mathbf{E}L_1 s}{L_1 s + R_1},$$

$$(\mathbf{Z}_{dp})_A = L_2 s + \frac{L_1 R_1 s}{L_1 s + R_1} = \frac{L_1 L_2 s^2 + R_1(L_1 + L_2)s}{L_1 s + R_1},$$

$$(\mathbf{Z}_{dp})_B = \frac{R_2}{R_2 C s + 1},$$

and

$$e = \mathbf{E}e^{st},\ v_3 = \mathbf{V}_3 e^{st}.$$

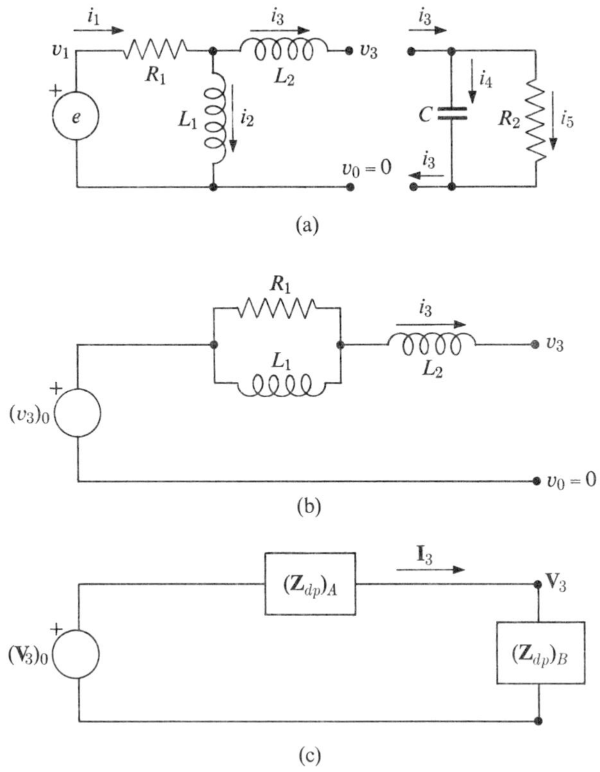

FIG. 14–10. Use of Thévenin's theorem to find the voltage at an internal point in a network. (a) Step-up pulse transformer network broken to expose v_3. (b) Thévenin equivalent network for left-hand portion of (a). (c) Equivalent network in terms of Thévenin source and impedances.

The amplitude $\mathbf{V}_3$ is immediately found from Fig. 14–10(c) and the above equations:

$$\mathbf{V}_3 = \frac{(\mathbf{V}_3)_0(\mathbf{Z}_{dp})_B}{(\mathbf{Z}_{dp})_A + (\mathbf{Z}_{dp})_B} = \frac{\mathbf{E}L_1 s}{(L_1 s + R_1)} \frac{1}{1 + (\mathbf{Z}_{dp})_A/(\mathbf{Z}_{dp})_B},$$

$$\frac{\mathbf{V}_3}{\mathbf{E}} = \frac{L_1 s}{L_1 s + R_1}\left[\frac{1}{1 + \dfrac{[L_1 L_2 s^2 + R_1(L_1 + L_2)s](R_2 C s + 1)}{(L_1 s + R_1)R_2}}\right],$$

$$\frac{\mathbf{V}_3}{\mathbf{E}} = \frac{R_2 L_1 s}{R_2 C L_1 L_2 s^3 + [R_1 R_2 C(L_1 + L_2) + L_1 L_2]s^2 + [R_1(L_1 + L_2) + R_2 L_1]s + R_1 R_2}.$$

This result agrees with the previous result, Eq. (14–35).

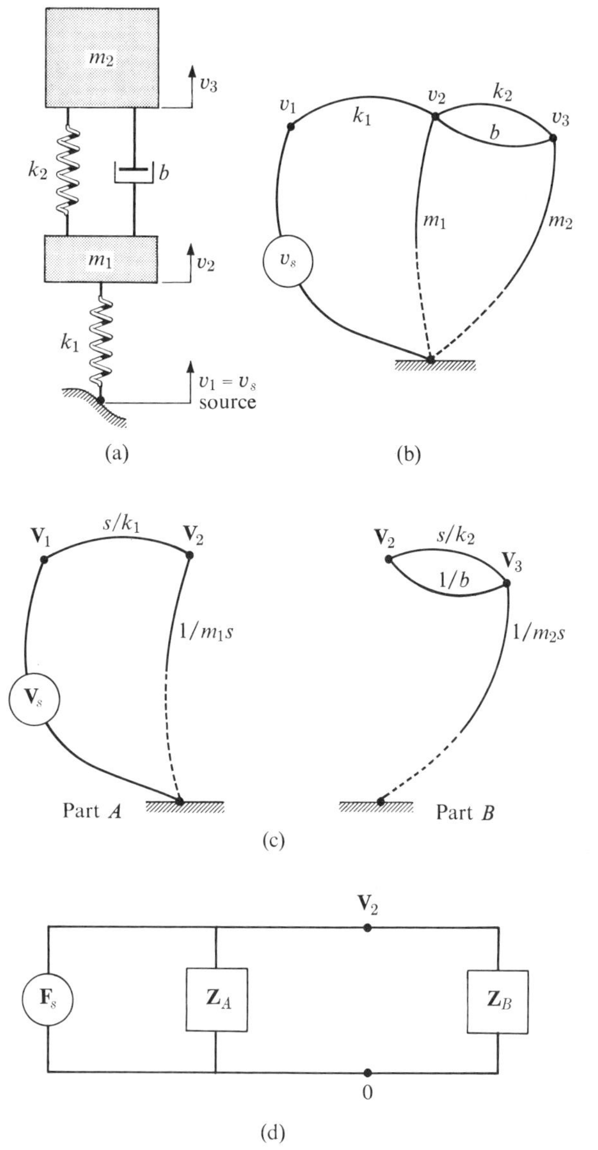

FIG. 14–11. Application of Norton's theorem to find the velocity at an internal point in a mechanical network. (a) System. (b) Linear graph. (c) System divided into two parts for application of Norton's theorem. (d) Norton equivalent.

b) *Norton equivalents.* The dual of Thévenin's theorem, attributed to E. L. Norton of the Bell Telephone Laboratories, and called *Norton's theorem,* may be stated as follows:

> Insofar as external or terminal characteristics are concerned, any two-terminal network containing ideal sources and elements is equivalent to a *through-variable source in parallel with the network in which all sources are set to zero.* The through-variable source has the instantaneous value of the through-variable which flows through the short-circuited terminals of the original network.

To illustrate Norton's theorem, let us consider the mechanical system of Fig. 14–11(a), which is a simplified model of an automobile (see Chapter 6). The velocity v_2 is to be found as a function of the input velocity v_s. Draw the system graph as in Fig. 14–11(b) and divide it into two parts at the point of interest, as shown in Fig. 14–11(c). Label each node with the amplitude of its across-variable (assuming exponential variations everywhere) and label each branch with its impedance. Part A can be replaced by a through-variable source $\mathbf{F}_s$, which is the short-circuit force for part A in parallel with the driving point impedance $\mathbf{Z}_A$ of part A. If $\mathbf{V}_2 = 0$ in Fig. 14–11(c), we have

$$\mathbf{F}_s = \frac{\mathbf{V}_1 - \mathbf{V}_2}{s/k_1} = \frac{k_{1s}\mathbf{V}_s}{s}.$$

The driving-point impedance is found by setting $\mathbf{V}_s = 0$, thus putting m_1 and k_1 in parallel at the $\mathbf{V}_2$ node:

$$\mathbf{Z}_A = \frac{(s/k_1)(1/m_1s)}{(s/k_1) + (1/m_1s)} = \frac{s}{m_1s^2 + k_1}.$$

The driving-point impedance for part B at $\mathbf{V}_2$ is

$$\mathbf{Z}_B = \frac{1}{m_2s} + \frac{(s/k_2)(1/b)}{(s/k_2) + (1/b)} = \frac{m_2s^2 + bs + k_2}{m_2s(bs + k_2)}.$$

The Norton equivalent is shown in Fig. 14–11(d), from which $\mathbf{V}_2$ is found to be

$$\mathbf{V}_2 = \mathbf{F}_s\left[\frac{\mathbf{Z}_A\mathbf{Z}_B}{\mathbf{Z}_A + \mathbf{Z}_B}\right] = \frac{k_1\mathbf{V}_s}{s}\left[\frac{\mathbf{Z}_A\mathbf{Z}_B}{\mathbf{Z}_A + \mathbf{Z}_B}\right],$$

$$\frac{\mathbf{V}_2}{\mathbf{V}_s} = \frac{k_1m_2s^2 + k_1bs + k_1k_2}{m_1m_2s^4 + (m_1 + m_2)bs^3 + (m_1k_2 + m_2k_2 + k_1m_2)s + k_1k_2}.$$

The differential equation for v_2 follows directly from this result.

14–4.3 Source Transformations

We have seen that any two-terminal network of ideal elements can be represented by either a Thévenin equivalent (across-variable source in series with the network impedance) or a Norton equivalent (through-variable source in parallel with the network impedance). Therefore these two representations must be equivalent to each other and one must be transformable into the other. To show how this transformation may be accomplished, let us take the Thévenin equivalent of any network as shown in Fig. 14–12(a), where $\mathbf{V}_s$ is the amplitude of the open-circuit across-variable $(\mathbf{V}_{mn})_o$ and $\mathbf{Z}$ is the network impedance as seen at terminals m-n when the sources are zero. Now form the Norton equivalent of this Thévenin network, giving Fig. 14–12(b). The network impedance is still $\mathbf{Z}$ but it now appears in parallel with a through-variable source $\mathbf{F}_s$. This source is the short-circuit value of $\mathbf{F}_q$ in Fig. 14–12(a), i.e., $\mathbf{F}_q$ when $\mathbf{V}_{mn} = 0$:

$$\mathbf{F}_s = \frac{\mathbf{V}_s}{\mathbf{Z}}.$$

Obviously, if we had started with the Norton equivalent, the Thévenin equivalent could have been found by letting $\mathbf{V}_s = \mathbf{Z}\mathbf{F}_s$.

The use of the above source transformation is illustrated by the electrical circuit of Fig. 14–13(a), where the voltage v is to be found. Form the Norton equivalent of e_s in series with R, then combine R and

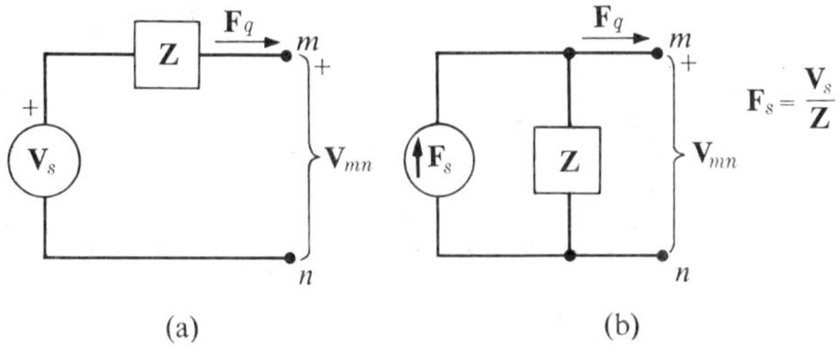

FIG. 14–12. Transformation from Thévenin equivalent to Norton equivalent. (a) Thévenin equivalent for any network. (b) Norton equivalent of Thévenin equivalent.

L in parallel as in Fig. 14–13(b). Next transform to a Thévenin equivalent to give the series network of Fig. 14–13(c). The solution for $\mathbf{V}$ is then, in terms of the symbols in Fig. 14–13,

$$\mathbf{V} = \frac{\mathbf{E}_{sA}2Ls}{2Ls + (2RLs + R^2)/(R + Ls)}$$

$$= \frac{\mathbf{E}_s}{R}\left(\frac{RLs}{R + Ls}\right)\frac{2Ls}{2Ls + (2RLs + R^2)/(R + Ls)},$$

$$\frac{\mathbf{V}}{\mathbf{E}_s} = \frac{2L^2s^2}{2L^2s^2 + 4RLs + R^2},$$

or

$$2L^2\frac{d^2v}{dt^2} + 4RL\frac{dv}{dt} + R^2v = 2L^2\frac{d^2e_s}{dt^2}.$$

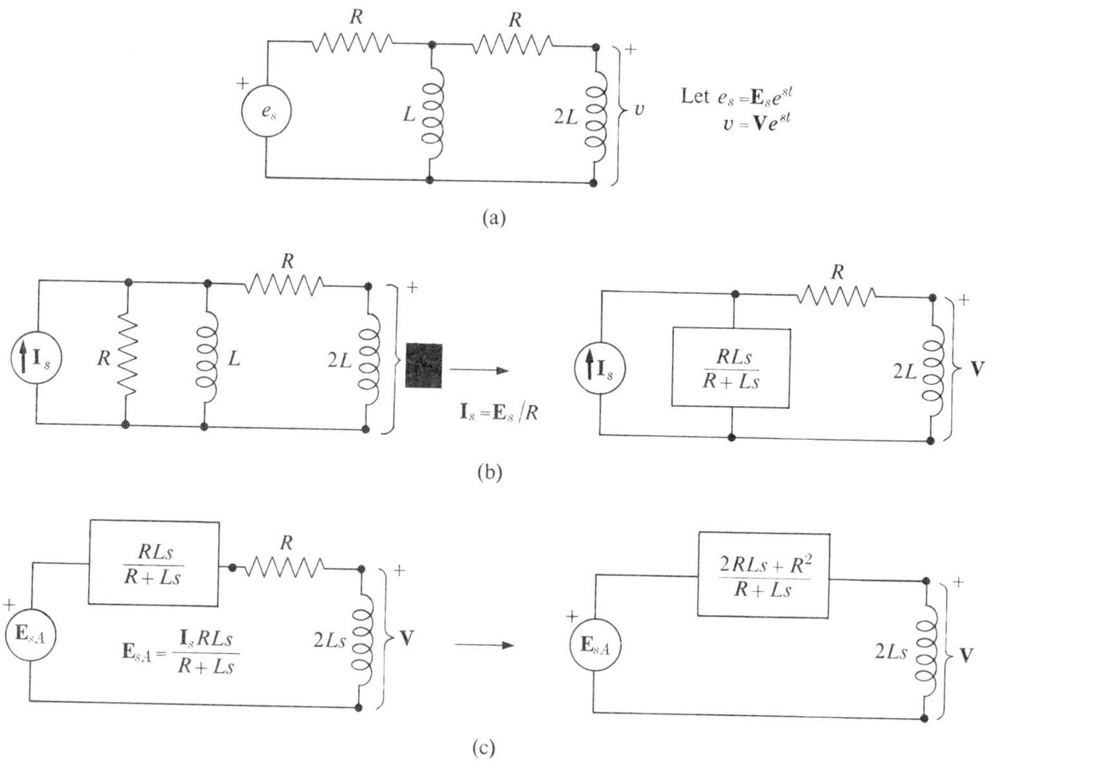

FIG. 14–13. Use of source transformations in solving for a network variable. (a) Electric circuit. (b) Step 1: Norton equivalent and combination of parallel impedances. (c) Step 2: Thévenin equivalent and combination of series impedances.

14–5. SUMMARY

The techniques described in this chapter simplify and systematize the analysis of systems of ideal elements by the use of impedance concepts.

The loop and node formulation methods yield sets of simultaneous algebraic equations in s which are conveniently expressed in matrix form. The system function and the differential equation for any output is easily found by solving the simultaneous equations by the method of determinants.

When only a few of the system variables must be determined, the techniques of impedance combination, Thévenin and Norton equivalents, and source transformations are particularly useful.

REFERENCES

1. Z. Hennyey, *Linear Electric Circuits*. Reading, Mass.: Addison-Wesley Publishing Company, Inc., 1962.
2. A. M. Tropper, *Matrix Theory for Electrical Engineers*. Reading, Mass.: Addison-Wesley Publishing Company, Inc., 1962.
3. L. de Pian, *Linear Active Network Theory*. Englewood Cliffs: Prentice-Hall, Inc., 1962.
4. E. Guillemin, *Theory of Linear Physical Systems*. New York: John Wiley and Sons, Inc., 1963.

PROBLEMS

14–1. Find a set of equations in matrix form that represents the system shown in Fig. 14–14 by using either the loop or node method of analysis.

14–2. For the electrical system shown in Fig. 14–15:

a) Draw the linear graph.

b) Draw the mechanical translational analog.

c) Draw the mechanical rotational analog.

d) Draw the lumped-fluid-system analog.

e) Find the system function between i and v if $R_1 = R$, $R_2 = 2R$, $C_1 = C$, and $C_2 = C/2$.

f) For $i = I \sin \omega t$ find the amplitude and phase of v in the steady state.

g) If $i = \text{const} = I$, what is the steady-state value of v?

14–3. Using *both* the loop and node methods, obtain the required set of equations in matrix form for the system shown in Fig. 14–16.

14–4. Using the node method, formulate the equations for the system shown in Fig. 14–17 and solve for the differential equation relating v to the inputs V and F.

14–5. For the system shown in Fig. 14–18, which is a model of a three-car train engaging a buffer as it comes to rest in a railroad station, use the node method to formulate the necessary equations in matrix form. Assume that the buffer is equivalent to a spring k_3 and damper b_3 connected end-to-end, and neglect friction between the track and the wheels.

14–6. Find a set of equations in matrix form for the fluid system shown in Fig. 14–19.

14–7. Solve for the driving-point impedance of the circuit (Fig. 14–20), using impedance combinations and determine the transfer function relating v to i.

14–8. In the system of Fig. 14–21, use impedance combinations to find the differential equation relating v_1 to F.

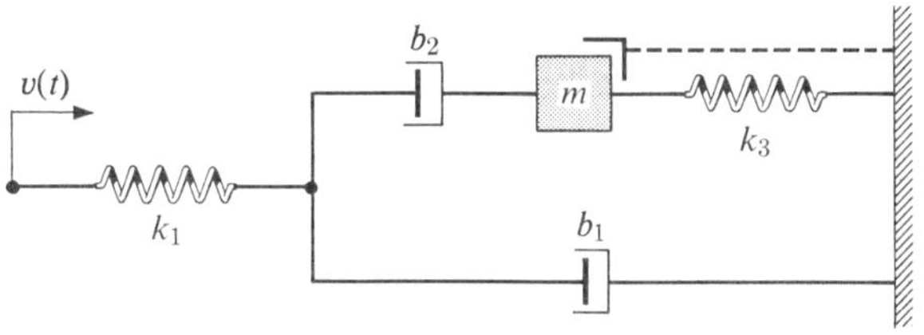

FIGURE 14–14

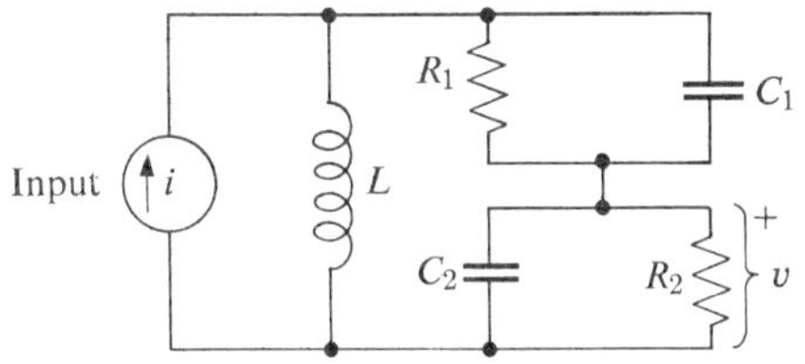

FIGURE 14–15

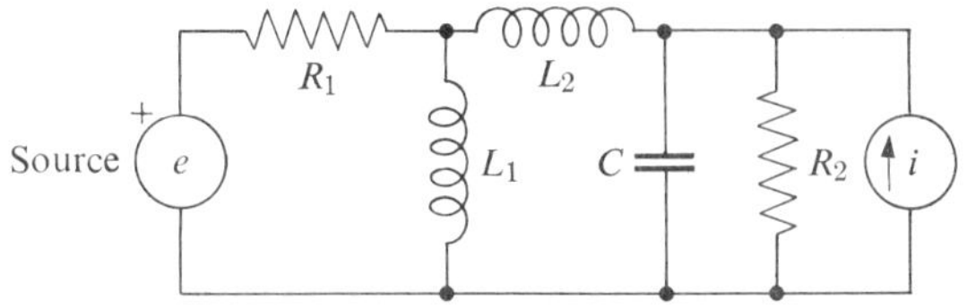

FIGURE 14–16

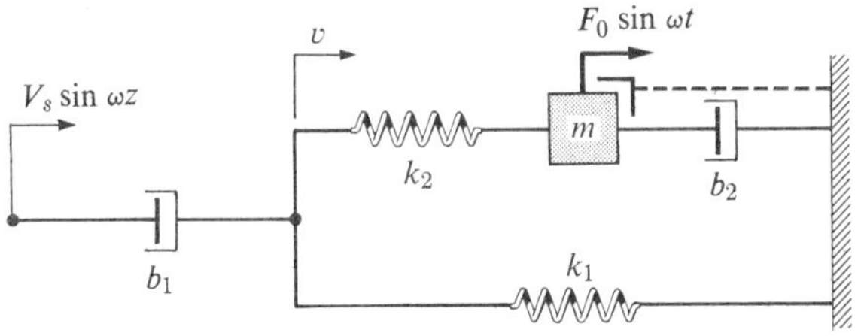

FIGURE 14–17

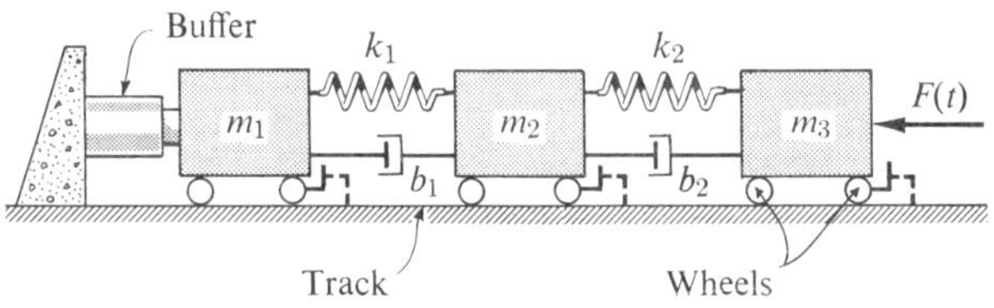

FIGURE 14–18

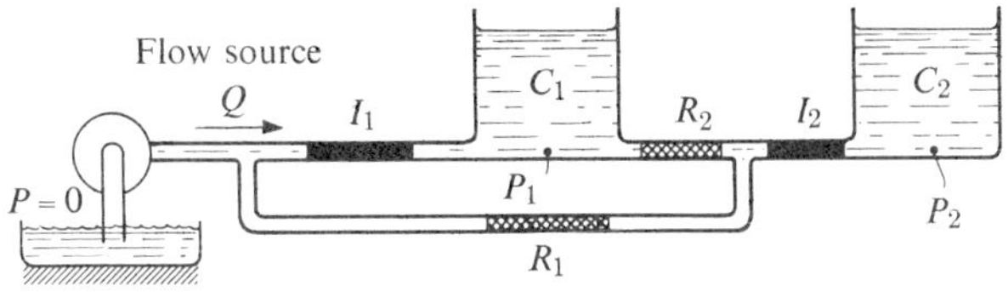

FIGURE 14–19

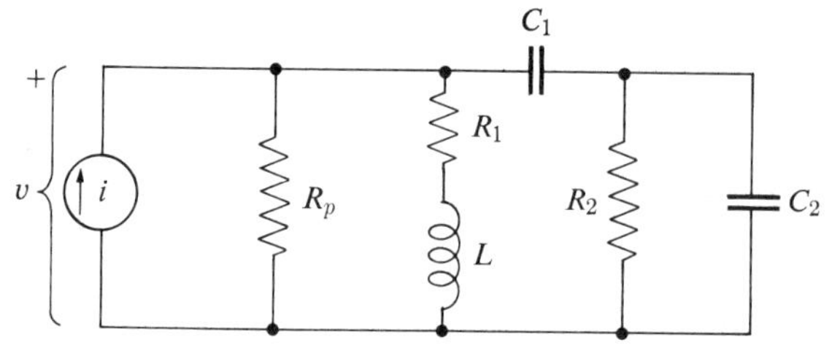

FIGURE 14–20

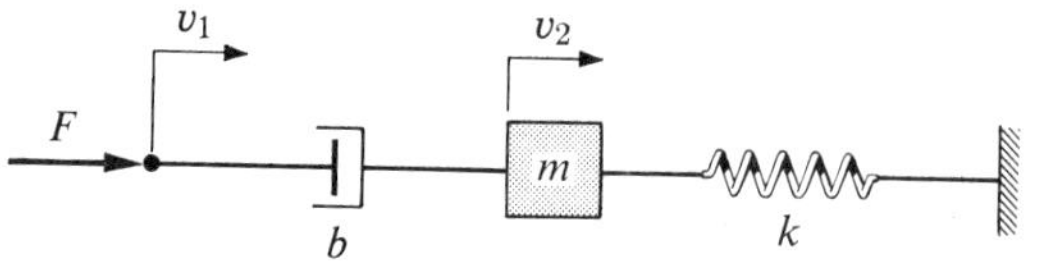

FIGURE 14–21

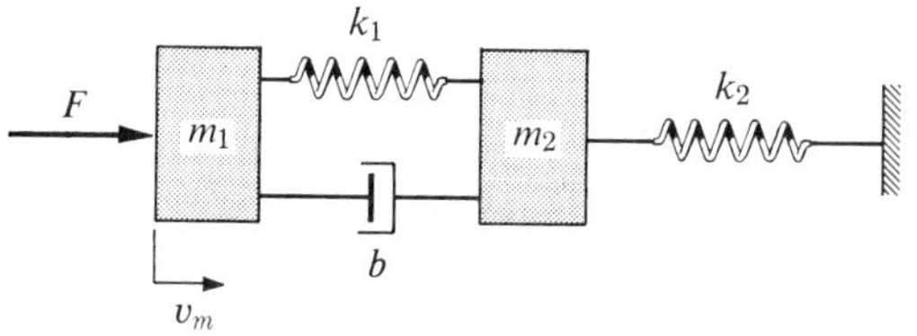

FIGURE 14–22

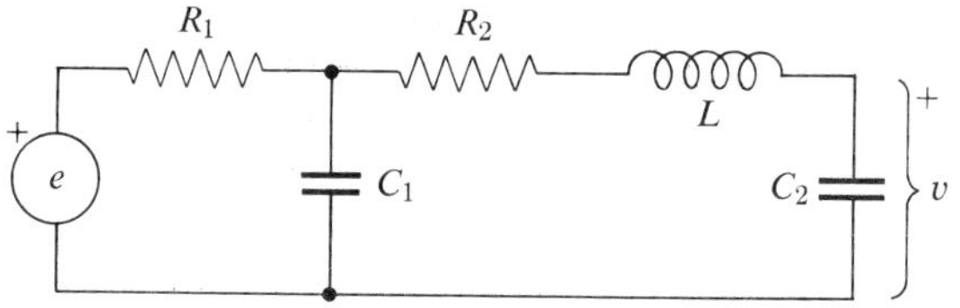

FIGURE 14–23

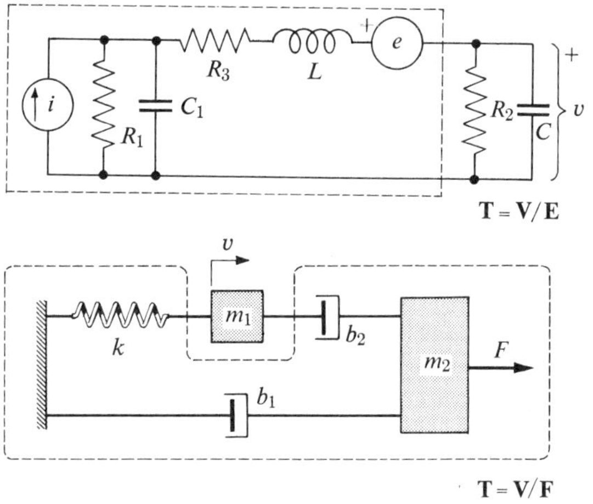

FIGURE 14–24

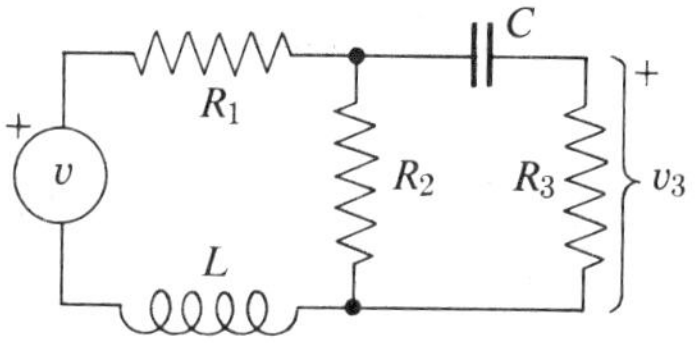

FIGURE 14–25

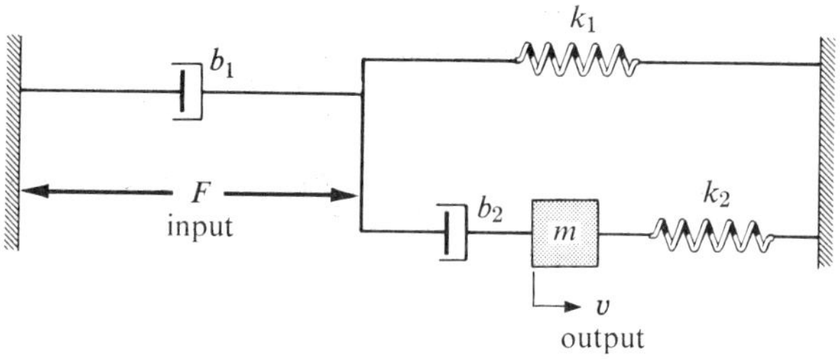

FIGURE 14–26

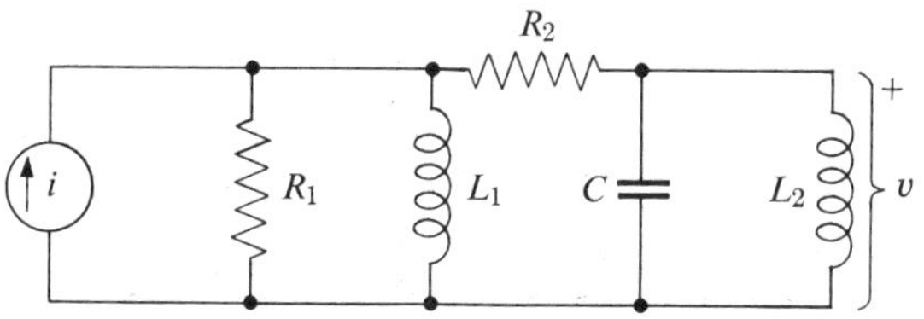

FIGURE 14–27

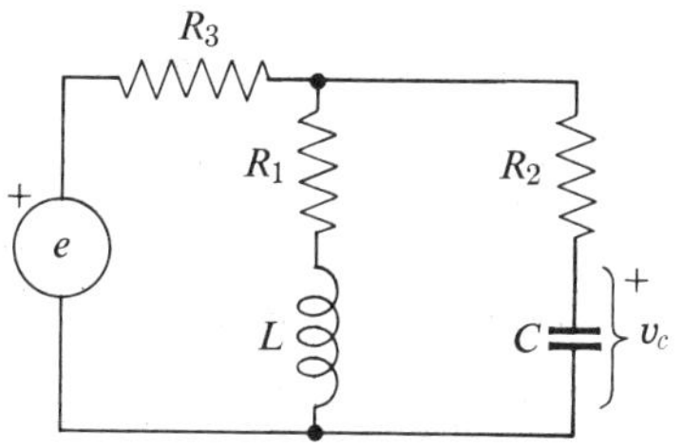

FIGURE 14–28

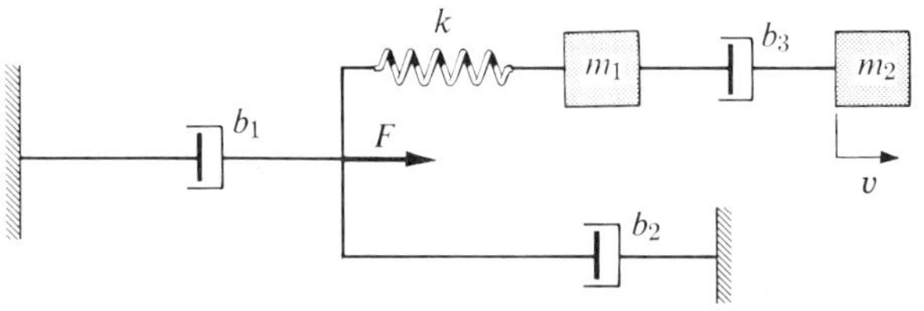

FIGURE 14–29

14–9. Using impedance methods, find the system function $\mathbf{V}_m/\mathbf{F}$ for the system shown in Fig. 14–22.

14–10. Find the transfer function $\mathbf{V}/\mathbf{E}$ for the system shown in Fig. 14–23, using either Thévenin or Norton equivalents.

14–11. In Problem 14–6, solve for the transfer function relating P_1 to Q by using Thévenin's theorem.

14–12. In Problem 14–8 find the differential equation relating v_2 to F by using Norton's theorem.

14–13. Find (a) Thévenin's equivalent and (b) Norton's equivalent for the subsystems enclosed in dashed lines (Fig. 14–24). Use either of these equivalents to find the required transfer functions.

14–14. Find the system function $\mathbf{V}_3/\mathbf{V}$ for the network shown in Fig. 14–25.

14–15. Using the method of source transformation, find the transfer function $\mathbf{V}/\mathbf{F}$ for the system shown in Fig. 14–26.

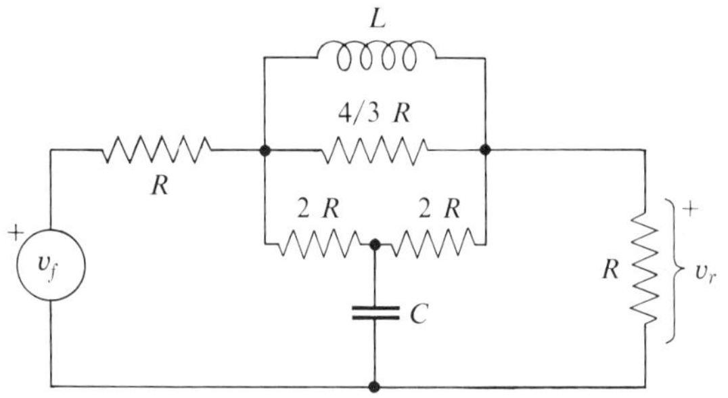

FIGURE 14–30

14–16. Using the method of source transformation, obtain the transfer function $\mathbf{V}/\mathbf{I}$ for the network shown in Fig. 14–27.

14–17. Use source transformations to find the differential equation for v_c (Fig. 14–28).

14–18. Determine the system function for v when F is the input to the system in Fig. 14–29.

14–19. The circuit shown in Fig. 14–30 is called a "bridged-T" filter.

a) Find the differential equation relating v_r to v_f.

b) If $R^2 = L/C$, $C = 100\,\mu\mu$f, $R = 10{,}000$ ohms, find the poles and zeros of the system function.

c) For a steady sinusoidal input voltage $v_f = V \sin \omega t$, sketch the amplitude and phase of the output sinusoid as a function of ω. Comment on the effectiveness of this circuit as a filter.

15 SINUSOIDAL STEADY-STATE ANALYSIS

15–1. INTRODUCTION

When the sinusoidal excitation function was discussed in Chapter 10, it was shown how periodic functions of time can be expressed by an equivalent series of sinusoidal functions. The role of the exponential function in describing and in working with sinusoidal functions was discussed in both Chapters 9 and 10. Then the forced part of the response of a linear system to exponential inputs which was discussed in Chapter 13 was used as a basis for the development of simplified techniques in system analysis in Chapter 14.

Of all the possible input or excitation functions, the sinusoids are by far the most widely encountered and used. A great majority of the electrical-power generation, transmission, and utilization systems throughout the world operate with sinusoidally varying voltages and currents, so that almost all electrical devices such as lights, motors, relays, solenoids, appliances, and electronic systems operate with alternating current (ac) power. Direct current (dc) power is normally used only for very high voltage distribution systems and for small mobile or standby systems containing storage batteries. Mechanical unbalance in rotating machinery such as gas turbines, internal-combustion engines, machine tool drives, and electric motors results in the generation of sinusoidal forces which excite sinusoidal or periodic mechanical vibrations. Ocean waves which cause rolling and pitching of ships at sea are often nearly sinusoidal. Most communication systems rely on the transmission of signals over long distances by means of high-frequency sinusoidal waves.

Furthermore, the use of a steady sinusoidal input offers a means of finding dynamic-response characteristics of a real system with a series of steady-state tests. Most of the dynamic-response specifications of control-system and communication-system components are given in terms of frequency-response characteristics (i.e., with steady sinusoidal inputs).

In this chapter we shall deal exclusively with the determination of the steady-state response of linear systems to sinusoidal inputs, often called *frequency-response analysis*. Thus, in the sections which follow, it is always assumed that the transient or natural part of the response has decayed to zero so that only the forced part of the response remains.

15–2. THE TRANSFER FUNCTION FOR SINUSOIDAL EXCITATION

The most general differential equation which relates an output or response quantity $x_r(t)$ to an input or driving function $x_f(t)$ for a system composed of lumped linear elements is of the form

$$\left[a_n \frac{d^n}{dt^n} + a_{n-1} \frac{d^{n-1}}{dt^{n-1}} + \cdots + a_0\right] x_r = \left[b_m \frac{d^m}{dt^m} + \cdots + b_0\right] x_f, \tag{15–1}$$

where the a_n and b_m coefficients are constants. For most physical systems $m \leq n$.

When the forcing function is exponential,

$$x_f = \mathbf{X}_f e^{st}, \tag{15–2}$$

it has been shown that the forced part of the response x_r is also an exponential,

$$x_r = \mathbf{X}_r e^{st}. \tag{15–3}$$

When these expressions (15–2) and (15–3) are substituted into the differential equation (15–1), the factor e^{st} may be canceled, leaving the *system function* or *transfer function* $\mathbf{T}(s)$:

$$\mathbf{T}(s) = \frac{\mathbf{X}_r}{\mathbf{X}_f} = \frac{b_m s^m + b_{m-1}s^{m-1} + \cdots + b_0}{a_n s^n + a_{n-1}s^{n-1} + \cdots + a_0}. \tag{15–4}$$

Although this transfer function is derived only for the forced part of the response to exponential inputs, it contains all the information necessary to obtain the transient response also, since the *poles* of the transfer function (values of s for which the denominator of $\mathbf{T}(s)$ vanishes) are the roots of the system characteristic equation.

The system function (15–4) may be factored to give the following alternative form:

$$\mathbf{T}(s) = \frac{b_m}{a_n}\frac{(s - r_{10})(s - r_{20})\cdots(s - r_{m0})}{(s - r_{1p})(s - r_{2p})\cdots(s - r_{np})}, \tag{15–5}$$

where $r_{10}, \ldots, r_{m0}$ are the *m zeros* and $r_{1p}, \ldots, r_{np}$ are the *n poles* of the system function. Note that the constant factor b_m/a_n plus the values of the zeros and poles completely specify the system function.

When the input $x_f(t)$ is a sinusoid, which is the case of interest here, s is a purely imaginary quantity, $s = j\omega$. Following the method described in Chapter 13, "Response of Linear Systems," we see that the input sinusoid

$$x_f = A \sin(\omega t + \alpha) \tag{15–6}$$

may be represented by

$$x_f = \mathrm{Im}[\mathbf{X}_f e^{j\omega t}], \tag{15–7}$$

where $\mathbf{X}_f$ is a complex number,

$$\mathbf{X}_f \triangleq A e^{j\alpha}, \tag{15–8}$$

and A is the (real) amplitude of the sinusoid. The reader should verify that Eqs. (15–6) and (15–7) are equivalent when the definition (15–8) is employed.

Any general sinusoid (including, for instance, a cosine) can therefore be represented by (15–7), providing that the complex number $\mathbf{X}_f$ is properly chosen. Alternatively, the input could be expressed as

$$x_f = \mathrm{Re}[\mathbf{X}_f e^{j\omega t}], \tag{15–9}$$

in which case, if

$$\mathbf{X}_f \triangleq B e^{j\theta}, \tag{15–10}$$

$$x_f = B \cos(\omega t + \theta). \tag{15–11}$$

In the following discussion the representation (15–7) in terms of imaginary parts will be used.

It has been shown in Chapter 13 that the response of a linear system x_r to an input given in Eq. (15–7) is given by

$$x_r = \mathrm{Im}[\mathbf{X}_r e^{j\omega t}]. \tag{15–12}$$

When Eq. (15–4) is substituted into (15–12),

$$x_r = \mathrm{Im}[\mathbf{X}_f \mathbf{T} e^{j\omega t}]. \tag{15–13}$$

Since $s = j\omega$ for sinusoidal excitation, the transfer function becomes

$$\begin{aligned} \mathbf{T}(j\omega) &= \frac{b_m(j\omega)^m + \cdots + b_0}{a_n(j\omega)^n + \cdots + a_0} \\ &= \frac{(B_i)j + B_r}{(A_i)j + A_r}. \end{aligned} \tag{15–14}$$

In Eq. (15–14), the constants B_i and A_i are the sums of the *odd* powers of ω multiplied by their corresponding coefficients, and the constants B_r and A_r are the sums of the *even* powers of $j\omega$ times their coefficients. For example, let us suppose that the transfer function is second order. Then

$$\mathbf{T}(j\omega) = \frac{b_2(j\omega)^2 + b_1(j\omega) + b_0}{a_2(j\omega)^2 + a_1(j\omega) + a_0} = \frac{(b_1\omega)j + (b_0 - b_2\omega^2)}{(a_1\omega)j + (a_0 - a_2\omega^2)}, \tag{15–15}$$

and, comparing (15–15) with (15–14), we see that

$$B_i = b_1\omega, \qquad A_i = a_1\omega, \tag{15–16a, b}$$

$$B_r = b_0 - b_2\omega^2, \qquad A_r = a_0 - a_2\omega^2. \tag{15–16c, d}$$

For a higher-order system the transfer function would be of the following form:

$$\mathbf{T}(j\omega) = \frac{(b_0 - b_2\omega^2 + b_4\omega^4 - \cdots) + j(b_1\omega - b_3\omega^3 + b_5\omega^5 - \cdots)}{(a_0 - a_2\omega^2 + a_4\omega^4 - \cdots) + j(a_1\omega - a_3\omega^3 + a_5\omega^5 - \cdots)}. \tag{15–17}$$

From Eq. (15–14) it may be seen that regardless of the order or complexity of the transfer function for sinusoidal excitation, it can always be expressed as a simple ratio of two complex numbers. Furthermore, Eq. (15–14) can be further reduced to a single complex number,

$$\mathbf{T}(j\omega) = M(\omega)e^{j\phi(\omega)}, \tag{15–18}$$

where $M(\omega) = \text{magnitude} = |\mathbf{T}(j\omega)|$,

$$M(\omega) = \sqrt{B_i^2 + B_r^2}/\sqrt{A_i^2 + A_r^2}, \tag{15–19}$$

and

$$\phi(\omega) = \text{phase angle} = \text{angle of } \mathbf{T}(j\omega) \quad \text{or} \quad \angle\mathbf{T}(j\omega) \tag{15–20}$$

or

$$\phi = \tan^{-1}(B_i/B_r) - \tan^{-1}(A_i/A_r).$$

Thus the response x_r may be written, by combining Eqs. (15–13) and (15–18), as

$$x_r = \text{Im}[\mathbf{X}_f M(\omega)e^{j(\omega t+\phi)}], \tag{15–21}$$

or from (15–8),

$$x_r = \text{Im}[Ae^{j\alpha}M(\omega)e^{j(\omega t+\phi)}] = \text{Im}[AM(\omega)e^{j(\omega t+\alpha+\phi)}], \tag{15–22}$$

$$x_r = AM(\omega)\sin(\omega t + \alpha + \phi) = X_r \sin(\omega t + \alpha + \phi),$$

where $X_r = |\mathbf{X}_r|$. Thus, in summary, if a system having a transfer function $\mathbf{T}(s)$ is excited by a sinusoid,

$$x_f = A\sin(\omega t + \alpha). \tag{15–23}$$

The steady-state response x_r is also a sinusoid having an amplitude $AM(\omega) = X_r$ and shifted in time by a phase angle ϕ:

$$x_r = AM(\omega)\sin(\omega t + \alpha + \phi) = X_r\sin(\omega t + \alpha + \phi). \tag{15–24}$$

The magnitude $M(\omega)$ is the absolute value of the frequency-response transfer function $\mathbf{T}(j\omega)$ and the phase angle $\phi(\omega)$ is the phase angle of the transfer function, Eqs. (15–18), (15–19), and (15–20).

Figure 15–1 shows the x_f and x_r of Eqs. (15–23) and (15–24) as time functions generated by the imaginary part of phasors rotating with speed ω in the complex plane. The detailed procedure for determining the amplitude- and phase-response characteristics of dynamic systems will be illustrated in the following sections.

15–2.1 Response of a First-Order System to Sinusoidal Excitation

The basic concepts of amplitude and phase angle of $\mathbf{T}(j\omega)$ developed above and the use of pole and zero phasors described in Section 13–4.4 may now be employed to reveal the frequency-response characteristics of linear systems. Let us consider first the case of first-order systems.

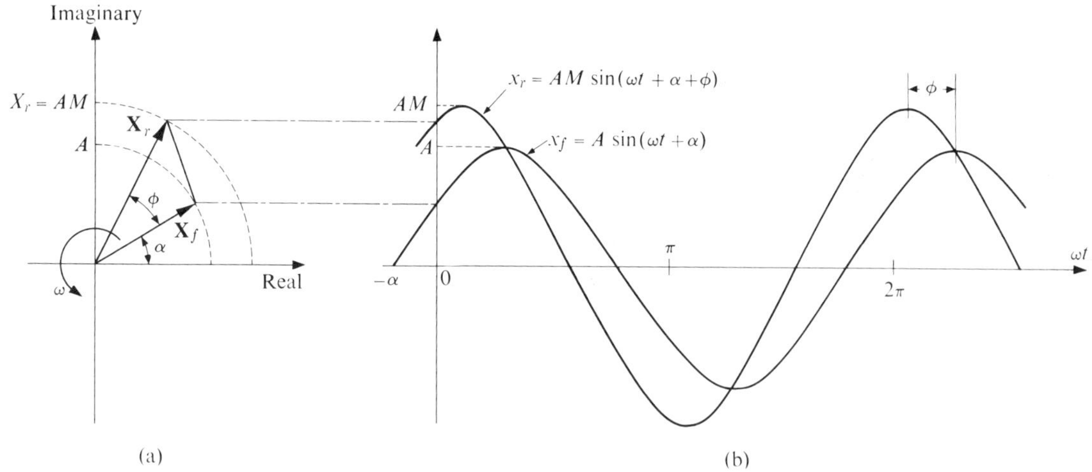

FIG. 15–1. Input and response sinusoids for a linear dynamic system.

The most general first-order linear system is represented by the following differential equation:

$$a_1 \frac{dx_r}{dt} + a_0 x_r = b_1 \frac{dx_f}{dt} + b_0 x_f. \qquad (15\text{–}25)$$

Let

$$\tau_1 = a_1/a_0, \qquad \tau_2 = b_1/b_0, \qquad K = b_0/a_0. \qquad (15\text{–}26)$$

Assume an exciting function,

$$x_f = A \sin \omega t = \operatorname{Im}[Ae^{j\omega t}]. \qquad (15\text{–}27)$$

The transfer function for this sinusoidal input is, from Eqs. (15–25), (15–26), and (15–27),

$$\mathbf{T}(j\omega) = \frac{\mathbf{X}_r}{\mathbf{X}_f} = \frac{\mathbf{X}_r}{A} = K\frac{(j\omega\tau_2 + 1)}{(j\omega\tau_1 + 1)}. \qquad (15\text{–}28)$$

This function may be written in the form

$$\mathbf{T}(j\omega) = K\mathbf{G}(j\omega) = K\frac{\tau_2}{\tau_1}\frac{(j\omega + 1/\tau_2)}{(j\omega + 1/\tau_1)} = K\frac{\tau_2 \mathbf{U}_0}{\tau_1 \mathbf{U}_p}. \qquad (15\text{–}29)$$

This function has a single real zero r_0 at $-1/\tau_2$ and a single real pole r_p at $-1/\tau_1$. That is,

$$\mathbf{U}_0 = (j\omega + 1/\tau_2) = (\sqrt{1 + \tau_2^2\omega^2}/\tau_2)e^{j\phi_0}, \qquad (15\text{–}30)$$

$$\mathbf{U}_p = (j\omega + 1/\tau_1) = (\sqrt{1 + \tau_1^2\omega^2}/\tau_1)e^{j\phi_p}. \qquad (15\text{–}31)$$

Using Eqs. (15–29), (15–30), and (15–31), we obtain the magnitude and phase angle of $\mathbf{T}(j\omega)$:

$$M(\omega) = K\sqrt{\frac{(1 + \tau_2^2\omega^2)}{(1 + \tau_1^2\omega^2)}} = \frac{|\mathbf{X}_r|}{A} = \frac{X_r}{A}, \qquad (15\text{–}32)$$

$$\phi(\omega) = \phi_0 - \phi_p = \tan^{-1}(\omega\tau_2) - \tan^{-1}(\omega\tau_1). \qquad (15\text{–}33)$$

The reader should note the ease with which this result was obtained by the use of the exponential-response method and complex numbers as compared with the method of direct solution in terms of sines and cosines used in Chapter 5, "Analysis of Elementary Dynamic Systems."

The results (15–32) and (15–33) also may be obtained graphically. Figure 15–2 shows the single *zero* (indicated by a dot in a circle) and the single *pole* (indicated by a cross in a circle) in the complex *s*-plane. The factors $\mathbf{U}_0$ and $\mathbf{U}_p$ described by Eqs. (15–30) and (15–31) are obtained graphically as in Chapter 13 by the drawing of phasors from the zero at $-1/\tau_2$ and the pole at $-1/\tau_1$ to the point $j\omega$ corresponding to the frequency of the input excitation:

$$M(\omega) = \frac{b_1}{a_1}\frac{|\mathbf{U}_0|}{|\mathbf{U}_p|}, \qquad (15\text{–}34)$$

$$\phi(\omega) = \phi_0 - \phi_p. \qquad (15\text{–}35)$$

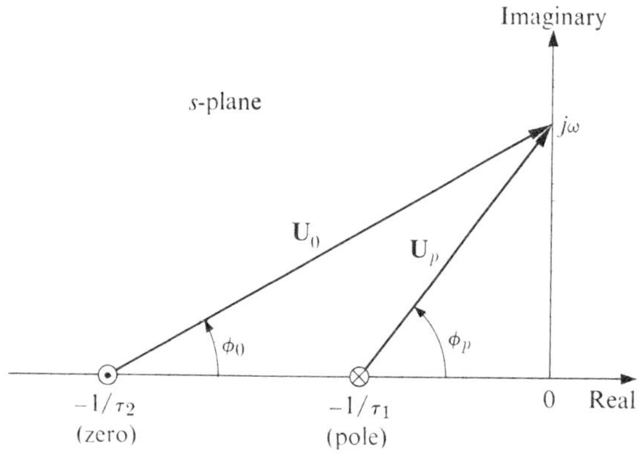

FIG. 15-2. Graphical construction of factors in a single-order system.

A good qualitative idea of the frequency response of the single-order system can be obtained from inspection of the diagram of Fig. 15-2, with the aid of Eqs. (15-34), and (15-35). When $\omega = 0$, ϕ_0 and ϕ_p are both zero, and hence $\phi = 0$. The magnitude M is then $(b_1/a_1)(\tau_1/\tau_2) = b_0/a_0$. For very large frequencies, $\omega \to \infty$, ϕ_0 and ϕ_p approach $\pi/2$, and again $\phi = 0$. The lengths of $\mathbf{U}_0$ and $\mathbf{U}_p$ approach very large and equal values, so that M approaches b_1/a_1. For frequencies near zero, the magnitudes do not vary much from their zero-frequency values and the amplitude response $M(\omega)$ is nearly constant or "flat" for frequencies $\omega \ll 1/\tau_1$. In this region, the phase angles ϕ_0 and ϕ_p are approximately

$$\phi_0 \approx \omega\tau_2; \qquad \phi_p \approx \omega\tau_1 \tag{15-36}$$

or

$$\phi_1 \approx \omega(\tau_2 - \tau_1); \qquad \omega \text{ small.} \tag{15-37}$$

Thus the low-frequency phase angle $\phi(\omega = \text{small})$ is proportional to frequency. Figure 15-3 is a sketch showing this qualitative behavior, where τ_2 has been assumed to be less than τ_1. The reader should construct a similar sketch for $\tau_2 > \tau_1$. Reference to Eqs. (15-32) and (15-33) will verify the correctness of the curves shown in Fig. 15-3. For the input function of amplitude A (Eq. 15-27), the response x_{r1} at ω_1 (see Fig. 15-3) for $\tau_1 > \tau_2$ is given by

$$x_{r1} = AM_1 \sin(\omega_1 t + \phi_1). \tag{15-38}$$

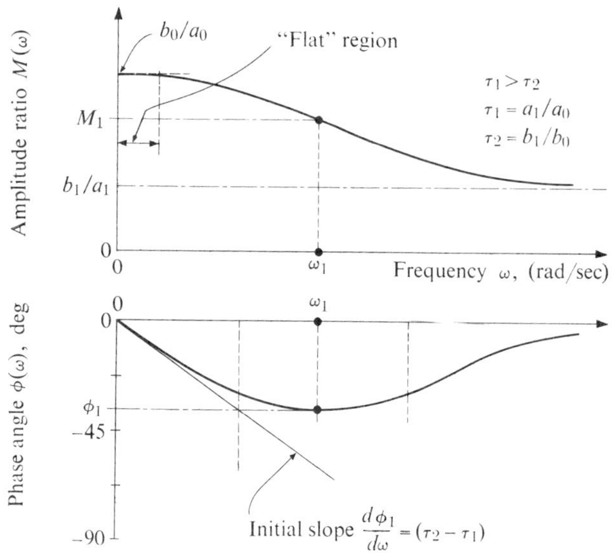

FIG. 15-3. Qualitative amplitude and phase characteristics for a general first-order system

$$T(j\omega) = (b_1 j\omega + b_0)/(a_1 j\omega + a_0).$$

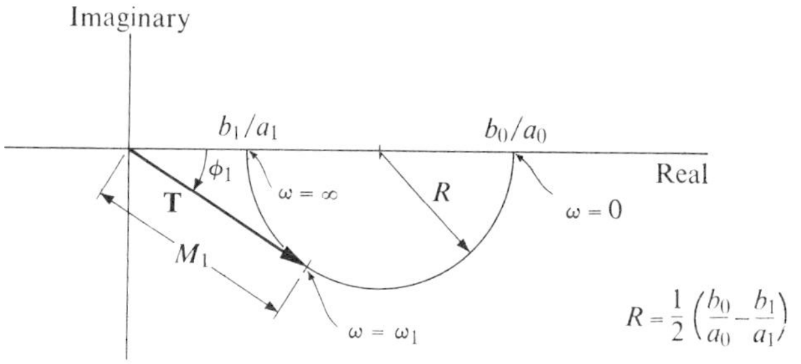

FIG. 15-4. Polar plot of $\mathbf{T}(j\omega) = (b_1 j\omega + b_0)/(a_1 j\omega + a_0)$ when $b_0 a_0 > b_1 a_1$.

Another way to portray the frequency response is by means of a single curve in the complex $\mathbf{T}(j\omega)$ plane. Because polar coordinates of magnitude and phase angle are frequently used instead of real and imaginary cartesian coordinates (see Chapter 9), this type of plot is often called a *polar plot*. The polar plot corresponding to the curves shown in Fig. 15-3 is given in Fig. 15-4. Actually the transfer function $\mathbf{T}(j\omega)$ for a system described by Eq. (15-25) can be considered to be made up of the two basic transfer functions $\mathbf{U}_0$ and $1/\mathbf{U}_p$ (Fig. 15-5).

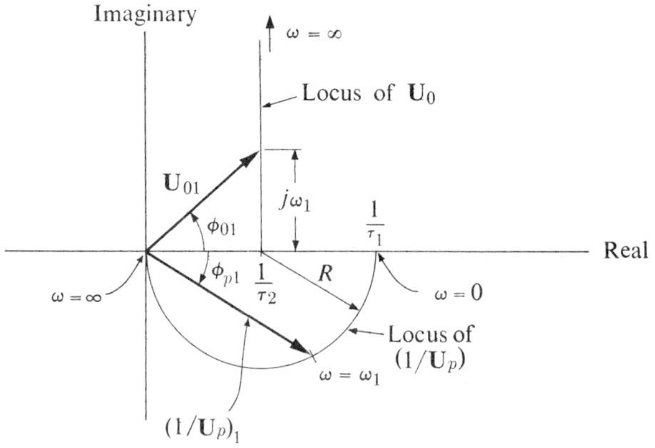

FIG. 15–5. Polar plots of $\mathbf{U}_0 = j\omega + 1/\tau_2$ and $1/\mathbf{U}_p = 1/(j\omega + 1/\tau_1)$.

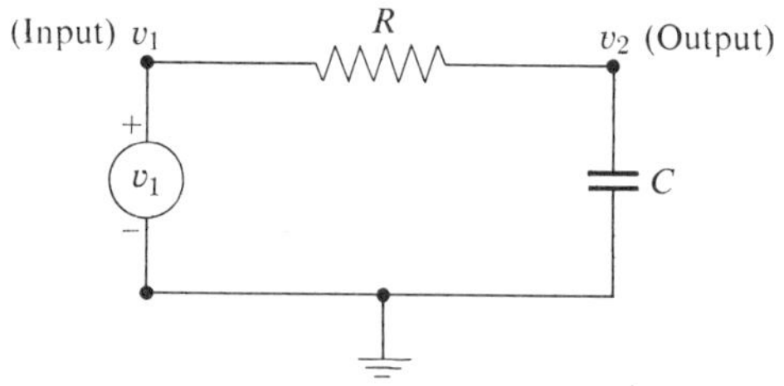

FIG. 15–6. Simple *RC* circuit.

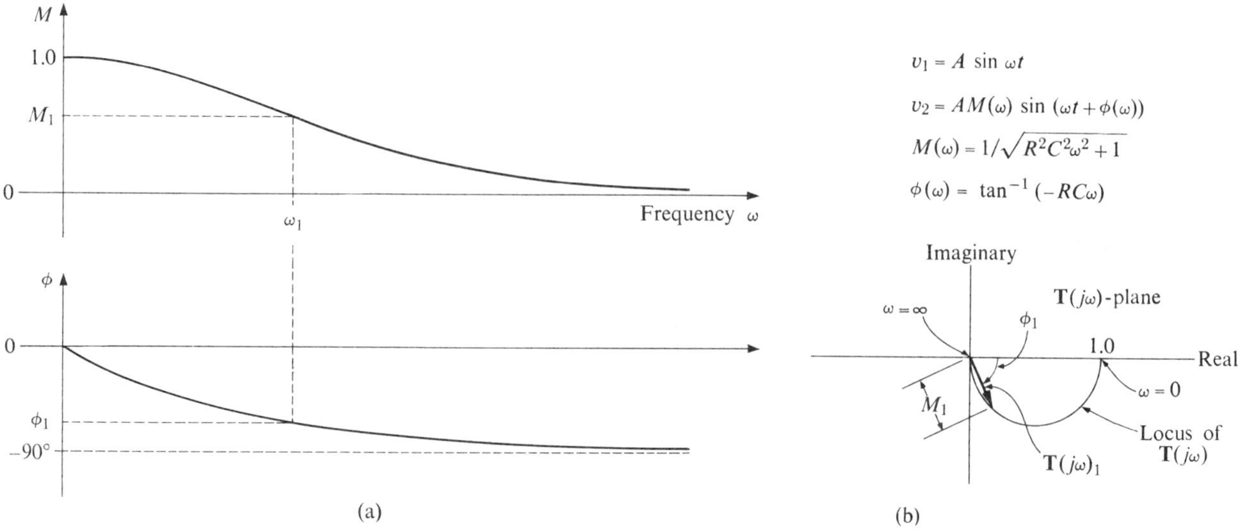

FIG. 15–7. Frequency response characteristics of *RC* circuit shown in Fig. 15–6.

Example. Illustrate the frequency-response characteristics of the *RC* circuit shown in Fig. 15–6. The input is

$$v_1 = A \sin \omega t.$$

The differential equation relating v_2 to v_1 is first found:

$$v_1 = v_2 + iR = v_2 + CR\, dv_2/dt,$$

or

$$(CRs + 1)\mathbf{V}_2 = \mathbf{V}_1, \tag{15–39}$$

so that

$$\mathbf{T}(j\omega) = \frac{\mathbf{V}_2}{\mathbf{V}_1} = \frac{1}{RCj\omega + 1} = \frac{1}{\tau \mathbf{U}_p},$$

where $\tau = RC$. This is similar to the $(1/\mathbf{U}_p)$ function shown in Fig. 15–5, and the two methods of representing the frequency-response characteristics of this specific system are shown in Fig. 15–7. The output is, of course, then given by

$$v_2 = AM \sin (\omega t + \phi). \tag{15–40}$$

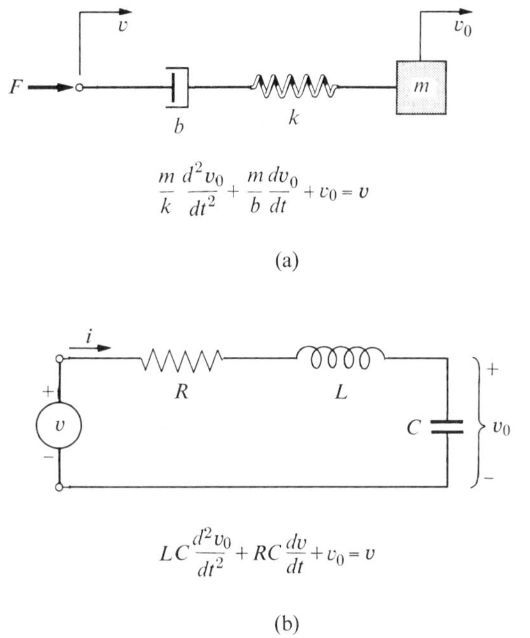

FIG. 15-8. Typical second-order systems.

(a)
$$\frac{m}{k}\frac{d^2 v_0}{dt^2} + \frac{m}{b}\frac{dv_0}{dt} + v_0 = v$$

(b)
$$LC\frac{d^2 v_0}{dt^2} + RC\frac{dv}{dt} + v_0 = v$$

15-2.2 Response of Typical Second-Order Systems

Two typical dynamic systems which are described by second-order transfer functions are shown in Fig. 15-8. The differential equations for both of these systems are of the form

$$\left[a_2 \frac{d^2}{dt^2} + a_1 \frac{d}{dt} + a_0 \right] x_r = b_0 x_f, \quad (15\text{-}41)$$

where $x_f = v$ and $x_r = v_0$ for both systems. This equation does not represent the most general second-order system since the right-hand side contains no derivative terms. However, a study of this particular equation will reveal several important characteristics exhibited by second-order dynamic systems when excited by steady sinusoids.

The transfer function for sinusoidal excitation is obtained by replacing d^n/dt^n by $(j\omega)^n$, x_f by $\mathbf{X}_f$, and x_r by $\mathbf{X}_r$ in Eq. (15-41):

$$\mathbf{T}(j\omega) = \frac{\mathbf{X}_r}{\mathbf{X}_f} = \frac{b_0}{(a_0 - a_2\omega^2) + ja_1\omega}. \quad (15\text{-}42)$$

This function may be written in the following non-dimensional form:

$$\mathbf{T}(j\omega) = K\mathbf{G}(j\omega) = \frac{K}{[1 - (\omega^2/\omega_n^2)] + 2\zeta(\omega/\omega_n)j},$$

(15-43)

where

$K = b_0/a_0 = |\mathbf{T}(j\omega)|_{\omega=0}$ = zero frequency magnitude,

$\omega_n = \sqrt{a_0/a_2}$ = undamped natural frequency,

$\zeta = \dfrac{a_1}{2\sqrt{a_0 a_2}}$ = damping ratio.

The function $\mathbf{G}(j\omega)$ has an absolute value or magnitude of unity at zero frequency, and it is the ratio of the output phasor at a given frequency ω to the output phasor at zero frequency. The significance of the natural frequency ω_n and damping ratio ζ has been discussed in connection with transient response, in Chapter 12.

Following the terminology of Eq. (15-43), we see that the magnitude M and phase ϕ of the transfer function are

$$M = \frac{b_0}{a_0}|\mathbf{G}(j\omega)| = \frac{b_0/a_0}{\sqrt{[1 - (\omega^2/\omega_n^2)]^2 + [2\zeta(\omega/\omega_n)]^2}},$$

(15-44)

$$\phi(j\omega) = -\tan^{-1}\left[\frac{2\zeta(\omega/\omega_n)}{1 - (\omega^2/\omega_n^2)}\right] = \angle \mathbf{G}(j\omega). \quad (15\text{-}45)$$

Consider first the case when the damping ratio ζ is zero. The magnitude of $\mathbf{G}(j\omega)$ is then

$$|\mathbf{G}(j\omega)| = \left(M\frac{a_0}{b_0}\right)_{\zeta=0} = \frac{1}{|1 - \omega^2/\omega_n^2|}. \quad (15\text{-}46)$$

At low frequencies, $\omega \ll \omega_n$, the magnitude of $\mathbf{G}$ is approximately unity. Under these conditions, the amplitude of the output sinusoid is proportional to the amplitude of the input sinusoid, that is, the response curve is "flat." When $\omega = \omega_n$, $|\mathbf{G}(j\omega)|$ becomes infinite. This is a *resonance* condition which always occurs when s becomes equal to a pole of the system function. In the sinusoidal case this can occur only when the poles of the system function or transfer function are purely imaginary, since s is purely imaginary for sinusoidal excitations. For very large exciting frequencies, $\omega \gg \omega_n$, the amplitude drops off as $(\omega_n/\omega)^2$. Figure (15–9) shows a sketch of the amplitude behavior just described.

The phase relationship for zero damping ratio is given by Eq. (15–45). Suppose that ζ is very small. Then at $\omega \ll \omega_n$, ϕ approaches zero. For large values of ω/ω_n, ϕ approaches $-\pi$. In the limit as ζ tends to zero, ϕ is zero for $\omega < \omega_n$ and $-\pi$ for $\omega > \omega_n$.

When the damping ratio is greater than zero, the magnitude of the transfer function will no longer become infinite since the denominator of Eq. (15–44) cannot vanish under any circumstances. For small ζ, the magnitude curve has the shape shown in Fig. (15–9). The maximum or peak value of M can be found by setting the derivative of M with respect to frequency ratio equal to zero. From Eq. (15–44),

$$\frac{dM}{d(\omega/\omega_n)} = \frac{(b_0/a_0)[-4(1 - \omega^2/\omega_n^2)\omega/\omega_n + 8\zeta^2\omega/\omega_n]}{-2[(1 - \omega^2/\omega_n^2)^2 + (2\zeta\omega/\omega_n)^2]^{3/2}} = 0.$$

Therefore the value of ω/ω_n at the peak M, which is called the *resonant frequency ratio*, is

$$\left(\frac{\omega}{\omega_n}\right)_p = \sqrt{1 - 2\zeta^2}. \qquad (15\text{–}47)$$

(Recall from Chapter 12 that the damped natural frequency of a second-order system is $\omega_d = \omega_n\sqrt{1 - \zeta^2}$.) A resonant peak exists in the frequency-response curve only for $\zeta < 1/\sqrt{2}$. When $\zeta > 1/\sqrt{2}$, the magnitude function drops off continuously with increasing frequency ratio. By substituting Eq. (15–47) into Eq. (15–44), we find the peak magnitude M_p:

$$M_p = \frac{b_0/a_0}{2\zeta\sqrt{1 - \zeta^2}}, \quad \text{when } \zeta < 1/\sqrt{2},$$
$$M_p = 1, \quad \text{when } \zeta > 1/\sqrt{2}. \qquad (15\text{–}48)$$

According to Eq. (15–45) the phase angle for finite ζ varies continuously from zero at zero frequency to

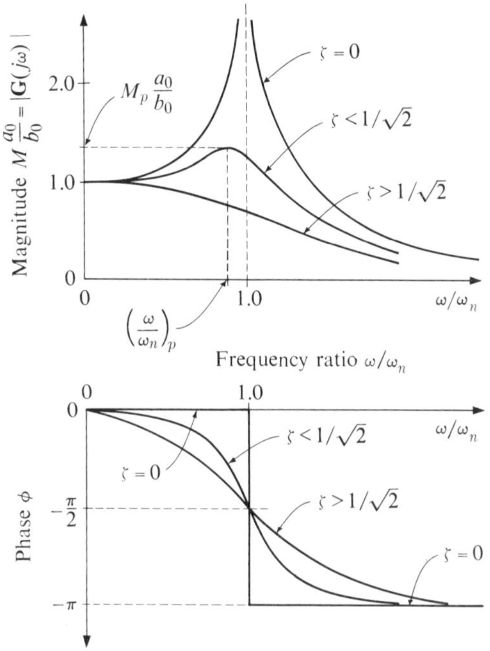

FIG. 15–9. Amplitude and phase plots for a second-order system.

$-180°$ at infinite frequency. In all cases, regardless of ζ, the phase angle is $-90°$ when $\omega = \omega_n$; that is, when ω equals the *undamped* natural frequency of the system. Representative curves of magnitude and phase for $\zeta > 0$ are sketched in Fig. 15–9.

Many electronic communication systems contain highly underdamped circuits; for example, most oscillators and many filter networks are designed to be as free from damping as possible. Certain types of mechanical vibration absorbers require spring-mass systems with very small damping to perform properly. A figure of merit or "quality" for a system which should possess low damping is the ratio of the system output amplitude at resonance to the amplitude at zero frequency. This quantity is called the Q of the system. From Eq. (15–48),

$$Q = M_p/(b_0/a_0) = 1/2\zeta\sqrt{1 - \zeta^2}, \qquad (15\text{–}49)$$

or, for small ζ,

$$Q \cong 1/2\zeta. \qquad (15\text{–}50)$$

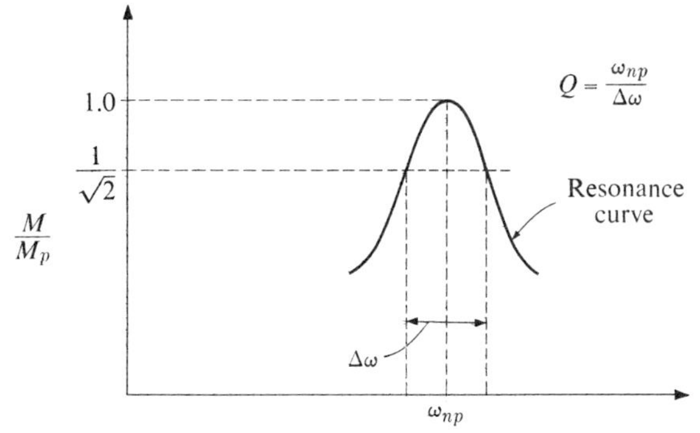

FIG. 15-10. Resonance curve illustrating the definition of Q.

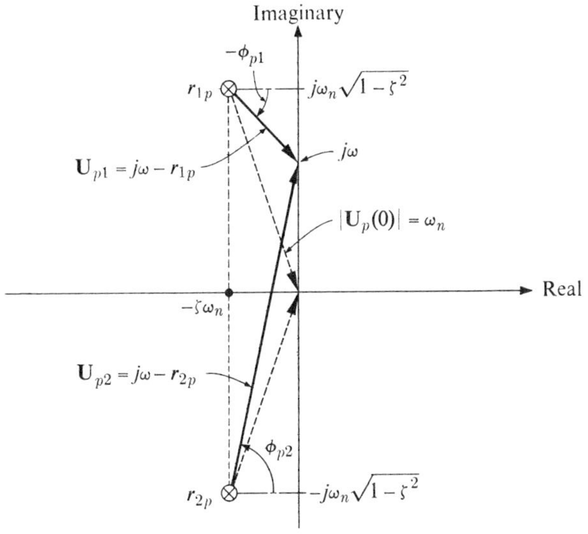

FIG. 15-11. Representation of underdamped second-order system in the complex plane.

A "high-Q" system is therefore one with small damping. In power and control systems, by contrast with communications systems, high Q is usually undesirable since it would encourage the buildup of large vibrations at the resonant frequency or frequencies.

In some systems the zero-frequency amplitude may be zero. This is true, for instance, of the systems in Fig. 15-8 if F or i are taken as inputs. In such a case an alternative way to define Q is in terms of the sharpness of the resonant peak. It can be shown (see Problem 15-18) that for small damping, Q^{-1} is the ratio between the width of the amplitude curve between the two points at $M = (1/\sqrt{2})M_p$ and the resonant frequency. This is illustrated in Fig. 15-10. The points where $M = (1/\sqrt{2})M_p$ are called the *half-power points* because power will normally be a function of amplitude squared (M^2).* The power at these points is half the power at resonance ($M = M_p$):

$$Q = \omega_{np}/\Delta\omega = 1/2\zeta. \tag{15-51}$$

The width of the response curve $\Delta\omega$ decreases as the Q increases. The Q is a measure of the so-called *selectivity* of the system. It is the property which permits a radio receiver to be tuned to one particular frequency and to reject all other frequencies. The tuning dial on the receiver varies a resonant LC circuit in the radio so that ω_{np} coincides with the broadcasting frequency of a particular radio station.

The frequency-response characteristics of a second-order system described by the transfer function (15-42) may be visualized by plotting the transfer-function factors in the complex plane. Following the method outlined in Section 13-4.4, we must first find the factors of the system function. From Eq. 15-41, we see that the system function in terms of s is

$$\mathbf{T}(s) = \frac{b_0/a_0}{(s^2/\omega_n^2 + 2\zeta s/\omega_n + 1)}, \tag{15-52}$$

where ω_n and ζ are as previously defined. The poles of this function are

$$r_{1p}, r_{2p} = -\zeta\omega_n \pm j\omega_n\sqrt{1 - \zeta^2}. \tag{15-53}$$

If $\zeta < 1$, these poles are complex conjugates with negative real parts. There are no zeros of (15-52). The poles are plotted in the complex plane, as shown in Fig. 15-11.

The transfer function may now be represented by vectors drawn from the poles r_{1p} and r_{2p} to the value of $j\omega$ corresponding to the particular forcing frequency (Fig. 15-11). Then

$$\mathbf{T}(j\omega) = \frac{(b_0/a_0)(\omega_n^2)}{(j\omega - r_{1p})(j\omega - r_{2p})}. \tag{15-54}$$

---

* The power which a resistive load would absorb depends on the square of the current in the resistance.

The reader should substitute the values of r_{1p} and r_{2p} from (15–53) into (15–54) and verify that (15–54) becomes identical to the previously obtained transfer function (15–43). Let

$$\mathbf{U}_{p1} = j\omega - r_{1p}, \qquad \mathbf{U}_{p2} = j\omega - r_{2p}. \tag{15–55}$$

The magnitude of $\mathbf{T}(j\omega)$ is then

$$M(\omega) = \frac{b_0/a_0(\omega_n^2)}{|\mathbf{U}_{p1}|\,|\mathbf{U}_{p2}|}. \tag{15–56}$$

Let ϕ_{p1} and ϕ_{p2} be the angles which the phasors $\mathbf{U}_{p1}$ and $\mathbf{U}_{p2}$ make with the real axis in Fig. 15–11. Then the phase ϕ of $\mathbf{T}(j\omega)$ is

$$\phi = -\phi_{p1} - \phi_{p2}. \tag{15–57}$$

From the geometry of Fig. 15–11, we see that when $\omega = 0$, $|\mathbf{U}_{p1}| = |\mathbf{U}_{p2}| = \omega_n$, and $\phi_{p2} = -\phi_{p1}$. Thus, from Eqs. (15–56) and (15–57),

$$M(\omega = 0) = b_0/a_0 = K \quad \text{and} \quad \phi(\omega = 0) = 0.$$

As ω increases, $|\mathbf{U}_{p1}|$ decreases and $|\mathbf{U}_{p2}|$ increases. If ζ is small so that r_{p1} lies close to the imaginary axis, $|\mathbf{U}_{p1}|$ will become very small when ω is close to ω_n, and therefore M will become large. This is the resonant condition. At this point, $\phi_{p1} \approx 0$ and $\phi_{p2} \approx 90°$ for small ζ, and $\phi \approx -90°$. As ω increases beyond ω_n, $|\mathbf{U}_{p1}|$ increases and approaches infinity, as does $|\mathbf{U}_{p2}|$ also. Hence M approaches zero. For large ω, $\mathbf{U}_{p1}$ and $\mathbf{U}_{p2}$ approach angles of $+90°$, and therefore ϕ approaches $-180°$. These qualitative conclusions are verified by the response plots in Fig. 15–9.

Generalized frequency-response charts for simple first- and second-order systems. Figures 15–12(a) and (b), pp. 370 and 371, are response charts giving amplitude ratio $|\mathbf{G}(j\omega)|$ and phase $\phi(\omega)$ for nondimensionalized first- and second-order transfer functions. The first-order function is described by the equation

$$\mathbf{G}(j\omega) = \frac{1}{j\tau\omega + 1}, \tag{15–58}$$

where τ is the time constant and ω is the exciting frequency. The second-order function is given by

$$\mathbf{G}(j\omega) = \frac{1}{-\omega^2/\omega_n^2 + 2j\zeta\omega/\omega_n + 1}, \tag{15–59}$$

where ζ is the damping ratio and ω_n is the undamped natural frequency.

15–2.3 Frequency Response of Higher-Order Systems

One of the advantages of steady sinusoidal analysis is the relative simplicity with which it can be applied to higher-order systems. It is not necessary to find roots of high-order polynomials or to evaluate batteries of undetermined coefficients. In each case it is necessary only to evaluate the real and imaginary terms in the numerator and denominator of the transfer function (Eq. 15–17) and then to find M and ϕ as functions of the excitation frequency.

An excellent example for applying frequency-response analysis is the vibration absorber discussed in Chapter 8 and illustrated in Fig. 15–13. The overall differential equation which was formulated for this system is of fourth order. In this discussion we shall assume that the damping effects of b_1 and b_2 are so small that they may be neglected except for the fact that infinitesimal amounts would be needed to damp out the transients occurring after a sinusoidal input was first applied.

When the damping terms b_1 and b_2 are neglected in Eq. (8–34), the sinusoidal transfer function relating the phasor for the velocity of the vibrating machine mass to the phasor for the velocity of the vibrating floor is

$$\mathbf{T}(j\omega) = \frac{\mathbf{V}_1}{\mathbf{V}} = \frac{2k_1m_2(j\omega)^2 + 2k_1k_2}{m_1m_2(j\omega)^4 + [k_2m_1 + m_2(k_2 + 2k_1)](j\omega)^2 + 2k_1k_2}. \tag{15–60}$$

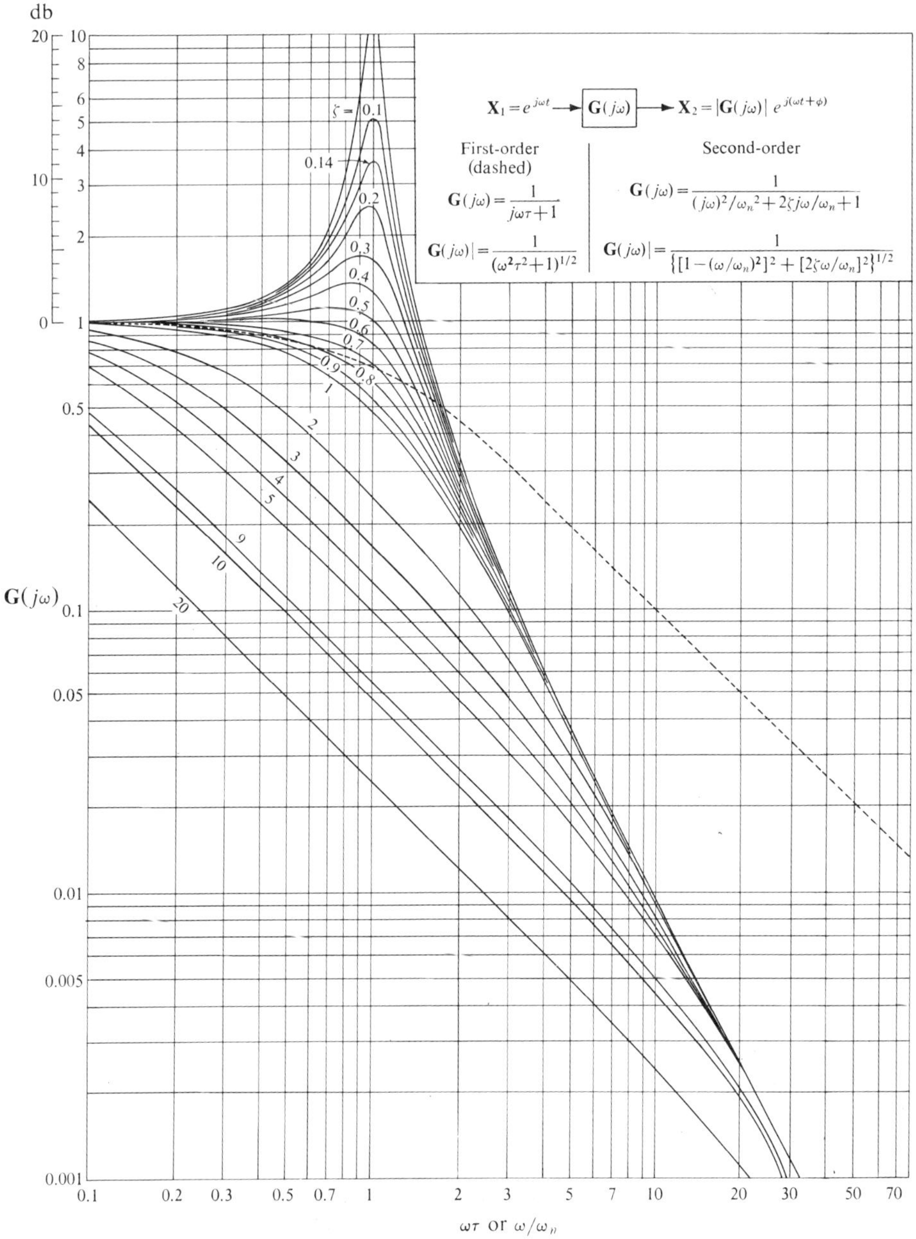

FIG. 15–12(a). Logarithmic display of amplitude of simple first- and second-order systems.

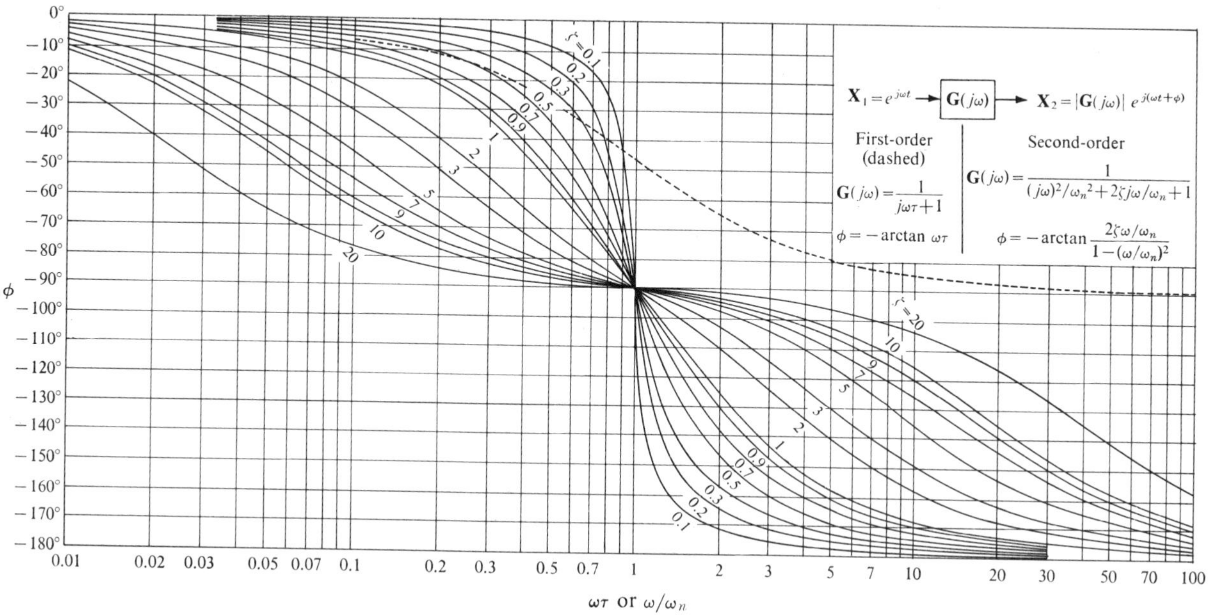

FIG. 15–12(b). Logarithmic display of phase angle of simple first- and second-order systems.

From the point of view of the number of system parameters to be dealt with, this is still a rather formidable looking equation. However, here we wish to examine only one aspect of the system, namely the means by which a mass-spring absorber (that is, the m_2, k_2 assembly) can act to reduce the vibration of a given machine (that is, the m_1, k_1 system) when the excitation frequency is at or near the natural frequency of the machine.

The sinusoidal transfer function relating $\mathbf{V}_1$ to $\mathbf{V}$ for the machine without the vibration absorber is readily obtained from Eq. (15–60) by letting m_2 and k_2 be equal to zero. When this is done, we obtain for the machine with no absorber,

$$\mathbf{T}_{na}(j\omega) = \frac{2k_1k_2}{k_2m_1(j\omega)^2 + 2k_1k_2} = \frac{2k_1}{m_1(j\omega)^2 + 2k_1}. \tag{15–61}$$

Note that it was necessary to remove m_2 from the transfer function first and then to allow the remaining terms with k_2 in them to have k_2 eliminated by cancellation from both numerator and denominator.

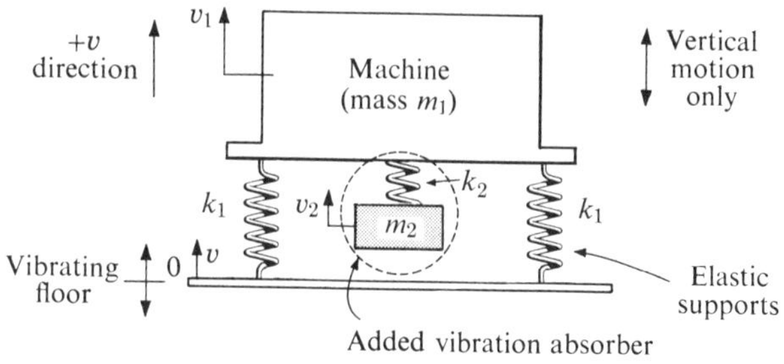

FIG. 15–13. Schematic diagram of vibration absorber identifying all elements.

Simply setting m_2 and k_2 both equal to zero in Eq. (15–60) would have led to an expression of the form 0/0, which, of course, would have been meaningless.

The transfer function given in Eq. (15–61) is the same as that discussed in the preceding section with $\zeta = 0$. The natural frequency of the machine alone, then, is

$$\omega_{n1} = \sqrt{2k_1/m_1}. \tag{15–62}$$

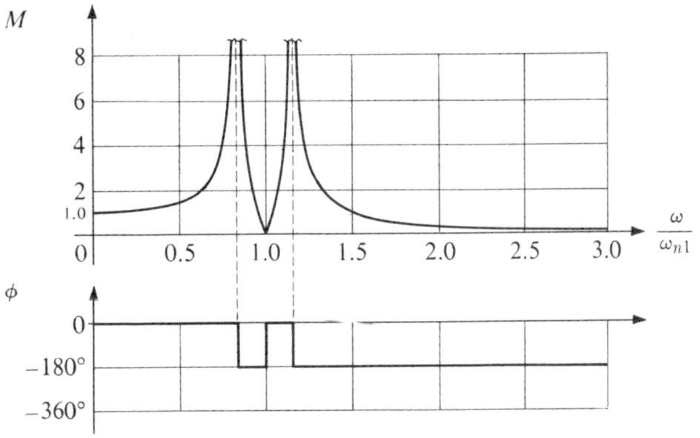

FIG. 15–14. Amplitude and phase vs. frequency of vibrating machine having tuned absorber, $m_1 = 10\,m_2$.

This is the frequency of excitation which would cause the machine to vibrate with excessive amplitude (theoretically infinite amplitude as shown in Fig. 15–9). Obviously, it would be a logical move to try to change the natural frequency of the machine if it happened that an unavoidable floor vibration were to exist at this frequency. However, it might not be feasible to change k_1 or m_1. This is a situation in which the vibration absorber can be very effective, because if we look at the transfer function (15–60) with the absorber acting, we see that it should be possible to make the numerator zero! (This will be fine if the denominator does not become zero at the same time.) Setting the numerator equal to zero and using an excitation frequency equal to the natural frequency of the machine alone, we get

$$\begin{aligned} 2k_1m_2(-2k_1/m_1) + 2k_1k_2 &= 0, \\ k_2/m_2 &= 2k_1/m_1. \end{aligned} \tag{15–63}$$

This says that the natural frequency $\omega_{n2} = \sqrt{k_2/m_2}$ of the absorber mass-spring assembly (when acting by itself) should be precisely equal to the natural frequency ω_{n1} of the machine (when acting by itself). (A quick examination of the denominator reveals that having $k_2m_1 = 2k_1m_2$ would not cause the denominator to become zero when $\omega^2 = 2k_1/m_1$.)

Now we may determine M and ϕ_1 for the machine with an absorber tuned to its own (individually acting) natural frequency by substituting $2k_1m_2/m_1$ for k_2 in Eq. (15–60):

$$\mathbf{T}(j\omega) = \frac{2k_1m_1(j\omega)^2 + 4k_1^2}{m_1^2(j\omega)^4 + (4k_1m_1 + 2k_1m_2)(j\omega)^2 + 4k_1^2}, \tag{15–64}$$

$$M(\omega) = \frac{4k_1^2 - 2k_1m_1\omega^2}{m_1^2\omega^4 + 4k_1^2 - (4k_1m_1 + 2k_1m_2)\omega^2}, \tag{15–65}$$

$$\begin{aligned} \phi(\omega) &= 0, && 0 < \omega < \omega_{r1}, \\ &= -\pi, && \omega_{r1} < \omega < \omega_{r2}, \\ &= 0, && \omega_{r2} < \omega < \omega_{r3}, \\ &= -\pi, && \omega_{r3} < \omega < \infty. \end{aligned} \tag{15–66}$$

where ω_{r1} and ω_{r3} are values of ω for which the denominator of (15–65) is zero, and ω_{r2} is the value of ω for which the numerator is zero.

The amplitude ratio M and phase angle ϕ are sketched as functions of ω in Fig. 15–14 for a system with $m_1 = 10m_2$. Now it is seen that adding the tuned vibration absorber has indeed changed the natural frequency of the system. In fact it has resulted in a system with two natural frequencies, one below and the other above the original natural frequency of the machine. And when the floor vibrates at the old natural frequency of the machine, the machine will not vibrate at all!

It is interesting to go one step further and explore what is happening to the absorber. The transfer function for the absorber, relating $\mathbf{V}_2$ to $\mathbf{V}_1$ is, by analogy with Eq. (15–61),

$$\mathbf{T}_a(j\omega) = \frac{k_2}{m_2(j\omega)^2 + k_2}. \tag{15–67}$$

Employing $k_2 = 2k_1m_2/m_1$ (the parameters to tune the absorber), we obtain

$$\mathbf{T}_a(j\omega) = \frac{2k_1}{m_1(j\omega)^2 + 2k_1},$$

which is precisely the same as Eq. (15–61).

Now we may combine Eqs. (15–64) and (15–67) to obtain the transfer function relating $\mathbf{V}_2$ to $\mathbf{V}$,

$$\frac{\mathbf{V}_1}{\mathbf{V}} \frac{\mathbf{V}_2}{\mathbf{V}_1} = \frac{\mathbf{V}_2}{\mathbf{V}} = \mathbf{T}_2(j\omega),$$

or

$$\mathbf{T}_2(j\omega) = \mathbf{T}(j\omega)\mathbf{T}_a(j\omega),$$

$$\mathbf{T}_2(j\omega) = \frac{4k_1^2}{m_1^2(j\omega)^4 + (4k_1m_1 + 2k_1m_2)(j\omega)^2 + 4k_1^2}. \tag{15-68}$$

The expressions for $M_2(\omega)$ and $\phi_2(\omega)$ are

$$M_2(\omega) = \frac{4k_1^2}{m_1^2\omega^4 + 4k_1^2 - (4k_1m_1 + 2k_1m_2)\omega^2}, \tag{15-69}$$

$$\begin{aligned}\phi_2(\omega) &= 0, & 0 < \omega < \omega_{r1}, \\ &= -\pi, & \omega_{r1} < \omega < \omega_{r3}, \\ &= -2\pi, & \omega_{r3} < \omega < \infty.\end{aligned} \tag{15-70}$$

and M_2 and ϕ_2 are sketched as functions of frequency in Fig. 15–15. Note that $\mathbf{V}_2$ is 180° out of phase with $\mathbf{V}$ when $\omega = \sqrt{2k_1/m_1}$. This is so because m_2 is vibrating in such a way that it precisely opposes the sinusoidally varying forces in the springs k_1 at this frequency.

15–3. LOGARITHMIC TECHNIQUES—BODE PLOTS

When the transfer function of a linear system consists of a number of factors which are combined by multiplication or addition,

$$\begin{aligned}\mathbf{T}(j\omega) &= k\mathbf{U}_a \times \mathbf{U}_b \div \mathbf{U}_c \\ &= [|k\mathbf{U}_a| \times |\mathbf{U}_b| \div |\mathbf{U}_c|]e^{j(\phi_a+\phi_b-\phi_c)},\end{aligned} \tag{15-71}$$

the logarithm of the magnitude $M(\omega)$ is given by

$$\log M = \log k + \log|\mathbf{U}_a| + \log|\mathbf{U}_b| - \log|\mathbf{U}_c|, \tag{15-72}$$

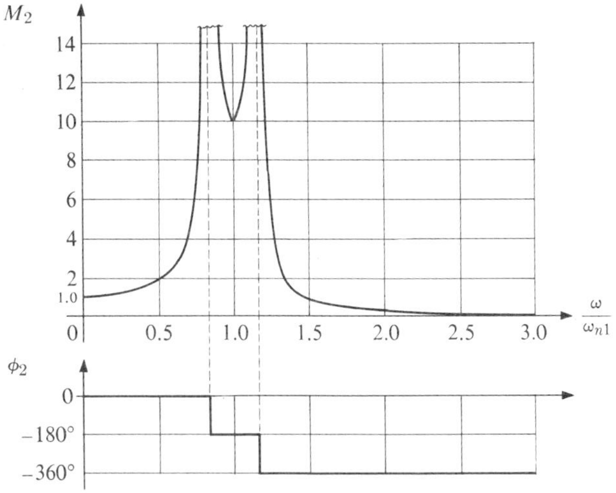

FIG. 15–15. Amplitude and phase vs. frequency of absorber mass.

and the phase angle (ϕ) is given by

$$\phi = \phi_a + \phi_b - \phi_c. \tag{15-73}$$

This means that when the magnitudes of the factors are expressed as logarithms, the magnitude of the combined factors can be found by simple addition.

Thus for a linear system which has a transfer function of the form

$$\mathbf{T}(j\omega) = \frac{b_m(j\omega - r_{01})\cdots(j\omega - r_{0m})}{b_n(j\omega - r_{p1})\cdots(j\omega - r_{pn})}, \tag{15-74}$$

the magnitude $M(\omega)$ is given by

$$M(\omega) = \frac{b_m|j\omega - r_{01}|\cdots|j\omega - r_{0m}|}{b_n|j\omega - r_{p1}|\cdots|j\omega - r_{pn}|} \tag{15-75}$$

and its logarithm is given by

$$\log M = \log(b_m/a_n) \\ + \sum_{i=1}^{m}\log|j\omega - r_{0i}| - \sum_{i=1}^{n}\log|j\omega - r_{pi}|, \tag{15-76}$$

and the logarithm of the magnitude of $\mathbf{T}(j\omega)$ is

obtained simply by adding the logs of the individual phasors associated with the zeroes and poles of $\mathbf{T}(j\omega)$. Similarly, the phase angle $\phi(\omega)$ is given by

$$\phi(\omega) = \sum_m \phi_{0m} - \sum_n \phi_{pn}. \tag{15–77}$$

It is common practice, especially in the fields of electronics, automatic control, and acoustics, to express the logarithms of magnitude functions in units known as *decibels* (db). One decibel is equal to 20 times the logarithm to the base 10 of the magnitude. Thus Eq. (15–76) can be expressed in decibels by multiplying each term by 20. The terminology log-magnitude (Lm) will be used to indicate the logarithm expressed in decibels:

$$\mathrm{Lm}(M) = 20\log_{10}(M) = \mathrm{db}. \tag{15–78}$$

If in Eq. (15–78), M is *doubled*, its increase is called an *octave*.* That is, $2M$ is one octave above M. From Eq. (15–78)

$$\begin{aligned}\mathrm{Lm}(2M) &= \mathrm{Lm}(M) + 20\log_{10}(2),\\ &= \mathrm{Lm}(M) + 6.02.\end{aligned} \tag{15–79}$$

Therefore the Lm of a quantity increases by 6.02 db per octave increase in the quantity.

Similarly, it may be shown that an increase of a *decade* or a factor of 10 corresponds to an increase of 20 db.

15–3.1 Log-Magnitude and Phase for a First-Order System

To illustrate the foregoing method of representing the steady amplitude and phase response, let us consider the general first-order system discussed in Section 15–2.1. The magnitude M for this system given earlier by Eq. (15–32) may be expressed also by

$$M(\omega) = (b_0/a_0)\sqrt{(\tau_2^2\omega^2 + 1)/(\tau_1^2\omega^2 + 1)}. \tag{15–80}$$

* Terminology used in work with musical scales.

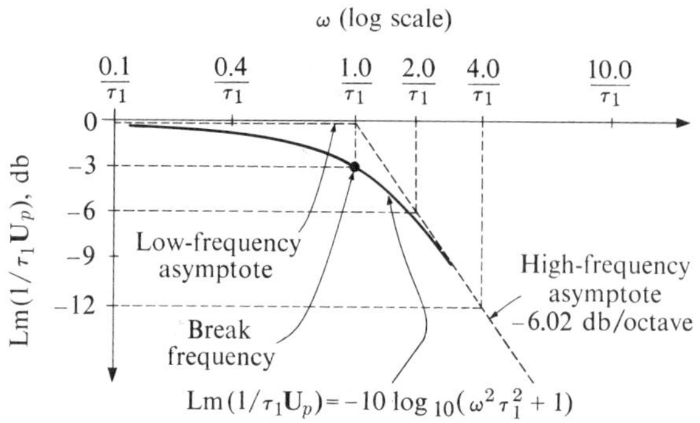

FIG. 15–16. Log-magnitude vs. frequency for a denominator factor of a transfer function.

The log-magnitude of this function is

$$\mathrm{Lm}(M) = 20\log_{10}(b_0/a_0) + 10\log_{10}(\omega^2\tau_2^2 + 1) - 10\log_{10}(\omega^2\tau_1^2 + 1). \tag{15–81}$$

Let

$$\mathrm{Lm}(1/\tau_1\mathbf{U}_p) = -10\log_{10}(\omega^2\tau_1^2 + 1) \tag{15–82}$$

and

$$\mathrm{Lm}(\tau_2\mathbf{U}_0) = 10\log_{10}(\omega_2\tau_2^2 + 1). \tag{15–83}$$

Consider first Eq. (15–82). For *low* frequencies, $\omega \ll 1/\tau_1$, $\mathrm{Lm}(1/\tau_1\mathbf{U}_p)$ is approximately

$$\mathrm{Lm}(1/\tau_1\mathbf{U}_p)|_{\omega\to 0} = 0. \tag{15–84}$$

This is the low-frequency asymptote of $\mathrm{Lm}(1/\tau_1\mathbf{U}_p)$. At *high* frequencies, $\omega \gg 1/\tau_1$, Eq. (15–82) gives

$$\mathrm{Lm}(1/\tau_1\mathbf{U}_p)|_{\omega\to\infty} = -20\log(\omega\tau_1). \tag{15–85}$$

On logarithmic coordinates, $\mathrm{Lm}(1/\tau_1\mathbf{U}_p)$ asymptotically approaches this straight line at high frequencies. This line passes through zero db when $\omega = 1/\tau_1$, and has a slope of -6.02 db per octave of ω (as shown in Eq. 15–79). The two asymptotes, Eqs. (15–84) and (15–85), are shown in Fig. 15–16. At any value of ω, the actual value of $\mathrm{Lm}(1/\tau_1\mathbf{U}_p)$ can be calculated from Eq. (15–82) or it may be read directly from the first-order system plot shown dashed in Fig. 15–12. Note

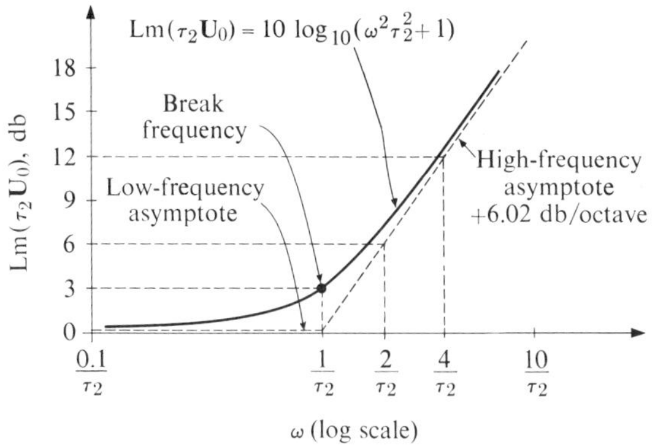

FIG. 15–17. Log-magnitude vs. frequency for a numerator factor of a transfer function.

in particular that when $\omega = 1/\tau_1$, the magnitude of $(1/\tau_1\mathbf{U}_p) = \sqrt{2}/2$ and $\mathrm{Lm}(1/\tau_1\mathbf{U}_p) = -3.01$ db. A rather good approximation to the curve of $\mathrm{Lm}(1/\tau_1\mathbf{U}_p)$ may be obtained by fairing in a curve using the -3.01 db point and the low- and high-frequency asymptotes as guides.

Figure 15–16 shows the complete $\mathrm{Lm}(1/\tau_1\mathbf{U}_p)$ curve vs. $\log(\omega)$. The frequency at which $\omega = 1/\tau_1$ is called the "break-point" for the first-order factor; that is, the point at which the asymptote "breaks" from a slope of zero to a slope of -6.02 db per octave. (On a straight log-log plot, the slope is -1.)

The log-magnitude for the numerator factor $\mathbf{U}_0$ of Eq. (15–29) will now be determined. From Eq. (15–83), we see that $\mathrm{Lm}(\tau_2\mathbf{U}_0)$ differs from $\mathrm{Lm}(1/\tau_1\mathbf{U}_p)$ only in that the break frequency of the $\mathrm{Lm}(\tau_2\mathbf{U}_0)$ asymptote occurs at $1/\tau_2$ and the slope of the high-frequency asymptote is $+6.02$ db/decade. Figure 15–17 is a sketch of $\mathrm{Lm}(\tau_2\mathbf{U}_0)$ showing the asymptotes and the break frequency.

The curve of $\mathrm{Lm}(\tau_2\mathbf{U}_0)$ is *identical* to the curve for $\mathrm{Lm}(1/\tau_1\mathbf{U}_p)$ except for sign and for a shift along the ω-axis, provided that equal scales are used for ω and $\mathrm{Lm}(\tau_2\mathbf{U}_0)$. In fact, celluloid templates having the shape of $\mathrm{Lm}(\tau_2\mathbf{U}_0)$ to standard scales can be purchased from drafting supply houses for use in constructing plots of first-order factors.

The general first-order system of Eq. (15–29) can now be assembled from its component parts simply

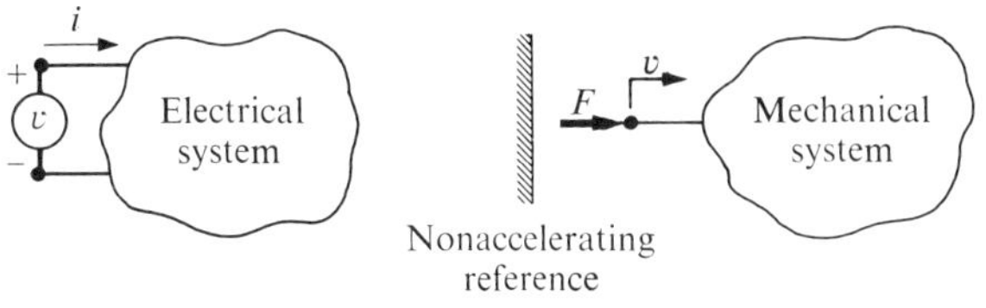

FIG. 15–18. Power inputs to electrical and mechanical systems.

by adding the log magnitudes and the phase angles of the components. This means that it is relatively simple to assemble the parts of a higher-order system *if its poles and zeros are known* or if a number of components have been determined by some other means, such as by experimental measurements.

Many engineers prefer not to work with decibel scales, so that frequently the log plots will be found on ordinary log scales (e.g., Figs. 15–12a and 15–12b).

These graphs of $\mathrm{Lm}(M)$ are often called *Bode plots* after H. W. Bode, for his work on the dynamic performance of electronic feedback amplifiers.

15–4. POWER RELATIONSHIPS IN THE SINUSOIDAL STEADY STATE

15–4.1 Definitions

At a terminal, or energy port, of an electrical system such as shown in Fig. 15–18, the power $\mathcal{P}$ which flows into the system at any given instant of time is given by the product of the current i and the voltage v. Similarly the power input to a mechanical system is the product of the force F and the velocity of the point at which the force F acts. In general, the power is the product of the *through*-variable f and the *across*-variable v:

$$\mathcal{P} = fv. \tag{15–86}$$

15–4.2 Average, Reactive, and Total Power

To determine the power being delivered to a system under steady sinusoidal conditions, we need to either (a) measure the amplitude and phase of the voltage and current at that point in the system or (b) know either the voltage or the current and the driving-point

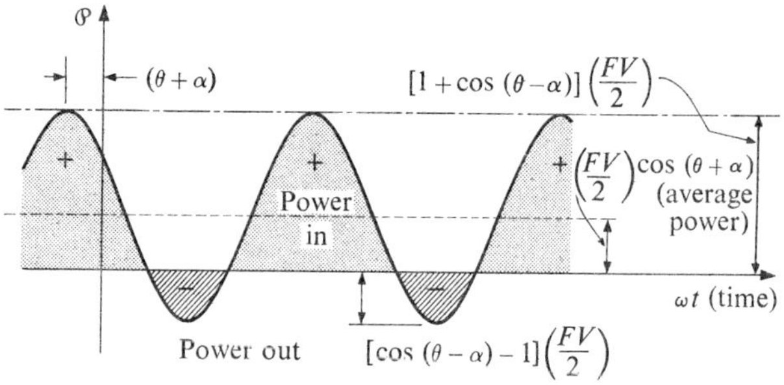

FIG. 15–19. Power transfer by sinusoidal through- and across-variables.

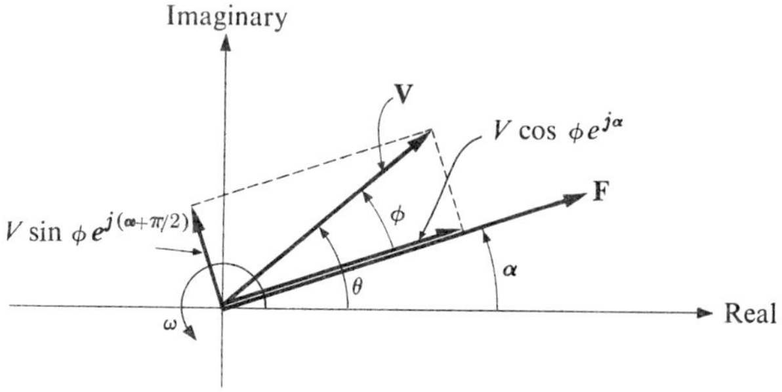

FIG. 15–20. Phasor representation of **V** and **F**.

impedance at that point. Here we shall describe f and v by means of exponential time functions,

$$f = \text{Re}[\mathbf{F}e^{j\omega t}], \tag{15–87}$$

$$v = \text{Re}[\mathbf{V}e^{j\omega t}], \tag{15–88}$$

where

$$\mathbf{F} = Fe^{j\alpha}, \tag{15–89}$$

$$\mathbf{V} = Ve^{j\theta}. \tag{15–90}$$

We may now employ the results of Section 9–7 to obtain for the power,

$$\mathcal{P} = (FV/2)[\cos(\theta - \alpha) + \cos(2\omega t + \alpha + \theta)]. \tag{15–91}$$

The variation of power flow, $\mathcal{P}$, with time is shown in Fig. 15–19. It is seen that the instantaneous power can be negative part of the time, even when the average power $\mathcal{P}_a$ given by Eq. (15–92) is positive:

$$\mathcal{P}_a = (FV/2)\cos(\theta - \alpha). \tag{15–92}$$

It is also interesting to note that the power varies sinusoidally about its average value with a frequency which is twice the frequency of the f and v sinusoids.

The phasors **F** and **V** used in describing f and v are shown in Fig. 15–20. The angle ϕ between the phasors **F** and **V** is the angle by which the sinusoidally varying f lags behind the sinusoidally varying v. The phasors, which could rotate at angular velocity ω to generate f and v, are shown in the position which they would then occupy when $t = 0$. Since $\phi = \theta - \alpha$, we may express the average power $\mathcal{P}_a$ by

$$\mathcal{P}_a = (FV/2)\cos\phi. \tag{15–93}$$

The phasor **V** may be resolved into two components, $V\cos\phi e^{j\alpha}$ *in phase with* **F** and $V\sin\phi e^{j(\alpha+\pi/2)}$ *leading* **F** by $\pi/2$. Thus the average power $\mathcal{P}_a$ is equal to one-half the product of **F** and that component of **V** which is in phase with **F**.

The component of **V** normal to or *at quadrature* with **F** produces no average power. However, this component $V\sin\phi e^{j(\alpha+\pi/2)} = j(FV/2)\sin\phi e^{j\alpha}$ is of some interest since it is related to the magnitude of **F** required to achieve a given average power when **V** has a specified value. The product $(FV/2)\sin\phi$ is called the *reactive* or imaginary power, $\mathcal{Q}_a$:

$$\mathcal{Q}_a = \frac{FV\sin\phi}{2}. \tag{15–94}$$

The angle ϕ between **F** and **V** is called the *power-factor angle*, and $\cos\phi$ is called the *power factor*. When $\phi = 0$, $\cos\phi = 1$ and the power factor is unity. If $\phi = 90°$, the power factor is zero and no average power exists. If, in Fig. 15–21, where the case of $\alpha = 0$ is shown, ϕ is positive, the power factor is said to be *lagging* and if negative, *leading*.*

---

* The terms *lagging* and *leading* refer to whether the through-variable phasor lags or leads the across-variable phasor.

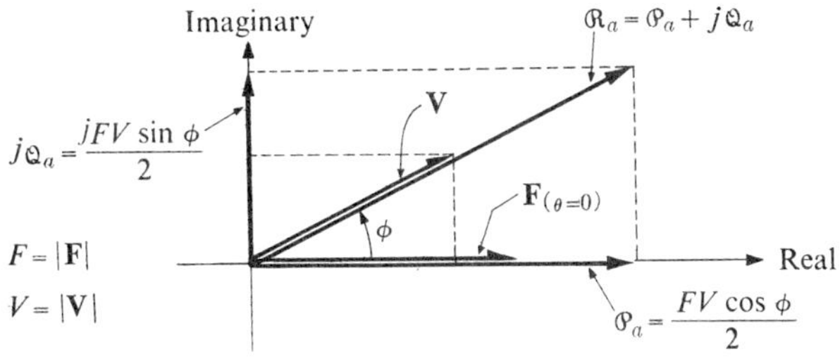

FIG. 15-21. Total, average, and reactive power diagram.

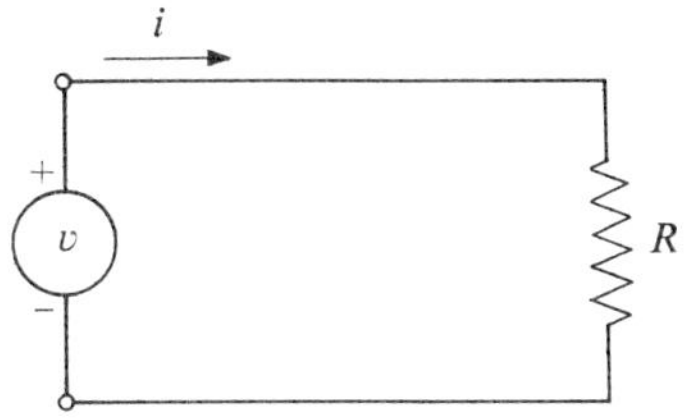

FIG. 15-22. Resistor excited by a voltage.

The *total power* $\mathcal{R}_a$ is defined as the vector sum of $j\mathcal{Q}_a$ and $\mathcal{P}_a$:

$$\begin{aligned}\mathcal{R}_a &= \mathcal{P}_a + j\mathcal{Q}_a \\ &= \frac{FV}{2}(\cos \phi + j \sin \phi) \\ &= \frac{FV}{2} e^{j\phi} = \frac{\mathbf{VF}}{2} = \frac{Fe^{j\theta}Ve^{-j\alpha}}{2}.\end{aligned} \tag{15-95}$$

Figure 15-21 shows Eq. (15-95) graphically. It should be borne in mind that total power $\mathcal{R}_a$ and reactive power $\mathcal{Q}_a$ are fictitious quantities which are useful primarily in visualizing the FV relationship but are not directly related to actual instantaneous or average power flow.

15-4.3 Root-Mean-Square or Effective Values

Let us consider the resistor shown in Fig. 15-22 which is being excited by a voltage v, producing a current flow i into the resistor.

If the voltage is *constant*, $v = v_0$, the current will also be constant, $i = i_0 = v_0/R$, and the power $\mathcal{P}$ will be

$$\mathcal{P} = vi = v_0 i_0 = v_0^2/R = \text{const.} \tag{15-96}$$

If, on the other hand, v is a sinusoidal voltage,

$$v = \text{Im}(\mathbf{V}e^{j\omega t}), \tag{15-97}$$

the current will be sinusoidal also:

$$i = \frac{v}{R} = \text{Im}\left(\frac{\mathbf{V}}{R} e^{j\omega t}\right). \tag{15-98}$$

From Eq. (15-93), the average power is

$$\mathcal{P}_a = \frac{1}{2}\frac{V^2}{R}, \tag{15-99}$$

since $\mathbf{I} = \mathbf{V}/R$ and $\phi = 0$.

Equation (15-99) may be made analogous to the relation for power delivered by *constant* currents and voltages by defining the *root-mean-square* (rms) or effective value of a sinusoid. Define the rms value of the sinusoid $v(t)$ as V_e, where

$$V_e = \sqrt{\frac{1}{T}\int_0^T v^2\,dt} \quad \text{or} \quad \sqrt{\frac{1}{2\pi}\int_0^{2\pi} v^2\,d(\omega t)}. \tag{15-100}$$

If $v = V \sin \omega t$, Eq. (15-100) gives

$$V_e = V/\sqrt{2}. \tag{15-101}$$

Thus the rms value of a sinusoid is $1/\sqrt{2}$ or 0.707 times the maximum value of the sine wave. If Eq. (15-101) is used to replace V by the rms value V_e, Eq. (15-99) becomes

$$\mathcal{P}_a = V_e^2/R, \tag{15-102}$$

which is of the same form as the equation for power delivered by a *constant* voltage of magnitude V_e.

The equations developed for power in Section 15-4.2 may be rewritten in terms of effective or rms values. In Eqs. (15-93), (15-94), and (15-95), let

$$F_e = F/\sqrt{2}, \qquad V_e = V/\sqrt{2}. \tag{15-103}$$

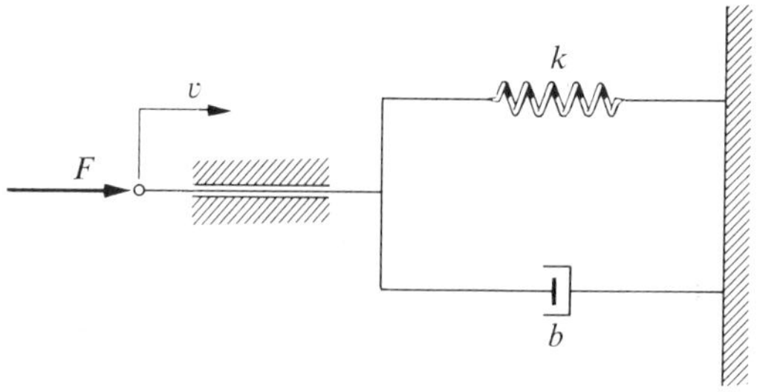

FIG. 15–23. Parallel spring dashpot system.

Then

$$\mathcal{P}_a = F_e V_e \cos\phi,$$
$$\mathcal{Q}_a = F_e V_e \sin\phi, \qquad (15\text{–}104)$$
$$\mathcal{R}_a = F_e V_e e^{j\Phi} = \mathcal{P}_a + j\mathcal{Q}_a.$$

15–4.4 Examples of Power Calculation

Equations (15–104) relate power to the through- and across-variables F_e and V_e. However, **F** and **V** and therefore F_e and V_e, are related by the driving point impedance $\mathbf{Z}_d$. Since ϕ has been defined as the phase angle by which leads **F**, the *admittance* may be written as

$$\mathbf{Y}_d = \frac{\mathbf{F}}{\mathbf{V}} = Y_d e^{-j\phi}, \qquad (15\text{–}105)$$

and the *impedance* as

$$\mathbf{Z}_d = \frac{\mathbf{V}}{\mathbf{F}} = Z_d e^{j\phi}. \qquad (15\text{–}106)$$

Caution should be exercised in assuming from Eqs. (15–106) and (15–105) that $\mathbf{Z}_d = 1/\mathbf{Y}_d$ at all times, since the system transfer function may be different when the input is v than when the input is f.

Equations (15–105) and (15–106) also determine the relationships between the effective values V_e and F_e:

$$F_e/V_e = Y_d, \qquad V_e/F_e = Z_d. \qquad (15\text{–}107)$$

Equations (15–104) may be written in terms of F_e and Z_d or V_e and Y_d as follows:

$$\mathcal{P}_a = F_e^2 Z_d \cos\phi, \qquad (15\text{–}108)$$

$$\mathcal{P}_a = V_e^2 Y_d \cos\phi. \qquad (15\text{–}109)$$

Example. The simple parallel spring-dashpot system shown in Fig. 15–23 is an approximate model of an automobile suspension system (spring and shock absorber) in which a sinusoidal force is applied due to periodic irregularity in the road surface. The differential equation relating F and v for this system is

$$b\,dv/dt + kv = dF/dt. \qquad (15\text{–}110)$$

Assuming a sinusoidal excitation,

$$F = \text{Im}[\mathbf{F}e^{j\omega t}], \qquad (15\text{–}111)$$

we see that the steady-state response of v will be

$$v = \text{Im}[\mathbf{V}e^{j\omega t}] \qquad (15\text{–}112)$$

and **V** is related to **F** by the sinusoidal system function or the driving-point impedance $\mathbf{Z}_d$,

$$\frac{\mathbf{V}}{\mathbf{F}} = \mathbf{Z}_d(j\omega) = \frac{j\omega}{bj\omega + k} = \frac{1}{b - j(k/\omega)}. \qquad (15\text{–}113)$$

The following numerical data are provided:

$k = 500$ lb/in., $\qquad \omega = 10$ rad/sec.

$b = 100$ lb sec/in., $\qquad V_e = 20$ in/sec (rms).

Find

a) the average power delivered to the system,

b) the reactive power delivered to the system,

c) the total power delivered to the system.

The *driving point impedance* $\mathbf{Z}_d$ may be expressed in polar form:

$$\mathbf{Z}_d = Z_d e^{j\phi} = \sqrt{\omega^2/(k^2 + b^2\omega^2)}\, e^{j\phi}. \qquad (15\text{–}114)$$

Thus

$$\phi = -\tan^{-1}\frac{-k}{b\omega}$$
$$= \tan^{-1}\left[\frac{500 \text{ lb/in.}}{(100 \text{ lb sec/in.}) \times (10 \text{ rad/sec})}\right]$$
$$= \tan^{-1} 0.5,$$
$$\phi = 26.6°,$$

and

$$Z_d = \sqrt{100/(25 \times 10^4 + 100 \times 10^4)}$$
$$= 1/111.6 = 8.95 \times 10^{-3} \text{ in/lb sec}.$$

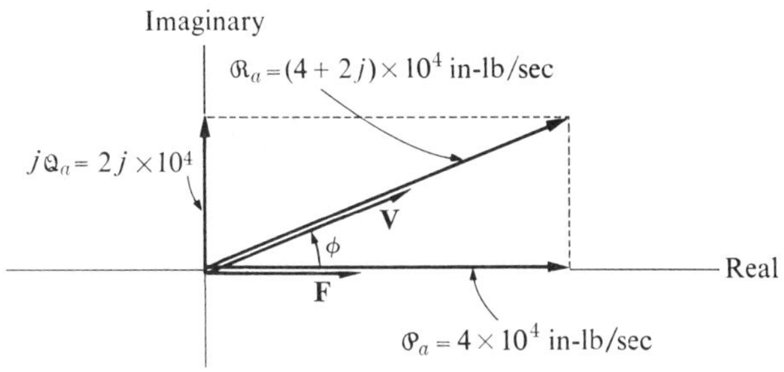

FIG. 15–24. Force, velocity, and power phasors for the system of Fig. 15–23.

The *admittance* for this system is

$$\mathbf{Y}_d = (1/Z_d)e^{-j\phi}, \qquad Y_d = 111.6 \text{ lb-sec/in.} \tag{15–115}$$

From Eq. (15–109) the average power $\mathcal{P}_a$ is given as

$$\begin{aligned}\mathcal{P}_a &= V_e^2(111.6)\cos 26.6^\circ \\ &= (20 \text{ in/sec})^2 \times (111.6 \text{ lb-sec/in}) \times (0.895) \\ &= 4 \times 10^4 \text{ in-lb/sec.}\end{aligned}$$

From Eq. (15–107), the rms value of the force is

$$F_e = V_eY_d = (20) \times (111.6) = 2.23 \times 10^3 \text{(rms).}$$

The average power could also be computed from Eq. (15–104):

$$\begin{aligned}\mathcal{P}_a &= F_eV_e\cos\phi = (2.23 \times 10^3) \times (20) \times (0.895) \\ &= 4 \times 10^4 \text{ in-lb/sec.}\end{aligned}$$

The reactive power $\mathcal{Q}_a$ is

$$\begin{aligned}\mathcal{Q}_a &= F_eV_e\sin\phi = (2.23 \times 10^3) \times (20) \times (0.447) \\ &= 2 \times 10^4 \text{ in-lb/sec,}\end{aligned}$$

and the total power is

$$\mathcal{R}_a = (4 + 2j) \times 10^4 \text{ in-lb/sec.}$$

Figure 15–24 shows the phasors representing **F**, **V**, $\mathcal{P}_a$, $j\mathcal{Q}_a$, and $\mathcal{R}_a$.

In Fig. 15–24, the power factor $\cos\phi$ is lagging. That is, **F** lags behind **V**.

15–5. STEADY-STATE PHASOR DIAGRAMS

Phasor diagrams showing the relationships between variables in various parts of physical systems provide a graphical representation of the behavior of these variables, and they often are useful as tools for system analysis.

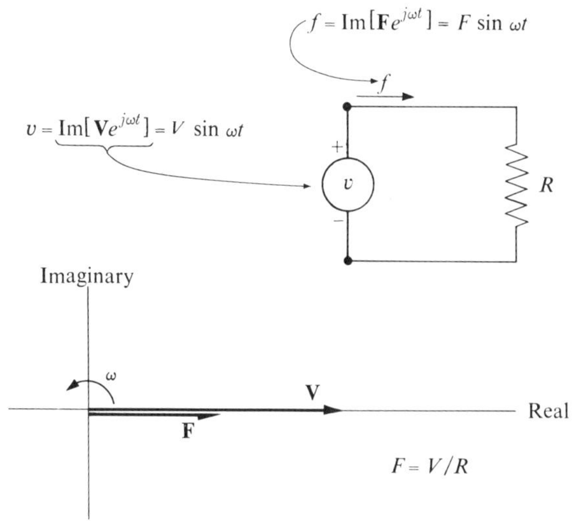

FIG. 15–25. Phasor diagram for ideal *D*-type element (resistance).

15–5.1 Phasor Relations for Ideal Elements

The simplest case of sinusoidal excitation occurs when a single ideal element is subjected to a sinusoidal input, producing a sinusoidal output. For example, let us consider an ideal resistance driven by a sinusoidal across-variable v:

$$v = \text{Im}(\mathbf{V}e^{j\omega t}), \tag{15–116}$$

where **V** is assumed real. That is, **V** will be taken as a reference phasor which lies along the real axis, as shown in Fig. 15–25.

The through-variable f then will be of the form

$$f = \text{Im}(\mathbf{F}e^{j\omega t}). \tag{15–117}$$

The sinusoidal transfer function for the ideal resistance is

$$\mathbf{F}/\mathbf{V} = 1/R. \tag{15–118}$$

By comparing this equation with (15–106), we see that the impedance $\mathbf{Z}_d$ at the input terminals of the resistance is

$$\mathbf{Z}_d = R.$$

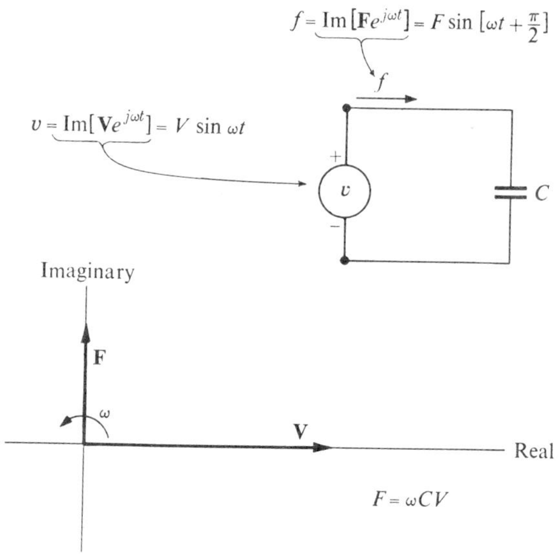

FIG. 15-26. Phasor diagram for ideal *A*-type element (capacitance).

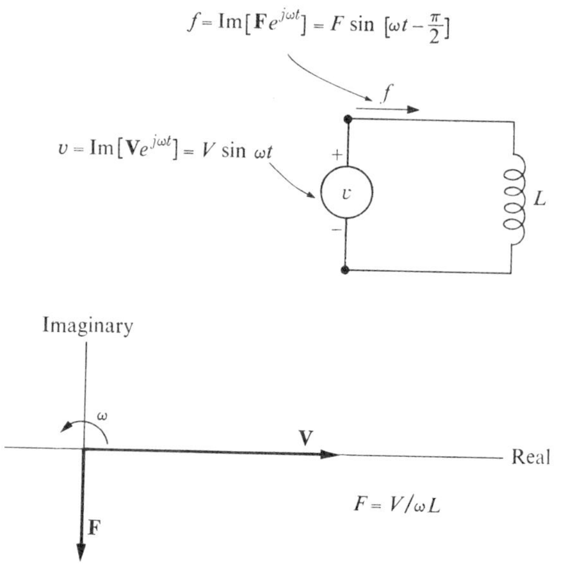

FIG. 15-27. Phasor diagram for an ideal *T*-type element (inductance).

The through-variable is therefore

$$\mathbf{F} = \mathbf{V}/\mathbf{Z}_d = \mathbf{V}/R. \tag{15-119}$$

This result is shown in Fig. 15-25. Thus the through-variable of the resistance is in phase with the across-variable.

An ideal capacitance excited by an across-variable, Eq. (15-116), will develop a through-variable given by Eq. (15-117), and it will have a sinusoidal transfer function:

$$\mathbf{F}/\mathbf{V} = Cj\omega. \tag{15-120}$$

Thus the impedance of an ideal capacitance is

$$\mathbf{Z}_d = 1/j\omega C = (1/\omega C)e^{-j\pi/2}. \tag{15-121}$$

The through-variable phasor is then

$$\mathbf{F} = \mathbf{V}\omega C e^{j\pi/2}. \tag{15-122}$$

Figure 15-26 shows the phase relation between the through-variable and the across-variable. Here the through-variable *leads* the across-variable by 90°.

The *capacitive reactance* X_c is defined as

$$X_c = 1/\omega C, \tag{15-123}$$

so that the impedance of an ideal capacitance is

$$\mathbf{Z}_d = 1/j\omega C = -j/\omega C = -jX_c. \tag{15-124}$$

An ideal inductance excited by an across-variable will have a sinusoidal transfer function,

$$\mathbf{F}/\mathbf{V} = 1/Lj\omega. \tag{15-125}$$

Thus the *inductive reactance* X_l, of an ideal inductance is $X_l = \omega L$ and its impedance is

$$\mathbf{Z}_d = j\omega L = jX_l = \omega L e^{j\pi/2}. \tag{15-126}$$

The through-variable phasor is then

$$\mathbf{F} = \mathbf{V}\left(\frac{1}{\omega L}\right) e^{-j\pi/2}. \tag{15-127}$$

Figure 15-27 shows the phase relation between the through-variable and the across-variable. Here the through-variable *lags* the across-variable by 90°.

15–5.2 Phasor Diagrams for Complete Systems

To illustrate the use of phasor diagrams for complete systems, let us consider the series electric circuit shown in Fig. 15–28.

The input to the circuit is a voltage v,

$$v = V \sin \omega t = \sqrt{2}\, V_e \sin \omega t, \tag{15–128}$$

where $V_e = 100$ v, rms. The parameter values are $R = 100$ ohms, $L = 0.2$ h, $C = 10\,\mu$f, and $\omega = 1000$ rad/sec. The current i and the voltage v_2 are to be determined.

The continuity equation requires that the current through each element be the same, and the compatibility or loop equation requires that v be the sum of the voltage drops across each element. Let

$$v = \text{Im}(\mathbf{V}e^{j\omega t}). \tag{15–129}$$

Then the voltage loop equation will be

$$v = v_2 + (v_3 - v_2) + (v - v_3),$$

$$\text{Im}[\mathbf{V}e^{j\omega t}] = \text{Im}[\mathbf{V}_2 e^{j\omega t}] + \text{Im}[(\mathbf{V}_3 - \mathbf{V}_2)e^{j\omega t}] + \text{Im}[(\mathbf{V} - \mathbf{V}_3)e^{j\omega t}]. \tag{15–130}$$

The imaginary operators may be discarded in (15–130) since the system is linear. Let $\mathbf{I}$ equal the complex amplitude of the current:

$$i = \text{Im}(\mathbf{I}e^{j\omega t}). \tag{15–131}$$

Then equation (15–130) can be rewritten using the impedance relationships developed in Section 15–5.1 for the ideal elements:

$$\mathbf{V} = \underbrace{\mathbf{I}/j\omega C}_{\text{(capacitance)}} + \underbrace{\mathbf{I}j\omega L}_{\text{(inductance)}} + \underbrace{\mathbf{I}R}_{\text{(resistance)}}. \tag{15–132}$$

If $\mathbf{I}$ is taken as a reference phasor located at the real axis for $t = 0$, Eq. (15–132) is represented graphically as shown in Fig. 15–29. The real and imaginary axes are not explicitly drawn in Fig. 15–29. From the geometry of Fig. 15–29 it can be seen that

$$\mathbf{V} = (100 + j100)\mathbf{I} = 100\sqrt{2}\, e^{j\pi/4}\mathbf{I}, \tag{15–133}$$

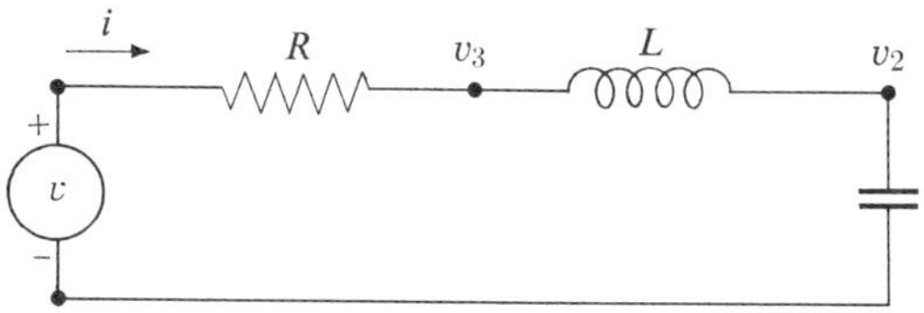

FIG. 15–28. Series electric circuit.

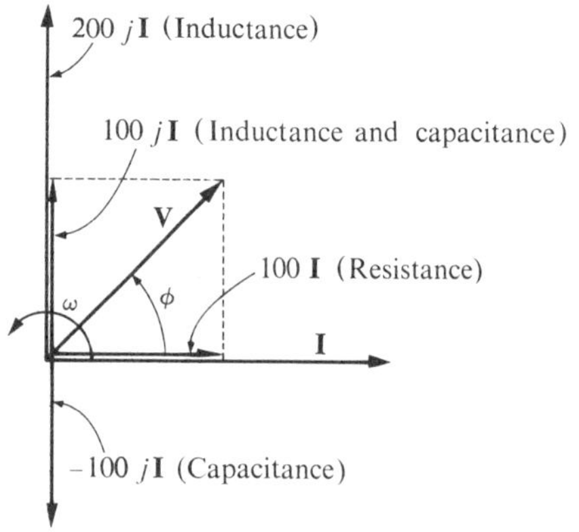

FIG. 15–29. Phasor diagram for circuit of Fig. 15–28.

where $\phi = \pi/4$ is the phase angle between $\mathbf{V}$ and $\mathbf{I}$. In this case the current $\mathbf{I}$ lags behind the voltage $\mathbf{V}$. Since the voltage $\mathbf{V}$ is known, $\mathbf{V} = \sqrt{2}\, V_e e^{j0}$, the current phasor is

$$\mathbf{I} = Ie^{-j\pi/4} = \frac{\mathbf{V}}{100\sqrt{2}} e^{-j\pi/4} = \frac{V_e}{100} e^{-j\pi/4} = 1.0e^{-j\pi/4} \tag{15–134}$$

or the rms current I_e is

$$I_e = I/\sqrt{2} = 0.707 \text{ amp, rms.}$$

The instantaneous current is given by Eq. (15–131):

$$i = \text{Im}(\mathbf{I}e^{j\omega t}) = \frac{V_e}{100} \sin(\omega t - \pi/4). \tag{15–135}$$

The voltage v_2 across the capacitor is represented by $-j/\omega C$ as shown in Fig. 15–29. From Eq. (15–134),

$$\mathbf{V}_2 = \frac{\mathbf{I}}{j\omega C} = \frac{\mathbf{I}}{\omega C} e^{-j\pi/2} = \frac{V_e}{100\omega C} e^{-j3\pi/4},$$

$$\mathbf{V}_2 = \frac{V_e}{1} e^{-j3\pi/4}. \tag{15–136}$$

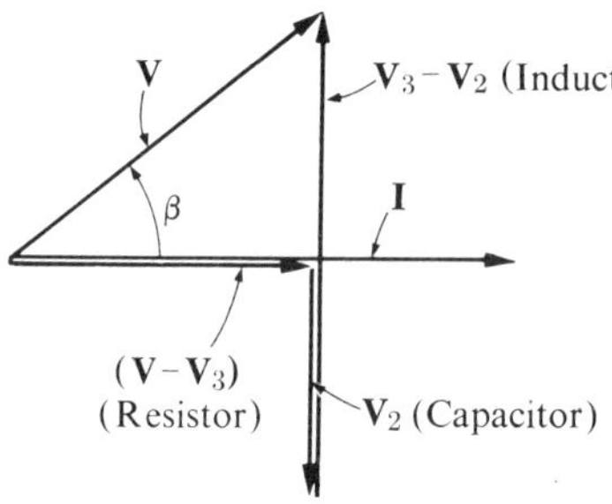

FIG. 15–30. Alternative phasor diagram for circuit of Fig. 15–28.

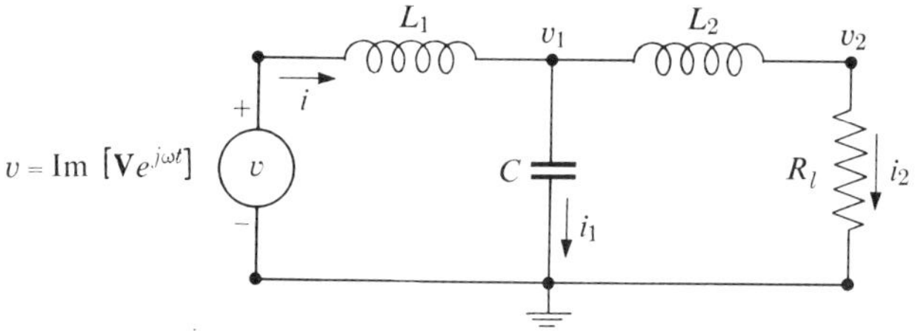

FIG. 15–31. Series-parallel electric circuit.

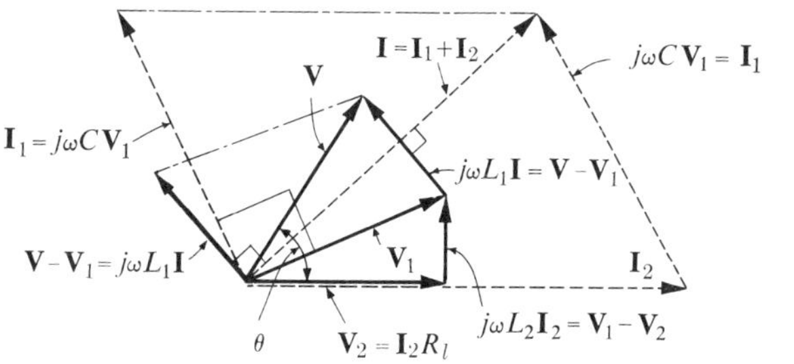

FIG. 15–32. Phasor diagram for the circuit of Fig. 15–31.

Therefore

$$v_2 = \text{Im}(\mathbf{V}_2e^{j\omega t}) = V_e \sin(\omega t - 3\pi/4). \tag{15–137}$$

The phasor diagram, Fig. 15–29, could have been constructed by adding the phasors in a vector fashion, as shown in Fig. 15–30.

The average power $\mathcal{P}_a$ transferred to the circuit at the input terminals can be computed by multiplying the rms voltage V_e by that part of the rms current which is in phase with $\mathbf{V}$. Thus

$$\begin{aligned}\mathcal{P}_a &= V_eI_e\cos\beta \\ &= (100)(0.707)(\cos\pi/4) = 50 \text{ watts.}\end{aligned} \tag{15–138}$$

Since power is dissipated only in the resistor, $\mathcal{P}_a$ should equal the resistor power:

$$(\mathcal{P}_a)_{\text{resistor}} = I_e^2R = (0.707)^2 100 = 50 \text{ watts.} \tag{15–139}$$

In conclusion of this discussion of phasor diagrams, a slightly more complex system will be discussed. Figure 15–31 shows a series-parallel electric circuit which is excited by a steady sinusoidal voltage source:

$$v = \text{Im}(\mathbf{V}e^{j\omega t}) = V\sin\omega t. \tag{15–140}$$

The voltage v_2 across the load resistor R_l and the average power absorbed by the circuit are to be found.

To construct a phasor diagram for this circuit without using trial and error methods, we must work backward from the output v_2 to the input. Accordingly, let us assume that $i_2 = \text{Im}(\mathbf{I}_2e^{j\omega t})$ is known and let $\mathbf{I}_2$ be a reference phasor, drawn horizontal in Fig. 15–32. The voltage v_2 is in phase with the current i_2 and its phasor $\mathbf{V}_2 = \mathbf{I}_2R_l$, as shown in Fig. 15–32. Since the current $\mathbf{I}_2$ also flows through the inductor L_2, the voltage difference across the inductor is

$$\mathbf{V}_1 - \mathbf{V}_2 = j\omega L_2\mathbf{I}_2. \tag{15–141}$$

This voltage leads $\mathbf{I}_2$ by 90°. When $\mathbf{V}_1 - \mathbf{V}_2$ is added to the voltage $\mathbf{V}_2$, the phasor representing $\mathbf{V}_1$ is obtained as shown in Fig. 15–32. The current $\mathbf{I}_1$ through C is determined by $\mathbf{V}_1$,

$$\mathbf{I}_1 = j\omega C\mathbf{V}_1, \tag{15–142}$$

and thus $\mathbf{I}_1$ leads $\mathbf{V}_1$ by 90°. The current $\mathbf{I}$ flowing through the inductor L_1 must be the sum of $\mathbf{I}_1$ and $\mathbf{I}_2$, by continuity. Therefore $\mathbf{I}$ is constructed in Fig. 15–32 by adding the phasors $\mathbf{I}_1$ and $\mathbf{I}_2$. The voltage $\mathbf{V} - \mathbf{V}_1$ across L_1 is

$$\mathbf{V} - \mathbf{V}_1 = j\omega L_1\mathbf{I}, \tag{15–143}$$

so that $\mathbf{V} - \mathbf{V}_1$ leads $\mathbf{I}$ by 90°. The phasor representing $\mathbf{V}$, which is actually the input function, is obtained in Fig. 15–32 by adding $(\mathbf{V} - \mathbf{V}_1)$ to $\mathbf{V}_1$. Therefore, $\mathbf{V}_2$ is a phasor which lags behind $\mathbf{V}$ by the angle θ. The magnitude of $\mathbf{V}_2$ relative to $\mathbf{V}$ can be obtained by carrying through the construction of Fig. 15–32 graphically to scale, or by working out the geometry analytically. It is suggested that the reader carry through the procedure outlined to find v_2 as a function of time for the parameters $C = 0.1\ \mu$f, $R_l = 100$ ohms, $L_1 = L_2 = 0.1$ h, $\omega = 100$ rad/sec, and $v = 10 \sin(\omega t)$ v.

REFERENCES

1. R. Oldenburger, "Frequency-Response Data Presentation Standards and Design Criteria," *Transactions ASME* **76,** 1155–1169 (November 1954).
2. C. S. Draper, W. McKay, and S. Lees, *Instrument Engineering*, Vol. 2. New York: McGraw-Hill Book Company, Inc., 1953 (pp. 286–317).

PROBLEMS

15–1. The network shown in Fig. 15–33 is used as a filter to prevent the high-frequency noise content of an input signal v_1 from getting through to the input of a cathode-ray oscilloscope. In other words, it is desired to observe the low frequencies without the noise on the scope.

a) For the following values of C and L, find the value of R for a network damping ratio of 0.8:

$$C = 0.1\ \mu\text{f}\ (10^{-7}\ \text{f}), \qquad L = 0.6\ \text{h}.$$

b) If the high-frequency noise portion of the input v_1 consists of a 20,000-cps sinusoid, what fraction (magnitude) of this noise portion of the input signal will appear at the input to the scope (v_2)?

c) How well does a 200-cps sinusoidal signal get through to the scope? That is, what is the magnitude of v_2 relative to v_1, and how much phase shift occurs between v_2 and v_1 at this frequency?

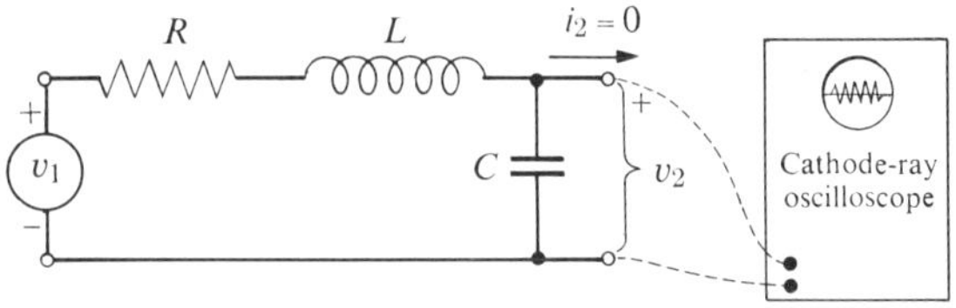

FIGURE 15–33

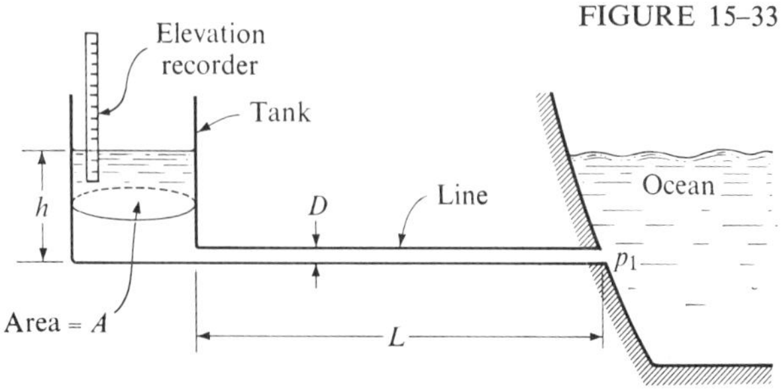

FIGURE 15–34

15–2. The system shown in Fig. 15–34 is used to record sea-level elevations and is called a *tide recorder*. Fluctuations in the level of the ocean surface are transmitted through the line to the tank where the water surface elevation is recorded by an electrical or mechanical recorder. It is usually desired that the tide recorder filter out high-frequency variations in the water surface, such as waves, and pass the low-frequency variations due to diurnal tides.

In a particular application, it is required that the variations in the tank level h due to waves be less than 1% of the variation due to tides and that the tide amplitudes be correct to 1% accuracy. It is known that the variation in p_1 due to waves has an amplitude of 2.5 psi (5 psi peak-to-peak) and a period of 12 sec; the variation due to tides has an amplitude of 3.75 psi (7.5 psi peak-to-peak) and a period of 12 hr.

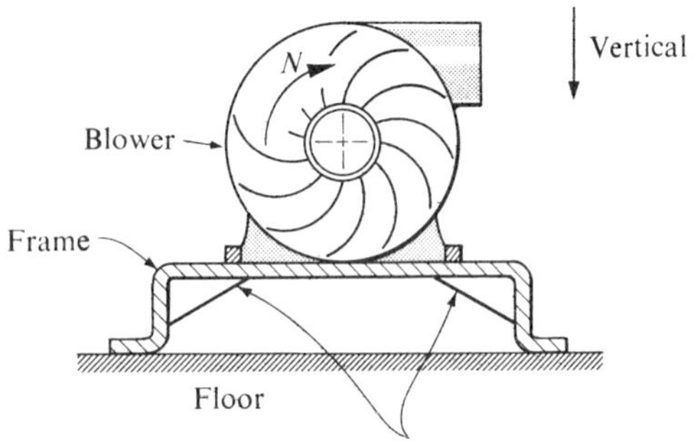

FIGURE 15–35

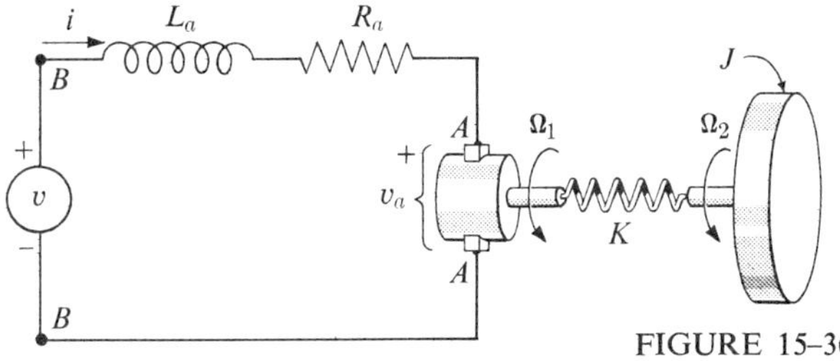

FIGURE 15–36

a) Determine whether the system will function properly if its geometry and the fluid properties are:

$L = 240$ in., $A = 144$ in^2, $D = 2$ in,
μ = viscosity = 1.74×10^{-6} lb-sec/in^2,
ρ = density = 10^{-4} lb-sec^2/in^4.

b) State (qualitatively) the effect on the variations of h due to the waves when each of the following modifications is made individually:

1) an increase in L,

2) an increase in D,

3) an increase in A.

15–3. A large blower for a building ventilating system is mounted on a steel frame which is rigidly connected to the floor of the building (Fig. 15–35). Due to unbalance in the rotor of the blower, the blower is subject to a sinusoidally varying vertical load at the speed of rotation ω. This speed is 1750 rpm. The brackets of the steel frame are flexible and in fact the base deflects vertically 0.026 in. under the weight of the blower, which is 500 lb.

Your assistant has designed the blower frame, and after the blower was installed and turned on, the blower was found to vibrate violently in the vertical direction. Your assistant has designed some braces, shown in Fig. 15–35, which he says will increase the vertical stiffness of the base by about a factor of 2, and thus reduce the vibration of the blower by about the same factor. He wishes you to approve this modification to the base. Would you accept his proposal? (Analyze his proposed modification for its effect on the vibration. Assume that his stiffness calculations of the brace are correct.) If you disapprove, what alternative would you suggest to him?

15–4. a) Find amplitude and phase angle for

$$\mathbf{G}(j\omega) = \frac{10}{(j\omega)(1 + j\omega)(2 + j\omega)} = Ae^{j\phi},$$

$$\mathbf{G}(j\omega) = \frac{10(1 + j\omega/2)}{(1 + j\omega)(1 + j\omega/4)} = Ae^{j\phi}.$$

b) Plot A and ϕ on log-log graph paper for various values of ω, that is, $\omega = 0.01, 0.1, 0.3, 1, 2, 3, 4$, and 10.

15–5. The schematic diagram in Fig. 15–36 shows an armature-controlled dc-motor with a fixed field being used to drive a rotational spring-inertia load. The motor armature circuit is being driven by a voltage source v. The torque and voltage equations for the motor, based on 100% electromechanical energy conversion, are

$$T = k_t i, \qquad v_a = k_g \Omega_1.$$

a) Find the driving-point impedance of the motor armature with its mechanical load as seen at the points A-A (that is, $\mathbf{V}_a/\mathbf{I}$).

b) Find the driving point impedance of the entire system as seen at the points B-B (that is, $\mathbf{V}/\mathbf{I}$).

c) Find the transfer function (system function) relating $\mathbf{\Omega}_2$ to $\mathbf{V}$.

d) Sketch the amplitude and phase of $\mathbf{\Omega}_2/\mathbf{V}$ as functions of forcing frequency ω by means of hand-drawn Bode plots (i.e., log frequency plots), using the following data:

$L_a = 0.05$ h, $K = 10{,}000$ in-lb/rad,
$R_a = 1.0$ ohm, $J = 25.0$ lb-in-sec^2,

Motor: rated power = 1.0 hp,
rated speed = 1200 rpm,
rated current = 10.0 amp.

15–6. A motor and generator are connected by a transmission line which, for the frequency of interest, can be represented by the lumped linear circuit elements shown in

Fig. 15–37. When a certain steady mechanical load is applied to the motor, the motor current and voltage are found to be

$$v_m = 141 \sin (120\pi t) \text{ volts},$$
$$i_m = 1.00 \sin (120\pi t - \pi/4) \text{ amp}.$$

For the steady sinusoidal operating condition, find the following quantities.

a) The average and instantaneous power delivered to the motor.

b) Draw a phasor diagram indicating the amplitudes and phases of the currents and voltages in each of the lumped elements and of the voltage and current at the generator terminals.

c) Find the load power factor as seen at the generator terminals, and compute the average power supplied by the generator.

d) What could you do (e.g., what additional linear elements could you add between the generator and the line) to increase the power factor obtained in part c) and thus reduce the amount of maximum current which the generator must supply for a given amplitude of the terminal voltage v_g?

15–7. Find the amplitude and phase of the voltage v_2 for the system shown in Fig. 15–38. The conditions are:

$v_1 = 100 \sin 120\pi$ (60 cps), $R_1 = 10$ ohms,
$C = 20\,\mu$f, $R_2 = 2$ ohms, $L = 2$ h.

15–8. A steam turbine which is used to drive a compressor in an oxygen plant runs at a constant (regulated) speed of 37,700 rpm. The turbine weighs 15,440 lb. When the turbine was installed, it was found that the turbine vibrated very violently during operation; the vibration could be easily detected in another building $\frac{1}{4}$ mi away. It was found that some unbalance was present in the rotor which was exciting a vibration at the frequency of shaft rotation. Since the machine and mount were assembled and installed, it was impractical to modify either the machine or the mount in any major way. You have been asked to design a system to reduce or eliminate these vibrations. You should think of a vibration absorber such as shown in Fig. 15–39.

When you measure the vertical and transverse stiffness of the turbine assembly on its mount, you find values of 4×10^5 lb/in. and 40×10^5 lb/in., respectively. Analyze the behavior of the turbine plus absorber in the vertical plane to determine the amount of force transmitted to the floor. Design the absorber, i.e., specify k_a and m_a. For practical purposes, m_a should weigh less than 1000 lb and k_a should be greater than 2500 lb/in. Comment on the influence of transverse and angular motion of the turbine mount on the effectiveness of the absorber.

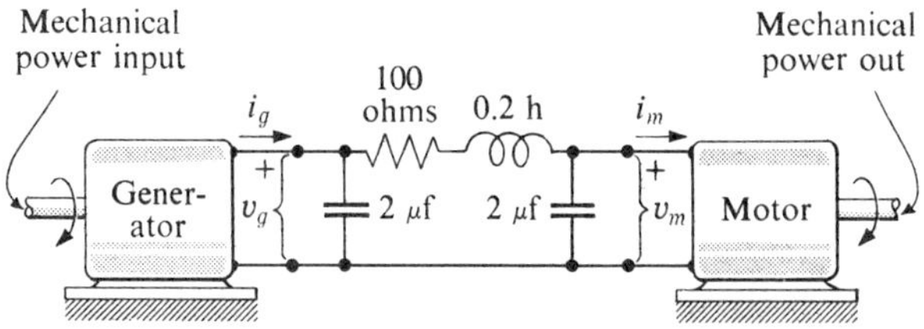

FIGURE 15–37

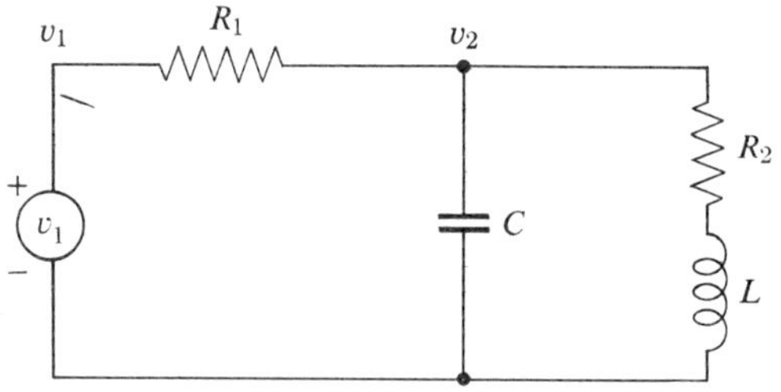

FIGURE 15–38

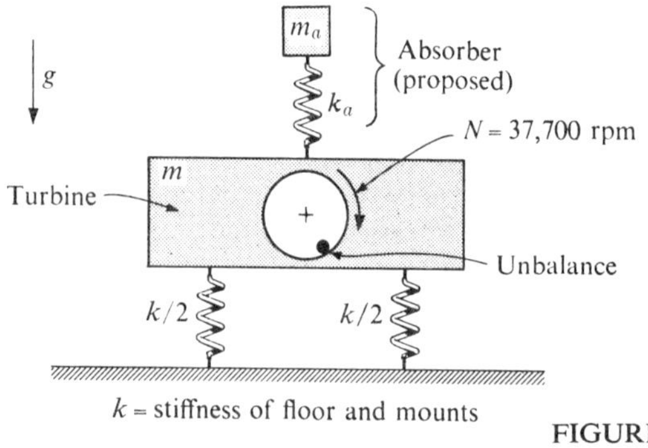

FIGURE 15–39

15–9. There is available in the laboratory a vibration analyzer, which is a device for measuring vibration amplitude and frequency. This device includes a vibration pickup to be held on the vibrating object in question and an electronic "black-box" console containing meters, amplifiers, etc. The analyzer operates over a frequency range of 500 to 40,000 cpm (note *per minute*). The upper

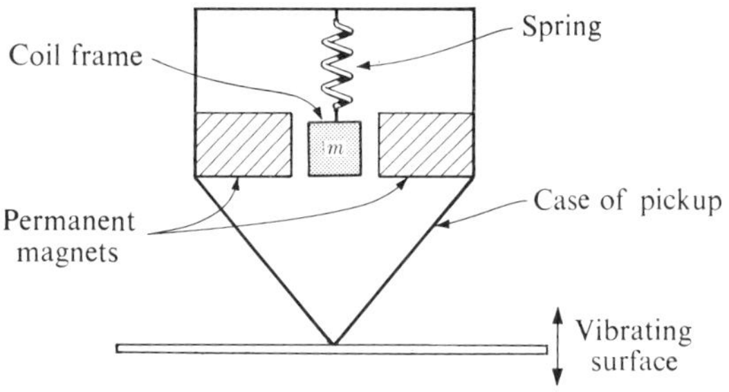

FIGURE 15–40

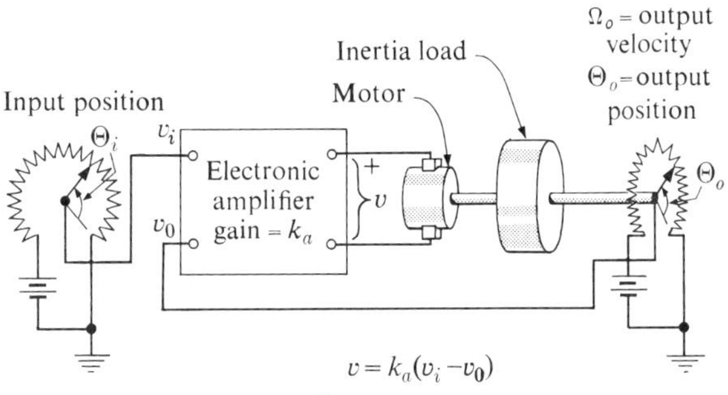

FIGURE 15–41

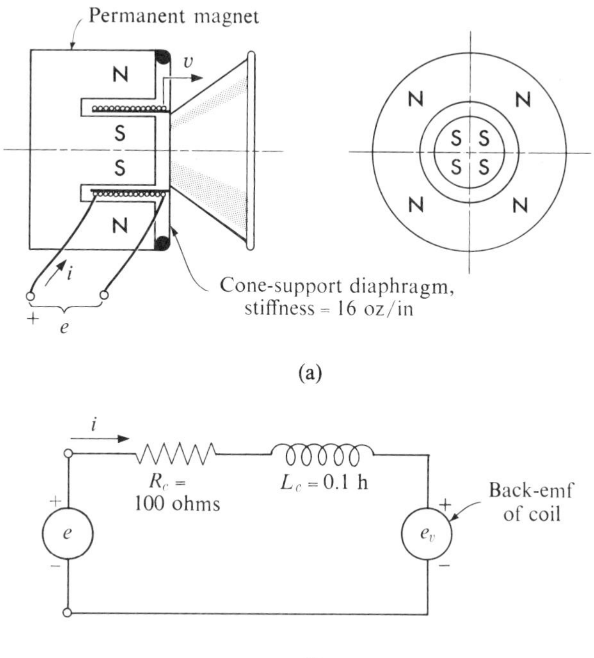

FIGURE 15–42

frequency may be extended by using the same pickup with different electronic consoles. Two other models have upper frequencies of 80,000 and 160,000 cpm. The over-all accuracy is 3% of full scale reading. The device has scale ranges of 0.0001, 0.001, 0.010, and 0.100 in. peak-to-peak, full scale.

The instruction book for this instrument says that . . . "the pickup is of the velocity type and consists of a coil moving in a permanent magnet field . . . . The velocity signal is electronically integrated, amplified, and rectified in the console, and the resultant displacement indication is presented on an amplitude meter."

Let us suppose that the pickup may be represented as shown in Fig. 15–40. You are asked to determine the necessary design parameters of the system so that the instrument will operate as the manufacturer states. The particular part of the design we are concerned with is the mechanical system of the spring and coil frame. It should be quite clear that this system will give an output voltage proportional to the velocity of the coil with respect to the permanent magnet. Your job is to design the system so that this velocity is related to the vibration velocity in the correct fashion for the indicated frequencies.

If necessary, you may assume that the vibrations are sinusoidal in time. Determine relative magnitudes for the spring constant and coil frame mass, and other pertinent quantities.

15–10. A simple feedback control scheme is to be employed to position a large inertia J in response to a reference input position Θ_i as shown in Fig. 15–41. The load inertia is much greater than that of the motor rotor to which it is connected by a very stiff shaft.

The electronic amplifier is assumed to be ideal and has a gain of k_a, that is, its output voltage is k_a times its input voltage regardless of the output current.

The motor with its inertia load has the transfer function:

$$\mathbf{G}(s) = \frac{\mathbf{\Omega}_o}{\mathbf{V}} = \frac{K_m}{\tau s + 1}.$$

The angular position of the load is sensed by means of the potentiometer connected as shown in Fig. 15–41.

a) Determine the system function between input position Θ_i and output position Θ_o.

b) Sketch the step- and frequency-response characteristics for the following parameter values:

$k_a = 200$ v/v,

potentiometer constant $= 1$ v/deg,

$\tau_m =$ time constant of motor with inertia load $= 1$ sec,

$K_m =$ steady-state motor gain $= 0.010$ deg/sec per volt.

15–11. Figure 15–42(a) shows a voice-coil type of loudspeaker. A current i through the coil interacts with the radial magnetic field of the permanent magnet and produces on the coil an axial force F_e which is proportional to the current i:

$$F_e = K_a i, \qquad K_a = 0.05 \text{ oz/ma}.$$

Any axial velocity v of the coil in the magnetic field will generate a counter-emf e_v in the coil:

$$e_v = k_v v.$$

The electric circuit for the coil is shown in Fig. 15–42(b). The mass of the moving elements of the speaker, including the mass of air which is moved by the speaker cone, weighs 0.2 oz. Assuming that the input voltage is sinusoidal with amplitude of 1 v, plot the approximate amplitude of the speaker cone displacement vs. frequency from 10 to 5000 cps. Comment on the quality of this device as a loudspeaker.

15–12. Figure 15–43 shows a proposed instrument for measuring the acceleration vs. time of "hot rod" cars during drag races. The differential equation relating scale reading y to case acceleration a is

$$m(d^2y/dt^2) + b(dy/dt) + ky = ma.$$

A Fourier analysis of typical auto acceleration vs. time curves obtained by other methods shows that an accelerometer which will measure up to $a_{\max} = 2g = 772$ in/sec^2 and which will give an output scale reading y approximately proportional to acceleration a for steady sinusoidal accelerations at frequencies from zero to 12 cps will be acceptable.

One of your junior engineers has designed a unit having mass m which weighs 6.44 lb, $k = 1.6$ lb/in., $b = 0.0058$ lb-sec/in., and has asked you to approve the drawings for manufacture.

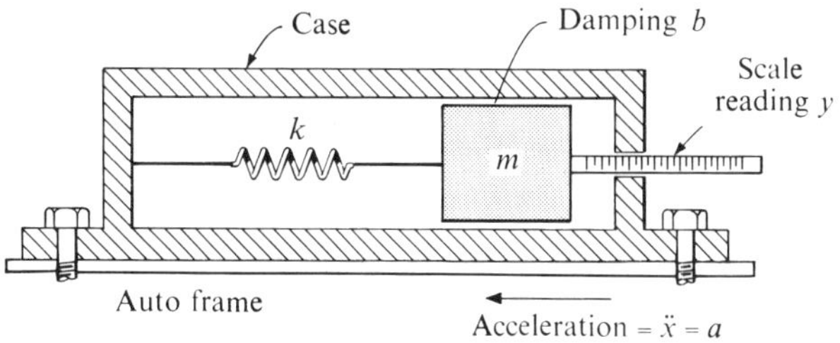

FIGURE 15–43

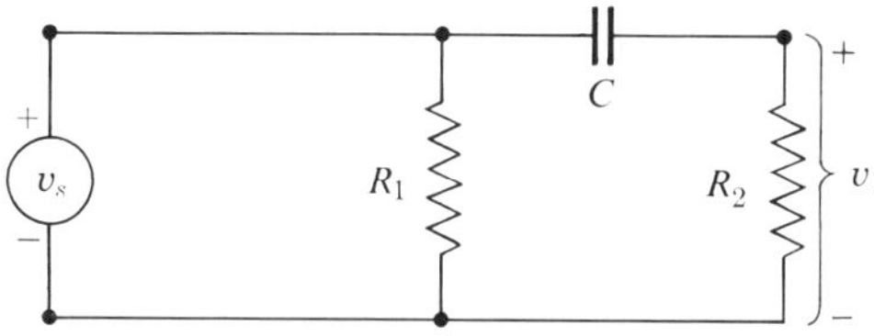

FIGURE 15–44

a) Calculate the full-scale deflection y for

$$a = a_{\max} = \text{const.}$$

b) For the steady sinusoidal acceleration $a = a_0 \sin \omega t = \text{Im}[\mathbf{A}e^{j\omega t}]$, sketch the amplitude of y with respect to a, $(|\mathbf{Y}/\mathbf{A}|)$, for

$$0 < \omega < 12 \text{ cps}.$$

c) Estimate the maximum error between the amplitude of the output scale reading y of the instrument and the acceleration amplitude a_o.

d) Recommend proper changes in the design to achieve acceptable performance. How will these changes affect the answer to part a)?

15–13. The circuit shown in Fig. 15–44 is intended to act as a "high-pass" filter for sinusoidal variations of v_s. That is, for low frequencies only a small fraction of v_s should appear at v_1.

a) Assuming that $v_s = V_o \sin \omega t$, find the differential equation for v_1.

b) Consider the long-term response of v_1 (i.e., for a time after the transient has died out). Find v_1 as a function of time for this case.

c) For $\omega = 1/(R_2C)$, find the amplitude V_1 of the voltage v_1 and the phase shift between v_1 and v_s.

d) Over what range of frequencies, will the filter "pass" v_s? (For what values of ω will $v_1 \approx v_s$?)

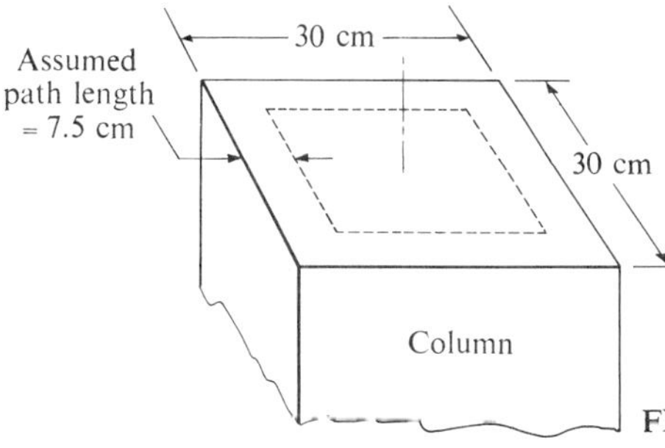

FIGURE 15–45

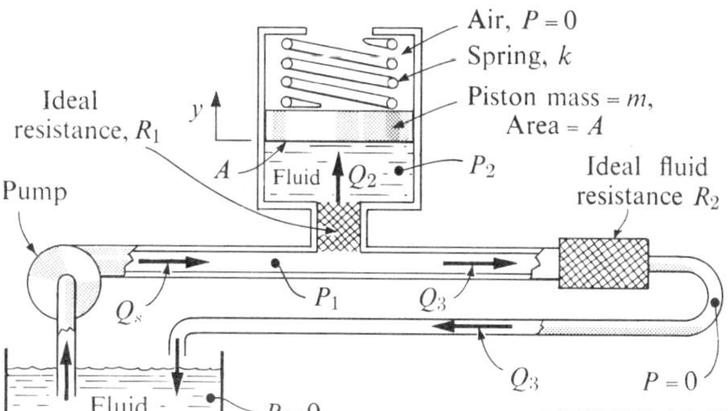

FIGURE 15–46

15–14. To study the influence of the daily temperature cycle on the free-standing columns of modern buildings, the column is modeled by a two-parameter system consisting of a thermal resistor and a thermal capacitor in series.* For a concrete column 30 × 30 cm (Fig. 15–45), assume for the heat transmission a representative path length of 7.5 cm and a net effective area normal to the heat flow of half the vertical surface area; for heat storage the whole column volume is to be counted. Use a sinusoidal daily temperature variation with an amplitude of 10°C (20°C peak-to-peak).

The thermal properties of concrete are:

Conductivity,	0.0022 cal/cm^2-sec-(°C/cm),
Specific heat,	0.20 cal/gr-°C,
Thermal capacitance,	0.50 cal/cm^3-°C.

a) Draw the system graph and determine the differential equation relating the column temperature to the atmospheric temperature.

b) Determine the parameters to be used for the thermal resistance and thermal capacitance per centimeter of column height. Watch the units.

* End effects are negligible so that heat flows in and out only through the vertical sides of the column.

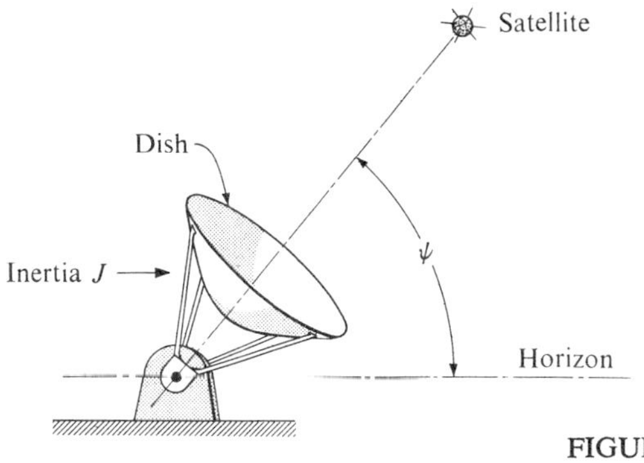

FIGURE 15–47

c) For the given atmospheric temperature variation, find the maximum daily variation of column temperature and the time of its occurrence (assuming maximum atmospheric temperature to occur at noon). [*Hint:* If you have solved part b) correctly, the system time constant equals 7.1 hr.]

15–15. Figure 15–46 shows a hydraulic "filter." The source Q_s is a rotary hydraulic piston pump which produces a flow having pulsations about an average value. A mathematical function which approximates Q_s has been found to be

$$Q_s = Q_0(1 + \epsilon \sin \omega t),$$

where ϵ is of the order of 0.1. The load which the pump drives might be a hydraulic motor or some other device, but assume that this load acts as an ideal hydraulic resistance R_2. The fluid may be assumed to be incompressible.

The hydraulic filter is intended to remove the pulsation $Q_o \epsilon \sin \omega t$ from the flow so that Q_3 into the load will be practically constant. For the following, neglect gravity and friction between piston and cylinder.

a) Formulate a lumped-parameter model of the above system. Draw a schematic diagram using equivalent masses, springs, and dashpots *or* using equivalent inductances, capacitors, and resistors.

b) Derive the differential equation relating Q_3 to Q_s.

c) Find the transfer function relating the sinusoidal part of Q_s to the sinusoidal part of Q_3 (i.e., remove steady components by subtraction).

d) For the given $Q_s = Q_0(1 + \epsilon \sin \omega t)$, calculate the value of Q_3 when $\omega = 0$ and when $\omega = \infty$.

e) As a designer of this filter system, state your recommendations for the relationships between R_1, R_2, ω, A, m, and k for the pulsation in Q_3 to be only 10% of the pulsation in Q_s. Assume that R_2 and ω are known constants.

15–16. A tracking radar is being designed to follow the motion of a satellite as shown in Fig. 15–47. An automatic control system is desired to position the elevation angle ψ of the radar dish in proportion to an input or command voltage v. A *servomotor* is available which is represented by the following equation, in which the dish inertia is proportional to τ_s:

$$\tau_s(d^2\psi/dt^2) + (d\psi/dt) = K_v v_1,$$

where v_1 is the voltage applied to the motor.

It is proposed to measure ψ by an electrical device such that

$$v_\psi = k_\psi \psi,$$

and to let v_1 be proportional to the error $v - v_\psi$, where v is the input voltage:

$$v_1 = K(v - v_\psi),$$

thus feeding back the position signal to achieve a closed-loop control system.

a) Find the differential equation relating ψ to input v.

b) Find the system "steady-state gain" $= (\psi/v)_{t\to\infty}$.

c) Find the undamped natural frequency and damping ratio of the complete system.

d) Derive an expression for the value of K which will produce the fastest response to a step change in v without overshoot of the steady-state, $(\psi)_{t\to\infty}$. Estimate the time T for ψ to reach 90% of $(\psi)_{t\to\infty}$.

15–17. A cam-driven element in a knitting machine is modeled as shown in Fig. 15–48. As the cam rotates at constant speed ω, it tends to move the cam follower horizontally back and forth with motion

$$x_1 = \epsilon_0 \sin \omega t$$

so long as the force on the cam (force in k) is compressive. If this force tends to become tensile, the follower will leave the cam surface and the resulting mass motion will cause the machine to "drop stitches." To prevent this from occurring, a constant force F_0 is applied to the mass as shown.

a) Let the compressive force between the cam and spring equal F_c. Derive the differential equation relating the *force difference* $(F_0 - F_c)$ to the input motion $x_1(t)$.

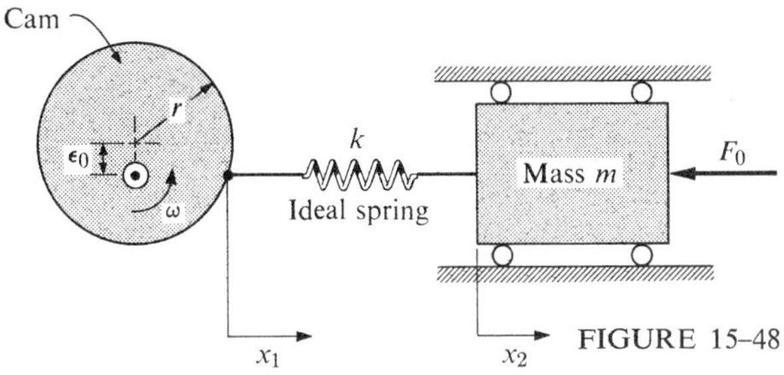

FIGURE 15–48

b) Find the steady-state amplitude of the sinusoidal variation of $(F_0 - F_c)$ resulting from $x_1(t)$.

c) Determine algebraically the lowest speed ω at which the follower leaves the cam surface (i.e., when F_c is zero at some time during the sinusoidal cycle).

d) Assuming that

$$m = (1/386)\ \text{lb-sec}^2/\text{in.}, \qquad k = 250\ \text{lb/in.},$$
$$\epsilon_0 = 0.25\ \text{in.}, \qquad F_0 = 10\ \text{lb},$$

compute the speed ω at which the follower will just leave the cam.

15–18. Consider a series RLC circuit similar to the circuit analyzed in Chapter 12 (Fig. 12–5) in which the resistance R is quite small so that the system is very resonant. The voltage v_{23} across the inductor L is to be used as the input signal to an electronic amplifier having an input impedance which is at least 1000 times as great as R, so that the power required to drive the amplifier input constitutes a negligible load on the RLC circuit.

a) For the RLC circuit show that:

$$\frac{\mathbf{V}_{23}(j\omega)}{\mathbf{V}_1(j\omega)} = \frac{LC(j\omega)^2}{LC(j\omega)^2 + RC(j\omega) + 1}.$$

b) Sketch the Bode plots of $\mathbf{V}_{23}/\mathbf{V}_1$ for this circuit with

$$R = 0.1\sqrt{L/C},$$

showing the value of the resonant peak, M_p, of the magnitude curve.

c) Determine the analytical expression for the frequency ω_{np} at which the resonant peak occurs.

d) Show that the ratio of the amplitude of v_{23} at resonance to the amplitude of v_{23} at $\omega \to \infty$ is the same as the ratio of the resonant peak frequency ω_{np} to the width of the magnitude curve at $M = M_p/\sqrt{2}$ (i.e., the difference between the frequencies at which $M = M_p/\sqrt{2}$).

e) Consider the input impedance of the amplifier picking up a signal from across L to be a pure resistance R_l, and show that the power delivered to the load resistance R_l at resonance is twice the power delivered to it when the excitation frequency is such that $M = M_p/\sqrt{2}$.

The above steps, if successfully completed, demonstrate the validity of the half-power-point definition of the quality Q of such a circuit given in Section 15–2.2.

15–19. If you did not do Problem 10–10 earlier, do it now. It fits in well with this chapter.

15–20. Determine the amplitude and phase of the first three Fourier series components of the output of a system described by the differential equation

$$5d^2x_r/dt^2 + 3dx_r/dt + 4x_r = x_f$$

when the input is a sawtooth wave, one period of which is given by

$$x_f = 10t, \qquad \text{for} \qquad -0.1 < t < 0.1 \text{ sec},$$

and sketch the resulting output waveform (composed of only these three components) as a function of time.

16 LAPLACE TRANSFORM METHODS

16-1. TRANSFORMATIONS IN SYSTEM ANALYSIS

Frequently an analytical problem can be simplified by a change in viewpoint called a *transformation*. Transformations may involve only simple changes of coordinates, or they may be very complicated and consist of the substitution of one function for another. Undoubtedly one of the most common transformations familiar to the reader is the logarithm, which transforms the multiplication of two numbers into a simpler process of adding exponents. By using available tables, we *transform* into logarithms, the numbers to be multiplied, and then we add the logarithms. Then we determine the *inverse transformation* or antilogarithm from the tables, thus finding the desired product.

In our earlier discussion of exponential forcing functions, we found that when the input is of the form $x_f = \mathbf{X}_f e^{st}$, the output is of the form $x_r = \mathbf{X}_r e^{st}$, where the two amplitudes $\mathbf{X}_f$ and $\mathbf{X}_r$ are related by the *algebraic* system function $\mathbf{T}(s)$. We can consider this result as deriving from an *exponential transformation*. Let us consider, for example, the second-order linear equation

$$a_2 \frac{d^2 x_r}{dt^2} + a_1 \frac{dx_r}{dt} + a_0 x_r = b_0 \mathbf{X}_f e^{st}. \tag{16-1}$$

Note that if x_r could be replaced by a new function such that x_r and all its derivatives contained the common factor e^{st}, this time factor would cancel from the equation, possibly resulting in a simpler equation.

Therefore, let us try the *transformation*

$$\mathbf{X}_r = e^{-st} x_r(t), \tag{16-2}$$

which transforms x_r into $\mathbf{X}_r$, where $\mathbf{X}_r$ is assumed to be independent of time. When (16-2) is solved for x_r and substituted into (16-1), $\mathbf{X}_r$ is found to be equal to the familiar function of s:

$$\mathbf{X}_r(s) = \left[\frac{b_0}{a_2 s^2 + a_1 s + a_0}\right] \mathbf{X}_f = \mathbf{T}(s)\mathbf{X}_f, \tag{16-3}$$

where $\mathbf{T}(s)$ is the system function. The time function $x_r(t)$ is found by applying the *inverse* of (16-2):

$$x_r(t) = \mathbf{X}_r(s) e^{st} = \mathbf{T}(s)\mathbf{X}_f e^{st}. \tag{16-4}$$

We know that the above transformation is imperfect, however, because the time function (16-4) is not the complete solution of Eq. (16-1) but only its particular integral. In addition, if the forcing function is not exponential, the transformation does not work at all because the e^{st} factor and thus the time variation does not drop out of the transformed equation. The Laplace transform overcomes this limitation of the exponential transformation and allows the *total* response to be found for any input function.

16–2. THE LAPLACE TRANSFORM

a) *Definition*. Consider the nth-order differential equation

$$a_n \frac{d^n x_r}{dt^n} + a_{n-1} \frac{d^{n-1} x_r}{dt^{n-1}} + \cdots + a_0 x_r = b_0 f(t). \quad (16\text{–}5)$$

If the transform (16–2) is applied to (16–5) and the result is divided by e^{st}, a factor e^{-st} will remain on the right-hand side of the resulting equation. Note, however, that if this equation were *integrated* with respect to time over a definite time interval, then the time variable would be removed and a function of the new variable s would remain. Therefore, in place of (16–2) let us define a new transformation called the *Laplace transform*. Since we are usually interested in the system behavior from $t = 0$, when the status of the system is known, to any arbitrarily large time, we will define the limits of the time integration to be $t = 0$ to $t = \infty$. The Laplace transform $\mathbf{F}(s)$ of any function $f(t)$ is by definition

$$\mathbf{F}(s) = \mathcal{L}\{f(t)\} = \int_{t=0}^{t=\infty} e^{-st} f(t)\, dt. \quad (16\text{–}6)$$

Sudden changes, sometimes of infinite magnitude, may occur in the system variables at $t = 0$, as we have seen. Therefore, in using Eq. (16–6), we must state whether $t = 0^-$ or $t = 0^+$ is intended as the lower limit in the integration.

For example, if $f(t)$ is a unit impulse

$$f(t) = u_i(t),$$

$f(t)$ is zero for $t \geqq 0^+$. Hence

$$\mathcal{L}_{0^+}\{u_i(t)\} = \int_{0^+}^{\infty} u_i(t) e^{-st}\, dt = 0.$$

On the other hand, if t is taken from 0^- in the integration,

$$\mathcal{L}_{0^-}\{u_i(t)\} = \int_{0^-}^{\infty} u_i(t) e^{-st}\, dt = \int_{0^-}^{0^+} u_i(t)\, dt = 1.$$

If the function to be transformed is finite between $t = 0^-$ and $t = 0^+$, then either limit will give the same Laplace transform, provided that the integral is properly evaluated. Consider the unit step

$$f(t) = u_s(t),$$

which is discontinuous at $t = 0$. For a limit $t = 0^+$,

$$\mathcal{L}_{0^+}\{u_s(t)\} = \int_{0^+}^{\infty} u_s(t) e^{-st}\, dt = \int_{0^+}^{\infty} e^{-st}\, dt = \frac{1}{s},$$

provided that $\mathrm{Re}(s) > 0$. For $t = 0^-$ as the lower limit,

$$\mathcal{L}_{0^-}\{u_s(t)\} = \int_{0^-}^{0^+} u_s(t) e^{-st}\, dt + \int_{0^+}^{\infty} e^{-st}\, dt = \frac{1}{s},$$

since the integrand of the integral from 0^- to 0^+ is finite. In general, therefore,

$$\mathcal{L}\{u_s(t)\} = 1/s. \quad (16\text{–}7)$$

If the system differential equation to be Laplace transformed does not change between $t = 0^-$ and 0^+, it is advantageous to take $t = 0^-$ as the limit in (16–6) so that discontinuities in $f(t)$ are included in the transformation. This will be done in the following development. The transform of a unit impulse will then be unity:

$$\mathcal{L}\{u_i(t)\} = 1. \quad (16\text{–}8)$$

If the system equation changes at $t = 0$ due to a change in system structure, the initial conditions must be found at $t = 0^+$ and the transformation (16–6) applied from $t = 0^+$ to $t = \infty$. In this situation, if an input function also occurs simultaneously at $t = 0$, its effect must be included in evaluating the initial conditions at $t = 0^+$.

b) *Convergence*. For a function to possess a Laplace transform, the defining integral (16–6) must converge. A complete discussion of the conditions for convergence cannot be presented here [1, 2]. However, convergence is usually assured for functions encountered in realistic system models. In general the integral will converge if $f(t)$ has only a finite number of finite discontinuities and increases more slowly than an exponential $e^{\sigma t}$ as $t \to \infty$. For example, the function $f(t) = e^{t^2}$ will cause $\mathcal{L}\{f(t)\}$ to diverge at $t = \infty$, and hence this function is not Laplace transformable.

c) *Elementary properties.* Several useful properties of the Laplace transform can be derived directly from its definition. The reader may easily verify that Laplace transformation is a *linear operation.* That is, if $f_1(t)$ and $f_2(t)$ are time functions which each have convergent transforms, and if a is any constant,

$$\mathcal{L}\{f_1(t) + f_2(t)\} = \mathcal{L}\{f_1(t)\} + \mathcal{L}\{f_2(t)\}, \tag{16–9}$$

$$\mathcal{L}\{af(t)\} = a\mathcal{L}\{f(t)\}. \tag{16–10}$$

The reader should note that $\mathcal{L}\{f_1(t)f_2(t)\}$ is *not* equal to $\mathcal{L}\{f_1(t)\} \cdot \mathcal{L}\{f_2(t)\}$.

The transform of the derivative of a function $f(t)$ whose transform is $F(s)$ can be found from (16–6):

$$\mathcal{L}\left\{\frac{df(t)}{dt}\right\} = \int_0^\infty e^{-st}\frac{df}{dt}\,dt.$$

Let us assume that $f(t)$ is *continuous** in the interval $t = 0^+$ to $t = \infty$, and integrate by parts, letting

$$u = e^{-st}, \quad du = -se^{-st}, \quad dv = df, \quad \text{and} \quad v = f(t),$$

$$\mathcal{L}\{df/dt\} = \int_0^\infty u\,dv = [e^{-st}f(t)]_0^\infty + s\int_0^\infty e^{-st}f(t)\,dt.$$

Therefore, if Re $(s) > 0$ and

$$\lim_{t\to\infty} \{e^{-st}f(t)\} = 0,$$

$$\mathcal{L}\{df/dt\} = s\mathcal{L}\{f(t)\} - f(0) = sF(s) - f(0). \tag{16–11}$$

* Problem 16–8 deals with the case where $f(t)$ contains discontinuities.

Note that $f(0)$ must correspond to the lower limit in the defining integral. We will take this to be 0^- unless otherwise specified.

Equation (16–11) may be applied successively to yield, for a function $f(t)$ whose first $n - 1$ derivatives are continuous from 0^+ to ∞,

$$\mathcal{L}\left\{\frac{d^n f(t)}{dt^n}\right\} = s^n\mathcal{L}\{f(t)\} - s^{n-1}f(0) - s^{n-2}\frac{df(0)}{dt} \cdots - \frac{d^{n-1}f(0)}{dt^{n-1}}. \tag{16–12}$$

Thus, for a function which originates at $t = 0$, differentiation n times transforms into multiplication by s^n.

d) *Useful transform pairs.* The transform $F(s)$ of any time function $f(t)$ for which the defining integral converges can be found by application of (16–6). The function $f(t)$ and its transform $F(s)$ are called a *transform pair.* We have already determined the transforms of a unit impulse (which was unity) and of a unit step.

Suppose that the transform of

$$f(t) = \sin \omega t$$

is required. By definition,

$$\mathcal{L}\{\sin \omega t\} = \int_0^\infty e^{-st} \sin \omega t\,dt = F(s).$$

Since $\sin \omega t = (e^{j\omega t} - e^{-j\omega t})/2j$,

$$F(s) = \frac{1}{2j}\int_0^\infty [e^{(-s+j\omega)t} - e^{-(s+j\omega)t}]\,dt.$$

Let $\alpha_1 = s + j\omega$, and $\alpha_2 = s - j\omega$. Then

$$F(s) = \frac{1}{2j}\int_0^\infty (e^{-\alpha_2 t} - e^{-\alpha_1 t})\,dt = \frac{1}{2j}\left[\frac{1}{\alpha_1} - \frac{1}{\alpha_2}\right] = \frac{2\omega}{2(s^2 + \omega^2)},$$

first order: $a_1 \frac{dy}{dt} + a_0 y = x$
Second order: $a_2 \frac{d^2y}{dt^2} + a_1 \frac{dy}{dt} + a_0 y = x$

TABLE 16–1

Short Table of Laplace Transform Pairs

$\mathcal{L}\{f(t)\} = \mathbf{F}(s) = \int_{0-}^{\infty} e^{-st} f(t)\, dt$

$\frac{d^2y}{dt^2} = s^2Y + sY(0) - Y'(0)$

$\frac{dy}{dt} = sY - Y(0)$; $y = Y$

	Function $f(t)$	Transform $\mathbf{F}(s)$	
1	$u_i(t)$	1	Unit impulse
2	$u_s(t)$	$\frac{1}{s}$	Unit step
3	$u_r(t)$	$\frac{1}{s^2}$	Unit ramp
4	e^{-at}	$\frac{1}{s + a}$	Exponential
5	$\sin \omega t$	$\frac{\omega}{s^2 + \omega^2}$	Sine
6	$\cos \omega t$	$\frac{s}{s^2 + \omega^2}$	Cosine
7	$\frac{1}{(n-1)!} t^{n-1} e^{-at}$	$\frac{1}{(s + a)^n}$; n = positive integer	Repeated roots
8	$\frac{e^{-\zeta\omega_n t}}{\omega_n\sqrt{1 - \zeta^2}} \sin \omega_n\sqrt{1 - \zeta^2}\, t$	$\frac{1}{s^2 + 2\zeta\omega_n s + \omega_n^2}$	Damped sine
9	$\frac{df(t)}{dt}$	$s\mathbf{F}(s) - f(0^-)$; $\mathbf{F}(s) = \mathcal{L}\{f(t)\}$	First derivative
10	$\frac{d^n f(t)}{dt^n}$	$s^n\mathbf{F}(s) - s^{n-1}f(0^-) - s^{n-2}\frac{df(0^-)}{dt} \cdots - \frac{d^{n-1}f(0^-)}{dt^{n-1}}$	nth derivative
11	$\int_0^t f(t)\, dt$	$\frac{1}{s}\mathbf{F}(s)$	Integration
12	$\int_0^t \cdots \int_0^t f(t)\, dt^n$	$\frac{1}{s^n}\mathbf{F}(s)$	n integrations
13	$\int_0^t f_1(\tau)f_2(t - \tau)\, d\tau$	$\mathbf{F}_1(s)\mathbf{F}_2(s)$; $\mathbf{F}_1(s) = \mathcal{L}\{f_1(t)\}$, $\mathbf{F}_2(s) = \mathcal{L}\{f_2(t)\}$	Convolution
14	$f(t - \tau)u_s(t - \tau)$	$e^{-\tau s}\mathbf{F}(s)$	Delay
15	$f(at)$	$\frac{1}{a}\mathbf{F}\left(\frac{s}{a}\right)$	Scale change
16	$e^{-at}f(t)$	$\mathbf{F}(s + a)$	Exponential attenuation
17	$tf(t)$	$-\frac{d}{ds}\{\mathbf{F}(s)\}$	Time multiplication

provided that Re $(s) > 0$. Thus,

$$\mathcal{L}\{\sin \omega t\} = \frac{\omega}{s^2 + \omega^2}.$$

Table 16–1 is a short table of Laplace transform pairs derived in this manner. The first eight entries are transform pairs for specific functions. The remaining pairs in the table are so-called *operational* transform pairs since they determine the effect on $\mathbf{F}(s)$ produced by certain mathematical operations on $f(t)$. Extensive tables of Laplace transforms may be found in References 1, 2, and 3.

16–3. TRANSFORMATION OF LINEAR DIFFERENTIAL EQUATIONS

The defining integral and the operational relationships developed in Section 16–2 permit the Laplace transform of any linear differential equation with constant coefficients to be determined. To illustrate, let us consider the second-order equation

$$a_2 \frac{d^2x_r}{dt^2} + a_1 \frac{dx_r}{dt} + a_0 x_r = b_0 x_f(t). \tag{16–13}$$

The transformed equation is, if we assume the equation to be valid at $t = 0^-$ so the initial conditions at $t = 0^-$ can be used in the transform,

$$a_2 s^2 \mathbf{X}_r(s) - a_2 s x_r(0^-) - a_2 \frac{dx_r}{dt}(0^-) + a_1 s \mathbf{X}_r(s) - a_1 x_r(0^-) + a_0 \mathbf{X}_r(s) = b_0 \mathbf{X}_f(s), \tag{16–14}$$

where

$$\mathbf{X}_r(s) = \mathcal{L}\{x_r(t)\} = \int_{0^-}^{\infty} e^{-st} x_r(t)\, dt = \text{transform of the response},$$

$$\mathbf{X}_f(s) = \mathcal{L}\{x_f(t)\} = \int_{0^-}^{\infty} e^{-st} x_f(t)\, dt = \text{transform of the forcing function}.$$

Equation (16–14) may be solved for x_r and written in the form

$$\underbrace{\mathbf{X}_r(s)}_{\substack{\text{transform}\\ \text{of}\\ \text{response}}} = \underbrace{\frac{b_0 \mathbf{X}_f(s)}{a_2 s^2 + a_1 s + a_0}}_{\substack{\text{transform of}\\ \text{response due to}\\ \text{the forcing function}}} + \underbrace{\frac{(a_2 s + a_1) x_r(0^-)}{a_2 s^2 + a_1 s + a_0}}_{\substack{\text{transform of}\\ \text{response due}\\ \text{to } x_r(0^-)}} + \underbrace{\frac{a_2\, dx_r(0^-)/dt}{a_2 s^2 + a_1 s + a_0}}_{\substack{\text{transform of}\\ \text{response due}\\ \text{to } dx_r(0^-)/dt}}. \tag{16–15}$$

Thus the transform of the response is the sum of terms due to the forcing function and the initial conditions at $t = 0^-$. This is not surprising since we proved in Chapter 11 that the response itself must consist of a sum of such individual responses.

The first term of Eq. (16–15) in fact corresponds to the second term in Eq. (11–16), and the second and third term of Eq. (16–15) to the first term in Eq. (11–16). We will see that application of the inverse Laplace transform to the terms in (16–15) will permit the system time response to be found directly. *The inverse of the first term in* (16–15) *will contain both the particular integral and the complementary function corresponding to the forcing function* $x_f(t)$. Similarly, the inverse of the second term is the complete response due to the initial condition $x_r(0^-)$, and so on.

Although Eq. (16–15) applies to a second-order equation, the conclusions just reached clearly hold for an nth-order equation.

16–4. THE RELATIONSHIP BETWEEN LAPLACE TRANSFORMS AND SYSTEM FUNCTIONS

If the initial conditions are zero so that only the input function $x_f(t)$ is present in (16–15), the ratio of $\mathbf{X}_r$ to $\mathbf{X}_f$ will be recognized as the system function derived previously for exponential forcing functions. This is a general result for any linear system.

If the initial conditions are zero, i.e., the system is initially at rest, the system function $\mathbf{T}(s)$ *as derived earlier in this book is the ratio of the Laplace transform of the response to the Laplace transform of the forcing function:*

$$\frac{\mathcal{L}\{x_r(t)\}}{\mathcal{L}\{x_f(t)\}} = \frac{\mathbf{X}_r(s)}{\mathbf{X}_f(s)} = \mathbf{T}(s). \tag{16–16}$$

If more than one forcing function is present, superposition can be used to find the response transform when all forcing functions act together.

The definition of the system or transfer function as the ratio of output to input Laplace transforms when the initial conditions are zero will be found extensively in more advanced literature on the subject of system dynamics and automatic control.

When nonzero initial conditions are present, additional system or transfer functions exist which give the Laplace transforms of the responses which are due to various initial conditions divided by the magnitudes of the respective initial conditions. These transfer functions can also be found by impedance methods if impulsive sources which establish the initial conditions are added to the system. This technique will be illustrated in Section 16–7.

It should be noted that in formulating the transformed equations for a system, it is often more convenient to apply the Laplace transform immediately to the path, vertex, and elemental equations. One reason for this is that the initial conditions may be more evident when the equations are formulated in this way. The resulting algebraic equations can then be combined and solved for the transformed variables of interest. When the initial conditions are zero, this procedure is *identical* to the impedance method developed in Chapters 13 and 14.

16–5. INVERSE LAPLACE TRANSFORMATION

The inverse of the transform of the output function is the response time function. Therefore, a time function whose transform is $\mathbf{X}_r(s)$ is required. There are several ways to find this function, the simplest and most direct being by

a) *Use of tables of transform pairs and the operational properties.* Although transform tables such as Table 16–1 are developed by transforming functions of time, they may also be used to find inverse transforms. The inverse will be unique except possibly at points where the time function is discontinuous. Whenever the function is continuous, the time function and its inverse are uniquely related.*

Frequently the transform of interest will be the ratio of two polynomials, in which case a *partial fraction expansion* can be made to facilitate inversion. Only proper fractions can be expanded by the following method. Therefore, if the order of the numerator of the transform is greater than or equal to the order of the denominator, divide the numerator by the denominator to obtain a polynomial in s plus a proper fraction. For example,

$$\underbrace{\frac{s^3+s^2+s+2}{s^2+s+1}}_{\text{improper fraction}} = \underbrace{s-1}_{\text{polynomial}} + \underbrace{\frac{2s+3}{s^2+2s+1}}_{\text{proper fraction}}.$$

The polynomial is inverted by using Relations 1 and 10 of Table 16–1. In the above case $s - 1$ inverts to give a unit doublet (derivative of an impulse) minus a unit impulse.

The method for expanding the proper fraction will be described by example. First the denominator of the fraction must be factored to find its roots and the fraction written in the form, for example, of

$$\mathbf{X}(s) = \frac{s+a}{s(s+b)^2(s+c)} = \frac{C_1}{s} + \frac{C_2}{(s+b)^2} + \frac{C_3}{s+b} + \frac{C_4}{s+c}. \tag{16–17}$$

This function has simple poles at $s = 0$ and $s = -c$, and has a double pole at $s = -b$. For

* For further discussion of this point, see Reference 2, Chapter 3.

each simple pole $-p_i$, a term $C_i/(s + p_i)$ should be placed in the partial fraction expansion, and for each multiple pole $-p_r$ of order q, a series of terms

$$\frac{C_a}{(s + p_r)^q} + \frac{C_{a+1}}{(s + p_r)^{q-1}} + \cdots + \frac{C_{q+a}}{s + p_r} \qquad (16\text{–}18)$$

should be included.

The constant for each simple pole can be found by multiplying $\mathbf{X}(s)$ by $(s + p_i)$ and letting $s \to -p_i$. Thus in the above example,

$$C_1 = \lim_{s \to 0} [s\mathbf{X}(s)] = \frac{a}{b^2 c},$$

$$C_4 = \lim_{s \to -c} [(s + c)\mathbf{X}(s)] = \frac{a - c}{-c(b - c)^2}.$$

The coefficients for a set of multiple poles are found by multiplying $\mathbf{X}(s)$ by $(s + p_r)^q$, taking successive derivatives with respect to s, and then letting $s \to -p_r$. In Eq. (16–18), therefore,

$$C_a = \lim_{s \to -p_r} [(s + p_r)^q \mathbf{X}(s)],$$

$$C_{a+1} = \lim_{s \to -p_r} \left[\frac{d}{ds} \{(s + p_r)^q \mathbf{X}(s)\}\right], \qquad (16\text{–}19)$$

$$\vdots$$

$$C_{a+m} = \lim_{s \to -p_r} \left[\frac{1}{m!} \frac{d^m}{ds^m} \{(s + p_r)^q \mathbf{X}(s)\}\right].$$

For our example, Eq. (16–17),

$$C_2 = \lim_{s \to -b} [(s + b)^2 \mathbf{X}(s)] = \frac{a - b}{-b(c - b)},$$

$$C_3 = \lim_{s \to -b} \left[\frac{d}{ds} \{(s + b)^2 \mathbf{X}(s)\}\right] = \frac{b^2 - 2ab + ac}{-b^2(c - b)^2}.$$

Once the C's in the expansion are known, Relations 2, 4, and 7 from Table 16–1 may be used to find the inverse transform. For Eq. (16–17),

$$x(t) = \mathcal{L}^{-1}[\mathbf{X}(s)] = C_1 + C_2 t e^{-bt} + C_3 e^{-bt} + C_4 e^{-ct}, \quad t > 0.$$

Often the operational Relations 9 through 17 in Table 16–1 can be used to find inverse transforms. For example, if $\mathbf{X}(s) = [1/s(s + a)]$, we can use Relation 11. Let $\mathbf{F}(s) = 1/(s + a)$ with inverse $f(t) = e^{-at}$. Then

$$\frac{1}{s}\mathbf{F}(s) = \frac{1}{s(s + a)},$$

$$\mathcal{L}^{-1}[\mathbf{X}(s)] = \int_0^t e^{-at}\, dt = \frac{1}{a}(1 - e^{-at}) = x(t).$$

Division of a transform by s is equivalent to integration of the inverse from 0 to t. Similarly, multiplication by s is equivalent to time differentiation of the inverse. Relation 6 in Table 16–1 can be derived from Relation 5 by using this property.

A second useful way of finding inverse transforms is provided by Relation 13 of Table 16–1, called the *convolution integral* or the *composition product.*

b) *Inversion by convolution.* For any system described by linear differential equations having constant coefficients, the transfer functions will be algebraic functions having the form of the ratio of two polynomials in s; see Eq. (16–16). Such functions can always be inverted by the partial-fraction method. However, the transform of the forcing function will not always be composed of polynomials, and we may be faced with a transform which cannot be expanded in partial fractions. In this case we note that any response transform

$$\mathbf{X}_r(s) = \mathbf{T}(s)\mathbf{X}_f(s) = \mathbf{F}_1(s)\mathbf{F}_2(s)$$

is the product of two transforms, one of which is the transfer function and the other of which is the transform of the forcing function. Notice that if the forcing function is a *unit impulse*, $x_f(t) = u_i(t)$, then its transform is unity. *Thus the transfer function is the Laplace transform of the response to a unit impulse.* Since $\mathbf{T}(s)$ is the ratio of polynomials, the unit impulse response $x_{\text{impulse}}(t)$ can be readily found.

$$\mathcal{L}^{-1}[\mathbf{T}(s)] = x_{\text{impulse}}(t).$$

Relation 14, the convolution integral, then allows us to obtain the response $x_r(t)$ *for any other forcing function* $x_f(t)$:

$$x_r(t) = \int_0^t x_f(\tau)x_{\text{impulse}}(t - \tau)\,d\tau. \tag{16–20}$$

This is an extremely important relation in linear system analysis. It tells us that *the response of a system to an impulse completely characterizes the system.* In Eq. (16–20) the response x_r may be thought of as a weighted replica of the input x_f and hence the impulse response is sometimes referred to as the system *weighting function.*

Note in Eq. (16–20) that $x(t - \tau)$ is defined to be zero for $t < \tau$.

c) *Inversion by explicit integration.* An explicit formula can be developed for the inverse transform [1],

$$\mathsf{f}(t) = \mathcal{L}^{-1}[\mathbf{F}(s)] = \frac{1}{2\pi j}\int_{c-j\infty}^{c+j\infty} e^{st}\mathbf{F}(s)\,ds, \tag{16–21}$$

where c is a real constant lying in the region where the transform $\mathbf{F}(s)$ is convergent. Because s is a complex variable, the use of the above equation is considerably more difficult than the use of the direct transform (16–6) which is integrated with respect to the real variable t. However, in those cases where the indirect methods (a), (b), and (c) fail, the inversion integral (16–21) must be used.

16–6. INITIAL AND FINAL VALUE THEOREMS

If the Laplace transform (16–6) is applied between 0^- and ∞, it is possible for the response $x_r(t)$ to be different at $t = 0^+$ from its value at $t = 0^-$. The initial value theorem permits $x_r(0^+)$ to be found from the Laplace transform $\mathbf{X}_r(s)$.

a) *Initial-value theorem.* From Relation 9 of Table 16–1,

$$s\mathbf{F}(s) - \mathsf{f}(0^-) = \mathcal{L}\left\{\frac{d\mathsf{f}(t)}{dt}\right\} = \int_{0^-}^{\infty} e^{-st}\,d\mathsf{f},$$

$$s(\mathbf{F}s) - \mathsf{f}(0^-) = \int_{0^-}^{0^+} e^{-st}\,d\mathsf{f} + \int_{0^+}^{\infty} e^{-st}\,d\mathsf{f}.$$

Since $e^{-st} = 1$ between 0^- and 0^+,

$$s\mathbf{F}(s) - \mathsf{f}(0^-) = \mathsf{f}(0^+) - \mathsf{f}(0^-) + \int_{0^+}^{\infty} e^{-st}\,d\mathsf{f}.$$

Now if we take the limit as $s \to \infty$ and assume that $\mathsf{f}(t)$ is finite from 0^+ to ∞ and has a Laplace transform

$$\lim_{s\to\infty}\int_{0^+}^{\infty} e^{-st}\,d\mathsf{f} = 0,$$

and we obtain

$$\mathsf{f}(0^+) = \lim_{s\to\infty}[s\mathbf{F}(s)]. \tag{16–22}$$

This result permits $\mathsf{f}(0^+)$ to be found directly from $\mathbf{F}(s)$.

b) *Final-value theorem.* If all the poles of $\mathbf{F}(s)$ lie in the left half of the s-plane, the value of $\mathsf{f}(t)$ as $t \to \infty$ can also be found from $\mathbf{F}(s)$. Again from Relation 9,

$$s\mathbf{F}(s) - \mathsf{f}(0^-) = \int_{0^-}^{\infty} e^{-st}\,d\mathsf{f}.$$

Now take the limit as $s \to 0$:

$$\lim_{s\to 0}[s\mathbf{F}(s) - \mathsf{f}(0^-)] = \mathsf{f}(\infty) - \mathsf{f}(0^-),$$

$$\mathsf{f}(\infty) = \lim_{s\to 0}[s\mathbf{F}(s)]. \tag{16–23}$$

If $\mathbf{F}(s)$ has poles on the imaginary axis or in the right half-plane, the response function will contain oscillating or exponentially increasing time functions, respectively, and $\mathsf{f}(\infty)$ will not have a constant limit. Equation (16–23) is then inapplicable.

Note that for a step input function, application of the initial- and final-value theorems is equivalent to letting $s = \infty$ and $s = 0$, respectively, in the system function $\mathbf{T}(s)$, since $\mathcal{L}\{u_s(t)\} = 1/s$.

16–7. EXAMPLE

To illustrate the use of Laplace transforms in system analysis, we will consider the mechanical network of Fig. 16–1. We wish to find $v_3(t)$ when the input v is a step of magnitude V and the mass has an initial velocity V_3. The values of the parameters and inputs are given in Fig. 16–1.

The transform $\mathbf{V}_3$ of v_3 can be derived by writing the appropriate path, vertex, and elemental equations for the system, taking their Laplace transforms and solving the resulting algebraic equations in s for $\mathbf{V}_3$. The more convenient and efficient impedance methods of Chapter 14 could be used directly to derive $\mathbf{V}_3$ if the initial conditions $v_3(0^-)$ and $f_1(0^-)$ were zero, since in that case the system function would be the ratio of $\mathbf{V}_3$ to the transform of v. So far as system behavior for $t > 0$ is concerned, it is always possible to transform a system with nonzero initial conditions into an equivalent system having zero initial conditions by adding impulsive across-variable sources in series with all T-type energy storers, and impulsive through-variable sources in parallel with all A-type energy storers. These sources are zero except at $t = 0$, at which time they are infinite. Their integrated across- and through-variables establish the required initial through- and across-variables, respectively, in the energy storers at $t = 0$.

For the system of Fig. 16–1, let us suppose that the spring had an initial force at $t = 0^-$ of $f_1(0^-)$ and the mass an initial velocity of $v_3(0^-)$. The linear graph of Fig. 16–1(b) can be modified as shown in Fig. 16–2 to include an impulsive velocity source v_k in series with the spring to establish $f_1(0^-)$ and an impulsive force source f_m in parallel with the mass to establish $v_3(0^-)$. The system is then assumed at rest with no initial conditions for $t < 0$. Between $t = 0^-$ and $t = 0^+$ the impulse in v_k will produce only finite forces in the spring, and hence nodes b and c will not displace. The integrated v_k goes entirely to deform the spring. The force established should be $f_1(0^-)$,

$$f_1(0^-) = k\int_{0^-}^{0^+} v_k\,dt = kV_i\int_{0^-}^{0^+} u_i(t)\,dt = kV_i,$$

where V_i is the strength of the impulse and $u_i(t)$ is the unit impulse. Thus

$$V_i = \frac{f_1(0^-)}{k}. \tag{16–24}$$

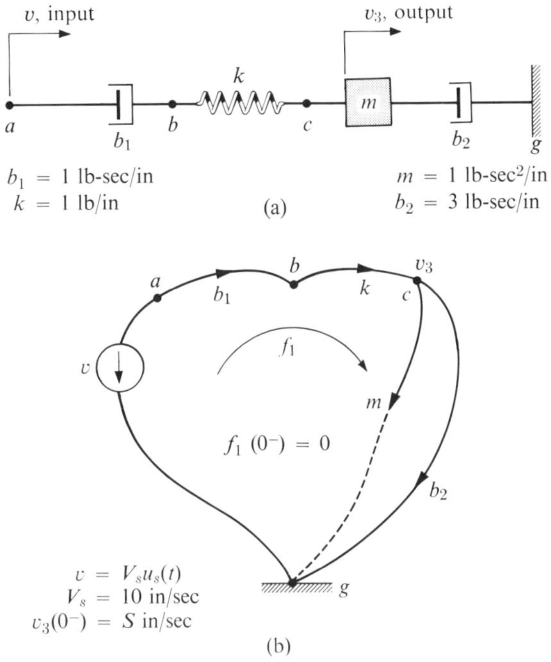

FIG. 16–1. Mechanical system and its graph.

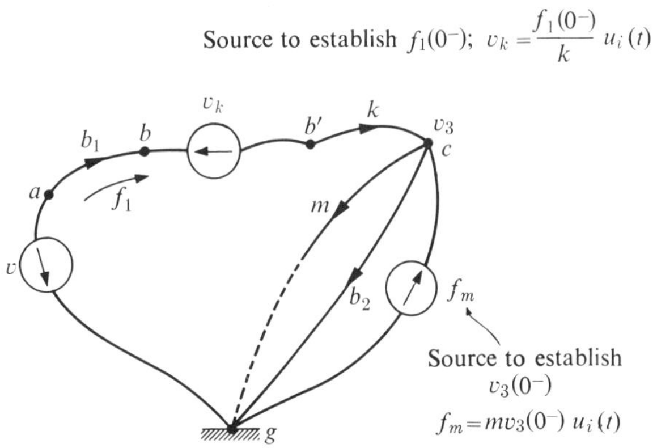

FIG. 16–2. Introduction of impulsive sources v_k and f_m to establish $f_1(0^-)$ and $v_3(0^-)$.

Similarly, the impulsive force f_m will act on the mass m between 0^- and 0^+, imparting velocity to it. During this time only finite forces will be established in the spring k and the damper b_2. These forces are negligible compared with f_m. Since we wish the mass velocity change to be $v_3(0^-)$,

$$v_3(0^-) = \frac{1}{m}\int_{0^-}^{0^+} f_m\,dt = \frac{F_i}{m}\int_{0^-}^{0^+} u_i(t)\,dt = \frac{F_i}{m}.$$

Thus

$$F_i = mv_3(0^-). \tag{16–25}$$

With the sources v_k and f_m present, we may now consider all initial conditions to be zero and apply the impedance methods of Chapter 14 to obtain $\mathbf{V}_3$. For the loop g-a-b-b'-c-g,

$$\mathbf{V}_3 = \mathbf{V} + \mathbf{V}_k - \mathbf{F}_1\left(\frac{1}{b_1} + \frac{s}{k}\right), \tag{16–26}$$

where the boldface symbols indicate transforms, $\mathbf{V}_3 = \mathcal{L}\{v_3\}$, etc. For node c,

$$\mathbf{F}_1 = \mathbf{V}_3(ms + b_2) - \mathbf{F}_m. \tag{16–27}$$

These equations combined give the result

$$\{b_1ms^2 + (b_1b_2 + mk)s + k(b_1 + b_2)\}\mathbf{V}_3 = b_1k(\mathbf{V} + \mathbf{V}_k) + (b_1s + k)\mathbf{F}_m. \tag{16–28}$$

When the transforms of Eqs. (16–24) and (16–25) are evaluated, Eq. (16–28) becomes

$$\mathbf{V}_3 = \frac{b_1kV_s/s + b_1f_1(0^-) + (b_1s + k)mv_3(0^-)}{b_1ms^2 + (b_1b_2 + mk)s + k(b_1 + b_2)}. \tag{16–29}$$

In the present problem, $f_1(0^-) = 0$ and $v_3(0^-) = V_3$. When the numerical values given in Fig. 16–1 are substituted, Eq. (16–29) becomes

$$\mathbf{V}_3 = \frac{10/s + 5(s + 1)}{s^2 + 4s + 4} = \frac{5s^2 + 5s + 10}{s(s + 2)^2}.$$

This transform is easily expanded by the method of Section 16–2.5(a):

$$\mathbf{V}_3 = \frac{5/2}{s} - \frac{10}{(s + 2)^2} + \frac{5/2}{(s + 2)}.$$

Relations 2, 4, and 7 of Table 16–1 permit the inverse to be found:

$$v_3(t) = \mathcal{L}^{-1}[\mathbf{V}_3] = \frac{5}{2} - 10te^{-2t} + \frac{5}{2}e^{-2t},$$

which is the desired response.

16–8. SUMMARY

The Laplace transformation permits the differential equations of linear systems to be transformed into algebraic equations which can be solved for the Laplace transforms of the desired response variables. The response variables are then found by determining the inverse transforms. It is not necessary to find the particular integral and the complementary function and then determine the arbitrary constants to satisfy the initial conditions; all this is automatically accomplished by the Laplace method.

For zero initial conditions the generalized impedance methods of Chapter 14 can be used directly to derive the Laplace transform of any output since it is given by the sum of the products of each system function with the transform of its respective input function. One can also include the initial conditions by means of impulsive sources, and then the impedance methods are applicable to any system. In other words, everything learned earlier in this book is directly applicable to Laplace methods; it is necessary only to take the inverse transform of the results derived by our previous methods.

The Laplace transform is primarily useful for

a) deriving theoretical results such as superposition, convolution, and initial- and final-value theorems,

b) solving for the response of high-order systems,

c) determining the response of distributed-parameter systems, which are described by partial differential equations (see Reference 1, Chapters 4, 7, and 8).

For simple first- and second-order systems the response can usually be found more simply by our earlier methods.

REFERENCES

1. R. V. Churchill, *Operational Mathematics*, 2nd Ed. New York: McGraw-Hill Book Company, Inc., 1958.
2. M. F. Gardner and J. L. Barnes, *Transients in Linear Systems*, Vol. 1. New York: John Wiley and Sons, Inc., 1952.
3. H. S. Carslaw and J. C. Jaeger, *Operational Methods in Applied Mathematics*. London: Oxford University Press, 1941.

PROBLEMS

16–1. Prove that the operator $\mathcal{L}\{\ \}$ is a linear operator as stated in Eqs. (16–9) and (16–10).

16–2. Not all functions of time possess Laplace transforms because the defining integral (16–6) may not exist. In some cases a function will have a transform over a limited region of the s-plane. The range of s over which the transform exists is called the *region of convergence* for the function. Find the region of the s-plane over which the following functions possess Laplace transforms:

a) e^{at}; $a > 0$ [*Ans.:* $s = \sigma \pm j\alpha$; $\sigma > a$],

b) te^{-at} [*Ans.:* $\sigma > 0$],

c) $\sin 3t$,

d) $\sinh 2t$.

16–3. By using Eq. (16–11) and the result derived in the text for $\mathcal{L}\{\sin \omega t\}$, derive $\mathcal{L}\{\cos \omega t\}$.

16–4. From the defining integral, find $\mathcal{L}\{e^{at}\}$. From this result and the property that $\mathcal{L}\{\ \}$ is a linear operator, verify the expression given in Table 16–1 for $\mathcal{L}\{\cos \omega t\}$.

16–5. Using the method of Problem 16–4, find $\mathcal{L}\{\cosh at\}$ and determine the values of s for which the transform exists.

16–6. Prove transform pair No. 8 in Table 16–1.

16–7. From the defining integral, prove that the transform of a delayed function

$$f(t-\tau) = 0, \qquad \text{for } t < \tau,$$
$$f(t-\tau) = f(t-\tau), \qquad \text{for } t > \tau$$

is $e^{-\tau s}\mathbf{F}(s)$, where $\mathbf{F}(s) = \mathcal{L}\{f(t)\}$. Thus multiplication of the transform of a function by $e^{-\tau s}$ delays the function by time τ.

16–8. For deriving Eq. (16–11), $f(t)$ was assumed to be continuous. Now assume that $f(t)$ contains a finite number n of finite discontinuities in the interval $t = 0^+$ to $t = \infty$. Show graphically how $f(t)$ may be decomposed into a continuous function plus a series of step functions located at the points of discontinuity $t_1, t_2, \ldots, t_n$. By using the delay theorem, No. 14 in Table 16–1, derive the transform of $f(t)$ for this case, letting $f_c(t)$ be the continuous part of $f(t)$.

16–9. Prove the convolution theorem, pair No. 13 in Table 16–1. [*Hint:* Use integration by parts.]

16–10. Prove pair No. 17 in Table 16–1.

16–11. Find the Laplace transform of the response $y(t)$ for the following equations:

$$\frac{dy}{dt} + 2y = \sin 6t, \qquad y(0^-) = 0;$$

$$\frac{dy}{dt} + 2y = \sin 6t, \qquad y(0^-) = 2;$$

$$\frac{dy}{dt} + 2y = \cos 6t, \qquad y(0^-) = 0.$$

Solve for $y(t)$ in each case.

16–12. Given the transform of $y(t)$:

$$\mathbf{Y}(s) = \frac{s+1}{s^2+s+1},$$

find $y(t)$.

16–13. Prove relations (16–19).

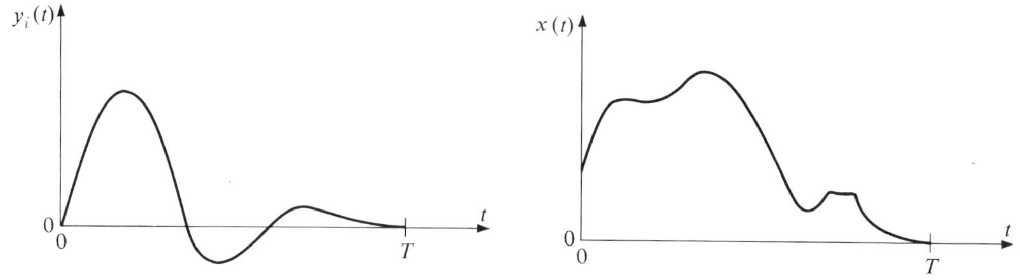

(a) Impulse response

(b) Input function

FIGURE 16–3

16–14. Find the inverse transform of the following:

$$\mathbf{Y}(s) = \frac{s + 2}{(s^2 + 4s + 4)(s + 1)^2}.$$

16–15. Obtain the Laplace transforms of the following:

a) t, b) e^{-nt}, c) $(1 - \cos at)/t$.

16–16. Obtain the inverse transforms of the following:

a) $$\frac{s^2 + 2}{s^3 - 2s^2 - 9s + 18} = \frac{s^2 + 2}{(s + 3)(s - 2)(s - 3)},$$

b) $$\frac{1}{s(s + a)(s + b)}.$$

16–17. Find

a) the Laplace tranform of $\mathsf{f}(t) = t(1 - \cos \omega t)$,

b) the inverse of

$$\mathbf{F}(s) = \frac{s + 1}{s^2 + 2s + 2}.$$

16–18. Find

a) $\mathsf{f}(t)$ if $\mathbf{F}(s) = \dfrac{3s + 4}{(s + 1)(s + 3)}$,

b) $\mathbf{F}(s)$ if $\mathsf{f}(t) = e^{-at}[\cos kt + \sin kt]$.

16–19. The transform of the response $y_i(t)$ of a system to a unit impulse is

$$\mathbf{Y}_i(s) = \frac{2}{s + 2}.$$

a) Find $y_i(t)$.

b) Using this result and the convolution theorem, find the response to an input function $x = 2 \sin 4t$.

16–20. The impulse response of a system $y_i(t)$ is shown in Fig. 16–3(a). An input function $x(t)$ is shown in Fig. 16–3(b). The analytical relationships for these functions are unknown (i.e., the functions may have been determined experimentally). Using the convolution theorem, develop a graphical method for determining the response $y(t)$ to the input $x(t)$. [*Hint:* In Relation No. 13 of Table 16–1, let

$$y(t) = \int_0^t x(\tau) y_i(t - \tau)\, d\tau.$$

Sketch $x(\tau)$ and $y_i(t - \tau)$ on the same graph. Integrate from $\tau = 0$ to $\tau = t$ to find y at time t. How long in terms of T shown in the sketches will the response $y(t)$ have a nonzero value; i.e., how long does the system "remember" the input $x(t)$?]

16–21. Find the initial and final value of the functions having the following Laplace transforms:

$$\mathbf{F}(s) = \frac{1}{s^2(s + \alpha)}, \qquad \mathbf{F}(s) = \frac{1}{s + a},$$

$$\mathbf{F}(s) = \frac{1}{(s^2 + \alpha^2)^2}, \qquad \mathbf{F}(s) = \frac{s + a}{s^2 + b^2}.$$

16–22. Using superposition and the delay theorem, find the transforms of the time functions shown in Figs. 16–4 and 16–5 for the

a) rectangular pulse [*Ans.:* $(1/s)(1 - e^{-Ts})$].

b) half-wave rectified sine wave
[*Ans.:* $(a/s^2 + a^2) \coth (\pi s/2a)$].

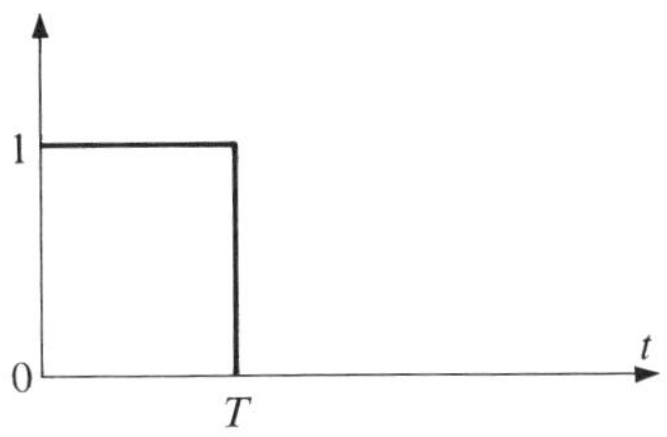

FIGURE 16–4

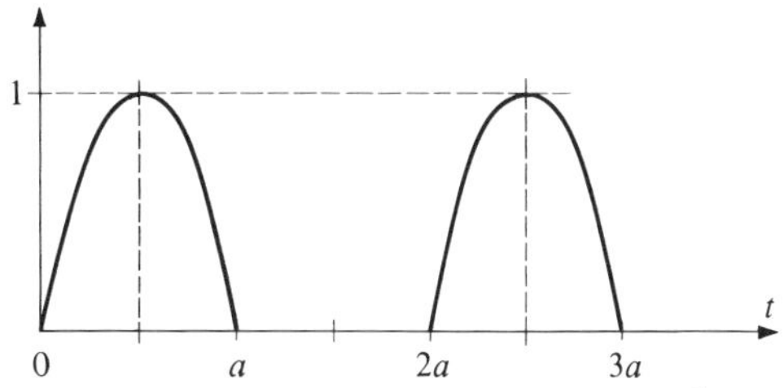

FIGURE 16–5

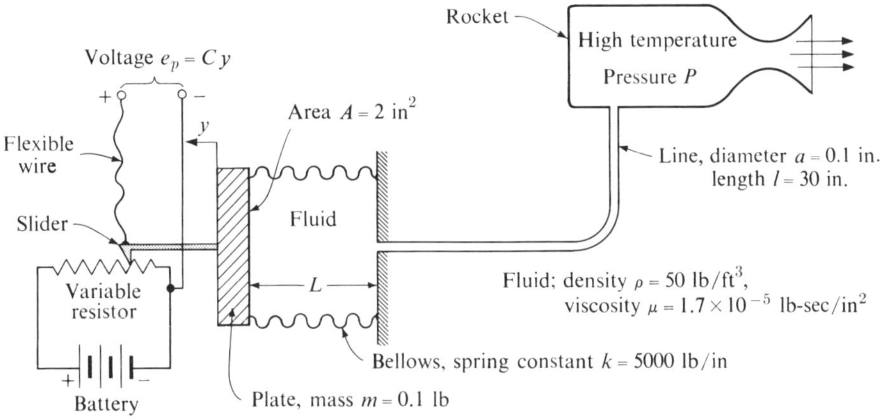

$L = 0.6$ in. $C = 100$ v/in

FIGURE 16–6

16–23. Using Laplace transforms, find the response of the system given to the stated inputs.

a) $\dfrac{d^2x}{dt^2} + \dfrac{dx}{dt} + x = y(t), \qquad y(t) = 10u_s(t),$

$x(0^-) = \dot{x}(0^-) = 0.$

b) Same as (a) except $y(t) = 0$; $x(0^-) = 10$, $\dot{x}(0^-) = 0$.

c) Same as (a) except $y(t) = 10 \cos t$.

d) Same as (a) except $y(t) = 10 \sin 2t$.

16–24. Find the unit impulse response of the following system:

$$\frac{d^2x}{dt^2} + 1.4\frac{dx}{dt} + x = y(t).$$

16–25. The scheme shown in Fig. 16–6 is proposed for measuring pressure variations in the combustion chamber of a large rocket motor. To isolate the pressure-measuring device from the high temperatures in and around the motor, a relatively long line is used as shown in the figure. To improve the dynamic characteristics of the pressure gage, this line and the other internal parts of the gage are filled with an essentially incompressible liquid having density ρ and viscosity μ. The pressure indicator consists of a chamber enclosed by a movable end plate and a cylindrical corrugated bellows which has a spring stiffness of k lb/in. As the pressure increases in the rocket chamber, fluid flows through the line into the bellows, causing the bellows to expand. The displacement y of the bellows is measured electrically by a variable resistor, resulting in a voltage e_p proportional to y.

a) Make a lumped-element model of this system using ideal elements, and draw the linear graph.

b) List the energy-storage elements which are independent and thereby predict the order of the system characteristic equation.

c) Compute the values of all elements in your model, using consistent units.

d) Derive the Laplace transform of the function relating e_p to $P(t)$, the pressure in the combustion chamber.

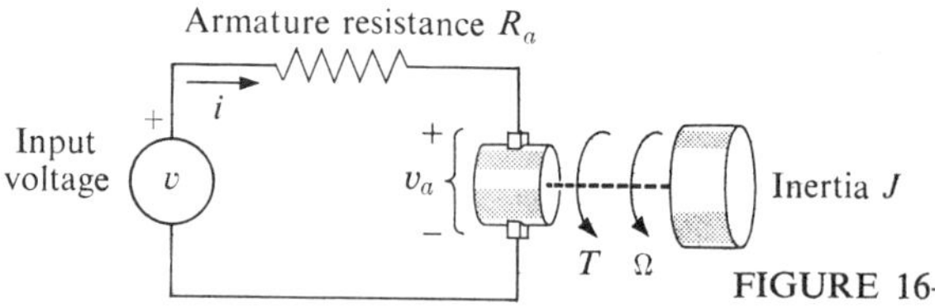

FIGURE 16-7

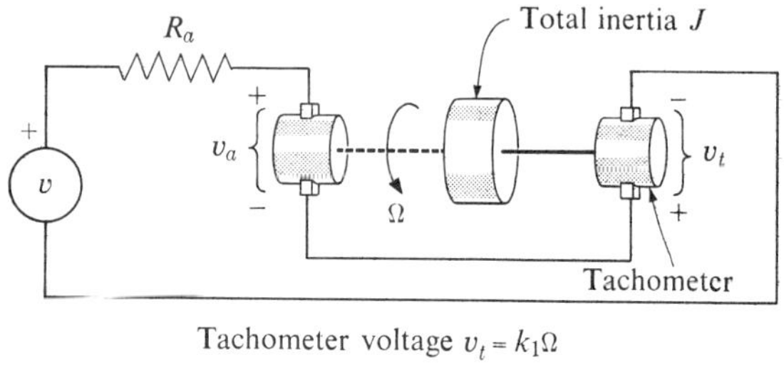

FIGURE 16-8

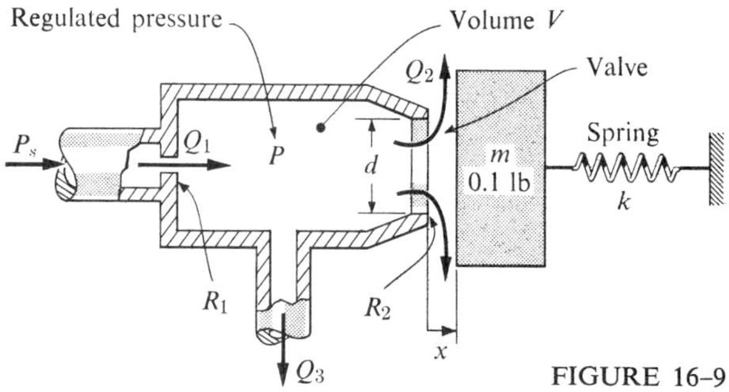

FIGURE 16-9

e) If the system is in a steady state with $P = 500$ psi when the pressure suddenly increases to 700 psi, determine $e_p(t)$ and sketch.

f) Comment on the importance of inertance and resistance in the line and of the mechanical spring and mass. How could the system response be improved?

16–26. In this problem we will consider the effect of negative feedback on the response of a system. Consider a motor which is to be used in a control system (Fig. 16–7). The inductance of the motor coils is assumed negligible and the motor drives an inertia load which has negligible damping.

$v_a = k_g\Omega =$ back emf of motor,

$T = k_t i =$ electromechanical torque,

$\Omega =$ rotor speed.

The motor has a field winding in which the current is held constant.

a) Determine the Laplace transform between v and Ω and solve for $\Omega(t)$ if v is a step input $V_0 u_s(t)$.

b) If a tachometer generator is used to feed back a voltage proportional to Ω as shown in Fig. 16–8, determine the new Laplace transform between v and Ω. Solve for the response $\Omega(t)$ in this case if $v = V_0 u_s(t)$ and compare with the results of (a).

c) The system gain, G, is defined as

$$G = \frac{\Omega \text{ steady-state}}{v} \text{ for constant } v.$$

Compare the gains and time constants for the two systems as functions of k_1.

d) If an amplifier with negligible time of response is inserted ahead of the motor, it is possible to maintain the same gain for (a) and (b). Determine the required amplification factor necessary to accomplish this and then compare the system without feedback with the one employing feedback. Can you state two or more reasons why feedback would be used in a system of this type?

16–27. Figure 16–9 shows a simplified model of a pressure regulating valve. When the pressure in the valve equals the desired pressure, $P = P_{\text{set}}$, $x = x_0 = 0.010$ in. and the spring force $F_s = P$ times the projected area $\pi d^2/4$. When P rises above P_{set}, the spring allows the valve to open further, thus permitting a larger flow Q_2 to occur. Taking additional flow Q_2 from the valve chamber tends to reduce P back to P_{set}.

The resistance to flow into the valve chamber is

$$R_1 = \frac{P_s - P}{Q_1} = \frac{100 \text{ lb-sec}}{\text{in}^5},$$

and the resistance to flow out of the valve chamber through the valve opening is

$$R_2 = \frac{P}{Q_2} = \frac{C_2}{\pi\, dx},$$

where

$$R_2 = \frac{100 \text{ lb-sec}}{\text{in}^5}$$

when $x = 0.010$ in., C_2 is a constant, and $d = 0.25$ in.

The volume of the valve chamber is $2\ \text{in}^3$ and the set pressure $P_{set} = 250$ psi. The bulk modulus β of the fluid is 100,000 psi.

a) Derive the differential equations relating P to Q_3 and x to Q_3.

b) Sketch the steady-state *regulation characteristics;*

$$P = f(Q_3) \qquad \text{for} \qquad 0 < Q_3 < 2\ \text{in}^3/\text{sec}$$

for $k = 0$, $k = \infty$, $k = 100$ lb/in. Which value of k gives the "best" regulation; i.e., keeps P closest to P_{set}?

c) Linearize the equations of part (a) where necessary for small variations in x and P. Find the Laplace transform relating P to Q_3.

d) Solve for $P(t)$ if Q_3 is a step change from $0.2\ \text{in}^3/\text{sec}$ to zero and sketch the result.

e) Discuss the results and recommend design criteria and modifications, if necessary, for this type of regulator.

Appendix COMMON UNITS AND CONVERSION FACTORS

Quantity	Metric (MKS) Unit	British Engineering System (BES) Unit	Conversion Factor*
Angular displacement	radians	radians	1; 1 rad = $180/\pi$ deg
Angular momentum	newton-meter-second = kilogram-meter^2/second n-m-sec = kg-m^2/sec.	pound-inch-sec; lb-in-sec.	8.8512
		slug-feet^2/second; slug-ft^2/sec.	0.7376
Capacitance (electrical)	farads = coulombs/volt f = c/v	same as MKS	1
Capacitance (fluid)	meters^5/newton; m^5/n	inches^5/pound; in^5/lb	4.218×10^8
Capacitance (thermal)	kilogram-calorie/degree C; kg-cal/°C	British thermal units/degree F Btu/°F	2.2046
Conductivity (electrical)	(ohm-meters)$^{-1}$; 1/(ohm-m)	(ohm-inches)$^{-1}$; 1/(ohm-in)	0.0254
Conductivity (thermal)	$\frac{\text{(kilogram-calorie)}}{\text{(second-meter-degree C)}}$; kg-cal/sec-m-°C	$\frac{\text{(British thermal unit)}}{\text{(second-inch-°F)}}$; Btu/sec-in-°F	0.0560
Charge	coulombs, c	same as MKS	1
Current	amperes = coulombs/second amp = c/sec	same as MKS	1
Density (mass)	kilograms/meter^3 = kg/m^3	slug/foot^3, slug/ft^3	0.001938
		pound-second^2/inch^4, lb-sec^2/in^4	0.936×10^{-7}
Displacement	meters, m	inches, in.	39.37
		feet, ft	3.2808

*Multiply quantity in MKS units by this factor to obtain quantity in BES units.

Quantity	Metric (MKS) Unit	British Engineering System (BES) Unit	Conversion Factor*
Energy or work	newton-meters, n-m = watt-seconds, watt-sec = joules, j	pound-inch, lb-in	8.8512
		pound-foot, lb-ft	0.7376
	kilogram-calorie, kg-cal (1 kg-cal = 4186 j)	British thermal unit, Btu	3.9685
	joule, j	British thermal unit, Btu	9.480×10^{-4}
Flux density	webers/meter^2 = w/m^2	lines/inch^2, l/in^2	6.45×10^4
Flux linkage	webers, w = volt-seconds, v-sec	lines, ℓ = maxwells, Φ	10^8
Force	newtons, n = $\frac{\text{kilogram-meter}}{\text{second}^2}$, kg-m/sec^2	pound, lb	0.2248
Frequency	radians/second = rad/sec or cycles/second, cps = Hertz, Hz	same as MKSA frequency in cps = (frequency in rad/sec)/2π	1
Heat	see Energy		
Inductance	henry, h = volt-second/ampere, v-sec/amp = webers/ampere = w/amp	same as MKS	1
Inertance (fluid)	newton-second^2/meter^5 = n-sec^2/m^5	pound-second^2/inch^5 = lb-sec^2/in^5	2.37×10^{-9}
Inertia (rotary)	kilogram-meter^2, kg-m^2 = newton-meter-second^2, n-m-sec^2	pound-inch-second^2 = lb-in-sec^2	8.8512
		slug-foot^2, slug-ft^2 = pound-foot-second^2, lb-ft-sec^2	0.7376
Magnetizing force	amperes/meter, amp/m	amperes/inch, amp/in	0.0254
Mass	kilograms, kg = newton-second^2/meter, n-sec^2/m	slug = pound-second^2/foot, lb-sec^2/ft	0.0684
		pound-second^2/inch, lb-sec^2/in	5.72×10^{-3}
		pound (mass)†	2.2046

*Multiply quantity in MKS units by this factor to obtain quantity in BES units.

†Note: Mass is sometimes expressed in units of pounds, termed *pounds mass*, to distinguish it from the unit of force, which is then termed *pounds force*. One pound mass is that amount of mass whose weight (a force) is one pound force when measured where the acceleration due to gravity is equal to the standard sea level value, 32.174 ft/sec^2.

Quantity	Metric (MKS) Unit	British Engineering System (BES) Unit	Conversion Factor*
Momentum	newton-seconds, n-sec	pound-seconds, lb-sec	0.2248
Permeability (magnetic)	webers/ampere-meter, w/amp-m	lines/ampere-inch, *l*/amp-in	2.54×10^6
Permittivity (electric)	coulomb/volt-meter, c/v-m = farads/meter, f/m	farads/inch, f/in	0.0254
Power	newton-meters/second, n-m/sec = watts = joules/second, j/sec	pound-inch/second, lb-in/sec	8.851
		pound-foot/second, lb-ft/sec	0.7376
		horsepower, hp (1 hp = 550 ft-lb/sec)	0.001341
Pressure	newtons/meter^2 = n/m^2	pound/inch^2, psi or lb/in^2	1.451×10^{-4}
		atmosphere, atm	9.88×10^{-6}
Pressure momentum	newton-second/meter^2, n-sec/m^2	pound-second/inch^2, lb-sec/in^2	1.451×10^{-4}
Resistance (electric)	ohms = volts/ampere, v/amp	same as MKS	1
Resistance (fluid)	newton-second/meter^5, n-sec/m^5	pound-second/inch^5, lb-sec/in^5	2.37×10^{-9}
Resistance (thermal)	°C-second/kilogram-calorie, °C-sec/kg-cal	°F-second/British thermal unit, °F-sec/Btu	0.453
Resistivity (electric)	ohm-meter, ohm-m	ohm-inches, ohm-in	39.37
Specific Heat	kilogram-calorie/kilogram-°C, kg-cal/kg-°C	British thermal unit/pound-°F, Btu/lb-°F	1
Temperature	degree Centigrade, °C	degree Farenheit, °F	(°F) = 32° + $\frac{9}{5}$ (°C)
Temperature (absolute)	degree Kelvin, °K (°K = °C + 273.16°)	degree Rankine, °R (°R – °F + 459.69°)	(°K) = $\frac{5}{9}$ (°R)
Torque	newton-meter, n-m	pound-inch, lb-in	8.8512
		pound-foot, lb-ft	0.7376
Voltage	volts, v = joules/coulomb, j/coul = webers/second, w/sec	same as MKS	1

*Multiply quantity in MKS units by this factor to obtain quantity in BES units.

Quantity	Metric (MKS) Unit	British Engineering System (BES) Unit	Conversion Factor*
Volume Flow Rate	meters^3/second, m^3/sec	inch^3/second, in^3/sec	6.102×10^4
		foot^3/minute, ft^3/min	2.119×10^3
		gallons/minute, gal/min (1 gal/min = 3.85 in^3/sec)	1.58×10^4
Weight	newtons, n	pounds, lb	0.2248
Weight density	newtons/meter^3, n/m^3	pounds/inch^3, lb/in^3	3.648×10^{-6}
Work	see Energy		

*Multiply quantity in MKS units by this factor to obtain quantity in BES units.

INDEX

INDEX

DIRECT METHOD

$$\tau \frac{dy}{dt} + y = x = A\cos\omega t \quad (1)$$

* get explicit form for $x(t,\omega)$
  $\rightarrow x(t,\omega) = A\cos\omega t$

* assume $y = B\cos(\omega t + \phi)$ $\rightarrow \frac{dy}{dt} = -B\omega\sin(\omega t + \phi)$
* differentiate $\rightarrow \dot{y}, \ddot{y}$ & plug into standard eq ($\ddot{y} + 2\zeta\omega_n\dot{y} + \omega_n^2 y = x$) / sub in to (1)